T0206145

Flammable Australia

The Fire Regimes and Bio

Fire is pivotal to the functioning of ecosystems in Australia, affecting the distribution and abundance of the continent's unique and highly diverse range of plants and animals. Conservation of this natural biodiversity therefore requires a good scientific understanding of processes involved in the action of fire on the landscape. This book provides a synthesis of current knowledge in this area and its application in contemporary land management. Central to the discussion is an exploration of the concept of the fire regime – the cumulative pattern of fires and their individual characteristics (fire type, frequency, intensity and season) – and its interactions with biodiversity. Contributions by 32 leading experts cover a broad sweep of topics, including prehistory, future climate change, fire behaviour, modelling of temporal and spatial patterns, plant and animal life cycles, case studies of major ecosystems, and management policies and systems. The result is a comprehensive treatment of this important ecological process set in the context of an entire continent.

ROSS BRADSTOCK is Principal Research Scientist in the Biodiversity Research Group of the New South Wales National Parks and Wildlife Service. He has studied the fire ecology of plant populations for nearly two decades and is currently engaged in studies of the nature and management of fire regimes in conservation reserves, including the application of policy and adaptive systems of fire management for biodiversity conservation. He has edited two books dealing with landscape and biodiversity conservation.

JANN WILLIAMS is Senior Fellow in the Department of Geospatial Science at Royal Melbourne Institute of Technology University. Her career to date has provided experience in both research and policy related to natural resource management. Her scientific interests include fire ecology and management, climate change impacts and environmental weeds, with a focus on tree-dominated systems. She is co-editor of *Eucalypt Ecology* (1997) and is currently President of the Ecological Society of Australia.

MALCOLM GILL has recently retired from the position of Senior Principal Research Scientist at the Centre for Plant Biodiversity Research, CSIRO Plant Industry. His research career has spanned many aspects of fire ecology, fire behaviour and fire management in a range of Australian ecosystems. He was the lead editor of the first anthology on Australian fire ecology, *Fire and the Australian Biota* (1981), and author of a variety of seminal papers on fire ecology which cemented the concept of the fire regime as a central tenet of the discipline.

Flammable Australia

The Fire Regimes and Biodiversity of a Continent

Edited by

ROSS A. BRADSTOCK
New South Wales National Parks and Wildlife Service

JANN E. WILLIAMS
Royal Melbourne Institute of Technology University

MALCOLM A. GILL
CSIRO Plant Industry

Tasmania

CAMBRIDGE
UNIVERSITY PRESS

CAMBRIDGE UNIVERSITY PRESS
Cambridge, New York, Melbourne, Madrid, Cape Town, Singapore,
São Paulo, Delhi, Dubai, Tokyo

Cambridge University Press
The Edinburgh Building, Cambridge CB2 8RU, UK

Published in the United States of America by Cambridge University Press, New York

www.cambridge.org
Information on this title: www.cambridge.org/9780521125314

First published 2002
This digitally printed version 2010

A catalogue record for this publication is available from the British Library

Library of Congress Cataloguing in Publication data

Flammable Australia : the fire regimes and biodiversity of a continent / edited by Ross
Bradstock, Jann Williams, Malcolm Gill.
     p. cm.
   Includes bibliographical references.
   ISBN 0 521 80591 0 (hardback)
    1. Fire ecology–Australia. 2. Biological diversity–Australia. I. Bradstock, R. A. (Ross
Andrew) II. Williams, Jann E. (Jann Elizabeth), 1961–   III. Gill, Malcolm, 1940–
QH197 .F563 2001
577.2–dc21   2001025479

ISBN 978-0-521-80591-9 Hardback
ISBN 978-0-521-12531-4 Paperback

Additional resources for this publication at www.cambridge.org/9780521125314

Contents

Plate section between pages 126 and 127*
*These plates are available for download in colour from
www.cambridge.org/9780521125314

Contributors

Grant E. Allan
Bushfires Council of the Northern Territory, PO Box 37346, Winnellie, Northern Territory 0821, Australia

Ross A. Bradstock
Biodiversity Research and Management Division, New South Wales National Parks and Wildlife Service, Box 1967, Hurstville, New South Wales 2220, Australia

Geoffrey J. Cary
Department of Forestry, The Australian National University, Canberra, Australian Capital Territory 0200, Australia

Wendy Catchpole
Department of Mathematics and Statistics, Australian Defence Force Academy, Northcott Drive, Canberra, Australian Capital Territory 2600, Australia

Peter C. Catling
Commonwealth Scientific and Industrial Research Organisation Sustainable Ecosystems, PO Box 84, Lyneham, Australian Capital Territory 2602, Australia.

James S. Clark
Department of Botany, Duke University, Durham, North Carolina 27708, USA

Janet S. Cohn
Biodiversity Research and Management Division, New South Wales National Parks and Wildlife Service, Box 1967, Hurstville, New South Wales 2220, Australia

Donna M. D'Costa
School of Geography and Environmental Science, Monash University, Clayton, Victoria 3168, Australia

Chris R. Dickman
Institute of Wildlife Research, School of Biological Sciences, University of Sydney, Sydney, New South Wales 2006, Australia

A. Malcolm Gill
Centre for Plant Biodiversity Research, Commonwealth Scientific and Industrial Research Organisation Plant Industry, GPO Box 1600, Canberra, Australian Capital Territory 2601, Australia

Anthony C. Grice
Commonwealth Scientific and Industrial Research Organisation Tropical Agriculture, Davies Laboratory, Private Bag, PO, Aitkenvale, Queensland 4814, Australia

Anthony D. Griffiths
Key Centre for Wildlife Management, Northern Territory University, Darwin Northern Territory 0909, Australia

Richard Hobbs
School of Environmental Science, Murdoch University, Murdoch, Western Australia 6150, Australia

Ken C. Hodgkinson
Commonwealth Scientific and Industrial Research Organisation Sustainable Ecosystems, GPO Box 284, Canberra, Australian Capital Territory 2601, Australia

David A. Keith
Biodiversity Research and Management Division, New South Wales National Parks and Wildlife Service, Box 1967, Hurstville, New South Wales 2220, Australia

A. Peter Kershaw
School of Geography and Environmental Science, Monash University, Clayton, Victoria 3168, Australia

Ian D. Lunt
The Johnstone Centre, Charles Sturt University, PO Box 789, Albury, New South Wales 2640, Australia

Michael A. McCarthy
School of Botany, University of Melbourne, Parkville, Victoria 3052, Australia

W. Lachlan McCaw
Western Australian Department of Conservation and Land Management, Brain St, Manjimup, Western Australia 6258, Australia

John W. Morgan
School of Botany, La Trobe University, Bundoora, Victoria 3083, Australia

James C. Noble
Commonwealth Scientific and Industrial Research Organisation Sustainable Ecosystems, GPO Box 284, Canberra, Australian Capital Territory 2601, Australia

Louise Rodgerson
Australian Flora and Fauna Research Centre, Department of Biological Sciences, University of Wollongong, Wollongong, New South Wales 2522, Australia

Jeremy Russell-Smith
Tropical Savannas Cooperative Research Centre, Centre for Indigenous Natural and Cultural Resource Management (Northern Territory University), and Bushfires Council of the Northern Territory, PO Box 1022, Palmerston, Northern Territory 0821, Australia

Richard I. Southgate
Parks and Wildlife Commission of the Northern Territory, PO Box 1046, Alice Springs, Northern Territory 0871, Australia

Peter Stanton
PO Redlynch, Queensland 4870, Australia

Elizabeth F. Sutherland
Biodiversity Research and Management Division New South Wales National Parks and Wildlife Service, PO Box 1967, Hurstville, New South Wales 2220, Australia

Robert J. Whelan
Australian Flora and Fauna Research Centre, Department of Biological Sciences, University of Wollongong, Wollongong, New South Wales 2522, Australia

Jann E. Williams
Department of Geospatial Science, Royal Melbourne Institute of Technology University, GPO Box 2476V, Melbourne, Victoria 3001, Australia

Richard J. Williams
Co-operative Research Centre for the Sustainable Development of Tropical Savannas, Tropical Ecosystems Research Centre, Commonwealth Scientific and Industrial Research Organisation Sustainable Ecosystems, PMB 44, Winnellie, Northern Territory 0822, Australia

John C. Z. Woinarski
Parks and Wildlife Commission of Northern Territory and Tropical Savannas Cooperative Research Centre, PO Box 496, Palmerston, Northern Territory 0831, Australia

Preface

Fires are a continual challenge to landscape managers and policy makers in many parts of the world. Landscape fires are of global significance through their effects on climate and climate change among many other aspects. The effects of fires on biodiversity also have global significance, because a loss of biodiversity within a continent, region or even local area may be a global loss, or can contribute to one. This book is about the effects of fires on the biodiversity of Australia, a continent with a large complement of unique flora and fauna. Many of the topics covered, however, will have relevance across the globe.

It is no longer just scientists who recognize that knowledge of the pattern of fires and their cumulative characteristics – the fire regime – is important if we are to understand and manage fire-prone landscapes. However, scientists are finding that they need to account for broad-scale temporal and spatial variation in fire regimes if they are to understand effects on biodiversity. Thus the level of complexity needed to elucidate the nature of fire regimes and their ecological effects is increasing. Consequently, emphases will be found in the chapters of this book that were unforeseen a decade or two ago, such as in *Fire and the Australian Biota*.[1]

The authors of this volume review the effects of fire regimes on ecosystems and landscapes from an Australian perspective. There are chapters on climate change, on the main Australian vegetation types and on aspects of management. Unfortunately, a planned chapter on contemporary Aboriginal management of fires was not completed.

All the chapters of this book have been peer-reviewed. We are extremely grateful to the following people for acting as reviewers in addition to the various chapter authors who also performed this task: Alan Andersen, Tony Auld, Tom Beer, Dick Braithwaite, Neil Burrows, Peter Clarke, Garry Cook, Rod Fensham, Gordon Friend, Lesley Head, Byron Lamont, John Ludwig, Keith McDougall, Helene Martin, David Morrison, Bob Parsons, John Pickard, Daniel Polakow, Hugh Possingham, Dick Rothermel and Wal Whalley.

The editors have greatly appreciated the assistance given to them by a Steering Group. Our thanks go to Kevin Tolhurst and Neil Burrows along with a number of chapter authors. We also wish to thank Belinda Pellow and Janet Cohn who provided crucial assistance with final editing of manuscripts.

All lead authors were invited to contribute to the book and we especially wish to thank Professor James Clark from the United States for his collaboration and his visits to Australia as part of this project. He was especially invited because of his international prominence in the fields of modelling of fire regimes, fire effects and palynology.

Financial support for Professor Clark's visits to Australia came from the Australasian Fire Authorities Council, CSIRO Plant Industry and Monash University. Donations from pertinent organisations meant that colour could be used in the book without increasing its price. The Tasmanian Fire Service and Tasmanian Parks and Wildlife Service are singled out for special thanks due to their emphatic support of this project which included financial contributions.

Finally, we emphasise that the opinions and conclusions expressed by the authors remain their responsibility. They do not necessarily represent those of the editors, their employers or the publisher. We trust, however, that the chapters and topics covered by this book will stimulate and inform readers.

Ross A. Bradstock
Jann E. Williams
A. Malcolm Gill

[1] Gill, A. M., Groves, R. H., and Noble, I. R. (eds) (1981) *Fire and the Australian Biota* (Australian Academy of Science: Canberra.)

Part I

Past and future

1

A history of fire in Australia

A. PETER KERSHAW, JAMES S. CLARK, A. MALCOLM GILL AND DONNA M. D'COSTA

Abstract

Over the last 20 years, the counting of charcoal in association with pollen in the construction of palaeoenvironmental records from swamp, lake and marine sediments has become routine. This has provided a substantial, although methodologically variable and geographically biased, data set with which to examine the history of burning on the Australian continent. Two generalised records from southeastern Australia demonstrate a general increase in burning within the later part of the Tertiary period in line with reduced precipitation, increased climatic variability and development or expansion of sclerophyll forest and heath vegetation. The one continuous record of vegetation and burning from the early part of the Quaternary suggests that this period may have experienced relatively stable environmental conditions. Increased fire activity is evident in the majority of records extending through at least the last glacial/interglacial cycle, particularly during drier glacials and during times of major climate change. A focus on southeastern Australia for the last 11 000 years (11 ka), a place and period containing the bulk of charcoal records, indicates high fire activity over the last few thousand years, with a major peak coinciding with the early phase of European settlement followed by reduction within recent decades to Early Holocene levels. It is demonstrated that climate has exerted the major control over both fire activity and vegetation change. There is a notable increase in fire activity centred on 40 ka before present (BP) which, in the absence of a major climate change around this time, is considered to most likely indicate early Aboriginal burning. The impact on the vegetation was largely to accelerate existing trends rather than cause a wholesale landscape change. It is difficult to separate the effects of climate and human-induced burning subsequent to this time until the arrival of Europeans.

Introduction

The first systematic examination of evidence for the role of fire in the historical development and maintenance of Australian vegetation was presented in *Fire and the Australian Biota* (Gill et al. 1981). The evidence was minimal. Kemp (1981) in her review of pre-Quaternary fire speculated on the vexed question of the extent to which the geological material 'fusain' was a product of biomass burning and on likely changes in fire activity through the Tertiary period based on vegetation/climate patterns. Most direct evidence for fire within the Cainozoic period derived from the brown coals of Victoria with the general acceptance that fusinite was largely charcoal derived from the burning of swamp vegetation, with lightning and spontaneous combustion being the major ignition agents. No data were available to determine whether or not these fires had any long-term effect on the character of the vegetation. In lieu of any direct fire–vegetation relationships, Kemp surmised that burning activity would have increased during the Tertiary and Quaternary periods in response to increasingly dry and variable climatic conditions.

Singh *et al.* (1981) focussed their paper 'Quaternary vegetation and fire history in Australia' on three recently produced pollen records incorporating measures of charcoal particles as direct evidence of past fires. All three records demonstrated some relationship between fire activity and climatic conditions but major features of the records related to the presumed impact of Aboriginal people on fire regimes and sustained vegetation change. The long record of

Lake George, near Canberra, subsequently extended and refined by Singh and Geissler (1985), provided what was considered to be a continuous record through the last four glacial/interglacial cycles, about the last 400 000 years (400 ka). The earlier part of this record showed a pattern of significant fire activity only during wetter and warmer interglacials which were dominated by Casuarinaceae with minor rainforest elements. It was considered that the cooler, drier glacials, dominated by herbaceous vegetation, were too open to support regular fires. This pattern changed at the beginning of stage 5 (the last interglacial, *c.* 125–130 ka BP – before present) with higher and more continuous burning. This was considered to be a result of an additional ignition source, the activities of Aboriginal people. There was a substantial replacement of Casuarinaceae by eucalypts and a reduction in rainforest elements during the last and present (Holocene) interglacials, and a general sustained increase in myrtaceous shrubs. The maintenance of high burning levels during the last glacial period was considered to be the result of an increase in sclerophyllous vegetation combined with persistent Aboriginal burning. Highest burning levels for the whole diagram were achieved near the top of the record, considered to be the result of activities of early European colonists, before a sharp decrease in recent years representing the implementation of a fire-exclusion policy.

A record from Lynch's Crater, from the Atherton Tableland, in the humid tropics region of northeast Queensland, covered the last glacial cycle, with complex rainforest dominant during the last interglacial under high precipitation levels and araucarian rainforest together with some sclerophyll forest or woodland prevalent during the first part of the last glacial period (Singh *et al.* 1981). Relatively low charcoal levels suggested infrequent burning, although fire activity was most evident during phases with higher sclerophyll representation, and also at times of major climate and vegetation change, presumably as a result of general environmental instability. A substantial increase in burning occurred around 38 ka BP, considered to be responsible for a gradual replacement, between 38 and 26 ka BP,

of 'fire sensitive' araucarian forest by sclerophyll communities, co-dominated by Casuarinaceae and *Eucalyptus,* and also for the extinction of a previously conspicuous member of swamp and riverine communities, the conifer *Dacrydium.* Aboriginal burning was hypothesised to be the major cause of change largely because it did not correspond with any time of known major climate change. Araucarian forest is now restricted, in northeastern Australia, to small, isolated and relatively fire-protected patches. High, although variable, burning levels have been maintained subsequently except for the mid Holocene when high precipitation allowed a re-expansion of complex rainforest within the area.

The much shorter sequence from Lashmar's Lagoon, on Kangaroo Island, South Australia, covering much of the Holocene period, showed consistently low charcoal values, suggesting low fire activity within surrounding *Casuarina stricta* woodlands, until there was some increase in burning around 5 ka BP, considered to be the result of a regional decrease in precipitation (Singh *et al.* 1981). Shortly afterwards, about 4.8 ka BP, there was a rapid replacement of the *Casuarina* woodland by eucalypt woodland, caused by the drier climate but probably facilitated by a short phase of higher burning activity indicated by a charcoal peak. Around 2.5 ka BP there was a major increase in charcoal particles, sustained until after the arrival of European people, but without any concomitant response in the vegetation. It was considered that this charcoal increase was the result of a shift from a regular low- intensity burning pattern maintained by Aborigines, to a pattern of infrequent but high-intensity fires with the disappearance of people from the island.

These three Quaternary records have been prominent in debate about the role of fire, and particularly the impact of people, in the evolution and dynamics of Australian vegetation for almost the last 20 years. Virtually all interpretations have been criticised. Most criticisms have been made on the Lake George record, largely because of the inferred presence of people at least 60 ka earlier than that revealed in the archaeological record (Flood 1995). It has also been proposed that the inferred time-scale is incorrect (Wright 1986) and that the greater 'fire sensitivity' of Casuarinaceae relative to *Eucalyptus* can be questioned (Ladd 1988). The most extreme opposing view to that of Singh *et al.* (1981) was presented by Horton (1982) who considered that Aboriginal people had little effect on existing fire regimes and that the pre-European settlement vegetation cover, as well as the composition and distribution of dependent fauna, would have been the same as it was regardless of whether people had been here or not. Bowman (1998) in the latest review of Aboriginal burning in relation to the Australian biota presents yet another examination of these records and, although he incorporates some more recent studies, is equivocal about the value of past vegetation and fire records for determination of the role of Aboriginal people in the development and maintenance of Australian landscapes.

Perhaps the largest overall problem with the Singh *et al.* (1981) paper was that it provided three stories which, although internally consistent, together appeared to contain major differences with respect to the timing of initial human impact and vegetation responses to fire generated by both natural and human causes. The time gap between the first evidence for inferred human burning at Lake George and Lynch's Crater (i.e. 90 ka) is substantial and, although originally explained by the likelihood that more open vegetation would be most readily settled and transformed by people, this interpretation was unconvincing. Similarly, arguments made for increased charcoal abundance on the grounds of both human presence and absence at different sites reduced the validity of the human-caused burning hypothesis in the minds of a number of critics. The consistency in the three records is that they show a general increase in charcoal through time and change from a less to more fire 'tolerant' vegetation, suggesting that fire has become an increasingly important component of the dynamics of the vegetation within the late Quaternary.

We are now in a position to make a more informed assessment of the nature of the fire record and its relationship to vegetation. There is now a more substantial database of over 100 pollen records with associated charcoal curves and additional information provided by elemental carbon analysis, fire scars on trees and identified soil charcoal. Greater understanding of charcoal–fire and charcoal–pollen relationships has been provided by experimental and fine resolution studies in Australia (R. L. Clark 1982, 1983) and particularly North America (e.g. J. S. Clark 1988; J. S. Clark and Hussey 1996). However, the data, in the main, are still crude both in terms of time resolution and accuracy and are biased towards certain parts of the continent. Following the example set by Singh *et al.* (1981), there is also a lack of consistency in methods of counting and portrayal of charcoal data. Counting methods range from point counting which provides a measure of the area of a microscope slide prepared from a pollen sample covered by charcoal, simple counting of charcoal particles above a certain size (usually 5, 10, 15 or 20 μm), to visual assessment of relative abundance. Quantitative measures of charcoal abundance are displayed variously in the form of concentration (e.g. $cm^2 cm^{-3}$, $mm^2 cm^{-3}$, particles cm^{-3}, particles g^{-1}), influx (e.g. $cm^2 yr^{-1}$, $cm^2 cm^{-2} yr^{-1}$) or as charcoal/pollen concentration or percentage ratios, with the pollen component being total pollen, total dryland pollen or an individual taxon such as *Eucalyptus*. This lack of consistency essentially prohibits the establishment of a quantitative, comparative measure of charcoal abundance and hence fire activity.

In line with the limitations of charcoal data for the determination of fire activity, information provided by pollen for the reconstruction of vegetation is similarly restricted. Of particular concern is our inability to identify pollen taxa to taxonomic levels which allow the recognition of differential responses to individual fires or to fire regimes. Consequently, we generalise the terminology applied to responses of individual species or species groups to higher taxonomic groupings and vegetation types. If communities or taxa are closely associated with high charcoal levels they are considered to be 'fire tolerant' relative to those which

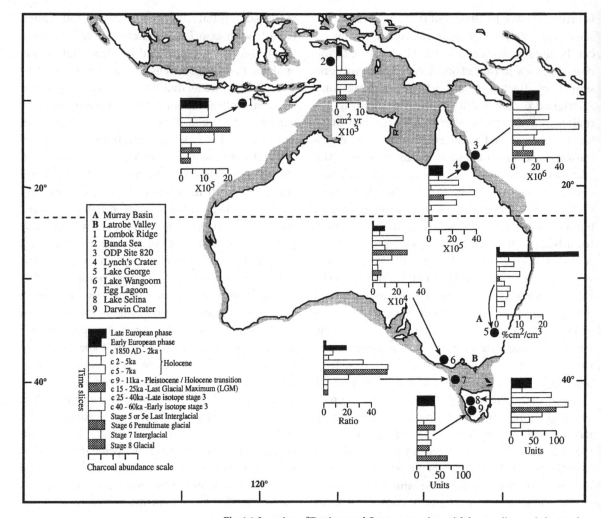

Fig. 1.1. Location of Tertiary and Quaternary sites with long pollen and charcoal records examined in this paper and relative abundance of charcoal for time slices for the Quaternary records. See Table 1.1 for sources and details of the different measures of charcoal.

experience low charcoal levels which are termed 'fire sensitive'.

Pre-Quaternary fire activity

The two major areas in Australia where the composition of vegetation in the Tertiary period has been reconstructed, the Murray River catchment and the Latrobe Valley (Fig. 1.1), have associated charcoal records. These records cover substantial portions of the major period of climate change experienced in

Australia, related to the continent's separation from Antarctica in the Eocene, about 40 million years ago (40 Ma BP), followed by its movement into lower latitudes. Although both study areas are in the southeastern part of the continent, they reflect some geographical variation along the present climatic gradient from moist coastal environments to the drier interior.

The generalised record from the Murray catchment (Martin 1990, 1991; Kershaw *et al.* 1994) (Fig. 1.2) shows the predominance of *Nothofagus* pollen,

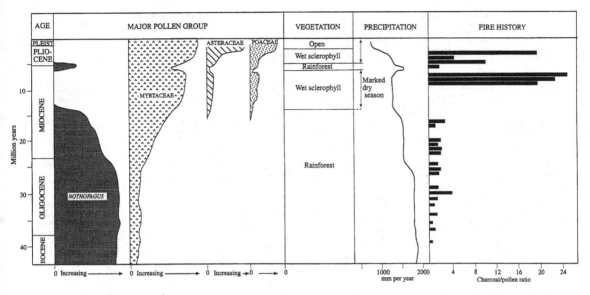

Fig. 1.2. Summary Cainozoic record of pollen and charcoal together with inferred vegetation and precipitation changes from the eastern part of the Murray Basin, N.S.W. Adapted from Martin (1990, 1991).

characteristic of Tertiary rainforest through most of Australia, from the Late Eocene and Oligocene periods. Charcoal levels indicate that fire was present, but probably very infrequent during the Eocene, and that fire activity was somewhat higher during the Oligocene. The steady increase in Myrtaceae from the Late Oligocene to the Early Miocene may indicate some opening of the rainforest canopy or, more likely in the absence of the herbaceous taxa Poaceae and Asteraceae, a change to warmer or more seasonal rainforest types. There is no evidence of an increase in burning during this period. The sharp decline in *Nothofagus* around 15 Ma BP in the Mid Miocene together with the presence of herbaceous taxa certainly indicates a reduction in rainforest and presence of more open canopied vegetation. It is inferred that the vegetation was wet sclerophyll (or tall open) forest dominated by Myrtaceae with a mix of rainforest taxa in the understorey. In the present Australian environment, the canopy of wet sclerophyll forests is dominated everywhere by the non-rainforest genus *Eucalyptus* (Gill and Catling, this volume), but in

these fossil assemblages other Myrtaceae, perhaps including genera of present-day rainforest margins such as *Tristania, Tristaniopsis* and *Syncarpia*, were conspicuous. Charcoal levels are very high in the Late Miocene samples and, although charcoal records are not available for Mid Miocene sequences, it may be assumed that high levels were achieved with the development of wet sclerophyll forest as intense fires are associated with this kind of vegetation today. Similar conditions continued through the Pliocene apart from a temporary re-expansion of rainforest during the Early Pliocene when burning was reduced to pre-wet-sclerophyll levels. The very Late Pliocene/Early Pleistocene witnessed a likely further opening of the canopy. The high levels of herbaceous taxa, combined with high values of Myrtaceae and lack of rainforest taxa, suggest the development of open-forest or woodland. Such vegetation types dominate the landscapes of these areas today.

The Murray catchment record provides data that support the proposition of Kemp (1981) that burning would have increased through the later

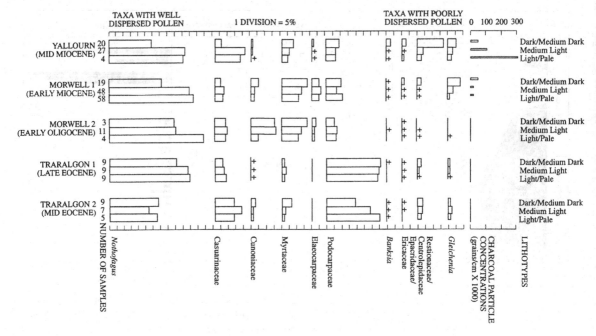

Fig. 1.3. Summary of Cainozoic pollen and charcoal records from the Latrobe Valley coal measures, western Victoria. Adapted from Kershaw *et al.* (1991).

part of the Cainozoic period in response to drier and more variable climatic conditions. There is certainly a good relationship between vegetation, inferred rainfall and charcoal although the record is too generalised to determine the degree to which fire was actively driving vegetation change rather than simply responding to climatically induced change. The suggested presence of wet sclerophyll forest is interesting because, in contemporary conditions, fire is considered essential for the maintenance of a mix of shade-intolerant eucalypts and shade-tolerant rainforest taxa. There is a great deal of speculation on the long-term stability of such communities and whether they can exist in the absence of fire generated in surrounding more open vegetation. The evidence from the Murray catchment is that wet sclerophyll, although different floristically from any vegetation type present today, can be maintained over long periods of time.

The Latrobe Valley provides a similar span of vegetation history to that of the Murray catchment but charcoal data are restricted to sequences derived from Eocene to Miocene coal seams which accumu-

lated as peat on the onshore section of the subsiding Gippsland Basin (Kershaw *et al.* 1991; Sluiter *et al.* 1995). The summary diagram (Fig. 1.3) indicates average pollen and charcoal values for samples from each major coal lithotype group within each coal seam. Although it has been traditionally thought that peat accumulation was largely continuous through the whole period, recent sequence stratigraphic analysis has suggested that phases of coal seam accumulation relate to major global sea-level highs and may only have occupied about 7.5 of the 26 Ma period represented by the coal measures (Holdgate *et al.* 1995). Furthermore, peat accumulation within each seam may have been restricted to sea-level highs of several thousand years duration within fourth-order eustatic cycles (Haq *et al.* 1987). Consequently the record may be very biased towards wetter periods which are most likely to accompany sea-level highs. Lithotype groups have been separated on the diagram because they represent different depositional environments within the coal-forming basin, although the actual nature of these environments is the source of debate.

Original palynological and plant macrofossil interpretation was that there was a lithotype colour gradient from light to dark representing hydroseral succession, initiated by periodic basin subsidence, from open water to swamp forest (Luly *et al.* 1980; Blackburn 1985; Kershaw *et al.* 1991). However, the predominantly lightening-upwards pattern identified from Markov chain analyses (Mackay *et al.* 1995) suggested the opposite. The presence of open water has also been questioned (Anderson and Mackay 1990). Sequence analysis suggests that the lithotypes may be more a reflection of climatic variation during sea-level highs than successional change in response to phases of subsidence (Sluiter *et al.* 1995).

In general terms, the taxa with well-dispersed pollen provide a regional picture of the vegetation, although many taxa will also have had a swamp forest representation. Coal formation began in the Mid Eocene after the Early Eocene global temperature peak. Temperatures continued to fall through the Eocene and this is reflected in the pollen by the increase in *Nothofagus* pollen. The decline in Podocarpaceae may be related to the general evolutionary replacement of gymnosperms by angiosperms. All taxa have strong rainforest affinities although there is some macrofossil evidence of the presence in Australia, and in the Latrobe Valley coals, of *Casuarina*, the sclerophyllous member of the Casuarinaceae, in the Miocene (Hill 1994), while components of the Myrtaceae, although not containing *Eucalyptus,* may also have been sclerophyllous. Within those taxa with poorly dispersed pollen, which must have derived from taxa growing within the swamp system, there is a general increase, particularly in the Morwell 1 and Yallourn seams. All are sclerophyll elements.

Throughout the whole recorded period the landscape was dominated by rainforest. However, the increase in light-demanding sclerophyll elements through time suggests that the vegetation was either becoming more open or was subject to increasing levels of disturbance or stress. The charcoal record is consistent with increased disturbance, being absent until the Early Miocene Morwell 1 seam and then increasing through to the

Mid Miocene. These changes relate to increased lithotype differentiation in the coals which is likely to have been primarily due to increased climatic variability at Milankovitch or sub-Milankovitch scales (i.e. 100 ka or less). The data are too coarse to indicate any role of fire in the initiation of vegetation change but, from a fine-resolution study of a section of the Morwell 1 seam, Blackburn and Sluiter (1994) concluded that burning occurred after a change to drier conditions and development of a more sclerophyllous vegetation and was clearly a response to, rather than a cause of, vegetation change.

Although there are major differences between the vegetation and depositional environments of the Murray Basin and Latrobe Valley sequences, there are sufficient similarities between them within their period of overlap to suggest synchronous broad regional changes in fire activity. Charcoal values were low until probably the Late Oligocene, suggesting infrequent or low intensity burning under a climate with low-variability and limited sclerophyll development. Higher charcoal levels from the Late Oligocene to Early Miocene indicate more intense and widespread fire activity under more variable climates. There is increased representation of sclerophyll vegetation in both swamp and dry land environments. The major change to high charcoal levels, perhaps indicating fire activity almost as great as that in the Pleistocene, occurred within the Mid Miocene, by about 16 Ma BP, accompanied by a highly variable climate and, although precipitation is estimated to have been about twice that of today, the representation of sclerophyll vegetation was substantial.

Pleistocene fire activity

The focus of Quaternary palynological study has continued to be on the construction of records extending from the present as far back into the past as possible and, as a result of the high cost of drilling economically unimportant Quaternary sediments and the discontinuous nature of Quaternary sedimentary sequences, there is very little information on the vegetation or fire history of the Early Pleistocene. The record of Martin (1990) suggests that there was a substantial reduction in precipitation at the beginning of the Pleistocene with a major expansion of open woodland and forest vegetation. This pattern is supported by evidence from the more refined and better dated early record of Lake George to the east (Kershaw *et al.* 1994) which shows the virtual disappearance of rainforest taxa and expansion of more open vegetation around 2.5 million years ago. Unfortunately there are no charcoal data for either record but the nature of the vegetation and the fact that fire had become a significant feature of the environment suggests that burning would have continued through the Pleistocene period. The only pollen and charcoal record that extends into the Early Pleistocene is from a marine core (ODP Site 820) off the coast of north-east Queensland adjacent to the Atherton Tableland (Kershaw *et al.* 1993) (Fig. 1.1). This record indicates that burning was regionally important from around 1.4 Ma, despite the pollen suggesting that the landscape was dominated by wet rainforest and drier araucarian rainforest, with restricted representation of sclerophyll vegetation through most of the recorded period. The extended record from Lake George (Singh and Geissler 1985) which, from the location of the Brunhes/Matuyama palaeomagnetic reversal, can be firmly dated to the base of the Mid Pleistocene (*c.* 800 ka BP) continues to show that burning was a feature of the environment. However, charcoal–vegetation relationships above this boundary appear to be different from those in the later part of the Pleistocene. Their elucidation is made difficult by discontinuous representation of pollen.

There are now 10 records, including those from Lynch's Crater and Lake George, which cover at least the last glacial/interglacial cycle. Their locations, along with summary charcoal records, are shown in Fig. 1.1 for all except Old Lake Coomboo Depression on Fraser Island (Longmore and Heijnis 1999) whose chronology is most uncertain. Some information on the nine selected records is included in the longer list of site records in Table 1.1. The charcoal summaries are averaged values for time slices selected to represent major identified past climatic phases apart from the most recent two slices which represent early and late phases of European occupation.

Time slices become shorter towards the present due to improved dating control and greater environmental understanding. However, it is only the more generalised Pleistocene record that will be considered in this section. Additional information on burning patterns is provided by the identification of major charcoal peaks in those records which are sufficiently refined to allow peak isolation (Fig. 1.4). Peaks during the Pleistocene are shown in relation to the SPECMAP marine oxygen isotope curve of Martinson *et al.* (1987) which provides a standard chronology for the late Quaternary. Positive $\delta^{18}O$ values relate to glacial periods when sea levels were low while negative $\delta^{18}O$ values indicate high sea-level, interglacial periods.

Almost all records show a general increase in charcoal abundance through time if climatic cycling at the Milankovitch scale is taken into account. In the Lynch's Crater, Banda Sea, Lake Wangoom and Egg Lagoon records, this increase is abrupt and takes place in the later part of isotope stage 3, centred on 30–40 ka BP. As has been noted, the charcoal increase at Lynch's Crater around 38 ka BP was accompanied by the beginning of the replacement of araucarian forest by eucalypt woodland and it has been hypothesised that increased burning was the cause of this sustained vegetation change. In the marine record of the Banda Sea, whose pollen is derived from both Indonesia and northern Australia, there is a marked reduction in eucalypts relative to grasses suggesting the development of a more open landscape. The impact of the increase in burning appears to have extended to Indonesia which experienced a marked and sustained reduction in Dipterocarpaceae, a canopy dominant of many Indonesian rainforest communities. Vegetation responses at the two other sites appear to have been minimal with some increase in grasses relative to daisies (Asteraceae) around Lake Wangoom within an open woodland landscape, and perhaps a sustained reduction in the extent of the rainforest and wet sclerophyll forest taxa *Phyllocladus* and *Pomaderris aspera* respectively with some epacrid heath development around Egg Lagoon.

The suggestion from Lake George that there was an earlier marked and sustained increase in burning with an associated alternation of the

Site Number	Site	Charcoal measure	Reference
1	Lombok Ridge	particles/cm³	Wang et al. (1999)
2	Banda Sea	particles/cm² per yr + mg/cm² per yr	van der Kaars et al. (2000)
3	ODP Site 820	particles/cm³	Moss (1999), Moss and Kershaw (2000)
4	Lynch's Crater	particles/cm³	Kershaw (1986)
5	Lake George	surface area%/unit volume of sediment	Singh et al. (1981), Singh and Geissler (1985)
6	Lake Wangoom	particles/cm³	Edney et al. (1990), Harle (1998)
7	Egg Lagoon	mm²/cm³	D'Costa (1997)
8	Lake Selina	not given	Colhoun et al. (1999)
9	Darwin Crater	not given	Colhoun and van de Geer (1988)
10	Burraga	cm.cm²	Dodson et al. (1994c)
11	Boggy Swamp	particles/cm² per yr	Dodson et al. (1986)
12	Butcher's Swamp	particles/cm² per yr	Dodson et al. (1986)
13	Black Swamp	particles/cm² per yr	Dodson et al. (1986)
14	Sapphire Swamp	particles/cm² per yr	Dodson et al. (1986)
15	Penrith Lakes	particle size abundance scale	Chalson (1989)
16	Warrimoo	particle size abundance scale	Chalson (1989)
17	King's Tableland	particle size abundance scale	Chalson (1989)
18	Ingar Swamp	particle size abundance scale	Chalson (1989)
19	Notts Swamp	particle size abundance scale	Chalson (1989)
20	Katoomba Swamp	particle size abundance scale	Chalson (1989)
21	Burralow Creek Swamp	particle size abundance scale	Chalson (1989)
22	Killalea Lagoon	charcoal/pollen ratio	Dodson et al. (1993)
23	Bondi Lake	charcoal/pollen ratio	Dodson et al. (1993)
24	Bega Swamp	mm²/ml	Green et al. (1988), G. Hope unpublished data
25	Rotten Swamp	cm²/cm² per yr	R. L. Clark (1986)
26	Club Lake	cm²/cm² per yr	Dodson et al. (1994a)
27	Tea Tree Swamp	charcoal/pollen ratio	Gell et al. (1993)
28	Lake Curlip	cm²/cm² per yr	Boon and Dodson (1992)
29	Hidden Swamp	particles/cm³	Hooley et al. (1980)
30	Loch Sport Swamp	particles/cm³	Hooley et al. (1980)
31	Lake Wellington	mm²/cm³	Reid (1989)
32	McKenzie Road Bog	charcoal/pollen ratio	Robertson (1986)
33	Greens Bush	mm²/cm³	Jenkins and Kershaw (1997)
34	Cranbourne Botanic Gardens	particles/cm³	Aitken and Kershaw (1993)

Table 1.1. (cont.)

Site Number	Site	Charcoal measure	Reference
35	Lake Mountain	charcoal/pollen ratio	McKenzie (1997)
36	Storm Creek	charcoal/pollen ratio	McKenzie (1997)
37	Tom Burns	charcoal/pollen ratio	McKenzie (1997)
38	Oaks Creek	charcoal/pollen ratio	McKenzie (1989)
39	Buxton	charcoal/pollen ratio	McKenzie (1989)
40	Powelltown	charcoal/pollen ratio	McKenzie (1989)
41	Lake Horden	particles/cm^3	Head and Stuart (1980)
42	Chapple Vale	charcoal/pollen ratio	McKenzie and Kershaw (1997)
43	Wyelangta	mm^2/cm^3	McKenzie and Kershaw (2000)
44	Aire Crossing	mm^2/cm^3	G. M. McKenzie unpublished data
45	West Basin	particles/cm^3	Gell et al. (1994)
46	Cobrico Swamp	mm^2/cm^3	Dodson et al. (1994b)
47	Lake Keilambete	cm^2/cm^3	Mooney (1997)
48	Northwest Crater, Tower Hill	particles/cm^3	D'Costa et al. (1989)
49	Main Lake, Tower Hill	particles/cm^3	D'Costa et al. (1989)
50	Lake Turangmoroke	charcoal/pollen ratio	Crowley and Kershaw (1994)
51	Jacka Lake	particles/cm^3	C. Greenwood unpublished data
52	Lake Tyrrell	mm^2/cm^3	Luly (1993)
53	Bridgewater Lake Core A	charcoal/pollen ratio	Head (1988)
54	Boomer Swamp	charcoal/pollen ratio	Head (1988)
55	Long Swamp	charcoal/pollen ratio	Head (1988)
56	Lashmar's Lagoon	particles/cm^3	Singh et al. (1981)
57	Lake Flannigan, King	mm^2/g	D'Costa (1997)
58	Killiecrankie, Flinders	charcoal/pollen ratio	Ladd et al. (1992)
59	Stockyard Swamp, Hunter	charcoal/pollen ratio	G. Hope unpublished data
60	Sundown Point, Hunter	charcoal/pollen ratio	G. Hope unpublished data
61	Lake Johnson	g	Anker (1991)
62	Poets Hill	charcoal/pollen ratio	Colhoun (1992)
63	Melaleuca Inlet	charcoal/pollen ratio	Thomas (1995)
64	Dublin Bog	charcoal/pollen ratio	Colhoun et al. (1991)
65	Den Plain 2	particles/cm^3	Moss (1994)
66	Ringarooma River	cm^2/cm^3	Dodson et al. (1998)
67	Big Heathy Swamp	charcoal/pollen ratio	Ellis and Thomas (1998)

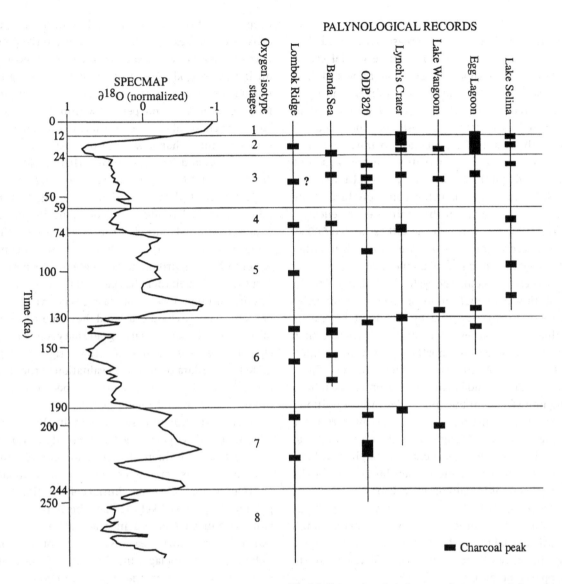

Fig. 1.4. Charcoal peaks in long continuous records from the Australian region in relation to the marine oxygen isotope (SPECMAP) record of Martinson *et al.* (1987).

vegetation receives some support from the ODP Site 820 where higher charcoal levels around 135 ka BP are closely associated with a sharp decline in *Araucaria* and some increase in the representation of *Eucalyptus*. Despite the proximity of Lynch's Crater to this site, there are no similar changes recorded at that time. However, the 38 ka BP event at Lynch's Crater appears to be recorded in the ODP site 820 record with a further decline in *Araucaria*

and further increase in *Eucalyptus* in association with increased charcoal levels. There are also sustained changes in vegetation around 175 ka BP, with a change in dominance from *Eucalyptus* to Poaceae in the northwestern Australian pollen component of the Lombok Ridge marine core and increased values for Poaceae and declines in ferns and palms within the ODP site 820 record, but neither has an associated charcoal event.

Darwin Crater (Tasmania) is the only record where there is a general decrease in averaged charcoal values through the last three glacial cycles, although there appears to be a major and sustained increase in herbs relative to rainforest and open-forest trees within what is estimated to be isotope stage 6. Jackson (1999) suggests that these changes were the result of an increase in burning which resulted in an ecological drift to a more open landscape and that the reduction in charcoal abundance was due to lesser quantities being produced from herbaceous vegetation than rainforest or eucalypt forest. However, through the full estimated 500 ka of the record, there are few overall trends in either the vegetation or charcoal abundance.

The effects of climatic cyclicity on charcoal representation are evident in most records where cycles are not totally obscured by trends of increasing values. Charcoal levels are higher in glacials than interglacials at all northern Australian sites, although values are extremely low and sporadic at Lynch's Crater, and have been interpreted as indicating more frequent burning under drier conditions. In southern Australia this pattern appears to be evident at Lake Wangoom and, surprisingly, at Darwin Crater, where less charcoal might have been expected under the more open landscapes of the glacial periods in light of the Jackson hypothesis. At Lake George, which exists under drier conditions than any of the other sites, lower charcoal values within glacial phases have been explained as due to the inability of the sparse cover of vegetation to support regular fires.

Another perspective on burning patterns is provided by the plots of charcoal peaks on Fig. 1.4 which may indicate times of high fire activity. There are concentrations of dates within the maxima of the last glaciation (stages 4 and 2) and the penultimate glacial (late stage 6), which could be interpreted as providing general support for the proposal that these drier periods were times of generally higher fire activity. However, it could be argued that the majority of these events are close to isotope stage boundaries and relate more to times of rapid climate change than extreme climatic conditions. This interpretation receives support from a high

proportion of the peaks which fall within interglacials. The major exception to this pattern is the presence of at least one charcoal peak within all records during the period 30–40 ka BP which, according to the marine isotope record, may have been one of the most climatically stable periods within the last 280 ka. Other charcoal peaks may relate to times of substantial climate change within isotope stages.

A somewhat different picture to that provided by Singh *et al.* (1981) on the pattern of burning and its impact on Australian vegetation is provided by this more substantial, though still geographically biased, data set. Limited data for the early part of the Quaternary indicate that fire has been a major component of the environment throughout this period and that, although there has been an increase in fire activity through at least the last glacial cycle with perhaps stepwise changes around the proposed dates of 130 and 40 ka BP, late Quaternary vegetation changes have been less dramatic than originally suggested. These data require a re-evaluation of the relative roles of climate and Aboriginal people in the production of fire and vegetation changes.

It is always difficult to separate totally the effects of climate and people on the fossil record as climate will have some influence on human activity. This influence is considered to be nowhere more apparent than in Australia where initial colonisation and possible impact is likely to have been related to climate-induced changes in sea-level and opportunities for short sea-crossings through the Indonesian Archipelago. In addition, any substantial human impact on the vegetation cover could lead to climate change as is evident with current greenhouse forcing. However, it seems unlikely that human impact could have been a significant factor in those trends and abrupt changes witnessed in the Lombok Ridge and ODP site 820 records that are evident as far back as 170 ka BP, well beyond the limit of the archaeological record. Furthermore, there is either no great correspondence between vegetation alterations and charcoal peaks or the initiation of vegetation change appears to have preceded charcoal peaks in earlier events, suggesting that fire may have been more a result than a cause of vegetation change. At Lake George, it was argued that the presence of people could not be detected until the beginning of the penultimate interglacial, some 130 ka BP, when the vegetation cover became sufficiently complete to carry fire. This argument cannot be extended to records from other sites situated within wetter environments. The general pattern of charcoal representation and/or inferred burning levels is that there has been strong climatic control over fire patterns with greater activity during drier glacials than interglacials (although the opposite may have been the case in environments similar to and drier than those around Lake George), with maximum burning during periods of substantial climate change.

The major exception to this pattern, at least in the Pleistocene, was during stage 3 when there were major sustained changes in the vegetation, particularly in northeastern and northern Australia and charcoal peaks in almost all other records at a time of relative climatic stability. The degree of synchroneity of these events is uncertain due to problems of dating so close to the limit of radiocarbon but they fall, almost certainly, within the period 45 to 30 ka BP. As people are known to have occupied most of the continent by this time, and as there is a close association between evidence for increased burning and vegetation change within the records, it might be fairly concluded that Aboriginal burning was the major causal factor and could well have impacted the whole of the Australian landscape, as well as some parts of Indonesia, within a short period of time.

The timing of presumed human impact is interesting in light of the archaeological debate over the actual timing of the arrival of people. Assuming that the colonisation process was rapid and impact was immediate, the data add support to the school of thought that accepts the validity of radiocarbon dates and that people arrived around 40 ka BP (Allen and Holdaway 1995) rather than the increasing number of sites dated by other methods including thermoluminescence (Roberts et al. 1990), optically stimulated luminescence (Roberts et al. 1993) and electron-spin resonance and uranium/thorium (Thorne et al. 1999) for a 50 to at least 60 ka BP arrival date. In fact there is a marked absence of charcoal peaks between 50 and 60 ka BP.

The data suggest that there has been a close relationship between climate and fire through the late Quaternary period although global climate, as reflected in the SPECMAP isotope curve, does not indicate any clear climatic trend through the last 280 ka. From about 40 ka BP, the directional change towards increased burning and more open vegetation can be attributed to an additional anthropogenic source of ignition but this appears to have resulted in an acceleration of an existing trend rather than the initiation of it. One climatic explanation for the trend is that the effectiveness of fire increased with a change to higher-amplitude glacial oscillations, which produced a great deal of vegetation instability within the 1 Ma to 700 ka (Shackleton et al. 1995). The effects of individual fires may have been cumulative through the process of ecological drift proposed by Jackson (1968). In his model, burning results in the repeated expansion of one community at the expense of another, in sequence, according to a shift in mean fire interval. If fires are sufficiently frequent or intense, then the vegetation change can be permanent. The process can be facilitated by the negative long-term effects of burning on the physical and chemical properties of soil, with alterations to soil moisture-holding capacity and fertility helping to maintain more open, sclerophyllous and hence more flammable vegetation. However, such a trend might be expected to have been global, which does not seem to have been the case. It is possible, however, that Australia has experienced a different late Quaternary climate history to much of the rest of the world. Isern et al. (1996) and Peerdeman et al. (1993) detected a change in oxygen isotope values of marine cores in the Coral Sea region some 500 to 250 ka BP which were interpreted as an average increase of about 4 °C in sea surface temperatures. Such an increase may be expected to have increased precipitation in northeastern Australia, due to increased evaporative power of the ocean, and perhaps cause a decrease rather than an increase in fire activity. However, if, as Isern et al. (1996) suggest, this temperature increase was related to the development of the West Pacific Warm Pool, then the production of a steep temperature gradient across the Pacific Ocean may have satisfied the prerequisites for the establishment of El Niño–Southern Oscillation (ENSO) variability, a major feature of Australia's climate. Even with higher precipitation, increased occurrence of droughts would have most likely allowed increased fire activity within the Australian landscape.

Holocene fire activity

A substantial number of palynological records with associated charcoal curves cover part or the whole of the Holocene period (i.e. the last 10 ka). The variability of charcoal representation within these records has led to the postulation of a whole range of suggestions about the relationship between fire, climate, various components of the vegetation and people. In order to determine and attempt to explain any consistent patterns of fire activity, a regional approach to data analysis has been adopted. Attention is focussed on southeastern Australia as this is where the majority of studies have been undertaken (Fig. 1.5).

From all available records, 58 were considered to have the resolution and dating control required for analysis of patterns in relation to the younger time slices identified in the longer records in the previous section. The location of these records, which in fact exclude the longer records from this region, are shown on Fig. 1.5 while some information about them is contained in Table 1.1. The 11–9 ka BP time slice represents the transition between the Pleistocene and the Holocene and covers a period of vegetation change in relation to increasing precipitation. The period 7–5 ka BP is identified as the Holocene precipitation peak, the time slice 4–2 ka BP incorporates a drier and perhaps cooler phase before a return to wetter conditions in the last 2 ka. The European time slices are identified on evidence for disturbance to the native vegetation and on weed pollen indicators. Chronological control is generally poor within the European period and although 1940, when fire-exclusion policies were introduced, was adopted as the nominal date for separation of older and younger European phases, only estimates of this date could be made for most records.

Because of differences in charcoal counting and presentation methods, and the variation in charcoal abundance related to the nature and size of the depositional basins from which the records were derived, it was considered that charcoal abundance measures could not be used to determine temporal and spatial patterns of fire activity. Instead, a simple ranking was undertaken initially of relative charcoal values in each time slice for each record. For example if a record covered all time slices the rankings would range from 6 for the time slice with the highest charcoal value to 1 with the lowest charcoal value, and if only the last three time slices were recorded (the minimum number in the data set) the rankings would range from 3 to 1. The average of the ranking values for time slices from all sites, and for subsets of sites surrounded by present-day wet forest (rainforest and tall open-forest), dry (open) forest, heath and woodland, were calculated and are shown on Fig. 1.6. Some sites were not included in the subsets either because the original vegetation cover was uncertain or the vegetation did not clearly fit into these groups. Conversely, a few sites which were surrounded by a mix of two community types were included in both relevant subsets. In addition to these summaries, times of identified charcoal peaks within any of the records from the data set, which may represent major fires or discrete times of intensive fire activity, are shown on Fig. 1.7.

The overall impression is that fire activity has been relatively constant over the Holocene period with greatest variation during the period of European occupation. The data suggest that burning increased during the early part of European settlement to levels higher than at any other time during the Holocene in all major vegetation types, a conclusion that was also reached by Singh *et al.* (1981). The increase was least within wet forest which may be due to contained areas being remote from major areas of land clearance and utilisation. This period was followed by a reduction in burning to present-day levels which are, on average, lower than at any time during the Holocene. However, levels in dry forest and heath environments appear still to be higher than those in the Early to Mid Holocene. The timings of the onset of high fire activity and the sub-

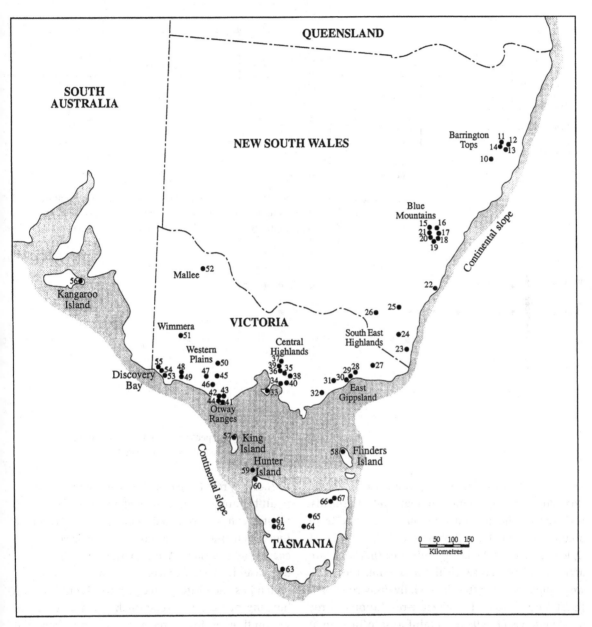

Fig. 1.5. Location of late Quaternary pollen and charcoal sites from south-eastern Australia examined in this chapter.

sequent transition from high to low levels of burning in the European occupation period are difficult to pin down due to a lack of absolute dating for most sites. Radiocarbon dates tend to be inaccurate or lack appropriate precision for these recent periods leaving ^{210}Pb dating as the only radiometric method which can be used, in addition to less

certain markers associated with historical events. Even with ^{210}Pb, accurate dating can only be achieved for the last 100 years and for certain records. It appears that the time of onset as well as the duration of high burning levels varied from place to place. Such variation is illustrated by fine resolution studies from McKenzie Road Bog, South Gippsland

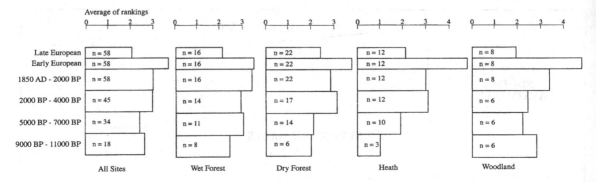

Fig. 1.6. Relative importance of burning in time slices from the Holocene deduced from individual site rankings of charcoal abundance for all site records, and separately for wet forest, dry forest, heath and woodland site records, in south-eastern Australia.

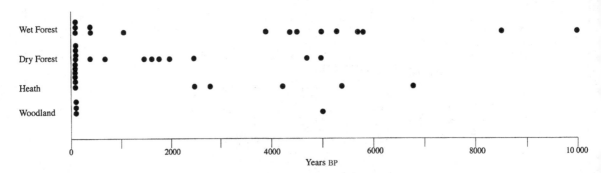

Fig. 1.7. Charcoal peaks recorded in the site records from south-eastern Australia shown in Fig. 1.5.

and Delegate River, East Gippsland. At McKenzie Road Bog, dated by detailed comparison with the historic record, the high burning phase occurred from about 1850 to 1890 (Robertson 1986) while at Delegate River, ^{210}Pb dating indicates that high fire activity did not cease until suppression measures were implemented after the socially disastrous 1939 fires (Gell *et al.* 1993). In some fine-resolution records (e.g. Dodson *et al.* 1998 from rainforest in Tasmania; C. Greenwood personal communication in the Victorian Wimmera) there is no evidence of an increase in burning since European arrival. The coarser-sampled records of the Barrington Tops region also show no increase in fire activity with the arrival of Europeans although, without absolute dating control, fire peaks considered to have occurred about 400 years ago might possibly represent early European burning.

The influence on the vegetation of alterations to

fire activity resulting from European activities is difficult to gauge as the bulk of changes within pollen records can be attributed to local or regional clearing which may or may not have involved fire. In coastal and western Victoria there was a general decrease in Casuarinaceae relative to *Eucalyptus* which has been interpreted as a result of European burning but no firm relationship can be detected within fine-resolution records. It is most likely that the casuarinas were selectively removed for firewood and other purposes. It has been suggested that disruption to the Aboriginal fire regime resulted in an increase in understorey shrubs relative to grasses within open-forest. However, the East Gippsland study of Gell *et al.* (1993) suggests exactly the opposite with an increase in grasses during the early European high burning phase and a decrease to present with the implementation of fire-control measures.

Within the pre-European part of the Holocene, there is a general increase in burning through time although there may have been a slight reduction in the Mid Holocene, between 7 and 5 ka BP, in response to an increase in rainfall and reduction in seasonality (Kershaw 1998). There was a significant increase in burning after 5 ka BP in all vegetation types, except the wet forests, and high fire activity was generally maintained or further increased after 2 ka BP. Increased burning is also suggested by the presence of charcoal peaks which centre on 5–4.5 ka and on 2 ka BP. The increase in burning may well have been a result of decreased precipitation after about 5 ka BP and high burning levels sustained by an onset or increase in activity of ENSO postulated for the region from this time (McGlone et al. 1992). A relationship with people may also be inferred from evidence of 'intensification' of occupation between 5 and 4 ka BP (Lourandos 1985). Although this concept has been hotly debated (see Head 1989; Dodson et al. 1992), a recent systematic study of the evidence for human occupation provides full support for both intensification and a high degree of synchroneity throughout Australia (Lourandos and David 2001). The relative importance of climate and human influence is difficult to assess but evidence of increased burning around this time from sites in New Zealand, where there is no suggestion of the presence of people, but where ENSO has a significant influence (McGlone et al. 1992; Ogden et al. 1998), suggests that climate was the major driving force.

The major burning response in the Mid Holocene was in heath and also open-forests which frequently contain a heath understorey. In general terms, this increase and the evidence for fire peaks around 5 and 2 ka BP provide regional support for the original Holocene fire record of R. L. Clark (1983) from such vegetation on Kangaroo Island in Singh et al. (1981). However, there is little support for the postulation of fire, as a result of Aboriginal burning, creating a replacement of Casuarinaceae woodlands by heath and eucalypt vegetation. Although Casuarina woodlands regionally declined, they did so at different times and generally without an obvious burning cause (Ladd 1988). In the Western Plains of Victoria,

their decline may have related to an increase in precipitation around 7 ka BP, while on coastal areas leaching of dune soils is likely to have favoured replacement by eucalypt and heath vegetation (Ladd et al. 1992). The development of eucalypt and heath vegetation would have favoured fire activity though, and this may help explain the more substantial increase in burning through the Holocene in these environments. The suggestion of R.L. Clark (1983) that a further increase in charcoal about 2.5 ka BP was the result of a change from regular low-intensity fires to less frequent high-intensity fires with the abandonment of Kangaroo Island by people is not supported by evidence from similarly 'abandoned' islands such as Flinders, King and Hunter. Burning did continue but at similar levels to those occurring previously.

General discussion and conclusions

The three scales of analysis of the pattern of burning on the Australian landscape, although limited in geographical spread and data type and consistency, have provided some quantitative basis for assessment of the history of fire in relation to environmental and vegetation change.

There is a general impression of an increase in fire activity through the late Cainozoic in association with drier and more climatically variable conditions and an increasingly fire 'tolerant' vegetation. The data, however, are insufficient to allow a determination of the degree to which fire has promoted the development and expansion of more open vegetation. Through most of the recorded period climate appears to have been the driving force of vegetation change with fire largely responding to vegetation changes. Although only crudely quantified, the Murray catchment records suggest that high levels of burning, in association with a wet sclerophyll forest, were achieved several million years ago, before the onset of the dramatic fluctuations in climate which have characterised the Quaternary and particularly the late Quaternary. The continuing drying trend has inhibited the preservation of a Quaternary palaeoecological record in this area, but it might be inferred that fire activity has been reduced with the establishment of a more open vegetation.

There were marked vegetation changes in mainland southeastern Australia around the Tertiary–Quaternary boundary with a substantial reduction in the extent of rainforest (Kershaw et al. 1994) but, despite the Murray catchment evidence for vegetation to attain and sustain high levels of fire activity, small patches of rainforest survived and managed to expand during brief, high-rainfall interglacial periods within the late Quaternary (D'Costa and Kershaw 1995; Harle et al. 1999). Even in Tasmania where it has been postulated that ecological drift, through vegetation feedbacks on fire and soil fertility, would have led to a progressive replacement of rainforest by more open vegetation (Jackson 1968), the net result on the vegetation around Darwin Crater over the last 500 ka has been a maintenance of the status quo.

In the humid tropics region of northeastern Australia there is little directional change in charcoal abundance or vegetation through much of the Quaternary period. There is a general increase in charcoal and a gradual replacement of the previously important araucarian forest by open sclerophyll forest and woodland during the later part of this period but the association between charcoal and vegetation change is not strong enough to demonstrate that fire was responsible for the initiation of this trend. It is likely that changes were climatically induced with droughts resulting from an increase in ENSO activity, superimposed on higher-amplitude Late Quaternary climatic cyclicity, being the critical features (Moss and Kershaw 2000). The evidence for some periodicity in ENSO activity from the Holocene records might help explain the stepwise decline of the araucarian forest.

The data suggest that the influence of human-induced burning on the development of the present vegetation cover might be much less than proposed by Singh et al. (1981) despite the fact that suggested times and patterns of charcoal increases have been recorded in additional records. The proposal of a human presence around Lake George by about 125 ka BP is not outlandish in light of sustained vegetation changes in other records at or before this

time, but, as mentioned, the recent evidence for altered climatic conditions in this part of the world provides, for the first time, an alternative explanatory mechanism. In addition, the earliest archaeological evidence for the arrival of people, although still debated in detail, has firmed for most of the continent at between 40 and 60 ka BP. On the other hand, evidence for some human impact, centred on about 40 ka BP has been strengthened by both archaeological and palaeoecological data. At this time there was no major global climate change and, unless the suggestion of Moss and Kershaw (1999) that there may have been a major phase of ENSO variability can be confirmed, a climatic cause, or predominantly climatic cause, can probably be discounted. In both the Lynch's Crater and ODP Site 820 records, the evidence indicates a response of vegetation to burning rather than a vegetation change resulting in higher fire activity. However, the ODP site 820 record demonstrates that the change represents an acceleration in an existing process rather than a singular event, while the degree of vegetation change proposed in other records for which there is a burning signal is much less than that in the climatically sensitive humid tropics region.

By Holocene time, people had been present for at least 30 ka and it might be expected that some balance had been achieved with the landscape. The data suggest that this was the case, with major vegetation changes and burning activity consistent with the inferred pattern of climate change. The relatively fire 'sensitive' species Allocasuarina verticillata (syn. Casuarina stricta) had survived to dominate the vegetation of much of coastal and inland Victoria during the Early Holocene and was then replaced by eucalypts and heath in the later Holocene most likely due to increased moisture levels and leached soils (or even salinity: Crowley 1994a, b) rather than fire as previously suggested by Singh et al. and others. There is also little support for alteration of fire regimes with the disappearance of people from continental islands during the Holocene. Fire activity has been higher in the later Holocene and, although it corresponds to an intensification of human occupation, is more parsimoniously explained by vegetation changes associated with drier and particularly more variable climatic conditions.

The presumed extinction of a species of Casuarinaceae in the Holocene of the Lake George record that dominated several interglacial periods prior to the penultimate interglacial was considered central to the argument for continued impact of human burning on the landscape. However, Crowley (1994a) suggested that increasing salinity levels may have resulted in the decline of Casuarinaceae and the there is always danger in assuming a similar vegetation response to different interglacial periods. Whitlock and Bartlein (1997) demonstrated the marked impact that different patterns of orbital solar forcing could have on vegetation from one interglacial to another and might explain the very different representation of Casuarinaceae, probably *Allocasuarina verticillata*, in the last three interglacials (Harle *et al.* 1999).

Acknowledgements

We thank Lesley Head and Helene Martin for very valuable comments on the manuscript and Sue Tomlins for drafting the text figures. Support for construction of long pollen and charcoal records in Australia was provided by Australian Research Council grants to APK.

References

Aitken, D., and Kershaw, A. P. (1993). Holocene vegetation and environmental history of Cranbourne Botanic Garden. *Proceedings of the Royal Society of Victoria* **105**, 67–80.

Allen, J., and Holdaway, S. (1995). The contamination of Pleistocene radiocarbon determinations in Australia. *Antiquity* **69**, 101–112.

Anker, S. A. (1991). Pollen and vegetation history from Lake Johnson, in western Tasmania. BSc. thesis, University of Newcastle.

Anderson, K. B., and Mackay, G. (1990). A review and re-interpretation of evidence concerning the origin of Victorian brown coal. *International Journal of Coal Geology* **16**, 327–347.

Blackburn, D. T. (1985). *Palaeobotany of the Yallourn and Morwell coal seams*, Palaeobotany project report 3. (State Electricity Commission: Victoria.)

Blackburn, D. T., and Sluiter, I. R. K. (1994). The Oligo-Miocene coal floras of southeastern Australia. In *The History of the Australian Vegetation: Cretaceous to Recent* (ed. R. S. Hill) pp. 328–367. (Cambridge University Press: Cambridge.)

Boon, S., and Dodson, J. R. (1992). Environmental response to land use at Lake Curlip, eastern Victoria. *Australian Geographical Studies* **30**, 206–221.

Bowman, D. M. J. S. (1998). Tansley Review no. 111. The impact of Aboriginal landscape burning on the Australian biota. *New Phytologist* **140**, 385–410.

Chalson, J. M. (1989). The Late Quaternary Vegetation and Climatic History of the Blue Mountains, N.S.W., Australia. PhD thesis, University of New South Wales, Sydney.

Clark, J. S. (1988). Particle motion and the theory of charcoal analysis: source area, transport, deposition, and sampling. *Quaternary Research* **30**, 81–91.

Clark, J. S., and Hussey, T. C. (1996). Estimating the mass of charcoal from sedimentary records: effects of particle size, morphology, and orientation. *The Holocene* **6**, 129–144.

Clark, R. L. (1982). Point count estimation of charcoal in pollen preparations and thin sections of sediments. *Pollen et Spores* **24**, 523–535.

Clark, R. L. (1983). Pollen and charcoal evidence for the effects of Aboriginal burning on the vegetation of Australia. *Archaeology in Oceania* **18**, 32–37.

Clark, R. L. (1986). The fire history of Rotten Swamp, ACT. Report to A.C.T. Parks and Conservation Service, Canberra.

Colhoun, E. A. (1992). Late Glacial and Holocene vegetation at Poets Hill Lake, Western Tasmania. *Australian Geographer* **23**, 11–23.

Colhoun, E. A., and van de Geer, G. (1988). Darwin Crater, the King and Linda Valleys. In *Cainozoic Vegetation of Tasmania*, Department of Geography Special Paper, pp. 30–71. (University of Newcastle: Newcastle.)

Colhoun, E. A, van de Geer, G., and Hannan, D. (1991). Late Glacial and Holocene vegetation history at Dublin Bog, North Central Tasmania. *Australian Geographical Studies* **29**, 337–354.

Colhoun, E. A., Pola, J. S., Barton, C. E., and Heijnis, H. (1999). Late Pleistocene vegetation and climate history of Lake Selina, western Tasmania. *Quaternary International* **57/58**, 5–23.

Crowley, G. M. (1994a). Groundwater rise, soil salinizations and the decline of *Casuarina* in southeastern Australia. *Australian Journal of Ecology* **19**, 417–424.

Crowley, G. M. (1994b). Quaternary soil salinity events and Australian vegetation history. *Quaternary Science Reviews* **13**, 15–22.

Crowley, G. M., and Kershaw, A. P. (1994). Late Quaternary environmental change and human impact around Lake Bolac, western Victoria, Australia. *Journal of Quaternary Science* **9**, 367–377.

D'Costa, D. M. (1997) The reconstruction of Quaternary vegetation and climate on King Island, Bass Strait, Australia. PhD thesis, Monash University, Melbourne.

D'Costa, D. M., and Kershaw, A. P. (1995). A late Quaternary pollen record from Lake Terang, Western Plains of Victoria, Australia. *Palaeogeography, Palaeoclimatology, Palaeoecology* **113**, 57–67.

D'Costa D. M., Edney, P. A., Kershaw, A. P., and De Deckker, P. (1989) Late Quaternary palaeoecology of Tower Hill, Victoria, Australia. *Journal of Biogeography* **16**, 461–482.

Dodson, J. R., Greenwood, P. W., and Jones, R. L. (1986). Holocene forest and wetland vegetation dynamics at Barrington Tops, New South Wales. *Journal of Biogeography* **13**, 561–585.

Dodson, J., Fullagar, R., and Head, L. (1992). Dynamics of environment and people in the forested crescents of temperate Australia. In *The Naive Lands: Prehistory and Environmental Change in Australia and the Southwest Pacific* (ed. J. Dodson) pp. 115–159. (Longman Cheshire: Melbourne.)

Dodson, J. R., McRae, V. M., Molloy, K., Roberts, F., and Smith, J. D. (1993). Late Holocene human impact on two coastal environments in NSW, Australia: a comparison of Aboriginal and European impacts. *Vegetation History and Archaeobotany* **2**, 89–100.

Dodson, J. R., de Salis, T., Myers, C. A., and Sharp, A. J. (1994a). A thousand years of environmental change and human impact in the alpine zone at Mt. Kosciusko, New South Wales. *Australian Geographer* **25**, 77–87.

Dodson, J. R., Frank, K., Fromme, M., Hickson, D., McRae, V., Mooney, S., and Smith, J. D. (1994b). Environmental systems and human impact at Cobrico Crater, south-

western Victoria. *Australian Geographical Studies* **32**, 27–40.

Dodson, J. R., Roberts, F. K., and de Salis, T. (1994c). Palaeoenvironments and human impact at Burraga Swamp in montane rainforest Barrington Tops National Park, New South Wales, Australia. *Australian Geographer* **25**, 161–169.

Dodson, J. R., Mitchell, F. J. G., Bögeholz, H., and Julian, N. (1998). Dynamics of temperate rainforest from fine resolution pollen analysis, Upper Ringarooma River, northeastern Tasmania. *Australian Journal of Ecology* **23**, 550–561.

Edney, P. A., Kershaw, A. P., and De Deckker, P. (1990). A Late Pleistocene and Holocene vegetation and environmental record from Lake Wangoom, Western Plains of Victoria, Australia. *Palaeogeography, Palaeoclimatology, Palaeoecology* **80**, 325–343.

Ellis, R. C., and Thomas, I. (1998). Pre-settlement and post-settlement vegetational change and probable Aboriginal influences in a highland forested area in Tasmania. In *Australia's Ever-Changing Forests*, Special Publication no.1 (eds. K. J. Frawley and N. Semple) pp. 63–86. (Department of Geography and Oceanography, Australian Defence Force Academy: Canberra.)

Flood, J. (1995). *Archaeology of the Dreamtime*, revd edn. (Angus and Robertson: Sydney.)

Gell, P., Stuart, I.-M., and Smith, D.J. (1993). The response of vegetation to changing fire regimes and human activity in East Gippsland, Victoria, Australia. *The Holocene* **3**, 150–160.

Gell, P. A., Barker, P. A., De Deckker, P., Last, W., and Jelicic, L. (1994). The Holocene history of West Basin Lake, Victoria, Australia: chemical changes based on fossil biota and sediment mineralogy. *Journal of Palaeolimnology* **12**, 235–258.

Gill, A. M., and Catling, P. C. (2001). Fire regimes and biodiversity of forested landscapes of southern Australia. In *Flammable Australia: The Fire Regimes and Biodiversity of a Continent* (eds. R. A. Bradstock, J. E. Williams and A. M. Gill) pp. 351–369. (Cambridge University Press: Cambridge.)

Gill, A. M., Groves, R. H., and Noble, I. R. (eds.) (1981). *Fire and the Australian Biota*. (Australian Academy of Science: Canberra.)

Green, D., Singh, G., Polach, H., Moss, D., Banks, J., and Geissler, E. A. (1988). A fine resolution palaeoecology and palaeoclimatology record from south-eastern Australia. *Journal of Ecology* **76**, 790–806.

Haq, B. U., Hardenbol, J., and Vail, P. R. (1987). Chronology of fluctuating sea levels since the Triassic. *Science* **235**, 1156–1167.

Harle, K. J. (1998). Patterns of vegetation and climate change in southwest Victoria over approximately the last 200 000 years. PhD thesis, Monash University, Melbourne.

Harle, K. J., Kershaw, A. P., and Heijnis, H. (1999). The contributions of uranium/thorium and marine palynology to the dating of the Lake Wangoom pollen record, Western Plains of Victoria, Australia. *Quaternary International* **57/58**, 25–34.

Head, L. (1988). Holocene vegetation, fire and environmental history of the Discovery Bay region, south-western Victoria. *Australian Journal of Ecology* **13**, 21–49.

Head, L. (1989). Prehistoric Aboriginal impacts on Australian vegetation: an assessment of the evidence. *Australian Geographer* **20**, 37–46.

Head, L., and Stuart, I.-M. F. (1980) Change in the Aire. Palaeoecology and Prehistory in the Aire Basin, southwestern Victoria. *Monash Publications in Geography* **24**, 1–102.

Hill, R. S. (1994). The history of selected Australian taxa. In *The History of the Australian Vegetation: Cretaceous to Recent* (ed. R. S. Hill) pp. 390–419. (Cambridge University Press: Cambridge.)

Holdgate, G. R., Kershaw, A. P., and Sluiter, I. R. K. (1995). Sequence stratigraphic analysis and the origins of Tertiary brown coal lithotypes, Latrobe Valley, Gippsland Basin, Australia. *International Journal of Coal Geology* **28**, 249–275.

Hooley, A. D., Southern, W., and Kershaw, A. P. (1980). Holocene vegetation and environments of Sperm Whale Head, Victoria, Australia. *Journal of Biogeography* **7**, 349–362.

Horton, D. R. (1982). The burning question: Aborigines, fire and Australian ecosystems. *Mankind* **13**, 237–251.

Isern, A. R., McKenzie, J. A., and Feary, D. A. (1996). The role of sea-surface temperature as a control on carbonate platform development in the western Coral Sea. *Palaeogeography, Palaeoclimatology, Palaeoecology* **124**, 247–272.

Jackson, W. D. (1968). Fire, air, water and earth – an elemental ecology of Tasmania. *Proceedings of the Ecological Society of Australia* **3**, 9–16.

Jackson, W. D. (1999). The Tasmanian legacy of man and fire. *Papers and Proceedings of the Royal Society of Tasmania* **133**, 1–14.

Jenkins, M. A., and Kershaw, A. P. (1997). A mid-late Holocene vegetation record from an interdunal swamp, Mornington Peninsula, Victoria. *Proceedings of the Royal Society of Victoria* **109**, 133–148.

Kemp, E. M. (1981). Pre-Quaternary fire in Australia. In *Fire and the Australian Biota* (eds. A. M. Gill, R. H. Groves and I. R. Noble) pp. 1–22. (Australian Academy of Science: Canberra.)

Kershaw, A. P. (1986). The last two glacial–interglacial cycles from northeastern Australia: implications for climate change and Aboriginal burning. *Nature* **322**, 47–49.

Kershaw, A. P. (1998). Estimates of regional climatic variation within southeastern mainland Australia since the Last Glacial Maximum from pollen data. *Palaeoclimates: Data and Modelling* **3**, 107–134.

Kershaw, A. P., Bolger, P., Sluiter, I. R. K., Baird, J., and Whitelaw, M. (1991). The origin and evolution of brown coal lithotypes in the Latrobe Valley, Victoria, Australia. *Journal of Coal Geology* **18**, 233–249.

Kershaw, A. P., McKenzie, G. M., and McMinn, A. (1993). A Quaternary vegetation history of northeastern Queensland from pollen analysis of ODP site 820. *Proceedings of the Ocean Drilling Program, Scientific Results* **133**, 107–114.

Kershaw, A. P., Martin, H. A., and McEwen Mason, J. R. C. (1994). The Neogene: a period of transition. In *The History of the Australian Vegetation: Cretaceous to Recent* (ed. R. S. Hill) pp. 299–327. (Cambridge University Press: Cambridge.)

Ladd, P. G. (1988). The status of Casuarinaceae in Australian forests. In *Australia's Ever-Changing Forests*. Special Publication no.1. (eds. K. J. Frawley and N. Semple) pp. 63–86. (Department of Geography and Oceanography, Australian Defence Force Academy: Canberra.)

Ladd, P. G., Orchiston, D. W., and Joyce, E. B. (1992). Holocene vegetation history of Flinders Island. *New Phytologist* **122**, 757–767.

Longmore, M. E., and Heijnis, H. (1999). Aridity in Australia: Pleistocene records of palaeohydrological change from the perched Lake sediments of Fraser Island, Queensland, Australia. *Quaternary International* **57/58**, 35–47.

Lourandos, H. (1985). Intensification and Australian prehistory. In *Prehistoric Hunter–Gatherers: The Emergence of Cultural Complexity* (eds. T. D. Price and J. A. Brown) pp. 385–423. (Academic Press: Orlando.)

Lourandos, H., and David, B. (2001). Long term archaeological and environmental trends: a comparison from Late Pleistocene–Holocene Australia. In *The Environmental and Cultural History and Dynamics of the Southeast Asian–Australasian Region* (eds. A. P. Kershaw, N. J. Tapper, B. David, P. Bishop and D. Penny) pp. 00–00. (Catena Verlag: Cremlingen, Germany.) (in press)

Luly, J. (1993). Holocene palaeoenvironments near Lake Tyrrell, semi-arid northwestern Victoria, Australia. *Journal of Biogeography* **20**, 587–598.

Luly, J., Sluiter, I. R. K., and Kershaw, A. P. (1980). Pollen studies of Tertiary brown coals: preliminary analyses of lithotypes within the Latrobe Valley, Victoria. *Monash Publications in Geography* **23**, 1–78.

Martin, H. A. (1990). Tertiary climate and phytogeography in southeastern Australia. *Review of Palaeobotany and Palynology* **65**, 47–55.

Martin, H. A. (1991). Tertiary stratigraphic palynology and palaeoclimate of the inland river systems in New South Wales. In *The Cenozoic of Australia: A Reappraisal of the Evidence*, Special Publication no. **18** (eds. M. A. J. Williams, P. De Deckker and A. P. Kershaw) pp. 181–194. (Geological Society of Australia: Sydney.)

Martinson, D. G., Pisias, N. G., Hays, J. D., Imbrie, J., Moore, T. C., and Shackleton, N. J. (1987). Age dating and orbital theory of the Ice Ages: development of a high resolution 0 to 300 000-year chronology. *Quaternary Research* **27**, 1–29.

Mackay, G. H., Attwood, D. H., Gaulton, R. J.,and George, A. M. (1995). The cyclical occurrence of brown coal lithotypes. State Electricity Commission of Victoria, Report no. SO/85/93, Victoria.

McGlone, M. S., Kershaw, A. P., and Markgraf, V. (1992). El Niño/Southern Oscillation climatic variability in Australasian and South American palaeoenvironmental records. In *El Niño: Historical and Palaeoclimatic Aspects of the Southern Oscillation* (eds. H. F. Diaz and V. Markgraf) pp. 435–462. (Cambridge University Press: Cambridge.)

McKenzie, G. M. (1989). Late Quaternary vegetation and climate in the central Highlands of Victoria with special reference to *Nothofagus cunninghamii* (Hook.) Oerst. rainforest. PhD thesis, Monash University, Melbourne.

McKenzie, G. M. (1997). The late Quaternary vegetation history of the south–central highlands of Victoria, Australia. I. Sites above 900m. *Australian Journal of Ecology* **22**, 19–36.

McKenzie, G. M., and Kershaw, A. P. (1997). A vegetation history and quantitative estimate of Holocene climate from Chapple Vale, in the Otway Region of Victoria, Australia. *Australian Journal of Botany* **45**, 565–581.

McKenzie, G. M., and Kershaw, A. P. (2000). The last glacial cycle from Wyelangta, the Otway region of Victoria, Australia. *Palaeogeography, Palaeoclimatology, Palaeoecology* **155**, 177–193.

Mooney, S. (1997). A fine resolution palaeoclimatic reconstruction of the last 2000 years, from Lake Keilambete, southeastern Australia. *The Holocene* **7**, 139–149.

Moss, P. T. (1994). Late Holocene environments of Den Plain, north-western Tasmania. BSc thesis, University of Melbourne, Melbourne.

Moss, P. T. (1999). Late Quaternary environments of the humid tropics of northeastern Australia. PhD. thesis, Monash Univeristy, Melbourne.

Moss, P. T., and Kershaw, A. P. (1999). Evidence from marine ODP Site 820 of fire/vegetation/climate patterns in the humid tropics of Australia over the last 250,000 years. In *Bushfire 99, Proceedings of the Australian Bushfire Conference*, Albury, Australia, July 1999, pp. 269–279.

Moss, P. T., and Kershaw, A. P. (2000). The last glacial cycle from the humid tropics of northeastern Australia: comparison of a terrestrial and a marine record. *Palaeogeography, Palaeoclimatology, Palaeoecology* **155**, 155–176.

Ogden, J., Basher, L., and McGlone, M. (1998). Fire, forest regeneration and links with early human habitation: evidence from New Zealand. *Annals of Botany* **81**, 687–696.

Peerdeman, F. M., Davies, P. J., and Chivas, A. R. (1993). The stable oxygen isotope signal in shallow-water, upper-slope sediments off the Great Barrier Reef (Hole 820A). *Proceedings of the Ocean Drilling Program, Scientific Results* **133**, 163–173.

Reid, M. (1989). Palaeoecological changes at Lake Wellington, Gippsland Lakes, Victoria, during the late Holocene: a study of the development of a coastal ecosystem. BSc. thesis, Monash University, Melbourne.

Roberts, R. G., Jones, R., and Smith, M. A. (1990). Thermoluminesence dating of a 50 000-year-old human occupation site in northern Australia. *Nature* **345**, 153–156.

Roberts, R. G., Jones, R., and Smith, M.A. (1993). Optical dating of Deaf Adder Gorge, Northern Territory indicates human occupation between 53 000 and 60 000 years ago. *Australian Archaeology* **37**, 58–59.

Robertson, M. (1986). Fire regimes and vegetation dynamics: a case study examining the potential of fine resolution palaeoecological techniques. BA thesis, Monash University, Melbourne.

Shackleton, N. J., Crowhurst, S., Hagelberg, T., Pisias, N., and Schneider, D. A. (1995). A new late Neogene timescale: applications to leg 138 sites. *Proceedings of Ocean Drilling Program, Scientific Results* **138**, 73–101.

Singh, G., and Geissler, E. A. (1985). Late Cainozoic history of fire, lake levels and climate at Lake George, New South Wales, Australia. *Philosophical Transactions of the Royal Society of London* **311**, 379–447.

Singh, G., Kershaw, A. P., and Clark, R. (1981).Quaternary vegetation and fire history in Australia. In *Fire and the Australian Biota* (eds. A .M. Gill, R. H. Groves and I. R. Noble) pp. 23–54 (Australian Academy of Science: Canberra.)

Sluiter, I. R. K., Kershaw, A. P., Holdgate, G. R., and Bulman, D. (1995). Biogeographic, ecological and stratigraphic relationships of Miocene brown coal floras, Latrobe Valley, Victoria, Australia. *International Journal of Coal Geology* **28**, 277–302.

Thomas, I. (1995). Where have all the forests gone? New pollen evidence from Melaleuca Inlet in southwestern Tasmania. *Monash Publications in Geography* **45**, 295–301.

Thorne, A., Grun, R., Mortimer, G., Spooner, N., Simpson, J. J., McCulloch, M., Taylor, L., and Curnoe, D. (1999). Australia's oldest human remains: age of the Lake Mungo 3 skeleton. *Journal of Human Evolution* **36**, 591–612.

van der Kaars, S., Wang, X., Kershaw, A. P., and Guichard, F. (2000). A late Quaternary palaeoecological record from the Banda Sea, Indonesia: patterns of vegetation, climate and biomass burning in Indonesia and northern Australia. *Palaeogeography, Palaeoclimatology, Palaeoecology* **155**, 135–153.

Wang, X., van der Kaars, S., Kershaw, A. P., and Bird, M. (1999). A record of fire, vegetation and climate through the last three glacial cycles from Lombok Ridge core G6–4, eastern Indian Ocean, Indonesia. *Palaeogeography, Palaeoclimatology, Palaeoecology* **147**, 241–256.

Whitlock, C., and Bartlein, P. J. (1997). Vegetation and climate change in northwest America during the past 125 kyr. *Nature* **388**, 57–61.

Wright, R. (1986). How old is zone F at Lake George? *Archaeology in Oceania* **21**, 138–139.

2

Importance of a changing climate for fire regimes in Australia

GEOFFREY J. CARY

Abstract

In recent times there has been considerable speculation about the sensitivity of fire regimes to the change in climate that may result from global warming. However, most of the discussion to date has not involved rigorous mechanistic analysis. This paper reviews the current state of our understanding of the effects that climate change may have on fire regimes in Australia and presents a new analysis for southeastern Australia. Climate-change scenarios for Australia are discussed and previous work on the effects of climate change on spatial patterns of annually summed Forest Fire Danger Index is reviewed. Various attempts at process-based modelling of fire regimes and the effects of climate change are then discussed before a detailed case study is presented for the Australian Capital Territory region, using output from the CSIRO DARLAM regional climate model and the FIRESCAPE fire regime model. The effect of a $2 \times CO_2$ climate was to increase the overall frequency of fire but not to alter the spatial pattern of fire frequency in the model system. More fires occurred in autumn under the changed climate compared with that modelled for the current climate. It is proposed that the increase in fire frequency is more likely to be a function of a reduction in the probability of fire extinguishment rather than significant increases in fire intensity at the other end of the fire behaviour spectrum. Finally, some consequences of predicted changes to fire regimes for biological diversity are briefly discussed.

Introduction

Ambient and antecedent meteorological conditions have important implications for the behaviour of forest and grassland fires (McArthur 1966a, 1967; Rothermel 1972; Luke and McArthur 1978; Sneeuwjagt and Peet 1985; Cheney *et al.* 1993, 1998). Temperature and relative humidity affect the moisture content of fine fuel, as does rainfall, while drought affects the moisture content of coarser fuels, deep litter beds and live vegetation (McArthur 1966b; Mount 1972; Burrows 1987). Wind speed, amongst other factors, determines flame angle and influences spotting distance, thereby also influencing the rate of spread of a fire, as well as fire shape (Peet 1967).

There is a close relationship between fire behaviour and the patterns of fire regime (defined by Gill 1975) that may characterise a particular site within a landscape. This is because a fire regime is a function of the many individual fire events that affect the site. The frequency and size of the individual fires that determine the fire regime are a function of ignition frequency and the rate at which fires spread across a landscape. For a constant spatial distribution and frequency of ignitions, the frequency of fire in sites across a landscape will be low if fires spread slowly, extinguish easily and are small. On the other hand, the frequency of fire will increase for the same landscape with the same spatial distribution and frequency of ignition if fires spread faster and burn more of the landscape before extinguishing. Other characteristics of the fire regime, particularly fire intensity, may also differ between the two scenarios.

These ideas are explored in this chapter. Initially, a brief overview of the science of climate change presents the range of greenhouse gas emissions and global warming scenarios for Australia. Then, various climate modelling approaches and current scenarios for climate change are investigated. The potential effects of these on distributions of fire danger, fire behaviour and fire regimes, primarily of forest and woodland systems, are reviewed from published studies and analysed using existing and new methodologies. Other effects of climate change, including potential changes to lightning-ignition patterns are also briefly addressed, whereas the importance of changed levels of biomass production (Pittock and Nix 1986) and decomposition are not.

No attempt is made to predict how a changing climate may affect fire regimes at a continental level because there is still a great deal of uncertainty associated with the scenarios of climate change and with aspects of generating fire regimes using process-based models. Rather, the concepts outlined provide a review of different methodologies that others may adopt as confidence in climate and fire modelling increases as a result of further research effort.

Climate change scenarios for Australia

The energy budget of the earth depends on the balance between incoming solar radiation and outgoing longwave radiation. About 30% of the energy from incoming solar radiation is scattered or reflected back into space by airborne particles, clouds or the earth's surface, while the remaining radiation results in warming of the earth's surface and atmosphere (Houghton et al. 1994). The energy balance of the earth is maintained by the emission of thermal longwave radiation, in the infrared part of the spectrum, from both the surface and the atmosphere (Trenberth et al. 1996).

Around 90% of the longwave radiation that is emitted near the surface of the earth is absorbed by the atmosphere, which in turn emits radiation either into space or back to the earth's surface. This occurs because of naturally occurring gases includ-ing water vapour, carbon dioxide (CO_2), ozone (O_3), methane (CH_4) and nitrous oxide (N_2O) in the atmosphere and because of clouds, all of which are capable of absorbing and emitting thermal radiation which originates from the earth's surface (Houghton et al. 1994). Therefore, the surface of the earth is warmer than it would be without the insulating effects of these gases. This is known as the natural greenhouse effect, and CO_2 and the other minor gases are referred to as greenhouse gases. Oxygen and nitrogen, which comprise about 99% of the atmosphere, are transparent to infrared radiation (Trenberth et al. 1996).

The enhanced greenhouse effect arises from an increase in the atmospheric concentration of greenhouse gases, particularly CO_2, which has resulted from the burning of fossil fuels and from other anthropogenic activities. The atmospheric concentration of CO_2 has increased from 280 ppm to about 356 ppm since the industrial revolution (Houghton et al. 1994; Manning et al. 1996) and is expected to double the pre-industrial revolution concentration within the next 50 to 100 years (Trenberth et al. 1996). This will tend to force an increase in global surface temperatures because it will result in a reduction of outgoing radiation which means the surface and troposphere must warm to restore a radiative balance (Houghton et al. 1994). The overall increase in global mean temperature, known as the climate sensitivity, that would arise from instantaneously doubling the atmospheric CO_2 concentration and allowing the climate to come to a new equilibrium is between 1.5 °C and 4.5 °C (CSIRO 1996).

Aerosol emissions are also important, largely because they have a negative radiative forcing effect on the earth's energy balance since they can scatter solar radiation back into space. They also act as nuclei on which cloud droplets form and hence they influence the formation, lifetime and radiative properties of clouds (Houghton et al. 1994; Trenberth et al. 1996).

The Intergovernmental Panel on Climate Change (IPCC) has produced six greenhouse gas emission scenarios (Leggett et al. 1992) (Fig. 2.1). These scenarios are non-interventionist in that they assume no

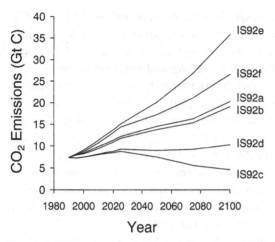

Fig. 2.1. IPCC IS92 scenarios for CO_2 emissions. From CSIRO (1996). (Figure reproduced by permission of CSIRO.)

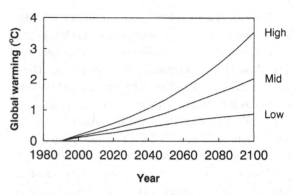

Fig. 2.2. IPCC (Houghton *et al.* 1996) global warming scenarios: high (emission scenario IS92e and climate sensitivity of 4.5 °C); mid (emission scenario IS92a and climate sensitivity of 2.5 °C); and low (emission scenario IS92c and climate sensitivity of 1.5 °C). From CSIRO (1996). (Figure reproduced by permission of CSIRO.)

governmental policy intervention of greenhouse-gas emissions, although scenario IS92c has emission levels and some input assumptions that are more characteristic of an interventionist rather than a non-interventionist scenario (Alcamo *et al.* 1994). IS92a represents a moderate scenario with global population reaching 11.3 billion by 2100, economic growth averaging 2.3% per annum between 1990 and 2100 and a mix of non-renewable and renewable energy sources being used. IS92e represents moderate population growth, high economic growth, high fossil fuel availability and the eventual phase-out of nuclear power. IS92c assumes that population first rises then declines by the middle of next century, that economic growth is low and that there are severe constraints on fossil fuel use.

IPCC have produced global warming scenarios for 1990 to 2100 based on the ranges of greenhouse gas and aerosol emission scenarios and climate sensitivities (Houghton *et al.* 1996). They range from an increase in average global temperature of 0.8 °C to 4.5 °C, depending on the greenhouse gas emission scenario and climate sensitivity adopted. Figure 2.2 outlines the global warming scenarios which assume changing atmospheric aerosol concentrations beyond 1990. A similar set of scenarios with aerosol concentrations maintained at their 1990 levels was also produced (Houghton *et al.* 1996).

Using this information, the CSIRO Climate

Change Research Program has produced regional climate change scenarios for Australia which indicate the expected change in temperature and precipitation, per degree of global warming, using a number of General Circulation Models (GCMs). GCMs are the most powerful means of understanding the earth's present climate and for making predictions about past or future climates (Gates *et al.* 1996; Kattenberg *et al.* 1996), including the climates that might exist under the influence of increased greenhouse gases and aerosols.

The CSIRO (1996) regional climate scenarios were generated using experiments from two types of GCMs. Coupled atmosphere–ocean GCMs, which link full atmospheric and ocean models, were used for transient simulations where the amount of CO_2 was gradually increased to produce the temperature scenarios and one set of precipitation scenarios. Older slab–ocean GCMs, which do not have a representation of the deep ocean, were used for equilibrium simulations where the CO_2 concentration was instantaneously doubled and the climate allowed to come to a new equilibrium. They were used to produce a second set of precipitation scenarios that are less preferable on theoretical grounds but are retained by CSIRO (1996) because of the differences in the scenarios compared with the coupled GCMs. Also, there was uncertainty over the role of the ocean in producing some aspects of the

Table 2.1. *Scenarios of temperature change for regions within Australia*

Region	Local warming per degree of global warming (°C)	Warming in 2030 (°C)	Warming in 2070 (°C)
Northern Coast (north of 25° S)	0.9 to 1.3	0.3 to 1.0	0.6 to 2.7
Southern Coast (south of 25° S)	0.8 to 1.6	0.3 to 1.3	0.6 to 3.4
Inland	1.0 to 1.8	0.4 to 1.4	0.7 to 3.8

Source: CSIRO (1996).

results from coupled models and some conflicts between the results of coupled models and the observed warming trends during the 20th century (CSIRO 1996). Details of the GCMs can be found in Basher *et al.* (1998) and Watson *et al.* (1998). Whetton *et al.* (1996) also produced a set of climate change scenarios for Australia and New Zealand using five slab–ocean GCMs.

The range in the local warming per degree of global warming (Table 2.1) arises from differences between GCMs and the range in global warming scenarios (Fig. 2.2) (CSIRO 1996). Local warming scenarios for 2070 range from a 0.6 °C to 3.8 °C rise. The precipitation scenarios also display significant ranges with considerable differences between those from different model types (Tables 2.2 and 2.3). For example, the range for eastern Australia during summer is −10% to +10% by 2070 for the coupled models and 0% to +20% for the slab–ocean models. Similarly, for all but the southwest of Western Australia, the range for the remainder of the continent is −20% to 0% for the coupled models and +4% to +30% for the slab–ocean models. For the southwest of Western Australia, the comparison is −10% to +10% for the coupled models and +4% to +30% for the slab–ocean models.

There are three important points that can be drawn from the climate change scenarios of CSIRO (1996) and others (e.g. Whetton *et al.* 1996). Firstly, the models predict that there is the potential for the climate in all areas of Australia to change sufficiently to modify average fire behaviour. By 2070, average temperatures may rise by as much as between 2.7 °C and 3.6 °C and precipitation may

change by −20% to +30%. Alternatively, changes may be much smaller. Nevertheless, the potential for significant change remains and this provides the imperative for the type of analyses outlined in this chapter.

Secondly, there are aspects of weather that affect fire behaviour that are not commonly included in published climate change scenarios like those of CSIRO (1996). For example, Beer *et al.* (1988) identified the particular sensitivity of the Forest Fire Danger Index (FFDI) (McArthur 1967; Noble *et al.* 1980) to small changes in relative humidity. Wind speed and direction are also fundamental to fire behaviour and the pattern of burning in a landscape.

Thirdly, the predictions contain significant variations. The ranges in predictions from different models, and types of models, for regions within Australia provide clear evidence that at present there is a great deal of uncertainty associated with predictions of future climates. This is recognised by climate modellers who use the term 'scenario' instead of 'prediction'.

The confidence in the predictions is further decreased with some insights into the evaluation of the individual models. Giorgi *et al.* (1998) analysed the bias (difference between climate simulated under a $1 \times CO_2$ (control) scenario and observed climate) for nine different atmosphere–ocean GCMs using transient simulations for the Australian region. Their analysis included some of the models used to produce the CSIRO (1996) climate change scenarios for Australia. According to Giorgi *et al.* (1998), the biases can be used to assess the models

Table 2.2. *Scenarios of precipitation change in (a) winter and (b) summer for regions within Australia based on coupled GCMs*

Location	Response per degree of global warming	Change in 2030	Change in 2070	
(a) Winter				
Region A	−10% to 0%	−8% to 0%	−20% to 0%	
Region B	−5% to +5%	−4% to +4%	−10% to +10%	
Region C	0% to +10%	0% to +8%	0% to +20%	
(b) Summer				
Region A	−10% to 0%	−8% to 0%	−20% to 0%	
Region B	−5% to +5%	−4% to +4%	−10% to +10%	

Source: CSIRO (1996). Reproduced by permission of CSIRO.

Table 2.3. *Scenarios of precipitation change in (a) winter and (b) summer for regions within Australia based on slab–ocean GCMs*

Location	Response per degree of global warming	Change in 2030	Change in 2070	
(a) Winter				
Region A	−5% to 0%	−4% to 0%	−10% to 0%	
Region B	−2.5% to +2.5%	−2% to +2%	−5% to +5%	
Region C	0% to +5%	0% to +4%	0% to +10%	
(b) Summer				
Region A	+5% to +15%	+2% to +12%	+4% to +30%	
Region B	0% to +10%	0% to +8%	0% to +20%	

Source: CSIRO (1996). Reproduced by permission of CSIRO.

because 'it can generally be expected that the better the match between the control run and the observed climate, the higher the confidence in simulated change scenarios'. A lower bias translates into a better match.

The bias associated with the model of Gordon and O'Farrell (1997) is representative of the findings of Giorgi *et al.* (1998). Bias for temperature was smaller in the Australian region than for any region in the world, although it was around 80% of the predicted temperature changes between a $1 \times CO_2$ and a $2 \times CO_2$ simulation. This means that errors in prediction are in the same order of the predicted changes. The summer and winter precipitation biases were around 4.5 to 6 times greater than the predicted changes for a doubled CO_2 climate.

Biases for predictions around the world are too large to warrant a high level of confidence in simulated climate change scenarios (Giorgi *et al.* 1998). Nevertheless, unless there are systematic errors that are shared by all GCMs, the fact that they largely agree on positive changes to summer and winter temperatures indicates that there is some confidence that can be placed in aspects of model prediction. Increased resolution of GCMs, used in combination with recently developed high-resolution Regional Climate Models (RegCMs) and long (>10 year) simulation runs, are showing an increased level of accuracy (Girorgi *et al.* 1998). This arises partly from the inability of the relatively coarse-scale GCMs to capture aspects of complex topography and features such as storm-tracks and the jet stream core.

CSIRO have developed a one-way nested RegCM known as the Division of Atmospheric Research Limited Area Model (DARLAM). It is driven at its lateral boundaries by the CSIRO slab–ocean GCM and has a grid cell resolution of about 125 km per side, giving 442 grid cells over Australia. However, given the nesting of RegCMs within GCMs, the deficiencies of the GCMs will influence the results from the regional climate model and the caution placed on the predictions of climate change that were discussed above should also apply.

Therefore, given the high level of uncertainty that is currently associated with many aspects of climate change scenarios, any predictions regarding possible changes of fire regimes and subsequently of patterns of biodiversity are also at least equally uncertain. Rather than devote this chapter to generating predictions that will undoubtedly change with increased sophistication of climate and fire modelling, it is more profitable to investigate approaches to understanding how fire regimes might change with an altered climate, as well as to carry out some case studies using these approaches. A number of examples for around Australia are considered.

Sensitivity of fire behaviour and fire regimes to climate change

Analysis of the effect of climate change scenarios on fire occurrence and behaviour provides the initial step for determining the sensitivity of fire regimes because models of fire regimes are usually based on an understanding of fire spread and extinguishment. In this section, examples of both approaches are reviewed and a new case study is presented.

Sensitivity of distributions of fire danger indices and frequency of extreme events

Beer *et al.* (1988) analysed the effect of climate change on fire danger in Australia. Their analysis was based on the climate change scenario of Pittock and Nix (1986) and included an increase in the mean annual temperature of between 3.5°C and 4.3 °C, a 40% increase in summer rainfall and a 20% decrease in winter rainfall. It is worth noting the differences between these climate change scenarios and those published more recently (CSIRO 1996; Whetton *et al.* 1996). For example, recent scenarios have a maximum temperature increase of around 3.6 °C (Table 2.1) and summer rainfall changes ranging between −20% and +10% for coupled models (Tables 2.2 and 2.3). Beer *et al.* (1988) also analysed the effect of a 20% increase in wind speed and a 20% reduction in relative humidity because of the importance of these variables in determining fire behaviour (McArthur 1967). For example, of all the variables affecting fire behaviour, annually

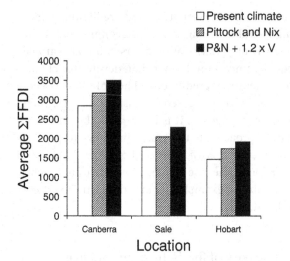

Fig. 2.3. Average ΣFFDI under observed climate (present) and the climate change scenario of Pittock and Nix (1986) and Pittock and Nix with wind speed increased by 20% (P&N + 1.2 × V) for Canberra, Sale and Hobart. Data from Beer *et al.* (1988).

summed FFDI (ΣFFDI) was the most sensitive to changes in daily relative humidity with a linear relationship of

$$\Sigma FFDI = -124.1\,(H_{min}) + 8603$$

where H_{min} is the annual average of the daily minimum relative humidity.

The climate change scenarios were applied to daily meteorological data collected between 1945 and 1986 from East Sale, Victoria. The temperature/rainfall scenarios increased the mean daily FFDI by around 10%, whereas the combined effect of those modifications plus the increase in wind speed and decrease in relative humidity resulted in an increase of 50% to 100% (Beer *et al.* 1988). Similar analyses (modifying temperature, precipitation and wind speed) were conducted for Canberra and Hobart and these indicated that, depending on the climate change scenario adopted, the average ΣFFDI increased by up to 35% (Fig. 2.3).

More recently, Beer and Williams (1995) analysed daily data produced by four- and nine-level GCMs (CSIRO4 with three-year runs and CSIRO9 with nine-year runs). Observed averages were calculated from 30 years of daily data and the details for determin-

ing screen temperatures, relative humidity and wind speed are given in Beer and Williams (1995). There were considerable discrepancies between the observed average daily weather variables for Sale, Victoria, and the modelled averages calculated using the GCMs with a $1 \times CO_2$ scenario, particularly for the nine-level model. The prevalence of bias in modelling of this type was discussed above and in Giorgi *et al.* (1998). These data gave an average ΣFFDI for Sale of 1773, 1958 and 3730 for the observed weather data and the simulated data from the four- and nine-level models respectively.

Doubling the CO_2 concentration in the GCM simulations resulted in the average ΣFFDI for Sale of 3035 and 4547 for the four- and nine-level models respectively. These resulted from increases in average temperature in both models and a decrease in average minimum relative humidity for the four-level model. The average wind speed showed little difference between the $1 \times CO_2$ and $2 \times CO_2$ scenarios or between model types (Beer and Williams 1995). Further, the relationship between annual average minimum relative humidity and the average ΣFFDI was reconfirmed. This provided a means of spatial extrapolation of average ΣFFDI to a continental scale (Fig. 2.4).

In a similar study, conducted for Russian and Canadian boreal forests, Stocks *et al.* (1998) predicted fire weather severity using the output from four GCMs. They concluded that a large increase in the area affected by extreme fire danger would be expected under a $2 \times CO_2$ scenario in both countries.

A variation on the use of averaged or summed FFDI to analyse the effect of climate change is to analyse the frequency of threshold events. Hennessy and Pittock (1995) produced maps, based on the climate change scenarios of CSIRO (1992), indicating how the frequency of days with a maximum temperature equal to or above 35 °C could change across Victoria in the year 2030 compared with the present (1995) climate. Each of the 12 sites used in the analysis displayed a range in the increase of days over the threshold because of the uncertainties associated with local warming scenarios. Three sites (Geelong, Omeo and Orbost) had lower-range scenarios of 0% increase in the number of days above

(a)

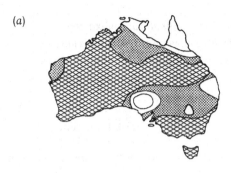

(b)

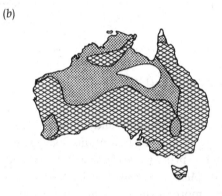

Fig. 2.4. Changes in ΣFFDI over Australia between the $1 \times CO_2$ and $2 \times CO_2$ climate simulations of the (a) CSIRO4 GCM, and (b) CSIRO9 GCM. The blank areas are regions of decreased ΣFFDI, the spotted regions are increases in ΣFFDI of between 0% and 10%, and areas that are cross-filled have ΣFFDI increasing by more than 10%. Reproduced from Beer, T., and Williams, A. (1995) Estimating Australian forest fire danger under conditions of doubled carbon dioxide concentrations. *Climatic Change* **29**, 169–188, Fig. 3, with kind permission from Kluwer Academic Publishers.

35 °C although the upper-range scenarios for these sites indicated significant increases in the number of days exceeding the threshold. Four sites (Ballarat, Bright, Mt Beauty and Omeo) had upper-range scenarios that saw the number of days exceeding the threshold double, at least.

By definition, an increase in the average FFDI will mean that there is a greater chance of fires starting, and once ignited, fires will have higher rates of spread, be more intense and more difficult to suppress (McArthur 1967). Similarly, Pinol et al. (1998) demonstrated that an increase in wildfire hazard indices between 1968 and 1994 in a number of prov-

inces in a Mediterranean locality of northeastern Spain was correlated with an increase in the number of fires. The increasing wildfire hazard reflected an increasing mean daily maximum temperature and a decreasing mean daily minimum relative humidity. Furthermore, Thompson et al. (1998) speculated that increases in the fire weather index (Van Wagner 1987) would probably result in decreased inter-fire intervals and larger fires as was observed in a number of northern hemisphere studies.

However, for a comprehensive insight into how patterns of fire regimes may change, more information than that contained in distributions of threshold events and variables such as FFDI and ΣFFDI are required. Knowledge of other processes and different methodologies is needed.

Sensitivity of fire regimes: simulation of fire regimes with climate change

As discussed, fire behaviour is one important factor in determining the pattern of fire regimes in a landscape. Others include patterns of fuel accumulation and ignition probability (Baisan and Swetnam 1990); topographic characteristics such as elevation (Kirkpatrick 1977), slope and aspect (Barker 1991); and the arrangement of these in the landscape relative to areas with high ignition probability (Stocker 1966; Gilbert 1959). Thus, the concept of the neighbourhood, particularly the ignition neighbourhood (Cary 1997), of a site is also important. Further, certain aspects of the weather, other than those incorporated into a fire danger index and including wind direction and the succession of particular types of days, will affect how a fire burns across a landscape. This information, along with that of fire spread, can be incorporated into models that generate fire regimes as their primary output (Clark 1990; Gardner et al. 1996; Keane et al. 1996; Cary 1997, 1998). These have been used to assess the effect of climate change on patterns of fire regimes (Clark 1990; Gardner et al. 1996; Cary and Banks 1999) in much the same way that the indices of fire danger were used in the previous section. Baker et al. (1991) proposed a general modelling framework for analysing the effect of climate change on disturbances in a landscape; however their model was still in the development stage and was not used in a climate change case study.

The FOREST model (Clark 1990) determines the probability of fire for any given month based on water balances, live and dead fuel dynamics, and lightning incidence, for separate stands of mixed conifer forest in northwestern Minnesota. The model predicted that, in the absence of fire suppression, there would have been a 20% to 40% increase in fire frequency in these forests in the warmer–drier 20th century compared to the 19th century. This would have resulted from changes to both the water balance and fuel load dynamics of the sites, although the time required for the build-up of fuels limited the extent to which increased moisture deficits could increase fire frequency in the model system.

Gardner *et al.* (1996) used the EMBYR fire model to investigate the effect of both wetter and drier climates on the number and size of fires and the fire return interval for a 90 000-ha section of the Yellowstone National Park. They used Thousand Hour time lag Fuel Moisture (THFM) as the driving variable for generating auto-correlated weather sequences for 1000-year simulations. THFM is the percentage moisture content of large ($\geq$7.6 cm) fuel particles that require approximately 1000 hours to respond to precipitation, relative humidity and temperature (Deeming *et al.* 1977). The method for generating weather sequences from the mean maximum and minimum yearly THFM is described in McCarthy and Cary (this volume). Climate change was simulated by modifying the mean maximum and minimum yearly THFM to generate a wetter and drier climate, compared to a nominal climate which was based on historical data (Table 2.4).

Gardner *et al.* (1996) found that the wetter scenario had 20% fewer fires than the nominal scenario while the drier scenario had 30% more. Further, the wetter scenario had a longer average return interval but had more frequent extreme fire events, a function of the formation of larger, more contiguous patches of highly flammable mature forests. These changes in fire regimes resulted in significant differences in the distribution of age classes of lodgepole pine (*Pinus contorta* Loudon var. *latifolia*

Table 2.4. *Mean minimum and maximum THFM percentiles used for generating autocorrelated weather sequences for the analysis of the effect of three weather scenarios on the characteristic of fires in Yellowstone National Park, Wyoming*

	Weather scenario		
	Wetter	Nominal[a]	Drier
Minimum	0.25	0.2	0.15
Maximum	0.80	0.75	0.70

Note:
[a] The nominal scenario is based on historical observations.
Source: Gardner *et al.* (1996).

Engelm.) forest (Romme and Despain 1989). The oldest age class (which contains lodgepole pine, Engelmann spruce (*Picea engelmanni* Parry) and subalpine fir (*Abies lasiocarpa* (Hook.) Nutt.)) occupied one-third more of the landscape under the wetter scenario compared with the drier. Thus they conclude that small changes in weather can produce major shifts in the age structure and spatial arrangement of the forest.

Cary and Banks (1999) demonstrated the use of fire regime simulators for studying the effects of climate change in the Australian Capital Territory (ACT) region of southeastern Australia, using the CSIRO (1992) climate change scenarios for Australia. Their analysis used the FIRESCAPE model (Cary 1997, 1998; McCarthy and Cary this volume), which generates fire regimes in spatially complex landscapes, to determine the effect of increasing daily temperatures by 2 °C and summer rainfall amounts by 20%. They hypothesised no change in the frequency of low-intensity fires (<500 kW m^{-1}) but an increase in the frequency of higher-intensity fires. For example, the frequency of cells predicted to burn with a fire line intensity between 2000 and 2500 kW m^{-1} was more than doubled for the changed climate compared with the normal climate scenario. This resulted from the cancelling effects of higher temperature and increased rainfall, a situation that did not occur during periods of drought when the higher-intensity fires tended to occur. As will be discussed below, the result for the frequency of low-intensity fires would be quite different if the range of both temperature and rainfall changes from more recent scenarios (CSIRO 1996) had been adopted and if the effect of relative humidity was also included.

At the landscape level, Cary and Banks (1999) found that fire frequency would increase significantly in around half their study landscape under the altered climate, while almost all of the remaining cells showing no significant change in fire frequency. Very few cells showed a decrease according to the cell-by-cell t-test comparison (see Fig. 2.5).

The sensitivity of fire regimes to climate change: a case study from a topographically complex landscape

The process-based fire regime simulations discussed so far represent fairly straightforward attempts at modelling the effect of climate change. Only one analysis, using simple and superseded climate change scenarios, has been conducted for an Australian landscape. No attempt has been made to assess the relationship between fire regime and modelled change in the average ΣFFDI or the occurrence of extreme events. The remainder of this chapter addresses these issues with a detailed case study from the ACT region.

The analyses are undertaken for a real landscape. More general results might have been produced using a hypothetical landscape and a number of different climates from around Australia. However, in many respects the 'real' landscape used in the following analyses is really somewhat hypothetical because of a number of important simplifying assumptions that have been made. These include the assumption of continuous forest and woodland vegetation cover, and the absence of water bodies or any anthropogenic structures and modifications in the landscape. Similarly, there is no anthropogenic ignition or suppression of fire. On the other hand, the landscape should be of general interest because

it is large enough (around 1 million hectares) to include a number of different land systems including mountain ranges, valley systems and plain areas, as well as the interfaces between them.

The unmodified (current) climate used in this analysis is that relevant to the ACT region. It is relevant for some areas of Australia but less so for others. Nevertheless, it represents a real climate system that may potentially be altered in the course of global climate change. As mentioned, the scenarios developed here are not meant to be definitive and will have a very short life span for reasons outlined above. A nationwide analysis of fire regimes under current climate is in itself a difficult enough task, let alone prediction of how they might be altered with climate change. Instead, this case study is intended to represent the current state of our ability to analyse the potential sensitivity of fire regimes to climate change. Its purpose is to promote discussion about the potential impacts of climate change amongst the Australian fire and meteorological research community.

There are a number of ways of incorporating information from GCMs into models that simulate fire danger, fire behaviour and fire regimes. One approach is to generate weather sequences at the appropriate temporal scale (say daily) and with different CO_2 concentrations, and generate predictions of fire behaviour and fire regimes under the different scenarios. This was the approach of Beer and Williams (1995). An alternative is to quantify changes to the average values of individual meteorological variables and use them to modify sequences of observed weather (Beer et al. 1988), or sequences from weather generation models that are increasingly being used for fire regime modelling (Gardner et al. 1996; Cary 1998). Each method has its own inherent advantages and disadvantages. The former method is attractive because it allows the modeller to incorporate the effect that climate change may have on the correlations between individual meteorological variables on any one day and also from one day to the next. On the other hand, simply modifying observed weather sequences assumes that the relationships between variables will be maintained into the future. However, using the output from the climate models directly may have an even greater drawback related to any biases that exist in the models (Giorgi *et al.* 1998). Therefore assuming that if a bias exists it will be shared by weather sequences produced under both $1 \times CO_2$ and $2 \times CO_2$ scenarios the differences between the two weather sequences may be more meaningful than the weather sequences themselves. This is especially the case if the response of a particular process to a change in the climate is nonlinear as may be the case with fire regimes.

As an example, consider some simulated climate data recently produced and supplied by the CSIRO Division of Atmospheric Research. Twenty years of daily data were supplied from $1 \times CO_2$ and $2 \times CO_2$ simulations using the regional climate model (DARLAM) nested in the CSIRO9 GCM. The experiments were conducted for a large number of grid cells across the Australian continent; however, the data presented here are representative of the grid cell which includes Canberra, ACT (Fig. 2.6a–d).

In the DARLAM experiment, the combination of a very high climate sensitivity (4.4 °C) and an equilibrium-type experiment resulted in a very large increase in the average monthly temperatures (Fig. 6a) for the $2 \times CO_2$ scenario compared with $1 \times CO_2$. Accordingly, the effects of climate change were scaled to produce three scenarios for a changed climate corresponding to the range of increases in temperature (0.6 °C to 3.4 °C) that are to be realistically expected for the study region by 2070 (Table 2.1). Thus, the 'small' change in climate scenario was defined as one having an average temperature increase of 0.6 °C with all other variables equivalently scaled to represent the same fraction of the full effect of climate change that 0.6 °C does for maximum temperature. The 'large' change in climate scenario was similarly scaled so that the increase in maximum temperature was scaled to 3.4 °C while the 'moderate' scenario was scaled so that the average increase in maximum temperature was 2 °C. This assumes that the effects of climate change are directly scalable in the fashion adopted here. The scenario of present climate ('no change') is based on the long-term averages for Canberra, ACT.

The effects of climate change were used to modify the daily weather produced by the FIRE-SCAPE model which adopts a modification of the Richardson weather generator (Richardson 1981; Richardson and Wright 1984). This avoided the problem of limited runs of data. Presumably as computation of altered weather sequences become more accurate and efficient, the preferable method of including long sequences of simulated daily weather directly into fire regime models will become more common. The FIRESCAPE model has been described in McCarthy and Cary (this volume). In order to examine the effect of a changed climate on fire regimes, a number of FIRESCAPE simulations were performed. These involved simulating spatial patterns in fire regimes using synthetic weather based on the current climate and simulations under the three changed climate scenarios (small, moderate and large).

In the simulations, different weather variables were modified in different ways. The effects of climate change on maximum and minimum temperature were treated as additive so that the appropriate temperature increase for each scenario was simply added to the daily maximum and minimum temperature. Rainfall and relative humidity were modified according to a ratio multiplier that modified the variable according to the ratio of monthly average of the daily variable observed in the $2 \times CO_2$ experiment over that for $1 \times CO_2$ (Fig. 2.6b, c) and scaled to the appropriate scenario. Wind speed was not adjusted because there was little difference between the monthly average of the daily 10-m zonal and meridional wind speeds resulting from the $1 \times CO_2$ and $2 \times CO_2$ experiments, nor for the overall wind speed at 10 m (Fig. 2.6d).

Four separate 500-year FIRESCAPE simulations were conducted, one for the present climate and one for each of the climate change scenarios (small, moderate and large). Single simulation runs were used because the results from a single 500-year simulation is equivalent to the average derived from five separate 100-year simulations. For each simulation, a number of output variables were recorded for each 1-ha cell in the study landscape. These were: (i) the mean inter-fire interval (years); (ii) the mean fire line intensity (kW m^{-1}); and (iii) the proportion

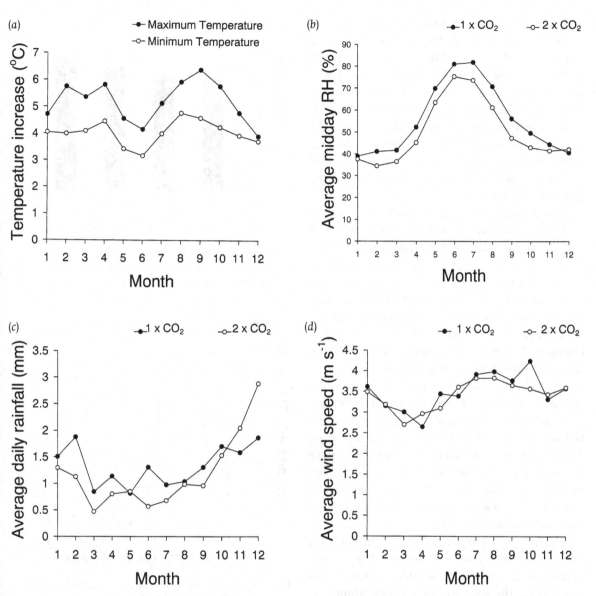

Fig. 2.6. Climatic averages for the Canberra region from $1 \times CO_2$ and $2 \times CO_2$ equilibrium experiments using the CSIRO DARLAM regional climate model nested in the CSIRO9 GCM. For (a) average daily maximum and minimum temperature the increase resulting from the $2 \times CO_2$ scenario compared with the $1 \times CO_2$ scenario is reported. The actual daily averages for the $1 \times CO_2$ and $2 \times CO_2$ scenarios are reported for (b) midday relative humidity (RH) (%); (c) average daily rainfall (mm); and (d) average daily wind speed (m s^{-1}). Climate sensitivity for the $2 \times CO_2$ simulation was 4.338 °C. Averages determined from 20 years of daily data supplied by the CSIRO Division of Atmospheric Research.

of times that the cell was burnt in each of the four seasons (summer = December–January–February, autumn = March–April–May, winter = June–July–August, spring = September–October–November). The average ΣFFDI was also calculated for the region based on the synthesised daily weather.

Changing the climate resulted in an increase in the average ΣFFDI (Fig. 2.7). The average ΣFFDI for the large change in climate scenario represents an increase of around 400 units or 20% over the average ΣFFDI for the normal climate scenario. This is similar to the percentage increase predicted by Beer and Williams (1995) for Sale, Victoria, using the CSIRO9 GCM which was the host model for the DARLAM simulations used in this study. It is worth noting that the average ΣFFDI for the normal climate scenario in Canberra is itself around 400 units lower than that observed between 1960 and 1998, largely because of small differences in the method used to calculated FFDI from the observed record and in the model system. These were predominantly related to some averaging of wind speed in the model system.

Climate change resulted in considerable changes in the overall frequency of fires in the study region but not necessarily to the spatial patterns in fire frequency (see Fig. 2.8a–d). The overall landscape average of the mean inter-fire interval of all cells in the study area decreased from around 42 years for the current climate to around 12 years for the large change in climate. The effect of the small change in climate was negligible, particularly when the variability of the average inter-fire interval is considered. These results represent a range in the increase of the number of fires experienced by each landscape cell from around two to three fires per century (no change) to eight or nine (large change) (Fig. 2.9). On the other hand, the landscape average of the mean fire line intensity for each cell increased by around 25% for the large change simulation, compared with the normal simulation, from 450 to 550 kW m^{-1} (Fig. 2.10).

A change in the seasonal distribution of fires is also predicted (Fig. 2.11). A decline of 0.1 in the proportion of cells predicted to burn in summer is observed for the large change scenario, compared with the normal scenario, while a larger increase (0.2 of all cells burnt) is predicted for autumn. No fires are expected in winter, even under the large change scenario. The effects of the other scenarios are intermediate.

Of all the fire regime variables, fire frequency was the most sensitive to a change in climate. These studies predict that for the moderate change scenario, average inter-fire intervals will be halved under a $2 \times CO_2$ climate. This prediction will change as a result of improved understanding and modelling of climate systems and better quantifications of

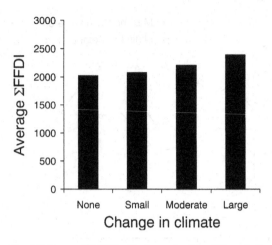

Fig. 2.7. Average ΣFFDI for Canberra for four different climate change scenarios: none (present climate) and a small, moderate and large change in climate.

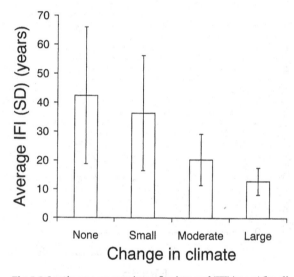

Fig. 2.9. Landscape average inter-fire interval (IFI) (years) for all cells in the ACT study region for four different climate change scenarios: none (present climate) and a small, moderate and large change in climate.

emission policies that affect emission scenarios. The relatively recent attempts at modelling fire regimes using models like FIRESCAPE also have considerable scope for an improved representation of the natural systems that they attempt to model. Therefore, the actual magnitude of the predicted changes is perhaps of less value than the insights that can be gained about the processes that are driving the predicted changes.

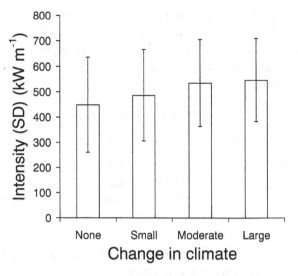

Fig. 2.10. Landscape average fire line intensity (kW m^{-1}) for all cells in the ACT study region for four different climate change scenarios: none (present climate) and a small, moderate and large change in climate.

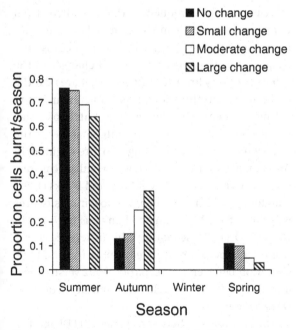

Fig. 2.11. Landscape average proportion of cells burnt per season in the ACT study region for four different climate change scenarios: none (present climate) and a small, moderate and large change in climate.

What then is responsible for the large predicted changes in inter-fire interval? Higher average fire intensity results from faster-spreading fires in the model system. The increase in the average rate of spread would contribute partially to the increased frequency of fire. The percentage of cells burnt in seasons other than summer was predicted to increase from 24% for the current climate to 36% for the large change scenario. Thus, similar to findings of Stocks *et al.* (1998) and Wotton and Flannigan (1993) for Russian and Canadian boreal forests, the fire season will be longer. This would contribute to an increase in fire frequency but again, probably not sufficiently to result fully in the large increase observed.

The increase in fire frequency must be attributed to a combination of these factors, as well as one other important mechanism – fire extinguishment. For model fires in FIRESCAPE, as is the case for real fires, there are three possible outcomes when a fire attempts to spread to an unburnt neighbouring cell. Firstly, the fire may successfully spread. Secondly, the fire may not spread but may not become extinguished. Rather, it remains as a very slow-moving fire or as glowing combustion that represents an important potential ignition source should the meteorological conditions become more conducive to the successful spread of the fire. Finally, the burning cell may extinguish and therefore never ignite the unburnt neighbour. For the current simulations, the extinguishment threshold was set at 80 kW m^{-1} and the lower limit for spreading a fire was set at 12 m h^{-1}.

The frequency of these possible outcomes was recorded during the fire regime simulations described above. Fires from a burning cell successfully spread to an unburnt neighbour in 69% of cases under a normal climate but this increased to 77% of cases for the large change scenario (Fig. 2.12). Conversely, the number of burning cells that extinguished without spreading to any neighbours decreased from 23% to 16%, representing a 30% decline in extinguishment. The number of smouldering events also declined slightly, also a function of an increase in successful spread events.

In order to demonstrate the importance of exting-

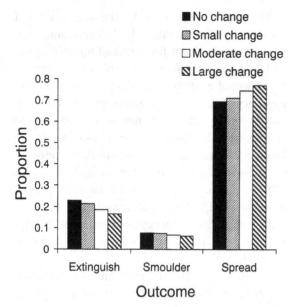

Fig. 2.12. Overall average proportion of cell-to-cell fire spread attempts that either extinguish, spread or do not extinguish but do not spread (smoulder) in the ACT study region for four different climate change scenarios: none (present climate) and a small, moderate and large change in climate.

uishment and the reason fire frequency is particularly sensitive to this process, consider a small, simple landscape comprising a row of adjacent landscape cells. A fire that is ignited at one end of the simple landscape will have a probability

$$p(B) = [1 - p(E)]^n$$

of spreading n cells along the landscape without extinguishing, where $p(E)$ is the probability of extinguishment for each of the climate change scenarios (Fig. 2.12) and n is the number of landscape cells away from the ignition point. Note that this assumes that the case of not spreading but not being extinguished is counted as a successful spread event.

$p(B)$ is very sensitive to $p(E)$. For example, the probability of an average fire burning 10 cells without extinguishing is double for the large change scenario compared to that representing no change (Table 2.5). The difference becomes greater as n becomes larger. Therefore, small changes in $p(E)$ have considerably greater effects on fire size. The effect would be greater for multi-dimensional

Table 2.5. *Probability p(B) that a fire will burn n cells along a simple linear landscape without extinguishing, for four different climate change scenarios: none (present climate) and a small, moderate and large change in climate*

n	Climate change scenario			
---	None	Small	Moderate	Large
10	0.0744	0.0904	0.1274	0.1631
20	0.0055	0.008	0.0162	0.0266
30	0.0004	0.0007	0.0020	0.0043

spread as is the case with model fires in FIRESCAPE and real fires in natural systems.

Rather than an increase in the maximum intensity of fires and a lengthened fire season, it is proposed that it is the reduction in extinguishment that is primarily responsible for the large predicted increase in fire frequency that resulted from climate change in the model system. Fire frequency is very sensitive to changes in climate because fire frequency is very sensitive to small changes in the intensity of very low-intensity fires that are near the extinguishment threshold. Therefore, research into extinguishment thresholds (see Catchpole this volume) may be fundamental for understanding the importance of climate change on fire regimes in Australian landscapes. These results contrast sharply with the findings of Cary and Banks (1999) who found that the frequency of low-intensity fires was not affected by climate change as outlined in the CSIRO (1992) scenarios. The difference arises from differences in postulated temperature and precipitation between the scenarios (1992 versus 1996) and the inclusion of relative humidity from the DARLAM data.

The relationship between average ΣFFDI and fire frequency (landscape average of inter-fire interval) was curvilinear (Fig. 2.13) as it would have to be because the interval between fires would have to drop to zero asymptotically as ΣFFDI tends to infinity. Thus, similar to the findings of Clark (1990), a lower limit on the inter-fire interval resulted from the need for fuels to accumulate in the model

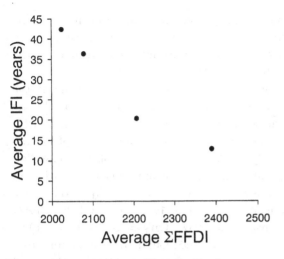

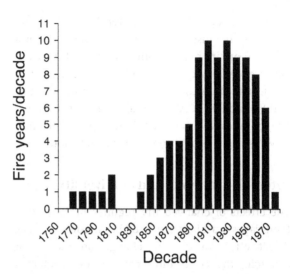

Fig. 2.13. Relationship between ΣFFDI and landscape average of inter-fire interval (IFI) (years) for the range of climate change scenarios in the ACT region study.

Fig. 2.14. Temporal trends in the total number of fires per decade determined from dendrochronological analysis of five stands of snow gum (*Eucalyptus pauciflora*) in the Brindabella Range, Australian Capital Territory. Data from Banks (1982).

system. Nevertheless, fire frequency in the model system was extremely sensitive to small changes in ΣFFDI. If the sensitivity is more general than for just the ACT region, potential increases in fire frequency might be expected for places like Sale, Victoria (Beer and Williams 1995) and elsewhere (Fig. 2.4) with climate change.

However, the realisation of these changes in the real world is likely to be less for one important reason. As discussed, the major difference between FIRESCAPE and the real world is the absence of anthropogenic effects in the model system. Humans are particularly good at modifying fire regimes. For example, dendrochronological studies (Banks 1982, 1997) have shown that fire frequency in the study area has already changed considerably as a result of human-induced changes to ignition patterns and changed management practices (Fig. 2.14). During the pre-European period, fire frequency was low but increased after European settlement in the local area in the 1840s until the 1970s when fire frequency had declined because of changed management.

Given this ability, the real change in fire regimes resulting from climate change will depend to what extent humans can modify the predicted changes in fire intensity, particularly in relation to a predicted decrease in the extinguishment of fires discussed

above. If the average probability of extinguishing in a $2 \times CO_2$ world is maintained at the current level then it is unlikely that even a large change in climate will result in particularly different fire regimes. On the other hand, if extinguishment levels decrease then changes to fire regimes may occur. The extent of change will depend on the extent to which the climate, which is driving the effect, also changes.

Other factors affecting fire regimes: alteration to lightning patterns

Lightning occurrence is an important determinant of fire regimes. Goldammer and Price (1998) studied how the global distribution of lightning frequency might be affected by a changed climate using the Goddard Institute for Space Studies GCM. They reported that the modelled frequency of lightning increased over all continental areas and doubled over most of Australia for a $2 \times CO_2$ climate simulation (with a 4.2 °C global warming) compared with the $1 \times CO_2$ simulation. Modelled cloud-to-ground lightning flashes increased more in the tropics and decreased with increasing latitude.

The effect of increased lightning activity on fire

regimes in Australia has received little attention, primarily because most impact studies have not focussed on fire regimes but have used FFDI or the occurrence of threshold events. There is the potential to use landscape-level fire regime models like EMBYR and FIRESCAPE to analyse the sensitivity of fire regimes to increased lightning occurrence. However this has not been undertaken to date.

Altered fire regimes and biodiversity

The effects of changed fire regimes in Australian landscapes must be considered in the context of the direct effect of climate change on biota. For example, the continuum concept of vegetation theory (Austin and Smith 1989) recognises the existence of a fundamental niche, which is the environmental space within which a particular species can survive and reproduce (Austin et al. 1990). Expression of the fundamental niche of species in the landscape will be altered by the direct effects of climate change, independently of its effect on fire regimes.

The fundamental niche of a species is reduced because of competition, herbivory, disease and aspects of metapopulation dynamics (Gleason 1926). This reduced environmental space is known as the species' realised niche and is expressed spatially as a species' geographic distribution. Basher et al. (1998) have reviewed studies into the effect that climate change may have on the geographic distribution of numerous animal and plant species in Australia. It is expected that some species may encounter a modification of their geographic distribution. Extinctions may occur in cases where species do not have the ability to migrate to new areas which are suitable for their survival, or if those suitable areas no longer exist. Fire regimes also affect the realised niche because disturbances, like fire, also influence the distribution of plant and animal species, and dynamics of plant populations, either independently from other factors or in interaction with them (Whelan and Main 1979; Anderson et al. 1987). A thorough treatment of the importance of fire regimes for plant and animal community dynamics and the dynamics of plant

and animal populations will be dealt with elsewhere; however, it is worth raising some issues to do with potential modifications of fire regime and the biodiversity in Australia.

The evidence suggests that fire frequency in the study area may increase as a result of climate change. Numerous empirical and theoretical studies have identified the importance of fire frequency as a determinant of plant population dynamics and plant community composition in southern, temperate Australia (Jackson 1968; Shugart and Noble 1981; Fox and Fox 1986; Nieuwenhuis 1987; Cary and Morrison 1995; Morrison et al. 1995; Gill 1997). Populations that exist near their fire frequency extinction threshold are at the most risk of substantial effects from climate change. Fire intensity has important effects on various life-cycle stages of plants (Purdie 1977; Gill and Groves 1980; Christensen et al. 1981; Auld 1986; Bradstock and Myerscough 1988; Bradstock and O'Connell 1988; Auld and O'Connell 1991); however predicted changes are small compared to those for fire frequency. Compared to variability in fire intensity (Fig. 2.10), any increase in intensity may not be biologically significant. Conversely, predicted changes to the proportion of fires occurring in autumn and spring are large. Several studies have identified the importance of post-fire conditions for seedling establishment (Bradstock and O'Connell 1988; Enright and Lamont 1989; Lamont et al. 1991; Burgman and Lamont 1992; Whelan 1995; Auld and Bradstock 1996) and survival of resprouting individuals (Daubenmire 1968). Nevertheless, variation in post-fire conditions due to the season of fire is probably relatively small compared with that arising from different years. Therefore, the main impacts of modified fire regimes are likely to be related to fire frequency. The magnitude of the impacts will remain largely unknown because of the uncertainty in a number of important processes related to global warming and to climate and fire modelling. Nevertheless, the potential for significant impacts remain.

Acknowledgements

Thanks to Janette Lindsay (Department of Geography, School of Resource Management and Environmental Science, ANU) for her ideas during early discussions about the writing of this chapter. Thanks to Professor Tom Beer and Dr David Keith who made some very helpful comments on an earlier draft. The CSIRO Division of Atmospheric Research is gratefully acknowledged for supplying the DARLAM simulated weather sequences and for permission to reproduce the CSIRO (1996) climate change scenarios for Australia. Thanks to Roger Jones and Kevin Hennessy of CSIRO (Division of Atmospheric Research) who between them supplied the raw DARLAM data required for the FIRESCAPE analysis, provided much explanation about the work of their organisation and also suggested a number of helpful references. Thanks to Rick McCrae (ACT Bureau of Emergency Services) who calculated and provided information on the FFDI for Canberra, ACT.

References

Alcamo, J., Bouwman, A., Edmonds, J., Grubler, A., Morita, T., and Sugandhy, A. (1994). An evaluation of the IPCC IS92 emission scenarios. In *Radiative Forcing of Climate Change and an Evaluation of the IPCC IS92 Emission Scenarios* (eds. J. T. Houghton, L. G. Meira Filho, J. Bruce, Hoesung Lee, B. A. Callander, E. Haites, N. Harris and K. Maskell) pp. 247–304. (Cambridge University Press: Cambridge.)

Anderson, L., Carlson, C. E., and Wakimoto, R. H. (1987). Forest fire frequency and western spruce budworm outbreaks in western Montana. *Forest Ecology and Management* 22, 251–260.

Auld, T. D. (1986). Population dynamics of the shrub *Acacia suaveolens* (Sm.) Willd.: fire and the transition to seedlings. *Australian Journal of Ecology* 11, 373–385.

Auld, T. D., and Bradstock, R. A. (1996). Soil temperatures after the passage of a fire: do they influence the germination of buried seeds? *Australian Journal of Ecology* 21, 106–109.

Auld, T. D., and O'Connell, M. A. (1991). Predicting patterns of post-fire germination in 35 eastern Australian Fabaceae. *Australian Journal of Ecology* 16, 53–70.

Austin, M. P., and Smith, T. M. (1989). A new model for the continuum concept. *Vegetatio* 83, 35–47.

Austin, M. P., Nicholls, A. O., and Margules, C. R. (1990). Measurement of the realised qualitative niche: environmental niches of five *Eucalyptus* species. *Ecological Monographs* 60, 161–177.

Baisan, C. H., and Swetnam, T. W. (1990). Fire history on a desert mountain range: Rincon Mountain wilderness, Arizona, U.S.A. *Canadian Journal of Forest Research* 20, 1559–1569.

Baker, W. L., Egbert, S. L., and Frazier, G. F. (1991). A spatial model for studying the effects of climatic change on the structure of landscapes subject to large disturbances. *Ecological Modelling* 56, 109–125.

Banks, J. C. G. (1982). The use of dendrochronology in the interpretation of the dynamics of the snow gum forest. PhD thesis, Australian National University, Canberra.

Banks, J. C. G. (1997). Trees the silent fire historians. *Bogong* 18, 9–12.

Barker, P. C. J. (1991). *Podocarpus lawrencei* (Hook. f.): population structure and fire history at Goonmirk Rocks, Victoria. *Australian Journal of Ecology* 16, 149–158.

Basher, R. E., Pittock, A. B., Bates, B., Done, T., Gifford, R. M., Howden, S. M., Sutherst, R., Warrick, R., Whetton, P., Whitehead, D., Williams, J. E., and Woodward, A. (1998). Australasia. In *The Regional Impacts of Climate Change: An Assessment of Vulnerability* (eds. R. T. Watson, M. C. Zinyowera and R. H. Moss) pp. 106–148. (Cambridge University Press: Cambridge.)

Beer, T., and Williams, A. (1995). Estimating Australian forest fire danger under conditions of doubled carbon dioxide concentrations. *Climatic Change* 29, 169–188.

Beer, T., Gill, A. M., and Moore, P. H. R. (1988). Australian bushfire danger under changing climate regimes. In *Greenhouse: Planning for Climate Change* (ed. G. I. Pearman) pp. 421–427. (CSIRO: Melbourne.)

Bradstock, R. A., and Myerscough, P. J. (1988). The survival and population responses to frequent fires of two woody resprouters *Banksia serrata* and *Isopogon anemonifolius*. *Australian Journal of Botany* 36, 415–431.

Bradstock, R. A., and O'Connell, M. A. (1988). Demography of woody plants in relation to fire: *Banksia ericifolia* L.f. and *Petrophile pulchella* (Schrad) R.Br. *Australian Journal of Ecology* 13, 505–518.

Burgman, M. A., and Lamont, B. B. (1992). A stochastic model for the viability of *Banksia cuneata* populations: environmental, demographic and genetic effects. *Journal of Applied Ecology* **29**, 719–727.

Burrows, N. D. (1987). The soil dryness index for use in fire control in the south-west of Western Australia. Western Australian Department of Conservation and Land Management, Technical Report no. 17, Perth.

Cary, G. J. (1997). Analysis of the effective spatial scale of neighbourhoods with respect to fire regimes in topographically complex landscapes. In *MODSIM 97: Proceedings of an International Congress on Modelling and Simulation* (eds. A. D. McDonald and M. McAleer) pp. 436–439. (Modelling and Simulation Society of Australia: Canberra.)

Cary, G. J. (1998). Predicting fire regimes and their ecological effects in spatially complex landscapes. PhD. thesis, Australian National University, Canberra.

Cary, G. J., and Banks, J. C. G. (1999). Fire regime sensitivity to global climate change: An Australian perspective. In *Advances in Global Change Research* (eds. J. L. Innes, M. M. Verstraete and M. Beniston) pp 233–246. (Kluwer Academic Publishers: Dordrecht and Boston.)

Cary, G. J., and Morrison, D. A. (1995). Effects of fire frequency on plant species composition of sandstone communities in the Sydney Region: Combinations of inter-fire intervals. *Australian Journal of Ecology* **20**, 418–426.

Catchpole, W. (2001). Fire properties and burn patterns in heterogeneous landscapes. In *Flammable Australia: The Fire Regimes and Biodiversity of a Continent* (eds. R. A. Bradstock, J. E. Williams and A. M. Gill) pp. 49–76. (Cambridge University Press: Cambridge.)

Cheney, N. P., Gould, J. S., and Catchpole, W. R. (1993). The influence of fuel, weather and fire shape variables on fire-spread in grasslands. *International Journal of Wildland Fire* **3**, 31–44.

Cheney, N. P., Gould, J. S., and Catchpole, W. R. (1998). Prediction of fire spread in grasslands. *International Journal of Wildland Fire* **8**, 1–13.

Christensen, P., Recher, H., and Hoare, J. (1981). Responses of open-forest (dry sclerophyll forests) to fire regimes. In *Fire and The Australian Biota* (eds. A. M. Gill, R. H. Groves and I. R. Noble) pp. 367–394. (Australian Academy of Science: Canberra.)

Clark, J. S. (1990). Twentieth-century climate change, fire suppression, and forest production and decomposition in northwestern Minnesota. *Canadian Journal of Forest Research* **20**, 219–232.

CSIRO (1992). *Climate Change Scenarios for the Australian Region* (Climate Impact Group, CSIRO Division of Atmospheric Research: Melbourne, Australia.)

CSIRO (1996) *Climate Change Scenarios for the Australian Region* (Climate Impact Group, CSIRO Division of Atmospheric Research: Melbourne, Australia.)

Daubenmire, R. (1968). Ecology of fire in grasslands. *Advances in Ecological Research* **5**, 209–266.

Deeming, J. E., Burgan, R. E., and Cohen, J. D. (1977). The National Fire Danger Rating System – 1978. U.S. Department of Agriculture, Forest Service, General Technical Report INT–39, Ogden, Utah.

Enright, N. J., and Lamont, B. B. (1989). Seed banks, fire season, safe sites, and seedling recruitment in five co-occuring *Banksia* species. *Journal of Ecology* **77**, 1111–1122.

Fox, M. D., and Fox, B. J. (1986). The effect of fire frequency on the structure and floristic composition of a woodland understorey. *Australian Journal of Ecology* **11**, 77–85.

Gardner, R. H., Hargrove, W. W., Turner, G. M., and Romme, W. H. (1996). Climate change, disturbances and landscape dynamics. In *Global Change and Terrestrial Ecosystems* (eds. B. Walker and W. Steffen) pp. 149–172. (Cambridge University Press: Cambridge.)

Gates, W. L., Henderson-Sellers, A., Boer, G. J., Folland, C. K., Kitoh, A., McAvaney, B. J., Semazzi, F., Smith, N., and Zeng, Q. C. (1996). Climate models: evaluation. In *Climate Change 1995: The Science of Climate Change* (eds. J. T. Houghton, L. G. Meira Filho, B. A. Callander, N. Harris, A. Kattenberg and K. Maskell) pp. 229–283. (Cambridge University Press: Cambridge.)

Gilbert, J. M. (1959). Forest succession in the Florentine Valley, Tasmania. *Proceedings of the Royal Society of Tasmania* **93**, 129–151.

Gill, A. M. (1975). Fire and the Australian flora: a review. *Australian Forestry* **38**, 4–25.

Gill, A. M. (1997). Eucalypts and fires: interdependent or independent? In *Eucalypt Ecology: Individuals to Ecosystems* (eds. J. E. Williams and J. C. Z. Woinarski) pp. 151–167. (Cambridge University Press: Cambridge.)

Gill, A. M., and Groves, R. H. (1980). Fire regimes in heathlands and their plant ecological effects. In

Ecosystems of the World, vol. 9B, *Heathlands and Related Shrublands* (ed. R.L. Specht) pp. 61–84. (Elsevier: Amsterdam.)

Giorgi, F., Meehl, G. A., Kattenberg, A., Grassl, H., Mitchell, J. F. B., Stouffer, R. J., Tokioka, T., Weaver, A. J., and Wigley, T. M. L. (1998). Simulation of regional climate change with global coupled climate models and regional modelling techniques. In *The Regional Impacts of Climate Change: An Assessment of Vulnerability* (eds. R. T. Watson, M. C. Zinyowera and R. H. Moss) pp. 427–438. (Cambridge University Press: Cambridge.)

Gleason, H. A. (1926). The individualistic concept of the plant association. *Bulletin of the Torrey Botanical Club* **53**, 7–26.

Goldammer, J. G. and Price, C. (1998). Potential impacts of climate change on fire regimes in the tropics based on MAGICC and a GISS GCM-derived lightning model. *Climatic Change* **39**, 273–296.

Gordon, H. B., and O'Farrell, S. P. (1997). Transient climate change in the CSIRO coupled model with dynamical sea-ice. *Monthly Weather Review* **125**, 875–907.

Hennessy, K. J., and Pittock, A. B. (1995). Greenhouse warming and threshold temperature events in Victoria, Australia. *International Journal of Climatology* **15**, 591–612.

Houghton, J. T., Meira Filho, L. G., Bruce, J., Lee, H., Callander, B. A., Haites, E., Harris, N., and Maskell, K. (eds.) (1994) *Climate Change 1994: Radiative Forcing of Climate Change and an Evaluation of the IPCC IS92 Emission Scenarios*. (Cambridge University Press: Cambridge.)

Houghton, J. T., Meira Filho, L. G., Callander, B. A., Harris, N., Kattenberg, A., and Maskell, K. (eds.) (1996) *Climate Change 1995: The Science of Climate Change*. (Cambridge University Press: Cambridge.)

Jackson, W. D. (1968). Fire, air, water and earth: an elemental ecology of Tasmania. *Proceedings of the Ecological Society of Australia* **3**, 9–16.

Kattenberg, A., Giorgi, F., Grassl, H., Meehl, G. A., Mitchell, J. F. B., Stouffer, R. J., Tokioka, T., Weaver, A. J., and Wigley, T. M. L. (1996). Climate models: projections of future climate. In *Climate Change 1995: The Science of Climate Change*. (eds. J. T. Houghton, L. G. Meira Filho, B. A. Callander, N. Harris, A. Kattenberg and K. Maskell) pp. 289–357. (Cambridge University Press: Cambridge.)

Keane, R. E., Morgan, P., and Running, S. W. (1996). FIRE-BGC: a mechanistic ecological process model for simulating fire succession on coniferous forest landscapes of the northern Rocky Mountains. U.S. Department of Agriculture, Forest Service, Intermountain Research Station, Research Paper INT-RP-484, Ogden, UT.

Kirkpatrick, J. B. (1977). Native vegetation of the west coast region of Tasmania. In *Landscape and Man*. (eds. M. R. Banks and J. B. Kirkpatrick) pp. 55–80. (Royal Society of Tasmania: Hobart.)

Lamont, B. B., Connell, S. W., and Bergl, S. M. (1991). Seedbank and population dynamics of *Banksia cuneata*: the role of time, fire, and moisture. *Botanical Gazette* **152**, 114–122.

Leggett, J., Pepper, W. J., and Swart, R. J. (1992). Emission scenarios for the IPCC: an update. In *Climate Change 1992: Supplementary Report to the IPCC Scientific Assessment* (eds. J. T. Houghton, B. A. Callander and S. K. Varney) pp. 69–95. (Cambridge University Press: Cambridge.)

Luke, R. H., and McArthur, A. G. (1978). *Bushfires in Australia*. (Australian Government Publishing Services: Canberra.)

Manning, M. R., Pearman, G. I., Etheridge, D. M., Fraser, P. J., Lowe, D. C., and Steele, L. P. (1996). The changing composition of the atmosphere. In *Greenhouse: Coping with Climate Change* (eds. W. J. Bouma, G. I. Pearman and M. R. Manning) pp. 3–26. (CSIRO Collingwood, Victoria.)

McArthur, A. G. (1966a). Weather and grassland fire behaviour. Commonwealth of Australia Forest and Timber Bureau, Leaflet no. 100, Canberra.

McArthur, A. G. (1966b). The application of a drought index system to Australian fire control. Forest Research Institute Publication, Forestry and Timber Bureau, Canberra.

McArthur, A. G. (1967). Fire behaviour in eucalypt forests. Commonwealth of Australia Forest and Timber Bureau, Leaflet no. 107, Canberra.

McCarthy, M. A., and Cary, G. J. (2001). Fire regimes in landscapes: models and realities. In *Flammable Australia: The Fire Regimes and Biodiversity of a Continent* (eds. R. A. Bradstock, J. E. Williams and A. M. Gill) pp. 77–93. (Cambridge University Press: Cambridge.)

Morrison, D. A., Cary, G. J., Pengelly, S. M., Ross, D. G., Mullins, B. J., Thomas, C. R., and Anderson, T. S. (1995). Effects of fire frequency on plant species composition of the sandstone communities in the Sydney region: inter-fire intervals and time-since-fire. *Australian Journal of Ecology* **20**, 239–247.

Mount, A. B. (1972). The derivation and testing of a soil dryness index using run-off data. Tasmanian Forestry Commission, Bulletin no. 4, Hobart.

Nieuwenhuis, A. (1987). The effect of fire frequency on the sclerophyll vegetation of the West Head, New South Wales. *Australian Journal of Ecology* **12**, 373–385.

Noble, I. R., Bary, G. A. V., and Gill, A. M. (1980). McArthur's fire-danger meters expressed as equations. *Australian Journal of Ecology* **5**, 201–203.

Peet, G. B. (1967). The shape of mild fires in jarrah forest. *Australian Forestry* **31**, 121–127.

Pinol, J., Terradas, J., and Lloret, F. (1998). Climate warming, wildfire hazard and wildfire occurrence in coastal eastern Spain. *Climatic Change* **38**, 345–357.

Pittock, A. B., and Nix, H. A. (1986). The effect of changing climate on Australian biomass production: a preliminary study. *Climatic Change* **8**, 243–255.

Purdie, R. W. (1977). Early stages of regeneration after burning in dry sclerophyll vegetation. II. Regeneration by seed germination. *Australian Journal of Botany* **25**, 35–46.

Richardson, C. W. (1981). Stochastic simulation of daily precipitation, temperature, and solar radiation. *Water Resources Research* **17**, 182–190.

Richardson, C. W., and Wright, D. A. (1984). WGEN: a model for generating daily weather variables. U.S. Department of Agriculture, ARS-8, Beltsville, Maryland.

Romme, W. H., and Despain, D. G. (1989). Historical perspectives of the Yellowstone fires of 1988. *Bioscience* **39**, 695–699.

Rothermel, R. C. (1972). A mathematical model for predicting fire spread in wildland fuels. U.S. Department of Agriculture , Forest Service, Research Paper INT-115, Ogden, UT.

Shugart, H. H., and Noble, I. R. (1981). A computer model of succession and fire response in the high-altitude *Eucalyptus* forest of the Brindabella Range, Australian Capital Territory. *Australian Journal of Ecology* **6**, 149–164.

Sneeuwjagt, R. J., and Peet, G. B. (1985). *Forest Fire Behaviour Tables for Western Australia*, 3rd edn. (Western Australian Department of Conservation and Land Management: Perth.)

Stocker, G. C. (1966). The effects of fires on vegetation in the Northern Territory. *Australian Forestry* **30**, 223–230.

Stocks, B. J., Fosberg, M. A., Lynham, T. J., Mearns, L., Wotton, B. M., Yang, Q., Jin, Z., Lawrence, K., Hartley, G. R., Mason, J. A., and McKenney, D.W. (1998). Climate change and forest fire potential in Russian and Canadian boreal forests. *Climatic Change* **38**, 1–13.

Thompson, I. D., Flannigan, M. D., Wotton, B. M., and Suffling, R. (1998). The effects of climate change on landscape diversity: an example in Ontario forests. *Environmental Monitoring and Assessment* **49**, 213–233.

Trenberth, K. E., Houghton, J. T., and Meira Filho, L. G. (1996). The climate system: an overview. In *Climate Change 1995: The Science of Climate Change* (eds. J. T. Houghton, L. G. Meira Filho, B. A. Callander, N. Harris, A. Kattenberg and K. Maskell) pp. 51–64. (Cambridge University Press: Cambridge.)

Van Wagner, C. E. (1987). Development and structure of the Canadian Forest Fire Weather Index System. Canadian Forest Service, Technical Report no. 35, Ottawa, Ontario.

Watson, R. T., Zinyowera, M. C., and Moss, R. H. (1998). Introduction. In *The Regional Impacts of Climate Change: An Assesment of Vulnerability* (eds. R. T. Watson, M. C. Zinyowera and R. H. Moss) pp. 19–28. (Cambridge University Press: Cambridge.)

Whelan, R. J. (1995). *The Ecology of Fire*. (Cambridge University Press: Cambridge.)

Whelan, R. J., and Main, A. R. (1979). Insect grazing and post-fire plant succession in south-west Australian woodland. *Australian Journal of Ecology* **4**, 387–398.

Whetton, P., Mullan, A. B., and Pittock, A. B. (1996). Climate-change scenarios for Australia and New Zealand. In *Greenhouse: Coping with Climate Change* (eds. W. J. Bouma, G. I. Pearman and M.R. Manning) pp. 145–170. (CSIRO: Collingwood, Victoria.)

Wotton, B. M., and Flannigan, M. D. (1993). Length of fire season in a changing climate. *Forestry Chronicle* **69**, 187.

Part II

Fire regimes and life histories

3

Fire properties and burn patterns in heterogeneous landscapes

WENDY CATCHPOLE

Abstract

Prediction of the imprint made by single fire, in terms of environmental conditions, topography, vegetation type and age, is a first step in modelling the vegetation pattern in a landscape after a period of years of exposure to fire. The expected imprint on the landscape can be characterised by the probability of a fire, the expected fire size and shape, the burning pattern within the fire and the pattern of associated spot fires. Each of these depends on fire behaviour characteristics such as the spread rates of head, flank and backfires, the intensity distribution within the fire and the probability of extinction. The first step in the process of determining the fire imprint is prediction of such fire behaviour in terms of the prevailing conditions. This chapter discusses the effect of vegetation, environmental and topographical conditions on fire behaviour, and the consequent effect on the fire imprint.

Introduction

Every fire leaves an imprint on the landscape. In mild conditions the area burned will be small, patches of unburnt material may be left within the fire boundary, and only surface vegetation will be killed. In severe conditions the burnt area will be large and all vegetation within the fire area may be destroyed. Prediction of a single fire imprint in terms of environmental conditions, topography, vegetation type and age is a first step in modelling the resultant vegetation pattern in a region after a period of years of exposure to fire.

The expected fire imprint can be characterised by the probability of a fire, the fire size and shape, the burning pattern within the fire and the pattern of associated spot fires. These depend on fire behaviour characteristics such as rate of spread, intensity (heat release per unit length of fireline) and the probabilities of ignition and extinction. The fire behaviour characteristics are determined by vegetation, as well as topographical and environmental conditions, which change spatially over the area where the fire burns, and/or temporally while the fire is alight. These then determine the final fire imprint. To predict this fire imprint we first need to predict fire behaviour in terms of the local prevailing conditions. We then need to model the variations in environmental conditions both spatially in terms of topography and vegetation, and temporally during the period of burning. Wind speed and dead fuel moisture content are major factors determining fire spread and intensity. Wind speed is stronger on ridges, funnels up valleys, and is reduced under forest canopies. Dead fuel moisture content varies with altitude and aspect, and is higher in shaded forests than open grassland. Both wind speed and moisture content vary diurnally and with changes in atmospheric conditions. All of these factors need to be taken into account when modelling fire behaviour on a landscape scale. This chapter concentrates primarily on the first step of predicting fire behaviour in terms of local environmental conditions, topography and vegetation with a view to predicting the fire imprint.

Ignition and extinction

The probability of a fire can be determined by the product of the probability of the ignition source and the probability of ignition and sustained burning from the source. Unplanned fires can be ignited naturally by lightning strikes or burning embers from another fire, by accident from sparking power lines, sparks from chain saws, escaped camp fires etc., or by arson. The probability of such occurrences varies spatially depending on land use and the propensity for lightning in an area. It varies temporally with weather conditions, season and weekday, and with critical conditions used to impose fire bans. It is essential to incorporate at least some of these components into a model of landscape fires. More human-caused ignition sources are likely around urban and recreational areas. Lightning tends to occur at higher elevations. If historical studies are used to determine ignition probability in terms of weather variables alone, the sources and ignition probabilities are combined, and care must be taken when importing the ignition model to an area where the source probability may be different.

For an ignition to take place the total heat energy from the ignition source must be sufficient to drive the moisture out of the fuel and raise the fuel temperature to ignition (around 320 °C). Sparks from chain saws have less chance of igniting fires than unattended camp fires because the heat source is small and not applied for long. Burning brands will only start spot fires if they remain burning for long enough to give off enough heat to ignite the fuel. Fine fuels with low moisture contents and low bulk densities (such as cured grass) are more easily ignited than thicker, closely packed, moister fuels, such as forest litter. Muraszew (1974) gave a model for ignition from burning brands based on heat transfer processes, but it has received little attention. Field studies in eucalypt forest indicate that when the litter moisture content is above 7% only large brands will be effective in starting new fires (Cheney and Bary 1969), but at moisture contents below 4% most flaming brands are capable of starting new fires.

Fires caused by lightning are common in much of mainland Australia. In the semi-arid regions of western Queensland, for example, Luke and McArthur (1978) reported that lightning caused 80% of all fires. Latham and Schlieter (1989), using simulated lightning strikes in no-wind conditions,

found that the duration rather than the intensity of the charge was important in determining ignition. They calculated the probability of ignition from a lightning storm from

$$\Pr(\text{ignition}) = \int \Pr(\text{ignition given the duration of strike} = t) f(t)\, dt,$$

and used a Weibull distribution for the distribution, $f(t)$, of strike duration. A similar model could be used for spot fire ignitions using brand size instead of lightning duration. Latham and Schlieter (1989) found that the probability of ignition in North American conifer litter depended on fuel type and surface moisture content. Surprisingly, they found that moisture content played a very small role in duff ignitions, and that duff depth was more influential.

Extensive laboratory work, using zero-wind fires, was done by Wilson (1985) who identified fuel surface area per surface area of ground and moisture content as the important variables determining self-extinction. Strictly though, because the experiments were done with a fixed ignition source (a line of burning alcohol), Wilson was testing the probability of ignition given the ignition source. The assumption that this gives the conditions for extinction is that enough heat is being given out from the ignition source to simulate the heat required for sustained burning if that were possible.

Wilson developed an extinction index to separate conditions of non-ignition, poor burning and sustained burning. This index is a strong function of fuel surface area and moisture content, but also depends weakly on fuel type and fuel temperature. Fuels with low moisture content and high surface area per surface area of fuel bed (through depth, packing ratio or particle size) have a high probability of sustained burning. Wilson did not explore the other extreme of high porosity (high surface area per volume of bed), which may limit combustion by limiting the available oxygen, nor did he extend the work to wind-aided fires. However, Wilson's work does go a long way towards an understanding of the factors affecting sustainability. Wilson (1990) gives the probability of burning as a function of the extinction index.

Table 3.1. *Extinction moisture contents for Australian vegetation*

Fuel type[a]	Wind speed (m s^{-1})[b]	Fuel stratum	Extinction moisture (%)	Reference
Grassland	≤3	elevated	20	Cheney and Sullivan (1997)
Grassland	>3	elevated	24	Cheney and Sullivan (1997)
Grassland	back fire	elevated	18	Cheney and Sullivan (1997)
Tropical savanna		elevated	35	Lacey *et al.* (1982)
Hummock grass	15 (gusts)	elevated	24	Gill *et al.* (1995)
Hummock grass	>1.5	elevated	35[c]	Burrows *et al.* (1991)
Mallee shrubland		shallow litter	8	McCaw (1998)
Forest		litter	20	Cheney (1981)
		litter	21	Burrows (1999a)
Forest	0–1.5	elevated	16	Buckley (1993)
Forest	0–1.5	surface litter	25	Buckley (1993)
Forest	0–1.5	litter and twigs	19	McCaw (1986)

Notes:

[a] For buttongrass moorland see Fig. 3.1.

[b] Approximate conversion from kmh^{-1}.

[c] Average moisture content of live and dead vegetation.

Some Australian prediction tables and guides to burning give the dead fuel moisture of extinction for the relevant vegetation type, or at least a guide to conditions that are too marginal for sustained burning. Suggested extinction moisture contents for Australian fuels are given in Table 3.1. It is probable that the moisture of extinction of forest fuels is higher than that required for successful ignition from a point source, such as a match or a spot fire, but this has not often been quantified. McCaw (1986) gives an example of this for fires in karri (*Eucalyptus diversicolor*) regrowth burning in a heavy layer of dead surface fuel. In this case a dead fuel moisture content of 15% was required for ignition, but once alight, the fires would sustain at a moisture content of 19%, and thus have the potential to burn steadily well into the evening and sometimes overnight.

The effect of wind on fuel ignition depends on the ignition source (Cheney and Sullivan 1997). Sparks may be blown out in high winds, while burning logs may be rekindled and re-ignite a fire.

Ellis (2000) found that ignition success by glowing firebrands was only possible in the presence of wind in laboratory studies in *Pinus radiata* litter. While wind speed may not always be vital to the ignition process, it is often necessary for sustained burning. Ignitions in windy conditions are hard to suppress, and there are generally more reported fires on windy days, even though the actual number of ignitions is not necessarily greater (Haines *et al.* 1973). Surprisingly though Krusel *et al.* (1993) found that wind speed was a poor predictor of severe fire occurrence in lightning fires in mallee shrubland.

Marsden-Smedley *et al.* (2001) did extensive work in the Tasmanian buttongrass moorlands (dominated by *Gymnoschoenus sphaerocephalus*) to determine conditions under which fires self-extinguish. As in Wilson's experiments the ignition source was a line of fire. The probability of sustained burning, p, was found to depend on wind speed, dead fuel moisture content and biomass. A model for p was given in the form

$$\ln[p/(1-p)] = a + bU - cM + eP,$$

where U is 2 m wind speed, M is moisture content and P is site productivity (coded 0 or 1). The probability of sustained burning, as a function of moisture content and wind speed, is shown in Fig. 3.1 for (a) low-productivity and (b) high-productivity moorland. Productivity depends on geology in the moorlands, and was used as a prediction variable for management convenience. Productivity is strongly related to both fuel load (weight of fuel per unit area) and continuity, which both contribute to the probability of sustained burning. Modelling using classification trees (Breiman *et al.* 1984) gave conditions for which fires would not sustain, and conditions for which they were marginal. Although some wind may be necessary for sustained burning in marginal conditions, very high winds speeds may actually extinguish fires in grasslands with relatively low biomass (Cheney and Sullivan 1997).

The fire danger indices for grassland and forest developed by McArthur (McArthur 1966, 1973, 1977), which depend on weather conditions and fuel flammability, are intended to indicate the probability of fire occurrence. The indices are strongly related to wind speed and the moisture content of the components of the fuel.

One might expect that in discontinuous fuels the probability of fire spread would be a function of wind speed. However, McCaw (1998) found that the moisture content of the surface litter was the governing factor for spread in discontinuous eucalypt mallee shrubland. Below 8% moisture content fires were non-sustaining at a range of wind speeds. On the other hand, Burrows *et al.* (1991) gave a threshold wind speed of 3–5 m s^{-1} (at a height of 2 m) for spread rate in the discontinuous hummock grasslands in the arid regions of Australia. Griffin and Allan (1984) reported a threshold of 1 m s^{-1} in similar vegetation. This threshold wind speed must, in fact, be a function of patch–gap distribution, patch size and height, and moisture content (Gill *et al.* 1995). Work on the conditions for fires to spread in terms of the patch–gap size and distribution, and environmental conditions (Bradstock and

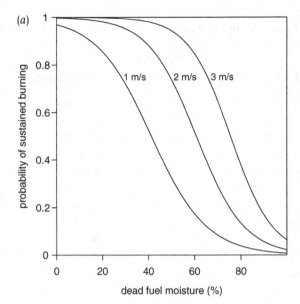

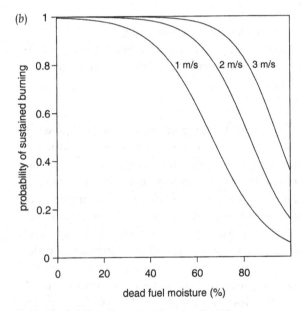

Fig. 3.1. Probability of sustained burning for (a) low-productivity and (b) high-productivity buttongrass moorland, in terms of dead fuel moisture content and wind speed. From Marsden-Smedley *et al.* (2001).

Gill 1993; Weber *et al.* 1995a; Mercer 1997) should help to quantify thresholds in these discontinuous fuels.

Burrows *et al.* (1991) noted that provided the fire

front width was greater than 20–30 m the fire would burn around small patches in hummock grassland and continue to spread. Thus the probability of spread was also dependent on fire size. Marsden-Smedley et al. (2001) report that under humid, low-wind conditions fire-line lengths of greater than 20 m were needed for sustained spread in buttongrass moorland.

Sustained ignition also depends on the percentage of dead fuel in the fuel array. If the percentage of dead fuel is less than 50% the grassland fire behaviour model developed by Cheney et al. (1998) predicts that grass fires are non-sustaining. Parrott and McDonald (1970) ignited a range of annual grass types with matches, and found that when the percentage of dead grass was 50% the probability of ignition by matches ranged from 0.5 to 0.7. Shrubland fires will burn with percentages of dead fuel as low as 30%, either because of the greater volatile content or the arrangement of live material above the dead material.

Young fuels often have poor continuity and not enough dead fuel to sustain low- to medium-intensity fires. For tropical savanna grassland Lacey et al. (1982) give required loads of 0.1 kg m^{-2} for fairly continuous grass growing after rain, and 0.15 kg m^{-2} in areas with bare ground between perennial tussocks. Burrows (1999a) found that fires in jarrah litter (*Eucalyptus marginata*) would not sustain when fuel loads were less than 0.35 kg m^{-2}. This would require at least a 2-year-old fuel. Marsden-Smedley and Catchpole (1995b) found that low–medium-intensity fires would not sustain in continuous moorlands of age less than 3 years. In heathland in New Zealand a low-intensity fire in 3-year-old heath only sustained when there was a continuous sedge understorey (G. Pearce personal communication). In the more discontinuous mallee shrubland fires are unlikely to spread well in fuels of less than 8 years old, unless conditions are severe (McCaw 1998). Intense wildfires however may be able to burn in sparse fuels that would not sustain lower-intensity fires. Under severe weather conditions fires may spread in shrublands 1 to 2 years following fire (Conroy 1996; Morrison et al. 1996). Fires in 3-year-old eucalypt fuels will burn

well in severe conditions (Buckley 1990; McCaw et al. 1992)

Fuel age determines biomass, dead-to-live ratio and continuity, all of which increase with age, and increase the probability of sustained fire spread. In analysis of fire regime data the number of fires of various fuel ages may be available from records and fire scars. From this the fire hazard rate (or equivalently the probability of an ignition) in vegetation of given age can be determined. Several researchers have proposed an increasing hazard rate with fuel age, or an asymptotic function that reflects that biomass and dead-to-live ratios may tend to level off with age (McCarthy and Cary, this volume).

Rate of spread

The McArthur forest and grassland meters (McArthur 1966, 1973, 1977) are widely used in eastern Australia to predict fire spread rate. In Western Australia accumulated fire behaviour knowledge has been succinctly condensed in the Forest Fire Behaviour Tables (FFBT) (Sneeuwjagt and Peet 1985). In the past decade prediction models for spread rate in other vegetation complexes have been developed, e.g. hummock grassland (Griffin and Allan 1984; Burrows et al. 1991), buttongrass moorland (Marsden-Smedley and Catchpole 1995b), mallee shrubland (McCaw 1998) and heathland (W. R. Catchpole et al. 1998a). Fire spread equations for grassland have been reformulated (Cheney and Sullivan 1997; Cheney et al. 1998). Further research has been done in other types of forest fuels (Cheney et al. 1992; Buckley 1993; Burrows 1999a,b). Project Vesta in Western Australia has been initiated to investigate fire behaviour in jarrah forest in more severe conditions than were previously used in the development of prediction tables, and to investigate the effect of fuel load on fire behaviour. The prediction equations that have been developed are fuel-type specific, and cannot easily be modified for other fuels. However, the accumulated knowledge of fire researchers in Australia and elsewhere has given us some understanding of the effects of fuel, environment and topography on fire spread.

Wind speed

Fire spread rate is strongly affected by wind speed. Increasing wind bends the flames over, and thus increases the radiative heat transfer. The convective heat transfer also increases due to the passage of turbulent hot gases over unburnt fuel. In low winds radiative transfer dominates, winds tend to be very variable in direction and strength, and spread rate is not much greater than in still air. Several researchers have suggested a threshold wind speed of 1–2.5 m s^{-1} (Fons 1946; Beer 1993; Cheney et al. 1998; Burrows 1999a). After this threshold, the convective process comes into play, and spread rate increases at least linearly with wind speed. At high wind speeds spotting in some vegetation types, or interaction with the atmosphere, may cause a greater than linear increase of spread rate as wind speed increases. Fig. 3.2 shows some of the prediction equations for spread rate as a function of wind speed (in the open at 10 m) that have been developed for different Australian fuel types, and the range of data (where known) used to formulate these equations. Predictions are for a temperature of 30 °C, humidity of 30% and drought factor 10 (see section on moisture content). The functions used range from somewhat less than linear to proportional to the square of the wind speed. This may be the result of fitting the same function in different data ranges where different heat transfer processes dominate, or of forcing predictions to agree with limited and unreliable wildfire data.

Spread rate over a landscape for low–moderate-intensity fires may be lower than that given by the prediction equations (Cheney and Sullivan 1997). At low to moderate wind speeds fires may be held up for brief periods by roads, gullies and natural firebreaks. At higher wind speeds flames would be long enough to bridge these gaps, or could spot over barriers.

In open heathland and grasslands the surface (2 m) wind speed is about two-thirds that of the 10-m wind speed (Marsden-Smedley and Catchpole 1995b; Cheney and Sullivan 1997; McCaw 1998;), while for hummock grasslands the ratio can be higher (Burrows and Van Didden 1991). In wood-

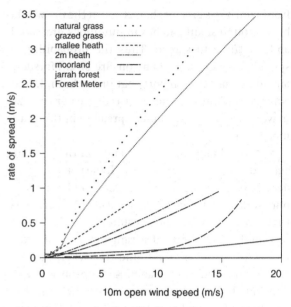

Fig. 3.2. Prediction equations for rate of spread as a function of wind speed (in the open at 10 m) for some Australian fuel types. Curves are given over the range of experimental and wildfire data used to develop the equations. Conditions: temperature 30 °C, humidity 30%, drought factor 10. Moisture content is estimated from the relevant prediction equations for each fuel type. Spread rate prediction equations are for (a) natural grass (Cheney et al. 1998), (b) grazed grass (Cheney et al. 1998), (c) mallee heath (McCaw 1998), (d) 2-m tall heath (W. R. Catchpole et al. 1998a), (e) mature buttongrass moorland (Marsden-Smedley and Catchpole 1995b), (f) jarrah forest (0.8 kg m^{-2}) (Sneeuwjagt and Peet 1985), (g) Forest Meter (1.25 kg m^{-2}) (McArthur 1973). Equations to fit the tables are from (f) Beck (1995) and (g) Noble et al. (1980).

lands and forests, the effective wind speed is reduced depending on tree height and density, canopy height, and shrub height and density. For woodland and open-forests with a grass understorey, Cheney and Sullivan (1997) give reduction factors in spread rate of 0.6 and 0.42 respectively. McArthur (1962) gives wind reduction factors (in terms of stand height) for eastern Australian forests. Reduction factors (in terms of canopy cover and topography) for Western Australian forests are tabulated in Sneeuwjagt and Peet (1985).

Moisture content

Fine dead fuel moisture content also has an important effect on rate of spread, especially when the humidity is low and the temperature is high. Australian and Canadian spread rate prediction equations have used a single, often exponential, decay function, $\Phi(M) = \exp(-kM)$, for the damping effect of fine dead fuel moisture content on rate of spread. In the USA the moisture effect is split into a cubic source and a linear sink term (Rothermel 1972). The Canadian function, based on Canadian data for pine litter and Australian data for eucalypt litter, has a function which approximates to exponential with $k = 0.13$, when M is given in percent dry weight. This is close to the damping factor $k = 0.11$ determined for grasslands by Cheney et al. (1998). This predicts a 25% increase in spread rate for a drop in moisture content of 2%, and a 50% increase for a drop of 4%. Figure 3.3 shows the relative effects of moisture content and wind speed on rate of spread in grasslands for the prediction model given in Cheney et al. (1998). McCaw (1998) found the moisture response in mallee shrubland to be similar to that of grass. Both grassland and mallee functions were generated from data on fires burned at moisture contents of less than 8%. In fires in buttongrass moorland, with higher moisture contents, Marsden-Smedley and Catchpole (1995b) found a much smaller damping factor ($k = 0.02$), and most of the effect of moisture was on fires with dead fuel moistures greater than 35%, where there was moisture on the fuel surface. W. R. Catchpole et al. (1998a) found no damping effect of moisture in heathland from eastern Australia and New Zealand (where moisture contents ranged from 10% to 30%). Some moisture effect should obviously be present, but the fuel may be responding more to the moisture content of the mixture of live and dead fuel, and thus may be relatively insensitive to changes in dead fuel moisture content. The effect of moisture content on rate of spread at lower moisture contents needs to be tested for heathland vegetation.

Dead fine fuel moisture content is not a direct input into the current McArthur meters, but is predicted from temperature and humidity within the meters. More recently fire researchers have incorporated moisture content directly into prediction equations, and later developed prediction equations for moisture content in the vegetation

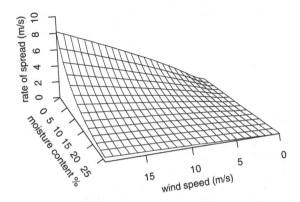

Fig. 3.3. Relative effects of dead fuel moisture content and wind speed on rate of spread for grasslands. From Cheney et al. (1998).

complex (Marsden-Smedley and Catchpole 1995b; McCaw 1998). Some of the prediction equations for moisture content used for Australian fuels are shown in Fig. 3.4, which shows the expected moisture content at an air temperature of 30 °C, as a function of humidity. It can be seen that the moisture contents of grass and forest litter are similar at 30 °C, and much lower than the moisture contents in the more shaded pine forests. The moisture content of the surface layer of forest litter is generally higher than that of the aerial fuel (Buckley 1993; Pook and Gill 1993), but is still closely influenced by temperature and humidity changes. In very open vegetation, such as mallee shrubland, the surface litter may be drier than the aerial fuel because of radiant heating from the ground (McCaw 1998). As well as air temperature and humidity, shading, season and time of day affect moisture content. For the same values of temperature and humidity McArthur (1967) notes that moisture contents may be 2%–3% lower in full sun than in shade, 1%–2% lower in summer than in spring or autumn, and 2%–3% lower in the afternoon, under adsorption conditions, than in the morning, under desorption conditions.

Fuels cannot respond immediately to changes in temperature and humidity so changes in dead fuel moisture content lag behind changes in meteorological conditions. Lag time and equilibrium moisture content (the moisture content attained under constant conditions) both depend on fuel particle

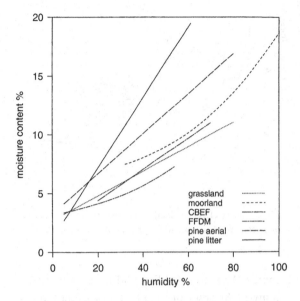

legend:
grassland ....
moorland - - - - -
CBEF ————
FFDM — - — -
pine aerial — - - —
pine litter ————

Fig. 3.4. Prediction equations for dead fuel moisture content (at a temperature of 30 °C) in terms of humidity, and their range of validity. Predictions are for (a) grassland Fire Danger Meter, Mark V (McArthur 1977), (b) buttongrass moorland (Marsden-Smedley and Catchpole (2001), (c) control burning of *Eucalyptus* forests (McArthur 1962), (d) fire behaviour in eucalypt forests (McArthur 1967), (e) pine plantation aerial dead fuel (Pook and Gill 1993), (f) pine plantation litter (Pook and Gill 1993). Equations to fit the tables are from (a) Noble *et al.* (1980), (c) Viney and Hatton (1989) and (d) Viney (1991).

size, density and weathering. Past variations in temperature, humidity and rainfall have been incorporated into moisture prediction equations using bookkeeping methods (e.g. Sneeuwjagt and Peet 1985; McCaw 1998; E.A. Catchpole *et al.* 2001). Fine fuels, such as grass, twigs and surface litter respond rapidly to changes in temperature and humidity, and generally have response times (to two-thirds of equilibrium moisture content) of an hour or two (Fosberg and Deeming 1971; Anderson 1990; E.A. Catchpole *et al.* 2001). Larger branches and logs may have response times ranging from several hours to several days.

After rain, the larger fuels and deep layer of the litter bed retain moisture, so there is less fuel available to burn. Forest fire behaviour prediction tables (FFBT) and meters assume that spread rate is proportional to available fine fuel load (McArthur 1967; Sneeuwjagt and Peet 1985), so that spread rate is

expected to decrease when the lower layer of litter and the larger branches are too wet to burn. This decrease in spread rate is modelled by decreasing the available fine fuel load. The FFBT use surface and profile moisture content to determine the 'available fuel', while the McArthur meter uses a 'drought factor'. This is a function of the number of days since rain, the amount of the last precipitation, and the Keetch–Byram drought index (KBDI), which indicates rainfall deficit (Keetch and Byram 1968). The drought factor in the Mark V Forest Meter has a maximum of 10, indicating that all fine fuel is available for burning. The equation fitted to the McArthur meter by Noble *et al.* (1980) allows the drought factor to exceed 10. If it can be shown that maximum drought effects occur when the drought factor is 10 a constraint should be put on the equation to reflect this. The drought factor indicates long-term moisture conditions, and may also reflect the lowered moisture contents of some of the live fuels, particularly when drought-stressed. A high drought index also indicates that few moist areas remain in the landscape to check fire spread (Cheney 1991).

Some fire authorities replace the KBDI with the soil dryness index (SDI) as an indicator of fire hazard. This was formulated by Mount (1972) as a replacement to the KBDI, which was developed in the southeast USA, and does not always reflect conditions in the more arid Australian climate. Burrows (1987) compared the performance of the SDI and KBDI in the southwest forests of Western Australia and found the SDI to be a superior index for fire control purposes.

Cheney *et al.* (1992), burning in eucalypt regrowth found no correlation between spread rate and drought index, and considered fires to be carried by the well-aerated trash fuel just above the forest litter. Buckley (1993) and Burrows (1999a, b) found no relationship between spread rate and fuel load in wind-aided forest fires. It thus seems likely that the larger forest fuels and deeper litter have only a minor effect on spread rate, and so a drought index (and its derived drought factor) is not relevant for spread rate prediction. The significance of drought indices in forest fuels may principally be in determining the probability of ignition, and the amount of large woody debris ignited and consumed (relating to suppression difficulty, tree damage and soil heating, and the risk of re-ignition at a later date). In grasslands, heathlands and moorlands burning is often possible a day or so after rain. In these open vegetation complexes, with little large fuel, the moisture content of the fine dead aerial fuels determines spread rate. Spread rates in mallee shrubland (McCaw 1998) and buttongrass moorland (Marsden-Smedley *et al.* 2001) were found to be independent of both SDI and KBDI.

Vegetation characteristics

It can be seen from Fig. 3.2 that vegetation type has a strong influence on spread rate. Fuel particle size, vegetation height, bulk density (mass of fuel per volume of fuel bed) and the percentage of dead fuel all differ markedly between vegetation types and influence spread rate and intensity.

Australian fire behaviour prediction tables and meters have been developed for specific vegetation complexes, and the only fuel structure characteristic that is included is fuel load. This has the disadvantage that the underlying prediction equations are not easily extended to slightly different conditions. It is useful here to summarise what is known about the effects of fuel structure on rate of spread.

In the Western Australian FFBT and the McArthur meter 'fine fuel' is assumed to be the major contributor to fire spread, but the definition of fine fuel varies from less than 6 mm to less than 10 mm. Larger fuels, which may burn in the firefront in severe conditions, can slow down spread rate due to the increased up-draught, which decreases radiation and convection ahead of the fire (E.A. Catchpole *et al.* 1993). Most vegetation has a range of particle sizes, weighted to the finer end of the scale, but the importance of the particle distribution in determining spread rate is not clear. In laboratory fires W.R. Catchpole *et al.* (1998b) found that particle size had no effect on spread rate once fuel diameters were less than 2 mm. In field studies Cheney *et al.* (1993) found no difference in spread rates between fires in tropical thick-stemmed grass (*Eriachne* sp.), and thin-stemmed tussock grass (*Themeda* sp.).

There is general agreement that spread rate increases with decreasing bulk density so that fires burn faster in well-aerated grass than in densely packed litter. In laboratory studies spread rate has been found to be inversely proportional to the square root of the bulk density (Carrier *et al.* 1991; W.R. Catchpole *et al.* 1998b).

If the fuel bulk density is constant (which is a reasonable assumption for forest litter; see Cheney 1981) a linear increase in spread rate with fuel load implies a linear increase with fuel bed depth. Using laboratory and field experiments in jarrah forest litter, Burrows (1999a, b) found a linear relationship between fuel load and spread rate in zero-wind fires, but could find no relationship for wind-aided fires. Burrows' experiments were done in single-aged fuel. With age litter load increases, but so do the amount of surface fuel, the understorey height and the percentage of dead fuel. All these could tend to increase spread rate. This is currently under investigation in Project Vesta, through comparison of simultaneous experimental fires in fuels of differing ages.

Unlike the McArthur Mark III grassland meter (McArthur 1966), the Mark V (McArthur 1977) included a linear effect of load on fire spread. To resolve the conflict between the meters Cheney *et al.* (1993) did extensive experimentation in tropical grassland in the Northern Territory and found no clear effect of load (or height). Cutting the grass to half the original height reduced the spread rate by a factor of 0.8, but this changed the bulk density and overall particle size, and was not a clear indication of a load or fuel depth effect. This led to a prediction model separating grazed and ungrazed pasture (Cheney *et al.* 1998). In shrub and heathlands spread rate has been related to the square root of fuel height (W.R. Catchpole *et al.* 1998a), but this could be due to lowered bulk density, senescence or increased load with increasing height (and therefore age). Laboratory studies show little effect of height on spread rate in wind-aided fires (Wolff *et al.* 1991; W.R. Catchpole *et al.* 1998b), but the range of heights used was small compared to the range in natural vegetation.

Cheney *et al.* (1992) introduced the concept of

'near surface fuel', the layer of predominantly horizontal aerated trash fuel above the forest litter. In *Eucalyptus sieberi* regrowth in coastal Victoria they found spread rate to depend on the continuity (and possibly height) of this layer. Fuel continuity is almost certainly an important factor in determining fire spread in most vegetation types, but it is difficult to quantify. In the discontinuous spinifex hummock grass communities continuity (or patchiness) has been incorporated into spread rate equations (Griffin and Allan 1984; Burrows *et al.* 1991). Burrows *et al.* (1991) found that Griffin and Allan's model considerably overpredicted spread rate in spinifex fuels in the Gibson Desert. Burrows *et al.* (1991) suggest that this may be due to the greater average gap size in their study area. A model being developed by Mercer (1997), in which the probability of spread depends on flame dimensions and patch distribution, shows promise as a prediction model for spread rate in these discontinuous fuels.

Continuity, biomass and dead-to-live fuel ratio are greater in older than in younger vegetation, and average live fuel moisture tends to be lower, as a smaller proportion of the shrub foliage is new growth. All these tend to result in greater spread rate as age increases. However the increase of these variables with age levels off eventually and so the effect of age on spread rate should also level off. Figure 3.5 shows the predicted effect of age on spread rate in buttongrass moorland (Marsden-Smedley and Catchpole 1995b) and jarrah forest (FFBT), respectively. Note that the predictions for jarrah do not take account of the effect of age (through shrub height and density) on fuel moisture content and surface wind speed.

Curing

The process of grasslands drying out and dying in the summer, or the dry season in the tropics, is known in Australia as 'curing'. Completely dead grass is 100% cured. The influence of live grass at a high fuel moisture content has a damping effect on fire spread rate. The damping functions fitted to the Mark III and Mark V grassland meters (Noble *et al.* 1980) are considerably different (see Fig. 3.6). Cheney *et al.* (1998) found that both of these func-

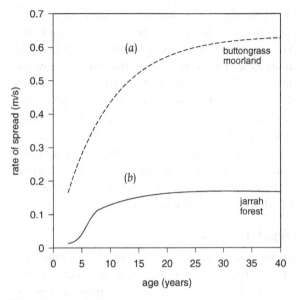

Fig. 3.5. Predicted effect of age on rate of spread in (a) buttongrass moorland (Marsden-Smedley and Catchpole 1995b) and (b) jarrah forest; fuel accumulation effect from Beck (1995), and shrub height dependence on age from Burrows (1994). Conditions: temperature 30 °C, humidity 30%, wind speed (10 m open) 10 m s^{-1}.

tions overestimated the effect of curing above a curing of 80%, and adjusted the curing function to reflect this. The latest curing damping function is a logistic function given by

$$\Phi(C) = 1.12 / \{1 + 59.2 \exp[-0.124(C - 50)]\},$$

where C is the degree of curing in percent (see Fig. 3.6). There is little supporting data for any of these functions, except the negligible effect of curing above 80% in the Northern Territory grass fires discussed in Cheney *et al.* (1993, 1998). Luke and McArthur (1978) suggest increased flammability when grasslands first cure due to increased volatiles, but they do not quantify the effect.

Satellite imagery now being using to estimate percentage dead fuel (i.e. percentage cured) cannot easily distinguish between (a) drying out in summer, (b) new growth after rain and (c) patchiness in the landscape caused by wetter gullies. The curing factor above is more relevant to (a), although the minimal effect on spread rate in $\Phi(C)$ at high curing could reflect the effect of a fire burning over

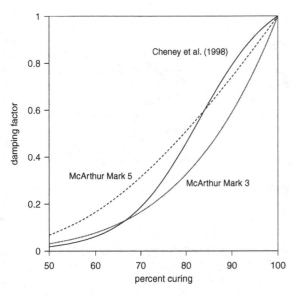

Fig. 3.6. Prediction equations for the damping effect of grassland curing on rate of spread.

the top of new growth in some of the fires in Cheney *et al.* (1993). On a landscape scale, the curing function relates to an average spread rate, taking account of the fact that the fire may be slowed down in less cured grass. In practice the spread rate will remain high in the cured plains, and the fire may well be extinguished in the gullies unless spotting occurs.

In heathlands and shrublands the percentage of dead fuel varies with age and drought stress. Age-related spread equations (e.g. Marsden-Smedley and Catchpole's (1995b) equation for moorland) incorporate the effect of increased senescence with age.

No direct use of the moisture content of the live fuel is made for predicting fire spread in Australian fuels, unlike in the USA where live fuel moisture content is directly input into the BEHAVE fire prediction equations (Andrews 1986). In the McArthur grassland meters curing acts as a surrogate for live fuel moisture content. In shrubland and moorland fire behaviour experiments no effect of live fuel moisture on spread rate has been detected (Marsden-Smedley and Catchpole 1995b; W.R. Catchpole *et al.* 1998a; McCaw 1998). However, in conditions of drought stress, when live fuel moisture contents are very low, fire behaviour may well

be affected. In forests, there can be more variation of live fuel moisture with season, for example when bracken (*Pteridium esculentum*) cures out. Some of the effect of lowered live fuel moisture on spread rate is probably reflected in the drought index in the McArthur forest meter.

Slope

Increased radiative and convective heat transfer in fires on slopes causes a major increase in spread rate. The McArthur forest meter predicts that a slope of 10° will double the spread rate of a forest fire, and a slope of 20° will quadruple the spread rate. Moreover the effects of slope and wind are multiplicative, so that the slope effect is equally strong in high winds as in low winds. Noble *et al.* (1980) fitted the equation

$$R = R_0 \exp(0.0693\,\theta)$$

to the McArthur forest meter (where R and R_0 are the spread rates on slope and flat ground respectively, and θ is the slope angle). This relationship is shown in Fig. 3.7. It also applies to downhill slopes, but eddy winds on lee slopes can complicate fire behaviour downslope (Cheney and Sullivan 1997). Beck (2000) found that McArthur's relationship overestimated the reduction in spread rate in downslope fires in jarrah litter, and suggested an alternative relationship for downslope fires given by

$$R = R_0 (1 + 0.055\,\theta \exp(0.06437\,\theta)),$$

which reduces the spread rate to about 0.70 of the zero-slope spread rate at a slope of $-10°$, after which there is virtually no more effect. For large forest fires, where the spread rate is dominated by spotting, Luke and McArthur (1978) recommend that the slope correction to spread rate should not be used.

Cheney *et al.* (1992), Burrows (1999a) and Beck (2000) verified McArthur's relationship in low-wind fires on slopes below 25° in eucalypt regrowth fuels, jarrah litter in the laboratory and karri forest, respectively. In other Australian vegetation types, there is little good data for fires on slopes. The slope effect formulated by Van Wagner (1977b) for conifer forests is less than McArthur's slope effect, provided that the slope is less than about 40° (see Fig. 3.7).

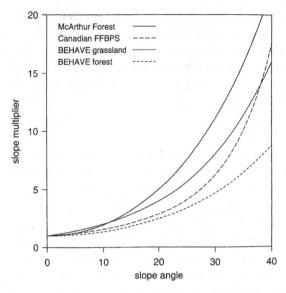

Fig. 3.7. The effect of slope on rate of spread as predicted by the McArthur forest meter, the Canadian FFBPS and the US BEHAVE system (for grassland and eucalypt forest litter). For eucalypt forest litter, the packing ratio = 0.085 (Beck 2000), and surface area to volume ratio = 5500 m^{-1} (Burrows 1994).

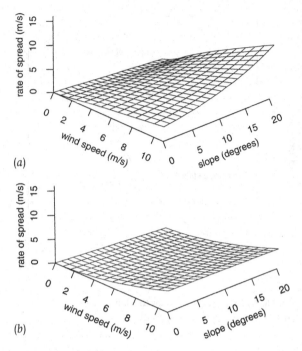

Fig. 3.8. The relative effect of wind speed and slope on rate of spread: (a) grassland fire behaviour model (Cheney and Sullivan 1997; Cheney et al. 1998), (b) US BEHAVE predictions for grassland (Rothermel 1972).

This slope effect is used by the Canadian Fire Behaviour Prediction System (Forestry Canada Fire Danger Group 1992), but is converted to an equivalent wind effect for use for fires not burning upslope. The BEHAVE system, developed in the USA based on Rothermel (1972), uses an additive wind and slope effect of the form

$$R = R_0 (\Phi_w + \Phi_s),$$

where Φ_w and Φ_s are the wind and slope effects respectively. The additive model results in the wind effect dominating the slope effect at higher wind speeds. Compare this with the multiplicative model used by McArthur (see Fig. 3.8). Rothermel's (1972) slope experiments were all done in zero wind. Weise and Biging (1997) tested the effect of wind and slope in a small open wind tunnel with wind speeds between −1 and 1 m s^{-1} and slopes between −17° and 17°. The McArthur equation for slope overestimates the slope effect in their data, particularly for backing fires. Weise and Biging's (1997) data also suggest that an additive model for slope may be more appropriate.

Several methods have been suggested for predicting fire spread on slopes not parallel to the wind (Pagni and Peterson 1973; Rothermel 1983; Gellie 1990; McAlpine et al. 1991). Weise and Biging (1997) found the best prediction methods to be Pagni and Peterson's physical model and an unpublished modification of BEHAVE suggested by F.A. Albini.

Surface wind speed increases up a slope. Because of its empirical nature this effect is built into the McArthur meter. McRae (1997) points out that using a wind–terrain model, which reflects this increase in wind speed, could cause overestimation of spread rate.

Rothermel (1972) showed that the slope effect for zero-wind fires increased with decreasing bulk density (compare the predicted slope effects for forest and grassland in Fig. 3.7). If this were true for wind-aided fires one would expect the slope effect to be greater in grass and shrub fires than in forest fires. Experimental fires on slopes in heath and shrubland are being carried out by New Zealand

Forest Research and by the European Commission funded Project Inflame in Portugal (Viegas 1999). Early results from New Zealand suggest that the slope effect is less than that predicted by the McArthur forest meter. Vegetation on slopes is often more sparse and discontinuous, broken by rocky outcrops. This tends to decrease spread rate, which would decrease the effect of slope.

Fire growth and headfire width

Fires from a point source increase their spread rate as they grow. At first the radiative transfer is low because the flames are drawn back into the centre of the fire. In fires where the convective up-draught is high compared to the wind speed, such as in forest fuels in dry conditions, fires may spread as backing-fires for a considerable time after ignition. Burrows (1994) notes that in jarrah forests in summer conditions a threshold surface wind speed of 1 m s^{-1} is needed to overcome this effect. When the influence of the convective up-draught is overcome the spread rate will increase due to the increase of radiative transfer, as the fire grows larger. However, even when the full effect of radiation is established fires still may continue to accelerate. The equation predicting fire growth in forest fires given by Cheney and Bary (1969) was

$$R = R_{ss} \exp(-a/t).$$

This equation describes a spread rate, R, which increases with time, t, but whose rate of increase becomes less as the fire tends to a quasi-steady rate of spread, R_{ss} ('quasi-steady' because spread rates are never constant in natural conditions). For forest fires Cheney and Bary give $a = 2.16$. This predicts that a fire will have reached 90% of its quasi-steady spread rate in 20 minutes. Low-intensity fires tend to reach a quasi-steady spread rate faster than high-intensity fires (McAlpine and Wakimoto 1991). Cheney and Gould (1995) found that low-intensity grass and woodland fires, lit from point sources, reached their maximum spread in 5–10 minutes, while medium- to high-intensity fires, in winds of above 4 m s^{-1}, took 12–45 minutes. Forest fires can take many hours to reach a quasi-steady spread rate (Cheney and Gould 1997). Consumption of progres-

sive vegetation layers allows greater penetration of the wind into the flaming zone, which could be a partial cause of the acceleration of forest fires. The time taken is less if there are many shifts in wind direction which cause the fire to spread with a broad front and increased spread rate (Cheney and Gould 1995). Quasi-steady state is reached when the convection column is fully developed and damps out minor fluctuations in the wind direction (Cheney 1981).

Cheney and Gould (1995) gave some quantification of the effect of fire-line width on rate of spread. They found that in grassland, and woodland fires with a grass understorey, fire growth was dependent on the effective headfire width (the width across the headfire) and the wind speed. The equation used (given in Cheney et al. 1998) was of the form

$$R = R_{ss}\{1 - \exp[-(b + cU)/W]\},$$

where U is the 2-m wind speed and W is the effective headfire width. At low wind speeds quasi-steady spread rate is achieved faster for the same headfire width. Woodland fires were found to need a wider front to achieve the same percentage of steady-state spread rate as grassland fires. Cheney and Gould (1995) hypothesised that this was due to the greater influence of thermal up-draughts in the woodlands. The fraction of quasi-steady rate of spread achieved in grassland as a function of wind speed and headfire width is shown in Fig. 3.9. It seems likely that fire growth should be a function of the intensity of the fire compared to the headfire width, and thus should be a function of fuel load and moisture content as well as wind speed. Cheney and Gould did not have a large enough range of data to explore this, and it needs further research in other vegetation types.

Increase in spread rate by spotting

Spotting occurs when flaming bark or twigs from forest or shrubland fires are lofted into the air and ignite the fuel ahead of the fire. There is little quantitative data on the effect of spotting on spread rate. Short-range spotting of 5–10 metres in front of the fire may have little impact on spread rate, as the

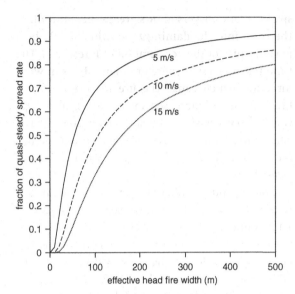

Fig. 3.9. Fraction of quasi-steady spread rate in terms of effective head fire widths at surface wind speeds of 5, 10 and 15 m s^{-1} in grasslands (Cheney *et al.* 1998).

spots are soon overrun by the main fire front (Cheney and Bary 1969). It does however make suppression very difficult. Long-range spotting, which can be 2–10 km in front of a forest fire, causes new fires, and has no effect on spread rate unless conditions are extreme and the fires join up.

Medium-range spotting of up to 2 km from medium- to high-intensity fires in long unburnt forests with fibrous bark, such as stringy-barked eucalypts, can increase spread rate significantly. Cheney and Bary (1969) report a three- to five-fold increase in spread rate caused by spotting, which is not reflected in the McArthur forest meter. When the main fire coalesces with the spot fires there are pulses of stronger intensity as the fires run together, due to increased radiative and convective activity (Cheney and Bary 1969). During these pulses more firebrands are lofted, and the fire moves forward in surges. In severe weather conditions this may cause firestorms capable of snapping or uprooting trees, as in the 1967 Hobart fires (McArthur and Cheney 1967).

Prolific short-range spotting from suspended dead material is common in shrublands. In discontinuous mallee communities spotting from euca-

lypt clumps may be a major fire-spread mechanism (Bradstock and Gill 1993). In this case, where spread is limited by clump interspacing, even short-distance spotting can bridge gaps, and either initiate spread or increase spread rate.

Atmospheric interactions

Byram (1959) introduced an energy criterion dependent on the ratio of the rate of conversion of thermal to kinetic energy in the convection column to the rate of horizontal flow of kinetic energy through unit vertical area. This is generally referred to as 'the power of the fire (P_f) divided by the power of the wind (P_w)', or Byram's convection number. Unpublished case studies by Byram showed that when P_f/P_w is greater or equal to 1 for at least 300 m above a fire, the fire tends to exhibit erratic behaviour, and 'blow up', with increasing spread rates due to strong down-draughts, and fire whirlwinds. Rothermel (1991) related P_f/P_w to changes in the appearance of the convection column, and used it as a predictor for 'blow-up' fires in North American coniferous forests. Possibly because of the lower crown loads, leading to lower intensities in eucalypt forests, this has not received much attention in Australia, but case studies by Burrows (1984), Buckley (1990) and Tolhurst and Chatto (1999) suggest that 'blow-up' conditions can be a feature of Australian forest fires.

Another possible predictor of 'blow up' fires is the Haines Index for atmospheric stability (Haines 1988) which Haines related to fire severity in North American conifer forests. Similar research done by Bally (1995) in Australia shows that the Haines Index gives additional predictive power over the McArthur Forest Fire Danger Index to predict fire activity. Better understanding of the interaction of a large fire with the atmosphere should evolve from recent work by Clark *et al.* (1996) who examined the effect of adding a heat source to a meteorological meso-scale model of the atmosphere.

Intensity

The fire-line intensity (heat release rate per unit fire-line length) is an important predictor of the effects of a fire, and of its suppression potential. Byram (1959) defined fire intensity as

$$I = h\, w_a R,$$

where h is the heat released by unit mass of fuel (the heat of combustion), w_a is the fuel involved in combustion (known as 'available fuel') and R is the spread rate. Byram (1959) primarily intended that intensity should be a measure of the heat given out in the convection column, and for this the heat released and fuel consumed in flaming combustion (rather than total combustion) should be used. The fuel consumed by flaming combustion should relate more directly to flame and scorch height. In practice though it is only possible to measure the total fuel consumed after the fire, and intensity measures have generally been based on this. It is, however, possible to obtain an estimate of the fuel consumed in flaming combustion by using size class limits based on mass loss studies in the laboratory (see, for example, Burrows 2001).

Byram (1959) suggests an adjustment to the heat of combustion for the heat that is lost to the convection column due to radiation to the ground and atmosphere, but a good measure of this is difficult to obtain. The heat of combustion is also reduced slightly by moisture content, due to losses through evaporation and desorption. Some heat energy also remains in unburnt residues. Byram referred to the net heat released into the convection column as the 'heat yield'. Luke and McArthur (1978) suggest an average value of 16 000 kJ kg^{-1} for Australian fuels.

While intensity is important in determining some fire effects it only tells part of the story. A fast-moving grassfire in light fuel can be as intense as a much slower-moving fire burning through heavy scrub, but its impact on overstorey woodland is much less. Other factors like flame residence time and total heat load (the product of residence time and intensity) may be more appropriate for determining the impact of the fire on vegetation. Cheney (1990) recommended that fire intensity only be used to compare fires in structurally similar fuel types.

Available fuel

Prediction of the percentage of available fuel is necessary for prediction of intensity. Young trees, thick twigs and branches, bark, and deep litter will burn in medium- and high-intensity fires, but not in low-intensity fires. Thus there tends to be a feedback loop of the available fuel with increasing intensity. To avoid this loop predictions of available fuel are needed in terms of fuel and environmental variables.

In laboratory experiments in eucalypt jarrah litter Burrows (2001) found that over a range of diameters (up to 75 mm) about 85% of the fuel was consumed in the fire. However, the amount burned by flaming combustion was highly dependent on fuel diameter, and ranged from 75% for leaves and fine fuel to 40% for 75-mm diameter branches. In low–moderate-intensity fires (<3000 kW m^{-1}) in thinning slash in karri forest McCaw et al. (1997) found that on average 95% of fine aerial fuel, 85% of 6–25 mm fuel and about 55% of litter and larger fuels were consumed. Total fuel consumption of all but the fine aerial fuels was inversely related to profile (whole litter) moisture content (and also SDI). Tables of available fuel for Western Australian forest litter are given in Sneeuwjagt and Peet (1985) in terms of surface and profile moisture contents. Available fuel from understorey shrubs is given in terms of shrub structure, height, density and fire intensity class. In the McArthur forest meter the drought factor is a direct predictor of available fuel. A drought factor of 10 indicates that all fuel is available to be burned. Albini and Reinhardt (1995) formulated a model of fuel consumption for large fuels in coniferous forests, but this would need calibration for Australian forest types.

In shrublands much of the dead fuel is below 6 mm, and is consumed even in low-intensity fires. Estimates of the maximum size of live twigs that are consumed vary from 2 to 9 mm. Whight and Bradstock (1999) have shown that the maximum tip diameter remaining after a fire increases with intensity and decreases with height above the ground, but a predictive equation has not yet been determined.

Moderate-intensity fires in shrubland and forest may kill, but not consume, live shrub and small tree material, which may remain standing for several years before it is knocked down and rots. This then increases the potential available fuel for the next fire season. This may not be a problem as the area may take two or three years before it is sufficiently recovered to burn, but if grasses invade, and the area becomes flammable, the old shrub material could add considerably to the available fuel. Patchy burning will result in more fuel being available for a subsequent burn in the same area. This happens for example in moorland where burns at high moisture contents may leave the lower stratum of the moorland unburned. This then dies, and leaves a flammable area of thatch that can remain for several years (Marsden-Smedley and Catchpole 1995a).

Flame length

Flame length has been modelled in terms of intensity (Byram 1959; Thomas 1971), and has sometimes been used to predict intensity (Alexander 1982). It is an indicator of suppression potential, as is intensity. Byram's and Thomas's equations for flame length take the form $L = a I^b$, where L is flame length (m) and I is Byram's intensity (kW m^{-1}). Byram's equation (for which $a = 0.0775$ and $b = 0.46$) appears to give better predictions in surface fuels (Albini 1976). Thomas's equation ($a = 0.0266$, $b = 0.67$), which gives higher predictions, has been used for estimating crown fire flame length in coniferous forests (Rothermel 1991). It is probable that the constants are in fact fuel-dependent, but flame lengths are difficult to quantify without video analysis (e.g. McMahon et al. 1986), and much of the variation in reported observations is probably in the methods of observation and criteria used. In shrubland fires Byram's equation has been found to underpredict slightly at medium–high intensities (Van Wilgen et al. 1985; W.R. Catchpole et al. 1998a), but predictions would be adequate for most purposes. For fires in jarrah forest Burrows (1999b) found that it underpredicted badly.

Flame length is critical to spread rate in discontinuous fuels such as mallee and hummock grasslands (Allan and Southgate, this volume; Bradstock

and Cohn, this volume). Bradstock and Gill (1993) found that flame length was strongly related to clump height and diameter, both in mallee and spinifex clumps, and determined that longer flames resulted when spinifex burned under mallee clumps causing crowning in the eucalypts.

Fire shape

Headfires are driven by radiative and convective heat transfer from flames leaning towards the unburnt fuel, and are strongly affected by variations in wind speed. Flank fires are partly driven by radiation from the smaller flames along the flank, made upright or curving inwards by the convection column. There is also intermittent convective transfer in turbulent and variable winds. Thus flank spread may increase with wind speed, but the wind has less influence on flank spread than it does on headfire spread. Flank spread, particularly in grassland fires, can be very much dominated by variable winds, which change the direction of fire spread. At the back of the fire the flames bend away from the unburnt fuel, and the fire spreads slowly by radiation and flame contact, at more or less the spread rate in zero-wind. Backfire spread rates are equal to or marginally less than zero-wind rates of spread (Beck 2000). Cheney and Sullivan (1997) state that backfire spread rate in grassfires is unaffected by wind speed, but decreases with increasing dead fuel moisture content from about 0.015 m s^{-1} at 3% to extinction at 18%. Fires in homogeneous fuel, terrain and environmental conditions form roughly elliptical shapes because the headfire spread rate is much faster than the back and flank spread rates. As a result of slower spread rates in the acceleration period of the fire the shape can be more like a teardrop. Changes in wind direction and terrain over the duration of the fire cause many fires to have irregular shapes.

To characterise fire shape the ratio of length to width has been used. On a landscape scale the effect of slope is not so important, and length-to-width ratios tend to be dominated by wind effects. McArthur (1966) gave a relationship between length-to-width ratio and wind speed obtained for

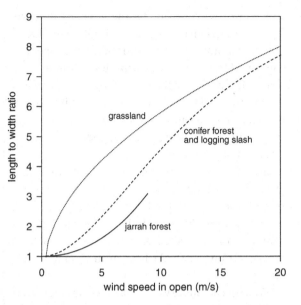

Fig. 3.10. Relationships between length-to-width ratio and wind speed for different vegetation types: grassland from McArthur (1966); conifer forest and logging slash from Forestry Canada Fire Danger Group (1992); jarrah forest from Beck (2000).

wildfires in grassland. This is shown in Fig. 3.10, together with a relationship determined for fires in conifer forests in Canada, and a relationship for laboratory fires in jarrah forest fuels from Burrows (1994) (modified by Beck 2000 to include wildfire data). Grassfires tend to be narrower than forest fires for the same 10 m open wind speed, but much of the difference can be attributed to the reduction in wind speed in forests. In hummock grasslands Burrows *et al.* (1991) report long narrow tongues with little or no back and flank spread because fires in this discontinuous fuel will not sustain without wind.

For fires in non-homogeneous conditions Anderson (1983) suggested characterising the fire by a shape index determined by dividing the outer perimeter of the fire by the perimeter of a circle with an area equal to that enclosed by the outer perimeter. Eberhart and Woodward (1987) found that shape index increased with fire size for fires in the Canadian boreal forests.

Simulations of the effect of variable wind on fire shape show that variability in wind direction can cause teardrop shapes or lemniscates (Richards 1999). Cheney *et al.* (1993) found that grassfires with wide heads had significantly greater spread rates, and related this to more variable winds. Down-draughts tend to blow the fire sideways and this gives rise to wider fires than might be expected. The width of a fire decreases with increasing consistency in wind direction (constancy). Constancy is defined as the vector mean wind divided by the mean wind speed regardless of direction (Maher and Lee 1977). Constancy can be as low as 30%–40% in summer atmospheric conditions, and this results in wide fires. Days with severe fire-weather in Australia are often characterised by strong winds. The passage of a cold front on such days can alter wind direction by 90°. This results in wide fires with greatly increased spread rates, and causes a dangerous situation for firefighters.

Fire size

Fire size is a function of spread rate, the probability of containment (especially by initial attack) and the weather pattern during the fire. If fire-weather is severe, there may be many ignitions and resources will be stretched, so more big fires develop, not just because the conditions are extreme, but because initial attack is impossible. In severe weather conditions fires may be impossible to contain, and may burn large areas in the course of a single day (see Cheney 1976). In the evening, when conditions become less severe, fires will die down and suppression may be possible, but long burnout times in large forest fuels may enable a fire to smoulder all night and build up again the next day, particularly in inaccessible regions.

Containment

Some fires can be attacked early, and so do not grow large. Whether containment is possible will depend on the remoteness of the fire, the priorities assigned for the region, and the intensity and spread rate of the fire. Different estimates of the intensity at which fires can be contained have been made in the literature but they vary considerably, from 2000 to 4000 kW m^{-1}, partly due to the type of vegetation

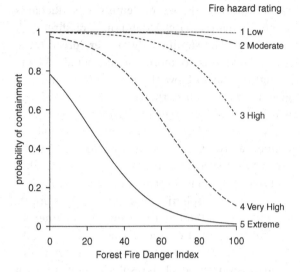

Fig. 3.11. Probability of successful initial attack in Victorian forests as a function of Forest Fire Danger Index for different overall fuel hazard ratings. After McCarthy and Tolhurst (1998).

considered and partly due the resources supposed available for containment. An empirically based probabilistic model for Victorian dry sclerophyll forests developed by McCarthy and Tolhurst (1998) splits the effects of environment and fuel (including the effect of bark hazard and thus spotting potential). It relates the probability of successful initial attack to the Forest Fire Danger Index (McArthur 1967) and the overall fuel hazard rating (McCarthy *et al.* 1998). The function used is

$$\ln[\,p_c/(1-p_c)\,] = a - bH - c\,\text{FFDI},$$

where p_c is the probability of successful initial attack, H is the overall fuel hazard rating, and FFDI is the Forest Fire Danger Index. The hazard rating takes values 1–5 denoting low (1) through to extreme (5). The function is graphed in Fig. 3.11, as a function of FFDI, over the range of hazard ratings. This idea could be extended to other forest types, and possibly to shrubland.

Horizontal patterns

The intensity within a fire in homogeneous conditions varies over the area of the fire, depending on whether the vegetation is burned by the head, flank or backfire. An expression for the intensity in terms of position was given by E.A. Catchpole *et al.* (1982), under the assumption of an elliptical fire shape. The diagram of the intensity distribution, given in their paper, clearly shows the large difference between head, flanks and back. While this is a stylised representation it does show that a great range of intensities will have been experienced even when conditions are uniform.

Intensity variations are seen in all fires, and relate to variations in wind speed, moisture content, fuel load and to barriers such as rock outcrops. Variability may be induced by the lighting technique in prescribed burns where 20%–30% of the vegetation, particularly in gullies, may be left unburnt.

Wind speed is more variable at lower average wind speeds. In summer, in New South Wales, wind speed variation has been found to be around 100% for winds of about 3 m s^{-1}, decreasing to about 40% for higher wind speeds (M. Speer personal communication). Variations in the wind speed cause variations in spread rate and hence variations in intensity. Richards and Walberg (1998) showed by simulation that variable wind speed is more important than variable fuel load in producing patterns in intensity in North American conifer fuels. Intensity tends to be more variable in forest and woodland than open vegetation. Intermittent up-draughts from the convection column cause lulls in the fire, then when the convection column breaks down the fire surges ahead causing marked variations in intensity.

In shrublands fuels may vary within a small area from discontinuous to continuous, and over a wide range of heights, causing large variations in intensity from both the effects on spread rate and available fuel. Based on the heathland fire behaviour model (W.R. Catchpole *et al.* 1998a) intensity in 2-m tall heathland would be about three times the intensity in 1-m tall heathland. Fuel moisture variations occur due to topography and vegetative shading and this also results in variations in intensity. In mallee shrubland, for example, the model developed by McCaw (1998) predicts that a decrease of 3% in moisture content would result in an increase in intensity of about 30%.

In low-intensity fires burning may be patchy. Hobbs and Atkins (1988) found that the variations in temperature within fires in shrubby woodlands were greater if the fire was less intense. Unburnt patches may be left within the fire boundary. Quantification of these patches, important in determining fire effects, is needed. It may then be possible to predict patchiness in terms of fuel and environmental variables. Eberhart and Woodward (1987) suggested (a) percentage burnt area per area enclosed by the outer perimeter of the fires, (b) median island size-class and (c) number of islands per 100 ha as indices. Percentage of burnt area at various distances from residual vegetation, important for seed dispersal and grazing, was also considered. Eberhart and Woodward (1987) showed that in the Canadian boreal forests the proportion of area unburned in a fire and also the median island area increased with fire size. Number of islands per 100 ha was greatest in medium-sized fires. This may be because larger-sized fires were associated with more extreme conditions.

Even in relatively homogeneous fuels, the perimeter of a fire is of a fractal nature due to variations in wind speed and direction. Grassfires often form several separate heads, which may later rejoin, or one may grow narrow and extinguish. When the fire eventually extinguishes this causes a fractured perimeter. McAlpine and Wotton (1993) found that the perimeter length of large fires could be two to four times that given by assuming an elliptical fire shape. (Exactly how much more was dependent on the length scale used to measure the perimeter.) Quantification of perimeter length using the shape or edge indices suggested by Eberhart and Woodward (1987) may be useful, for studies in post-grazing fire effects, for example. The shape index has been discussed. The edge index is given by determining the whole fire perimeter (including the edges of the inner islands) and dividing by the perimeter of a circle which has area equal to that of the total fire area. Edge index increased with fire size in Eberhart and Woodward's studies in the Canadian boreal forests.

Vertical patterns

In stratified vegetation, like shrubland and forests, fires may burn through the lower strata and leave the upper stratum either untouched or only partially damaged. This occurs, for example, in low-intensity prescribed burns in forests where the live crown of the overstorey is untouched, or in fires that burn through the litter in old shrubland. In rarer cases fires may burn through the upper stratum above a lower wetter stratum, such as when fires in shrubland or moorland burn over wet ground (Cheney and Sullivan 1997). Within a fire vertical patterning may occur, depending on fire intensity. In some areas only ground fuels and low bark may be burned. In other areas crown scorching or complete crown consumption may occur. To consider what determines the vertical distribution of burning and consequently vegetation mortality, fires can be conveniently divided into ground, surface and crown fires.

Ground fires

Ground fires that creep along in the peat or deep duff layer are not common in mainland Australia as decomposition rates are fast, and the humus layer is thin (Cheney 1981). They may occur in peat beds or deep organic soils that have dried out in prolonged droughts. In Tasmania where conditions are colder, and decomposition rates are slower, ground fires are more common. Ground fires seldom start without surface fires, which are their ignition source, but may persist for days after the surface fire has been extinguished. If environmental conditions become more severe they have the potential to flare up and become surface fires.

Surface fires

Surface fires burn in grassland, forest litter and understorey, or shrubland. The effect of the fire on vegetation mortality depends on what temperature the vegetation is exposed to, the length of the exposure time, and nature and extent of the insulating layer (such as bark). Mercer and Weber (1994) and Weber et al. (1995b) modelled the flame temperature distribution at all heights in the flame and plume above the flame, but for simplicity the temperature distribution is usually characterised by the flame height, scorch height and flame residence time at the base of the flame. Scorch height is more important than flame height in determining the impact of the fire and the vertical extent of the fire imprint.

The residence time of a fire determines cambium damage and root mortality. It is also important in determining seed release, and soil temperature. Residence time can be split into flaming combustion time and glowing combustion time (where the cellulose products burn by glowing combustion with little flame). Most plant deaths occur in the flaming region. In the region of glowing combustion behind the fire front the deeper layers of litter, large branches and logs burn, causing damage to nearby cambium and to roots. Flaming time is given by D/R, where R is the spread rate and D the flame depth. For individual branches and logs it is principally a function of surface area to volume ratio and moisture content (Anderson 1969; Cheney 1981). For continuous fuel beds it is a strong function of available fuel load and moisture content (Burrows 1999b; W.R. Catchpole unpublished data). Typical values of about 5 seconds in light grassland and 50 seconds, or more, in forest litter have been reported. Glowing combustion time is much longer, and can even last days in some vegetation types, such as tussock grassland (Cheney and Sullivan 1997).

Tree crown damage can be related to scorch height. Even if the scorch height is too low to kill the crowns, stems may suffer injury, which can later lead to insect attack. However if the fire is sufficiently intense to actually burn in the crowns, much more serious damage ensues, and mortality is high. Complete crown scorch can be expected in most forest types even at moderate intensities of $500-3000$ kW m^{-1} (Cheney 1981). Several authors have given simple models of scorch height for forest or woodland (e.g. Cheney et al. 1992; Buckley 1993; Burrows 1994; Williams et al. 1998). These predict scorch height as a function of spread rate (and air temperature), flame height or intensity. Scorch height has been shown to vary with season and with drought factor (Burrows 1994; Tolhurst 1995). This probably reflects the fact that leaves are more resist-ant to damage at the higher live fuel moisture contents associated with spring burns. Van Wagner (1973) gave a physically based model of scorch height in terms of fire-line intensity and wind speed, but Burrows (1994) and Gould et al. (1997) found that this model underpredicted scorch height in Australian forest types. Gould et al. (1997) developed a different physical model, calibrated for eucalypt regrowth forest, which predicted better than Van Wagner's model, but it needs further development to produce accurate predictions for any forest type. In these physically based models scorch height increases strongly with intensity and weakly with ambient temperature, but decreases with wind speed as a result of the tilting flames. A dependence on live fuel moisture content, or drought index, should be considered in further development of these models.

Crown fires

Van Wagner (1977a) and Alexander (1998) developed models for the initiation of crown fires in coniferous forests. They are based on two criteria. Firstly, the surface fire must be intense enough, the foliar moisture content low enough and the flame residence time long enough (in Alexander's model) so there is sufficient heating for the crowns of the trees to ignite. Secondly, the crown bulk density must be great enough for the fires to spread. Because the rough bark of some eucalypts can enable fire to reach the crowns so easily in Australian forests the crown initiation model is probably too simplistic for Australian fuel types. On the other hand, the conditions for critical spread may be more easily calibrated for use in eucalypt forests.

In Australian forests crown fires can be expected in low forest types at surface-fire intensities of $3000-7000$ kW m^{-1}, and in most forest types at intensities above 7000 kW m^{-1} (Cheney 1981). Crowning has been related to bark hazard, which is a function of fuel age (in the sense of when the bark of the tree last burned) and tree type (rough or smooth bark). A photographic guide to fire hazard (that relates to crowning potential) was given by McCarthy et al. (1998) for Victorian forests.

Dead fuel moisture content affects crowning

potential by affecting flame height, and the ability of the fire to ascend the bark into the crowns. McArthur (1967) states that below a moisture content of 7% crown fire development is possible, and the strong convective circulation thus created instigates spotting. He also notes that crown fires are fairly frequent when the moisture content is below 4% in medium fuels of 7–8 tons per acre (1.6–1.8 kg m^{-2}). Live fuel moisture, which may vary seasonally in eucalypt species (McCaw 1998), may affect spread rate in the crowns. Very strong winds actually limit crowning because the flame angle is reduced (Luke and McArthur 1978). None of these effects has yet been properly quantified.

Little work has been done to determine crowning conditions for old shrublands and mallee/shrub communities that have distinct crown layers. McCaw (1998) gives a provisional flame length of 5 m needed for crowning in a 3–4-m mature mallee/shrub community, which corresponds to a rate of spread of about 0.4 m s^{-1} and fire intensity of 8500 kW m^{-1}.

Spot-fire patterns

Spot fires that result in a distinct imprint away from the main fire are caused by medium- to long range-spotting. Spotting of over 2 km can be expected from low forest types when intensities are between 3000 and 7000 kW m^{-1} and crowning occurs. In very high intensity fires of over 7000 kW m^{-1} much larger spotting distances can be expected. The McArthur forest meter predicts spotting distance in terms of the Forest Fire Danger Index and fuel load. For example, in a forest with a fuel load of 2 kg m^{-2}, the meter predicts spotting of about 4 km when the FFDI is 50. This implies an intensity of over 11 000 kW m^{-1}. The likelihood of long-distance spotting depends heavily on the dominant forest species. McCarthy et al. (1998) categorise bark hazard from low to extreme depending on the type of tree and build-up of bark since the last fire, but this does not necessarily give a guide to long-range spotting. Candle-bark species are more likely to generate long-range spot fires than stringy-bark species (Cheney and Bary 1969). Long candle-bark steamers can cause spotfires 20–30 km ahead of the main fire. The brands are light, aerodynamic, and have a long flaming time. Spotting is more likely in late summer, or after a drought period, when the candle bark is shed from the trees. Long-distance spotting occurs when the fire has a large convection column so the brands are lofted high, and the upper winds are strong enough to carry the brands some kilometres ahead of the fire front.

Albini (1983) modelled the spotting process for fuels in the USA to give a prediction for maximum spotting distance. Morris (1987) shows this to be primarily a function of fire-line intensity and wind speed. For Australian conditions, the greater availability of bark, longer particle burnout times, and more aerodynamic shapes would require the model to be greatly modified.

There is no guide either from the Australian literature or from Albini's (1983) model as to how many spots there will be. Fire maps of intense wildfires generally show clusters of several spots caused by surges in fire intensity and strong convective updraughts. Spots fan out in front of the main fire due to variability in the wind direction. In some cases upper winds, which tend to be in a westerly direction in severe fire danger conditions in eastern Australia, may carry the spot fires in a different direction from the main fire (Luke and McArthur 1978). Cheney and Bary (1969) note that firebrands are likely to be thrown to the left of the direction of fire spread because of the tendency of the wind velocity vector to advance in an anticlockwise direction with increasing height. Topography also affects spot-fire distribution. On the windward side of ridges the fire intensity tends to be high due to slope and wind effects. At the top of the ridge turbulence and reverse flow cause firebrand lofting, and the brands may be carried to the next ridge where new spot fires form (Cheney and Bary 1969). In severe conditions fire may spread from ridge top to ridge top carried by the spotting process.

Fire danger indices

The fire behaviour information required for determining a wildfire imprint is essentially contained in spread rate, intensity and ignition/extinction probability. These are all highly interdependent. Fire danger indices, such as the McArthur Forest Fire Danger Index (FFDI) and Grassland Fire Danger Index (GFDI), are indices of weather conditions and fuel flammability, and as such are indicators of ignition probability and spread rate in average fuels with no slope. Because variations in intensity are governed primarily by variations in spread rate the fire danger indices are also indicators of average intensity to be expected in a given fuel type. Studies have shown that annual numbers of fires and the fire size attained are highly correlated with such fire danger indices (e.g. Gill *et al.* 1987; Pinol *et al.* 1998). McCarthy and Cary (this volume) and Cary (this volume) show how fire danger indices can be used in modelling expected fire regimes.

Summary

The idea of simulating fire patterns to increase our understanding of fire regimes is fairly recent. As simulation models become more complex they will require more detailed information on fire behaviour. The past decade has seen much research and experimentation that has increased our ability to predict rate of spread. Much less emphasis has been placed on other aspects of fire behaviour, such as the probability of ignition, the conditions under which a fire will self-extinguish, the criteria for crowning and the expected spot-fire pattern. Even less emphasis has been placed on predicting the horizontal burn pattern within a fire, which could be critical to survival of some species and may be important for seed dispersal.

High-resolution meteorological models are improving predictions of wind speed in terms of topography, and wind reduction factors have been developed to allow for canopy cover. However an area where information is scarce is the prediction of dead fine fuel moisture content in terms of vegetation type, overstorey cover and topography.

If we are to gain understanding from fire regime modelling of the effects of different fire management practices and of global changes in weather, information of this kind is urgently needed.

References

Albini, F. A. (1976). Computer-based models of wildland fire behavior: a user's manual. U.S. Department of Agriculture Forest Service, Intermountain Forest and Range Experimental Station, Ogden, UT.

Albini, F. A. (1983). Potential spotting distance from wind-driven surface fires. U.S. Department of Agriculture Forest Service, Intermountain Forest and Range Experimental Station, Research Paper INT-309, Ogden, UT.

Albini, F. A., and Reinhardt, E. D. (1995). Modeling ignition and burning rate of large woody natural fuels. *International Journal of Wildland Fire* 5, 81–91.

Alexander, M. E. (1982). Calculating and interpreting forest fire intensities. *Canadian Journal of Botany* 60, 349–357.

Alexander, M. E. (1998). Crown fire thresholds in exotic pine plantations of Australasia. PhD thesis, Australian National University, Canberra.

Allan, G., and Southgate. R. (2001). Fire regimes in the spinifex landscapes of Australia. In *Flammable Australia: The Fire Regimes and Biodiversity of a Continent* (eds. R. A. Bradstock, J. E. Williams and A. M. Gill) pp. 145–176. (Cambridge University Press: Cambridge.)

Anderson, H. E. (1969). Heat transfer and fire spread. U.S. Department of Agriculture Forest Service, Intermountain Forest and Range Experimental Station, Research Paper INT-69, Ogden, UT.

Anderson, H. E. (1983). Predicting wind-driven wildland fire size and shape. U.S. Department of Agriculture Forest Service, Intermountain Forest and Range Experimental Station, General Technical Report INT-305, Ogden, UT.

Anderson, H. E. (1990). Moisture diffusivity and response time in fine forest fuels. *Canadian Journal of Forest Research* 20, 315–325.

Andrews, P. L. (1986). BEHAVE: fire behavior prediction and fuel modeling system: BURN subsystem, Part 1. U.S. Department of Agriculture Forest Service, Intermountain Forest and Range Experimental Station, General Technical Report INT-131, Ogden, UT.

Bally, J. (1995). The Haines Index as a predictor of fire activity in Tasmania. In *Proceedings of Bushfire 95* (Parks and Wildlife Service: Hobart, Tasmania.)

Beck, J. A. (1995). Equations for the forest fire behaviour tables for Western Australia. *CALMScience* 3, 325–348.

Beck, J. A. (2000). Towards an operational geographic information and modelling system for fire management in Western Australia. PhD thesis, Curtin University, Western Australia.

Beer, T. (1993). The speed of a fire front and its dependence on wind speed. *International Journal of Wildland Fire* 3, 193–202.

Bradstock, R. A., and Cohn, J. S. (2001). Fires regimes and biodiversity in semi-arid mallee ecosystems. In *Flammable Australia: The Fire Regimes and Biodiversity of a Continent* (eds. R. A. Bradstock, J. E. Williams and A. M. Gill) pp. 238–258. (Cambridge University Press: Cambridge.)

Bradstock, R. A., and Gill, A. M. (1993). Fire in semi-arid, mallee shrublands: size of flames from discrete fuel arrays and their role in the spread of fire. *International Journal of Wildland Fire* 3, 3–12.

Breiman, L., Friedman, J. H., Olshen, R. A., and Stone, C. J. (1984). *Classification and Regression Trees*. (Wadsworth and Brooke/Cole: Monterey, CA.)

Buckley, A. J. (1990). Fire behaviour and fuel reduction burning. Bemm river wildfire, October 1988. Department of Conservation and Natural Resources, Research Report no. 28, Victoria.

Buckley, A. J. (1993). Fuel reducing regrowth forests with a wiregrass fuel type: fire behaviour guide and prescriptions. Department of Conservation and Natural Resources, Research Report no. 40, Victoria.

Burrows, N. D. (1984). Predicting blow-up fires in the jarrah forest. Forestry Department of Western Australia, Technical Paper no. 12, Perth.

Burrows, N. D. (1987). The Soil Dryness Index for use in fire control in the south-west of Western Australia. Western Australian Conservation and Land Management, Technical Report no. 17, Perth.

Burrows, N. D. (1994). Experimental development of a fire management model for jarrah (*Eucalyptus marginata* Donn ex Sm.) forest. PhD thesis, Australian National University, Canberra.

Burrows, N. D. (1999a). Fire behaviour in jarrah forest fuels: 1. Laboratory experiments. *CALMScience* 3, 31–56.

Burrows, N. D. (1999b). Fire behaviour in jarrah forest fuels: 2. Field experiments. *CALMScience* 3, 57–84.

Burrows, N. D. (2001). Combustion rate of eucalypt forest fuel particles. *International Journal of Wildland Fire*. (in press)

Burrows, N. D., and Van Didden, G. (1991). Patch burning desert nature reserves in Western Australia using aircraft. *International Journal of Wildland Fire* 1, 49–55.

Burrows, N. D., Ward, B., and Robinson, A. (1991). Fire behaviour in spinifex fuels on the Gibson Desert Nature Reserve, Western Australia. *Journal of Arid Environments* 20, 189–204.

Byram, G. M. (1959). Combustion of forest fuels. In *Forest Fire: Control and Use*. (ed. K. P. Davis) pp. 61–80. (McGraw-Hill: New York.)

Carrier, G. F., Fendell, F. E. and Wolff, M. F. (1991). Wind-aided fire-spread across arrays of discrete fuel elements. I. Theory. *Combustion Science and Technology* 75, 31–51.

Cary, G. J. (2001). Importance of a changing climate for fire regimes in Australia. In *Flammable Australia: The Fire Regimes and Biodiversity of a Continent* (eds. R. A. Bradstock, J. E. Williams and A. M. Gill) pp. 26–46. (Cambridge University Press: Cambridge.)

Catchpole, E. A., de Mestre, N. J., and Gill, A. M. (1982). Intensity of a fire at its perimeter. *Australian Forest Research* 12, 47–54.

Catchpole, E. A., Catchpole, W. R., and Rothermel, R. C. (1993). Fire behaviour experiments in mixed fuel complexes. *International Journal of Wildland Fire* 3, 45–57.

Catchpole, E. A., Catchpole, W. R., Viney, N., McCaw, W. L., and Marsden-Smedley, J. B. (2001). Estimating equilibrium moisture content and response time in the field. *International Journal of Wildland Fire*. (in press)

Catchpole, W. R., Bradstock, R. A., Choate, J., Fogarty, L. G., Gellie, N., McCarthy, G. J., McCaw, W. L., Marsden-Smedley, J. B., and Pearce, G. (1998a). Co-operative development of equations for heathland fire behaviour. In *Proceedings of the 3rd International Conference of Forest Fire Research and 14th Conference of Fire and Forest Meteorology* (ed. D. X. Viegas) pp. 631–645. (Univesity of Coimbra: Portugal.)

Catchpole, W. R., Catchpole, E. A., Rothermel, R. C., Morris, G. A., Butler, B. W., and Latham, D. J. (1998b). Rate of spread of free-burning fires in woody fuels in a wind tunnel. *Combustion Science and Technology* 131, 1–37.

Cheney, N. P. (1976) Fire disasters in Australia, 1945–75. *Australian Forestry* 39, 245–268.

Cheney, N. P. (1981). Fire behaviour. In *Fire and the Australian Biota* (eds. A. M. Gill, R. H. Groves and I. R.

Noble) pp. 151–175. (Australian Academy of Science: Canberra.)

Cheney, N. P. (1990) Quantifying bushfires. *Mathematical and Computer Modelling* 13, 9–15.

Cheney, N. P. (1991) Models used for fire danger rating in Australia. In *Proceedings of the Conference on Bushfire Modelling and Fire Danger Rating Systems*, 11–12 July 1988 (eds. N. P. Cheney and A. M. Gill) pp. 19–28. (CSIRO Division of Forestry: Canberra.)

Cheney, N. P., and Bary, G. A. V. (1969). The propagation of mass conflagrations in a standing eucalypt forest by the spotting process. The Technical Co-operation Program Mass Fire Symposium, Defence Standards Laboratory, Melbourne.

Cheney, N. P., and Gould, J. S. (1995). Fire growth to a quasi-steady rate of forward spread. *International Journal of Wildland Fire* 5, 237–247.

Cheney, N. P., and Gould, J. S. (1997). Letter: Fire growth and acceleration.' *International Journal of Wildland Fire* 7, 1–5.

Cheney, N. P., and Sullivan, A. (1997). *Grassfires, Fuel, Weather and Fire Behaviour*. (CSIRO: Melbourne.)

Cheney, N. P, Gould, J. S., and Knight, I. (1992). A prescribed burning guide for young regrowth forests of silvertop ash. Forest Commission New South Wales, Research Report no. 16, Sydney.

Cheney, N. P., Gould, J. S., and Catchpole, W. R. (1993). The influence of fuel, weather and fire shape variables on fire-spread in grasslands. *International Journal of Wildland Fire* 3, 31–44.

Cheney, N. P., Gould, J. S., and Catchpole, W. R. (1998). Prediction of fire spread in grasslands. *International Journal of Wildland Fire* 8, 1–13.

Clark, T. L., Jenkins, M. A., Coen, J. L., and Packham, D. R. (1996). A coupled atmosphere–fire model: role of the convective Froude number and dynamic fingering at the fireline. *International Journal of Wildland Fire* 6, 177–190.

Conroy, B. (1996). Fuel management strategies for the Sydney region. In *The Burning Question: Fire Management in New South Wales* (ed. J. Ross) pp. 73–83. (University of New England: Armidale.)

Eberhardt, K. E., and Woodward, P. M. (1987). Distribution of residual vegetation associated with large fires in Alberta. *Canadian Journal of Forest Research* 17, 1207–1212.

Ellis, P. (2000). The aerodynamics and combustion characteristics of eucalypt bark: a firebrand study. PhD thesis, Australian National University, Canberra.

Fons, W. L. (1946). Analysis of fire spread in light forest fuels. *Journal of Agricultural Research* 72, 93–121.

Forestry Canada Fire Danger Group (1992). Development and structure of the Canadian forest fire behavior prediction system. Forestry Canada Information, Report ST-X-3, Ottawa.

Fosberg, M. A., and Deeming, J. E. (1971). Derivation of the 1- and 10-hour timelag fuel moisture calculations for fire-danger rating. U.S. Department of Agriculture Forest Service, Rocky Mountain Forest and Range Experimental Station, Research Note RM-233, Fort Collins, CO.

Gellie, N. J. H. (1990). Improving models with PREPLAN: a description of the current system and what is needed. *Mathematical and Computer Modelling* 13, 27–36.

Gill, A. M., Christian, K. R., Moore, P. H. R., and Forrester, R. I. (1987). Bushfire incidence, fire hazard and fuel reduction burning. *Australian Journal of Ecology* 12, 299–306.

Gill, A. M., Burrows, N. D., and Bradstock, R. A. (1995). Fire modelling and fire weather in an Australian desert. *CALMScience Supplement* 4, 29–34.

Gould, J. S., Knight, I., and Sullivan, A. L. (1997). Physical modelling of leaf scorch height from prescribed fires in young *Eucalyptus sieberi* regrowth forests in south-eastern Australia. *International Journal of Wildland Fire* 7, 2–20.

Griffin, G. F., and Allan, G. E. (1984). Fire behaviour. In *Anticipating the Inevitable: A Patch Burn Strategy for Fire Management at Uluru (Ayers Rock–Mt Olga) National Park* (ed. E. C. Saxon) pp. 55–58. (CSIRO: Melbourne.)

Haines, D. A. (1988). A lower atmosphere severity index for wildland fire. *U.S. National Weather Digest* 13, 23–27.

Haines, D. A., Main, W. A., and Crosby, J. S. (1973). Forest fires in Missouri. U.S. Department of Agriculture Forest Service, North Central Forest Experimental Station, Research Paper NC-87, St Paul, MN.

Hatton, T. J., and Viney, N. R. (1989). Assessment of existing fine fuel moisture models applied to *Eucalyptus* litter. *Australian Forestry* 52, 82–93.

Hobbs, R. J., and Atkins, L. (1988). Spatial variability of experimental fires in south-west Western Australia. *Australian Journal of Ecology* 13, 295–299.

Keetch, J. J., and Byram, G. M. (1968). A drought index for

forest fire control. U.S. Department of Agriculture Forest Service, Southeastern Forest Experimental Station, Research Paper SE-38, Asheville, NC.

Krusel, N., Packham, D., and Tapper, N. (1993). Wildfire activity in the mallee shrubland of Victoria, Australia. *International Journal of Wildland Fire* 3, 217–228.

Lacey, C. J., Walker, J., and Noble, I. R. (1982). Fire in Australian tropical savannas. In *Ecological Studies*, vol. 42, *Ecology of Tropical Savannas*. (eds. B. J. Huntley and B. H. Walker) pp. 246–272. (Springer-Verlag: Berlin.)

Latham, D. J., and Schlieter, J. A. (1989). Ignition probabilities of wildland fuels based on simulated lightning discharges. U.S. Department of Agriculture Forest Service, Intermountain Forest and Range Experimental Station, Research Report INT-411, Ogden, UT.

Luke, R. H., and McArthur, A. G. (1978). *Bushfires in Australia*. (Australian Government Publishing Service: Canberra.)

Maher, J. V., and Lee, D. M. (1977). *Upper Air Statistics, Australia, Surface to 5 mb, 1957 to 1975*. (Australian Government Publishing Service: Canberra.)

Marsden-Smedley, J. B., and Catchpole, W. R. (1995a). Fire behaviour modelling in Tasmanian buttongrass moorlands. I. Fire characteristics. *International Journal of Wildland Fire* 5, 203–214.

Marsden-Smedley, J. B., and Catchpole, W. R. (1995b). Fire behaviour modelling in Tasmanian buttongrass moorlands. II. Fire behaviour. *International Journal of Wildland Fire* 5, 215–228.

Marsden-Smedley, J. B., and Catchpole, W. R. (2001). Fire behaviour modelling in Tasmanian buttongrass moorlands. Fuel moisture. *International Journal of Wildland Fire*. (in press)

Marsden-Smedley, J. B., Catchpole W.R., and Pyrke, A. (2001). Fire behaviour modelling in Tasmanian buttongrass moorlands: sustaining versus non-sustaining fires. *International Journal of Wildland Fire*. (in press)

McAlpine, R. S., and Wakimoto, R. W. (1991). The acceleration of fire from point source to equilibrium spread. *Forest Science* 37, 1314–1337.

McAlpine, R. S., and Wotton, B. M. (1993). The use of fractal dimensions to improve wildland fire perimeter prediction. *Canadian Journal of Forest Research* 23, 1073–1077.

McAlpine, R. S., Lawson, B. D., and Taylor, E. (1991). Fire spread across a slope. In *Proceedings of the 11th Conference of Fire and Forest Meteorology*, Missoula, MT. (eds. P. L. Andrews and D. F. Potts) pp. 218–225. (SAF Publishing: Bethesda, MD.)

McArthur, A. G. (1962). Control burning in eucalypt forests. Forest Research Institute, Forest and Timber Bureau of Australia, Leaflet no. 80, Canberra.

McArthur, A. G. (1966). Weather and grassland fire behaviour. Forest Research Institute, Forest and Timber Bureau of Australia, Leaflet no. 100, Canberra.

McArthur, A. G. (1967). Fire behaviour in eucalypt forests. Forest Research Institute, Forest and Timber Bureau of Australia, Leaflet no. 107, Canberra.

McArthur, A. G. (1973). Forest fire danger meter, Mark V. Forest Research Institute, Forest and Timber Bureau of Australia, Canberra.

McArthur, A. G. (1977). Grassland fire danger meter Mark V. In *CSIRO Division of Forest Research Annual Report 1976–1977* (CSIRO: Canberra.)

McArthur, A. G., and Cheney, N. P. (1967). *Report on the Southern Tasmanian Bushfires of 7 February 1967*. (Government Printer: Hobart.)

McCarthy, M. A., and Cary, G. J. (2001). Fire regimes in landscapes: models and realities. In *Flammable Australia: The Fire Regimes and Biodiversity of a Continent*. (eds. R. A. Bradstock, J. E. Williams and A. M. Gill) pp. 77–93. (Cambridge University Press: Cambridge.)

McCarthy, G. J., and Tolhurst, K. G. (1998). Effectiveness of fire-fighting first attack operations, DNRE Victoria, 1991/92–1994/95. Department of Natural Resources and Environment, Research Report no. 45, Victoria.

McCarthy, G. J., Tolhurst, K. G., and Chatto, K. (1998). Fire hazard guide. Department of Natural Resources and Environment, Research Report no. 47, Victoria.

McCaw, W. L. (1986). Behaviour and short term effects of two fires in regenerated karri (*Eucalyptus diversicolor*) forest. Department of Conservation and Land Management, Western Australia, Technical Report no. 9, Perth.

McCaw, W. L. (1998). Predicting fire spread in Western Australian mallee-heath shrubland. PhD thesis, University College, UNSW, Canberra.

McCaw, W. L., Simpson, G., and Mair, G. (1992). Extreme wildfire behaviour in 3-year-old fuels in a Western Australian mixed *Eucalyptus* forest. *Australian Forestry* 55, 107–117.

McCaw, W. L., Smith, R., and Neal, J. (1997). Prescribed burning of thinning slash in regrowth stands of karri (*Eucalyptus diversicolor*). 1. Fire characteristics, fuel consumption and tree damage. *International Journal of Wildland Fire* **7**, 29–40.

McMahon, C. K., Adkins, C. W., and Rogers, S. L. (1986). A video image analysis system for measuring fire behavior. *Fire Management* **47**, 10–15.

McRae, R. (1997). Considerations on operational wildfire spread modelling. In *Bushfire 97: Proceedings of the Australasian Bushfire Conference* (eds. B. J. McKaige, R. J. Williams and W. M. Waggitt) pp. 288–293. (CSIRO: Darwin.)

Mercer, G. N. (1997). Simulation of fire spread in discrete fuels using flame contact. In *Bushfire 97: Proceedings of the Australasian Bushfire Conference* (eds. B. J. McKaige, R. J. Williams and W. M. Waggitt) pp. 302. (CSIRO: Darwin.)

Mercer, G. N., and Weber, R. O. (1994). Plumes above fires in a cross wind. *International Journal of Wildland Fire* **4**, 201–207.

Morris, G. A. (1987). A simple method for computing spotting distances for wind-driven surface fires. U.S. Department of Agriculture Forest Service, Intermountain Forest and Range Experimental Station, Research Note INT-374, Ogden, UT.

Morrison, D. A., Buckney, R. T., Bewick, B. J., and Cary, G. J. (1996). Conservation conflicts over burning bush in south-eastern Australia. *Biological Conservation* **76**, 167–175.

Mount, A. B. (1972). The derivation and testing of a soil dryness index using run-off data. Tasmania Forest Communications, Bulletin no. 4, Tasmania Forestry Commision, Hobart.

Muraszew, A. (1974). Firebrand phenomena. The Aerospace Corporation, Aerospace Report ATR-74(8165–01)-1, El Segundo, CA.

Noble, I. R., Bary, G. A. V., and Gill, A. M. (1980). McArthur's fire-danger meters expressed as equations. *Australian Journal of Ecology* **5**, 201–203.

Pagni, P. J., and Peterson, T. G. (1973). Flame spread through porous fuels. In *14th Symposium (International) on Combustion.* pp. 1099–1106. (Combustion Institute: Pittsburgh, PA.)

Parrott, R. T., and McDonald, C. M. (1970). Ignitability of swards of various annual species. *Australian Journal of Experimental Agriculture and Animal Husbandry* **10**, 76–83.

Pinol, J., Terradas, J., and Lloret, F. (1998). Climate warming, wildfire hazard and wildfire occurrence in coastal eastern Spain. *Climate Change* **38**, 345–357.

Pook, E. W., and Gill, A. M. (1993). Variations of live and dead fine fuel moisture in *Pinus radiata* plantations of the Australian Capital Territory. *International Journal of Wildland Fire* **3**, 155–168.

Richards, G. D. (1998). The effect of wind direction variations on perimeter length, shape and rate of spread. In *Proceedings of the 13th Conference of Fire and Forest Meteorology*, Lorne, Australia, pp. 359–363. (International Association of Wildland Fire.)

Richards, G. D., and Walberg, R. (1998). The computer simulation of crown fire streets.In *Proceedings of the 3rd International Conference of Forest Fire Research and 14th Conference of Fire and Forest Meteorology* (ed. D. X. Viegas) pp 435–440. (University of Coimbra, Portugal.)

Rothermel, R. C. (1972). A mathematical model for predicting fire spread in wildland fuels. U.S. Department of Agriculture Forest Service, Intermountain Forest and Range Experimental Station, Research Paper INT-115, Ogden, UT.

Rothermel, R. C. (1983). How to predict the spread and intensity of forest and range fires. U.S. Department of Agriculture Forest Service, Intermountain Forest and Range Experimental Station, General Technical Report INT-143, Ogden, UT.

Rothermel, R. C. (1991). Predicting behavior and size of crown fires in the northern Rocky Mountains. U.S. Department of Agriculture Forest Service, Intermountain Forest and Range Experimental Station, Research Paper INT-438, Ogden, UT.

Sneeuwjagt, R. J., and Peet, G. B. (1985). *Forest Fire Behaviour Tables for Western Australia.* (Department of Conservation and Land Management: Perth.)

Thomas, P. H. (1971). Rates of spread of some wind-driven fires. *Forestry* **44**, 155–175.

Tolhurst, K. G. (1995). Fire from a flora, fauna and soil perspective: sensible heat measurement. *CALMScience Supplement* **4**, 45–88.

Tolhurst, K. G., and Chatto, K. (1998). Behaviour and threat of a plume-driven bushfire in West Central Victoria, Australia. In *Proceedings of the 13th Conference on Fire and Forest Meteorology*, Lorne, Australia, pp. 321–331. (International Association of Wildland Fire.)

Van Wagner, C. E. (1973). Height of crown scorch in forest fires. *Canadian Journal of Forest Research* 3, 373–378.

Van Wagner, C. E. (1977a). Conditions for the start and spread of crown fires. *Canadian Journal of Forest Research* 7, 23–24.

Van Wagner, C.E. (1977b). Effect of slope on fire spread rate. Environment Cananda, Canadian Forest Service, Bi-Monthly Research Notes no. 33, 7–8, Ottawa.

Van Wilgen, B. W., Le Maitre, D. C., and Kruger, F. J. (1985). Fire behaviour in South African fynbos (macchia) vegetation and predictions from Rothermel's fire model. *Journal of Applied Ecology* 22, 207–216.

Veigas, D. X. (1999). GESTOSA 98: shrubland experimental fire general report. Inflame Project, ENV4–CT98–0700, Coimbra, Portugal.

Viney, N. R. (1991). A review of fine fuel moisture modelling. *International Journal of Wildland Fire* 1, 215–234.

Weber, R. O., Mercer, G. N., Mahon, G. M., and Catchpole, W. R. (1995a). Predicting maximum inter-hummock distance for fire spread. In *Bushfire '95: Australian Bushfire Conference* (Parks and Wildlife Service: Hobart.)

Weber, R. O., Gill, A. M., Lyons, P. R. A., Moore, P. R. H., Bradstock, R. A., and Mercer, G. N. (1995b). Modelling wildland fire temperatures. *CALMScience Supplement* 4, 23–26.

Weige, D. R., and Biging, G. S. (1997). A qualitative comparison of fire spread models incorporating wind and slope effects. *Forest Science* 43, 170–180.

Whight, S., and Bradstock, R. A. (1999). Indices of fire characteristics in sandstone heath near Sydney, Australia. *International Journal of Wildland Fire* 9, 145–153.

Williams, R. J., Gill, A. M., and Moore, P. H. R. (1998). Seasonal changes in fire behaviour in a tropical savanna in northern Australia. *International Journal of Wildland Fire* 8, 227–239.

Wilson, R. A. (1985). Observations of extinction and marginal burning states in free burning porous fuel beds. *Combustion Science and Technology* 44, 179–193.

Wilson, R. A. (1990). Reexamination of Rothermel's fire spread equations in no-wind and no-slope conditions. U.S. Department of Agriculture Forest Service, Intermountain Forest and Range Experimental Station, Research Paper INT-434, Ogden, UT.

Wolff, M. F., Carrier, G. F., and Fendell, F. E. (1991). Wind-aided fire-spread across arrays of discrete fuel elements. II. Experiment. *Combustion Science and Technology* 75, 261–289.

4

Fire regimes in landscapes: models and realities

MICHAEL A. MCCARTHY AND GEOFFREY J. CARY

Abstract

Spatial properties of fire regimes may have important influences on ecological systems. Several different landscape models have been developed to help determine how fire regimes vary spatially and temporally. Mechanistic models rely on simulating the spread of fires across an array of cells representing the landscape. The spread of fires across a landscape may be simulated with simple cellular automata, or rely on more complex process-based models. In contrast, footprint models use statistical descriptions of fires to (essentially) place fires in the landscape. The occurrence, size and shape of fires may be drawn from a probability distribution. Alternatively, stochastic fires may be modelled with appropriate spatial correlation to ensure that points close together are more likely to burn in the same year than points far apart. Comparisons between the different models using data on fire sizes and shapes are limited by the availability of suitable data. Hazard functions, describing how the probability of fire at a point changes with the time since the last fire, may be used to compare the different models. For the most basic models, it is proposed that the hazard (instantaneous risk of fire) changes in proportion to the state (e.g. amount and moisture content) of the fuel. In contrast, abrupt non-linear relationships between fuel and hazard may occur in the mechanistic models. In these models, there is often a threshold fuel load above which the rate of spread increases abruptly. This abrupt non-linearity makes parameter estimation of simple cellular automata difficult, and their use is most valuable for theoretical rather than applied purposes. The process-based models avoid the problem of parameter estimation by using models of weather and fire behaviour to control fire ignition and spread. Details of Cary's FIRESCAPE model are described to provide an example of such process-based models. Hazard functions have rarely been determined from records of real fires, and there is only a little evidence to indicate strong non-linear relationships. Once determined, hazard functions may be used to predict the age structure of landscapes that are subject to random and systematic disturbances.

Introduction

Fire regimes have integral roles in ecosystems throughout the world (Naveh 1975; Gill *et al.* 1981; Booysen and Tainton 1984; Goldammer 1990; Johnson 1992; Pyne *et al.* 1996). Gill (1975) proposed that effects of fire regimes on ecosystems could be described in terms of the time between fires, the time since the last fire, the intensity of the fire, the type of fuel burnt and the season of occurrence. Spatial aspects may also contribute to the influence of fire regimes on ecosystems, through influences on processes such as movement of organisms between burnt and unburnt areas, and the mix of different vegetation types in adjacent areas (Gill 1997). These factors need to be considered if appropriate fire regimes are to be applied in managed fireprone ecosystems. Further, fire regimes are characterised by variability. Fires do not always occur at the same interval, at the same intensity, or at the same time of year. Such variability can have important influences on ecosystems (Cary and Morrison 1995; McCarthy and Burgman 1995; D.A. Morrison *et al.* 1995; Clark 1996; Gill and McCarthy 1998; McCarthy and Lindenmayer 1998). Spatial variability in the influence of fire can contribute to particular patterns in landscapes, such as the development of multi-aged mountain ash forest (McCarthy and

Lindenmayer 1998). Such variability should be accounted for when describing and prescribing fire regimes (Gill and McCarthy 1998).

Spatial considerations are being increasingly seen as important in ecological studies (Kareiva 1994), through influences on processes such as local population dynamics (Allee 1938; McCarthy 1997), spread of organisms (Turchin 1998) and species interactions (Hassell and Wilson 1997). Spatial effects of fires may interact with patterns of the abundance and distribution of organisms, and with physical features of landscapes such as climate and topography (Cary 1998). The ability to predict the influence of fire regimes in landscapes will depend on the accuracy of models that are developed for this purpose. These will be limited by the level of complexity that can be included, which will ultimately be limited by the level of understanding and knowledge. Our aim in this paper is to review different theoretical approaches to studying fire regimes in landscapes, to highlight and develop links between these approaches, and to discuss how the models contribute to understanding the effects of fires in landscapes.

Point-based fire regime models

Prior to extension to entire landscapes, it is necessary to consider models of fire at points in landscapes to provide a framework for the review. If one considers a point in a landscape, it is not possible to be sure when it will next be burnt. In areas that experience frequent fire, we would know that there is a high chance of the point burning in the next few years, but we can not be sure of the exact time. The uncertainty means that the occurrence of fire is best described in terms of probability, and variation in intervals between fires may be described with probability distributions. Such distributions indicate the likelihood of different fire intervals occurring. The chance of the occurrence of fire can be considered in terms of how the instantaneous risk of fire changes with the time since the last fire. The instantaneous risk of fire is termed the hazard (Johnson and Gutsell 1994). The concept of hazard is easiest to understand in discrete time; in this case

it is the proportion of areas of a given age that burn in the next time step. In continuous time, there is a mathematical relationship between the hazard and the probability distribution of fire intervals, such that one defines the other (Johnson and Gutsell 1994). Mathematically, this relationship is given by

$$F(t) = 1 - \exp[-H(t)],$$

where $F(t)$ is the cumulative probability function of fire intervals (t), and $H(t)$ is the integral of the hazard function $h(t)$, which describes how the instantaneous probability of fire changes with time since the last fire. The cumulative probability function of fire intervals $F(t)$ is the probability of a fire occurring before time t since the last fire. The probability density function of fire intervals is $f(t)$, the derivative of $F(t)$.

The simplest type of point-based fire model is the case in which the hazard does not change with time since the last fire. In this case, the distribution of fire intervals is described by an exponential probability distribution (Van Wagner 1978; Johnson and Gutsell 1994). The mean fire interval is equal to the inverse of the hazard. Mathematically, the hazard is equal to $h(t) = c$, $H(t) = ct$, $F(t) = 1 - \exp(-ct)$, and the probability density of the fire intervals is equal to

$$f(t) = c\exp(-ct).$$

These functions are shown graphically in Fig. 4.1.

McCarthy et al. (2001) suggested that for many vegetation communities, hazard may change with fuel load and vegetation succession. Thus, where annual grasses provide the majority of the fuel, the exponential model may be appropriate on an annual time scale. However, in forest communities where fuels accumulate over several years, different models for the hazard function $h(t)$ may be appropriate. Changes in the composition and quantity of fuel may mean that the hazard would increase or decrease with the time since the last fire. The most common approach to modelling such changes is to use the Weibull model (Johnson and Gutsell 1994) in which hazard is equal to (Fig. 4.2)

$$h(t) = ct^b.$$

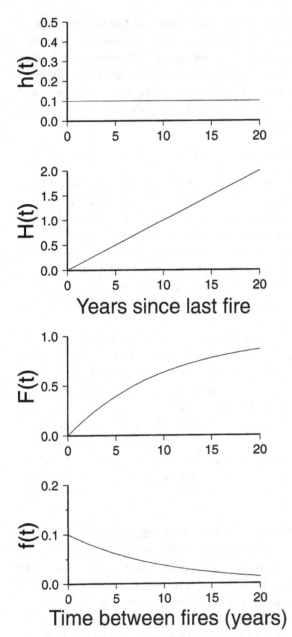

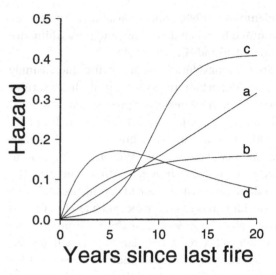

Fig. 4.2. Hazard functions varying with time since the last fire, illustrating the (a) Weibull, (b) Olson, (c) logistic and (d) moisture models. From McCarthy *et al.* (2001).

Fig. 4.1. The relationship between the hazard ($h(t) = 0.1$), the cumulative hazard ($H(t) = 0.1t$), the cumulative probability ($F(t) = 1 - \exp(-0.1t)$) and the probability density function ($f(t) = 0.1\exp(-0.1t)$) for the exponential model of fire intervals.

The hazard increases with time since the last fire when *b* is greater than zero, and decreases with time since the last fire when *b* is less than zero. When *b* is equal to zero, this model is equivalent to the exponential model in which the hazard is con-

stant. Clark (1989) provides an example of how these models of the fire interval distribution can be used to assist ecological interpretation. He fitted Weibull models to fire interval data from northwest Minnesota. The data were derived from fire scars on red pine (*Pinus resinosa*) trees and from stratigraphic charcoal. For all time periods examined and for both types of data, the estimated value of parameter *b* in the Weibull model was greater than 0, indicating that flammability tended to increase with the time since the last fire. Clark (1989) interpreted the parameters of the model in terms of the past climate. Periods with shorter fire intervals were more likely to be hot and dry.

The age structure of a landscape (the distribution of ages as measured by the time since the last fire) that is exposed to fires will depend on the fire interval distribution. Simply put, the age of the landscape will increase as fire intervals increase. This age structure is analogous to the 'shifting mosaic steady state' of Bormann and Likens (1979). A mathematical expression formalises the relationship between the fire interval distribution and the expected age structure of a landscape (Van Wagner 1978; Johnson and Gutsell 1994):

$$a(t) = [1 - F(t)]/m,$$

where $a(t)$ is the probability density function of age structure, m is the mean fire interval and $F(t)$ is defined as above. The age structure here is the probability density function describing the distribution of ages across the forest. In the case of the exponential model of fire intervals, the mean of the landscape age structure (i.e., the average age of a forest in which trees are killed by fire) will equal the mean fire interval ($1/c$).

The expected age structure gives the point-based models an implicit spatial component, and they can be extended to consider the mix of vegetation types in a landscape. For example, Jackson (1968) presented a conceptual model using hazard functions to explain distributions of vegetation types in Tasmania. Henderson and Wilkins (1975) used semi-Markov modelling to formalise the conceptual model of Jackson (1968).

Van Wagner (1978) used the exponential model and a graphical analysis to estimate the mean fire interval from data on the age structure of landscapes. The observed landscape age structure will only conform to the theoretical expectation $a(t)$ when the landscape is large enough such that random variation in the area burnt each year is small (Boychuk et al. 1997). This rarely occurs in forests, which typically experience large fires during infrequent dry periods. For example, Baker (1989) demonstrated that the age structure varied widely over time in the Boundary Waters Canoe Area, an area larger than 400 000 ha and more than 80 times the average fire size. Similarly, Cumming et al. (1996) were unable to identify a shifting mosaic steady state in boreal mixed-wood forests of Alberta, Canada, even in areas up to 73 600 km².

Reed (1994, 1997) formalised Van Wagner's approach by using the statistical method of maximum likelihood estimation, and provided solutions for the Weibull model (see also Reed et al. 1998). Additionally, Reed (1994) introduced a method to account for the fact that landscapes invariably deviate from the theoretical expectation because the area burnt varies each year.

The age structure of a landscape will also differ from the above model if the probability of fire varies across the landscape. Such spatial variation can be

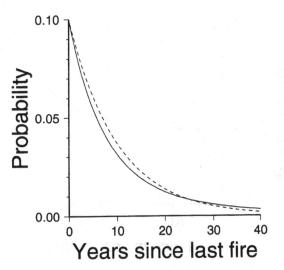

Fig. 4.3. Comparison of the expected age class distribution across a landscape subject to random fires for the cases where the probability of fire is equal to 0.1 in all places (dashed line) and where the probability of fire varies uniformly between 0.01 and 0.19 (solid line). In both cases, the interval between fires averaged across the landscape is 10 years, but the age structure of the landscape differs.

incorporated into the model by describing fire occurrence for each point in the landscape using one of the models discussed previously, and then representing the landscape as an aggregation of points with different parameter values. The landscape model can be constructed once the probability distribution of the parameter values has been defined. If variation in the hazard across the landscape is described by the probability density function $s(v)$, and the hazard is constant with time since the last fire at points (the exponential model), the expected age structure of the landscape would equal

$$a(t) = \int_0^\infty s(v)v e^{-vt} dv.$$

For example, if the hazard is uniformly distributed across the landscape between the values of y and z, $s(v)$ would equal $1/(z-y)$, and the age structure would equal

$$a(t) = \int_y^z v e^{-vt}/(z-y)\, dv,$$

which may be solved to give (Fig. 4.3)

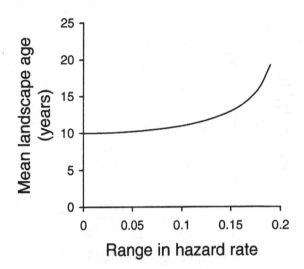

Fig. 4.4. Mean age of a landscape (as measured by the time since the last fire) versus the range of hazard rates. In all cases the average hazard is 0.1, and the probability distribution of hazard rates varies uniformly across the given range. The case shown in Fig. 4.3. is equivalent to a range in hazard rate of 0.18, which has a mean age of 16.4 years.

$$a(t) = [y e^{(z-y)t} - z + (e^{(z-y)t} - 1)/t] e^{-zt}/t/(z-y).$$

The mean of this distribution is $\ln(z/y)/(z-y)$. This mean is greater than $2/(z+y)$, which is the mean interval between fires averaged across all points in the landscape. Thus, variation in the hazard across landscapes will tend to increase the average age of the landscape (Fig. 4.4). Such spatial variation in the hazard will contribute to errors when using the methods of Reed (1994, 1997) and co-authors (Reed *et al.* 1998) to estimate fire intervals from data on age structure. Such methods are only applicable to areas in which the probability of fire is relatively uniform.

The parameters of the fire interval models may also be estimated using data on fire intervals. Maximum likelihood estimation is an established statistical method that has been used in reliability studies (Crowder 1991) and is applicable to studies of fire intervals (Johnson and Gutsell 1994). The method estimates the parameters of the fire interval models such that the observed intervals are most likely to occur. When comparing the fit of different models, the simplest model is selected unless a more complex model (with one or more additional parameters) provides a significantly better fit to the data. This criterion for selecting the best model is based on the difference in the log–likelihood values, but is only appropriate when the models are nested (Crowder 1991). Models are nested when the simpler model is a reduced parameter version of the more complex one (e.g. the exponential and Weibull models). Data in which the start or end of the interval is unknown can be included when using maximum likelihood estimation. In such cases, the interval is known to be at least a certain number of years long. In reliability studies, such data are referred to as being censored (Crowder 1991).

Hazard functions of the Weibull form are commonly used in reliability studies (Crowder 1991), and their use has been carried over into studies of fire intervals. McCarthy *et al.* (2001) suggested that hazard functions other than the Weibull might be more appropriate, depending on the ecosystem being studied. They suggested that the hazard function should reflect the underlying ecological processes of the system, such as changes in the vegetation structure and the fuel with time since the last fire. A number of functions were proposed (McCarthy *et al.* 2001). In the Olson model, the hazard changes with the fuel load as predicted by the Olson (1963) equation (Fig. 4.2). The Olson model is similar to a Weibull model with increasing hazard, although the hazard approaches an asymptote in the Olson model while it is unbounded in the Weibull model. In the logistic model, the hazard increases slowly at first before reaching an asymptote (Fig. 4.2). The logistic model has been suggested independently by Li *et al.* (1997). It would be suitable in circumstances where the hazard increases slowly immediately after a fire before rising to an asymptote. In the moisture model, the hazard first increases and then declines; this may be suitable in cases where mesic vegetation (e.g. rainforest) becomes established in a process of succession (Fig. 4.2).

Few authors have derived hazard functions directly from fire data, so it is still unclear which forms are applicable in real ecosystems. Lang (1997) derived hazard functions from maps of fires in jarrah (*Eucalyptus marginata*) forest in southwest

Western Australia. Prior to the introduction of prescribed fires, the hazard function increased towards an asymptote with increasing time since fire, in a manner similar to the fuel accumulation curve. During the more recent era of prescribed burning, the hazard function has exhibited a distinct peak that corresponds to the approximate interval between planned fires. The data of Lang (1997) suggest that, in the absence of prescribed fire, the hazard function for jarrah forest may be approximated by an Olson curve.

Gill *et al.* (2001) derived hazard functions for three different regions of Kakadu National Park in northern Australia using maps of fires derived from remotely sensed data. In the lowland areas, which are dominated by grassy woodlands, a constant hazard was expected. However, their data indicated that the hazard tended to decline with time since the last fire. It was not clear whether such a result was due to changes in the flammability of the fuel, or to heterogeneity in the chances of burning. Such heterogeneity would contribute to an apparent decline in the hazard because those areas that are less likely to burn would tend to remain unburnt for an extended period and become older. These older areas would have a lower probability of fire than younger areas and cause an apparent negative relationship between age and hazard, even though there is not a decline in hazard with age for individual points. Heterogeneity in the probability of fire across landscapes was identified in Cary's (1997, 1998) FIRESCAPE model that incorporated the effect of a site's neighbourhood on the fire regime.

In many cases, landscapes are subject to both prescribed disturbance and unplanned fires. Examples of these are mountain ash forests that are subject to stand-replacing fires and are logged on rotations of 50–80 years, and dry sclerophyll forests throughout Australia that are burnt by prescribed fire and unplanned fire. In these cases, the age structure of the forests, as measured by the time since the last disturbance, will be influenced by both random and deterministic factors. The models of age structure described above may be simply modified to accommodate the addition of deterministic disturbance, by truncating the expected age-class distribution

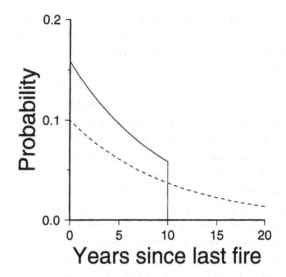

Fig. 4.5. Expected age-class distribution across a landscape subject to random disturbance with a mean interval of 10 years. The effect of including an additional deterministic disturbance 10 years after any previous disturbance (solid line) is compared to the case where there is no such deterministic disturbance (dashed line).

above the prescribed rotation, and redistributing the associated area to the younger age classes:

$$a(t) = [1 - F(t)]/m/F(T), \qquad 0 < t < T,$$

where T is the rotation length of the deterministic disturbance, and m is the mean interval between unplanned disturbances, as defined above. We have made the simplifying assumption that prescribed fires are entirely deterministic, with no areas escaping fires because of 'patchy' burns. Incorporating probabilistic prescribed fires at regular intervals would be straightforward by appropriate modification of the hazard function.

Thus, in the presence of random disturbance with a mean interval of $1/c$ (exponential model, $F(t) = 1 - e^{-ct}$), and prescribed disturbance when the age exceeds T years, the expected age structure is (Fig. 4.5)

$$a(t) = ce^{-ct}/(1 - e^{-cT}), \qquad 0 < t < T.$$

The combined effects of random and systematic disturbance may influence the abundance of older elements of the landscape (e.g. hollow-bearing trees in mountain ash forests) and the yield of products that depend on forest age (e.g. wood and water). Thus, the point-based fire models can be modified to accommodate the combined influence of random and systematic processes on the overall fire regime. The addition of systematic disturbance has the effect of truncating the age-class distribution that would arise under random disturbance.

The point-based models indicate how deterministic and stochastic disturbances combine to influence the resulting age structure of a landscape, but they do not explicitly answer the question of how to manage a landscape to maintain a particular age structure. Richards et al. (1999) used a relatively simple model of landscape disturbance to answer such a question. They examined fire management strategies for maintaining a diversity of age classes in a nature reserve, where the management options included lighting fires in different age classes, controlling fires, or doing nothing. The optimal strategy for maintaining a diversity of age classes in the model was solved using stochastic dynamic programming. An interesting result was that active management (either lighting or controlling fires) was often the optimal strategy even when the landscape was already in a desirable state (Richards et al. 1999).

The expected age structure gives the point-based models an implicit spatial component. The models emphasise the relationship between the fire regime and the resulting age structure of the landscape, and provide a theoretical basis for analysing the more complex spatially explicit models of fire regimes.

Spatial fire regime models

The occurrence of fire in a landscape may be thought of as the occurrence of fire across a set of points, and most spatial models of fire are represented in this way. McCarthy and Gill (1997) classified spatial fire models as being either mechanistic or footprint models, depending on how the fires were simulated. Mechanistic models simulate ignition and spread of fires, while footprint models rely on a statistical representation of fire sizes and shapes.

Cellular automata models

Mechanistic fire models are often based on cellular automata (von Neumann 1966) in which landscapes are represented by an aggregation of cells (Turner et al. 1989; Gardner and O'Neill 1991; Green 1994; Bradstock et al. 1998). Fires are spread using either site percolation or bond percolation (Gardner and O'Neill 1991). Bond percolation involves random ignition of cells, and probabilistic spread between cells. Site percolation is similar although spread is deterministic, while the chance of a cell being susceptible to fire is probabilistic. These two methods produce different distributions of fire sizes and different patterns of unburnt vegetation (McCarthy and Gill 1997). There is a complex non-linear relationship between the parameters of these models and the annual risk of fire within each cell in the landscape (McCarthy and Gill 1997). Fires may spread extensively if the density of flammable cells exceeds a certain threshold, yet be quite contained when below this threshold (Turner et al. 1989; Green 1994). Such extreme non-linearity makes it difficult to estimate the model parameters using formal statistical methods such as maximum likelihood estimation. Additionally, landscape fire models may be very sensitive to the input parameters. Nevertheless, such observations are of fundamental interest, because they demonstrate that small changes in conditions may cause abrupt changes in fire behaviour. Such 'phase transitions' are a feature of complex systems and may be observed in real fire behaviour (Turner et al. 1989; Green 1994).

Ratz (1995) developed a spatial fire model for boreal forest using a cellular automaton. Bond percolation was used, with the ignition probability derived from empirical data on the density of ignitions. One set of simulations was performed in which the probability of spread was independent of age. Other simulations were performed in which the probability of fire spreading to a cell depended on the time since the last fire in the cell. The probabilities of spread were varied to generate different mean fire intervals. Ratz (1995) demonstrated that the simulations in which the probability of spread increased with time since the last fire produced a more complex landscape structure than the simulations in which the probability of spread was constant. The former simulations produced landscape patterns that better matched the observed landscape structure of boreal forests (Ratz 1995). In itself, this does not indicate that the hazard increases with time since fire in boreal forests because other spatial models of fire produce different patterns of burnt and unburnt areas (McCarthy and Gill 1997).

Bradstock *et al.* (1998) extended the use of cellular automata models of fire spread to consider the interaction between prescribed and unplanned fires and the persistence of plant species. The model incorporated temporal variation in the probability of fire, and used a form of bond percolation to simulate fire spread. The parameters of the model were derived from a combination of empirical observations and a model of fire spread. The results indicated that controlling the size of unplanned fires with prescribed burning was incompatible with conservation of plant species, particularly obligate seeders.

Simple cellular automata models are usually used for theoretical purposes or examining the consequences of general prescriptions, rather than for making predictions in specific landscapes. In these circumstances, spatial variation in the parameters is often ignored, although temporal changes in the risk of ignition may be considered (Ratz 1995; Bradstock *et al.* 1998). Their application to specific case studies is limited by difficulties of parameter estimation caused by the strong non-linear relationship between fire properties and the parameters. More recently, process-based models on real landscapes have been developed that attempt to increase realism by incorporating spatial variation in the ignition and spread of fires. These use models of physical processes (e.g. ignition by lightning, and fire spread as a function of weather, vegetation and topography) to parameterise the probabilities required by the cellular automata. In contrast to the simple cellular automata, such models are usually applied to real landscapes for specific purposes.

Process-based simulation models

Process-based fire simulation models can be classified according to whether they simulate fire events or fire regimes. Fire event simulators (e.g. *Sirofire*, Coleman and Sullivan 1996) predict the shape of an individual fire as it grows over the course of a few hours or days. Fire regime simulators model the occurrence of the many fires that characterise a fire regime over much longer periods, typically hundreds of years. Fire regime simulators can be further divided according to whether they simulate the fire regime for an individual point or across a landscape with variation in the fire regime from one point to another. Examples of point-based fire regime simulators are, or are included in, BRIND (Shugart and Noble 1981), FIRESUM (Keane *et al.* 1989), FOREST (Clark 1990) and VAFS/STANDSIM (Roberts 1996). The simulation of fire regimes at the level of the landscape is a comparatively recent area of research and is the focus of this section.

FIRE-BGC (Keane *et al.* 1996), EMBYR (Gardner *et al.* 1996) and ELFM (Wu *et al.* 1996) are three examples of process-based landscape-level fire regime simulators. FIRE-BGC is a mechanistic ecological process model for simulating fire succession on coniferous forest landscapes of the northern Rocky Mountains. It was constructed by combining the gap-replacement model (FIRESUM) (Keane *et al.* 1989) with FOREST-BGC, an ecosystem process model that calculates the carbon, water and nitrogen cycles through a forest ecosystem (Running and Coughlan 1988). Fires, and subsequently fire regimes, are generated in three stages. Firstly, the location of ignition points is generated by computing a probability of fire occurrence for each cell in the landscape using the Weibull probability density distribution (Johnson and Van Wagner 1985) which is adjusted to account for the size of the fire. Fire-free intervals and average fire size must be determined from fire history data. Secondly, the rate of spread and intensity of a fire spreading from a burning cell to a neighbour is determined using the FARSITE spatial fire model (Keane *et al.* 1996; Finney 1998). FARSITE predicts fire intensity and rates of spread across the landscape using the FIREMOD approach of Albini

(1976) and the principles of Rothermel (1972). Fuel dynamics are modelled from the rates of production and decomposition in a number of fuel classes including duff, litter and three classes of wood of increasingly larger dimensions. Fuel class quantities are reduced by fire according to a consumption factor.

The fire regime implementation of EMBYR links a spatial model of fire spread with stochastic weather and ignition models. It employs a semi-empirical approach to spreading fires, based on the models of Rothermel (1972). In the study of Gardner et al. (1996), sequences of daily weather are produced by firstly generating an autocorrelated Brownian motion, with displacements produced by Gaussian noise, which is then normalised between randomly selected minimum and maximum values of thousand hour lag fuel moisture (THFM). Daily weather records are then matched to this randomly generated THFM. Gardner et al. (1996) state that this approach, of selecting the entire weather record based on a single randomly selected variable, resulted in consistent relationships between fuel moisture, wind speed and direction, and fire ignitions. Fire start probabilities are drawn from a probability surface developed from historical records.

Wu et al. (1996) modelled fire regimes in the Florida Everglades by stochastic simulation of ignition by lightning, fire spread and stochastic extinguishment on a cellular landscape. Fuel load was modelled as a function of vegetation type and time since the last fire, and fuel moisture depended on the periodicity and length of intervals between rain events. Fire spread was simulated as a function of wind speed, wind direction, fuel load and fuel moisture, based on the work of Rothermel (1972) and subsequent modifications. Daily weather data were input from meteorological records.

These models (Gardner et al. 1996; Keane et al. 1996; Wu et al. 1996) are at the forefront of fire regime simulation in landscapes. However, they are of limited use in Australian systems because of their reliance on the semi-empirical firespread algorithms originally developed by Rothermel (1972). In Australia, these algorithms are not widely used and hence are poorly understood compared to the empirical approach to predicting fire behaviour (McArthur 1967) that has been routinely used for many decades. Another limitation is the generation of daily weather in FIRE-BGC where many of the basic factors affecting fire behaviour, including wind speed and fuel moisture content, are preset by the user and are constant throughout the simulation of forest history. In a largely parallel research project, a new fire regime simulator (FIRESCAPE) that addresses these issues has been developed for Australian landscapes (Cary 1998).

Description of the FIRESCAPE model

In FIRESCAPE, spatial patterns in fire regimes are synthesised by simulating individual fire events that are combined over time into patterns of fire frequency, fire intensity and the season of fire occurrence (Cary 1998). It has been implemented for around 900 000 ha of the Australian Capital Territory (ACT) region of Australia. The study area has varying topography and is represented by square cells of 100 m to a side (Fig. 4.6). When there are no fires burning in the model system, FIRESCAPE operates on a daily time step that changes to hourly whenever a fire ignites. At present, ignition probability depends on daily weather because ignitions are lightning-caused. Anthropogenic ignitions can be included but this would require further model development.

Daily weather is generated by a modified version of the Richardson-type stochastic climate generator. Long-term daily weather observations can be used, however, it may be 'desirable to generate synthetic sequences of weather data based on the stochastic structure of the meteorological process' (Richardson 1981) for multiple simulation runs. Daily weather is generated in three stages. Initially, a Markov chain with only two states, wet or dry, is used to generate sequences of wet and dry days. Next, the amount of precipitation on wet days is generated using a truncated power of the normal model (Hutchinson 1995). The remaining weather variables are then simulated so that serial correlations within a variable and cross correlations between variables are maintained (Matalas 1967; Richardson 1981).

The Richardson weather model was parameterised for Canberra (ACT) using 16 years of daily rainfall data from 1978 to 1993 and 11 years of other daily weather data from 1978 to 1988. The simulated weather had a frequency distribution of Forest Fire Danger Index (McArthur 1967) that was similar to observed data both on a yearly basis and in the summer months. Of equal importance, the simulated weather had a similar frequency distribution of different length runs of low and moderate fire danger days as can be generated from the observed weather (Cary 1998).

Ignition locations are generated from an empirical model of lightning strike. The probability of ignition is positively associated with the macro-scale elevation at the broad spatial scale, primarily reflecting the effect of mountain ranges on storm occurrence. It is also positively associated with the magnitude of the meso-scale elevation residual (the difference between the elevation of a site and the average elevation measured at a broader spatial scale). These findings are developed from the data of McRae (1992) and are consistent with the current understanding of atmospheric electricity and lightning occurrence. They reflect patterns found in similar studies in Yosemite and Sequoia National Parks, California, USA (Vankat 1983).

The spread of fire from a burning cell to its immediate neighbours is based on the elliptical fire spread model (Van Wagner 1969). For the sake of parsimony, the spread of fires is modelled on the same system of grid cells as the other data required by the model. The fire line propagates by moving from one fixed point to another fixed point after the appropriate amount of time has elapsed.

Headfire rate of spread is determined for each individual burning cell using the equation form of McArthur's Forest Fire Danger Meter (McArthur 1967). Hourly meteorology, fuel load and drought factor are calculated for individual cells as required. The drought factor combines information on short-term rainfall patterns with longer-term dryness from the Soil Dryness Index (SDI) (Mount 1972). The SDI is determined from the daily data and takes into account the effect of landscape variation on soil moisture dynamics via the modification of daily evaporation as a function of solar radiation budgets.

Fuel loads are modelled using the empirical approach first proposed by Olson (1963). It has been found to describe adequately the pattern of litter accumulation in a number of Australian systems including sub-alpine eucalypt (Raison et al. 1983) and open eucalypt forest (Fox et al. 1979). Fire line intensity (kW m^{-1}) (Byram 1959) is calculated for the spread of fire from one cell to the next for characterising this aspect of the fire regime and for determining the extinguishment of the individual fire events.

The basic output of the model is a set of maps that represent the spatial variation in components of the fire regime (Fig. 4.7).

Verification of FIRESCAPE

A rigorous test of the fire regime patterns produced by FIRESCAPE would require a thorough knowledge of long-term fire regimes across the landscape. Although some aspects of this type of data are available for some areas of Australia (Cary and Morrison 1995), comprehensive data do not exist for the ACT region study area. Nevertheless, output can be compared with a general knowledge on fire regimes in the study area and related systems. A comparison of this nature is presented in Table 4.1.

There is some agreement between the model output and the minimum inter-fire intervals observed in dry sclerophyll forests. This reflects the overriding importance of rates of fuel accumulation. Also, there is general agreement between the predicted and observed intensity of fires. The predicted average intensity of fires burning during the warmer parts of the day is lower than the range identified for intense forest fires. This may occur because some fires occurring during the middle of the day can be of a low intensity, so the predicted average would be expected to be lower than the published observations. The theoretical maximum fire line intensity proposed by Trevitt (1994) is greater than the maximum intensity predicted by FIRESCAPE but this is also expected because the proposal must be based on an extreme fire danger rating and fuel loads in the order of 40 tonnes ha^{-1}. This combination is rare in the ACT region, particularly for montane areas.

Table 4.1. *Reported values of fire regime parameters for forested systems (general and ACT-specific) compared with the corresponding values derived from output of the ACT implementation of FIRESCAPE*

Parameter	Observed value	Reference	FIRESCAPE
General observations			
Fire frequency	Minimum inter-fire interval in dry sclerophyll forest is 3 to 4 years	Christensen et al. (1981)	Minimum inter-fire interval is 2 years
Fire intensity	Intense forest fires have a fire line intensity of between 2000 and 4000 kW m^{-1}	Christensen et al. (1981)	Average fire line intensity from 1100 to 1500 hours is 1800 kW m^{-1}
Fire intensity	Theoretical maximum fire line intensity is 100 000 kW m^{-1}	Trevitt (1994)	Maximum fire line intensity is around 45 000 kW m^{-1}
Observations specific to fire regimes in the ACT			
Fire frequency	Average inter-fire interval for the Cotter Catchment is around 50 years	Shugart and Noble (1981)	Average inter-fire interval is 50 years
Fire frequency	Inter-fire interval on plains is greater than that in the mountains in the ACT	Pryor (1939)	Average inter-fire interval on plains is 27 years and in mountains is 50 years
Fire season	Fire season is week 45 to week 11 of year. Fire season is week 50 to week 8 of year. Fire season is week 48 to week 11 of year	Pryor (1939) Walker (1981) Luke and McArthur (1978)	In FIRESCAPE, 76% of fires occur in summer (weeks 49 to 8), 14% in autumn, (weeks 9 to 22), 0.1% in winter (weeks 23 to 36), and 8.9% in spring (weeks 37 to 48)

The model also performed well in two out of three comparisons with more specific observations from the ACT region. Firstly, output was in close agreement with the assumed fire frequency nominated by Shugart and Noble (1981), based on the findings of Costin (1954), for the mountainous regions of the ACT. Also, the majority of cells burnt were done so in those parts of the year nominated as fire seasons by various authors (Table 4.1). These findings indicate that the FIRESCAPE model is capable of generating sensible fire regimes through modelling based on lower-level processes.

On the other hand, the inability of the model to reproduce the relative frequency of fire in the mountains versus the plains in the ACT demonstrates a possible shortcoming, particularly given that the emphasis is on spatial variation in fire regimes. Possible explanations for this include: (1) the poor representation or absence of important processes in the model; (2) the spatial and temporal span of the observations made by Pryor (1939); and (3) the effect of anthropogenic burning in the real system, before the observations were made by Pryor (1939), which is not included in the model system.

Footprint models

Spatial models of fires may be based on probability distributions, with the area burnt being a random variable drawn from a specified probability distribution. For example, Green (1989) simulated fires in landscapes by drawing the number of fires from a Poisson distribution and the area of each fire from a geometric distribution. Fires were elliptical in shape and were located randomly in the landscape. Other authors have drawn the size of fires from a probability distribution, ignited a random point for each fire and spread it randomly (Baker et al. 1991). Such models may be considered a hybrid of a footprint model (the area of fires is drawn from a probability distribution) and a mechanistic model (the shape of the fire is determined by spreading the fire).

The spatial fire models that have been discussed above attempt to randomly simulate the incidence of fire at points, while simultaneously modelling the shape of fires. It is reasonable to expect points close together to be more likely to burn in the same fire than points far apart. This dependence may be described in terms of the strength of the correlation between the incidence of events. Points that are close together will be highly correlated, while the correlation in the incidence of fire between points far apart will be small. Thus, McCarthy and Lindenmayer (1998) simulated fires as a set of spatially correlated events, in which both the annual probability of fire and the correlation in the incidence of fire were specified. The correlated fire model does not explicitly model the spread of fire, so it is of the footprint type. The mountain ash (*Eucalyptus regnans*) forest studied by McCarthy and Lindenmayer (1998) was represented as an aggregation of cells. They used the distance between cells to predict the strength of spatial correlation, with cells close together being more likely to burn in the same year than cells far apart. The relationship between the correlation in the incidence of fire and distance between cells was estimated from a fire that occurred in the study area. Although McCarthy and Lindenmayer (1998) assumed that the annual risk of fire was the same for all points in their landscape, the correlated fire model can accommodate spatial variation in the risk of fire. The correlated fire model is based on constructing a matrix that specifies the correlation between the incidence of fire for all pairs of cells. This correlation matrix may become prohibitively large for most computers when the number of cells exceeds a few thousand.

McCarthy and Lindenmayer (1998) used the correlated fire model to simulate the development of multi-aged mountain ash forest. Multi-aged forest occurs where young forest is adjacent to old forest, providing a combination of important habitat elements for arboreal marsupial species (Lindenmayer 1996). In the 3-ha areas that were modelled by McCarthy and Lindenmayer (1998), the predicted prevalence of multi-aged forest closely matched the observed prevalence in the 373 sites that were surveyed. The model helped to demonstrate that the incidence of multi-aged forest depended on the spatial scale at which the forest was examined. Additionally, it demonstrated that development and loss of multi-aged mountain ash forest is a dynamic process depending on the fire history of the previous centuries.

The correlated fire model has also been used in simulating the metapopulation dynamics of the greater glider (*Petauroides volans*) and Leadbeater's possum (*Gymnobelideus leadbeateri*), two species of arboreal marsupial that inhabit mountain ash forests (McCarthy and Lindenmayer 1999, 2000). Fire influences the habitat quality of forest for both these species. The models simulated both partial disturbance by fire within patches, and spatial correlation in the incidence of fire between patches. The correlated fire model was able to simulate both these aspects without modelling disturbance in the intervening areas. Previous metapopulation models have typically ignored partial disturbance within patches, requiring the implicit assumption that small and large patches are uniformly affected by disturbances (McCarthy and Lindenmayer 1999). The models demonstrated that positive correlations in the incidence of fire between patches produced higher risks of extinction, while allowing partial disturbance of patches decreased the predicted risk of extinction. Landscape fire models provide a mechanism for linking important spatial processes with metapopulation models, an important avenue for further development of metapopulation modelling (Wiens 1997).

Comparing FIRESCAPE to point-based models

Process-based simulators of fire regimes will generate fire regimes with implicit properties, while the statistically based models require these properties as input. As discussed above, McCarthy et al. (2001) suggested a number of different ways in which hazard may change with time since fire, depending on the properties of the vegetation and fuel. The simulated hazard function of FIRESCAPE (Cary 1998) provides a test of this idea. In FIRESCAPE, fuel load is the only factor to change systematically with time since fire, with other factors remaining unchanged (e.g. topographic position) or varying randomly (e.g. weather). The reasoning of McCarthy et al. (2001) would predict that changes in the prob-

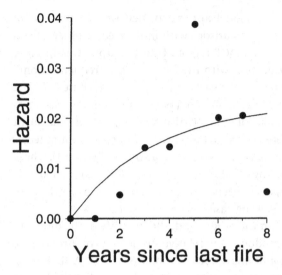

Fig. 4.8. Hazard versus time since the last fire for an arbitrary cell in the FIRESCAPE model (dots). The fitted line is the best-fitting Olson model with a fuel accumulation parameter of 0.3, $h(t) = 0.023\,(1-e^{-0.3t})$. For all cells examined, the hazard at year 1 was less than that predicted by the Olson model.

ability of fire in FIRESCAPE should reflect changes in the abundance of fuel. The fuel accumulation curve in FIRESCAPE followed the Olson (1963) equation with the accumulation parameter equal to 0.3. FIRESCAPE was used to simulate the fire regime for a period of 10000 years, and the hazard, as a function of time since the last fire, was determined for a set of 20 cells in different parts of the landscape. For each time period since fire (0, 1, 2, . . . years), the hazard was calculated as the proportion of times that a cell burnt within the next 1-year period. This derived hazard function was then compared to the shape of the fuel accumulation curve.

In all cases, the hazard function derived from the FIRESCAPE model had a form that was different from the fuel accumulation curve. For all 20 cells, the hazard function indicated that fires did not occur one year after a previous fire. A typical example of the hazard functions is shown in Fig. 4.8. There is an apparent non-linear relationship between the amount of fuel (and therefore the rate of spread in FIRESCAPE) and the probability of fire. When the fuel load is low, the probability of fire occurrence is disproportionately low before increasing rapidly. A function of this form is more akin to the logistic model, although the fuel accumulates according to the Olson model. This non-linear relationship is likely to be due to the same properties that contribute to non-linearity of cellular-automata models (Turner *et al.* 1989; Gardner and O'Neill 1991; Green 1994; McCarthy and Gill 1997). Low fuel loads would effectively mean that the probability of fire spreading between cells would be reduced. Such reductions have the effect of reducing the chance of cells burning, but such changes may be distinctly non-linear, with the chance of fire remaining relatively low until the probability of spread is greater than a certain threshold (Turner *et al.* 1989; Gardner and O'Neill 1991; Green 1994; McCarthy and Gill 1997). It is not clear at this time if the non-linear relationship between fuel load and the probability of fire occurs in reality. It may simply be an artefact of the discrete representation of space in cellular models. Using data derived from mapped fires in jarrah forest, Lang (1997) found that the hazard function was similar in form to the fuel accumulation curve, suggesting that a non-linear relationship did not exist in his study area. Similarly, there is only a little evidence in other parts of the world for a delayed increase in hazard after fire (Baker 1989; Gill and McCarthy 1998).

Discussion

The point-based fire models and the correlated fire model are statistical representations of the system. The fire interval distribution is a required input parameter of these models that could be estimated using methods such as those of Reed (1994, 1997) and colleagues (Reed *et al.* 1998). In the mechanistic models, the mean fire interval for each cell arises as a consequence of the other input parameters. For the process-based models (Wu *et al.* 1996; Cary 1998), the mean fire interval is also modified by the topography and climatic conditions. Such spatial realism requires considerable information about factors influencing the spread of fires. The predictions of these models can be tested, with the appropriate test depending on the purpose of the model. Wu *et al.* (1996) tested their model by comparing the predicted mean fire size, annual area burnt and the

largest fire size. Cary (1998) demonstrated several features of the fire regime were reasonable, while others appeared to be different from observations (Table 4.1). Cary (1998) also found that the predicted fire regime could help to explain the distribution of plant species in the landscape, which was the original purpose of the model. The properties that are most likely to be important are the size and shape of fires (McCarthy and Gill 1997) and how the hazard changes with time since fire.

As implemented by McCarthy and Lindenmayer (1998), the correlated fire model did not account for spatial or temporal variation in the probability of fire. Neither did they consider how the strength of correlation might vary with the direction between cells, as a consequence, for example, of the prevailing wind. All these factors could be included in the model without major modification. However, it would be difficult to incorporate particular details about fire behaviour, such as unidirectional spotting in advance of the firefront. Such processes would require a very complex pattern of correlations between cells, and it may not be possible to estimate the appropriate values. Therefore, the correlated fire model is suitable for modelling fires in relatively small areas of up to a few thousand hectares. The computational requirements of this method also mean that it is suited to these smaller scales. In larger areas where wind direction is likely to have an important influence on the pattern of burnt and unburnt vegetation, the process-based models (e.g. Wu *et al.* 1996) would be more appropriate. The advantage of the correlated fire model is that when only small areas are considered, the spread of fires from external areas may be considered implicitly rather than explicitly, with consequent reductions in model complexity.

The different spatial models generate different spatial patterns. Such spatial patterns may have important influences on the response of the system. There are few data from real systems on patterns of fires, so it is difficult to know what are reasonable models (McCarthy and Gill 1997). Tests such as comparing the mean predictions (Wu *et al.* 1996) are useful, but the predicted variation in fires may be equally important (McCarthy and Burgman 1995;

McCarthy and Lindenmayer 1998). Attempts to model the persistence of animal and plant populations in fire-prone landscapes may depend on the assumed patterns of fires (Lindenmayer and Possingham 1994; Bradstock *et al.* 1996; McCarthy and Lindenmayer 1999). In many population models, the responses to disturbance in large and small patches are identical, with a homogeneous and uniform effect. However, large patches are more likely to experience partial disturbance than small patches (Seagle and Shugart 1985; Eberhart and Woodward 1987; Morrison *et al.* 1992), and risks faced by species that are sensitive to disturbance will tend to be smaller in large patches (McCarthy and Lindenmayer 1998, 1999). Thus, the spatial patterns predicted by the different fire models may influence predictions of the effect of fire regimes on species.

A diversity of approaches has been used to model fire regimes in landscapes, reflecting the purposes of the models and the experiences of the model-builder. The different models may be compared by examining variations in the size and shape of fire, and how the hazard changes over time. There is only limited knowledge about these aspects of fire regimes because of a paucity of historical information about fire regimes. Maintaining accurate and consistent maps of fires would help to redress this lack of data. Contemporary fire regimes are often characterised by fire intervals that are long with respect to periods of record. This paucity of data means that it is often necessary to average across space or across time in order to describe fire regimes. Models of fire regimes in landscapes provide a means of integrating these data to retain information about how fire regimes may vary spatially and temporally.

Acknowledgements

The meteorological data for Canberra were supplied by the Australian Bureau of Meteorology and the Commonwealth Scientific and Industrial Research Organisation. The development of the FIRESCAPE model was partly supported by an Australian Postgraduate Research Award. Professor Ian Noble, Dr Malcolm Gill and Dr John Gallant are gratefully acknowledged for their discussion during its development. We are grateful to Dr Ross Bradstock and Dr Malcolm Gill for discussions about landscape fire models, and to Professor Hugh Possingham and Professor Jim Clark for comments that helped us to improve the manuscript. The work was completed whilst M. A. McCarthy worked at the Department of Applied and Molecular Ecology, University of Adelaide, Adelaide and the Centre for Resource and Environmental Studies, The Australian National University, Canberra.

References

Albini, F. A. (1976). Computer-based models of wildland fire behaviour: a user's manual. U.S. Department of Agriculture Forest Service, Intermountain Forest and Range Experiment Station, Ogden, UT.

Allee, W. C. (1938). *The Social Life of Animals.* (Heinemann: London.)

Baker, W. L. (1989). Landscape ecology and nature reserve design in the Boundary Waters Canoe Area. *Ecology* **70**, 23–35.

Baker, W. L., Egbert, S. L., and Frazier, G. F. (1991). A spatial model for studying the effects of climatic change on the structure of landscapes subject to large disturbances. *Ecological Modelling* **56**, 109–125.

Booysen, P. de V., and Tainton, N. M. (eds.) (1984). *Ecological Effects of Fire in South African Ecosystems.* (Springer-Verlag: New York.)

Bormann, F. H., and Likens, G. E. (1979). Catastrophic disturbance and the steady state in northern hardwood forests. *American Scientist* **67**, 660–669.

Boychuk, D., Perera, A. H., Ter-Mikaelian, M. T., Martell, D. L., and Li, C. (1997). Modelling the effect of spatial scale and correlated fire disturbances on forest age distribution. *Ecological Modelling* **95**, 145–164.

Bradstock, R. A., Bedward, M., Scott, J., and Keith, D. A. (1996). Simulation of the effect of spatial and temporal variation in fire regimes on the population viability of a *Banksia* species. *Conservation Biology* **10**, 776–784.

Bradstock, R. A., Bedward, M., Kenny, B. J., and Scott, J. (1998). Spatially-explicit simulation of the effect of prescribed burning on fire regimes and plant extinctions in shrublands typical of south-eastern Australia. *Biological Conservation* **86**, 83–95.

Byram, G. M. (1959). Combustion of forest fuels. In *Forest Fire: Control and Use*. (ed. K. P. Davis) pp. 61–89. (McGraw-Hill: New York.)

Cary, G. J. (1997). Analysis of the effective spatial scale of neighbourhoods with respect to fire regimes in topographically complex landscapes. In *Proceedings of the International Congress on Modelling and Simulation*. (eds. A. D. MacDonald and M. McAleer) pp. 434–439. (University of Tasmania: Hobart.)

Cary, G. J. (1998). Predicting fire regimes and their ecological effects in spatially complex landscapes. PhD thesis, The Australian National University, Canberra.

Cary, G. J., and Morrison, D. A. (1995). Effects of fire frequency on plant species composition of sandstone communities in the Sydney region: combinations of inter-fire intervals. *Australian Journal of Ecology* **20**, 418–426.

Christensen, P., Recher, H., and Hoare, J. (1981). Responses of open-forest (dry sclerophyll forests) to fire regimes. In *Fire and the Australian Biota*. (eds. A. M. Gill, R. H. Groves and I. R. Noble) pp. 367–393. (Australian Academy of Science: Canberra.)

Clark, J. S. (1989). Ecological disturbance as a renewal process: theory and application to fire history. *Oikos* **56**, 17–30.

Clark, J. S. (1990). Twentieth-century climate change, fire suppression, and forest production and decomposition in northwestern Minnesota. *Canadian Journal of Forest Research* **20**, 219–232.

Clark, J. S. (1996). Testing disturbance theory with long-term data: alternative life-history solutions to the distribution of events. *American Naturalist* **148**, 976–996.

Coleman, J. R., and Sullivan, A. L. (1996). A real-time computer application for the prediction of fire spread across Australian landscapes. *Simulation* **67**, 230–240.

Costin, A. B. (1954). A study of the ecosystems of the Monaro region of New South Wales with special reference to soil erosion. Soil Conservation Service of New South Wales, Sydney.

Crowder, M. J. (1991). *Statistical Analysis of Reliability Data*. (Chapman and Hall: London.)

Cumming, S. G., Burton, P. J., and Klinkenberg, B. (1996). Boreal mixed-wood forests may have no representative areas: some implications for reserve design. *Ecography* **19**, 162–180.

Eberhart, K. E., and Woodward, P. M. (1987). Distribution of residual vegetation associated with large fires in Alberta. *Canadian Journal for Forest Research* **17**, 1207–1212.

Finney, M. A. (1998). FARSITE fire area simulator: model development and evaluation. U.S. Department of Agriculture Forest Service, Rocky Mountain Forest and Range Experiment Station, Research Paper RMRS-RP-4, Ogden, UT.

Fox, B. J., Fox, M. D., and McKay, G. M. (1979). Litter accumulation after fire in a eucalypt forest. *Australian Journal of Botany* **27**, 157–165.

Gardner, R. H., and O'Neill, R. V. (1991). Pattern, process, and predictability: the use of neutral models for landscape analysis. In *Quantitative Methods in Landscape Ecology*. (eds. M. G. Turner and R. H. Gardner) pp. 289–307. (Springer-Verlag: New York.)

Gardner, R. H., Hargrove, W. W., Turner, G. M., and Romme, W. H. (1996). Climate Change, disturbances and landscape dynamics. In *Global Change and Terrestrial Ecosystems*. (eds. B. Walker and W. Steffen) pp. 149–172. (Cambridge University Press: Cambridge.)

Gill, A. M. (1975). Fire and the Australian flora: a review. *Australian Forestry* **38**, 4–25.

Gill, A. M. (1997). From 'mosaic burning' to the understanding of the fire-dynamics of landscapes. In *Bushfire 97: Proceedings of the Australasian Bushfire Conference*. (eds. B. J. McKaige, R. J. Williams and W. M. Waggitt) pp. 89–94. (CSIRO: Darwin.)

Gill, A. M., and McCarthy, M. A. (1998). Intervals between prescribed fires in Australia: what intrinsic variation should apply? *Biological Conservation* **85**, 161–169.

Gill, A. M., Groves, R. H., and Noble, I. R. (eds.) (1981). *Fire and the Australian Biota*. (Australian Academy of Science: Canberra.)

Gill, A. M., Ryan, P. G., Moore, P. H. R., and Gibson, M. (2001). Fire regimes of World Heritage Kakadu National Park, Australia. *Austral Ecology*. (in press)

Goldammer, J. G. (ed.) (1990). *Fire in the Tropical Biota: Ecosystem Processes and Global Change*. (Springer-Verlag: Berlin.)

Green, D. G. (1989). Simulated effects of fire, dispersal and spatial pattern on competition within forest mosaics. *Vegetatio* **82**, 139–153.

Green, D. G. (1994). Connectivity and complexity in ecological systems. *Pacific Conservation Biology* **1**, 194–200.

Hassell, M. P., and Wilson, H. B. (1997). The dynamics of spatially distributed host-parasitoid systems. In *Spatial Ecology: the Role of Space in Population Dynamics and Interspecfic Interactions*. (eds. D. Tilman and P. Kareiva) pp. 75–110. (Princeton University Press: Princeton.)

Henderson, W., and Wilkins, C. W. (1975). The interaction of bushfires and vegetation. *Search* 6, 130–133.

Hutchinson, M. F. (1995). Stochastic space–time weather-models from ground-based data. *Agricultural and Forest Meteorology* 73, 237–264.

Jackson, W. D. (1968). Fire, air, water and earth – an elemental ecology of Tasmania. *Proceedings of the Ecological Society of Australia* 3, 9–16.

Johnson, E. A. (1992). *Fire and Vegetation Dynamics: Studies from the North American Boreal Forest.* (Cambridge University Press: Cambridge.)

Johnson, E. A., and Gutsell, S. L. (1994). Fire frequency models, methods and interpretations. *Advances in Ecological Research* 25, 239–287.

Johnson, E. A., and Van Wagner, C. E. (1985). The theory and use of two fire history models. *Canadian Journal of Forest Research* 15, 214–220.

Kareiva, P. (1994). Space: the final frontier for ecological theory. *Ecology* 75, 1.

Keane, R. E., Arno, S. F., and Brown, J. K. (1989). FIRESUM: an ecological process model for fire succession in western coniferous forests. U.S. Department of Agriculture Forest Service, Intermountain Research Station, Research Paper INT-266, Ogden, UT.

Keane, R. E., Morgan, P., and Running, S. W. (1996). FIRE-BGC: a mechanistic ecological process model for simulating fire succession on coniferous forest landscapes of the northern Rocky Mountains. U.S. Department of Agriculture Forest Service, Intermountain Research Station, Research Paper INT-RP-484, Ogden, UT.

Lang, S. (1997). Burning the bush: a spatio-temporal analysis of jarrah forest fire regimes. BSc. thesis, The Australian National University, Canberra.

Li, C., Ter-Mikaelian, M., and Perera, A. (1997). Temporal fire disturbance patterns on a forest landscape. *Ecological Modelling* 99, 137–150.

Lindenmayer, D. B. (1996). *Wildlife and Woodchips*. (University of New South Wales Press: Sydney.)

Lindenmayer, D. B., and Possingham, H. P. (1994). The risk of extinction: ranking management options for Leadbeater's possum using population viability analysis. Centre for Resource and Environmental Studies, Australian National University, Canberra.

Luke, R. H., and McArthur, A. G. (1978). *Bushfires in Australia*. (Australian Government Publishing Services: Canberra.)

Matalas, N. C. (1967). Mathematical assessment of synthetic hydrology. *Water Resources Research* 3, 937–945.

McArthur, A. G. (1967). Fire behaviour in eucalypt forests. Commonwealth of Australia Forest and Timber Bureau, Leaflet no.107, Canberra.

McCarthy, M. A. (1997). The Allee effect, finding mates and theoretical models. *Ecological Modelling* 103, 99–102.

McCarthy, M. A., and Burgman, M. A. (1995). Coping with uncertainty in forest wildlife planning. *Forest Ecology and Management* 74, 23–36.

McCarthy, M. A., and Gill, A. M. (1997). Fire modelling and biodiversity. In *Frontiers in Ecology* (eds. N. Klomp and I. Lunt) pp. 79–88. (Elsevier Science: Oxford.)

McCarthy, M. A., and Lindenmayer, D. B. (1998). Multi-age mountain ash forests, wildlife conservation and timber harvesting. *Forest Ecology and Management* 104, 43–56.

McCarthy, M. A., and Lindenmayer, D. B. (1999). Incorporating metapopulation dynamics of greater gliders into reserve design in disturbed landscapes. *Ecology* 80, 651–667.

McCarthy, M. A., and Lindenmayer, D. B. (2000). Spatially-correlated extinction in a metapopulation of Leadbeater's possum. *Biodiversity and Conservation* 9, 47–63.

McCarthy, M. A., Gill, A. M., and Bradstock, R. A. (2001). Theoretical fire interval distributions. *International Journal of Wildland Fire* 10, 73–77.

McRae, R. H. D. (1992). Prediction of areas prone to lightning ignition. *International Journal of Wildland Fire* 2, 123–130.

Morrison, D. A., Cary, G. J., Pengelly, S. M., Ross, D. G., Mullins, B. J., Thomas, C. R., and Anderson, T. S. (1995). Effects of fire frequency on plant species composition of sandstone communities in the Sydney region: inter-fire interval and time-since-fire. *Australian Journal of Ecology* 20, 239–247.

Morrison, M. L., Marcot, B. G., and Mannan, R. W. (1992).

Wildlife Habitat Relationships: Concepts and Applications. (University of Wisconsin Press: Madison.)

Mount, A. B. (1972). The derivation and testing of a soil dryness index using run-off data. Tasmanian Forestry Commission Bulletin no. 4, Hobart.

Naveh, Z. (1975). The evolutionary significance of fire in the Mediterranean region. *Vegetatio* **29**, 199–208.

Olson, J. S. (1963). Energy balance and the balance of producers and decomposers in ecological systems. *Ecology* **44**, 322–331.

Pryor, L. D. (1939). The bush fire problem in the Australian Capital Territory. *Australian Forestry* **4**, 33–38.

Pyne, S. J., Andrews, P. L., and Laven, R. D. (1996). *Introduction to Wildland Fire.* (Wiley: New York.)

Raison, R. J., Woods, P. V., and Khanna, P. K. (1983). Dynamics of fine fuels in recurrently burnt eucalypt forests. *Australian Forestry* **46**, 294–302.

Ratz, A. (1995). Long-term spatial patterns created by fire: a model oriented towards boreal forests. *International Journal of Wildland Fire* **5**, 25–34.

Reed, W. J. (1994). Estimating the historic probability of stand-replacement fire using the age-class distribution of undisturbed forest. *Forest Science* **40**, 104–119.

Reed, W. J. (1997). Estimating historical forest-fire frequencies from time-since-last-fire-sample data. *IMA Journal of Mathematics Applied in Medicine and Biology* **14**, 71–83.

Reed, W. J., Larsen, C. P. S., Johnson, E. A., and Macdonald, G. M. (1998). Estimation of temporal variations in historical fire frequency from time-since-fire map data. *Forest Science* **44**, 465–475.

Richards, S. A., Possingham, H. P., and Tizard, J. (1999). Optimal fire management for maintaining community diversity. *Ecological Applications* **9**, 880–892.

Richardson, C. W. (1981). Stochastic simulation of daily precipitation, temperature, and solar radiation. *Water Resources Research* **17**, 182–190.

Roberts, D. W. (1996). Modelling forest dynamics with vital attributes and fuzzy systems theory. *Ecological Modelling* **90**, 161–173.

Rothermel, R. C. (1972). A mathematical model for predicting fire spread in wildland fuels. U.S. Department of Agriculture Forest Service, Intermountain Forest and Range Experimental Station, Research Paper INT-115, Ogden, UT.

Running, S. W., and Coughlan, J. W. (1988). A general model of forest ecosystem processes for regional applications. I. Hydrologic balance, canopy gas exchange and primary production processes. *Ecological Modelling* **42**, 125–154.

Seagle, S. W., and Shugart, H. H. (1985). Faunal species richness and turnover on dynamic landscapes: a simulation study. *Journal of Biogeography* **15**, 759–774.

Shugart, H. H., and Noble, I. R. (1981). A computer model of succession and fire response in the high-altitude Eucalyptus forest of the Brindabella Range, Australian Capital Territory. *Australian Journal of Ecology* **6**, 149–164.

Trevitt, C. (1994). Developing a national bushfire strategy. *Search* **25**, 126–127.

Turchin, P. (1998). *Quantitative Analysis of Movement: Measuring and Monitoring Population Redistribution in Animals and Plants.* (Sinauer: Sunderland, MA.)

Turner, M. G., Gardner, R. H., Dale, V. H., and O'Neill, R. V. (1989). Predicting the spread of disturbance across heterogeneous landscapes. *Oikos* **55**, 121–129.

Van Wagner, C. E. (1969). A simple fire-growth model. *Forestry Chronicle* **45**, 103–104.

Van Wagner, C. E. (1978). Age-class distribution and the forest fire cycle. *Canadian Journal of Forest Research* **8**, 220–227.

Vankat, J. L. (1983). General patterns of lightning ignitions in Sequoia National Park, California. In *Proceedings: Symposium and Workshop on Wilderness Fires*, pp. 408–411. U.S. Department of Agriculture, General Technical Report INT-182, Missoula, MT.

von Neumann, J. (1966). *Theory of Self-Replicating Automata.* (University of Illinois: Urbana.)

Walker, J. (1981). Fuel dynamics in Australian vegetation. In *Fire and the Australian Biota* (eds. A. M. Gill, R. H. Groves and I. R. Noble) pp. 101–127. (Australian Academy of Science: Canberra.)

Wiens, J. A. (1997) Metapopulation dynamics and landscape ecology. In *Metapopulation Biology: Ecology, Genetics, and Evolution* (eds. I. A. Hanski and M. E. Gilpin) pp. 43–62. (Academic Press: San Diego.)

Wu, Y. G., Sklar, F. H., Gopu, K., and Rutchey, K. (1996). Fire simulations in the Everglades landscapes using parallel programming. *Ecological Modelling* **93**, 113–124.

5

Critical life cycles of plants and animals: developing a process-based understanding of population changes in fire-prone landscapes

ROBERT J. WHELAN, LOUISE RODGERSON, CHRIS R. DICKMAN,
AND ELIZABETH F. SUTHERLAND

Abstract
Studies of fire responses of plants and animals in Australian ecosystems have been accumulating rapidly in the past 20 years. Many patterns have been identified, ranging from local elimination of a species by a single fire to enhancement of population size or extent by fire, the latter amounting to almost complete dependence on fire for persistence. There is still insufficient information on fire response patterns to be able to identify clear associations between particular fire responses and fire regimes – partly because of insufficient replication of fire studies on particular species, and partly because patterns have been shown to vary over space and time, and are affected by variation in characteristics of the organism, the fire, the landscape and the climate. We conclude that, except at a very coarse level, fire responses will be site-specific, making it almost impossible to predict the effects of a particular fire (or sequence of fires) on any species, based on any number of studies of fire-response patterns at other sites and at other times. Instead of seeking general fire-response patterns, we argue that the focus should be on using field experiments to understand how the life-cycle *processes* that produce various patterns of fire response (mortality in fire, recolonisation, survival and establishment of individuals after fire, reproduction and population growth) and the factors that mediate them (e.g. habitat quality, nest sites, nutrient and food availability, predation) respond to different fire, landscape and climatic characteristics. This approach represents a major challenge for fire ecology and we present some case studies to illustrate how such field experiments may be conducted. An understanding of the way in which important processes respond to fire can then be combined with knowledge of an organism's life cycle at the site and knowledge of the 'environmental templet' (fire history, landscape characteristics and climate) to predict the effects of a particular fire or fire regime.

Introduction

Will this fire cause local extinction of a particular organism? Will this series of fires cause a population decline? Will we need to burn this area deliberately to prevent a particular species from going extinct? How often? What fire regime should we prescribe in the recovery plan for this endangered species? Questions such as these are being asked frequently by conservation biologists, environmentalists and land managers. Satisfactory answers are not usually known, at least for particular situations.

Many patterns in the presence and absence of organisms and in the details of population change in relation to fires have been described. A single fire may eliminate some species temporarily, while other species may suffer virtually no mortality. Populations of some species may decline substantially soon after fire, while other species display population increases. Some species can increase in abundance after a single fire yet disappear after two fires in close succession. There are reports of a particular species declining in abundance after a fire in one area but increasing or maintaining numbers after another fire. How consistent are these patterns across species, and over a number of fires for a given species? What factors may explain the variation?

Over the last 25 years of fire ecology, three important principles have emerged in attempts to explain and predict the effects of fire on biota. Firstly, fires have both direct and indirect effects on individual organisms. Direct effects result from the fire itself causing emigration and/or mortality. Indirect effects result from the fire causing changes to some things (such as resources) which, in turn, affect the biota. Thus, a fire may cause rapid emigration or mortality either by direct effects on organisms (lethal temperatures, smoke, anoxia) or by altering the habitat to make it inappropriate for the organism's survival.

Secondly, the responses of a population of organisms to a particular fire will be affected both by the fire event itself and by processes occurring after the fire ('interval-related processes'). A clear example of this is where a population is maintained despite the passage of fire only to decline in the weeks and months afterwards as individuals are driven out or killed by starvation or by predation, and/or fail to reproduce.

Thirdly, the effects of fire on the biota result from the *fire regime*, not just a single fire in isolation (Gill 1975). Thus, the responses of an individual organism, population or community will depend on the starting state of a community, determined by the past history of the site (e.g. time since the last fire), as well as on the characteristics of the particular fire (which include season, intensity, type of fire, patchiness, post-fire climatic conditions etc.). Fire regime characteristics are explored in more detail in McCarthy and Cary (this volume).

The impacts of a particular fire on plants and animals clearly result from a complex interaction of many factors: the characteristics of the particular fire, the antecedent fire regime, the characteristics of the habitat, the pre- and post-fire climatic conditions, and the biology of the organisms. Population changes following fire, or through a sequence of fires, will be determined by the fates of individual organisms (mortality of adults, post-fire mortality,

reproductive rate and recruitment rate, emigration and immigration) and these fates are likely to be age- or stage-specific. Hence, life cycles of organisms are central to the study of fire effects. An understanding of the basic biology of organisms is fundamental to the prediction of their responses to fire (Newsome and Catling 1983).

The patterns of fire responses in the biota are many, they are complex, and they appear to be difficult to predict. Can we find generalisations about the responses of organisms to particular fire events? Do we have the ability to predict population and community responses to individual fires and to fire regimes? How can land managers choose particular fire management actions which will produce a desired outcome? At about the time of the publication of *Fire and the Australian Biota* (Gill *et al.* 1981), John Harper published an influential paper entitled: 'After description' (Harper 1982). He argued that 'The search for generality may sacrifice both realism and precision . . . and lead the ecologist to large-scale survey which runs the risk of yielding results that are only trite and superficial.' This 'trilemma' can be visualised as an ecologists' 'fire triangle' (Fig. 5.1), in which many of our studies represent a balance between precision (e.g. detailed, tightly controlled experiments, perhaps in the laboratory) and realism (perhaps a field study with more variability and less replication). Most studies are therefore in the bottom portion of the triangle. In this paper, we ask how the pinnacle of 'generality' or 'predictive power' might be achieved without sacrificing 'precision' or 'realism' at the base of the fire triangle. This challenge of achieving precision, realism and generality in ecology was well explored by Hairston (1989).

In this chapter, we argue that the key to solving this trilemma is to seek to understand ecological *processes* in relation to fire. Rather than solely describing the population or community changes that occur fire after fire, attention should be focussed on the *causes* of the changes. Understanding causes of patterns requires experimentation (Harper 1982; Connell 1983; Hairston 1989; Underwood 1998) and experiments designed to uncover processes have been relatively rare in fire

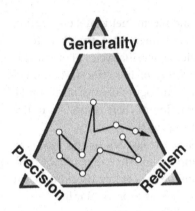

Fig. 5.1. The ecologists' 'fire triangle', based on Harper (1982) and Hairston (1989), illustrating the trade-offs between precision and realism of fire studies and the achieving of generality or predictive power.

ecology studies. Indeed, even studies of patterns in population responses have often lacked replication of fire events. The emphasis of this chapter is therefore on the need for more experimental studies in fire ecology. We first describe briefly a number of patterns of fire population responses that appear in the fire literature. We summarise a range of processes that have been invoked to explain the patterns that have been observed, and then examine these processes, focussing on studies that have used an experimental approach, for vertebrates, invertebrates and plants.

Patterns

Population responses to a single fire: vertebrates and vascular plants

Patterns of response to fire vary enormously across species and community types. Measures of population sizes at different times after a fire or, more commonly, in a number of sites of a particular vegetation type that are different ages since fire (a 'chronosequence' study), reveal many possible fire response patterns (Fig. 5.2).

A null pattern of response in which populations remain unchanged after a fire event (Fig. 5.2a) has been observed or inferred for saxicolous lizards (velvet geckoes *Oedura lesueurii*, leaf-tailed geckoes *Phyllurus platurus*) and their invertebrate prey (C. Dickman

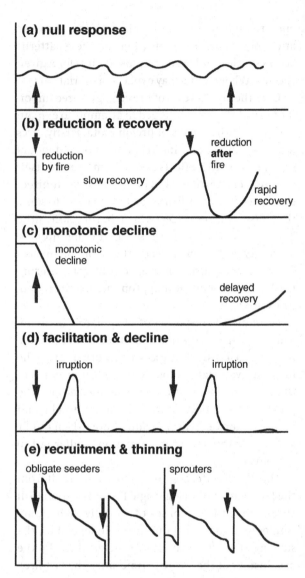

Fig. 5.2. Schematic diagram illustrating a range of fire response patterns populations may exhibit over time. (a) Null response – population size remains unchanged in response to a fire; (b) reduction and recovery – population size declines soon after the fire and remains low for some period followed by recovery (fast vs. slow recovery); (c) monotonic decline in population size, leading to local extinction, perhaps with eventual recovery; (d) facilitation and decline – population size increases following fire and then declines; (e) recruitment and thinning – population size dramatically decreases immediately after fire, then rapidly increases, followed by gradual decline, for obligate seeders (left) and sprouters (right).

personal observation). Some frogs may also maintain populations unchanged in the post-fire environment (Bamford 1992; Hannah and Smith 1995). Null patterns of response to fire have been reported surprisingly often in many vertebrate species (e.g. Bamford 1986; Baker *et al.* 1997; Recher 1997) and may occur more commonly than is often appreciated. Some plant species also appear to exhibit a 'null response', where established plants of a long-lived species survive by sprouting (Purdie (1977) described these species as 'fire-resistant decreasers') and recruitment after fire balances mortality.

A second pattern of response to a fire is a population declining during or soon after the fire, and remaining at low density for some time before recovering (Fig. 5.2b). The decline may be immediate or delayed, and the recovery period could range from rapid to prolonged. This pattern has been observed most often in small mammals and birds, and infrequently in reptiles and amphibians (Friend 1993). Among small mammals, the insectivorous *Antechinus swainsonii*, *A. stuartii* and *A. agilis* usually decline in abundance during or soon after fire, and may not begin to recover for at least 3 years. In contrast to the relatively rapid recovery of *Antechinus* species, *Rattus fuscipes* (bush rat) and *R. lutreolus* (swamp rat), in a range of studies, appear to take 6 years or more to achieve their former densities (Fox 1982, 1990). Populations of other species, such as the eastern bristlebird *Dasyornis brachypterus* (Baker 1997) and the heath rat *Pseudomys shortridgei* (Cockburn 1978; Baynes *et al.* 1987) may take longer still. Among plants, many rainforest species present in wet sclerophyll forest in eastern Australia show this pattern. Established plants are eliminated by high-intensity fire and invade slowly thereafter (Ashton 1981a).

A monotonic decline to local extinction after a fire, with no immediate recovery (Fig. 5.2c) may occur for some vertebrates. In an isolated bushland park in Perth, Western Australia, for example, Recher (1997) recorded losses of the pallid cuckoo *Cuculus pallidus* and fan-tailed cuckoo *C. pyrrhophanus* following an extensive wildfire. Other species were still declining 6 years after the fire. Few studies have been conducted for long enough to observe the

time of recovery of such populations, so it is possible that this pattern could simply be a longer-term version of that shown in Fig. 5.2b.

In contrast to the patterns of stasis or post-fire decline described so far, a fourth pattern is represented by a post-fire increase in abundance followed by a decline (Fig. 5.2d). Among the most dramatic examples of this pattern are the transient increases in numbers of many species of birds that follow fires to exploit temporarily increased prey availability (McCulloch 1966; Main 1981; Braithwaite and Estbergs 1987), although many of these situations may simply represent local concentrations of wide-ranging species at a fire, rather than real increases in local population size within the area affected by fire. Populations of the introduced house mouse *Mus domesticus* were undetected prior to the 1972 wildfire at Nadgee Nature Reserve, New South Wales, but irrupted soon afterwards (Newsome *et al.* 1975). This pattern has been observed for the house mouse in many areas. Native rodents (e.g. *Pseudomys* spp.) and some marsupials (e.g. *Sminthopsis* spp.) display a similar irruption, though the period of large population size is more prolonged than for the house mouse (Cockburn *et al.* 1981; Fox 1982, 1990; Lunney *et al.* 1987). Similar post-fire population peaks have been described for several species of birds (Woinarski and Recher 1997), but seldom for reptiles or frogs (Friend 1993). Some plant species may also display this fire response pattern. 'Irruptive behaviour' is seen in post-fire annuals and ephemerals, as described for Western Australian kwongan vegetation (Bell *et al.* 1984) and the wet–dry tropics (Stocker 1966; Gill 1981a), although this life history appears to be rare in many Australian fire-prone ecosystems (Hopkins and Griffin 1984; Whelan 1995). A medium time-scale example of this pattern is seen in obligate seeding shrubs which have a limited life span but fire-stimulated germination (e.g. the acacias: Auld 1987). Over a very long time-scale, *Eucalyptus* populations in mixed forest (e.g. *E. regnans*, *E. delegatensis*, *E. nitens*) increase after a high intensity fire and eventually disappear as rainforest species such as *Nothofagus* attain dominance (see Howard and Ashton 1973, cited in Ashton 1981a).

For many plant species, population changes may appear to be much more abrupt than for vertebrates, partly because assessments of population sizes typically exclude seeds. Thus, for some species, population size (excluding seeds) declines to zero as an immediate result of a fire, only to increase almost instantaneously to a very high density as a result of mass germination, followed by a persistent population decline (Fig. 5.2e). Many of the 'obligate seeder' species exemplify this fire response pattern (Specht *et al.* 1958; Auld 1987; Whelan 1995). Where some adult plants survive a fire, either because the fire is patchy or because the species is capable of sprouting, population size does not decline to zero as a result of fire, but recruitment may produce the same pattern of rapid increase and subsequent decline.

Invertebrates: a special case?

The patterns of fire responses exhibited by invertebrates are perhaps more variable and even more difficult to detect than those presented above for vertebrates and plants. The lack of clearly defined patterns may be attributable to a number of different factors which we explore here. Invertebrates deserve far more attention than they have received to date in fire ecology. They are numerous, diverse, they certainly play important functional roles in ecosystems, and the high densities of many taxa make them ideal subjects for replicated fire studies because adequate replication can be managed at scales that are ecologically meaningful. Despite this, our knowledge of patterns of fire responses of invertebrates has not progressed very much beyond the situation presented by Campbell and Tanton (1981). We therefore consider that it is worthwhile to explore the special problems of invertebrates in more detail.

Firstly, invertebrates are a tremendously diverse group that exhibit a wide range of life histories and morphologies, and inhabit many different types of habitats. Therefore, when considered at broad taxonomic levels (i.e. phylum, class, order) it is hardly surprising that patterns in fire responses have not emerged. Friend and Williams (1996) studied spiders, and also insects belonging to the orders Coleoptera, Diptera, Hemiptera, Hymenoptera and Orthoptera, and found that variation in abundances was unrelated to fire. When a finer taxonomic resolution was used (e.g. morphospecies of Coleoptera), some fire-related responses became apparent.

Secondly, invertebrate populations change both temporally (e.g. seasonally and annually) and spatially (e.g. different habitats, different regions). This is exemplified by a study by Greenslade and Rosser (1984) in which incorporation of monthly pre-fire samples allowed the substantial seasonal variation in abundance of five taxa to be distinguished from any effects of burning. It is essential that this background variation be taken into account when attempting to interpret patterns relating to fire responses.

Thirdly, key attributes of the fire regime (e.g. intensity) potentially influence the pattern of fire responses, and are likely to vary between different fires. Delettre (1994) attributed changes in chironomid species assemblages between sites in Brittany, France, partly to differences in fire intensity. Unfortunately, these attributes are rarely recorded in studies of invertebrates and fire and have not been studied in manipulative experiments.

Fourthly, a variety of different sampling protocols has been used to estimate invertebrate diversity and abundance, including pitfall trapping, sweep netting and litter sampling. Given that measurements made by different techniques are often not comparable (Southwood 1966; Friend 1995; Whelan 1995), caution must be used when comparing studies that have used different sampling protocols. Hindmarsh and Majer (1977), for example, found that extraction of invertebrates from litter and sampling by pitfall traps yielded different patterns of relative abundances of invertebrates among orders in sites of different post-fire age (Fig. 5.3).

Finally, it is essential that experiments are carefully designed, incorporating pre- and post-fire sampling of control and burnt sites with sufficient replication, especially given these different sources of variability. Unfortunately, very few studies have been conducted which fit these requirements (Friend 1995; Whelan 1995) and we are still some way from establishing the generality or otherwise

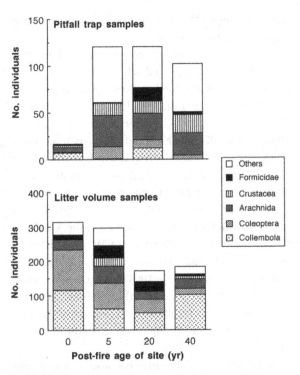

Fig. 5.3. Numbers of invertebrates of five different orders recorded in sites of four post-fire ages: 0 (immediately after fire), 5, 20 and 40 years after the last fire in southwest Australian eucalypt forest. Numbers were estimated by counts of animals in fixed volumes of litter (top) and by pitfall trap sampling (bottom). Modified from Hindmarsh and Majer (1977), after Whelan (1995), with permission of Cambridge University Press.

of fire-response patterns exhibited by different invertebrate taxa.

Given this state of knowledge of patterns in fire responses of invertebrates, it is essential that consistencies in fire responses, if they do exist, be identified. In a review of invertebrates in relation to fire in Australia, Friend (1995) identified Araneae, Lepidoptera, Isopoda, Blattodea, Thysanura and possibly Diptera as taxa that are most likely to display consistent responses to fire. Whilst it would appear sensible to focus on the fire responses of these taxa, the majority of studies investigating invertebrates and fire have focussed on ants, Orthoptera and Collembola.

Despite these problems some patterns have emerged. Many invertebrate taxa appear to decline after fire and then recover quite quickly (Whelan

1995; Friend and Williams 1996). For example, in a grassy woodland in New South Wales, invertebrate abundance declined during the first year post-fire and recovery had occurred by the second year (Greenslade 1997). Similar findings were reported for the total abundance of insects, and for individual insect taxa, in an Illinois sand prairie (Anderson *et al.* 1989). Ants were an exception to this pattern as they increased significantly after fire but had returned to abundance levels recorded on unburnt plots by the second year after fire. Numerous other studies, both in Australia (e.g. O'Dowd and Gill 1984; Tap and Whelan 1984; O'Dowd 1985; York 1996, 1999) and elsewhere (e.g. Ahlgren 1974) have reported similar findings for ants. These reported increases in ant abundance after fire may be due to increased ant activity rather than an increase in the size of ant populations, either by immigration or recruitment. Experiments designed to distinguish between these alternatives have yet to be conducted.

Many studies have found no change in invertebrate abundance after fire (e.g. Abbott *et al.* 1985; Collett and Neumann 1995). Collett and Neumann (1995) compared abundance of beetles between control and burnt sites, before and after fire in eucalypt forest in Victoria, Australia. They did not find significant changes in total beetle abundance, or in the abundances of Nitidulidae, Leiodidae and 28 minor families, after burning.

Variations in fire response patterns

Several of the population-level fire responses described above may be displayed by the range of organisms present in a community after a particular fire. For example, Fox and Fox (1987) summarised fire-response patterns for the small mammal species of a eucalypt forest and two heathland communities at Myall Lakes, New South Wales (Fig. 5.4). These studies reveal low numbers immediately after fire with rapid increases and declines for *Sminthopsis murina* and *Pseudomys* spp. (cf. Fig. 5.2d) alongside steady increases for the rats (*Rattus fuscipes* and *R. lutreolus*) (cf. Fig. 5.2b).

One explanation for inter-specific differences is obviously variation in life cycles and other characteristics as they relate to fire. The rats and *Pseudomys*

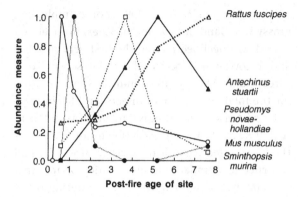

Fig. 5.4. Array of different fire-response patterns in small mammal community at Myall Lakes, New South Wales. For each species, numbers are expressed as a proportion of the maximum numbers recorded for that species during the span of the study. Modified from Fox and McKay (1981), after Whelan (1995), with permission of Cambridge University Press.

differ in reproductive rate, for example. Similarly, of two co-occurring congeneric plant species, one may display the pattern illustrated in the obligate-seeder part of Fig. 5.2e because adults die in the fire, whereas the resprouter would display the other pattern illustrated in Fig. 5.2e. Of more interest, however, is the fact that fire-response patterns can vary among sites and among different fires even within a single species. Further, life-history characteristics can also vary within species. It is important to explore the extent of this sort of variation, for it poses particular challenges to the ability of fire ecologists to come up with both precision and generality (see Fig. 5.1).

A close examination of the fire-response patterns in Fig. 5.4, for those small mammal species that occur in both the forest and the heathland sites, reveals differences between the habitats. *Pseudomys novaehollandiae*, for example, displayed an irruptive pattern in forest (see Fox and McKay 1981) but a slower increase and slower decline in the dry heath. What might have caused these differences: the nature of the fires themselves; different fire histories of the sites; different post-fire changes in food, cover, predation or competition; different landscape characteristics (e.g. topographic heterogene-

ity)? Are these differences between habitats consistent among sites or fires?

Patterns of decline and recovery in *Antechinus* and *Rattus* have been reported with some consistency, based mostly on detailed studies from three sites in southeastern Australia: Nadgee Nature Reserve (Newsome *et al.* 1975; Recher *et al.* 1975; Catling 1986) and Myall Lakes National Park (Fox 1982, 1983, 1990; Fox *et al.* 1985) in New South Wales, and the Otway Ranges in Victoria (Wilson *et al.* 1990; Wilson 1996). Other patterns of response for these species have been described (Whelan *et al.* 1996; Sutherland 1998) in which *Antechinus stuartii* numbers did not decline to near zero after the fire. What is different between the Nadgee wildfire of 1972 and the Royal National Park wildfire of 1994? Do the fire-response patterns differ because of different fires, different landscape characteristics, different post-fire processes such as predation, food availability, cover etc.?

Similar variations in fire responses are apparent in studies of birds (e.g. Loyn 1997; Woinarski and Recher 1997). Recher (unpublished; cited in Woinarski and Recher 1997) found that populations of most heathland bird species recovered rapidly after the 1972 wildfire at Nadgee. In contrast, Roberts (1970) reported that many heathland bird species had disappeared or were still in low numbers 2.5 years after a heath fire. Fire responses of ground parrots (*Pezoporus wallicus*) varied between two heath types in Victoria (Meredith *et al.* 1984), based on an extensive chronosequence study, with numbers greatest in 5-year-old sites in shrub–heathland and in 15-year-old sites in 'graminoid' heathland. In contrast, MacFarland (1991) reported a post-fire population peak at 5–8 years in graminoid heathlands and sedgelands in southeast Queensland. Densities of ground parrots had not yet started to decline after 17 years, in a longitudinal study after a fire at one sedgeland site in Barren Grounds Nature Reserve in New South Wales (Whelan and Baker 1999). What fire, habitat and climate variables might explain these differences in fire responses?

As time passes after fire in some eucalypt-dominated tall open-forests in eastern mainland Australia and in Tasmania, vertebrate- and wind-dispersed rainforest species encroach (Ashton 1981a, b). However, the precise nature of this pattern is apparently highly variable. Ashton (1981a) suggested that in some areas (e.g. the Atherton Plateau) the invasion of rainforest species is very rapid. In the southeast, invasion and increase in abundance of a temperate rainforest tree species (*Nothofagus cunninghamii*) within *Eucalyptus delegatensis*, *E. nitens* and *E. regnans* tall open-forests are frequently observed although they sometimes occur slowly. Further, in some places, *Nothofagus* is absent from this 'invasion', being replaced by other rainforest species (e.g. *Elaeocarpus*). Why does rainforest invasion occur more slowly at some sites than others?

While much of the variation in fire responses within species, illustrated in the above examples, may be attributable to variation in fires, climate, habitat characteristics and other environmental factors, studies on some plants suggest that life-history characteristics may vary from site to site. For example, Rice and Westoby (1999) found seven species of *Triodia* across Australia that had both populations that were killed by fire and populations that resprouted after fire. Another three species were apparently always killed by fire and another five species only resprouted. Similarly, the degree of bradyspory (also called 'serotiny') in species of *Banksia* has been found to vary substantially among sites in southeastern New South Wales (Whelan *et al.* 1998) and also along a latitudinal gradient in southwestern Australia (Cowling and Lamont 1985). Ingwersen (1977) reported that *Banksia ericifolia* (the archetypal obligate seeder!) resprouts at coastal headland sites at Jervis Bay, whereas plants are killed by fire in most other locations.

Sequence of fires

The above discussion about fire-response patterns relates to descriptions or inferences about changes after a single fire event. There are few studies which examine the population responses of species in response to a particular sequence of fires, especially with animals. One study of invertebrates that has incorporated a sequence of fires into their experimental design is that of Blanche *et al.* (2001) who examined the effect of dry season burns over a 7-year period on the abundance and diversity of beetles in the wet–dry tropics of northern Australia. Details of this study are discussed below.

The effects of sequences of fires have been most closely examined for obligate-seeder plant species, because it is their life cycle which is likely to be most at risk with increased fire frequency (Keith 1996). A gross indication of this is the finding by Nieuwenhuis (1987) that obligate seeder species collectively showed lower cover in sites that were frequently burnt compared to rarely burnt sites of Hawkesbury Sandstone vegetation near Sydney. Morrison *et al.* (1995) and Cary and Morrison (1995) investigated this in more detail and found that resprouting (cf. obligate seeding), nature of seed bank (soil-stored vs. canopy-stored), and length of the juvenile period (time from germination to first fruit set) were important correlates of species maintenance through consecutive fires. Various features of fires, particularly length of the shortest inter-fire interval, interacted with these features of plants to influence the species richness of the plant community. For example, shorter inter-fire intervals were associated with reduced mean species richness per sample plot (Fig. 5.5).

Processes that might explain the patterns

There are two important features of the patterns summarised above. Firstly, not many of the patterns have been studied in any detail. Methodologies vary, some methodologies are flawed or questionable, and there is little or no replication among fires or sites for any one species. Secondly, in those species for which there is some replication across fires or sites, there seems to be a lot of variation in observed or inferred fire response.

Perhaps some of this variation may be explained by differences in sampling techniques. However, the studies reviewed above suggest roles for one or more of the following factors in causing variation in fire response patterns in particular situations: fire history, fire characteristics, habitat characteristics, biotic interactions (e.g. predation, herbivory, competition), climate, and site-to-site variation in an

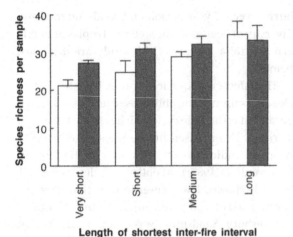

Fig. 5.5. Effects of different fire histories (specifically the length of the shortest inter-fire interval over a 28-year period) on mean species richness (±s.e.) per 100 m² sample in Hawkesbury Sandstone vegetation near Sydney. Obligate seeder species are open bars and sprouters are stippled bars. Data from Morrison *et al.* (1995).

organism's life history. Drawing generalisations or making precise predictions about fire effects for any particular fire will be problematic. Instead, if we understood the processes which may produce the patterns we do see, it might be feasible to make better predictions in particular circumstances.

It is these processes and the factors which affect them that form that basis of various classification schemes designed to predict or describe population and community responses to fires. Perhaps the most frequently cited schemes are the 'vital attributes' of plants devised by Noble and Slatyer (1980) (Table 5.1), and the fire response 'key' presented by Gill (1981b). Noble and Slatyer's scheme identifies functional groups of plants and uses them to predict the response of a population to fire. The functional groups are based on the following processes:

(1) means of arrival in the burnt site – dispersal from off-site sources or various forms of persistence of propagules through the fire;
(2) means of persistence within the site – effects of fire on mortality of adult and juvenile plants, and retention vs. loss of reproductive capacity by surviving plants;

(3) patterns of establishment of new plants – whether successful recruitment is confined to the immediate post-fire period, continues in the absence of further fire or requires unburnt vegetation; and
(4) three 'critical' time periods – time to reach reproductive maturity, time for all plants to die out in a site and time for all propagules to die out in a site.

Classification schemes such as this might allow prediction of fire responses at a relatively coarse level, for example, by indicating which functional groups of organisms might be most susceptible to local extinction as a result of extreme fire regimes, such as high frequency fire. Keith (1996), for example, used the following four life-cycle processes to identify the fire regimes that are most likely to cause population decline or local extinction in particular functional groups: (1) death of established plants and seeds, (2) failure of seed release or germination, (3) interruption of growth and maturation, and (4) failure of seed production (Table 5.2).

There does not appear to be a similar set of 'vital attributes' presented for animals in fire-prone regions. However, various authors have identified features such as ability to find refugia (local mobility in the face of fire), long-distance dispersal ability, breadth of diet (or ability to switch diet after fire), susceptibility to predation and competition, dependence on particular habitat, and intrinsic rate of population increase as factors which can determine the fire responses of mammals or birds (Fox 1982; Higgs and Fox 1993; Newsome and Catling 1983; Woinarski and Recher 1997).

We have divided the various possible processes which might contribute to fire response patterns into four main categories, to account for the fact that the attributes of organisms operate by interaction with features of the environment. These categories are: (A) the organism's life cycle, (B) fire characteristics, (C) landscape characteristics, and (D) climate (Fig. 5.6). In this schematic representation, we collapse 'vital attributes' (Noble and Slatyer 1980) and 'life-cycle processes' (Keith 1996) into four key processes: (1) mortality caused by the fire, (2)

Table 5.1. *Summary of 'vital attributes' classification, indicating categories of dispersal, establishment and persistence, and the ecological outcomes of each category*

Means of arrival or persistence	Population outcome
Dispersal from off-site sources	
D	Post-fire site occupied by juveniles derived from propagules continually dispersed from off-site sources, irrespective of whether pre-fire site is occupied by juveniles, adults or propagules, or is unoccupied
Persistence of on-site propagules through fire	
S	Post-fire site occupied by juveniles derived from propagule pool that persisted on site through the fire; propagules are long-lived and persist after standing plants senesce and die
G	As for S, but propagule pool exhausted by fire and hence populations in the juvenile phase are eliminated by fire
C	As for S, but propagule pool exhausted by fire and short-lived (i.e. do not persist beyond senescence and death of adults); hence populations in juvenile or propagule phase eliminated by fire, while those in adult phase converted to juvenile phase
Persistence of on-site vegetative material through fire	
V	Juveniles and adults survive fire, but adults lose reproductive capability and functionally return to juvenile phase; hence post-fire site occupied only by juvenile phase
U	Juveniles and adults survive fire, and adults retain reproductive capability; hence post-fire site occupied by both juveniles and adults if these were present in pre-fire population
W	As for U, but juveniles unable to persist through fire (i.e. development of fire resistance takes longer than maturation); hence, populations in adult phase remain in adult phase after fire, but those in juvenile phase are eliminated by fire
C	As for V, but juveniles unable to persist through fire (i.e. development of fire resistance takes longer than maturation); hence populations in juvenile or propagule phase eliminated by fire, while those in adult phase converted to juvenile phase
Multiple methods of arrival or persistence	
Δ	Propagules dispersed from off-site sources and adults persist through fire and retain their reproductive capability (i.e. D and U, or D and W)
Σ	Long-lived propagules and adults persist on site and retain their reproductive capability after fire (i.e. S and U, S and W, or G and U)
Γ	Long-lived propagules persist, but exhausted after fire and adults persist and retain their reproductive capability after fire (i.e. G and W)

Table 5.1. (*cont.*)

Means of arrival or persistence	Population outcome
Pattern of establishment	
T	Able to establish and grow to maturity immediately after fire when there is little competition for resources, and continue establishing and growing indefinitely afterwards despite increased competition (includes gap-phase recruitment).
I	Able to establish and grow to maturity only in the period immediately after fire; unable to continue establishing and growing as time elapses since fire and there is increased competition for resources
R	Unable to establish and grow to maturity immediately after disturbance; establishment and growth requires conditions that only become available after some time has elapsed since fire.

Source: Noble and Slayter (1980).

Table 5.2. *Fire-driven mechanisms of plant population decline and extinction*[a]

Life-cycle processes	Characteristics of fire regime	Mechanism of decline/extinction
1. Death of standing plants and seeds	(a) High-intensity fires	Heat-caused death of vital tissues above ground
	(b) Fires consuming extreme quantities of ground fuel	Heat-caused death of vital tissues below ground
	(c) Peat fires	As for 1(b)
	(d) High-frequency fires	Progressive depletion of bud banks, starch reserves and/or structural weakening Competition with opportunistic exotic species
	(e) Low-frequency fires	Progressive decline of standing plants and seed banks through senescence Competition with community dominants
2. Failure of seed release and/or germination	(a) Low-frequency fires	Low rate of recruitment due to infrequent recruitment events
	(b) Low-intensity fires	Low rate of recruitment due to release of few seeds
	(c) Fires with short residence time	Low rate of recruitment due to poor germination following poor soil heating

Table 5.2. (*cont.*)

Life-cycle processes	Characteristics of fire regime	Mechanism of decline/extinction
	(d) Fires resulting in poor penetration of smoke-derivatives in to soil	Low rate of recruitment due to poor smoke-induced germination
3. Failure of seedling establishment	(a) Low-frequency fires	High mortality of seedlings appearing between fires (competition, predation, disease)
	(b) Fires followed by dry conditions	High post-fire mortality of seedlings by desiccation
	(c) Fires consuming all organic material in soil	Physical changes to soil reducing germination conditions
	(d) Small or patchy fires	High levels of post-fire herbivory on seedlings
4. Interruption of growth and maturation	(a) High-frequency fires	Fire-induced death of pre-reproductive, juvenile plants Fire-induced death of juvenile plants that have not yet developed resistance
5. Failure of seed production	(a) Low-frequency fires	Low rate of recruitment due to infrequent flowering cues
	(b) Autumn/winter fires	Poor seed production due to 'suboptimal' flowering cues
	(c) Low-frequency, small or patchy, low-intensity fires	Low seed availability caused by high predation rates (failure of predator satiation)

Note:
[a] Keith (1996) used this information to predict which life-history types are most likely to be affected in each case.
Source: Modified from Keith (1996), with permission of the Linnean Society of New South Wales.

recolonisation, (3) survival and establishment of individuals after fire, and (4) post-fire reproduction and population growth. These key processes and the range of biotic factors that control them are summarised in Table 5.3.

For a particular organism, fires (and fire regimes) (B) will affect these key processes either directly (e.g. mortality in a fire) or indirectly (e.g. by affecting the habitat structure which increases risk of predation which causes population decline). In addition, other environmental attributes will interact with fire to affect organisms: attributes of the landscape (C), and of the climate (D) (see Table 5.4). For example, good local mobility may permit an animal to survive a low-intensity fire but not a fast-moving firefront. Recolonisation of burnt areas may be rapid after a patchy fire but not after an extensive, 'clean' fire. Post-fire population growth of plants may be rapid if there are good post-fire rains but slow if there is a prolonged dry period.

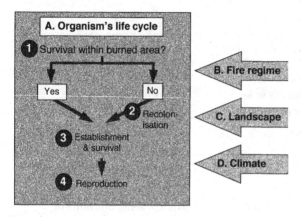

Fig. 5.6. Schematic diagram of the various processes which may contribute to determining the patterns of population change in relation to fire, divided into four categories: (A) attributes of the organism, (B) attributes of the fire and fire regime, (C) attributes of the landscape, and (D) climatic characteristics.

In the following sections, we examine how the four key life-cycle processes can be affected by features of fire, landscape and climate, and, in particular, how fires might affect the biotic factors and interactions which control these processes.

Vertebrates

Mortality caused by fire

No studies have yet reliably quantified the mortality of Australian vertebrates during fire. Mortality has been assumed in a number of studies (e.g. Fox 1978) when individuals present before a fire were not present after. However, this loss does not distinguish between mortality and emigration, nor does it allow for possible changes in habitat use that may lead to altered trappability or ease of observation. Mortality can be quantified only by following the fate of individuals through a fire event.

Features of an animal's social or individual behaviour may lead to differential risk of mortality during fire. Within a species, adults or dominant individuals may have a better chance of survival than transient individuals. It could be predicted that individuals with established territories containing nest sites in burrows, tree hollows or rock crevices are more likely to be protected from fire.

Transient individuals could be more vulnerable as they lack site familiarity and do not find suitable protection as fire approaches. Lawrence (1966) studied nest characteristics that were likely to provide protection during a fire. The 'minimum survival qualities' of a burrow were identified as being greater than 12 cm depth, with several surface openings and insulation of dry mineral soil. Nests in deep rock crevices, under logs or in tree trunks insulated by living bark were also protected from heat and other fire effects (see also Fox 1978).

Individual behaviour in the face of fire may influence survival prospects. Anecdotal reports suggest that there are body-size differences in animal behaviour during a bushfire. Kangaroos and other large and/or highly mobile animals either run or fly ahead of the firefront or double back through it (Christensen 1980). Small vertebrates may find shelter in burrows, rock crevices or insulated tree hollows, as suggested by observations of lizards and snakes actively seeking shelter in burrows in advance of an approaching firefront (Main 1981). Medium-sized mammals appear to be at higher risk as they are less mobile and are less able to find shelter. Hollow-dwelling species may find protection if sufficiently insulated, but species such as Koalas (*Phascolarctos cinereus*) which snooze in exposed forks of trees, ring-tailed possums (*Pseudocheiris peregrinus*) that make nests of dried leaves, and terrestrial birds such as the rufous (*Atrichornis rufescens*) and noisy scrub birds (*A. clamosus*), which are poor fliers and are associated with dense shrub vegetation, may be at risk of injury or death.

Further opportunities for survival during fire may be offered by the landscape. Animals can take advantage of features not usually utilised. Whelan *et al.* (1996), for example, found that *Antechinus stuartii* nest in rock crevices in the absence of tree hollows. Similarly, Braithwaite (1987) suggested that lizards in the frequently burnt habitats of the wet–dry tropics of Australia make use of unusual forms of shelter compared to those from habitats that experience lower frequency fires such as in central Australia. Topography of the land can also influence survival prospects through the differential effects of fire between moist gullies and

Table 5.3. *Summary of life-cycle processes referred to in section A of Fig. 5.6 and the various biological and environmental factors that might affect them*

Process	Factors affecting the process
1. Impact of fire on mortality of individuals	Characteristics of the organism (e.g. sprouter or seeder plants) interacting with characteristics of the fire (intensity, time since last fire, season); Behaviour in the face of fire (for animals); Opportunities for survival offered by the habitat/landscape
2. Recolonisation from within/outside burned area	Organism's characteristics, determining dispersal distances; Proximity of unburned refugia from which to recolonise; Presence/absence of a particular habitat state required by the organism; Interaction of season of dispersal and season of fire
3. Post-fire survival inside the burned area	Presence of adequate resources (food, light, water, nutrients), themselves affected by post-fire climate; Predation/herbivory levels; Presence of mutualists (e.g. pollinators, seed dispersers, mycorrhizal fungi)
4. Growth, reproduction and population increase	Right climatic conditions for germination and recruitment; Rate of growth and maturation (e.g. juvenile period in plants) determined by interaction of characteristics of the organism and post-fire climatic conditions; Suitable habitat, food availability etc

Table 5.4. *The environmental context for the factors which in turn control the processes of mortality, colonisation, post-fire survival, establishment, reproduction and population increase*

Environmental 'attributes'	Features of each attribute which can affect factors listed in Table 5.3 and thence control the processes
(B) Attributes of fire and fire regime	Intensity and other aspects of fire behaviour (type, back vs. head, patchy vs. clean) Initial conditions (e.g. time and climate since last fire) Season of burning
(C) Attributes of the landscape	Presence, size and location of features that act as refugia Features of the landscape that affect fire behaviour (e.g. slope, aspect, topographic discontinuities)
(D) Attributes of the climate	Leading up to a fire – effects on population structure at the time of fire After fire – determining plant recruitment and growth, hence animal habitat

drainage lines, and the more exposed and drier mid-slope and ridge-top areas (Lunney *et al.* 1987; Whelan 1995).

Quantifying mortality, especially in relation to high-intensity wildfires, is obviously challenging. Studies of the fates of individuals are sorely needed, with animals tagged before fire. Radio telemetry studies may be used in some situations to determine movements during and after fire. These approaches can be used to compare the effects of different characteristics of fires (e.g. large, high-intensity vs. low-intensity, patchy fires). Griffiths (1995) provided a good example of the sort of study that is needed; frillneck lizards (*Chlamydosaurus kingii*) in eucalypt woodland in Kapalga (Northern Territory) were captured, had radio transmitters attached and individuals were monitored before, during and after experimental fires in early dry season (low intensity) and late dry season (high intensity). Mortality was negligible in the early dry-season fires, during which most lizards remained in the same tree. In contrast, mortality was substantial in the later fires, during which lizards moved to find shelter in large trees and termite mounds, from which they emerged after about 30 min.

Recolonisation

Studies of recolonisation after fire suffer similar problems to those investigating mortality: it is difficult to know the fates of individuals. It is, however, important to know whether individuals appearing after a fire are colonists from outside the burned area or previously occurring individuals simply emerging from refugia. Further, individual tagging studies are necessary to distinguish between breeding up within an area and ongoing colonisation.

Individuals surviving in unburnt refugia are able to disperse into burnt areas after fire. Studies of rodents have demonstrated that it may often be transient and/or immature individuals that recolonise burnt areas first (Leonard 1974; Christensen and Kimber 1975; Friend 1979; White 1992). Animals may also exhibit post-fire habitat shifts until burnt areas recover favourable conditions (Fox 1983). The rock-rat (*Zyzomys woodwardi*), for example, moved from closed forest where the ground cover and leaf litter

had been largely removed by fire, to a nearby area of structurally complex rocky scree slope (Begg *et al.* 1981).

Dispersal from refugia into burnt areas depends not only on the availability of suitable resources, but also on the degree of mobility. Among birds, species that readily cross open spaces or fly above the canopy are likely to disperse further than ground and shrub-layer species that usually maintain discrete home ranges. However, it is difficult to draw generalisations on the degree of mobility of a species dependent on body size or any other single factor.

Survival of individuals after fire

Successful recolonisation requires that individuals survive in the burnt area and this requires that the critical resources for a species must be available. For vertebrate species, the critical resource requirements are usually identified as food and cover. This is based on several studies, largely on small mammals and birds and, to a smaller extent, on lizards (Pianka 1996), where correlations have been drawn between species abundance and various vegetation measures (e.g. Christensen and Kimber 1975; Fox 1981, 1982; Wilson *et al.* 1990). Some researchers have further implied a direct relationship between cover and food availability (e.g. Masters 1993; Reid *et al.* 1993). In cases where the habitat requirements of a species include resources that are reduced by a fire, it may appear reasonable to draw conclusions about the effect that a fire will have on the population. There is, however, little direct experimental evidence to link the post-fire abundance of any species with resource availability. In addition, some recent experimental studies on vertebrates (e.g. Monamy 1998; Sutherland 1998) have demonstrated that the relationships between post-fire resource availability and population recovery are complex.

Sutherland (1998) conducted a replicated, manipulative experiment to examine the influence of food and nest site availability on small mammals after a bushfire in coastal eucalypt forest near Sydney. Twelve sites, burnt 1 year previously, were used and the following treatments were applied (three replicate sites per treatment): food supplementation only, nest addition only, both food supplementation and nest addition, and no manipulation. In addition, four unburnt sites were used for comparison. The experiment was monitored for 26 months.

In the early period of the experiment, the bush rat (*Rattus fuscipes*) and the brown antechinus (*A. stuartii*) increased in abundance in response to food supplementation. However, this response was not maintained for either species over the 26-month duration of the study. The rat species showed increased juvenile recruitment into sites with additional resources, and the brown antechinus showed reduced dispersal of juveniles away from sites with higher resource availability. Interestingly, however, neither of these factors led to long-term increases in population abundance (Sutherland 1998).

In another controlled and replicated field experiment, Monamy (1998) investigated the effect of reducing plant cover on small mammals in coastal wet heath at Myall Lakes. Decreased cover was designed to simulate one feature of a post-fire environment. *Pseudomys gracilicaudatus* and *Rattus lutreolus* declined rapidly on experimental plots (in contrast to control plots). *Pseudomys gracilicaudatus* recovered to control levels within 6 months of the manipulation and *R. lutreolus* after 1.5 years, after vegetation cover had been restored. In contrast, no experimental effects were found for *P. novaehollandiae* or *Mus domesticus*, two species that often exploit the earliest stages of disturbed heath. These differences in the timing of recovery among species provide strong support for a prediction of the habitat accommodation model of Fox (1982) that species should re-enter an area during the phase in vegetation succession in which their habitat requirements are met.

A range of non-manipulative studies provides some further indications of the processes affecting vertebrate populations after fire. A fire may expose alternative food sources that are not available in unburnt habitats. Knob-tailed geckoes (*Nephurus stellatus*) exploit surface-active insects in newly burnt areas, achieving high local abundances due to this more accessible food resource (Dickman *et al.* 1991). Similarly, sooty owls (*Tyto tenebricosa*) which usually

feed on a range of arboreal and terrestrial mammals, have demonstrated a post-fire prey-shift to a diet almost entirely consisting of the terrestrial bush rats (*R. fuscipes*) despite arboreal mammals still being present in the foraging area (Loyn *et al.* 1986). This was attributed to the removal of terrestrial ground cover used by *R. fuscipes*, while tree hollows required by arboreal species remained after fire despite the removal of the canopy cover. Predation risk may also increase with more extensive foraging activity resulting from reduced food availability (Begg *et al.* 1981).

Predation pressures may also change in the post-fire environment though the impact of predation on post-fire survival is often assumed but rarely quantified. Using radio telemetry and extensive trapping and tagging, Christensen (1980) found that, after fire, almost half the individuals in a population of woylies (*Bettongia penicillata*) were killed by predators. Further, Newsome *et al.* (1983) provided evidence that high predation by dingoes upon macropods suppressed macropod population recovery for 2 years after a fire. Post-fire nest predation limited population recovery of fairy wrens (*Malurus* spp.) (Brooker and Brooker 1994).

Intra-specific and inter-specific competition for unburnt refugia may occur following a fire, which could lead to differential survival. In one study, female *Rattus lutreolus velutinus* competitively exclude males from high-quality habitat with dense ground cover during the non-breeding season. During the breeding season, however, males expand their ranges into the high quality female habitat. This reduces inter-specific competition and allows the sympatric *Pseudomys higginsi* also to expand into sites of higher quality (Monamy and Fox 1999). Intra-specific differences in patterns of post-fire survival of male and female white-browed scrub wrens (*Sericornis frontalis*) have also been recorded (Baker *et al.* 1997).

It is possible to predict features of an animal's physiology or life-history strategy that increase the likelihood of mortality after fire. For example, species adapted to xeric conditions that are able to tolerate high temperatures and reduced water availability may be able to tolerate the exposed conditions of the post-fire environment more readily than other species (Haim and Izhaki 1994).

The use of species habitat requirements to derive models of disturbance impact and post-disturbance succession have been attempted only rarely for vertebrates in Australia. Newsome and Catling (1983), for example, used the house mouse (*Mus domesticus*) to develop models of abundance in relation to food and nesting resources and predation. These models were then compared to actual post- fire population data for native mammals after a fire. The responses of each species fitted different models, both in the short- and long-term, indicating that complexities arise from differences in life histories, body size, longevity, diet and shelter requirements (Newsome and Catling 1983).

Growth, reproduction and population change

Different life-history strategies will affect species ability for population recovery after fire. Even if the necessary resources for breeding are available, different breeding strategies will allow some species a faster rate of population growth than others. Some vertebrates are able to breed when conditions are broadly favourable, whereas others are restricted by any combination of seasonal, behavioural, physiological, nutritional, or environmental factors.

Friend (1993) reviewed the impact of fire on small mammals, reptiles and amphibians by comparing reproductive patterns against reported post-fire population responses. For small mammals, the trends were quite marked. Species common in the early post-fire successional stages were found to have opportunistic (e.g. *Mus domesticus*), or temporally dynamic breeding strategies (e.g. *Isoodon obesulus*), and/or to exhibit high reproductive potential (e.g. *Pseudomys* spp.). Mid-successional species were generally polyoestrous, with short gestation and weaning periods (e.g. *Sminthopsis murina*). Species associated most commonly with late-successional habitat were generally monoestrous with seasonal breeding patterns (e.g. *Antechinus stuartii*), or polyoestrous but limited by restricted seasonal activity (e.g. *Tarsipes rostratus*). With reptiles and amphibians, however, diet, shelter requirements and foraging patterns seem to determine post-fire population

responses, rather than reproductive strategy (Friend 1993).

It is possible that post-fire population responses could be estimated by considering the reproductive strategy and habitat requirements of a species. Yom-Tov (1987) showed that 'old-endemic' Australian birds have smaller clutches than 'new-endemics', but breed more often. Thus, they could show different responses to fire although this does not appear to have been tested experimentally. Plasticity in resource requirements or reproductive strategies could indicate a lower susceptibility to fire. For example, James (1991) found dietary shifts in *Ctenotus* skinks in response to drought or wet conditions. Similarly, Dickman *et al.* (1999) found shifts in abundance of *Ctenophorus* dragons in relation to vegetation growth and rainfall, that could resemble post-fire patterns of vegetation recovery.

It is possible that, if resources are available, reproductive strategies may produce a more rapid build-up of populations. Silky mouse, (*P. apodemoides*) populations can respond rapidly to habitat disturbance due to a female-biased sex ratio and promiscuous breeding strategy (Cockburn 1981). Other species, such as *R. lutreolus* (Catling 1986) can breed at lower body weights. Rapid population increases are possible for rodents in particular. Many rodent species breed within weeks or months of birth and produce several young at a time.

There are dangers, however, in making generalisations about the factors which might determine vertebrate responses to fire. Braithwaite (1995) emphasised variability both among and within species in fire responses. Different lizard species survived and prospered to differing degrees following fires at various times within the dry season in the seasonal tropical savannas. Further, the effects of different fire treatments (early, mid, late dry season) on small mammals varied from one year to the next, suggesting that populations are affected by an interaction between climate and fire regime.

Although a particular species may be considered to prefer early, mid or late successional regrowth, many individuals will actually survive, reproduce and be at high abundance during different successional stages. In the small mammal community at

Myall Lakes (New South Wales), *R. fuscipes* and *A. stuartii* were regarded as mid- to late-successional species (Fox 1982, 1983). In very similar habitat in Ku-ring-gai Chase (New South Wales), and in the absence of *Pseudomys*, these species were the dominant small mammals during the early successional period (Sutherland 1998). There will be site- and community-specific factors that drive patterns of vertebrate response to fire. Management decisions, therefore, must be made on relevant data from the location in question.

Invertebrates

Given the less-than-ideal state of the data relating to the fire-response patterns exhibited by invertebrates, it is not surprising that our understanding of the processes generating these patterns is poor. Very few studies (e.g. Tap 1996; Blanche *et al.* 2001) have conducted manipulative experiments to test specific hypotheses to explain fire-response patterns. However, some insight into processes can be gained from studies that have correlated fire-response patterns with ecological parameters, such as life-history traits and climatic variables. There are also several manipulative experimental studies that were designed to test other hypotheses, but do provide some indirect evidence relating to fire-response processes. The lack of suitable studies requires that examples be drawn from other regions of the world, in addition to Australian studies.

Mortality caused by fire

Few studies have examined the direct effects of fire on invertebrates but there is considerable anecdotal evidence that some invertebrates can escape fire (Gandar 1982; Warren et al. 1987; Bock and Bock 1991). Highly mobile species may escape if the speed and intensity of the fire are not too great (Gandar 1982; Gillon 1972). Gillon (1972) observed acridid grasshoppers flying away from the firefront in African savannas and estimated a survival rate of 88%. In contrast, Bock and Bock (1991) found large numbers of the charred corpses of a large, flightless grasshopper (*Dactylotum variegatum*) after fire in Arizona grasslands.

Mobility is not the only mechanism that can be used to escape the direct effects of fire. Animals that spend at least part of their life cycle in the soil, or in other protected refugia, may survive the fire (Gandar 1982; Warren *et al.* 1987; Delettre 1994). Survival of chironomid larvae that live in the litter layer or the first centimetres of soil in Brittany appeared to be related to fire temperature (Delettre 1994). In tall-grass prairies, increased abundance after fire of certain species of grasshoppers (in Kansas: Knutson and Campbell 1976; Evans 1984, 1985) and coral hairstreak butterflies (*Satyrium titus*) (in Illinois: Swengel 1996) may be attributed to these species existing in the soil as eggs during the fire season.

In Western Australia, the differential survival of wood-eating versus litter-harvesting termites after fire appeared to be associated with their nesting habits (Abensperg-Traun and Milewski 1995). Wood-eating termites nest on the soil surface in wood that is destroyed during fire, whilst litter-harvesting species nest in hard, clay mounds that appear to protect the termites from incineration. Species which have periods of aestivation or hibernation in protected locations, such as in the soil, may also be protected from fire, as was found in Australia for mites (Wallace 1961) and spiders (Main 1987).

Clearly, survival during dormancy periods is likely to be linked to fire season. In the studies conducted by Wallace (1961) and Main (1987), survival was recorded when the fire season coincided with the period of dormancy. Other attributes of the fire may also be important in determining survival during the fire. Patchy fires will offer more refugia from which to escape incineration than more complete burns. Further, we can predict that the interaction between the distribution and size of unburnt patches and the mobility of animals will be crucial in determining the probability of survival.

Whilst some organisms clearly can survive the direct effects of fire, the majority of organisms probably die. In a Canadian boreal forest, when unburnt and burnt plots were compared immediately after fire (1 hour), a 95.5% reduction in the abundance of invertebrates in the burned plots was found (Paquin and Coderre 1997). The most likely explanation for this is increased mortality in burnt plots, but care must be taken when interpreting declines in numbers following fire since this could be due to increased mortality, or alternatively, to emigration to unburnt patches in the face of fire. Carefully designed experiments are needed to distinguish between these alternative explanations.

Recolonisation

Whilst some invertebrates, in some situations, may survive the direct effects of fire, many invertebrates probably have to recolonise burnt patches from unburnt refugia. Several types of unburnt refugia may be exploited by invertebrates, the most obvious being unburnt patches of the habitat, therefore, the patchiness of the fire is of key importance. Other potential refugia include dense crowns of plants (Whelan *et al.* 1980; Main 1981; Gandar 1982; D. Keith and M. Tozer unpublished data), thick layers of leaf litter, such as that adjacent to large logs (Andrew *et al.* 2000), thick bark on tree trunks, and the soil under rocks and in burrows (Main 1981; Warren *et al.* 1987). Fire intensity will influence the abundance and/or effectiveness of these refugia.

The dispersal ability of a species may play a role in determining whether it can recolonise burnt patches, and how quickly recolonisation occurs. Swengel (1996) suggested that the powerful and wide-ranging flight of regal fritillary butterflies (*Speyeria idalia*) might explain their abundance after fire in Illinois prairie. In contrast, poweshiek skipperling butterflies (*Oarisma poweshiek*) are poor fliers and occurred in relatively low abundance after fire compared to pre-fire levels (Swengel 1996).

When attempting to predict the ability of invertebrates to recolonise from unburnt patches, attributes of the fire, such as intensity, extent and season of burn, must be considered. The spatial distribution of unburnt patches, which is likely to be determined in part by fire intensity, may greatly influence the rate of recolonisation. If unburnt patches are scarce, recolonisation is likely to be slower relative to recolonisation from abundant patches. In addition, small unburnt patches are unlikely to provide as great a diversity of invertebrates as larger patches. The extent of the fire also influences recolonisation rates. In Western Australian woodland, grasshopper populations recovered more quickly after fires that covered a small area than those which burnt a larger area (Whelan and Main 1979). Presumably small burnt patches are more easily recolonised than large patches. The season of fire may also influence the recolonisation potential of a species, if the timing of dispersal occurs at an inappropriate time relative to the fire season. Experiments measuring the interaction between the dispersal abilities of invertebrates and key attributes of fire, at a range of spatial scales, would greatly improve our understanding of recolonisation processes after fire.

Survival of adults after fire

The post-fire environment offers a vastly altered landscape relative to the pre-fire situation. Whether individuals have survived the direct effects of a fire, or whether they have recolonised a burned area from unburnt refugia, they must survive in this altered landscape. A variety of mechanisms (e.g. changes to habitat structure, food availability and microclimate) have been suggested as important in determining whether or not invertebrates survive in the post-fire landscape. The two studies that we emphasise here are Blanche *et al.* (2001) and Tap (1996) since they used manipulative experiments to test specific hypotheses.

Blanche *et al.* (2001) examined the fire response of beetles as part of the landscape-scale, replicated experiment at Kapalga Research Station in the wet–dry tropics of northern Australia. The design consisted of three replicate plots (15–20 km^2) for each of three fire regimes (early dry-season burn, late dry-season burn and unburnt). Ground-active beetles were sampled twice a year (middle of dry season and early in wet season) over a seven-year period using pitfall traps. The results are complex and reveal an interesting interaction between the season of sampling, the timing of fire and the amount of rainfall prior to sampling. In the wet season, numbers of individuals, species and families of ground-active beetle were significantly reduced following fires late in the dry season. This reduction was detected only when substantial amounts of rainfall occurred prior to sampling. An interaction was also detected, between fire intensity and amount of rainfall prior to sampling, that affected the abundance of several common species and families.

Tap (1996) manipulated vegetation cover to determine what effect this would have on macro-arthropod abundance after burning. The design consisted of eight replicate plots of each of the eight treatments (1) burnt only; (2) burnt then vacuumed; (3) burnt, vacuumed then covered with cuttings; (4) scorched; (5) slashed then cuttings left; (6) slashed then cuttings removed; (7) slashed then mown; (8) no treatment. Macro-arthropods were sampled with pitfall traps on four occasions during a 1-year period after the initiation of treatments. Here too the findings are complex, though overall most macro-arthropods were trapped in the 'burnt then vacu-umed' treatment and fewest in the control plots. Ants accounted for 90% of all macro-arthropods caught and it appears that the presence or absence of cover as well as burning were important in determining their abundance.

The paucity of other experimental studies limits our understanding of the processes important to post-fire survival of invertebrates. However, there is a variety of studies that have correlated fire-response patterns with ecological parameters that suggest factors worthy of further investigations. These include the availability of food items, changes in microclimatic variables and increased predation pressures.

Fire may remove some plant species whilst promoting the growth of others (Whelan 1995). Specialist invertebrate herbivores are likely to be more sensitive to these changes than generalist herbivores. Swengel (1996) reported that all the fritillary butterfly species she studied in Illinois prairie declined in the first growing season after burning, which corresponded with a decrease in their food source, violets. Bock and Bock (1991) provided a similar explanation for the decline of Gomphocerinae grasshoppers in Arizona grasslands during the first summer post-fire, and attributed the survival of the herb-feeding grasshopper *Melanoplus gladstoni* to the increased availability of herbs after the burn. In Australia, massive seed fall

often occurs after fire and several studies have reported that increased availability of seeds after fire leads to large increases in the abundance of seed-removing ants (Ashton 1979; O'Dowd and Gill 1984; Wellington and Noble 1985a, b; Andersen 1987, 1988). This provides one possible explanation for the increased abundance of ants observed after fire in many Australian habitats.

The availability of resources after fire is likely to be greatly influenced by attributes of the fire itself. Both the intensity and season of fire may play an important role in determining the abundance and diversity of plants after fire and these in turn may influence the invertebrate herbivores which feed upon them, as has been suggested for grasshoppers (Knutson and Campbell 1976).

Survival following a fire may also be affected by changes in environmental parameters, such as surface temperature, soil moisture and insolation. In one study the survival of chironomids in the post-fire environment appeared related to soil moisture, as larvae require moisture for growth and pupation (Delettre 1994). In contrast, Lussenhop (1976) concluded that soil moisture was relatively unimportant compared to soil temperature to the survival of arthropods in a post-fire prairie. Temperature has also been implicated in the survival of termites after fire, in that high surface temperatures on exposed soils resulted in a reduced time available for foraging (Gandar 1982).

Predator populations are likely to change during the post-fire period and this may in turn influence the invertebrates they prey upon. The increased abundance of grasshoppers in a Kansas tall-grass prairie during several years post-fire was attributed, in part, to the decreased abundance of their predators, including quail (Knutson and Campbell 1976). A different explanation was provided for the increased abundance of grasshoppers in South African savanna by Gandar (1982); darkening of the cuticles of some grasshoppers, which developed as a result of the darker, burnt background, camouflaged them despite the increased abundance of predators such as insectivorous birds. The increased abundance of predators after fire has frequently been observed though the actual impact of these increases have not been quantified. Gillon (1972) found a variety of invertebrate prey in the stomach contents of raptors feeding immediately after fire, and it appeared that these invertebrates were captured as they attempted to escape the fire.

Growth, reproduction and population change

Studies of invertebrates after fire have focussed on changes in abundance of individuals and neglected the life-history processes integral in driving these changes, growth and reproduction. By examining growth and reproduction and subsequent population changes in invertebrates after fire we will gain much-needed insight into the longer-term consequences of fire on invertebrate populations. Clearly, the availability of resources is going to be a key determinant of whether individuals can grow and reproduce, and ultimately whether populations will increase or decline. However, it is essential these processes are examined in the context of the fire regime, and attributes of the landscape and climate, if we are to fully appreciate the impact of fire on invertebrate populations. Again, we advocate the way to move forward is through the use of experimental studies.

Plants

Mortality caused by fire

Mortality of individual plants clearly depends on anatomical characteristics of the plant (presence of a lignotuber; presence of protected meristems on trunk and branches), characteristics of a fire, and climatic variables. Plants may be classified as 'sprouters' if a significant number (perhaps >66%; Gill and Bradstock 1992) of established plants can survive after 100% canopy scorch (Gill 1981a, b). Variation in mortality of adults is obvious among tree and shrub species, after a particular fire in any plant community. Closely related, congeneric obligate-seeder and sprouter shrubs can occur side-by-side in many eastern Australian heathlands, woodlands and forests: e.g. *Banksia ericifolia* and *B. cunninghamii* (obligate-seeders) with *B. spinulosa* (sprouter) (R. Whelan personal observation); *B. marginata* (obligate seeder) with *B. paludosa, B. oblongifo-*

lia and *B. serrata* (sprouters) (Keith 1996; R. Whelan personal observation); *Epacris obtusifolia* (obligate seeder) with *E. paludosa* (sprouter) (Keith 1996). Even among long-lived forest trees, *Eucalyptus regnans* (obligate seeder) and *E. obliqua* (sprouter) co-occur in many wet-sclerophyll forest areas (Ashton 1981a).

Among the species that are reasonably classified as sprouters, the ability to survive a particular fire varies among individuals. The size of a plant may partially explain this, as plants with larger trunks, thicker bark (e.g. *Eucalyptus oreades*: Glasby *et al.* 1988) and taller canopies (above scorch height: Morrison 1995; Keith 1996) are more likely to survive. Burrows (1985) found that older (perhaps less vigorous) trees of *Banksia grandis* in southwest Australian jarrah forest suffered greater mortality in fires than smaller trees. Williams (1995) examined tree mortality in *Eucalyptus miniata* – *E. tetrodonta* open-forest in the Kapalga experiment and found significant effects of fire intensity and tree species. Percent mortality was much greater for trees <6 m tall (*c.* 12%) than for 18-m trees (<2%). Likewise, Bradstock and Myerscough (1988) found that survival of *Banksia serrata* plants was related to the interaction between plant size and fire intensity.

The time between fires may affect survival of sprouter species too. Even though lignotubers may contain dormant buds that are protected from the heat of fire, repeated defoliation (e.g. by burning) may deplete starch reserves (Bowen and Pate 1993) or the 'bank' of meristems (Zammit 1988). Starch reserves and meristems may be replenished rapidly, so it may require very frequent fires to cause death. Noble (1982, 1984, 1989) conducted some elegant field experiments to examine the effects of frequency and season of burning on survival of mallee (*Eucalyptus incrassata*) in western New South Wales, which are models for much-needed studies of factors causing variation in mortality rates of established plants. These studies involved burning plots of mallee vegetation in different seasons (spring, autumn and winter) and at different frequencies. A single autumn fire caused little mallee mortality (Noble 1982) but a second fire a year after the first reduced populations to 10% of starting numbers. This study involved addition of cereal straw to the burnt plots in an attempt to separate the effects of season and intensity. Further, the burning treatments were paralleled with cutting treatments (in which mallee stems were cut by axe). Both the cutting and burning treatments revealed a substantial effect of season, with virtually no mortality as a result of burning or cutting in two consecutive springs but substantial mortality after two autumn treatments.

One feature of population dynamics which has been little studied in relation to fire, but is likely to have a very significant impact on the likelihood of local extinction as a result of an inappropriate fire regime, is patchiness of fires. Even the most intense fires may leave patches of vegetation unburnt (e.g. Nadgee Nature Reserve: Fox 1978; Royal National Park: Whelan *et al.* 1996). These patches provide opportunities for survival in the landscape even if established plants are killed and seed banks have been depleted in the areas which were burnt. Short-distance seed dispersal may, therefore, determine spread of populations from these 'refugia' and explain the patterns of expansion and contraction of clumps of obligate-seeder species, as described by Keith (1995) for *Banksia ericifolia*.

Recolonisation

Recolonisation by seeds originating from outside the burnt area (vital attribute 'D' of Noble and Slatyer (1980); Table 5.1) appears to be rare in most fire-prone Australian ecosystems (Whelan 1986); seeds of most species probably originate within the burnt area (Purdie 1977) from soil-stored or canopy-stored seed banks. In contrast, some weed species appear to lack a long-lived, dormant seed bank and post-fire populations are founded by long-distance seed dispersal (e.g. *Epilobium angustifolium*: Myerscough 1980; Whelan 1995). Hobbs and Atkins (1990) noted that abundances of weedy species in recently burnt areas of Swan Coastal Plain woodland were greater after an autumn fire than a spring fire, so season of burning could influence the amount of long-distance dispersal that results in successful establishment.

The invasion of *Eucalyptus* forest by mesophytic species may be analogous to the long-distance dispersal of weed species, because many rainforest species lack seed dormancy and do not retain viability for extended periods. An important question here is: to what extent does dispersal from beyond the boundaries of a burnt area determine the timing of appearance of these mesophytic species?

It is worth noting that recolonisation from outside the perimeter of a burnt area may be required if an inappropriate fire regime (or other disturbance regime) causes local extinction of a population. Obligate seeder species with canopy-stored seed banks are likely to be the most at risk (vital attribute 'G' in Table 5.1), due to short inter-fire periods, because all adults die and all available seeds germinate or die in the first fire and a second fire occurs before a new seed bank has developed (see Keith 1995, 1996). Recolonisation in such situations will be an episodic process, because significant seed release and dispersal will not occur until the next fire burns the mosaic landscape and releases the canopy-stored seed bank from the plants in the formerly unburnt refugia. Dispersal distances may be quite short (of the order of 40–50 m) for a range of species (Whelan 1995, p. 96; Hammill *et al.* 1998).

Survival after fire

Given that there can be good survival of established plants and many species have protected seed banks, post-fire processes are probably more significant than the direct effects of a fire. Climate and herbivory in the post-fire period may both have a significant impact on survival of established plants. Experimental studies are required to test the effects of these processes on mortality of plants which survive a fire. Leigh and Holgate (1979) used exclosure cages to show that mortality rate approached 80% over 5 years after fire for *Daviesia mimosoides* plants that were grazed and burnt, whereas it was only 40% for plants protected from grazing. More experimental studies such as these are needed if we are to recognise the effects of herbivory, which are often not obvious in plant populations.

Growth, reproduction and population change

This is perhaps the area in which there has been most demographic work done on plant populations, partly because the germination and establishment phase of a plant's life cycle is most amenable to the 1-year and 3-year studies of honours and postgraduate students! It is also partly because this phase is critical to population persistence of obligate seeder species and the phase at which these species are most sensitive to external factors (Keith 1996).

Many plant species in fire-prone ecosystems are stimulated to flower by fire (see Gill 1981b), but the effects of time since last fire and season and intensity of fire are less well known. Taylor *et al.* (1998) quantified flowering of *Xanthorrhoea fulva* in six sites which provided comparisons of spring and summer fires, each of which had a short time span (3.75–5.25 yr) and a long time span (9.3–16.9 yr) since the previous fire, and long unburnt sites. There was no flowering in the long unburnt sites, and the proportion of crowns flowering was greater after the summer fires than the spring fires (Fig. 5.7). Interestingly, better flowering occurred in the site with the short fire interval, but it is difficult to assess the cause of this as fire intensity was greater in the summer fire than the spring fire. It is difficult to separate an intensity effect from a season effect, even with site manipulation (see Noble 1984).

Whelan *et al.* (1996) speculated that the stimulation of *Xanthorrhoea* flowering in the winter immediately after the 1994 fire at Royal National Park may have been due to the timing of this fire (summer). A fire later in the year might have caused a delay in the flowering response until the second winter. The data collected by Taylor *et al.* (1998) supported this for *X. fulva*; flowering after the summer fire was in the next flowering season, but it was delayed until the second flowering season after the spring fire.

The density of recruits is likely to be affected by the time since the last fire, if seedling numbers are linked to seed supply. For species with long-lived seed banks, the numbers of viable seeds available for germination may be expected to be low immediately after a fire, because post-fire germination depletes the seed bank until the first reproduction starts to add viable seeds to the seed bank. Thereafter, seed 'availability' might be expected to increase with each year's reproduction (Brown and Whelan 1999). However, annual inputs to the seed

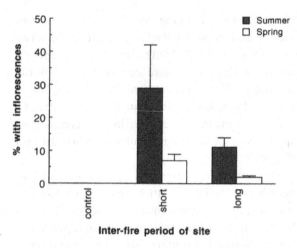

Fig. 5.7. Percent (±s.e.) of *Xanthorrhoea fulva* crowns that had flowered by three years after a summer fire (January 1991 – solid bars) and a spring fire (October 1994 – stippled bars) in sites that had previously been burnt either a short or a long time ago. Control sites were last burnt in 1988. After Taylor *et al.* (1998).

bank may start to decline after some years. In a study which followed a number of sites in a chronosequence for a 3-year period, Auld and Myerscough (1986) showed that fruiting success of *Acacia suaveolens* plants declined with time since fire. Vaughton (1998) found that seed densities of *Grevillea macleayana* (formerly *G. barklyana*) in the soil were greatest in sites that were 7–16 years since fire and lower in both young and old sites.

All else being equal, plants producing the most seeds and having an early age of first reproduction should show the greatest population growth after fire. Life history can clearly constrain seed supply, and the contrast between obligate seeder and sprouter species illustrates this. Numbers of seeds produced per plant are generally greater for obligate seeder species than for congeneric sprouters (Cowling *et al.* 1987; Meney *et al.* 1997). Further, Benwell (1998) found that six species of shrubs in northeastern New South Wales showed ecotypic variation in sprouting across a number of study sites. Seedling densities were generally highest in sites where the plants were acting as obligate seeders than in sprouting sites.

High seed numbers may not necessarily translate

into higher seedling numbers and greater recruitment. Plots of survivorship for species in seasonal environments typically reveal substantial mortality associated with summer (Whelan and Main 1979; Wellington and Noble 1985a), though Purdie (1977) warned that the magnitude of various factors causing seedling mortality (insect attack, water stress, fungal infection) can vary among species, sites, burnt and unburnt areas and season of the year. Experiments are required to assess the impact of these factors on population dynamics. A good model is a study by Wellington and Noble (1985a) of seedling survivorship in the mallee eucalypt, E. incrassata. They manipulated herbivory by caging plots containing high seedling densities, showing that (in that system at that time) herbivory had no effect on seedling survivorship. They applied nutrient- and water-additions, alone and in combination, to small plots of seedlings at high and at low density, showing that water and nutrient availability were both limiting survival, but only in high-density plots. In contrast, Leigh and Holgate (1979) found that herbivore exclusion had a marked effect on seedling survivorship. The impact of post-fire herbivory on seedling survival and hence on recruitment to plant populations is an important issue that has received too little attention in Australian systems. Intensity, season and frequency of burning, and especially patchiness of the burnt landscape, all have the potential to affect numbers of herbivores and hence cause marked changes in the post-fire plant community (see Whelan 1995).

The potential for season of burning to have an impact on recruitment was shown by Hobbs and Atkins (1990) in a comparison of two sites, one burnt in September 1984 and the other in March 1985. They noted an absence of seedlings after the spring fire, but successful establishment of seedlings of many species after the autumn fire. This study was conducted in areas that had been burned previously for another study; an approach that is typical of many fire studies. Replicated manipulative experiments are needed to separate site effects from the factor of interest (e.g. season of burning). Clark (1988) compared the effects of spring and autumn burning in the Sydney region by allocating four

small plots (25 m²) to one of the following burning treatments: unburnt, spring burn, autumn burn. The burnt plots were reburnt 6 years later (maintaining the spring and autumn comparison). In general, recovery (measured by both numbers of species and numbers of individuals) was better in autumn after the first burns, supporting the Hobbs and Atkins (1990) result, but better in spring after the second burns. However, different species showed different patterns. The second fires led to a greater reduction in species per plot than the first, especially in the autumn plots. Regression analyses suggested that temperature and rainfall may explain some of this variation more than the season of burning per se. Further experiments (e.g. by watering plots) are needed to isolate the proximate factors determining recruitment after spring versus autumn fires.

Conclusions

Many studies investigating the responses of populations to fire have tended to focus on describing patterns. Whilst many of these studies have speculated as to the processes generating these patterns, few studies have specifically examined the underlying mechanisms that produce particular fire-response patterns. In order to improve our understanding of fire responses of plant and animal populations a fundamentally different approach must be taken, and this understanding will only come by conducting manipulative experiments.

Even if replicated experimental studies do start to improve our knowledge of the range of fire-response patterns, the evidence presented above indicates that our ability to generalise or make accurate predictions about fire responses of organisms will be limited by strong site-specific and fire-specific factors. This leaves land managers in a difficult situation, especially when faced with the challenge of predicting the response of endangered or threatened species to applied fires. As stated by Williams et al. (1994), fire prescriptions will not necessarily be 'portable' from one site to another, because fire responses will vary.

We have argued that the best approach to making predictions about fire effects is to understand the processes that determine fire responses in a range of situations and apply this knowledge to the particular case. The state of our understanding of processes is relatively poor, so replicated, manipulative fire experiments are much needed. Knowledge of a couple of key attributes of the organism (e.g. dispersal ability, intrinsic rate of reproduction) can be combined with any knowledge that can be gleaned about survival during fire, survival in the post-fire environment, and population growth, and the factors which determine these processes, to make a prediction of likely fire effects. Refining these predictions will require a two-pronged approach. Firstly, experimental studies are required to improve the state of knowledge of the effects of different aspects of fire regime, climate and landscape on the key life-cycle processes. Secondly, fire management can be set up as experimental (with adequate replication), and simple monitoring can test whether the predicted fire response eventuated (Williams *et al.* 1994; Whelan and Baker 1999).

For example, Keith *et al.* (this volume) used current knowledge of dispersal ability (both local, in the face of fire, and long-distance), reproductive rate, susceptibility to predators and competitors, breadth of diet, and breadth of habitat association to make predictions of fire responses of two heathland bird species and two mammal species. On the basis of the existing information, different predictions might be made for different types of fire (short fire interval vs. long; patchy vs. 'clean' burn etc.). Appropriate design of management fires in areas containing these species would enable testing of the predicted fire response, with a minimal amount of monitoring. Detailed manipulative experiments are needed to further clarify the real habitat associations, susceptibility to predation, diet breadth and survival abilities of these species, and how they respond to different types of fires.

Acknowledgements

We thank numerous colleagues for contributing to the development of this book chapter: special thanks to Ross Bradstock, David Keith, Jack Baker, Mike Dunlop and Jann Williams.

References

Abbott, I., Van Heurck, P., and Wong, L. (1985). Responses to long-term fire exclusion: physical, chemical and faunal features of litter and soil in a Western Australian forest. *Australian Forestry* **47**, 237–242.

Abensperg-Traun, M., and Milewski, A. V. (1995). Abundance and diversity of termites (Isoptera) in unburnt versus burnt vegetation at the Barrens in Mediterranean Western Australia. *Australian Journal of Ecology* **20**, 413–417.

Ahlgren, I. F. (1974). The effects of fire on soil organisms. In *Fire and Ecosystems* (eds. T. T. Kozlowski and C. E. Ahlgren) pp. 47–72. (Academic Press: New York.)

Andersen, A. N. (1987). Effects of seed predation by ants on seedling densities at a woodland site in S. E. Australia. *Oikos* **48**, 171–174.

Andersen, A. N. (1988). Immediate and longer term effects of fire on seed predation by ants in sclerophyllous vegetation in south-eastern Australia. *Australian Journal of Ecology* **13**, 285–293.

Anderson, R. C., Leahy, T., and Dhillion, S. S. (1989). Numbers and biomass of selected insect groups on burned and unburned sand prairie. *American Midland Naturalist* **122**, 151–162.

Andrew, N., Rodgerson, L., and York, A. (2000). The use of leaf litter associated with logs as refuges from fire by ants. *Australian Journal of Ecology* **25**, 99–107.

Ashton, D. H. (1979). Seed harvesting by ants in forests of *Eucalyptus regnans* F. Muell. in Central Victoria. *Australian Journal of Ecology* **4**, 265–277.

Ashton, D. H. (1981a). Fire in tall open-forests (wet sclerophyll forests). In *Fire and the Australian Biota* (eds. A. M. Gill, R. H. Groves and I. R. Noble) pp. 339–366. (Australian Academy of Science: Canberra.)

Ashton, D. H. (1981b). The ecology of the boundary between *Eucalyptus regnans* F. Muell. and *E. obliqua* L'Herit. in Victoria. *Proceedings of the Ecological Society of Australia* **11**, 74–94.

Auld, T. D. (1987). Population dynamics of the shrub *Acacia suaveolens* (Sm.) Willd.: survivorship throughout the life cycle, a synthesis. *Australian Journal of Ecology* **12**, 139–151.

Auld, T. D., and Myerscough, P. J. (1986). Population dynamics of the shrub *Acacia suaveolens* (Sm.) Willd.: seed production and pre-dispersal seed predation. *Australian Journal of Ecology* **11**, 219–234.

Baker, G. B., Dettmann, E. B., and Wilson, S. J. (1997). Fire and its impact on avian population dynamics. *Pacific Conservation Biology* 3, 206–212.

Baker, J. (1997). The decline, response to fire, status and management of the Eastern Bristlebird. *Pacific Conservation Biology* 3, 235–243.

Bamford, M. J. (1986). The dynamics of small vertebrates in relation to fire in *Banksia* woodland near Perth, Western Australia. PhD thesis, Murdoch University.

Bamford, M. J. (1992). The impact of fire and increasing time after fire upon *Heleioporus eyrei, Limnodynastes dorsalis* and *Myobatrachus gouldii* (Anura: Leptodactylidae) in Banksia woodland near Perth, Western Australia. *Wildlife Research* 19, 169–178.

Baynes, A., Chapman, A., and Lynam, A. J. (1987). The rediscovery, after 56 years, of the Heath Rat *Pseudomys shortridgei* (Thomas, 1907) (Rodentia: Muridae) in Western Australia. *Records of the Western Australian Museum* 13, 319–322.

Begg, R. J., Martin, K. C., and Price, N. F. (1981). The small mammals of Little Nourlangie Rock, N.T. V. The effects of fire. *Australian Wildlife Research* 8, 515–527.

Bell, D. T., Hopkins, A. J. M., and Pate, J. S. (1984). Fire in the Kwongan. In *Kwongan: Plant Life on the Sandplain* (eds. J. S. Pate and J. S. Beard) pp. 178–204. (University of Western Australia Press: Nedlands.)

Benwell, A. S. (1998). Post-fire seedling recruitment in coastal heathland in relation to regeneration strategy and habitat. *Australian Journal of Botany* 46, 75–101.

Blanche, K. R., Andersen, A., and Ludwig, J. A. (2001). Rainfall-contingent detection of fire impacts: responses of beetles to experimental fire regimes in a tropical savanna. *Ecological Applications*. (in press)

Bock, C. E., and Bock, J. H. (1991). Response of grasshoppers (Orthoptera: Acrididae) to wildfire in a south-eastern Arizona grassland. *American Midland Naturalist* 125, 162–167.

Bowen, B. J., and Pate, J. S. (1993). The significance of root starch in post-fire shoot recovery of the resprouter *Stirlingia latifolia* R.Br. (Proteaceae). *Annals of Botany* 72, 7–16.

Bradstock, R. A., and Myerscough, P. J. (1988). The survival and population response to frequent fires of two woody resprouters *Banksia serrata* and *Isopogon anemonifolius*. *Australian Journal of Botany* 36, 415–431.

Braithwaite, R. W. (1987). Effects of fire regimes on lizards in the wet–dry tropics of Australia. *Journal of Tropical Ecology* 3, 265–275.

Braithwaite, R. W. (1995). A healthy savanna, endangered mammals and Aboriginal burning. In *Country in Flames*. (ed. D. B. Rose) pp. 91–102. (Biodiversity Unit of Department of Environment, Sport and Territories and Northern Australia Research Unit, Australian National University: Canberra and Darwin.)

Braithwaite, R. W., and Estbergs, J. A. (1987). Firebirds of the Top End. *Australian Natural History* 22, 298–302.

Brooker, L. C., and Brooker, M. G. (1994). A model for the effects of fire and fragmentation on the population viability of the Splendid Fairy-wren. *Pacific Conservation Biology* 1, 344–358.

Brown, C. L., and Whelan, R. J. (1999). Seasonal occurrence of fire and availability of germinable seeds in *Hakea sericea* and *Petrophile sessilis*. *Journal of Ecology* 87, 932–941.

Burrows, N. D. (1985). Reducing the abundance of *Banksia grandis* in the Jarrah forest by the use of controlled fire. *Australian Forestry* 48, 63–70.

Campbell, A. J., and Tanton, M. T. (1981). Effects of fire on the invertebrate fauna of soil and litter of a eucalypt forest. In *Fire and the Australian Biota* (eds. A. M. Gill, R. H. Groves and I. R. Noble.) pp. 215–241. (Australian Academy of Science: Canberra.)

Cary, G. J., and Morrison, D. A. (1995). Effects of fire frequency on plant species composition of sandstone communities in the Sydney region: combination of inter-fire intervals. *Australian Journal of Ecology* 20, 418–426.

Catling, P. C. (1986). *Rattus lutreolus*, colonizer of heathland after fire in the absence of *Pseudomys* species? *Australian Wildlife Research* 13, 127–139.

Christensen, P. E. S. (1980). The biology of *Bettongia penicillata* Gray, 1837, and *Macropus eugenii* (Demarest, 1817) in relation to fire. Forests Department of Western Australia, Bulletin 91, Perth.

Christensen, P. E. S., and Kimber, P. C. (1975). Effect of prescribed burning on the flora and fauna of south-west Australian forests. *Proceedings of the Ecological Society of Australia* 9, 85–106.

Clark, S. S. (1988). Effects of hazard-reduction burning on populations of understorey plant species on Hawkesbury sandstone. *Australian Journal of Ecology* 13, 473–484.

Cockburn, A. (1978). The distribution of *Pseudomys shortridgei* (Muridae: Rodentia) and its relevance to that of other heathland *Pseudomys*. *Australian Wildlife Research* 5, 213–219.

Cockburn, A. (1981). Population processes of the Silky Desert Mouse *Pseudomys apodemoides* (Rodentia), in mature heathlands. *Australian Journal of Zoology* 8, 499–514.

Cockburn, A., Braithwaite, R. W., and Lee, A. K. (1981). The response of the Heath Rat, *Pseudomys shortridgei*, to pyric succession: a temporally dynamic life-history strategy. *Journal of Animal Ecology* 50, 649–666.

Collett, N. G., and Neumann, F. G. (1995). Effects of two spring prescribed fires on epigeal Coleoptera in dry sclerophyll eucalypt forest in Victoria, Australia. *Forest Ecology and Management* 76, 69–85.

Connell, J. H. (1983). On the prevalence and relative importance of interspecific competition: evidence from field experiments. *American Naturalist* 122, 661–698.

Cowling, R. M., and Lamont, B. B. (1985). Variation in serotiny of three *Banksia* species along a climatic gradient. *Australian Journal of Ecology* 10, 345–350.

Cowling, R. M., Lamont, B. B., and Pierce, S. M. (1987). Seed bank dynamics of four co-occurring *Banksia* species. *Journal of Ecology* 75, 289–302.

Delettre, Y. R. (1994). Fire disturbance of a chironomid (Diptera) community on heathlands. *Journal of Applied Ecology* 31, 560–570.

Dickman, C. R., Henry-Hall, N. J., Lloyd, H., and Romanow, K. A. (1991). A survey of the terrestrial vertebrate fauna of Mount Walton, western Goldfields, Western Australia. *Western Australian Naturalist* 18, 200–206.

Dickman, C. R., Letnic, M., and Mahon, P. S. (1999). Population dynamics of two species of dragon lizards in arid Australia: the effects of rainfall. *Oecologia* 119, 357–366.

Evans, E. W. (1984). Fire as a natural disturbance to grasshopper assemblages of tallgrass prairie. *Oikos* 43, 9–16.

Evans, E. W. (1985). Community dynamics of prairie grasshoppers subjected to periodic fire: predictable trajectories or random walks in time? *Oikos* 52, 283–292.

Fox, A. M. (1978). The '72 fire of Nadgee Nature Reserve. *Parks and Wildlife* 2, 5–24.

Fox, B. J. (1981). The influence of disturbance (fire, mining) on ant and small mammal species diversity in Australian heathland. In *Proceedings of the Symposium on Dynamics and Management of Mediterranean-Type Ecosystems*, pp. 213–219. U.S. Department of Agriculture Forest Service, General Technical Report PSW-58, Albany, CA.

Fox, B. J. (1982). Fire and mammalian secondary succession in an Australian coastal heath. *Ecology* 63, 1332–1341.

Fox, B. J. (1983). Mammal species diversity in Australian heathlands: the importance of pyric succession and habitat diversity. In *Mediterranean-Type Ecosystems: The Role of Nutrients* (eds. F. J. Kruger, D. T. Mitchell and J. U. M. Jarvis) pp. 473–489. (Springer-Verlag: Berlin.)

Fox, B. J. (1990). Changes in the structure of mammal communities over successional time scales. *Oikos* 59, 321–329.

Fox, B. J., and McKay, G. M. (1981). Small mammal responses to pyric successional changes in eucalypt forest. *Australian Journal of Ecology* 6, 29–41.

Fox, B. J., Quinn, R. D., and Breytenbach, J. (1985). A comparison of small mammal succession following fire in shrublands of Australia, California and South Africa. *Proceedings of the Ecological Society of Australia* 14, 179–197.

Fox, M. D., and Fox, B. J. (1987). The role of fire in the scleromorphic forests and shrublands of eastern Australia. In *The Role of Fire in Ecological Systems* (ed. L. Trabaud) pp. 23–48. (SPB Academic: The Hague.)

Friend, G. R. (1979). The response of small mammals to clearing and burning of eucalypt forest in south-eastern Australia. *Australian Wildlife Research* 6, 151–163.

Friend, G. R. (1993). Impact of fire on small vertebrates in mallee woodlands and heathlands of temperate Australia: a review. *Biological Conservation* 65, 99–114.

Friend, G. R. (1995). Fire and invertebrates: a review of research methodology and the predictability of post-fire response patterns. *CALMScience Supplement* 4, 165–174.

Friend, G. R., and Williams, M. R. (1996). Impact of fire on invertebrate communities in mallee-heath shrublands of southwestern Australia. *Pacific Conservation Biology* 2, 244–267.

Gandar, M. V. (1982). Description of a fire and its effects in the Nylsvley Nature Reserve: a synthesis report. *South African National Scientific Reports Series* **63**, 1–39.

Gill, A. M. (1975). Fire and the Australian flora: a review. *Australian Forestry* **38**, 4–25.

Gill, A. M. (1981a). Coping with fire. In *The Biology of Australian Plants* (eds. J. S. Pate and A. J. McComb) pp. 65–87. (University of Western Australia Press: Nedlands.)

Gill, A. M. (1981b). Adaptive responses of Australian vascular plant species to fires. In *Fire and the Australian Biota* (eds. A. M. Gill, R. H. Groves and I. R. Noble) pp. 243–272. (Australian Academy of Science: Canberra.)

Gill, A. M. and Bradstock, R. A. (1992). A national register for the fire responses of plant species. *Cunninghamia* **2**, 653–660.

Gill, A. M., Groves, R. H., and Noble, I. R. (eds.) (1981). *Fire and the Australian Biota*. (Australian Academy of Science: Canberra.)

Gillon, Y. (1972). The effect of bushfire on the principal acridid species of an Ivory Coast savanna. *Proceedings of the Annual Tall Timbers Fire Ecology Conferences* **11**, 419–471.

Glasby, P., Selkirk, P. M., Adamson, D., Downing, A. J., and Selkirk, D. R. (1988). Blue Mountains Ash (*Eucalyptus oreades* R. T. Baker) in the western Blue Mountains. *Proceedings of the Linnean Society of New South Wales* **110**, 141–158.

Greenslade, P. (1997). Short term effects of a prescribed burn on invertebrates in grassy woodland in south-eastern Australia. *Memoirs of the Museum of Victoria* **56**, 305–312.

Greenslade, P., and Rosser, G. (1984). Fire and soil-surface insects in the Mount Lofty Ranges, South Australia. In *Medecos IV* (ed. B. Dell.) pp. 63–64. (Botany Department, University of Western Australia: Nedlands.)

Griffiths, A. D. (1995). We like our lizards frilled not grilled: the short-term effects of fire on frillneck lizards in the top end. In *Country in Flames*. (ed. D. B. Rose) pp. 87–90. (Biodiversity Unit of Department of Environment, Sport and Territories and Northern Australia Research Unit, Australian National University: Canberra and Darwin.)

Haim, A., and Izhaki, I. (1994). Changes in rodent community during recovery from fire; relevance to conservation. *Biodiversity and Conservation* **3**, 573–585.

Hairston, N. (1989). *Ecological Experiments: Purpose, Design and Execution.* (Cambridge University Press: Cambridge.)

Hammill, K., Bradstock, R. A., and Allaway, W. G. (1998). Post-fire seed dispersal and species reestablishment in proteaceous heath. *Australian Journal of Botany* **46**, 407–419.

Hannah, D. S., and Smith, G. C. (1995). Effects of prescribed burning on herptiles in southeastern Queensland. *Memoirs of the Queensland Museum* **38**, 529–531.

Harper, J. L. (1982). After description. In *The Plant Community as a Working Mechanism* (ed. E. I. Newman) pp. 11–25. (Blackwell Scientific Publications: Oxford.)

Higgs, P., and Fox, B. J. (1993). Interspecific competition: a mechanism for rodent succession after fire in wet heathland. *Australian Journal of Ecology* **18**, 193–201.

Hindmarsh, R., and Majer, J. D. (1977). Food requirements of the Mardo (*Antechinus flavipes* Waterhouse) and the effect of fire on Mardo abundance. Forests Department of Western Australia, Research Paper no. 31, Perth.

Hobbs, R. J., and Atkins, L. (1990). Fire-related dynamics of a *Banksia* woodland in south-western Western Australia. *Australian Journal of Botany* **38**, 97–110.

Hopkins, A. J. M., and Griffin, E. A. (1984) Floristic patterns. In *Kwongan: Plant Life on the Sandplain* (eds. J. S. Pate and J. S. Beard) pp. 69–83. (University of Western Australia Press: Nedlands.)

Ingwersen, F. (1977). Vegetation development after fire in Jervis Bay Territory. MSc thesis, Australian National University, Canberra.

James, C. D. (1991). Temporal variation in diets and trophic partitioning by coexisting lizards (*Ctenotus*: Scincidae) in central Australia. *Oecologia* **85**, 553–561.

Keith, D. A. (1995). Mosaics in Sydney heathland vegetation. *CALMScience Supplement* **4**, 199–206.

Keith, D. A. (1996). Fire-driven extinction of plant populations: a synthesis of theory and review of evidence from Australian vegetation. *Proceedings of the Linnean Society of New South Wales* **116**, 37–78.

Keith, D.A., McCaw, W. L., and Whelan, R. J. (2001). Fire regimes in Australian heathlands and their effect on plants and animals. In *Flammable Australia: The Fire Regimes and Biodiversity of a Continent* (eds. R. A. Bradstock, J. E. Williams and A. M. Gill) pp. 199–237. (Cambridge University Press: Cambridge.)

Knutson, H., and Campbell, J. B. (1976). Relationships of grasshoppers (Acrididae) to burning, grazing and range sites of native tallgrass prairie in Kansas. In *Proceedings of Tall Timbers Conference on Ecological Animal Control by Habitat Management*, no. 6, pp. 107–120, Tallahassee, FL.

Lawrence, G. E. (1966). Ecology of vertebrate animals in relation to chaparral fire in the Sierra Nevada foothills. *Ecology* **47**, 278–291.

Leigh, J. H., and Holgate, M. D. (1979). Responses of understorey of forests and woodlands of the southern tablelands to grazing and burning. *Australian Journal of Ecology* **4**, 25–45.

Leonard, B. V. (1974). The effects of burning on forest litter fauna in eucalypt forests. In *Proceedings of the 3rd Fire Ecology Symposium*, pp. 43–49. (Monash University: Melbourne.)

Loyn, R. H. (1997). Effects of an extensive wildfire on birds in far eastern Victoria. *Pacific Conservation Biology* **3**, 221–234.

Loyn, R. H., Traill, B. J., and Triggs, B. E. (1986). Prey of Sooty Owls in East Gippsland before and after fire. *Victorian Naturalist* **103**, 147–149.

Lunney, D., Cullis, B., and Eby, P. (1987). Effects of logging and fire on small mammals in Mumbulla State Forest, near Bega, New South Wales. *Australian Wildlife Research* **14**, 163–181.

Lussenhop, J. (1976). Soil arthropod response to prairie burning. *Ecology* **57**, 88–98.

MacFarland, D. C. (1991). The biology of the Ground Parrot, *Pezoporus wallicus*, in Queensland. III. Distribution and abundance. *Wildlife Research* **18**, 199–213.

Main, A. R. (1981). Fire tolerance of heathland animals. In *Heathlands and Related Shrublands of the World*, vol. 9B, *Analytical Studies* (ed. R. L. Specht) pp. 85–90. (Elsevier: Amsterdam.)

Main, B. Y. (1987). Ecological disturbance and conservation of spiders: implications for biogeographic relics in southwestern Australia. In *The role of Invertebrates in Conservation and Biological Survey* (ed. J. D. Majer) pp. 89–97. (Western Australian Department of Conservation and Land Management: Perth.)

Masters, P. (1993). The effects of fire-driven succession and rainfall on small mammals in spinifex grassland at Uluru National Park, Northern Territory. *Wildlife Research* **20**, 803–813.

McCarthy, M. A., and Cary, G. J. (2001). Fire regimes of landscapes: models and realities. In *Flammable Australia: The Fire Regimes and Biodiversity of a Continent* (eds. R. A. Bradstock, J. E. Williams and A. M. Gill) pp. 77–93. (Cambridge University Press: Cambridge.)

McCulloch, E. M. (1966). Swifts and bushfires. *Emu* **65**, 290.

Meney, K. A., Dixon, K. W., and Pate, J. S. (1997). Reproductive potential of obligate seeder and resprouter herbaceous perennial monocots (Restionaceae, Anarthiaceae, Ecdeiocoleaceae) from south-western Western Australia. *Australian Journal of Botany* **45**, 771–782.

Meredith, C. W., Gilmore, A. M., and Isles, A. C. (1984). The Ground Parrot (*Pezoporus wallicus* Kerr) in southeastern Australia, a fire adapted species? *Australian Journal of Ecology* **9**, 367–380.

Monamy, V. (1998). Native rodent distributions along successional gradients: the role of plant cover in structuring a small mammal community. PhD thesis, University of New South Wales, Sydney.

Monamy, V., and Fox, B. J. (1999). Habitat selection by female *Rattus lutreolus* drives asymmetric competition and coexistence with *Pseudomys higginsi*. *Journal of Mammalogy* **80**, 232–242.

Morrison, D. A. (1995). Some effects of low-intensity fires on co-occurring small trees in the Sydney region. *Proceedings of the Linnean Society of New South Wales* **115**, 103–109.

Morrison, D. A., Cary, G. J., Penjelly, S. M., Ross, D. G., Mullins, B. G., Thomas, C. R., and Anderson, T. S. (1995). Effects of fire frequency on plant species composition of sandstone communities in the Sydney region: interfire interval and time since fire. *Australian Journal of Ecology* **20**, 239–247.

Myerscough, P. J. (1980). Biological flora of the British Isles: *Epilobium angustifolium* L. *Journal of Ecology* **68**, 1047–1074.

Newsome, A. E., and Catling, P. C. (1983). Animal demography in relation to fire and shortage of food: Indicative models. In *Mediterranean-Type Ecosystems: The Role of Nutrients* (eds. F. J. Kruger, D. T. Mitchell and J. U. M. Jarvis) pp. 490–505. (Springer-Verlag: Berlin.)

Newsome, A. E., McIlroy, J., and Catling, P. C. (1975). The effects of an extensive wildfire on populations of twenty ground vertebrates in south-east Australia. *Proceedings of the Ecological Society of Australia* **9**, 107–123.

Newsome, A. E., Catling, P. C., and Corbett, L. K. (1983). The feeding ecology of the Dingo. II. Dietary and numerical relationships with fluctuating prey populations in South-eastern Australia. *Australian Journal of Ecology* **8**, 345–366.

Nieuwenhuis, A. (1987). The effect of fire frequency on the sclerophyll vegetation of the West Head, New South Wales. *Australian Journal of Ecology* **12**, 373–385.

Noble, I. R., and Slatyer, R. O. (1980). The use of vital attributes to predict successional changes in plant communities subject to recurrent disturbances. *Vegetatio* **43**, 2–21.

Noble, J. C. (1982). The significance of fire in the biology and evolutionary ecology of mallee *Eucalyptus* populations. In *Evolution of the Flora and Fauna of Arid Australia* (eds. W. R. Barker and P. J. M. Greenslade) pp. 153–159. (Peacock Publications: Adelaide.)

Noble, J. C. (1984). Fire in mallee (*Eucalyptus* spp.) shrublands of southwestern New South Wales. In *Medecos IV*. (ed. B. Dell) pp. 130–131. (Botany Department, University of Western Australia: Nedlands.)

Noble, J. C. (1989). Fire regimes and their influence on herbage and mallee coppice dynamics. In *Mediterranean Landscapes in Australia: Mallee Ecosystems and their Management* (eds. J. C. Noble and R. A. Bradstock) pp. 168–180. (CSIRO: Melbourne.)

O'Dowd, D. J. (1985). Effect of low-intensity fire on the activity of ants and other invertebrates at Belair Recreation Park, South Australia. In *Soil and Litter Invertebrates of Australian Mediterranean-Type Ecosystems* (eds. P. Greenslade and J. D. Majer) pp. 73–76. (Western Australian Institute of Technology: Perth.)

O'Dowd, D. J., and Gill, A. M. (1984). Predator satiation and site alteration following fire: mass reproduction of alpine ash (*Eucalyptus delegatensis*) in southeastern Australia. *Ecology* **65**, 1052–1066.

Paquin, P., and Coderre, D. (1997). Deforestation and fire impact on edaphic insect larvae and other macroarthropods. *Environmental Entomology* **26**, 21–30.

Pianka, E. C. (1996). Long-term changes in lizard assemblages in the Great Victorian Desert: dynamic habitat mosaics in response to wildfires. In *Long-Term Studies of Vertebrate Communities* (eds. M. L. Cody and J. A. Smallwood) pp. 191–215. (Academic Press: New York.)

Purdie, R. W. (1977). Early stages of regeneration after burning in dry sclerophyll vegetation. II. Regeneration by seed germination. *Australian Journal of Botany* **25**, 35–46.

Recher, H. F. (1997). Impact of wildfire on the avifauna of Kings Park, Perth, Western Australia. *Wildlife Research* **24**, 745–761.

Recher, H. F., Lunney, D., and Posamentier, H. (1975). A grand natural experiment: the Nadgee wildfire. *Australian Natural History* **18**, 154–163.

Reid, J. R. W., Kerle, J. A., and Baker, L. (1993). Mammals. In *Uluru Fauna: The Distribution and Abundance of Vertebrate Fauna of Uluru (Ayers Rock – Mount Olga) National Park, N.T* (eds. J. R. W. Reid, J. A. Kerle and S. R. Morton) pp. 69–78. (Australian National Parks and Wildlife Service: Canberra.)

Rice, B., and Westoby, M. (1999). Regeneration after fire in *Triodia* R. Br. *Australian Journal of Ecology* **24**, 563–572.

Roberts, P. E. (1970). Some effects of a bushfire on heathland birdlife. *Proceedings of the Royal Zoological Society of New South Wales* **89**, 40–43.

Southwood, T. R. E. (1966). *Ecological Methods: With Particular Reference to the Study of Insect Populations.* (Methuen: London.)

Specht, R. L., Rayson, P., and Jackman, M. E. (1958). Dark Island heath (Ninety Mile Plain, South Australia). VI. Pyric succession changes in composition, coverage, dry weight and mineral nutrient status. *Australian Journal of Botany* **6**, 59–88.

Stocker, G. C. (1966). Effects of fires on vegetation in the Northern Territory. *Australian Forestry* **30**, 223–230.

Sutherland, E. F. (1998). Fire, resource limitation and small mammal populations in coastal eucalypt forest. PhD thesis, University of Sydney.

Swengel, A. B. (1996). Effects of fire and hay management on abundance of prairie butterflies. *Biological Conservation* **76**, 73–85.

Tap, P. M. (1996). Arthropods and fire: studies in a southeast Australian heathland. PhD thesis, University of Wollongong, New South Wales.

Tap, P. M., and Whelan, R. J. (1984). The effect of fire on populations of heathland invertebrates. In *Medecos IV*. (ed. B. Dell) pp. 147–148. (Botany Department, University of Western Australia: Nedlands.)

Taylor, J. E., Monamy, V., and Fox, B. J. (1998). Flowering of *Xanthorrhoea fulva*: the effect of fire and clipping. *Australian Journal of Botany* **46**, 241–251.

Underwood, A. J. (1998). *Experiments in Ecology.* (Cambridge University Press: Cambridge.)

Vaughton, G. (1998). Soil seed bank dynamics in a rare, obligate-seeding shrub, *Grevillea barklyana* (Proteaceae). *Australian Journal of Ecology* **23**, 375–384.

Wallace, W. R. (1961). Fire in the jarrah forest environment. *Journal of the Royal Society of Western Australia* **49**, 33–44.

Warren, S. D., Scifres, C. J., and Teel, P. D. (1987). Response of grassland arthropods to burning: a review. *Agriculture, Ecosystems and Environment* **19**, 105–130.

Wellington, A. B., and Noble, I. R. (1985a). Post-fire recruitment and mortality in a population of the mallee *Eucalyptus incrassata* in semi-arid south-eastern Australia. *Journal of Ecology* **73**, 645–656.

Wellington, A. B., and Noble, I. R. (1985b). Seed dynamics and factors limiting recruitment of the mallee *Eucalyptus incrassata* in semi-arid south-eastern Australia. *Journal of Ecology* **73**, 657–666.

Whelan, R. J. (1986). Seed dispersal in relation to fire. In *Seed Dispersal.* (ed. D. R. Murray) pp. 237–271. (Academic Press: Sydney.)

Whelan, R. J. (1995). *The Ecology of Fire.* (Cambridge University Press: Cambridge.)

Whelan, R. J., and Baker, J. R. (1999). Fire in Australia: coping with variation in ecological effects of fire. In *Protecting the Environment, Land, Life and Property: Bushfire Management Conference Proceedings* (eds. F. Sutton, J. Keats, J. Dowling and C. Doig) pp. 71–79. (Nature Conservation Council: Sydney.)

Whelan, R. J., and Main, A. R. (1979). Insect grazing and post-fire plant succession in south-west Australian woodland. *Australian Journal of Ecology* **4**, 387–398.

Whelan, R. J., Langedyk, W., and Pashby, A. S. (1980). The effects of wildfire on arthropod populations in jarrah–*Banksia* woodland. *Western Australian Naturalist* **14**, 214–220.

Whelan, R. J., Ward, S., Hogbin, P., and Wasley, J. (1996). Responses of heathland *Antechinus stuartii* to the Royal National Park wildfire in 1994. *Proceedings of the Linnean Society of New South Wales* **116**, 97–108.

Whelan, R. J., de Jong, N., and von der Burg, S. (1998). Variation in bradyspory and seedling recruitment without fire among populations of *Banksia serrata* (Proteaceae). *Australian Journal of Ecology* **23**, 121–128.

White, N. A. (1992). The effects of prescribed burning on small mammals. In *Fire Research in Rural Queensland* (ed. B. R. Roberts) pp. 467–482. (University of Southern Queensland: Toowoomba.)

Williams, J. E., Whelan, R. J., and Gill, A. M. (1994). Fire and environmental heterogeneity in southern temperate forest ecosystems: implications for management. *Australian Journal of Botany* **42**, 125–137.

Williams, R. J. (1995). Tree mortality in relation to fire intensity in a tropical savanna of the Kapalga region, Northern Territory, Australia. *CALMScience Supplement* **4**, 77–82.

Wilson, B. A. (1996). Fire effects on vertebrate fauna and implications for fire management and conservation. In *Fire and Biodiversity: The Effects and Effectiveness of Fire Management*, Biodiversity Series, Paper no. 8 (ed. J. R. Merrick) pp. 131–147. (Biodiversity Unit, Department of the Environment, Sport and Territories: Canberra.)

Wilson, B. A., Robertson, D., Moloney, D. J., Newell, G. R., and Laidlaw, W. S. (1990). Factors affecting small mammal distribution and abundance in the eastern Otway Ranges, Victoria. *Proceedings of the Ecological Society of Australia* **16**, 379–396.

Woinarski, J. C. Z., and Recher, H. F. (1997). Impact and response: a review of the effects of fire on the Australian avifauna. *Pacific Conservation Biology* **3**, 183–205.

Yom-Tov, Y. (1987). The reproductive rates of Australian passerines. *Australian Wildlife Research* **14**, 319–330.

York, A. (1996). Long-term effects of fuel reduction burning on invertebrates in a dry sclerophyll forest. In *Fire and Biodiversity: The Effects and Effectiveness of Fire Management*, Biodiversity Series, Paper no. 8 (ed. J. R. Merrick) pp. 163–181. (Biodiversity Unit, Department of the Environment, Sport and Territories: Canberra.)

York, A. (1999). Long-term effects of repeated prescribed burning on forest invertebrates: management implications for the conservation of biodiversity. In *Australia's Biodiversity – Responses to Fire: Plants, Birds and Invertebrates*, Biodiversity Technical Paper no. 1 (eds. A. M. Gill, J. C. Z. Woinarski and A. York) pp. 181–266. (Department of the Environment and Heritage: Canberra.)

Zammit, C. (1988). Dynamics of resprouting in the lignotuberous shrub *Banksia oblongifolia*. *Australian Journal of Ecology* **13**, 311–320.

6

Spatial variability in fire regimes: its effects on recent and past vegetation

JAMES S. CLARK, A. MALCOLM GILL AND A. PETER KERSHAW

Abstract

Spatial variability in fire regimes produces heterogeneous vegetation and coexistence of life history strategies. Models used to analyse and predict effects of fire frequency typically assume a homogeneous stochastic process that applies everywhere on the landscape. Although these models assume a landscape mosaic of patch ages, each patch having a unique fire history, the distribution of fire intervals is assumed to be everywhere the same. Here we develop a tractable model that accommodates a range (continuous mixture) of fire intervals to understand how landscape heterogeneity in disturbance regime affects the mix of life histories that might coexist. Rather than assume one average interval that applies to the landscape as a whole, we adopt an underlying density of average intervals fitted to the fractions of the landscape experiencing different average intervals. The result is a continuous mixture of fire intervals that applies to the heterogeneous landscape as a whole. Application of the model to mapped landscapes in Tasmania suggests that the model provides a guide to the range of woodland and forest types expected under a heterogeneous fire regime. Examples of 'mixed' assemblages from fossil pollen records suggest that the degree of landscape heterogeneity in fire regimes changed with shifts in regional climate.

Introduction

Past and present Australian diversity depends on spatial variability in fire regimes. Fire constrains the range of life history attributes found along gradients from grassland to rainforest (Jackson 1968;

Brown and Podger 1982; Ellis 1985; Webb and Tracey 1994). The high diversity of vegetation types observed in close proximity reflects, in part, the patchiness of burns. While many variables influence vegetation heterogeneity including soils, hydrology and climate, fire is especially effective at selecting for particular functional types. The high spatial variability in fire regimes that characterizes modern sclerophyll/rainforest mosaics is frequently invoked when attempting to explain 'mixed' assemblages of the past, i.e. those containing populations that are both tolerant of ('resprouters' or 'sprouters') or sensitive to ('seeders') fire. Climate changes following the end of the Pleistocene are attended by contraction of sclerophyll and expansion of rainforests in northeast Queensland (Kershaw 1983, 1992; Walker and Chen 1987) and expansion of sclerophyll at the expense of steppe in Southeastern Australia (Kershaw 1998).

Models provide a tool for the analysis of fire interval effects and guidance on how changes in fire regimes might impact past and present vegetation. Tractable models currently in use quantify the relationship between fire return times and life history for a uniformly stochastic landscape (i.e. where the fire regime is everywhere the same). This approach has provided good predictions for the life histories expected for treefall gaps (Clark 1991a) and fire (Clark 1991b, 1996) in eastern North America. The importance of spatial variability for past and present Australian landscapes suggests the need for a more general approach.

Here we present a stochastic disturbance model based on renewal theory (Cox 1962) to explore how variability affects persistence of different life histories. Like previous efforts, our approach integrates life history information with fire interval data to predict reproductive success. We extend these efforts by incorporating landscape variability in fire regimes. We apply the model to a modern landscape for which published data allow us to characterise landscape variability. This analysis provides us with estimates of the range of life histories we might expect to occupy landscapes having homogeneous vs. spatially variable fire regimes. Published data permit comparison of model and data. We then turn to several examples from the palaeorecord of vegetation change and point out how the variability we incorporate in our analysis might help us understand past vegetation change. With pollen evidence for vegetation types, we use disturbance models to suggest ranges of fire intervals that are compatible with life histories corresponding to those vegetation types.

A model for integrating disturbance and life history

Rationale

To aid our interpretation of past fire regimes, we develop a simple model that captures the classic notion of disturbance consequences for life history. The underlying assumption is one of disturbances causing both mortality and, for some species, opportunity for recruitment. Whether or not disturbance mediates persistence depends on timing relative to life history schedules. Frequent disturbance can eliminate species that require extended fire-free periods for maturation. Conversely, rare disturbance results in local extinction of species that depend on disturbance for recruitment (Comins and Noble 1985). There is thus an 'intermediate' frequency that optimises the odds that species can persist (Jackson 1968; Heinselman 1973; Clark 1991b; Huston 1994) and that maximises diversity across landscapes of species with differing (but overlapping) schedules of life history and rates of colonisation and exclusion (Connell 1978; Huston 1979; Caswell and Cohen 1991).

Obviously, fire return interval is not the only aspect of fire regimes that affects vegetation. Fire intensity determines mortality of canopy species in forests and woodlands. Although low-intensity surface fires may not kill large trees, they may eliminate seedlings and have lasting impact on dynamics of understorey vegetation. Peat or duff fires have residual effects that can influence composition for many years. Seasonality of burns determines mortality and post-fire recruitment. While many studies recognise the importance of these many components of fire regimes (Gill 1975), data needed to incorporate them all in models rarely exist. In this chapter we focus on fire return time, because it is a key element of fire regimes, and data are frequently available from satellite imagery (Russell-Smith *et al.* 1997), tree rings (Clark 1990; Swetnam *et al.* 1999), sediment charcoal (Clark 1990), or stand age structure (Johnson 1979; Johnson and Van Wagner 1985).

Models based on renewal theory (e.g. Cox 1962) provide a means for quantifying the relationships between life history and disturbance frequency (Clark 1991b, 1996). Renewal theory began as an analysis of collections of objects that are 'self-renewing'. Common applications include problems involving failure and replacement; renewal theory is readily extended to disturbances that 'reset' vegetation on mosaics of landscape patches. The components include a density of disturbance events and life history schedules describing survivorship and fecundity. This framework accommodates dispersal among disturbed patches, although approximations provide ways to simplify some aspects of this effect (Clark 1991a; Clark and Ji 1995). The approach is sufficiently flexible to abet species having fundamentally different mortality responses to disturbance. The summary that follows borrows from previous, more in-depth, treatments of Clark (1991a, 1996) but extends these earlier analyses to the problem of disturbance regimes that vary as a mixture.

Disturbance intervals vary, representing the realisation of some underlying stochastic process. Previous analyses employ models that assume the process to be 'uniform' and focus on how changes in disturbance risk with time since the last disturbance control the distribution of patch ages across landscapes (Van Wagner 1978; Clark 1991b; Gill *et al.*

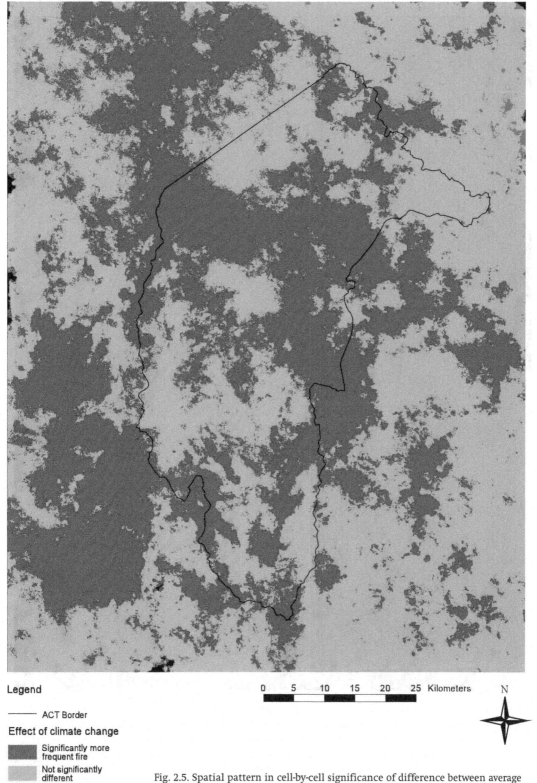

Legend

—— ACT Border

Effect of climate change

- Significantly more frequent fire
- Not significantly different
- Significantly less frequent fire
- No Data

Fig. 2.5. Spatial pattern in cell-by-cell significance of difference between average inter-fire interval simulated for current and changed climate from the CSIRO (1992) Climate Change Scenarios for the Australian Region. Modified from Cary and Banks (1999).

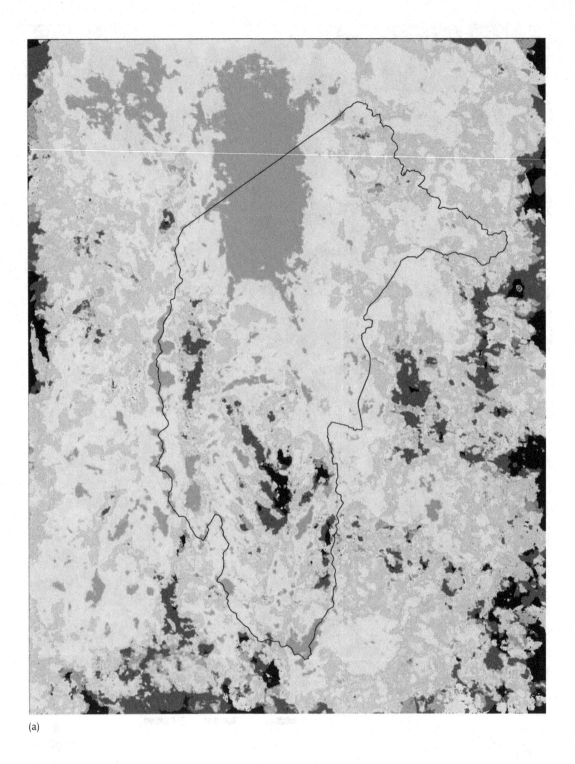

(a)

Fig. 2.8. Spatial patterns of average inter-fire interval (IFI) (years) predicted for the Australian Capital Territory region for four different climate change scenarios: (a) none (present climate); (b) small change in climate; (c) moderate change in climate; and (d) large change in climate.

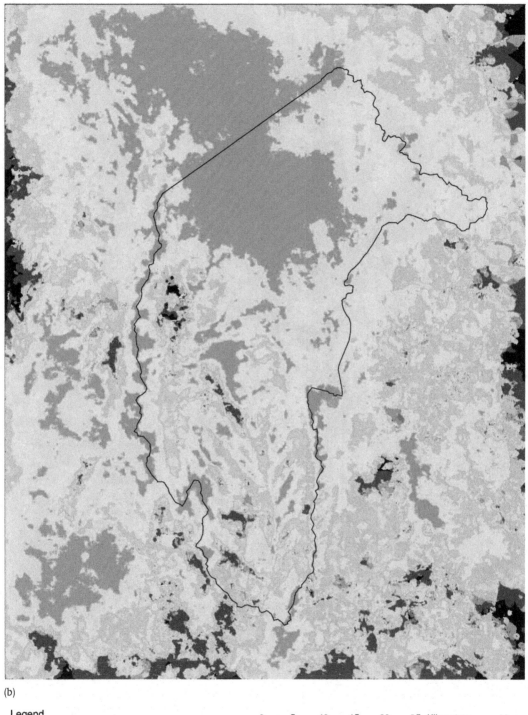

(b)

ACT Border

Average Interval (years)

0 - 10	30 - 40	70 - 80
10 - 20	40 - 50	80 - 90
20 - 30	50 - 60	> 90
	60 - 70	No Data

0   5   10   15   20   25   Kilometers

N

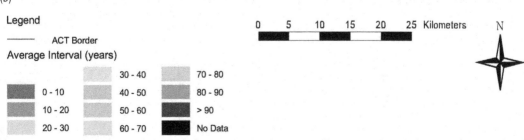

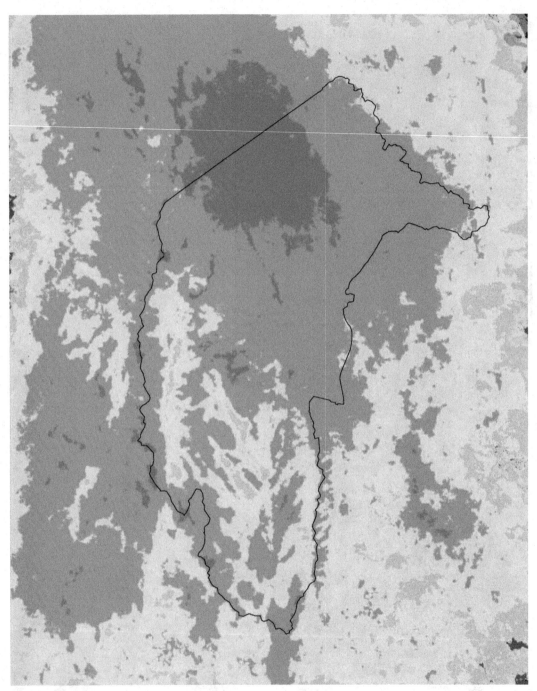

(c)

(d)

Legend

——— ACT Border

Average Interval (years)

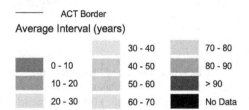

0   5   10   15   20   25   Kilometers

N

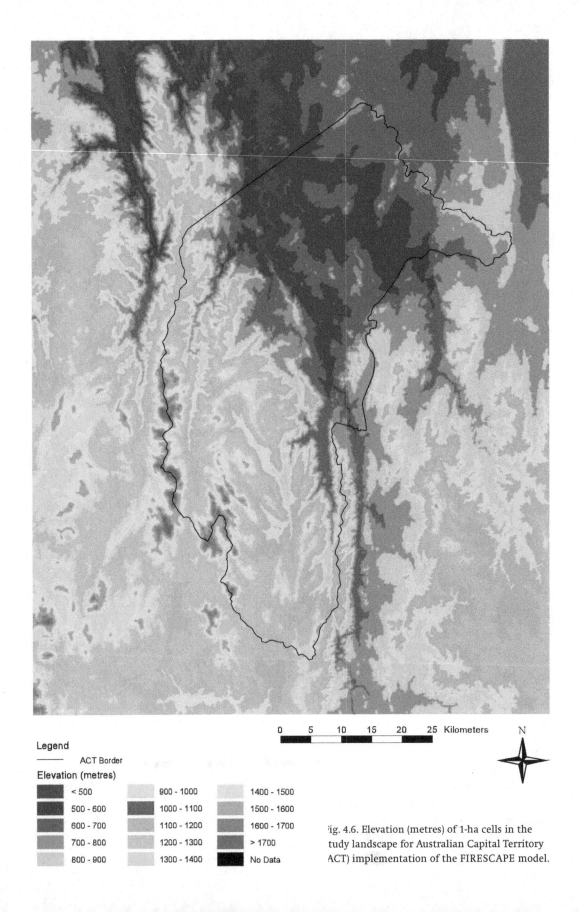

Legend

———   ACT Border

Elevation (metres)

▓	< 500	▓	900 - 1000	▓	1400 - 1500
▓	500 - 600	▓	1000 - 1100	▓	1500 - 1600
▓	600 - 700	▓	1100 - 1200	▓	1600 - 1700
▓	700 - 800	▓	1200 - 1300	▓	> 1700
▓	800 - 900	▓	1300 - 1400	▓	No Data

Fig. 4.6. Elevation (metres) of 1-ha cells in the
study landscape for Australian Capital Territory
(ACT) implementation of the FIRESCAPE model.

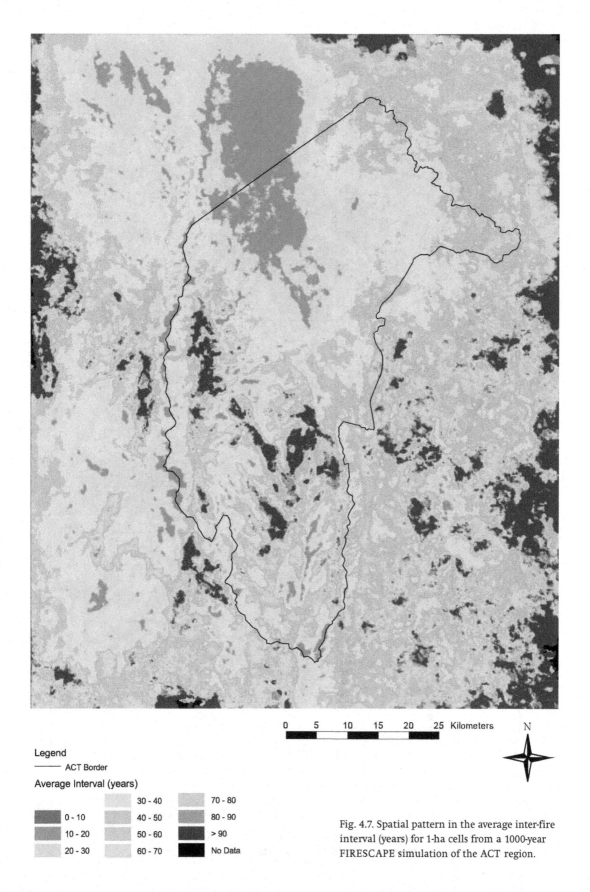

0   5   10   15   20   25   Kilometers   N

Average Interval (years)

	30 - 40	70 - 80
0 - 10	40 - 50	80 - 90
10 - 20	50 - 60	> 90
20 - 30	60 - 70	No Data

Fig. 4.7. Spatial pattern in the average inter-fire interval (years) for 1-ha cells from a 1000-year FIRESCAPE simulation of the ACT region.

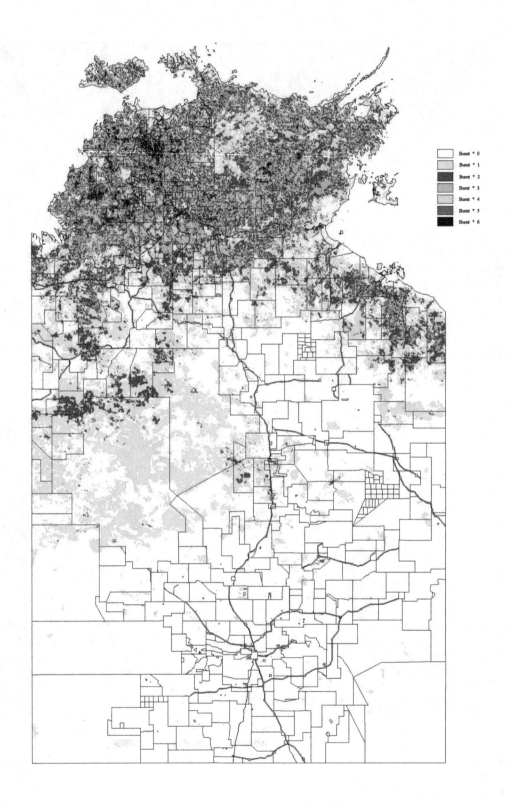

Fig. 7.4. Fire frequency in the Northern Territory derived from AVHRR (Advanced Very High Resolution Radiometer) fire history data, 1993 to 1998.

Fig. 8.2. Species-rich, annually burnt, *Themeda triandra*-dominated grassland in western Victoria. Note the abundance of bare ground and low levels of grass biomass.

Fig. 8.3. A dense, closed sward of *Themeda triandra* in an undisturbed, species-poor grassland in the Australian Capital Territory. Compare grass biomass levels and the amount of bare ground against the frequently burnt remnant in Fig. 8.2.

(a)

(b)

Fig. 12.2. Typical savanna vistas. (a) Mesic savanna on loamy sand (tenosol) at
Kapalga research station, NT; (b) semi-arid savanna on red earth (kandosol) at
Victoria River Research Station (Kidman Springs), NT; (c) treeless grassland plain
on cracking clay (vertosol) on the Barkly Tableland, NT; (d) escarpment ('stone
country') in Gregory National Park, NT.

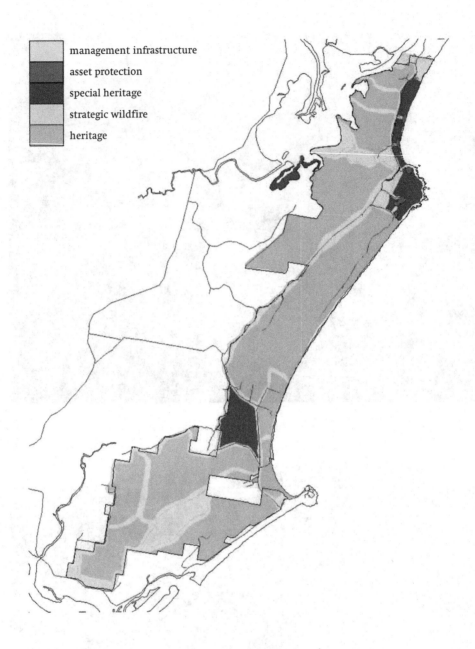

management infrastructure

asset protection

special heritage

strategic wildfire

heritage

Fig. 17.3. Fire management zones within Crowdy Bay National Park near Port Macquarie, New South Wales (from New South Wales National Parks and Wildlife Service 1997). Bushfire Adantage zones (1380 ha) contain access trails, watering points or fire breaks, Asset Protection zones (28 ha) contain recreational facilities and other functional structures, Strategic Wildfire control zones (1753 ha) are for prevention of fire spread to or from neighbouring settled areas, Special Heritage zones (548 ha) identify fire exclusion zones to protect cultural heritage, rainforest and wetlands and Heritage zones (5720 ha) are for conservation of fire-prone biodiversity. Each zone is characterised in the fire management plan by its habitats, threatened species, cultural resources and assets and has specific fire management objectives, strategies, actions and performance indicators.

2000) and the range of life histories that might persist (Clark 1996). Although a disturbance event is random, with a probability prescribed by an underlying risk, the relationship between risk and age is assumed to be deterministic and everywhere the same.

The usual assumption of a deterministic and unique risk function that applies everywhere is violated in all but the most uniform landscapes. Accumulation of fuels, fuel moisture and susceptibility to windthrow vary with topography (Romme 1982; Grimm 1983; Foster 1988; Clark 1990). For example, the mean interval of fires can vary dramatically over short distances, from frequent on upland sites to virtually non-existent in low, protected sites (Romme and Knight 1981; Brown and Podger 1982). Although stochastic, models that assume a uniform risk would predict extinction of disturbance-intolerant types that are, in fact, common in many landscapes characterised by high average disturbance frequency (e.g. Romme and Knight 1981; Webb and Tracey 1994).

The model we outline here recognizes the variability inherent in all disturbance regimes with an underlying distribution of disturbance risks that can be fitted to data. We begin by demonstrating how variable risk is incorporated in standard models and then analyse implications for life history.

A mixture of disturbance events

Landscapes are not homogeneous, so models of landscape or regional fire need explicit treatment of spatial variation. The appendix demonstrates how landscape variability in the fire interval affects the distribution of patch ages. Here a patch is defined as the area affected by a recent disturbance (e.g. a fire). For the examples discussed here we assume a Weibull density of events for a given landscape position, with scale parameter A (time) and shape parameter c (dimensionless). We further assume that A varies across a landscape, such that $1/A^c$ is gamma distributed with parameters α and θ. In Appendix 6.1 we derive the density of fire intervals across this landscape, where the scale parameter A varies continuously from place to place. The density of A was

selected for its flexible shape that might be parameterized to a range of landscapes and convenience (Appendix 6.1). For $c = 2$, A is distributed as inverse χ^2.

The composite density of fire intervals across a variable landscape is more leptokurtic than is the density for any given component of that landscape. The well-known exponential case illustrates the general result. If risk is independent of time since the last disturbance, then $c = 1$, and the density of intervals at a given landscape location is exponential (Fig. 6.1b). This density plots as a straight line on a semilog coordinate system, with slope $= -1/A$ (Fig. 6.1c). Fire ecologists use such plots to estimate the average fire interval and to infer whether risk may change with patch age, i.e. time since the last fire. The straight line suggests risk may be independent of patch age. A mode displaced from the origin indicates increase in risk with age (Fig. 6.1e). A positive second derivative (a slope that becomes less negative) indicates a decrease with age.

The mixture density that results when risk varies spatially demonstrates that the temporal interpretations based on spatially uniform landscapes do not apply. A density of disturbance intervals taken from this variable landscape has a fat tail, despite the fact that risk may be constant on any given portion of the landscape (Fig. 6.1c, f). Even with a constant (age-independent) risk on any part of the landscape, the mixture density has a shape that standard practice would interpret as decreasing risk with time since the last disturbance. Likewise, landscapes with increasing risk with time since the last disturbance have this effect masked by the high kurtosis that simply results from the fact that the density is a mixture (Fig. 6.1e, f).

The dramatic change in disturbance distributions caused by landscape variability in Fig. 6.1 are not the result of unrealistic parameter values. For these examples having mean rate parameters of $A = 20$ yr, the variable case has a standard deviation of $q = 20$. Thus, variability in disturbance process across landscapes has a profound effect on the distribution of events. The next section examines consequences for life history strategies.

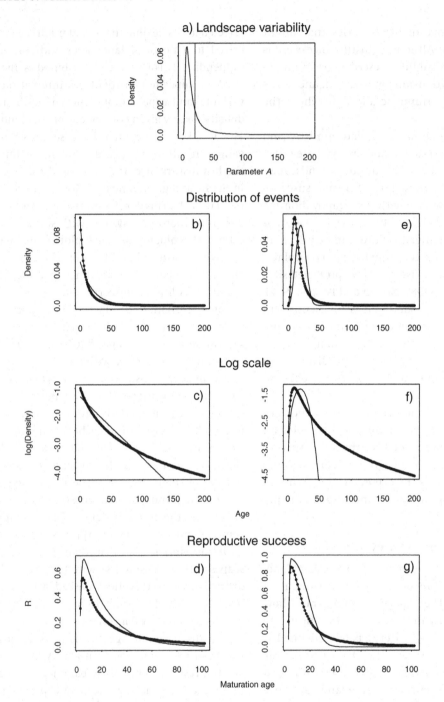

Fig. 6.1. A comparison of fire interval distributions and reproductive successes for homogeneous and variable landscapes. (a) The inverse χ^2 density of average fire intervals A across a landscape compared with the constant (mean = 20 yr) value (vertical line) used to illustrate a homogeneous landscape. (b, e) Densities of fire intervals for constant (b) and increasing (e) risk for homogeneous (thin line) and heterogeneous (dotted line) landscapes. Shape parameters are $c = 1$ and $c = 3$, respectively. Heterogeneity has the effect of increasing the variance, shifting the mass of the density to short and long values. (c, f) Same as b and e on a log scale. (d, g) Reproductive successes for life history types characterized by different maturation ages. Variability reduces the advantage of the strategy that is optimal at the mean interval of 20 yr and enhances success of the suboptimal types with delayed maturation (40 to 60 yr).

Table 6.1. *Life history parameters used in Fig 6.2*

Vegetation type	t_1	t_2	Mean fire interval (yr)
Heathland	5	50	10
Woodland	8	50	10
Open forest (eucalypt)	10	150	20
Tall open forest (eucalypt)	20	400	50
Rainforest	100	1000	300

Reproductive success and variability in time vs. space

For a given disturbance regime we could define an 'optimal' life history as the combined fecundity and survivorship schedule that maximises the probability of surviving and obtaining recruitment opportunities that occur following fire (Clark 1991b). Appendix 6.1 outlines the derivation of a 'relative net reproductive rate R', or the relative success of reproductive effort, in the context of a spatially varying disturbance regime. Net reproductive rate is used as an index of success, because it integrates how mortality risk and fecundity over the lifetime affect overall population growth rate. Population success depends on both survivorship and reproduction, and species can differ in their age-specific allocation to these activities. For instance, long-lived species having low reproductive rates may be more successful under many circumstances than short-lived species that produce copious seed. A species with low reproduction at any given age can still have net overall growth ($R_0 > 1$) by having occasional successful recruitment. Net reproductive rate combines both of these effects on population success. For a given spatial and temporal pattern of burns, there exists some value of R that is maximised with respect to maturation age for a given disturbance frequency. Clark (1991b, 1996) provides examples for uniform landscapes. For the examples that follow we use parameter estimates listed in Table 6.1. We apply the model to functional types in a general sense. Our reference to vegetation types, such as 'grassland', 'wattle', and so forth is intended to group life histories that share similar maturation ages and longevities.

The impact of variability on life history diversity (i.e. functional types having a range of maturation ages and longevities) can be viewed from the perspective of disturbance frequency. The range of life histories represented by savanna to rainforest communities segregates across a gradient of disturbance frequency (Fig. 6.2). Each life history type finds an optimum across this gradient that reflects its reproductive success for a uniform disturbance process, which occurs at $dR/dA = 0$,

$$A_{opt} = \frac{\chi\,(t_2^c - t_1^c)}{c\,(\ln t_2 - \ln t_1)}.$$

The disturbance gradient in Figs. 6.2a and b applies to a homogeneous landscape (eqn 6.1; see Appendix 6.1), where the timing of fire has two different levels of variability. Variability in return time of fires on a given piece of ground is determined by parameter c. If risk is constant ($c = 1$), variability is relatively high, and the distribution of events is exponential. Fig. 6.2a shows eqn 6.1 (Appendix 6.1) for $c = 1$, and a range of average intervals A. This high variability in return time results in broad overlap of life histories (eqn 6.5) across a range of average disturbance intervals (Fig. 6.2a). A range of life history schedules is successful at a given average interval, because the variability favours different life histories at different times and places. Thus, a homogeneous landscape with average return time of 50 yr ($\log 1/A = -1.7$) might support a mixture of grassland, woodland and eucalypt forest.

High values of c in eqn 6.1 mean that the probability of fire increases sharply at patch ages approaching the mean fire interval, and fire occurrence tends to be more 'periodic'. In other words

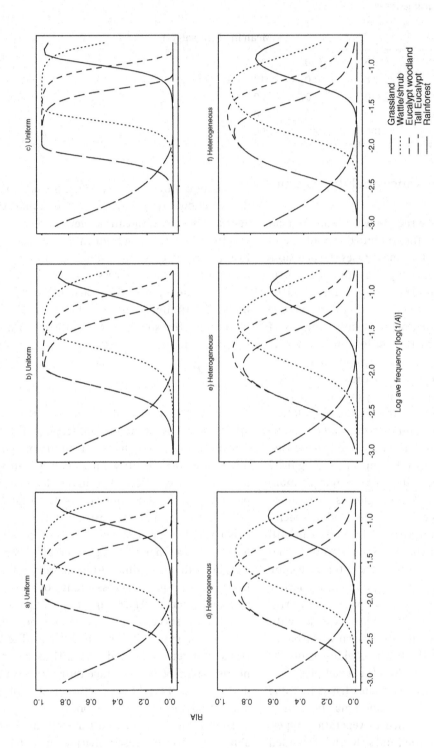

Fig. 6.2. Relative reproductive success across a gradient of disturbance frequencies at different levels of variability within patches ((a) to (c)) and across a landscape (the contrasts between (a) and (d) ($c = 1$), (b) and (e) ($c = 3$) and (c) and (f) ($c = 5$)). Both types of variability reduce maximum success of any one type while increasing success away from the respective optima. The most variable landscape (d) supports a broad diversity of types, whereas the least variable (c) would favour low diversity.

variability in fire return times within a patch is low. The low variability in fire return times provides strong advantage to the life history schedule that is best timed for disturbances that occur predictably at that age (eqn 6.5). Our analysis demonstrates that even a modest tendency for risk to increase with patch age has a large effect on reproductive success. Thus, the curves for life histories show greater reproductive success near their respective optima, but reproductive success drops off precipitously away from the optima (contrast Fig. 6.2a, b, c). With reduced variance of $c = 3$, there is greater segregation along a gradient of mean disturbance frequencies (Fig. 6.2b). At a mean interval of 50 yr, the simple model outlined here predicts dominance by woodland and eucalypt forest in a much less equitable mix (Fig. 6.2b, c) than does the variable, exponential case (Fig. 6.2a).

The timing of fires represents only one of two sources of variability that affects reproductive success. Landscape complexity that affects disturbance represents a second source of variation in reproductive success and is summarised by the coefficient of variation in A, i.e. parameter q (see Appendix 6.1). The expected reproductive success across variable landscapes differs substantially from the uniform case. Defining the 'optimal' life history as that which realises a higher reproductive success than all other life histories for a given disturbance regime (Appendix 6.1), we see that landscape variability diminishes the magnitude of the advantage for the optimal life history (Fig. 6.1d). For the average disturbance parameter of $A = 20$ in Fig. 6.1d, the optimal strategy has a maturation age of approximately 5 yr (eqn 6.5). Reproductive success is poor for shorter maturation times, because these short-lived species go locally extinct before the next disturbance, which occurs, on average, every 20 yr. Conversely, long-lived species are likely to be killed by disturbance before reaching maturation and, thus, are unable to take advantage of the recruitment opportunities that occur too early in life. The optimal maturation time of ~5 yr enjoys a substantial advantage over one of, say, 50 yr.

Variability in the disturbance regime makes for a more equitable distribution of life history types.

Variation tends to neutralise the advantage for the optimal life history, providing more equitable reproductive success for a broad range of life histories (eqn 6.6 and Fig. 6.1d). A single strategy does not achieve the reproductive success possible in a uniform landscape (the peak at 5 yr for the uniform case in Fig. 6.1d), but success of long-lived species is raised due to the increased possibility of extended fire-free intervals. This result does not mean that there is no longer an optimal strategy for any given portion of the landscape. Rather, the integrated success across this landscape of mixed disturbance regimes is more equitable than it is for any one portion. Variability means that the optimal strategy at the mean suffers to the benefit of those that are optimal under disturbance regimes that diverge from the mean. Low landscape variability promotes dominance by a single life history type.

Variability within vs. among landscape locations has some common and some disparate effects. Low variability both within (Fig. 6.2c, f) and among (Fig. 6.2a, b, c) patches produces maximum segregation of life history types across a gradient of disturbance frequencies. A given average frequency selectively promotes the life history that is best timed for that disturbance regime (the peak values in all examples from Fig. 6.2). High variability within a patch (Fig. 6.2a) or among patches (Fig. 6.2d) both promote overlap. For a given landscape average, a diversity of life history types is maintained due to the refuges provided by variability in disturbance regimes. Whereas a uniform disturbance process having a mean interval of 30 yr and low variability within patches (large c) predicts dominance by savanna and open woodland types (log $1/A = -1.5$ in Fig. 6.2c), variability within (Fig. 6.2a) or among (Fig. 6.2f) patches permits reproductive success by the spectrum of life history schedules. Variability across a landscape produces broad overlap in reproductive success curves at higher average intervals (Fig. 6.2d).

Application to modern landscapes

To demonstrate how the model is implemented on variable landscapes, we applied it to a forest/grassland mosaic in northeastern Tasmania. The

Table 6.2. *Parameters and predictions for the Ellis example*

Type	t_1	t_2	Average interval	Mt Maurice			Paradise Plains		
				% map	Variable	Uniform	% map	Variable	Uniform
Rainforest	100	1000	500	43.5	0.128	0.130	22.0	0.0811	0.0850
Tall Eucalypt	50	400	125	50.0	0.683	0.731	3.5	0.521	0.640
Eucalypt woodland	10	150	50	6.50	0.571	0.545	19.0	0.657	0.640
Wattle/shrub/grass	8	50	45	0	0.481	0.445	17.5	0.599	0.543
Grassland	3	10	10	0	0.116	0.0980	37.5	0.197	0.132
Spearman\(average interval)				160.2	0.662		117		
P value Cv (lambda)				84.8	0.223		177		

elevated plateau studied by Ellis (1985) supports a rich mosaic of vegetation types ranging from grasslands of *Poa labillarderi*, to *Eucalyptus delegatensis* forest and woodlands, to temperate rainforest, dominated by *Nothofagus cunninghamii*. The distributions of these vegetation types are controlled by topography, which determines moisture availability and frequency and severity of fire. Fire and vegetation history is not well understood, but successional patterns indicate dynamic responses to recent fires that include shifts in rainforest and woodland boundaries.

Ellis's (1985) surveys are the basis for vegetation maps for two areas, Mt Maurice and Paradise Plains. Vegetation is classified in 12 types, which we summarize here in five groups having similar life history schedules. We do not include Ellis's 'wetland' type. 'Rainforest', 'tall eucalypt', 'eucalypt woodland', 'wattle/shrub', and 'grassland' have different maturation ages and longevities summarized in Table 6.2. Today, the two areas have different representations of vegetation types. Mt Maurice is dominated by rainforest and tall eucalypt forest (Table 6.2). There is some eucalypt woodland, but grassland is essentially absent. The Paradise Plains are more complex, having a broad range of vegetation types.

From estimates of fire intervals and landscape variability we predicted reproductive success and compared that with life history types represented in Ellis's maps. For each of the $n = 5$ vegetation types we used estimates from Ellis and other sources (Jackson 1968) as basis for an average fire interval A_i, where i is the vegetation type. For simplicity, we assumed $c = 1$. Ellis provides the fractions of area occupied by each group p_i; which permit calculation of the landscape mean fire interval

$$\mu = \sum_{i=1}^{n} A_i p_i$$

and variance

$$Var[A] = \sum_{i=1}^{n} (A_i - \mu)^2 p_i.$$

These are the appropriate estimates of mean and variance for our purposes, because they represent the weighting as it affects reproductive success across a landscape. Parametric and non-parametric estimates of reproductive success were obtained for both landscapes. Parametric estimates were obtained by first fitting a gamma density for $1/A$ to frequency estimates for each portion of the landscape by maximising a log likelihood based on eqn 6.2 (see Appendix 6.1) weighted by landscape area,

$$\ln L = \sum_{i=1}^{n} p_i f(A_i) = \sum_{i=1}^{n} p_i [\alpha \ln(\theta) - \theta/A_i - (\alpha + 1) \ln(A_i) - \ln(\Gamma(\alpha))],$$

with respect to gamma parameters $\sigma \alpha$ and θ. The contribution of each type to this weighted likeli-

Mt Maurice

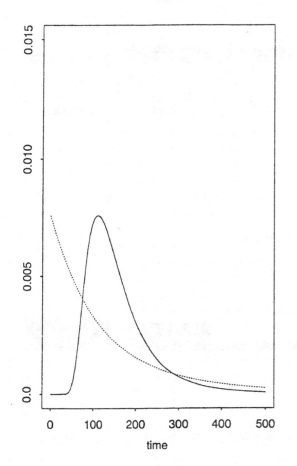

Paradise Plains

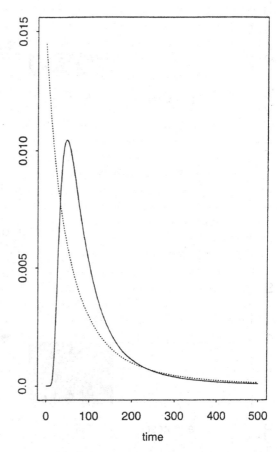

Fig. 6.3. The inverse χ^2 densities of A (solid lines) and fire intervals (dotted lines) for Mt Maurice and Paradise Plains.

hood is proportional to its representation on the landscape. Where data are sparse, parameters can be estimated by moment matching. Using the landscape weighted mean and variance of A, then α can be estimated from the coefficient of variation q, and θ from α and the expectation of A (see Appendix 6.1).

Model results illustrate the importance of landscape variability for vegetation types included in this analysis. Mt Maurice has a relatively low landscape average frequency estimate of $E[A] = 160$ and low variability, with a coefficient of variation of $q = 0.53$ (Fig. 6.3). The mean interval is optimal for maturation ages of 30 to 50 yr and, so, benefits forest trees. As demonstrated in the foregoing analysis, low vari-

ability provides little benefit to suboptimal life histories, such as shrub and grassland. The inconsequential role of variability is observed in the calculations of Table 6.2. Both 'variable' (i.e. that which incorporates landscape variance) and 'uniform' predict approximately the same reproductive successes for each type. The suboptimal wattle/shrub and grassland values are low for both estimates. The low variability that does exist benefits the short-lived functional types, which are not optimal at the mean interval of 160 yr (Fig. 6.4).

The Paradise Plains have a somewhat lower landscape average interval of $E[A] = 117$ yr and higher variability of $q = 1.51$. Figure 6.3 demonstrates how the

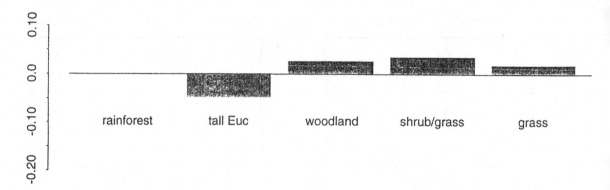

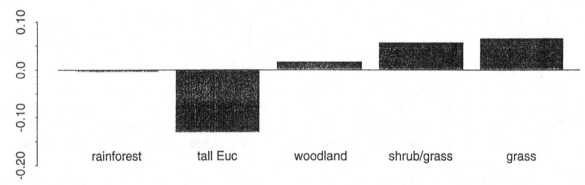

Fig. 6.4. The advantages conferred by variability, represented as the differences in predicted reproductive success on variable vs. homogeneous landscapes (Table 6.2). The long-lived types suffer under with increased variability, because they are optimal near the mean. The shorter-lived types rely on this variability.

probability density for Paradise Plains spans a broader range, thus providing for reproduction by more short-lived functional types (Table 6.2). The higher variability at Paradise Plains provides reproductive benefit to life histories that are suboptimal at the landscape average, and it reduces success of types that are optimal at the average. The average interval is best suited to tall eucalypt forest (Table 6.2). Variability means that this type is less successful across the landscape than it would be in a uniform landscape (Fig. 6.4). All other types realise

greater reproductive success with variability than they would in a uniform environment.

Where sufficient data exist, it is likely that a nonparametric approach would yield more accurate results. The gamma density for $1/A$ is flexible, but the continuous variability in risk it assumes may not best describe all landscapes. Because of the extreme variability in ignition and potential for spread on many landscapes, we expect that a nonparametric distribution of mean intervals A would be needed to capture inhomogeneous (including multimodal) distributions. Fortunately, the methodology we present here is readily adapted to a non-parametric distribution of risk variables.

Implications for past fire and vegetation

Palaeorecords provide our sole opportunity to examine consequences of climate changes that span decades to centuries at broad geographic scales. Unfortunately, the restricted view of past vegetation response provided by the fossil record limits its application to future climate change. Fossil pollen spectra from lake sediments represent a subset of the species present in surrounding vegetation, because species differ in pollen production and dispersal. Taxonomic resolution of pollen grains is coarse. There is little spatial information contained in pollen spectra that might be used to infer how species were patterned across landscapes. It is generally impossible to distinguish changes in the overall abundance of a taxon from changes in the distance at which it grows from the sample (core) location. Sediment charcoal data often display changes in abundance over time that are related to the overall importance of fire on the landscape. Data can be noisy, however, and so are difficult to interpret (R.L. Clark 1983; J.S. Clark *et al.* 1997). Most charcoal data do not provide clear evidence of fire frequency. Failing precise fire interval information and a detailed understanding of species present, it is generally impossible to interpret the role of fire in mediating effects of climate change.

In Australia, the obvious impact and complex effects of fire during past climate change, together with limitations of fossil charcoal records, have fostered considerable effort to infer past fire importance from pollen assemblages. Species are constrained by life history to a range of fire regimes; vegetation composition interpreted from fossil pollen helps to constrain the interpretation of fire regimes that could have prevailed in the past. Foresters and ecologists have long acknowledged that species requiring fire-free intervals for maturation are excluded from areas where fires occur at high frequency. Likewise, species that require fire

for recruitment are unable to persist where fires are sufficiently infrequent. The potential of the approach in view of few alternatives motivates analysis to refine the relationships between fire regime and vegetation composition. Improved quantitative methods could provide the basis for interpretations of past, present and future fire–vegetation interactions. The evidence for temporal changes in Australian fire regimes is provided in Chapter 1 (Kershaw *et al.*, this volume).

Pervasive evidence for mosaics of vegetation typical of long fire intervals (rainforest) and that in which shorter intervals were common (sclerophyll types) is a striking feature of the Australian palaeo-record of fire and vegetation. The juxtaposition of savanna and rainforest types in pollen diagrams emphasises a history of taxa with contrasting fire responses coexisting on complex landscapes. The Atherton Tableland, which today supports a mosaic of rainforest vegetation near the regional boundary with savanna, provides an example. Rough terrain and steep climate gradients make for a rich history of dynamics across glacial cycles. Pollen data from the region show a general pattern of savanna expansion during the most recent xeric glacial intervals with complex rainforest being more prominent during several interglacials (Kershaw 1983). Sediment charcoal accumulation is highest during glacial periods, especially the most recent one, when pollen of savanna types such as grasses and *Eucalyptus* are abundant with rainforest taxa *Podocarpus* and *Araucaria*. *Casuarina* is also prominent in these assemblages, and it is probably a sclerophyll element in north Queensland. At the height of the most recent glacial, (38 000 to 9 000 yr BP) rainforest disappears from the record, with an increase in charcoal and expansion of *Casuarina*, *Eucalyptus* and grasses. *Eucalyptus* charcoal in soils dated 12,000 to 27,000 yr BP indicates that present rainforest areas supported sclerophyll taxa (Hopkins *et al.* 1990). By contrast, charcoal accumulation tends to be low during interglacials, when rainforest expands. Charcoal accumulation is especially high after 40,000 yr BP, suggesting a link to human activities.

Spatial variability in fire and vegetation appears necessary to explain the past assemblages. For

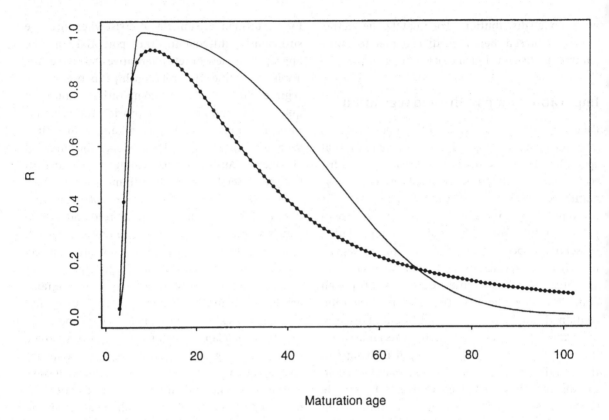

Fig. 6.5. Contrasting reproductive success for a mean interval of 50 yr for heterogeneous (dotted line) and homogeneous (solid line) landscapes.

instance, dry rainforest types (containing microphyll vine forest and *Araucaria*), which are rare both today and in previous interglacials, expanded at the beginning of the last glacial (approximately 78,000 to 38,000 yr BP) at Lynch's Crater. *Araucaria* survived in pre-settlement north Queensland in fire-protected patches, such as riparian zones and on rocky prominences. Kershaw (1992) notes that there might have been a mosaic of vegetation types separated by narrow ecotones during interglacials. He speculates that the ecotonal areas occupied by this vegetation type during glacial times became 'squeezed out' during interglacials, when savanna gave way directly to complex rainforest. If so, glacial intervals would have been times of low variability within landscape types (e.g. Fig. 6.2c) and rather discrete differences in fire regimes across landscape units. A distribution of mean fire intervals that is

bimodal might describe such a landscape better than eqn 6.2. The broad range of vegetation types suggested for interglacials is closer to the continuous mixture we analyse in Fig. 6.2d–f. Of course, fire was not the complete explanation for vegetation change across glacial–interglacial transitions. For instance, *Araucaria* forest dominates areas of southern Queensland today (e.g. the Bunya Mountains) where habitat variation is no greater than in north Queensland.

Unlike our example from Paradise Plains, the high fire frequency suggested by vegetation and charcoal at the height of the last glaciation (from 38,000 to 26,000 yr BP) is probably not characterised by the high landscape variability that would permit rainforest persistence. Although the average fire interval of ~ 100 yr at Paradise Plains is probably too short for rainforest, heterogeneity allows persistence of rainforest. The xeric climate of the full glacial suggests greater homogeneity, when pollen diagrams show little evidence for rainforest taxa. The collapse of *Podocarpus* and *Araucaria* populations and apparent extinction of the rainforest conifer *Dacrydium* with abrupt increases in charcoal during the full glacial suggests lack of landscape areas sufficiently protected from frequent fire. Here we expect that the distribution of average fire intervals had low variance. At an average interval of 50 yr, our analysis suggests extinction of rainforest on a uniform landscape, but persistence in a moderately variable one (Fig. 6.5). The non-synchronous expansion of rainforest across the Atherton Tableland in the early Holocene might reflect spread from refuges occupied when the Tableland was dominated by sclerophyll types (Walker and Chen 1987; Kershaw 1992). Different sites show this transition occurring between 8000 and 6000 yr BP.

At Lake George previous interglacials tend to have lower charcoal accumulation and higher percentages of *Casuarina* pollen percentages than do the present and penultimate interglacials (Singh and Geissler 1985). Thus, the demise of *Casuarina* and expansion of *Eucalyptus* may coincide with increased burning. But the relationship between charcoal accumulation and fossil pollen is inconsistent in the core. Because *Casuarina* includes species with a range of responses to fire, the trends in pollen and charcoal do not provide clear insights into the role of climate change, humans, fuels and fire. Whereas some *Casuarina* species tolerate frequent fire (e.g. *Casuarina torulosa* on the Atherton Tableland: Kellman 1986), others benefit from fire protection (Bowman 1998). This potential range of fire responses within a single pollen taxon confounds efforts to interpret how changes in temporal and spatial variability of fire occurrence changed in the past.

Our interpretations here do not exclude changes in fire regimes that may have resulted from human use of fire. If fire did increase with arrival of humans, it is still likely that heterogeneous fuels would have allowed for persistence of functional types that would be excluded in homogeneous landscapes.

Conclusions

Our approach represents a straightforward extension of traditional notions of life history and disturbance. The idea that responses to fire depend on how individuals time their reproductive activities is a pervasive one. By incorporating spatial variability we broaden this concept to better accommodate actual landscapes. Our minimal model retains the simplicity that permits analysis (Figs. 6.1, 6.2) and application to modern landscapes (Figs 6.3, 6.4, Table 6.2). For the Atherton Tableland example, it suggests times of high spatial variability in fire regimes (bimodal for the first half of the last glacial) and low variability (the last glacial maximum).

Our simple model complements more complex models that are implemented on spatially explicit landscapes. Although we limit our analysis to two types of landscape variability, our approach could accommodate other assumptions regarding variability. Complex models suffer from limited potential for analysis. By contrast, the simplicity of our model cannot accommodate details specific to individual landscapes, including the covariances in risk that can develop across patch boundaries, and complex patterns of landscape variability. On the other hand, if such covariances are weak, simple models may provide a rough guide to how variability may affect vegetation at the regional scale. Indeed, our simple model could even be parameterised from spatially explicit simulations in order to facilitate analysis of these complex models.

Appendix 6.1   Model for analysis of variability

Landscape variability in patch structure

Here we derive a landscape density of disturbance intervals under the assumption that risk varies continuously across this landscape. First, assume the landscape is homogeneous. Let a_i represent the event that the ith disturbance occurs at patch age $a = a_i$ yr after the last disturbance, and A be the mean of the density of disturbance intervals.

Further assume that the density of intervals for a given expected value A is a Weibull density

$$f(a|A) = \frac{c\chi a^{c-1}}{A^c} \exp\left[-\chi\left(\frac{a}{A}\right)^c\right], \tag{6.1}$$

where c is a shape parameter, and

$$\chi \equiv [\Gamma(1+1/c)]^c$$

is a composite parameter that simplifies notation. A constant risk results in an exponential density of intervals and represented is by the special case $c = \chi = 1$ (Fig. 6.1b).

Now suppose that the mean interval A can vary across the landscape, due to fuel structure, ignition probability, and so on. Now A is not fixed, and the density of disturbance intervals varies across the landscape, depending on the underlying variability in A. This more complex disturbance process can be analysed if we select a density for A that will permit us to solve for the marginal density of events a_i. We assume that $1/A^c$ is distributed as gamma density, because it accommodates a range of variances, and it permits a solution for the marginal density of disturbance events. Then the marginal density of A itself is

$$f_A(A) = \frac{\theta^\alpha c \exp\left[-\dfrac{\theta}{A^c}\right]}{A^{c\alpha+1}\Gamma(\alpha)}, \tag{6.2}$$

where θ and α are parameters that derive from the gamma distribution of $1/A^c$. For $c = 2$, $f_A(A)$ is Invχ^2. The mean value for A across the whole landscape,

$$\bar{A} = \int_0^\infty Af_A(A)\, dA = \frac{\theta^{1/c}\Gamma(\alpha-1/c)}{\Gamma(\alpha)},$$

is expected to differ from that for a given landscape location having a particular value for A

$$E[a|A] = \int_0^\infty af(a|A)\, da.$$

The coefficient of variation (ratio of the standard deviation to the mean) is

$$q = \frac{\sqrt{\Gamma(\alpha)\Gamma(\alpha-2/c) - \Gamma^2(\alpha-1/c)}}{\Gamma(\alpha-1/c)}.$$

To describe landscape variability with a given value of q we can evaluate the foregoing expression for α and then determine

$$\theta = \left[\frac{\bar{A}\Gamma(\alpha)}{\Gamma(\alpha-1/c)}\right]^c.$$

For $c = 1$, these expressions collapse to $q = (\alpha-2)^{-1/2}$ and $\theta = A(\alpha-1)$.

There are other possibilities for both the density of events in time $f(a|A)$ and the density for the rate parameter $f_A(A)$. We focus on this pair, because both are flexible and, thus, might capture temporal and landscape variability, and because the pair permits a solution for the marginal density of events $f(a)$. The marginal density incorporating spatial variance in risk $f_A(A)$ is obtained by integrating over the variability in A,

$$f(a) = \int_0^\infty f(a|A)f_A(A)\, dA = \frac{\alpha\theta^\alpha c\chi a^{c-1}}{(\theta + \chi a^c)^{\alpha+1}}. \tag{6.3}$$

Eqn 6.3 is the density of disturbance intervals across a landscape for which the mean interval for any given location varies according to eqn 6.2. The equilibrium density of patch ages is

$$v(a) = \frac{1}{\mu}\int_a^\infty f(a')\, da' = \frac{1}{\mu}\left(1+\frac{\chi a^c}{\theta}\right)^{-\alpha},$$

where μ is the mean interval between disturbances for the landscape as a whole

$$\mu = \int_0^\infty af(a)\, da.$$

Variability and life history

The consequences of spatially varying disturbance processes for life histories can be analysed within the framework of existing theory (Clark 1991b, 1996). The analysis that follows contrasts the effects of disturbance for a uniform landscape with that of a variable landscape. As our index of life history consequences we use $R \equiv \dfrac{R_0}{\beta}$, which represents the relative success of reproductive effort. β is seed production realised by a tree that survives to t_2 yr, and R_0 is the net reproductive rate, which incorporates the mortality schedule and opportunities for recruitment. Using the approach outlined in Clark (1991b, 1996), for a species that is killed by fire and that requires fire for recruitment, the fraction of potential seed output β represented by R_0 is given by the integrated opportunities for recruitment during the reproductive interval (t_1, t_2),

$$R = \int_{t_1}^{t_2} f(a)\,da, \qquad (6.4)$$

because $f(a)$ is the product of the age-specific risk of fire and the probability of not being killed by fire by age a. For an unvarying disturbance risk (constant A) with a Weibull density eqn 6.4 has the solution

$$R = \exp\left[-\chi\left(\frac{t_1}{A}\right)^c\right] - \exp\left[-\chi\left(\frac{t_2}{A}\right)^c\right] \qquad (6.5)$$

(Clark 1991b).

Variability in disturbance regimes changes this result. For A distributed according to eqn 6.2, we solve the integral by introducing a new variable $z = \theta + \chi a^c$,

$$R = \int_{t_1}^{t_2} f(a)\,da = \alpha\theta^\alpha \int_{\theta + \chi t_1^c}^{\theta + \chi t_2^c} z^{-\alpha-1}dz$$

which permits the solution for reproductive success

$$R = \theta^\alpha\left[\frac{1}{(\theta + \chi t_1^c)^\alpha} - \frac{1}{(\theta + \chi t_2^c)^\alpha}\right]. \qquad (6.6)$$

As the limit is variability tends to zero, this result tends to eqn 6.5. We arrive at the same result by solving R for a spatially uniform disturbance process in eqn 6.5 and then distributing that reproductive success over a landscape experiencing the range of disturbance, represented by $f(A)$,

$$E[R] = \int_0^\infty R(A) \times f(A)dA.$$

For the case where the full distribution of A can only be approximated by a mean and variance, we obtain a non parametric approximation by expanding about the mean A

$$E[R] = R(\bar{A}) + \frac{Var[A]}{2}\frac{dR}{dA^2}\bigg|_{\bar{A}} + \cdots \qquad (6.7)$$

For our model this result is

$$E[R] = \exp\left[-\chi\left(\frac{t_1}{A}\right)^c\right] - \exp\left[-\chi\left(\frac{t_2}{A}\right)^c\right] + \frac{var[A]}{2} \qquad (6.8)$$

$$\left\{\left(\frac{c\chi t_1}{A^{c+1}}\right)^2 \exp\left[-\chi\left(\frac{t_1}{A}\right)^c\right] - \left(\frac{c\chi t_2}{A^{c+1}}\right)^2 \exp\left[-\chi\left(\frac{t_2}{A}\right)^c\right]\right\}.$$

Rainforest species that are intolerant of frequent crown-killing fires are accommodated by the same analysis but require an alternative treatment of reproductive success. Whereas fire-dependent species are recruited only following fire, rainforest species continue to be recruited during intervals between fires; they are killed by fire, but they are not dependent on fire for recruitment. For the simplest assumption of recruitment that does not change with patch age the integral for reproductive success is

$$R = B\int_{t_1}^{t_2}\int_a^\infty f(x)\,dx\,da = B\int_{t_1}^{t_2} S(a)da,$$

where B is a rate constant, and the inner integral is the probability of no fire occurring before age a. (Clark (1991b) analyses the more complex case of successional change following fire.) For an unvarying disturbance process the result

$$S(a) = \exp\left[-\chi\left(\frac{a}{A}\right)^c\right] \qquad (6.9)$$

reflects the fact that reproductive success for a rainforest life history is always maximised in the absence of fire ($A \to 8$). With variance in A the result is

$$S(a) = \left(1 + \frac{\chi a_1^c}{\theta}\right)^{-\alpha}. \qquad (6.10)$$

As for the life histories requiring disturbance for recruitment, this result tends to unity as the variance in A tends to zero.

References

Bowman, D. M. J. S. (1998). The impact of aboriginal landscape burning on the Australian biota. *New Phytologist* **140**, 385–410.

Brown, M. J., and Podger, F. D. (1982). Floristics and fire regimes of a vegetation sequence from sedgeland-heath to rainforest at Bathurst Harbour, Tasmania. *Australian Journal of Botany* 30, 659–676.

Caswell, H., and Cohen, J. E. (1991). Disturbance, interspecific interaction and diversity in metapopulations. In *Metapopulation Dynamics: Empirical and Theoretical Investigations*. (eds. M. Gilpin and I. Hanski) pp. 193–218. (Academic Press: London.)

Clark, J. S. (1990). Landscape interactions among nitrogen mineralization, species composition, and long-term fire frequency. *Biogeochemistry* 11, 1–22.

Clark, J. S. (1991a). Disturbance and tree life history on the shifting mosaic landscape. *Ecology* 72, 1102–1118.

Clark, J. S. (1991b). Disturbance and population structure on the shifting mosaic landscape. *Ecology* 72, 1119–1137.

Clark, J. S. (1996). Testing disturbance theory with long-term data: alternative life history solutions to the distribution of events. *American Naturalist* 148, 976–996.

Clark, J. S. and Ji, Y. (1995). Fecundity and dispersal in plant populations: implications for structure and diversity. *American Naturalist* 146, 72–111.

Clark, J. S., Stocks, B. J., Cachier, H., and Goldammer, J. G. (1997). *Sediment Records of Biomass Burning and Global Change*. (Springer Verlag: Berlin.)

Clark, R. L. (1983). Pollen and charcoal evidence for the effects of Aboriginal burning on the vegetation of Australia. *Archaeology in Oceania* 18, 32–37.

Comins, H. N., and Noble, I. R. (1985). Dispersal, variability and transient niches: species co-existence in a uniformly variable environment. *American Naturalist* 126, 706–723.

Connell, J. H. (1978). Diversity in tropical rainforests and coral reefs. *Science* 199, 1302–1310.

Cox, D. R. (1962). *Renewal Theory*. (Chapman and Hall: London.)

Ellis, R. C. (1985). The relationship among eucalypt forest, grassland and rainforest in the highland area in north-eastern Tasmania. *Australian Journal of Ecology* 10, 297–314.

Foster, D. R. (1988). Disturbance history, community organization and vegetation dynamics of the old-growth Pisgah forest, south-western New Hampshire, U.S.A. *Journal of Ecology* 76, 105–134.

Gill, A. M. (1975). Fire and the Australian flora: a review. *Australian Forestry* 38, 4–25.

Gill, A. M., Ryan, P. G., Moore, P. H. R., and Gibson, M. (2000). Fire regimes of World Heritage Kakadu National Park, Australia. *Austral Ecology*. (in press)

Grimm, E. C. (1983). Chronology and dynamics of vegetation change in the prairie-woodland region of southern Minnesota, U.S.A. *New Phytologist* 93, 311–350.

Heinselman, M. L. (1973). Fire in the virgin forest of the Boundary Waters Canoe Area, Minnesota. *Quaternary Research* 3, 329–382.

Hopkins, M. S., Graham, A. W., Hewett, R., Ash, J., and Head, J. (1990). Evidence of late Pleistocene fires and eucalypt forest from a North-Queensland humid tropical rainforest site. *Australian Journal of Ecology* 15, 345–347.

Huston, M. A. (1979). A general hypothesis of species diversity. *American Naturalist* 113, 81–101.

Huston, M. A. (1994). *Biological Diversity*. (Cambridge University Press: Cambridge.)

Jackson, W. D. (1968). Fire, air, water and earth: an elemental ecology of Tasmania. *Proceedings of the Ecological Society of Australia* 3, 9–16.

Johnson, E. A. (1979). Fire recurrence in the subarctic and its implications for vegetation composition. *Canadian Journal of Botany* 57, 1374–1379.

Johnson, E. A., and Van Wagner, C. E. (1985). The theory and use of two fire history models. *Canadian Journal of Forest Research* 15, 214–220.

Kellman, M. (1986). Fire sensitivity of *Casuarina torulosa* in north Queensland, Australia. *Biotropica* 18, 107–110.

Kershaw, A. P. (1983). Holocene pollen diagram from Lynch's Crater, north-eastern Queensland, Australia. *New Phytologist* 94, 669–682.

Kershaw, A. P. (1992). The development of rainforest–savanna boundaries in tropical Australia. In *The Nature and Dynamics of the Forest–Savanna Boundary* (eds. P. A. Furley, J. Proctor and J. A. Ratter) pp. 255–271. (Chapman and Hall: London.)

Kershaw, A. P. (1998). Estimates of regional climatic variation within southeastern mainland Australia since the Last Glacial Maximum from pollen data. *Palaeoclimates: Data and Modelling* 3, 107–134.

Kershaw, A. P., Clark, J. S., and Gill, A. M. (2001). A history of fire in Australia. In *Flammable Australia: The Fire Regimes and Biodiversity of a Continent*. (eds. R. A. Bradstock, J. E.

Williams and A. M. Gill) pp. 3–25. (Cambridge University Press: Cambridge.)

Romme, W. H. (1982). Fire and landscape diversity in subalpine forests of Yellowstone National Park. *Ecological Monographs* **52**, 199–221.

Romme, W. H., and Knight, D. H. (1981). Fire frequency and subalpine forest succession along a topographic gradient in Wyoming. *Ecology* **62**, 319–326.

Russell-Smith, J., Ryan, P. G., and DuRieu, R. (1997). A Landsat MSS-derived fire history of Kakadu National Park, monsoonal northern Australia, 1980–1994: seasonal extent, frequency and patchiness. *Journal of Applied Ecology* **34**, 748–766.

Singh, G., and Geissler, E. A. (1985). Late Cenozoic history of fire, lake levels and climate at Lake George, New South Wales, Australia. *Philosophical Transactions of the Royal Society of London* **311**, 379–447.

Swetnam, T. J., Allen, C. D., and Betancourt, J. L. (1999). Applied historical ecology: using the past to manage the future. *Ecological Applications* **9**, 1189–1206.

Van Wagner, C. E. (1978). Age-class distribution and the forest fire cycle. *Canadian Journal of Forest Research* **8**, 220–227.

Walker, D., and Chen, Y. (1987). Palynological light on tropical rainforest dynamics. *Quaternary Science Reviews* **6**, 77–92.

Webb, L. J., and Tracey, J. G. (1994). The rainforests of northern Australia. In *Australian Vegetation* (ed. R. H. Groves) pp. 87–130. (Cambridge University Press: Cambridge.)

Part III

Ecosystems: grasslands

7

Fire regimes in the spinifex landscapes of Australia

GRANT E. ALLAN AND RICHARD I. SOUTHGATE

Abstract

Spinifex grasslands cover more than 25% of the Australian landscape. Spinifex is a generic term that includes three genera and over 60 species of perennial hummock grass. The grasses occur on soils that are low in nutrients and in climatic regions that are typically arid or semi-arid. Fuel accumulation is strongly linked to the post-fire cumulative rainfall and time since fire. The interval between fires can range from less than 3 years to more than 30 years. In the past, Aboriginal burning is thought to have reduced the size and increased the frequency of fires in spinifex grasslands. These days, fires can be extremely large, commonly covering many thousands of square kilometres. Data are presented that describe the frequency, extent and patchiness of fires in spinifex grasslands that extend along a climatic gradient from the Great Victoria Desert to Arnhem Land. Fire has a pronounced effect on the distribution of plants and animals. Although the evidence that a changed fire regime has lead to a decline in medium-sized mammals is equivocal, there is also strong evidence that the presence of mature spinifex is important to the maintenance of plant and animal species richness and diversity. Large fires in the spinifex ecosystems are contributing to a decline in the distribution of 'fire-sensitive' plants such as mulga (*Acacia aneura*) in certain areas. Further research needs to more clearly determine the size and frequency of fires that best suit animals and plants with particular life history characteristics, but active management programmes are required immediately. Remote sensing technology should be used to effectively monitor fire extent and pattern in the extensive spinifex-dominated landscapes and these data are needed to formulate fire management plans.

Introduction

Fires in spinifex grasslands rarely receive the attention of the general public. This may be because the fires are not considered catastrophic or the burnt areas are very remote and out of view. Nevertheless, fires in spinifex grasslands are inevitable and can be very large, frequently exceeding 10 000 km². Large fires are a significant issue of national importance in terms of biodiversity. Other important issues are associated with loss of property, and feed for domestic animals. There is uncertainty regarding the way to best manage fire frequency, fire size and fire patchiness in the spinifex grasslands to maintain biodiversity and 'fire-sensitive' plant and animal communities, protect pastoral activities and maintain cultural values.

This chapter describes the fire regimes in spinifex grasslands, provides a synthesis of the issues regarding fire, flora and fauna, and outlines methods used to monitor and manage an extensive landscape for conservation, primary production and cultural significance. New information is presented as regional comparisons of fire size, patchiness and frequency of burning. It also discusses the issues of patch burning for the conservation of threatened mammals, the perceived fiery battleground between spinifex and 'fire-sensitive' plants and the role of Aboriginal people and bushfire agencies in maintaining an appropriate fire regime.

Spinifex grasslands of Australia

Spinifex grasslands cover more than a quarter of the Australian continent. They are dominated by perennial grasses of the genera *Triodia* (Lazarides 1997) plus *Symplectrodia* and *Monodia* (Jacobs 1992). Following the revision of the genus *Triodia* (Lazarides 1997), which now encompasses the genera *Plectrachne*, there are 64 species of *Triodia*, two species of *Symplectrodia* (Lazarides 1984) and one species of *Monodia* (Jacobs 1985). The taxonomy used within this chapter follows Lazarides (1997) and provides a synonym for the first mention of each species affected by the revision.

All spinifex species have distinctive sclerophyllous leaves which form into large dome-shaped hummocks that may subside in the centre with age and form distinctive rings. Spinifex is frequently divided into 'hard' or 'soft' types based on leaf anatomy and amount of leaf resin. The leaves of the 'hard' species are generally more rigid and pungent and do not produce noticeable amounts of leaf resin (Jacobs 1992).

Spinifex communities are primarily restricted to the arid and semi-arid climatic regions and are associated with soils that are low in essential nutrients such as nitrogen and phosphorus (Winkworth 1967). They occur on both sandy and skeletal soils. Spinifex can also be an important understorey component on rocky substrates in tropical regions, such as the Arnhem Land Plateau of the Northern Territory (NT), well beyond the boundaries of the arid zone, with an average annual rainfall in excess of 1000 mm. In the eastern sub-humid region they occur as small, scattered areas in what are regarded as relict populations. A few species are widespread, but the majority have either localised distributions or occur in specialised habitats. The major alliances are:

- *Triodia pungens* and *T. schinzii* (syn. *Plectrachne schinzii*) form most of the spinifex communities of the sandplains and some sand ridges and sandier floodplains north and west of Alice Springs, plus the west coast region of Western Australia;

- *Triodia bitextura* (syn. *Plectrachne pungens*) occurs in the monsoonal climate regions of the north;
- *Triodia basedowii* is common southwest of Alice Springs, plus the Simpson Desert and channel country of Queensland, and vast areas of Western Australia, including the Great Victoria, Gibson and Great Sandy Deserts;
- *Triodia scariosa* (syn. *T. irritans*) occurs in the south, generally associated with mallee communities;
- *Triodia longiceps* is localised but occurs sporadically over large areas in western Queensland ranges and in the floodplains and channels of non-perennial watercourses;
- *Triodia wiseana* and allied varieties occur on rocky hills and slopes, in the Pilbara and Kimberley regions of Western Australia, and scattered areas between; and
- *Triodia mitchelli* and allied varieties and the related *T. marginata* occur in Queensland and northern New South Wales generally on sandplains or rocky hills in areas often regarded as marginally semi-arid (Jacobs 1984).

Despite these regional descriptions, Griffin (1991) notes that the distributions of the spinifex alliances have only been approximately mapped at different scales and levels of resolution. Although single species dominance was the norm, there was often a significant admixture of a second species.

Landscape changes associated with topography and soils frequently cause significant changes to plant communities. Small features within the spinifex landscapes often support 'fire-sensitive' species or communities (Latz 1995). The term 'fire-sensitive' is used to describe species or communities dominated by such species which are perceived to be disadvantaged by relatively frequent burning. Note that spinifex species may be formally defined as fire-sensitive according to the definition of Gill (1981) (i.e. plants are typically killed by a fire-sufficient to scorch the crown). Changing the scale of observation alters how readily 'fire-sensitive' and spinifex communities are identified and mapped. In a regional survey of central Australia, Perry *et al.* (1962) described and mapped seven of the 88 land systems as spinifex sandplains and dunefields. The survey described their component land units, which are non-spinifex vegetation communities that frequently include 'fire-sensitive' vegetation. The scale of their mapping precluded the delineation of these units. In a regional survey of the Simpson Desert region at a finer scale, Purdie (1984) recognised significant differences within the extensive areas of longitudinal sand dunes, based on size and density of dunes, plus vegetation and soil characteristics (Graetz *et al.* 1982; Purdie 1984). Twenty of the 32 land systems mapped and described by Purdie (1984) were part of the spinifex-dominated sandplains and dunefields which covered 79% of the survey area. Omitted from their maps were small areas of interdune red-earth plains, alluvial floodplains and watercourses, claypans, salt lakes and salinas, and lateritic rises and ridges, each with a variety of non-spinifex vegetation. Frequently these areas include 'fire-sensitive' species. Land unit surveys, which have a finer resolution and are capable of delineating these features, are generally restricted to smaller areas, frequently individual pastoral properties with areas less than 6000 km². Jessop and King (1997) described the land resources of New Crown Station, a pastoral property bordering the western edge of the Simpson Desert in the Northern Territory, on the border with South Australia. The spinifex-covered sandplains and dunefields were separated into five land units. In addition, inclusions within the land systems associated with small rocky outcrops, relict plains of mulga (*Acacia aneura*) and drainage lines were mapped as individual units. The above examples illustrate the textural richness and complexity of plant communities which may be threatened by wildfires from the surrounding spinifex communities and hint at the importance that the spatial characteristics of fire may play in affecting plant and animal communities.

Land tenure across the vast spinifex landscapes is variable between Australian states. It may be summarised into the four main land tenure categories of pastoral lands, Aboriginal lands, conservation areas and vacant crown land (Table 7.1 and Fig. 7.1). The important features are:

- the largest areas of spinifex occur in Western Australia (WA) and the Northern Territory (NT), there are moderate areas in South Australia (SA) and Queensland, and only small areas in New South Wales (NSW) and Victoria (which are generally considered part of the mallee lands);
- Aboriginal land and vacant crown land in WA, NT and SA have the highest proportion of spinifex;
- WA has a very large area of vacant crown land, whereas within the NT and SA, most of the vacant crown land has been converted to Aboriginal land over the past 25 years; and
- almost all the spinifex lands in Queensland and NSW are under pastoral tenure.

With the exception of the Karatha, Mt Tom Price and Port Hedland in the Pilbara region of WA, there are no cities or large towns within the landscapes dominated by spinifex and the human population density is very low.

Despite the low human population, spinifex grasslands form an arena where two cultures, with very different perspectives on fire, reside. Aboriginal people have traditionally been burners whereas European people have had a deep-seated urge to control and suppress fire. Whereas fire has been an integral and essential part of the landscape for traditional Aboriginal people, European inhabitants of Australia have considered fire as a fearful and threatening element of the landscape that needs to be fought and suppressed. There is uncertainty regarding the past and existing pattern of fire in spinifex grasslands and the attitudes of European and Aboriginal people are changing quite rapidly. Many younger Aboriginal people are adopting a more conservative European view of fire whereas many conservationists and a growing number of pastoralists recognise the importance of fire in the landscape to maintain productivity and diversity of flora and fauna. It is therefore critical that we more fully understand the consequences of different fire regimes before the experience and knowledge of Aboriginal people in burning the landscape is inevitably usurped by western influence.

There is currently much speculation about the

Table 7.1. *Spatial extent of spinifex in Australia by land tenure within each state*

	Western Australia	Northern Territory	South Australia	Queensland	New South Wales	Victoria	Total
Area of state (km² × 10³)	2518	1341	980	1727	797	226	7659
Area of spinifex (km² × 10³)	1083	700	162	131	16	7	2100
Proportion of state with spinifex (%)	43.0	52.2	16.6	7.6	2.0	2.9	27.4
Area of pastoral lands (km² × 10³)	1228	691	574	1586	711	154	4973
Pastoral lands with spinifex (%)[a]	25.9	37.2	6.2	7.8	2.1	0.4	15.1
Spinifex on pastoral lands (%)[b]	12.7	19.1	3.6	7.1	1.9	0.2	9.8
Area of Aboriginal lands (km² × 10³)	197	513	187	21	0	0	920
Aboriginal lands with spinifex (%)	67.7	68.5	44.8	0	4.6	0	61.8
Spinifex on Aboriginal lands (%)	5.3	26.2	8.6	0	0	0	7.4
Area of conservation area (km² × 10³)	153	38	203	53	38	30	530
Conservation areas with spinifex (%)	36.4	27.3	20.6	13.3	2.5	19.1	23.0
Spinifex on conservation areas (%)	2.2	0.8	4.3	0.4	0.1	2.6	1.6
Area of vacant crown land (km² × 10³)	924	94	10	54	42	38	1186
Vacant crown land with spinifex (%)	61.8	87.0	6.9	0.8	0.7	0.6	55.3
Spinifex on vacant crown land (%)	22.7	6.2	0.1	0	0	0.1	8.6

Notes:

[a] Calculated as the proportion of the area of pastoral lands with spinifex cover to the total area of pastoral lands (row 4).

[b] Calculated as the proportion of the area of pastoral lands with spinifex cover to the total area of the state (row 1). The sum of rows 6, 9, 12, 15 is equal to row 3, 'the proportion of state with spinifex'.

Source: Updated from Griffin (1984), using AUSLIG (1990) and AUSLIG (1993).

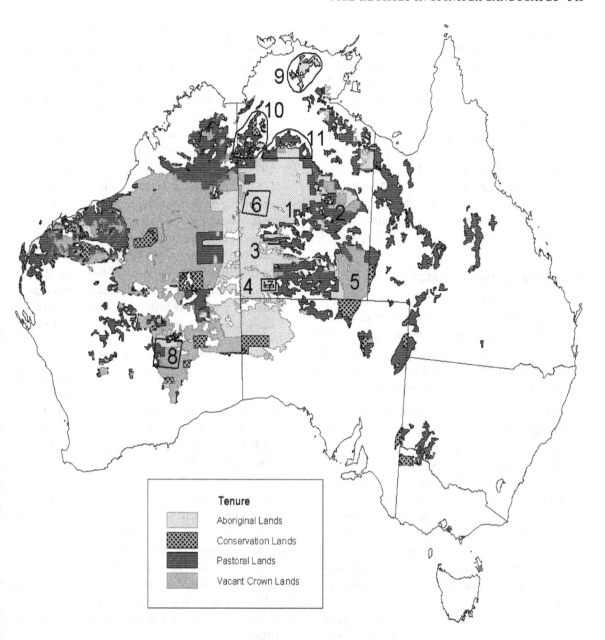

Fig. 7.1. Map of the distribution of spinifex grasslands in Australia, stratified by land tenure. Fire history database areas: (1) Tanami Desert; (2) Northeast central Australia; (3) Western Desert; (4) Southwest NT; (5) Simpson Desert; (6) Tanami Desert subregion; (7) Uluru National Park region; (8) Great Victoria Desert region; (9) Arnhem Land Plateau region; (10) Gregory National Park region; and (11) Northern Tanami region. Cadastre and tenure data sources: AUSLIG (1990) and AUSLIG (1993).

composition of vegetation and the pattern of fire prior to the arrival of Aboriginal people over 50 000 years ago and Europeans a little over 200 years ago (Flannery 1994; Latz 1995). It is argued that less scleroform vegetation existed before the arrival of Aboriginal people and the increased use of fire has resulted in an increase in 'fire-tolerant' species in the landscape. Because of the propensity of Aboriginal people to use fire, it has also been argued that the current fire pattern, especially in the spinifex grasslands, is very different from that at the time of European settlement (Bolton and Latz 1978; Latz and Griffin 1978; Kimber 1983; Latz 1995).

Fire in the spinifex grasslands

During the past 15 years several fire studies have collected appropriate spatial information to describe fire regimes in spinifex communities and make comparisons between regions of Australia. Several significant wildfires have occurred in the arid zone of Australia during the last 20 years (Fig. 7.2):

- In December 1983 a large wildfire west of Alice Springs burnt an area of approximately 20 000 km². In an unusual occurrence for central Australia, which was associated with a period of above average rainfall, a second fire in 1985 reburnt nearly 5000 km² of the previous fire area. The area is a mixed spinifex, mulga and open woodland vegetation complex (Wilson et al. 1990) although at the regional and national scales it is predominantly a spinifex grassland. No ecological studies have been undertaken in this area.
- In 1989/90 several wildfires followed 2 years of above average rainfall in the southwest of the NT and an area of approximately 5400 km² was burnt, much of which had been previously burnt by a 1976 wildfire. Control efforts were associated with two fires near the boundary of the pastoral properties and Aboriginal land. The other fires burned unattended. The high fuel loads in the region persisted and several other large wildfires in 1990 and 1991 burnt an area of 7250 km², including a portion of Uluru–Kata Tjuta National Park.

- In 1994 a series of wildfires burnt an area in excess of 100 000 km², in the Tanami and Great Sandy Desert regions of the NT and WA. The fires occurred after a period of few fires in the region and a rapid biomass accumulation following a tropical depression in February–March 1993 which nearly doubled the average annual rainfall in the northern Tanami region. In November 1994 control efforts to restrict the spread of the fire into pastoral and conservation lands was occurring on four separate fronts. The control teams were working independently and the shortest distance between two teams was just over 200 km. The greatest proportion of the fire burned unattended. These wildfires highlighted the need for better links between the development of spatial fire history databases and the implementation of land management actions.

Fuels in the spinifex grasslands

Spinifex is generally the major component of the fuel load, but there are significant regional and seasonal variations which influence the opportunity for fire. There is a consistent pattern of fuel development following a fire, throughout the spinifex ecosystems (Griffin 1992), but the rate of change is variable primarily in response to rainfall and time (Griffin 1984). Most spinifex is killed by fire and regeneration is from seed, which generally is initiated by the first effective rainfall after the fire. This may range from a period of weeks to years in the most arid regions of Australia. Once established, the individual perennial hummocks tend to accumulate biomass after major rainfall events and tend to lose little of it during the intervening dry periods (Griffin et al. 1990).

In the early stages of recovery, spinifex cover and biomass are very low and there is a high diversity of species, associated with reduced competition from spinifex and an increase in nutrients following the fire (Griffin 1991; Allan and Baker 1990). These recently burnt areas have insufficient fuel loads to carry a fire, and depending on the regional and seasonal conditions, this may persist for up to 15 years (Curry 1996). However, in high-rainfall periods in the arid zone, spinifex fuel habitat was observed to

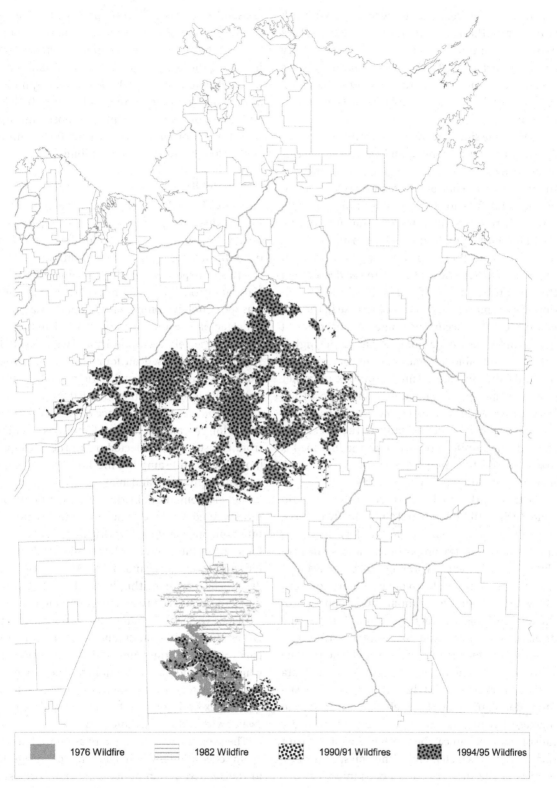

	1976 Wildfire		1982 Wildfire		1990/91 Wildfires		1994/95 Wildfires

Fig. 7.2. Map of significant wildfires in the Northern Territory.

accumulate sufficient fuel to burn twice within 3 years during the 1970s (Griffin *et al.* 1983; P.K. Latz personal communication; and the case history above). In such cases, *Triodia* species may not always constitute the bulk of the fuel. Short-lived grasses, such as *Aristida* species, can also produce large quantities of fuel.

As the time since fire increases, spinifex becomes dominant in both biomass and cover and species diversity declines. There are regional differences in the accumulation rate of spinifex and its proportion of the total fuel load between spinifex communities. There is a general pattern that follows the north-to-south rainfall gradient from the tropical savannas of northern Australia to the most arid region in southern Australia. There is a decrease in the accumulation rate of spinifex and the proportion of non-spinifex fuels and an increase in the variability of the occurrence and importance of non-spinifex fuels along this gradient. In the higher and more reliable rainfall regions of northern Australia, the fuel load is a mix of spinifex and other grasses. Some species of spinifex can resprout following fires, and the rate of growth of all species is rapid in response to the reliability of the wet season rainfall. The spinifex proportion of the fuel load is low with short fire intervals, but spinifex will dominate in areas with longer fire intervals.

Further south into the northern edges of the arid zone, such as the Tanami and Great Sandy Deserts, the spinifex proportion of the fuel load increases and the non-spinifex proportions increase in variability. During years of average to below-average rainfall, the non-spinifex biomass is very low, but following above-average rainfall, its proportion of the total fuel load can increase dramatically. In the drier portions of the arid zone that are still dominated by summer monsoonal rainfall, such as the southwest NT including the Uluru–Kata Tjuta National Park region and the Gibson Desert WA, spinifex domination of the fuel load continues to increase. Fire intervals are regulated by the recovery rate of spinifex, but can be shortened by significant rainfall events, which increase the non-spinifex proportion of the fuel load. The non-spinifex proportion remains small but its importance increases. It assists fire to spread across the gaps between the spinifex hummocks that would otherwise be too large. In the more southern regions of the arid zone, including the Great Victoria Desert, rainfall is more evenly distributed throughout the year. Spinifex is always the dominant proportion of the fuel load and non-spinifex components are mostly unimportant. The rate of spinifex growth and fuel accumulation is slow and fire intervals are long.

Fire behaviour

Fire spread usually takes place through a community of discrete spinifex hummocks separated by areas of bare ground. Griffin and Allan (1984) developed the first model for spinifex communities from a series of experimental fires at Uluru–Kata Tjuta National Park. The model to predict rate of spread was split into two components, a weather factor and a fuel factor. The weather factor was calculated from temperature, relative humidity and wind speed. The fuel factor was based on four factors, the proportions of spinifex cover and bare ground, a patchiness variable and fuel moisture. Patchiness was the ratio of the mean and variance of patch sizes of spinifex along a transect. The fuel factor was calculated from transect data collected using the wheelpoint (point intercept) sampling method. The associated computer program and manual could be run on hand-held computers in the field (Griffin 1985). It incorporated two fire behaviour models into the program, one developed by Griffin and Allan (1984) for spinifex fuels and another originally developed for continuous grasslands by McArthur (1977) and described as an algorithm by Noble *et al.* (1980) for grassland fuels. The two models provided the park manager with the option of using the program in either the spinifex landscapes that dominate the park or in the grassland and mulga woodlands which surround the two monoliths. The two models also worked together if data transects covered a mix of spinifex and non-spinifex areas, such as mulga swales within the sand dune areas.

The program was designed to assist park managers in scheduling fire management operations. Due to the relatively stable fuel conditions over short periods of time (months) within spinifex communities, the computer program provided the operator with the option of entering up to five different weather scenarios. The program predicted the rate of spread for each weather scenario along the actual transect of fuel conditions sampled. Therefore the manager could choose the best weather conditions for the desired fire management outcome, and schedule the burning activity when conditions were optimal. Despite the versatility, the model was not widely used or tested within parks in central Australia due to the idiosyncrasies of the hand-held computers available at the time. The opportunity to use the model on newer more reliable hand-held computers has never been realised.

Burrows *et al.* (1991) tested Griffin and Allan's (1984) model in the Gibson Desert region of Western Australia. The model overestimated the rate of spread of their fires, partially due to a higher level of patchiness within their fuels. Due to this patchiness, a threshold wind speed of 12–17 km h^{-1} was required before the flames would carry from one fuel patch to the next and the fire would spread. Their new rate of spread (ROS) model used the four parameters of wind speed, temperature, ground cover expressed as a ratio of spinifex cover to bare ground and fuel moisture.

A second fire behaviour model (Griffin and Allan 1993) was developed for the spinifex areas of central Australia following an expanded programme of experimental fires under a greater range of fuel loads and weather conditions. The model parameters were very similar to the Uluṟu model but the calculation of the fuel factor was based on airborne radiometer data. It was designed to operate as part of a programme of aerial control burning operations over the vast areas of spinifex grasslands. The model has not been used or tested due to insufficient efforts in the areas of technology transfer and the very limited use of aerial control burning in the southern portion of the NT.

In a review of rate of spread (ROS) models for spinifex, Gill *et al.* (1995) identified four factors critical to the ability to predict when discrete hummock-grass fuels will ignite, whether or not fires will spread and how fast they will travel once they do spread. The factors are:

- how to predict the moisture content of hummock grasses;
- how to characterise and predict fuel distributions as a function of time since fire or cumulative rainfall since fire;
- how to depict fire weather in arid lands, and
- how to parameterise the factors limiting fire spread in hummock-grass fuels.

Predicting the moisture content, or curing state, is a difficult issue for spinifex fuels. Cheney and Sullivan (1997) indicated that the curing state of a grassland – live fuel moisture content (FMC) – is expressed as the fraction of dead material in the sward. Measurements to calculate curing state should exclude dead fuel, rather than calculate fuel moisture content from a sample of the entire fuel load. It is not practicable to separate the live and dead fuel components of spinifex. The hummocks are a complex mix of live and dead fuels which contribute to its flammable nature. Burrows *et al.* (1991) sampled a profile through an entire hummock on the assumption that the proportions of live and dead material would be constant for a given spinifex species. Their results, which agreed with Griffin and Allan (1984), found that under normal daytime conditions in the desert, fuel moisture content has a relatively small effect on rate of spread compared to other variables, except when fuels were very moist (>30%). Live spinifex was difficult to ignite during the rare times when moisture content was in excess of 35%. The problem was also recognised by McArthur (1972), who observed that in the arid regions it was not the curing process that is the significant variable in fire behaviour. More important was whether there had been sufficient rain for growth to provide sufficient fuel to sustain the combustion process, especially in the inter-hummock areas. Griffin *et al.* (1983) used fire records from the Bushfires Council of the NT to quantify this process. The best relationship was found between area burnt and two or three years of cumulative antecedent rainfall.

Simulation models of discrete fuels have been used to parameterise the factors limiting fire spread in hummock-grass fuels (Bradstock and Gill 1993;

Weber *et al.* 1995; Mercer 1997). Further work is required to incorporate their flame length and flame residence time parameters into the existing ROS models and test the results in the field with both resinous and non-resinous species of spinifex.

Topography has not been considered in any of the spinifex ROS models. All the models were generated from experimental fires on the relatively flat landscapes of the arid zone sandplains. Although the spinifex-covered ranges are much less extensive, active fire management to provide protection of many fire-sensitive species is an important issue. Many species, such as *Acacia undoolyana*, *A. macdonnellensis* and *Callitris glaucophylla*, have distributions that are restricted to rocky ridges possibly as a refuge from fire (Bowman and Latz 1993; Duguid 1999). Many of these scenically spectacular portions of the landscape are now part of national parks or other conservation reserves where park managers have to implement fire management programs. These programs would benefit considerably from updated spinifex fire behaviour models with topographic factors.

Despite their availability, fire managers in central Australia do not use the rate of spread models. Experience and trial and error are used to guide the timing of both prescribed burning programmes and wildfire suppression activities. It is possibly the open nature of the landscape and the general absence of infrastructure, in comparison to the southern landscapes of Australia, that reduces the need for accurate knowledge of fire spread. Nevertheless there is a role for fire spread models in the vast areas of the spinifex landscapes to improve the timing of management programmes used to create patches within large areas burnt by previous wildfires.

Fire regimes

In the spinifex landscapes all fuels are above ground; fuel type (*sensu* Gill 1975) is uniformly a surface fuel. Fire intensity is most strongly affected by fuel load and by fire rate of spread. Local fuel loads can reach over 13 t ha^{-1} in the desert regions of WA (Burrows *et al.* 1991) and up to 40 t ha^{-1} on the sandstone escarpment regions of Kakadu National

Park (Russell-Smith *et al.* 1998). Quasi-steady-state fuel loads may be around 8–10 t ha^{-1} for studied sites (Burrows *et al.* 1991) but are lower (by inspection) elsewhere. Fuel accumulation rates vary within and between regions by spinifex species. There are slight differences between growth rates among the three dominant spinifex species on sandplain areas of central Australia with average annual rainfall between 200 and 450 mm (Griffin 1991). Similar differences were found between three other spinifex species in sandstone escarpments in Arnhem Land with average annual rainfall greater than 1000 mm (Russell-Smith *et al.* 1998). Fuel accumulation rates also vary between substrate types and landforms but information to quantify the observed differences is not available. Rates of spread vary widely. Maximum rates of spread ranged from 3.2 km h^{-1} (Griffin and Allan 1984) to 5.5 km h^{-1} (Burrows *et al.* 1991) to 7.2 km h^{-1} (Griffin and Allan 1993). Measured fire intensities reached 14 628 kW m^{-1} (Burrows *et al.* 1991) but this value would have been exceeded by several of the experimental fires by Griffin and Allan (1993). No studies have addressed the impact of fire intensity in spinifex landscapes.

Seasonality has been little studied but varies widely. In the tropical north fires occur in the long dry season from April to November with a bias toward the late dry season when diel (24 h) weather variability is lowest and fires can run for weeks (Gill *et al.* 1995; Russell-Smith *et al.* 1998). In central Australia most country is burnt during the summer months, October to January (Griffin *et al.* 1983). In the temperate region of spinifex landscapes, the summer–autumn period is when most fires occur. We know nothing of the effects of fire seasonality on the flora and fauna at any one locality.

Below we discuss how fire extent and patchiness has affected fire regimes, particularly fire frequency, in spinifex communities, using an extensive range of fire history data from different regions containing spinifex communities. It is appropriate to split fire extent into the two individual factors of fire size and fire patchiness. Fire size, patchiness and therefore frequency can be measured and estimated from satellite imagery. Regional comparisons can be made using spatial databases in a geographic information system (GIS). The numbers and size of fires create the area burnt per year, which in turn, is the determinant of fire interval on a yearly basis. The size distribution of fires may be called 'external patchiness' while the pattern of burnt and unburnt areas within a fire may be called 'internal patchiness' (Gill 1998). However, here 'fire patchiness' is defined as the heterogeneity of fire, or fuel, ages within a region, and it can be shown as patchiness on a time-since-fire map.

Databases

A central Australian fire history database was created in 1984 by visual interpretation of multi-temporal Landsat images and aerial photographs (Allan 1993). It covers the southern half of the NT, an area of 660 000 km² (36 1:250 000 topographic map sheets) and is an inventory of historic fire events, indicating the year of the last fire for each patch with an emphasis on the spinifex grasslands. The mapping of the fire events began with the most recent Landsat images and worked back through time to the oldest air photos. During the mid 1970s, when extensive fires were associated with the period of above-average rainfall, the time since fire had to be estimated by image interpretation due to a lack of available satellite images. A few patches in the southwest NT that were unburnt during the mid 1970s were dated to fires visible on the 1950 aerial photographs. A single update of the fire history map was made using Landsat images in 1985. Due to the high cost of Landsat imagery, subsequent annual updates used NOAA AVHRR (National Oceanic and Atmospheric Administration Advanced Very High Resolution Radiometer) satellite data for the period from 1986 to the present (Allan 1993). Despite the coarser 1 km² resolution, it was possible to map the significant regional fire events.

There are about 5000 patches within the original fire history mosaic for central Australia, and 2480 patches that represent individual fires after 1979 (Table 7.2). Most of the remaining patches have a fire age prior to 1979/80 but are not individual fires. The patches are either unburnt portions of a single fire age that has been broken up by more recent fires or

large areas representing numerous fire ages from the mid 1970s period. Some patches within the mosaic represent non-spinifex portions of the landscape, including salt lakes, mulga, non-spinifex plains and ranges.

The central Australian fire history database was divided into five regions for further analysis (Fig. 7.1). Each is dominated by spinifex grasslands but with different mean annual rainfall and landscape characteristics. The five regions, in order of decreasing mean annual rainfall, are the Tanami Desert, Northeast Central Australia, Western Desert, Southwest NT and Simpson Desert. All five regions are predominately Aboriginal land, but all include portions of pastoral land and conservation areas. The remaining area of central Australia is not dominated by spinifex and is primarily pastoral land. The Tanami Desert receives a mean annual rainfall of 431 mm, which has ranged from 145 mm to 913 mm over the past 20 years at Rabbit Flat (Jones and Weymouth 1997). The region is dominated by extensive sandplains, but includes a few large areas of sand dunes, palaeodrainage channels and associated salt lakes, and scattered very small rocky outcrops. Northeast Central Australia includes the Davenport Ranges in the west and the sandplain of the Wakaya Desert in the east, which is similar to the Tanami area, but is topographically distinct and without palaeodrainage channels. Barrow Creek/Kurundi receive a mean annual rainfall of 380 mm. The Western Desert is topographically similar to the Tanami, but receives less rainfall. The mean annual rainfall at Kintore is 321 mm, which has ranged from 68 mm to 884 mm since 1970 (Jones and Weymouth 1997). The Southwest NT is also topographically similar to the Tanami, including a mix of sandplains, dunefields and salt lakes but fewer distinct palaeodrainage channels. There are more rocky outcrops and associated areas of non-spinifex vegetation, including the significant features of Uluṟu and Kata Tjuṯa, and also the Petermann Ranges in the extreme southwest. Rainfall records from Uluṟu–Kata Tjuṯa National Park indicate an average annual rainfall of 300 mm, with a range from 111 mm to 776 mm since 1970 (Jones and Weymouth 1997). The Simpson Desert is dominated

Table 7.2. *Summary characteristics of the fire history of central Australia*

	Number of patches	Burnt (km²)	Burnt (%)	Largest patch burnt (km²)	Largest patch burnt (%)	Largest patch burnt (year)	Mean patch size (km²)	Number of patches per 100 km²
Tanami Desert (101 687 km²)								
All years	2428			4380	4.3	1979/80	41.9	2.39
1984/85	252	19 867	19.5	3512	3.5		78.8	
1983/84	168	15 109	14.9	3083	3		89.9	
1982/83	191	21 048	20.7	2985	2.9		110.2	
1981/82	46	2068	2	647	0.6		45	
1980/81	42	3119	3.1	1759	1.7		74.3	
1979/80	462	25 178	24.8	4380	4.3		54.5	
Northeast Central Australia (50 471 km²)								
All years	259			25 398	50.3	mid 1970s	194.9	0.51
1984/85	92	11 752	23.3	3403	6.7		127.7	
1983/84	8	89	0.2	41	0.08		11.1	
1982/83	9	730	1.4	564	1.1		81.1	
1981/82	1	61	0.1	61	0.1		60.8	
1980/81	0	0	0	0	0		0	
1979/80	22	3288	6.5	1197	2.4		149.5	
Western Desert (65 839 km²)								
All years	1365			19 217	29.2	1982/83	48.2	2.07
1984/85	247	14 087	21.4	2502	6.2		57	
1983/84	159	4296	6.5	270.8	0.4		27	
1982/83	238	23 493	35.7	19 217	29.2		98.7	
1981/82	34	1117	1.7	168.8	0.3		32.8	
1980/81	58	1122	1.7	294.6	0.4		19.3	
1979/80	46	420	0.6	69	0.1		9.1	
Southwest Northern Territory (58 878 km²)								
All years	439			18 088	30.7	mid 1970s	134.1	0.75
1984/85	79	6782	11.5	1453	2.5		85.8	
1983/84	50	11 762	20	5159	8.8		235.2	
1982/83	52	1611	2.7	831	1.4		31	
1981/82	1	16	0.03	16	0.03		16.1	
1980/81	29	395	0.7	78	0.13		13.6	
1979/80	1	7	0	7	0.01		7	
Simpson Desert (99 624 km²)								
All years	366			36 056	36.2	mid 1970s	272.2	0.37
1984/85	17	477	0.5	141	0.1		28.1	
1983/84	12	126	0.13	82	0.08		10.5	
1982/83	3	8	0.008	4	0.004		2.7	
1981/82	1	13	0.01	13	0.01		13	
1980/81	3	89	0.08	57	0.06		29.7	
1979/80	25	829	0.83	327	0.3		33.2	
Total (376 500 km²)								
All years	4857			36 056	9.6	mid 1970s	77.5	1.29

by long parallel sand dunes, which can extend for several hundred kilometres and range in height from 6 m to 15 m. The sand dunes are distinct from the other regions with a less vegetated, more mobile dune crest, dominated by sandhill canegrass (*Zygochloa paradoxa*). There are scattered rocky outcrops and a few linear ridges perpendicular to the southeast to northwest trending dunes. A relatively small area of sandplain occurs on the northern edge of the region. It is the driest area of central Australia. On its southern margin, Finke receives an annual rainfall of only 222 mm, but has experienced a range from 70 mm to 806 mm during the past 30 years. At Jervois on its northern margin the average annual rainfall increases to 312 mm (Jones and Weymouth 1997).

Fire history for these regions was complemented by a further set of databases, including those for other regions and two detailed sub-components of the central Australian regions (Fig. 7.1). A fire history database for the Tanami Desert Subregion covers an area of approximately 25 230 km² (Fig. 7.1). It is a subset from the central Australian fire history database but the time sequence was extended to 1996 using Landsat MSS images, rather than AVHRR images. It includes the area surrounding Rabbit Flat and its rainfall and landscape characteristics are the same as those of the Tanami Desert described above. Another additional database including Uluṟu–Kata Tjuṯa National Park covers an area of approximately 7000 km², for the period from 1950 to 1998 (Fig. 7.1). Its rainfall and landscape characteristics are the same as those of the Southwest NT region described above. The region's fires include a mix of both wildfires and management fires. Park management actively burn small strategic patches with the objective of maintaining a microcosm of the diversity of the surrounding region within the Park (Saxon 1984).

A fire history database was available for the Great Victoria Desert in WA (Fig. 7.1), covering an area of approximately 25 000 km². The fire history was created by Curry (1996) using Landsat MSS satellite data for the period from 1972 to 1994, although satellite data were not available or processed for each year. Spinifex (*Triodia basedowii*) with an overstorey of marble gums (*Eucalyptus gongylocarpa*) and *Acacia* and *Eremophila* shrubs (Pianka 1996) dominate this region. It also includes several salt lakes that are part of a palaeodrainage system. The average rainfall for the area is approximately 250 mm, with a mix of summer and winter rainfall events.

Finally, a fire history database for the Top End of the NT, north of 18° S, was created from a sequence of NOAA AVHRR satellite data from 1993 to 1998. Despite its short time frame, it covers a high rainfall zone with a short inter-fire period and provides an insight to fire in the northern semi-arid and tropical spinifex country. Three spinifex areas were selected from the fire history database using the 1:5 000 000 vegetation map of Australia (AUSLIG 1990). The areas represent the Arnhem Land Plateau area including a portion of Kakadu National Park, the Gregory National Park region and the Northern Tanami Desert region. In all three areas the coarseness of the regionalisation includes a mix of other communities more typical of the tropical savannas which can sustain a high frequency fire regime. The Arnhem Land area is a rugged dissected sandstone plateau with a mix of both *Symplectrodia* and *Triodia* spinifex. It is restricted to the area of Kakadu National Park and the adjacent Aboriginal land of the Arnhem Land region. The Gregory National Park region is primarily spinifex of the genus *Triodia* on sandstone uplands, but includes some areas of sandplain extending north from the Tanami Desert. It is a mix of land tenure, including National Park, Aboriginal reserve and pastoral lands. The Northern Tanami region lies to the south and east of the Gregory region and is predominantly *Triodia* species on sandplain within Aboriginal land. The region is a complex mix of vegetation communities with significant areas of Lancewood (*Acacia shirleyi*) and bulwaddy (*Macropteranthes kekwickii*) communities in a fine-grained mosaic. Both species are considered 'fire-sensitive' (Latz 1995). Although the species can recover from a fire, frequent fires will be detrimental and allow the spinifex communities to expand.

Fire size and extent

'Fire extent' is the area burnt by a single fire. It is dependent upon a number of factors including the continuity and homogeneity of fuel, fuel moisture content and weather conditions. Homogeneity is affected by features that will affect the growth of spinifex such as the habit of the individual species, the continuity of substrate and the antecedent rainfall pattern. Antecedent rainfall has been recognised as a factor related to area burnt in central Australia, especially if there has been sufficient rainfall to provide fuel to sustain combustion between the spinifex hummocks (McArthur 1972).

Very large fires have occurred in spinifex grasslands, but fire size, in general, shows significant variation between regions (Table 7.2) in central Australia. During the 6-year period from 1979/80 to 1984/85 there were 2480 individual fires mapped as part of the central Australia fire history database. The total area burnt was 176 565 km^2 and the average fire size was 71.2 km^2. The size of the fires ranged from 0.067 km^2 to 19 486.3 km^2. The Tanami and Western Desert regions have significantly higher numbers of fires and the smallest average fire size. The largest fire in the Tanami region was only one-quarter the size of the largest fire in the Western Desert region. There was a significant relationship ($r^2 = 0.73$) between the amount of rainfall in the preceding year and the area burnt per year. The relationship between rainfall and the number of fires was also significant but not as strong ($r^2 = 0.53$). Rainfall was represented as the total annual rainfall for central Australia from the Bureau of Meteorology ¼° raingrids (Jones and Weymouth 1997).

During the 6-year period, the total area burnt was highest for the three regions in the western part of central Australia (Table 7.2). There was 87 389 km^2 burnt in the Tanami, 44 535 km^2 in the Western Desert and 20 573 km^2 burnt in the Southwest NT region. This contrasts to only 15 920 km^2 burnt in the Northeast region and 1542 km^2 burnt in the Simpson Desert. The three western regions received above average rainfall during 1981/82, followed by average rainfall in 1982/83. In contrast both the Northeast region and the Simpson Desert had received several years of below-average rainfall until 1983/84, when above-average rainfall fell in the Northeast. During the next year, 1984/85, there

were 92 fires which burnt 11 752 km^2 in the Northeast, or 23.3% of the region (Table 7.2). The widespread nature of these fires can be explained by both the climatic patterns and previous fire patterns. The exceptionally wet years of the mid 1970s were followed by widespread fires throughout central Australia including this Northeast region. In the following 10 years spinifex began its slow recovery, through a period of average to below-average rainfall. Lack of fuel and the absence of human ignition sources, due to limited access, restricted the opportunities for fire in the region. The above-average rainfall in 1983/84 provided a surge of growth for both the spinifex (Griffin 1991) and the annual and perennial species that fill the gaps between the hummocks. The result was uniform fuel loads across a large area. It also coincided with a period of increased access to the region and the establishment of Aboriginal outstations.

Comparisons were made of the mean annual area burnt and its variance between regions in central Australia and elsewhere (Fig. 7.3). The Simpson and Great Victoria Desert are near the origin with low values of both mean proportion burnt per year and its variance. The mean and the variance increases with increasing rainfall for all five central Australian regions, using only the 1979 to 1985 period. The pattern is the same for the 25-year fire history databases in the Great Victoria, Uluru–Kata Tjuta and Tanami Desert subregion. The greatest separation occurs between the regions of the Top End of the NT. The variance is similar for both the Arnhem Land Plateau region and the Northern Tanami, but the mean proportion burnt per year is higher in the wetter Arnhem Land region.

Fire intervals

Fire interval is the time between successive fires at the same location. It is dependent on the locations, numbers and sizes of fires; i.e. locations and areas burnt each year. The frequency of fire across the entire area of the NT for the period 1993 to 1998 is shown in Fig. 7.4. A small proportion of the NT has an annual fire frequency and has been burnt six times in 6 years (see Williams *et al.* this

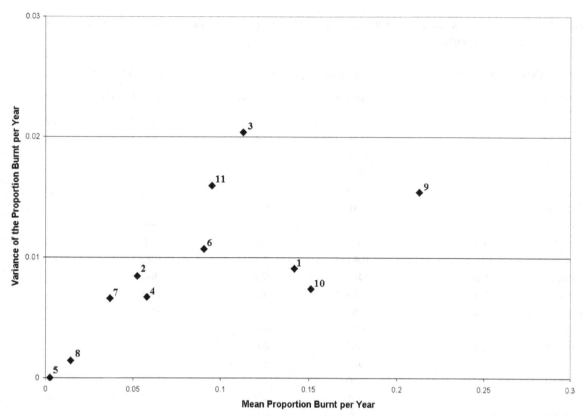

Fig. 7.3. Comparison of regional fire regimes. Fire history database areas: (1) Tanami Desert; (2) Northeast central Australia; (3) Western Desert; (4) Southwest NT; (5) Simpson Desert; (6) Tanami Desert subregion; (7) Uluṟu–Kata Tjuṯa National Park region; (8) Great Victoria Desert region; (9) Arnhem Land Plateau region; (10) Gregory National Park region; and (11) Northern Tanami region.

volume). Extending the time scale of fire history using the central Australian database shows that 41% of the NT has been burnt at least twice and has a known fire interval. As a complement to fire interval, fire return period, or fire cycle (Van Wagner 1978) is the time required to burn an area equivalent to the size of the region. It is the inverse of the average proportion of the region burnt per year if fires occur in a simple random pattern. The simple random pattern is assumed here.

Within central Australia for the period from 1979 to 1985, the average regional proportion burnt per year ranged from 14.2% in the Tanami to 0.25% in the Simpson Desert (Table 7.2). The fire return period is estimated as 7.0 years in the Tanami, 8.9 in the Western Desert and 17.2 in the Southwest region. These are logical values that follow the rainfall gradient in the western side of the southern NT. However, the fire return period for the Northeast central Australia region is 19 years and 385 years for the Simpson Desert. This highlights the high variability in both the Northeast region and the Simpson Desert and the need for longer datasets. As a comparison, extending the fire history database to 25 years for subsets of the regions (Table 7.3) changes the fire return period. The fire return period was 9 years in the Tanami Desert subregion, 32 years in the Uluṟu region and 77 years in the Great Victoria region (Table 7.3).

The data can also be expressed as numbers of

Table 7.3. *Comparison of annual areas burnt and number of times burnt for the Tanami Desert, Uluṟu–Kata Tjuṯa National Park and Great Victoria Desert regions*

Tanami Desert region		Uluṟu–Kata Tjuṯa N.P. region		Great Victoria Desert region[a]	
Time (years)	Burnt (%)	Time (years)	Burnt (%)	Time (years)	Burnt (%)
				Pre 1972	5.2
		1976	34.4		
		1977	0.0		
		1978	0.0		
		1979	0.0	1972–79	13.2
1979/80	35.0	1980	0.0	1980	0.1
1980/81	1.1	1981	0.0	1981	2.1
1981/82	1.5	1982	0.1	1982	6.4
1982/83	19.9	1983	2.4	1983	1.2
1983/84	20.6	1984	9.9	1984	1.0
1984/85	13.7	1985	3.8	1985	1.5
1985/86	1.3	1986	3.7	1986	0.9
		1987	0.3	1987	1.6
1986/88	2.4	1988	0.0		
		1989	6.5		
1988/90	0.5	1990	17.4		
		1991	3.9	1988–1991	1.8
1990/92	12.6	1992	0.6		
1992/93	0.9	1993	0.2		
1993/94	19.9	1994	0.3	1992–1994	0.0
1994/95	18.8	1995	0.0		
1995/96	5.7	1996	0.0		
		1997	0.4		
		1998	0.1		
Average proportion of region burnt per year					
1979–1994	11.0	1972–1998	3.2	1972–1994	1.3
Apparent fire return period (years)					
	9.1		31.6		76.9
Number of times burnt (%)					
0	0.6	0	14.2	0	65.7
1	16.9	1	55.0	1	33.5
2	47.4	2	30.0	2	0.8
3	33.7	3	0.8	3	0
4	1.3	4	0	4	0

Note:

[a] Data from Curry (1996).

times during a study interval that a pixel has burnt. In the Tanami subregion, about 33% of the sample area had been burnt three times in 25 years (Table 7.3). In the Uluṟu region 30% had been burnt twice, but in the Great Victoria Desert region 33.5% had only been burnt once (Table 7.3). There is a very high annual variability of area burnt, which reduces the significance of regional variations based on average values. For example, in the Uluṟu region, the annual variability for the period 1972–98 ranges from no fires for several years to 35.6% of the region burning in a single year. The average annual area burnt was 3.2%. No fire occurred in 10 of the 27 years of the sequence. There are differences in the proportion of fire ages between the Park and the surrounding region, however the proportion of the number of times burnt is similar.

The Uluṟu fire history dataset provides information on the minimum time and rainfall between fires. The first fire within the perimeter of the large area burnt in November 1976 was a wildfire in the eastern portion of the Park in November 1986. The fire burnt within a corridor of Triodia pungens along Britten-Jones creek, which would have received additional moisture associated with the palaeo-drainage channel (Jacobson et al. 1989). Griffin's (1991) models of spinifex recovery after fire indicate a more rapid response in T. pungens communities in comparison to T. basedowii communities that dominate the rest of the Uluṟu region. The cumulative rainfall for the 10-year period between the two fires was 3087 mm, or a rain*time value of 30870, i.e. number of years since last fire multiplied by cumulative rainfall in mm (Griffin et al. 1990). At the time of the 1986 wildfire, fire managers at Uluṟu were having difficulty burning areas of the 1976 fire age under less extreme weather conditions. The situation changed after 960 mm of rainfall in the 2 years between August 1987 and July 1989 (Masters 1993). By 1990 the 1976 fire age would carry a fire in moderate weather conditions and fire managers were able to burn more areas. By 1990 the cumulative rainfall had reached 4384 mm and the rain*time value was 61376 after 14 years. Numerous wildfires occurred within the region and by 1991 the fuel load in the 1976 fire age had reached a level that required

caution at all times. Masters (1991) measured an increase in spinifex cover in the Triodia basedowii community from 33% to 47% in the time between 1987 and 1990.

Fires occur in the non-spinifex portions of arid Australia. These fires are less frequent and the fuel is entirely from non-spinifex grasses and forbs, which is usually a mixture of annuals and perennials. Big fires in these areas can only occur about a year after a sequence of high rainfall events. Subsequently fuel loads diminish by either grazing or natural decomposition. Smaller fires can occur soon after single effective rainfall events. This was described for several habitat types by Griffin et al. (1983). In the Uluṟu region, a large proportion (71%) of non-spinifex country was burnt once (1972–98), mainly by the large wildfire in 1976 (cf. about 50% of spinifex communities). Only a very small proportion (3%) of non-spinifex was burnt again in the subsequent 22 years, by fires associated with the 1988–90 period of above-average rainfall, whereas about 30% of spinifex communities were burnt twice.

The fire history information in the Tanami Desert subregion indicates that small areas will carry a fire after 4–5 years and a minimum cumulative rainfall between 1676 and 1821 mm or a rain*time value (Griffin et al. 1990) between 6704 and 9105. The minimum value remains doubtful due to the 2-year interval between updates of the fire history. The average minimum time between fires is 5 years with an average cumulative rainfall of 2028 mm. This gives a rain*time value of 10139, which is considerably higher than the value of 6300 suggested by Griffin et al. (1990) as the minimum required for a patch to accumulate sufficient fuel to burn again under extreme summer conditions.

In the spinifex grasslands of the Top End of the NT, between latitudes 11° S and 18° S, the average fire interval, based on 6 years of data, is shorter than that for the desert regions as a result of the higher and less variable rainfall. In the Arnhem Land Plateau area only 20% of the region remained unburnt through 6 years, in contrast to 35% in the Gregory National Park region and 52% in the Northern Tanami. In Arnhem Land, the 1996 fires were the most extensive. In the Gregory region

Table 7.4. *Minimum interval between fires and mean cumulative rainfall for the spinifex landscapes of the Arnhem Land, Gregory National Park and Northern Tanami regions, using 1993 to 1998 AVHRR fire history*

| Minimum interval between fires (years) | Arnhem Land region | | | Gregory National Park region | | | Northern Tanami region | | |
| | % of region | Cumulative rainfall (mm) | | % of region | Cumulative rainfall (mm) | | % of region | Cumulative rainfall (mm) | |
		Mean	S.D.		Mean	S.D.		Mean	S.D.
1	13.9	1151	296	4.0	749	117	1.2	553	111
2	21.3	2411	549	14.7	1490	162	27.4	1252	121
3	11.0	3590	812	25.0	2261	226	6.0	1976	212
4	24.4	4719	1027	10.5	2967	290	1.6	2427	260
5	5.1	5950	1408	6.4	3740	325	6.4	3126	198
6	4.7	7145	1572	4.3	4157	457	5.7	3303	331
Unburnt	19.6	7015	1792	35.0	4359	524	51.6	3293	337

there was a more even distribution of fires, whereas in the Northern Tanami the large proportion burnt in 1994 was part of even larger fires (Fig. 7.2). For the 26% of Arnhem Land, which had been burnt twice, only 4% had a fire interval of 1 year, whereas 12.6% had an interval of 2 years and 9.5% had an interval of 3 years. In contrast, the areas burnt twice in the Gregory and Northern Tanami regions had similar patterns, both dominated by areas with a fire interval of 3 years. The lower rainfall and different spinifex species result in a longer interval to accumulate fuel load sufficient to carry another fire.

The fire history was also intersected with rainfall grids produced by the Bureau of Meteorology (Jones and Weymouth 1997) to check for regional differences between fire interval and cumulative rainfall (Table 7.4). For example, in Arnhem Land, 13.9% of the region burnt with a minimum time between fires of only 1 year and an average of 1151 mm of rainfall between fires. However over 50% of the region had a minimum inter-fire period between 2 and 4 years, which is comparable to the results from the Landsat-based fire history of the sandstone escarpment area of Kakadu National Park (Russell-Smith *et al.* 1998). The rates for fuel accumulation in different regional sandstone habitats typically are

sufficient to support intense late dry-season fires every 1 to 3 years, although in areas dominated only by spinifex the fuel accumulation is slightly slower and the interfire period is 2 to 4 years. In the Gregory region, 4% of the region burnt with a minimum interval of 1 year. However this probably represents an inaccuracy associated with the coarse resolution of the dataset. Visual interpretation of the patterns of the images show fires do not carry into areas burnt the previous year. About 15% of the region experienced at least two fires within 6 years. There was a minimum fire interval of 2 years and an average cumulative rainfall between fires of 1490 mm. In the Northern Tanami, about 27% of the region was burnt with a minimum fire interval of 2 years and an average of 1252 mm of cumulative rainfall (Table 7.4).

Fire patchiness

Patchiness of fire is equated here to be the heterogeneity of fire ages within a region. Knowledge of patchiness is important to the objectives of most fire management programmes, which are to reduce the occurrence of extensive wildfires. These principles and strategies, frequently termed patch burning, have been incorporated into the management ethos of most conservation areas and some pastoral lands in the arid zone. Achievement of this objective is linked to creating patches of low fuel loads, which will disrupt the spread of wildfires, restrict their overall extent and leave unburnt areas within their overall perimeter.

The patchiness of a single fire depends on the continuity and homogeneity of fuel, which in spinifex grasslands is related to the time since the last fire, and accumulation rate of fuel, both spinifex and non-spinifex, in association with preceding rainfall. The occurrence of a fire also requires an ignition source. In most seasons, areas of low fuel quantity, created by previous fires, frequently restrict the spread of a fire. The fire frequency information in central Australia indicates that the time-scale changes regionally. In the Tanami subregion, a second fire rarely occurs in areas with a fire age less than 5 years. In the Uluṟu region the time-scale is up to 10 years, whereas in the Great Victoria Desert region fire ages up to 20 years were observed to restrict the spread of subsequent fires (Curry 1996).

There are spatial and seasonal patterns in the two sources of fire ignition in spinifex grasslands, lightning and human ignition. Lightning is seasonal but randomly distributed and in central Australia occurs most frequently from October to January (Griffin et al. 1983). Human ignition of fires in spinifex is less seasonal and its distribution is linked to lines of travel, primarily current roads and tracks. It has been speculated that prior to European settlement, Aboriginal burning patterns would have been more dispersed across the landscape, although still aligned to popular travel or hunting routes (Griffin and Allan 1986).

In the 36 years after 1953 the size, distribution, frequency and intensity of fires in a part of the Western Desert in Western Australia changed dramatically (Burrows and Christensen 1990). In 1953 small patch burns dominated the fire mosaic but, by 1986, the area was dominated by a few large wildfires. The fine-scale patchiness of the fire mosaic disappeared after the nomadic lifestyle of the Aboriginal people in the region ended. Although the nomadic lifestyle of Aboriginal people in central Australia has also ended their influence on the patchiness of the fire mosaic was still obvious along the roads, tracks and outstations in the Western and Tanami Desert regions of the NT in 1985 (Allan and Griffin 1986).

There are obvious differences in the patchiness of the five central Australian fire regions, represented by the area histograms for each of the fire ages from the mid 1970s to 1995 (Fig. 7.5). The Tanami region has the most uniform distribution of fire ages in contrast to the highly skewed distributions in the Simpson Desert and the Northeast region. Data from the Tanami subregion indicate that this is due to the influence of fires since 1981, and in particular fires after 1990 burning within the extensive area burnt in 1979/80. In all regions the impact of the extensive fires which occurred during the mid 1970s still persists and these fires represent the largest area of each region. However, random models would predict the largest burnt area persisting to be of recent fire age (McCarthy and Cary, this volume).

For the spinifex grasslands of the Top End of the NT, patchiness can be represented by the time since the last fire within the 6-year fire history data (Table 7.5). The potential for the area to carry a fire is associated with the time since the last fire. For the Northern Tanami, 75% of the region has not been burnt for more than 5 years and the potential for a widespread wildfire would be increasing. In the Arnhem Land region the diversity of fire ages is more uniform. In the Gregory region 43% has a fire age greater than 5 years.

Influence of fire regimes on plants and animals

Despite the habitat being low in nutrients (Stafford Smith and Morton 1990), the world's richest reptile fauna (Pianka 1969; Morton and James 1988) and a large proportion of the Australian mammal and bird fauna occurred, until recently, in spinifex grasslands (Morton 1990; Reid and Fleming 1992). Most of the animals were small to medium-sized and less than 5 kg weight. Populations of the larger animals including emus (Dromaius novaehollandiae), euros (Macropus robustus) and red kangaroos (Macropus rufus) are still scattered throughout,

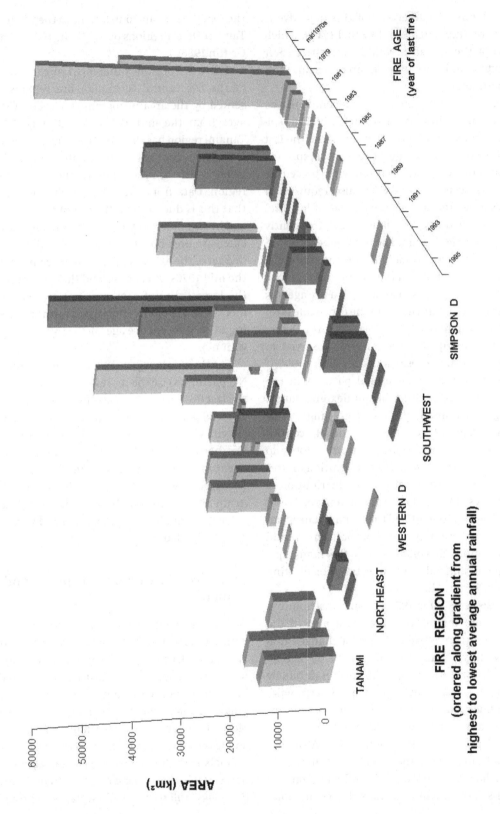

Fig. 7.5. Fire patchiness in central Australia fire history, as represented by area of land in annual classes of date of last fire, up to 1995.

Table 7.5. *Regional comparison of time since last fire for spinifex landscapes of the Arnhem Land, Gregory National Park and Northern Tanami regions, using 1993 to 1998 AVHRR fire history*

Fire age: year last burnt	Arnhem Land region (% of area)	Gregory N.P. region (% of area)	Northern Tanami region (% of area)
1998	15.8	13.8	6.3
1997	14.6	16.1	12.4
1996	36.1	10.2	2.3
1995	4.2	16.0	0.5
1994	7.5	9.0	26.7
1993	2.3	0.0	0.2
Pre 1993	19.6	35.0	51.6

although they occur more commonly in association with pockets of non-spinifex grasslands. A number of introduced feral species such as feral cats (*Felis catus*), foxes (*Vulpes vulpes*), rabbits (*Oryctolagus cuniculus*), camels (*Camelus dromederius*) and donkeys (*Equus asinus*) have demonstrated no difficulty in extending their range to include spinifex grasslands. Whereas more than half of the medium-sized mammal species have declined or become extinct in the last 50 years, the reptile and bird fauna has remained mostly intact (Morton 1990). However, it is predicted that components of the bird fauna will decline in the near future (Reid and Fleming 1992; Woinarski and Recher 1997).

It has been suggested that the loss of medium-sized mammal species from the arid and semi-arid part of Australia has been a consequence of a changed fire regime (Bolton and Latz 1978; Saxon 1983; Saxon and Dudzinski 1984; Burbidge *et al.* 1988; Johnson *et al.* 1989). It was recognised that mature spinifex provided shelter for medium-sized mammals and regenerating areas provided a greater variety and a more nutritious array of plants. Frequent fires, set by Aboriginal people, are believed to have produced a fine-grain mosaic of different ages. Movement of these people on to settlements and the loss of traditional lifestyles have resulted in this fire regime being replaced by large wildfires (Burrows and Christensen 1990). In other words, it has been proposed that habitat of a partic-

ular 'grain-size' was required for the survival of the medium-sized mammals and this had been altered substantially in the last 50–100 years. It has also been recognised that a fire regime of a particular grain-size is required for survival of a suite of threatened 'fire-sensitive' plants (Bowman and Latz 1993). Certain plant species are known to be easily killed by fire and unable to mature and set seed if the fire interval is too short. In addition, it is believed that the colonisation of burnt areas by spinifex has allowed fire to penetrate further into 'fire-sensitive' communities leading eventually to their elimination.

While observations of the spatial patterning of 'fire-sensitive' vegetation to avoid fire have been made (Bowman and Latz 1993), few studies have undertaken repetitive sampling to look at changes through time, or changes associated with different fire histories. Similarly, much has been written about the importance of the implementation of patch burning and a return to Aboriginal burning practice but few studies have tested the notion that fire grain-size has resulted in the loss of medium-sized animals in spinifex grasslands or the suite of 'fire-sensitive' plants. The following sections examine the evidence surrounding these issues and the prevailing notion that the patch size and juxtaposition may be important for many plant and animal species and, therefore, may necessitate consideration for effective management of biodiversity.

Plants

Change in plant composition and structure in spinifex communities following fire has been examined by Zimmer (1940), Burbidge (1943), Suijdendorp (1981), Jacobs (1984), Allan and Baker (1990) and Griffin (Griffin *et al.* 1990; Griffin 1992) primarily using space for time sampling. A number of other studies including Allan and Griffin (1986), Burrows and Christensen (1990), Masters (1991) and Southgate and Masters (1996) have reported on the successional changes in burnt patches as part of fauna surveys.

Following fire, the species able to resprout from meristems, roots or rhizomes regain some cover almost immediately. Given reasonable rainfall, the early stages in the succession are dominated by a suite of short-lived grasses and forbs. The richness of plant species on recently burnt areas can be two- to three-fold greater than in neighbouring patches of mature spinifex (Allan and Baker 1990). Some of the short-lived grasses and forbs such as *Solanum* spp., *Yakirra australiense, Dactyloctenium radulans* and *Portulaca* spp. produce abundant seed and fruit that feature prominently in the diet of medium-sized mammals (Southgate 1990; Lundie-Jenkins 1993) and, in the past, have provided important foods for traditional Aboriginal people (Latz 1995). Latz (1995) noted that the suite of plants germinating after fire will differ depending on the fire regime factors of fire intensity and season of fire. A response following a low-intensity winter fire in an area with primarily summer rainfall favours the re-establishment of perennial plants at the expense of the short-lived 'fire-weeds'.

With time and additional rainfall, spinifex and other perennials begin to replace short-lived species and vertical structural diversity increases as the woody plants emerge above the spinifex (Griffin 1992). Over long periods without fire, increased growth of perennial shrubs and trees can lead to the death of spinifex and a decrease in vegetative ground cover (Jacobs 1984). Spinifex hummocks characteristically die out in the centre first and eventually form rings 2–4 m in diameter. The senescence and decline in the dominance of spinifex can

result in a slight increase in species richness (Allan and Baker 1990).

The observed increase of species diversity after fire was the basis for Suijdendorp's (1967, 1981) recommendations for the use of fire to improve pasture in spinifex communities in the northwest of Western Australia. As a test for this widely reported recommendation, Holm and Allen (1988) measured changes to the nutritive value of soft spinifex (*Triodia pungens*) after fire. The common grass species that returned to the site after burning (post-fire increasers) were no more nutritious than soft spinifex and it seems that little is gained from manipulating spinifex pastures through burning if the aim is to encourage alternative grass species. Nevertheless, they recorded the typical short-term post-fire response associated with an increased species diversity and biomass of non-spinifex plants.

'Fire-sensitive' plant communities

Spinifex grasslands frequently include, or are adjacent to, areas of 'fire-sensitive' Acacia woodlands, such as mulga (*Acacia aneura*), lancewood (*A. shirleyi*) and the rare undoolyana wattle (*A. undoolyana*). Fires that commonly start from within spinifex communities can severely affect these species. Mulga has low tolerance to fire (Hodgkinson and Griffin 1982). Mature trees and seedlings readily succumb to moderately intense fire and generally do not resprout. Germination of mulga can be stimulated by fire, but very intense fires, frequently repeated fires or lack of follow-up rain to initiate seedling growth may deplete the seed supply. High-intensity wildfires, particularly at a frequency that prevents replenishment of the mulga seed store in the soil, can cause considerable long-term damage. These changes in vegetation may not occur gradually but at discrete periods in time, either in association with large rainfall events or droughts (Griffin and Friedel 1985). These periods are apparently crucial to proper management. Changes to the balance between spinifex and mulga communities have been observed in the Pilbara region of Western Australia (Start 1986). In the absence of Aboriginal burning, large areas of mulga woodland (including the associated animals and plants) are being replaced by spinifex-dominated communities. Subsequent work described the complex mosaic of spinifex and mulga (van Leeuwen *et al.* 1995) in the region and found that the perimeter of the mulga was being eroded by fires from the spinifex (van Etten 1988). The floristics between the two communities across the boundary were not distinctly different, but were qualified with lack of information on the fire regime and number of fires and age of communities. The recommendation of van Leeuwen *et al.* (1995) was for a more sensitive fire management programme to be developed to maintain a balance between the two communities.

Fire history data from both the Tanami subregion and the central Australian regional fire history have provided the opportunity to make observations of the impact of fire on isolated mulga communities in the region. During the initial compilation of the central Australian regional fire history isolated mulga communities were mapped due to their distinctive spectral characteristics on Landsat satellite images. Most patches were isolated, possibly remnants of more extensive mulga communities from long ago, which may have remained as a result of a fire regime in the past associated with Aboriginal activity. A few other patches represent the edges of mulga lands adjacent to the spinifex areas where their distribution and patchiness is also a function of different soil and landform characteristics. The subsequent updates of the fire history from 1984 to 1995 indicated that fire has had an impact on these areas. For instance, within the Tanami subregion the area of mulga mapped in 1984 was relatively insignificant. There were six patches covering a total of 8.2 km², or 0.03% of the 25 230 km² study area. After fires in 1984/85 the unburnt portion of Mulga was reduced to 5.3 km²; after fires in 1993/94 that unburnt portion was 4.6 km²; after fires in 1994/95 the unburnt portion was 2.6 km² and after fires in 1995/96 the unburnt portion was 2.2 km²

The situation was similar for the much larger Tanami region of the central Australian regional fire history. Within the 119 414km² area of the Tanami region there was 369 km² of mulga, or 0.3% of the area. Fires in 1984/85 burnt 37 km² of mulga. It was not possible to extend the information to 1995 because of the change from Landsat to NOAA AVHRR imagery and the associated loss of spatial accuracy. Nevertheless, the data provide a general indication that the remnant mulga patches are being burnt, rather than protected, by the current fire regime. Although mulga can recover from a fire, small patches are subject to encroachment by spinifex and their future is obviously uncertain. The story for undoolyana wattle and lancewood is similar to that of mulga. The large and conspicuous 'fire-sensitive' plants may not be the only species at risk. Numerous species persist in the spinifex communities only as seeds or dormant root-stock between fires (Griffin 1984). Ephemeral plant species richness may be declining because there are fewer fire patches and therefore fewer chances for plants to coincide with suitable rainfall conditions to allow seed set.

Animals

The distribution and abundance of animal species in spinifex communities has been widely studied. Several different approaches have been used, including general fauna surveys, fauna studies designed to determine response to fire events and autecological studies. Unfortunately the lack of information to describe the characteristics of the fire regime has restricted the fire descriptors to simple measures of fire age or time since last fire. Originally fauna surveys had the simple aim of capturing and identifying which animals occurred within a study area. Surveys were short-term and provided only limited information on population size and its fluctuations but did contribute information on the spatial variability of species occurrence. Fauna surveys have provided significant information on the diversity of species within the spinifex communities and comparisons between the earliest collections and more recent ones have highlighted the dramatic decline and extinction of many species. The Horn Expedition in 1894 and that of H.H. Finlayson in the 1930s were the last to collect several of the medium-sized mammals which have become extinct (Kerle and Fleming 1996). In the Great Victoria Desert studies in the 1960s recorded exceptionally high numbers of reptile species within spinifex communities and inspired subsequent studies in other desert regions of Australia (Pianka 1969). Surveys within the Simpson and Tanami Deserts by Gibson (1986) and Gibson and Cole (1988) and Great Sandy Desert by McKenzie and Youngson (1983) sampled mammals, birds and reptiles, to identify priority areas for conservation. The information from these surveys was not linked to spatial information on the fire regimes of the regions.

More recent studies have aimed to address specific questions associated with the impact of environmental parameters on community composition and abundance of threatened species of faunal assemblages (Masters 1993, 1996; Reid et al. 1993; Reid and Hobbs 1996; Southgate and Masters 1996; McAlpin 1997). Several clear trends have emerged from this work. Some mammal, bird and reptile species display a preference for early post-fire states and others for later states. Reptiles such as the central netted dragon (Ctenophorous nuchalis) and military dragon (C. isolepis) were more common on recently burnt habitat whereas species such as the desert skink (Ctenotus pianka) and jewelled gecko (Diplodactylus elderi) were more common on or confined to unburnt patches. Birds like the banded whiteface (Aphelocephala nigricincta) and crimson chat (Ephthianura tricolor) were abundant in recently burnt habitat whereas the striated grass wren (Amytornis striatus) and variegated fairy wren (Malurus lambert) were largely confined to areas of mature spinifex as was the dasyurid (Ningaui ridei). Some species such as the singing honey-eater (Lichenostomus virescens) and a fossorial skink (Lerista bipes) were found to be generalists and abundant across different aged fire states.

It was found that the richness of mammal and reptile species was generally greater in mature spinifex compared to recently burnt areas. Masters (1993, 1996) found that only two of eight mammal species and two of 40 reptile species were found in greater abundance on more recently burnt patches compared to the long unburnt spinifex. Reid et al. (1993) found 13 species of reptile and three bird species were confined to mature spinifex but none to recently burnt areas.

Rainfall has a highly significant and interactive effect on the response of animals to fire (Reid et al. 1993). Large rainfall events produced an explosive increase in the abundance of birds, both as a rapid invasion by mobile species and as a steady increase of resident species. Nomadic birds attained their highest abundance on the recently burnt patches after heavy rains that promoted vigorous growth of ephemeral plants. Small mammals also responded rapidly following big rains with a 10- to 100-fold increase in abundance and a two-fold increase in species richness. It is interesting that Mus domesticus, generally described as a colonising species and an invader of burnt country in the wetter parts of Australia, was unable to survive in spinifex grasslands except during the very wet years. It became more abundant in the mature spinifex communities than in recently burnt areas (Masters 1993) and was also the only species that declined significantly following fire in a small-mammal community that included Pseudomys hermannsburgensis, Notomys alexsis and three Sminthopsis species (Southgate and Masters 1996).

Overall, the changes brought about by exceptionally large rainfall events far exceed those produced by fire in isolation. However, the conclusion from these studies has been that patch burning in spinifex grasslands maximises species diversity of small mammals, reptiles and birds. The regenerating areas act as fire breaks and ensure that patches of mature spinifex remain. Reid et al. (1993) provided five rules for the use of fire in the management of fauna at Uluru. These are:

1. that the majority of the standing mulga and much of the regenerating 1976 mulga be protected from fire;

2. that the majority of 1976 aged spinifex be protected from fire at least until surveys to search for rare species are concluded;

3. that mallee–spinifex areas are protected from fire until they have been surveyed for rare species;

4. that surveys of these habitats be carried out to assess and map their habitat quality for wildlife, and that recommendations arising from the

research be formally written into the fire management strategy, and

5. that not withstanding the above points, the patch-burn strategy be vigorously pursued to promote landscape and faunal diversity.

Although the creation of a mosaic of fire ages has been advocated, the optimum patch size has not been clearly defined and is absent from prescriptions. The vegetation patches examined in these studies have varied in size from a few hectares to hundreds of square kilometers.

Medium-sized mammals and the fire-mosaic hypothesis

Few field studies have attempted to determine empirically whether the decline and extinction of medium-sized mammals was the result of a changed fire regime from that previously maintained by traditional Aboriginal burning practices. Short and Turner (1994) investigated the response of three medium-sized mammals to vegetation patterns that ranged from comparatively uniform to highly diverse spinifex grasslands on Barrow Island. Neither the scale of mosaic nor the disturbance associated with oil extraction and exploration activity were found to have any significant effect on the condition, numbers or reproductive status of the three medium-sized mammals considered, the brush-tailed possum (*Trichosurus vulpecula*), golden bandicoot (*Isoodon auratus*) and burrowing bettong (*Bettongia lesueur*). However, extension of these results to refute the fire-mosaic hypothesis is arguable mainly on the grounds that fire frequency (50–100 years) and the absence of predators on Barrow Island create conditions very different to those on the mainland.

Some autecological studies concerning remaining threatened species in spinifex grasslands on mainland Australia have aimed to identify the importance of fire in the distribution of these species and determine the patch or grain-size of fire regimes that produce most favourable habitat. Some studies have used GIS spatial datasets, including the central Australian fire history and other spatial habitat characteristics, to develop habitat

suitability models for specific species. Masters *et al.* (1997) concluded that habitat variables relating to the distribution of spinifex species were the most important influences on the distribution of mulgara (*Dasycercus cristicauda*) in the Tanami Desert of the NT. The species was most frequently encountered in *Triodia basedowii* but also occurred in *T. pungens* in the presence of *Melaleuca* spp. or rocky features. There was extremely low probability of encountering mulgara where *T. schinzii* was dominant. Spacing between hummocks and hummock form were important attributes and associated with the type of spinifex species. Fire age, as time since last fire, and fire patchiness were poor predictors of mulgara distribution mainly because mulgaras were found in big blocks of mature spinifex (mainly *T. basedowii*) and in small patches of mature spinifex within recently burnt blocks. It was evident that the resolution of the fire history database from AVHRR images was too coarse for mapping the fire size relevant to the 'response grain' (Wiens 1990) of this species. Satellite images with a finer resolution, such as Landsat, may be required to map suitable habitat.

A similar approach to Masters *et al.* (1997) was adopted to examine the importance of fire to the distribution of the bilby (*Macrotis lagotis*). The bilby was once very widespread on the Australian mainland and the pattern of its decline is more consistent with the spread of introduced predators (cats and foxes) and herbivores (rabbits and stock). At the present time there is little scope for management of predators and competitors within the NT and WA component of the species' current range due to the extent, remoteness and relative inaccessibility of the region. Fire management is the most practical management option available (Southgate *et al.* 1997). It is probable that the resolution of the fire history database from AVHRR images will be appropriate for studies of the bilby. It is larger and more mobile than the mulgara. Nevertheless, preliminary analysis indicates a poor relationship between bilby distribution and fire age *per se*. A stronger association between distribution and the categories of substrate was evident. Bilby sign was encountered far more frequently in drainage and laterite (or rock feature) substrates compared to sandplain and dunefield areas. However, fire may play a crucial role in allowing animals to recolonise drainage line and laterite habitats that are often isolated and only a minor component of the overall landscape.

The evidence in support of the patch-burn mosaic hypothesis is equivocal to date. Although there is little doubt that fire is important to the availability of food and shelter for many animals in spinifex grasslands, the signals from the remnant medium-sized mammal species with regard to fire age are indistinct. This is not surprising since the very persistence of the remnant species indicates their resilience to the changed fire regime. There is often a strong overriding effect caused by seasonal weather conditions and introduced predators and herbivores. Critical examination of the patch-burn mosaic hypothesis will require the use of species that have disappeared from a locality. It will need to determine whether reintroduced populations of these species can prosper in the absence of a fine-grained fire pattern where introduced predator and herbivore numbers are significantly reduced.

Fire management

Numerous land management agencies have adopted programmes for planning and implementing fire management activities. The principles or priorities for fire management on conservation areas are:

- to protect infrastructure, including historic and cultural sites;
- to maintain a full range of natural communities and species; and
- to minimise the risk of wildfires burning major portions of the reserve.

The implementation of these principles requires knowledge of the vegetation, its sensitivity to fire regimes, its current fire history and successional state and its spatial patterning. The best approach is to develop and maintain a GIS database to direct the formulation of annual fire action plans. This approach has been adopted and implemented on many conservation areas with varying degrees of commitment and success. There have been no attempts to extend this approach to the extensive

areas of spinifex-dominated sandplain and dune-fields. The development of new management systems must be scientifically sound, but also practical and tailored to what managers can realisitically achieve. The problems that must be overcome include difficulty of access, lack of spatial data associated with fire-sensitive vegetation and adjacent ecosystems, insufficient funds, the absence of a single responsible land manager, and the need to develop culturally appropriate methods which incorporate western ideas and objectives with Aboriginal aspirations and management philosophies.

A possible solution is the recommendation of Morton et al. (1995) for the integration of pastoral production and conservation within Australia's rangelands. Their approach is based on a model of the Australian landscape, as pockets of fertile areas within a predominantly infertile background. They suggest that society would need to financially assist land managers to achieve ecologically sustainable land management goals, especially on land with marginal economic value such as the spinifex grasslands. They recommend the mapping and classifying of the landscape into the three broad categories of resource-rich, resource-poor and poor with scattered patches of richness, and then to stratify the landscape into ecosystem types. Although their concept covers all of Australia's rangelands, it could be applied to the spinifex grasslands for the development of a strategy for fire management. Programmes that involve the least amount of intervention will be most likely to be initiated and sustained. The necessary path is to first identify areas where the fire pattern is naturally patchy. These frequently appear to coincide with resource-rich areas. The second step is to focus fire management activities in the surrounding areas to create areas of a fine-grained mosaic linking the resource-rich areas. Although the implementation of this approach may be expensive, more extensive fire management programmes must be initiated.

The information from many ecological studies has not been incorporated into fire management programmes. Frequently this is due to the lack of a spatial context because most ecological information is based on relatively site-specific locations. It should be the responsibility of the ecologist, not the land manager, to make the appropriate extrapolations to regional areas. In many areas action is required immediately and it is necessary to start using available data and improve the approach through time. In many areas this will require speculation about the location of resource-rich areas and past fire regimes and an assessment of current land use conflicts in order to initiate a management programme based on the current principles (e.g. protection of assets and fire-sensitive habitats) used in many conservation areas. The approach can be used at a range of spatial scales, from small-scale intensive management programmes up to a regional desert scale for the Western and Simpson Deserts. At the small-scale, Duguid and Schunke (1998) and Duguid (1999) formulated a management programme for *Acacia undoolyana* communities in the East Macdonnell Ranges to provide protection to the isolated pockets of the 'fire-sensitive' community. At the regional scale more general programmes to minimise the potential for extensive wildfires may have to consider differences across climate gradients and adjust the management and fire frequency time-scales. It will also be necessary to compile databases for large areas with minimal spatial data, following the approach of Masters *et al.* (1997) and incorporating data from a variety of scientific disciplines, especially geology. However, the implication that the coarse resolution of AVHRR data is insufficient poses a new challenge to land managers to maintain a finer resolution fire mosaic over the vast areas of the spinifex landscapes of Australia.

Effective fire management of the spinifex landscapes of Australia requires the integration of a diversity of information, including climatic, geomorphic, geologic, botanical, faunal, social and historical. For example, the use of geological information was important in building spatial datasets of the Tanami Desert (Masters *et al.* 1997). Additional geological information could include aeromagnetic and radiometric data (English 1998), night-time AVHRR images (Tapley 1990) and the hyperspectral (AVIRIS) datasets. Co-operative arrangements between the mining companies and ecologists could be arranged in order to obtain cheaper access to the geological data while providing ecological information to their management programmes that face the difficult issue of managing large areas of the landscape with few people and minimal opportunities for economic gain. A workshop at Uluru–Kata Tjuta National Park took this approach to bring together experts from a variety of fields to help contribute information for a new plan of management (Woodcock 1997).

Land tenure is another factor affecting the implementation and maintenance of scientifically determined fire management programmes. Most of the spinifex landscapes are Aboriginal tenure (1.22 million km^2) and the impacts of fires starting on Aboriginal lands and extending out into other land tenures and vegetation communities is a commonly expressed concern. Successful fire management programmes require:

- co-operation between fire management agencies and Aboriginal people, both individually and via the Land Councils;
- greater effort to achieve a mutual understanding of fire issues and appropriate compromises; and
- implementation of aerial control burning (ACB) programmes in spinifex country. ACB programmes within the extensive areas of spinifex sandplain and dunefields have been tested (Burrows and van Didden 1991). The programs must deal with the difficulty of scheduling operations at an ecologically and climatically appropriate time, the cost of the operation to cover large areas and a lack of support from Aboriginal people, who have expressed their desire to burn from the ground.

Further research and information is required on the definition of fire weather in arid lands including spinifex communities (Gill *et al.* 1995). More information on diel weather parameters is needed. Several studies on the climate, weather and fires in the NT (Love and Downey 1985; Tapper *et al.* 1993) have looked at regional patterns and general conditions but fire behaviour is linked to specific local weather conditions. There is a need for detailed continuous (10 to 30 minute) records of weather parameters in association with actual fires, especially controlled burns which rely on subtle diel changes in wind speed, temperature and relative humidity to restrict and suppress fires. As an example, fire managers need to know when subtle changes in seasonal conditions have crossed a threshold that will allow fires to continue through the night and suppression activities may be required.

The analysis of spatial patterning of vegetation moisture or curing state in the vegetation derived from satellite images needs more detailed calibration. Maps of natural fuel breaks and their persistence in time also need to be added to the spatial databases available for land managers when developing and implementing fire management programmes. Fuel loads in adjacent communities, specifically 'fire-sensitive' ones like mulga, can be used to control fire spread and may be critical to their long-term survival.

Bushfire agencies must play a key role in coordinating programmes and act as the watchdog to ensure the programmes appropriate to the seasonal conditions are implemented and that co-ordination across tenure boundaries is effective. Fire management must be widespread and will not be effective in isolation. The successful implementation of fire management must be based on documented strategies and programmes, which have an endorsement and commitment from all stakeholders. The adaptive management approach should be used to monitor the success of the program and provide the appropriate feedback to ensure its ongoing effectiveness. The monitoring programme must include a mix of spatial information on the fire regime in combination with biodiversity studies.

Acknowledgements

The authors thank the Bushfires Council of the Northern Territory, especially Russell Anderson and Jeremy Russell-Smith, for their support of this work. The quality of the final product was improved by the helpful comments and suggestions of the official reviewers, Malcolm Gill, Anne Kerle, Mike Fleming and Pip Masters. Ross Bradstock, Jann Williams and Malcolm Gill deserve thanks for the invitation to contribute and their confidence in our ability to meet their expectations and commitments. Most importantly, Coral Allan earned acknowledgement for her unfailing support and encouragement that ensured the completion of this chapter.

References

Allan, G. E. (1993). The Fire History of Central Australia. CSIRO/CCNT Bushfire Research Project, vol. 4. CSIRO/CCNT Technical Report, CSIRO, Alice Springs.

Allan, G.E., and Baker, L. (1990). Uluru (Ayers Rock–Mt Olga) National Park: an assessment of a fire management programme. *Proceedings of the Ecological Society of Australia* **16**, 215–220.

Allan, G. E., and Griffin, G. F. (1986). Fire ecology of the hummock grasslands of central Australia. In *Proceedings of the 4th Conference of the Australian Rangeland Society* pp. 126–129. (Armidale, NSW.)

AUSLIG (Australian Surveying and Land Information Group) (1990). *Vegetation Atlas of Australian Resources, Digital Dataset.* (AUSMAP, Department of Administrative Services: Canberra.)

AUSLIG (Australian Surveying and Land Information Group) (1993). *State and Northern Territory Boundary, Digital Dataset.* (AUSMAP, Department of Administrative Services: Canberra.)

Bolton, B. L., and Latz, P. K. (1978). The western hare-wallaby *Lagorchestes hirsutus* (Gould) (Macropodictae) in the Tanami Desert. *Australian Wildlife Research* **5**, 285–293.

Bowman, D. M. J. S., and Latz, P. K. (1993). Ecology of *Callitris glaucophylla* (Cypressaceae) on the Macdonnell Ranges, central Australia. *Australian Journal of Botany* **41**, 217–225.

Bradstock, R. A., and Gill, A. M. (1993). Fire in semi-arid, mallee shrublands: size of flames from discrete fuel arrays and their role in the spread of fire. *International Journal of Wildland Fire* **3**, 3–12.

Burbidge, A. A., Johnson, K. A., Fuller, P. J., and Southgate, R. I. (1988). Aboriginal knowledge of the mammals of the Central Deserts of Australia. *Australian Wildlife Research* **15**, 9–39.

Burbidge, N. T. (1943). Ecological succession observed during regeneration of *Triodia pungens* R.Br. after burning. *Journal of the Royal Society of Western Australia* **28**, 149–156.

Burrows, N. D., and Christensen, P. E. S. (1990). A survey of Aboriginal fire patterns in the Western Desert of Australia. In *Fire and the Environment: Ecological and Cultural Perspectives* (eds. S. C. Nodvin and T. A. Waldorp) pp. 297–305. U.S. Department of Agriculture Forest Service, General Technical Report SE-69. (Southeastern Forest Experimental Station: Asheville, NC.)

Burrows, N. D., and van Didden, G. (1991). Patch-burning desert nature reserves in Western Australia using aircraft. *International Journal of Wildland Fire* **1**, 49–55.

Burrows, N. D., Ward, B., and Robinson, A. (1991). Fire behaviour in Spinifex fuels on the Gibson Desert Nature Reserve, Western Australia. *Journal of Arid Environments* **20**, 189–204.

Cheney, N. P., and Sullivan, A. (1997). *Grassfires: Fuel, Weather and Fire Behaviour*. (CSIRO: Melbourne.)

Curry, J. C. (1996). A time series analysis of the spectral response of fires and vegetation regrowth in Landsat imagery of Australia's Great Victoria Desert: an initial analysis. MSc thesis, University of Texas at Austin, USA.

Duguid, A. (1999). Protecting *Acacia undoolyana* from wildfires: an example of off-park conservation from central Australia. In *Proceedings of the Australian Bushfire Conference*, 7–9 July 1999, Albury, Australia, pp. 127–131. (Charles Sturt University: Albury.)

Duguid, A., and Schunke, D. (1998). Final report on Project no. 290 *Acacia undoolyana* (Undoolya Wattle) Species Recovery Plan. Internal Report to Threatened Species and Communities Section, Environment Australia, Parks and Wildlife Commission of the Northern Territory, Alice Springs.

English, P. M. (1998). *Cainozoic Geology and Hydrogeology of Uluṟu–Kata Tjuṯa National Park: Geoscience for Land and Water Management*. (Australian Geological Survey Organisation Monograph: Canberra.)

Flannery, T. F. (1994). *The Future Eaters*. (Reed Books: Port Melbourne.)

Gibson, D. F. (1986). A biological survey of the Tanami Desert in the Northern Territory. Conservation Commission of the Northern Territory, Technical Report no. 30, Alice Springs.

Gibson, D. F., and Cole, J. R. (1988). A biological survey of the Northern Simpson Desert. Conservation Commission of the Northern Territory, Technical Report no. 40, Alice Springs.

Gill, A. M. (1975). Fire and the Australian flora: A review. *Australian Forestry* **38**, 4–25.

Gill, A. M. (1981). Adaptive responses of Australian vascular plant species to fires. In *Fire and the Australian Biota* (eds. A. M. Gill, R. H. Groves and I. R. Noble) pp. 243–272. (Australian Academy of Science: Canberra.)

Gill, A. M. (1998). An hierarchy of fire effects: impact of fire regimes on landscapes. In *Proceedings of 3rd International Conference of Forest Fire Research and 14th Conference of Fire and Forest Meteorology* (ed. D. X. Viegas) pp. 129–144. (Luso: Portugal.)

Gill, A. M., Burrows, N. D., and Bradstock, R. A. (1995). Fire modelling and fire weather in an Australian desert. *CALMScience Supplement* **4**, 29–34.

Graetz, R. D., Tongway, D. J., and Pech, R. P. (1982). An ecological classification of the lands comprising the southern Simpson Desert and its margins. CSIRO Rangelands Research Unit, Technical Memorandum 82/2, Deniliquin.

Griffin, G. F. (1984). Hummock grasslands. In *Management of Australia's Rangelands* (eds. G. N. Harrington, A. D. Wilson and M. D. Young) pp. 271–284. (CSIRO: Melbourne.)

Griffin, G. F. (1985). Manual for collection and analysis of data for fire behaviour predictions: applications of a fire management strategy at Uluṟu National Park. CSIRO Division of Wildlife Rangelands Research, Technical Memorandum no. 22, Alice Springs.

Griffin, G. F. (1991). Characteristics of three Spinifex alliances in central Australia. *Journal of Vegetation Science* **1**, 435–444.

Griffin, G. F. (1992). Will it burn – should it burn?: management of the Spinifex grasslands of inland Australia. In *Desertified Grasslands, Their Biology and Management*, Linnean Society Symposium Series no. 13 (ed. G. P. Chapman) pp. 63–76. (Academic Press: London.)

Griffin, G. F., and Allan, G. E. (1984). Fire behaviour. In *Anticipating the Inevitable: A Patch-Burn Strategy for Fire Management at Uluṟu (Ayers Rock–Mt Olga) National Park* (ed. E. C. Saxon) pp. 55–68. (CSIRO Australia: Melbourne.)

Griffin, G. F., and Allan, G. (1986). Fire and the management of Aboriginal owned lands in central Australia. In *Science and Technology for Aboriginal*

Development, Project Report no. 3 (eds B. D. Foran and B. Walker) (CSIRO Australia: Melbourne.)

Griffin, G. F., and Allan, G. (1993). Airborne monitoring system data and auxiliary information for the experimental fire sites in the hummock grassland of Central Australia. CSIRO/CCNT Bushfire Research Project, vol. 18. CSIRO/CCNT Technical Report, CSIRO, Alice Springs.

Griffin, G. F., and Friedel, M. H. (1985). Discontinuous change in central Australia: some implications of major ecological events for land management. *Journal of Arid Environments* **9**, 63–80.

Griffin, G. F., Price, N. F., and Portlock, H. F. (1983). Wildfires in the central Australian rangelands 1970–1980. *Journal of Environmental Management* **17**, 311–323.

Griffin, G. F., Morton, S. R., and Allan, G. E. (1990). Fire-created patch-dynamics for conservation management in the hummock grasslands of central Australia. In *Proceedings of the International Symposium on Grasslands Vegetation*, Huhhot, China (ed. Yang Hanxi) pp. 239–247. (Science Press: Beijing.)

Hodgkinson, K. C., and Griffin, G .F. (1982). Adaptation of shrub species to fires in the arid zone. In *Evolution of the Flora and Fauna of Arid Australia* (eds. W. R. Barker and P. J. M. Greenslade) pp. 145–52. (Peacock Publications: Frewville, SA.)

Holm, A. McR., and Allen, R. J. (1988). Seasonal changes in the nutritive value of grass species in Spinifex pastures of Western Australia. *Australian Rangeland Journal* **10**, 60–64.

Jacobs, S. W. L. (1984) Spinifex. In *Arid Australia*. (eds. H. G. Cogger and E. E. Cameron) pp. 131–142. (Australian Museum: Sydney.)

Jacobs, S. W. L. (1985). A new grass genus from Australia. *Kew Bulletin* **40**, 659–661.

Jacobs, S. W. L. (1992). Spinifex (*Triodia, Plectrachne, Symplectrodia*, and *Monodia*: Poaceae) in Australia. In *Desertified Grasslands, Their Biology and Management*, Linnean Society Symposium Series no. 13 (ed. G. P. Chapman) pp. 47–62. (Academic Press: London.)

Jacobson, G., Lau, G. J., McDonald, P. S., and Jankowski, J. (1989). Hydrogeology of the Lake Amadeus–Ayers Rock Region, Northern Territory. Bureau of Mineral Resources, Bulletin no. 230, Canberra.

Jessop, P. J., and King, D. (1997). The land resources of New Crown Station. Department of Lands Planning and Environment, Technical Memorandum no. TM 96/18, Alice Springs.

Johnson, K. A., Burbidge, A. A., and McKenzie, N. L. (1989). Australian macropods: status, causes of decline, and future research and management. In *Kangaroos, Wallabies and Rat-Kangaroos* (eds. I. Hume, G. Grigg and P. Jarman) pp. 641–657. (Surrey Beatty and Sons: Chipping Norton, NSW.)

Jones, D., and Weymouth, G. (1997). An Australian monthly rainfall dataset. Bureau of Meteorology, Technical Report no. 70, Melbourne.

Kerle, J. A., and Fleming, M. R. (1996). A history of vertebrate fauna observations in central Australia: their value for conservation. In *Exploring Central Australia: Society, the Environment and the 1894 Horn Expedition* (eds. S. R. Morton and D. J. Mulvaney) pp. 341–366. (Surrey Beatty and Sons: Chipping Norton, NSW.)

Kimber, R. (1983). Black lightning: Aborigines and fire in central Australia and the Western Desert. *Archaeology in Oceania* **18**, 38–45.

Latz, P. K. (1995). *Bushfires and Bushtucker: Aboriginal Plant Use in Central Australia*. (IAD Press: Alice Springs.)

Latz, P. K., and Griffin, G. F. (1978). Changes in Aboriginal land management in relation to fire and to food plants in central Australia. In *The Nutrition of Aborigines in Relation to the Ecosystem of Central Australia* (eds. B. S. Hetzel and H. J. Frith) pp. 77–85. (CSIRO: Melbourne.)

Lazarides, M. (1984). New taxa of tropical Australian grasses (Poaceae). *Nuytsia* **5**, 273–303.

Lazarides, M. (1997) A revision of *Triodia* including *Plectrachne* (Poaceae, Eragrostideae, Triodiinae). *Australian Systematic Botany* **10**, 381–489.

Love, G., and Downey, A. (1985). Fire weather in central Australia. In *Proceedings of Fire Weather Services Conference*, May, Adelaide, pp. 7.1–7.29. (Australian Bureau of Meteorology: Canberra.)

Lundie-Jenkins, G. (1993). Ecology of the Rufous Hare-wallaby, *Lagorchestes hirsutus* Gould (Marsupialia: Macropodidae), in the Tanami Desert, Northern Territory. I. Patterns of habitat use. *Wildlife Research* **20**, 457–476.

Masters, P. (1991). Fire-driven succession: the effects on lizards and small mammals in Spinifex grasslands. MAppSc thesis, Curtin University of Technology, Perth.

Masters, P. (1993). The effect of fire-driven succession and rainfall on small mammals in Spinifex grassland at Uluru National Park, Northern Territory. *Wildlife Research* **20**, 803–813.

Masters, P. (1996). The effect of fire-driven succession on reptiles in Spinifex grassland at Uluru National Park, Northern Territory. *Wildlife Research* **23**, 39–48.

Masters, P., Nano, T., Southgate, R., Allan, G., and Reid, J. (1997). *The Mulgara: Its Distribution in Relation to Landscape Type, Fire Age, Predators and Geology in the Tanami Desert. Report to Environment Australia* (Parks and Wildlife Commission NT: Alice Springs.)

McAlpin, S. (1997). Conservation of the Great Desert Skink, *Egernia kintorei*, at Uluru–Kata Tjuta National Park, N.T. Australian Nature Conservation Agency, Project Report no. DN67, Canberra.

McArthur, A. G. (1972). Fire control in arid and semi-arid lands of Australia. In *The Use of Trees and Shrubs in the Dry Country of Australia* (ed. N. Hall) pp. 488–516. (Australian Government Publishing Service: Canberra.)

McArthur, A. G. (1977). Grassland Fire Danger Meter Mark V. Country Fire Authority, Victoria.

McCarthy, M. A., and Cary, G. J. (2001). Fire regimes in landscapes: models and realities. In *Flammable Australia: The Fire Regimes and Biodiversity of a Continent* (eds. R. A. Bradstock, J. E. Williams and A. M. Gill) pp. 77–93. (Cambridge University Press: Cambridge.)

McKenzie, N. L., and Youngson, W. K. (1983). Mammals. In *Wildlife of the Great Sandy Desert, Western Australia* (eds. A. A. Burbidge and N. L. McKenzie) pp. 62–93. Wildlife Research Bulletin of Western Australia no. 12, Perth.

Mercer, G. (1997). Simulation of fire spread in discrete fuels using flame contact. In *Bushfire97: Proceedings of the Australasian Bushfire Conference* (eds. B. J. McKaige, R. J. Williams and W. M. Waggitt) poster presentation. (CSIRO: Darwin.)

Morton, S. R. (1990). The impact of European settlement on the vertebrate animals of arid Australia: a conceptual model. *Proceedings of the Ecological Society of Australia* **16**, 201–213.

Morton, S. R., and James, C. D. (1988). The diversity and abundance of lizards in arid Australia: a new hypothesis. *American Naturalist* **132**, 237–256.

Morton, S. R., Stafford Smith, D. M., Friedel, M. H., Griffin, G. F., and Pickup, G. (1995). The stewardship of arid Australia: ecology and landscape management. *Journal of Environmental Management* **43**, 195–217.

Noble, I. R., Bary, G. A. V., and Gill, A. M. (1980). McArthur's

fire danger meters expressed as equations. *Australian Journal of Ecology* **5**, 201–203.

Perry, R. A., Mabbutt, J. A., Litchfield, W. H., Quinlan, T., Lazarides, M., Jones, N. O., Slatyer, R. O., Stewart, G. A., Bateman, W., and Ryan, G. R. (1962). *General Report on Lands of the Alice Springs Area, Northern Territory, 1956–57,* Land Research Series no. 6. (CSIRO: Melbourne.)

Pianka, E. R. (1969). Habitat specificity, speciation, and species density in Australian desert lizards. *Ecology* **50**, 498–502.

Pianka, E. R. (1996). Long-term changes in lizard assemblages in the Great Victoria Desert: dynamic habitat mosaics in response to wildfires. In *Long-term Studies of Vertebrate Communities* (eds. M. L. Cody and J. A. Smallwood) pp. 191–215. (Academic Press: San Diego.)

Purdie, R. (1984). *Land Systems of the Simpson Desert Region*, Natural Resources Series no. 2, Division of Water and Land Resources. (CSIRO: Melbourne.)

Reid, J., and Fleming, M. R. (1992) The conservation statue of birds in arid Australia. *Rangelands Journal* **14**, 65–91.

Reid, J. R. W., and Hobbs, T. J. (eds.) (1996). *Monitoring the Vertebrate Fauna of Uluru National Park, Phase II: Final Report*, Consultancy Report to Australia Nature Conservation Agency. (CSIRO: Alice Springs.)

Reid, J. R. W., Kerle, J. A., and Morton, S. R. (eds.) (1993). *Uluru–Kata–Tjuta Fauna: The Distribution and Abundance of Vertebrate Fauna of Uluru (Ayers Rock–Mt Olga) National Park, NT*. (Australian National Parks and Wildlife Service: Canberra.)

Russell-Smith, J., Ryan, P. G., Klessa, D., Waight, G., and Harwood, R. (1998) Fire frequency and fire-sensitive vegetation of the sandstone Arnhem Plateau, monsoonal northern Australia: application of a remotely-sensed fire history. *Journal of Applied Ecology* **35**, 829–846.

Saxon, E. C. (1983). Mapping the habitats of rare animals in the Tanami Wildlife Sanctuary (Central Australia): an application of satellite imagery. *Biological Conservation* **27**, 243–257.

Saxon, E. C. (ed.) (1984). *Anticipating the Inevitable: A Patch-Burn Strategy for Fire Management at Uluru (Ayers Rock–Mt Olga) National Park*. (CSIRO: Melbourne.)

Saxon, E. C., and Dudzinski, M. L. (1984). Biological survey and reserve design by Landsat mapped ecoclines: a catastrophe theory approach. *Australian Journal of Ecology* **9**, 117–123.

Short, J., and Turner, B. (1994) A test of the vegetation

mosaic hypothesis: a hypothesis to explain the decline and extinction of Australian mammals. *Conservation Biology* **8**, 439–449.

Southgate, R. I. (1990). The distribution and abundance of the bilby. MSc thesis, Macquarie University, Sydney.

Southgate, R.I., and Masters, P. (1996). Fluctuations of rodent populations in response to rainfall and fire in a central Australian hummock grassland dominated by *Plectrachne schinzii*. *Wildlife Research* **23**, 289–303.

Southgate, R.I., Allan, G.E., Paltridge, R., Masters, P. and Nano, T. (1997). Management and monitoring of bilby populations with the application of landscape, rainfall and fire patterns: preliminary results. In *Bushfire97: Proceedings of the Australasian Bushfire Conference* (eds. B. J. McKaige, R. J. Williams and W. M. Waggitt) pp. 140–145. (CSIRO: Darwin.)

Stafford Smith, D. M., and Morton, S. R. (1990). A framework for the ecology of arid Australia. *Journal of Arid Environments* **18**, 255–278.

Start, A. N. (1986). Status and management of Mulga in the Pilbara region of Western Australia. In *The Mulga Lands* (ed. P. Sattler) pp. 136–138. (Royal Society of Queensland: Brisbane.)

Suijdendorp, H. (1967). A study of the influence of management practices on Spinifex (*Triodia pungens*) grazing. MSc thesis, University of Western Australia.

Suijdendorp, H. (1981). Responses of the hummock grasslands of northwestern Australia to fire. In *Fire and the Australian Biota* (eds. A. M. Gill, R. H. Groves and I. R. Noble) pp. 417–426. (Australian Academy of Science: Canberra.)

Tapley, I. J. (1990). Relation between landform and soil parameters and NOAA-AVHRR derived temperatures across palaeochannels in the Great Sandy Desert region of Western Australia. In *Proceedings of 5th Australasian Remote Sensing Conference*, pp. 928–941. (Australian Remote Sensing Conference Inc. Perth.)

Tapper, N. J., Garden, G., Gill, J., and Fernon, J. (1993). The climatology and meteorology of high fire danger in the Northern Territory. *Rangeland Journal* **15**, 339–351.

van Etten, E. J. B. (1988). Environmental factors affecting the boundary between mulga (*Acacia aneura*) and hummock grassland (*Triodia* spp.) communities. MAppSc thesis, Curtin University of Technology, Western Australia.

van Leeuwen, S. J., Start, A. N., Bromilow, R. B., and Fuller, P. J. (1995). Fire and the floristics of Mulga woodlands in the Hammersley ranges, Western Australia. In *Ecological Research and Management in the Mulgalands* (eds. M. J. Page and T. S. Beutel) pp. 169–175. (University of Queensland, Gatton College: Gatton.)

Van Wagner, C. E. (1978). Age-class distribution and the forest fire cycle. *Canadian Journal of Forest Research* **8**, 220–227.

Weber, R. O., Mercer, G. N., Mahon, G. J., Catchpole, W. R., and Gill, A. M. (1995). Predicting maximum inter-hummock distance for fire spread. In *Bushfire95: Proceedings of the Australasian Bushfire Conference*. (Parks and Wildlife Service: Hobart.)

Whelan, R. J. (1995). *The Ecology of Fire*. (Cambridge University Press: Cambridge.)

Wiens, J. A. (1990). On the use of 'grain' and 'grain size' in ecology. *Functional Ecology* **4**, 720.

Williams, R. J., Griffith, A. D., and Allan, G. E. (2001). Fire regimes and biodiversity in the savannas of northern Australia. In *Flammable Australia: The Fire Regimes and Biodiversity of a Continent* (eds. R. A. Bradstock, J. E. Williams and A. M. Gill) pp. 281–304. (Cambridge University Press: Cambridge.)

Wilson, B. A., Brocklehurst, P. S., Clark, M. J., and Dickinson, K. J. M. (1990). Vegetation survey of the Northern Territory, Australia. Conservation Commission of the Northern Territory, Technical Report no. 49, Darwin.

Winkworth, R. E. (1967). The composition of several arid Spinifex grasslands of central Australia in relation to rainfall, soil water relations, and nutrients. *Australian Journal of Botany* **15**, 107–130.

Woinarski, J. C. Z., and Recher, H. F. (1997). Impact and response: a review of the effects of fire on the Australian avifauna. *Pacific Conservation Biology* **3**, 183–205.

Woodcock, L. G. (ed.) (1997). *Proceedings of the 'Back to the Future' Natural Resources Research Workshop, 27–29 August 1997.'* Uluru-Kata Tjuta National Park (Australian Geological Survey Organisation: Canberra.)

Zimmer, W. J. (1940). Plant invasions in the Mallee. *Victorian Naturalist* **56**, 143–147.

8

The role of fire regimes in temperate lowland grasslands of southeastern Australia

IAN D. LUNT AND JOHN W. MORGAN

Abstract

The role of fire regimes in temperate lowland grasslands in Australia is discussed. Over 99% of native grasslands in most regions have been destroyed or highly modified for agriculture. Fire is excluded from most agricultural operations and is regularly used in only a small proportion of grassland remnants. Most of the grassland fire literature focusses on frequently burnt grasslands dominated by *Themeda triandra* in western Victoria, and little information is available on grasslands dominated by other species. *Themeda* grasslands tolerate frequent burning, and many remnants with a diverse flora are burnt at 1–3-year intervals. The fire ecology of *Themeda* grasslands is discussed in detail, and a conceptual model is developed which highlights the importance of disturbances such as fire for preventing competitive exclusion by the dominant grasses. Relatively little fire research has been conducted in Australian temperate grasslands compared to sclerophyllous ecosystems. Issues requiring further research include: experimental studies of ecosystem responses to particular fire regimes; identifying processes that promote seedling recruitment; and the effects of different fire regimes on grassland fauna and soil nutrient fluxes.

Introduction

Before European colonisation, natural and Aboriginal-lit fires may have played a pivotal role in controlling the ecological structure and functioning of the grassy plains of southeastern Australia. The importance of fire regimes in the original landscape will probably never be known, however, since the natural grasslands were rapidly and completely transformed to agricultural crops and pastures after settlement. Nowadays, biomass on the plains is controlled by introduced stock and mechanical harvesters, not by flames. Frequent burning of native grasslands is increasingly becoming an anachronistic oddity, confined to small, isolated sites on roadsides and rail easements where fragments of the original vegetation have been protected from landscape disturbance regimes such as those due to cropping and grazing. Many of these small fragments are critically important for the conservation of Australia's temperate grassland estate (Stuwe 1986; McDougall and Kirkpatrick 1994).

In this chapter, we review the effects of fire regimes on temperate, lowland (non-alpine) grassland communities in southeastern Australia. Whilst we focus on natural, treeless grasslands, we also include relevant information from secondary grasslands derived from other ecosystems such as grassy woodlands and shrublands. Our focus is on native vegetation, and grasslands composed predominantly of exotic species are excluded.

Compared to fire-prone ecosystems such as heathlands and wet sclerophyll forests, very little research has been undertaken on fire effects in temperate Australian grasslands. Since virtually all published information is on flora rather than fauna, and most published studies are from grasslands dominated by *Themeda triandra* (kangaroo grass), especially on the western basalt plains of Victoria, botanical studies from this ecosystem inevitably dominate this review.

Ecosystem distribution

Prior to European colonisation, natural grasslands occupied large areas of southeastern Australia, most notably on the basalt plains of western Victoria, in south-central South Australia (SA), and in the Riverina region of northern Victoria. The Riverina region of south-central New South Wales (NSW) is generally considered to have been originally dominated by woodlands of *Acacia pendula* and *Atriplex nummularia*, with both species being rapidly eliminated by heavy stock grazing to form extensive secondary grasslands (Moore 1953; Leigh and Noble 1972). However, little accurate information is available on original vegetation structure in the region, and it has been argued that at least some areas in NSW may always have been treeless grasslands (McDougall and Kirkpatrick 1994). Smaller areas of grassland occurred in the Wimmera and Gippsland in Victoria, on the NSW Southern Tablelands and Tasmanian Midlands (Fig. 8.1).

These grassland regions share a suite of similar climatic and geomorphological features including: relatively low rainfall (usually 400–600 mm mean annual rainfall, grading into drier, semi-arid ecosystems in the NSW Riverina and in South Australia); flat to gently undulating topography; and heavily textured, poorly drained soils of relatively high fertility. The combination of fertile soils on gentle slopes permitted widespread agricultural development after European colonisation, causing a dramatic decline in the extent and integrity of natural grasslands. Today, lowland temperate grasslands in all regions are threatened ecosystems. Most remnants are small, isolated, and managed for purposes other than biodiversity conservation (McDougall and Kirkpatrick 1994).

Plant composition

Natural temperate grasslands are typically dominated by perennial tussock grasses with inter-tussock forbs, smaller grasses and sedges (see Fig. 8.2). Dominant grass genera include *Themeda*, *Austrodanthonia* (formerly *Danthonia*), *Austrostipa* (formerly *Stipa*) and *Poa*. *Themeda triandra* or *Poa* species dominate remnants in areas of relatively high rainfall (e.g. southern Victoria, NSW Southern Tablelands and the Tasmanian Midlands), whereas *Austrodanthonia* and *Austrostipa* species co-dominate the Riverine plains and south-central SA

(with *Lomandra* species). *Austrodanthonia* and *Austrostipa* species are also widespread in higher-rainfall regions in dry, stony sites and where stock grazing has depleted the original dominant *Themeda triandra* (C. W. E. Moore 1953; R. M. Moore 1962; Willis 1964; Sharp 1995). Tussock grasses often contribute most of the above-ground (and presumably below-ground) biomass in undisturbed *Themeda*- and *Poa*-dominated communities (see Fig. 8.3, colour plate; McDougall 1989; Morgan and Lunt 1999). By contrast, biomass is usually considerably lower in *Austrodanthonia* and *Austrostipa* grasslands (Williams 1970; Williams and Roe 1975).

In all but the driest regions (i.e. NSW Riverina and south-central SA), most native species are herbaceous perennials (Costin 1954; Willis 1964; Benson 1994; McDougall and Kirkpatrick 1994; Tremont and McIntyre 1994). By contrast, native annuals (many shared by semi-arid ecosystems further inland) are abundant in south-central South Australia (Davies 1997) and the Riverina (Foreman 1996; Benson *et al.* 1997). In all regions, most perennial species are hemicryptophytes or geophytes which possess regenerative buds at or below the soil surface (McIntyre *et al.* 1995; Morgan 1998a). Indeed, almost 90% of species in frequently burnt *Themeda* grasslands in southern Victoria belong to one of these two life-form groups (Morgan 1999). In contrast to the predominance of perennial species in the native flora, many exotic species are annuals (therophytes). Exotic annual grasses and forbs are now abundant in most grassland remnants in all regions (McDougall and Kirkpatrick 1994).

Grassland fire regimes

Grassland burning regimes can conveniently be divided into three broad phases: (1) pre-European fire regimes; (2) traditional European regimes (late 1800s–1970s); and (3) recent burning practices (post-1980), which include the use of fire for conservation management.

Pre-European fire regimes

The nature of pre-European burning regimes in grasslands remains largely unknown. It is widely

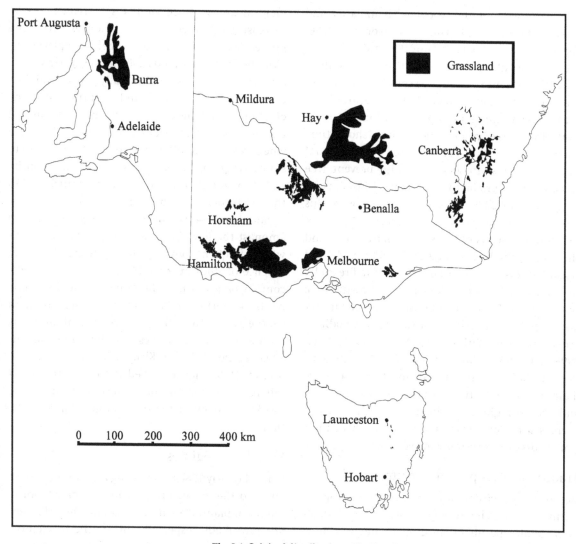

Fig. 8.1. Original distribution of lowland temperate grasslands in south-eastern Australia (based on Lunt *et al.* 1998 and Rehwinkel 1997). Shaded areas supported treeless grasslands at the time of European settlement except the NSW Riverina, which was originally dominated by woodlands of *Acacia pendula* and *Atriplex nummularia*. Both were rapidly eliminated by heavy stock grazing to form extensive secondary grasslands (Moore 1953; Leigh and Noble 1972). Most native vegetation in most grassland regions has now been transformed to agricultural crops and pastures.

assumed that grasslands were often burnt by fires ignited by Aborigines and lightning, but little information exists to support this belief, apart from many brief historical references which mention widespread burning by Aborigines.

Considerable debate has taken place in other con-

tinents over whether grassland areas were 'natural' vegetation formations or artefacts of the burning practices of indigenous peoples (e.g. Sauer 1950; Stewart 1956; Vogl 1974; Axelrod 1985). Several recent regional historical studies from southeastern Australia have demonstrated high fidelities between regional grassland and woodland boundaries and soil distributions (e.g. Fensham 1989; Foreman 1996; Fensham and Fairfax 1997; Lunt 1997a; Morcom and Westbrooke 1998). Consequently, it seems most likely that absence of trees was controlled by a combination of soil and regional climatic features, with pre-European fire regimes playing a minor role in controlling tree regeneration. Over short time periods, competition between tree seedlings and established grass tussocks may also prevent tree recruitment in productive grasslands (Fensham and Kirkpatrick 1992), but it is difficult to envisage this mechanism operating over millennia.

A number of recent invasions by native trees and shrubs into small grasslands on hillside 'balds' and coastal headlands have been attributed to fire exclusion since European settlement (e.g. Howitt 1890; Fensham and Fairfax 1996; Costello 1998). Similar processes have occurred in a variety of woodland ecosystems (e.g. Withers and Ashton 1977; Allen 1998; Lunt 1998). Thus, it is possible that frequent burning before European colonisation may have maintained small, treeless areas in mountainous and coastal regions and in some woodlands, but there is no evidence of large-scale tree or shrub invasion across extensive grassland ecosystems.

Traditional European fire regimes

Despite the paucity of information on pre-European burning regimes, it is widely assumed that the dominant landscape disturbance agent in grasslands has changed from burning under low grazing pressures before European colonisation, to grazing with little burning since (Groves and Williams 1981; Mack 1989). All temperate grassland regions in southeastern Australia are now utilised for stock grazing and other agricultural pursuits, including irrigation and cropping. In most temperate grassland regions, few grazing enterprises involve regular burning.

At local scales, however, many small sites in some regions have been burnt frequently for over a century as a means of fire protection for rural communities. Annual and biennial summer burning have traditionally been conducted by local fire brigades for fire control along roadsides in western

Victoria, at least since the 1940s (Morgan 1998a, b). Increasingly, however, these practices are being phased out and burning is being replaced by ploughed firebreaks and herbicides. Similarly, state railway authorities traditionally burnt rail-line easements (often annually in late spring) since the creation of rail-lines in the 19th century. Annual burning was a vital means of fire control in an era when steam trains often started fires from sparks in smoke. Ongoing rail closures (which have continued throughout 20th century) have contributed to the demise of rail-line burning.

Small rail reserves and some road easements are amongst the few places in most grassland regions where frequent burning has continued in the absence of stock grazing. Many of these sites now contain some of the highest quality grassland remnants, in terms of biodiversity conservation, in southeastern Australia, and they contain many endangered plant species (Scarlett and Parsons 1993; McDougall and Kirkpatrick 1994). Indeed, Scarlett (1994) has described the frequently burnt rail reserve remnants of western Victoria as the 'working model' of the original grassland vegetation.

Recent fire regimes

Since the early 1980s, increasing attention has been given to the use of fire as a management tool to maintain plant diversity in remnant *Themeda triandra* grasslands, especially in southern Victoria. Nowadays, two types of grassland are regularly burnt: (1) small, narrow rail and road easements, which historically have been burnt by railway staff or local fire brigades for fire control purposes; and (2) previously grazed (and rarely burnt) paddocks which are now being burnt to maintain grassland diversity by state conservation agencies. The different management authorities tend to use different fire regimes. Roadside remnants are often burnt early in summer (late December and January) after curing of the grasses, whilst conservation reserves are usually burnt in early autumn (late March and April) after the end of the fire control season. Burning regimes in conservation reserves have tended to be erratic, and conservation agencies have

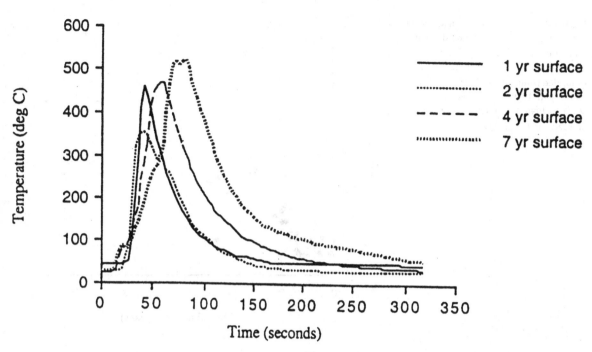

Fig. 8.4. Temperature at the soil surface (recorded at 3-s intervals) during grassland fires at sites that differed in the time since the last fire. Reproduced from Morgan (1999).

given a far lower commitment to implementing planned regimes than have rural fire brigades or railway agencies (Lunt and Morgan 1999a). Fragmentation of natural grasslands into small, isolated remnants has had inevitable consequences for grassland fire regimes. Contemporary fires in natural grasslands are small, tightly controlled, and usually intentionally lit. Landscape-scale wildfires very rarely occur.

Fire behaviour

Grassland management fires usually only burn small areas and are ignited when weather conditions are calm. Consequently, the rate of forward spread is slow and fire intensity is usually low. Recorded fire intensities from *Themeda triandra* grasslands range from 100 to 1200 kW m^{-1} (Groves 1974; Morgan 1999). By contrast, intensities up to 18 000 kW m^{-1} have been recorded from savanna fires in the Northern Territory (Williams *et al.* 1998).

Morgan (1999) described the characteristics of a number of management burns in *Themeda* grasslands in western Victoria. Fire intensity was found to be positively correlated with the rate of forward spread, but was not significantly correlated with fuel load, which indicates the importance of weather conditions in controlling fire behaviour. Maximum fire temperatures at ground level were highly variable, and ranged from 98 °C to 522 °C at the base of *Themeda* tussocks (Morgan 1999). Maximum temperatures were as high in annually burnt sites as in sites burnt 4 and 7 years previously (mirroring the similar fire intensities between sites), but temperatures were more variable and the duration of soil surface heating was much shorter at annually burnt sites. Temperatures at the soil surface exceeded 100 °C for less than 1 min in an annually burnt grassland, compared to 2–3 min in sites burnt less frequently (Fig. 8.4). All grassland fires were rapid events, and ambient soil temperatures returned within 5 min of the passage of the firefront (Morgan 1999). Fires had little effect on soil temperatures at 10 mm depth (Morgan 1999),

Fig. 8.5. Biomass accumulation (dry weight in t ha^{-1}) with increasing time since fire in temperate *Themeda triandra*-dominated grasslands in southeastern Australia. Compiled from data in Groves (1965, 1974), McDougall (1989), Lunt (1994, 1995), Morgan (1996, 1997a, 1998a, 1999) and Morgan and Lunt (1999).

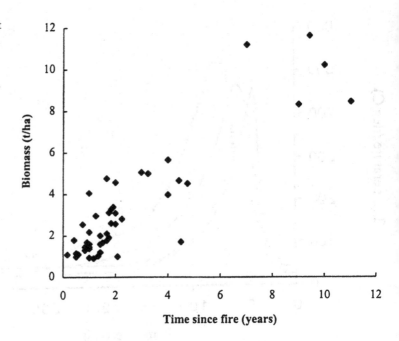

presumably because the fire front passed quickly (Groves 1974) and because temperature changes at depth are determined by surface fuel consumption (Bradstock and Auld 1995). Substantial and prolonged soil heating is more likely to occur in the months after a fire due to solar radiation absorption by the bare earth (Lunt 1995; Auld and Bradstock 1996; Morgan 1999).

Responses to single fires

Grassland structure

In grasslands dominated by large, long-lived, perennial tussock grasses such as *Themeda triandra* and *Poa* species, the biomass of the dominant grasses is at least one to two orders of magnitude greater than that of most inter-tussock species. As biomass accumulates after burning (or the absence of other disturbance regimes), the dominant grasses often contribute over 90% of community biomass (McDougall 1989; Tremont 1994).

The rate of biomass accumulation after burning is extremely variable in *Themeda triandra* grasslands in southeastern Australia (Fig. 8.5). Biomass levels generally increase up to about 10 years, but accumu-

lation rates in the first 2–5 years vary both between sites (according to site productivity) and within sites, according to seasonal moisture availability (Fig. 8.5). Biomass levels recorded 1 year after burning range from 0.9 to 4.0 t ha^{-1}, and after 2 years from 1.0 to 4.8 t ha^{-1} (Groves 1965; Lunt 1995; Morgan and Lunt 1999). A maximum level of 11.6 t ha^{-1} (mostly *Themeda*) was recorded from areas of the Laverton North Grassland Reserve (in western Melbourne) which were unburnt and ungrazed for 9–10 years (McDougall 1989).

Groves (1974) recorded a declining rate of biomass production from *Themeda triandra* as time progressed after burning, with rapid growth in the first 2 years, and considerably slower growth from 2 to 5 years. It was suggested that after 5 years, a steady state might be achieved with new growth being balanced by death and decomposition of old tissue (Groves 1974). More recent results, however, have questioned the stability of long-undisturbed *Themeda* swards. McDougall (1989) documented substantial differences in *Themeda* biomass and tussock densities between areas burnt 2 and 9 years previously. Tussock density in the long-unburnt area was only 25% of that in the recently burnt area, and most unburnt tussocks were poorly rooted, had a small basal area and were easily killed by trampling. McDougall (1989) predicted that long-unburnt *Themeda* tussocks would die, and that most litter would decompose after tussocks senesced.

This prognosis was confirmed by a recent chronosequence study which found that *Themeda* populations declined in the absence of burning or grazing. Morgan and Lunt (1999) compared *Themeda* biomass, cover, tiller and tussock densities between areas unburnt for varying periods, and documented a decline in tussock and tiller densities from 6 years after burning. By 11 years after fire, few live tillers or tussocks remained in the sward, and dead grass formed a thick layer of thatch over the soil surface. Almost 75% of *Themeda* tussocks died in this period. The cause of *Themeda* death was assumed to be self-shading by accumulated litter. Thus, instead of a steady state of balanced biomass production and litter decomposition, undisturbed *Themeda* swards may accumulate large quantities of litter which smother and kill tussocks long before substantial decomposition occurs. Thus, productive grasslands dominated by dense *Themeda* swards may not be sustainable over long periods of time unless accumulated grass litter is periodically removed by burning, grazing, slashing or some other mechanism. Before European colonisation, patches of senescent grassland presumably existed across the landscape, to be subsequently recolonised by propagules from nearby areas. Nowadays, the small size of most remnants makes such losses unsustainable, and areas of dead grass are quickly occupied by exotic species, against which a healthy *Themeda* sward provides considerable defence (Lunt and Morgan 1999b). In contrast to *Themeda triandra*, *Poa* species appear to be able to maintain their dominance without disturbance by burning or other forms of biomass removal. However, little is known of the fire ecology of these grasslands.

Few fire studies have been conducted in grasslands dominated by *Austrodanthonia* and *Austrostipa* species (Foreman 1996). *Austrodanthonia* and many *Austrostipa* species typically have less biomass and shorter life spans than *Themeda triandra* or *Poa* species (Williams 1970; Williams and Roe 1975). Consequently, litter accumulation and competitive exclusion occur much more slowly (if at all) in these low productivity grasslands, and biomass removal (by fire, grazing or other means) is not necessarily required in order to maintain plant diversity (Williams 1969). Williams (1969) recorded little accumulation of biomass after 16 years of grazing and burning exclusion in an *Austrodanthonia* grassland at Deniliquin (although changes in plant composition did occur during this period).

Botanical composition

Accumulating grass biomass leads to changes in the resources available to inter-tussock species. Closed grass swards control access to space, nutrients, water and light, creating adverse conditions for many associated species and consequent species poverty (Grubb 1977; Grime 1979). A strong, negative correlation exists between biomass levels in *Themeda* grasslands and the amount of photosynthetically active radiation which is transmitted through the grass canopy, with less than 10% of incident sunlight reaching the ground level beneath grasslands of 4 t ha^{-1} biomass (Fig. 8.6); this amount of biomass is commonly attained within 2–4 years of burning (Fig. 8.5). Dense grass litter presumably also influences soil moisture and nutrient levels, although little information is available from Australia (but see Wijesuriya and Hocking 1999). Grass biomass also affects fauna habitat structure (e.g. Baker-Gabb 1993; Osborne *et al.* 1995), including herbivore abundances. In a seed burial experiment, Watson (1995) found substantial seed predation (assumed to be by rodents and invertebrates) beneath a dense grass cover, but little seed predation in burnt, open areas. Similar results have been obtained outside Australia (e.g. Reader 1992).

Little information is available on the shade tolerance of most inter-tussock forbs, but indirect evidence suggests that ramets and genets of many species are intolerant of prolonged deep canopy shading (Stuwe and Parsons 1977; Scarlett and Parsons 1990; Morgan 1997a). For example, Scarlett and Parsons (1990) documented numerous declines of the endangered daisy *Rutidosis leptorrhynchoides*, as a result of grass competition in the absence of frequent burning. Subsequent experimental studies

Fig. 8.6. Relationship between biomass in *Themeda* grasslands and the proportion of photosynthetically active radiation above the grass layer which is transmitted through the grass canopy to ground level. Compiled from data in Lunt (1995) and Morgan (1996, 1997a, 1998e). All light measurements were made with a LiCor light meter with Quantum sensor.

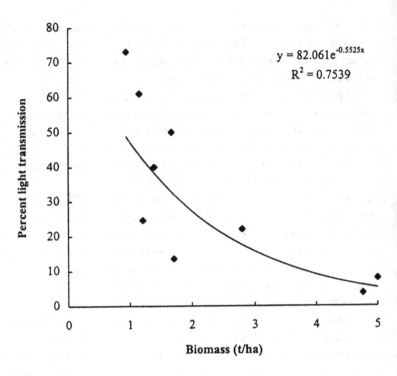

$$y = 82.061e^{-0.5525x}$$
$$R^2 = 0.7539$$

have shown that the species requires open conditions to recruit, grow and flower (Morgan 1995, 1997a). It would appear that high species richness can only be maintained at the small scale where grassland canopies are naturally sparse or frequently destroyed.

Effects of fires on demographic processes

Fires can affect plant populations at all stages of the plant life cycle. Consequently, an understanding of plant demographic processes and population biology is required to predict the impacts of different fire regimes in grasslands. In this section, we summarise some distinctive features of the grassland flora of southeastern Australia and describe the implications of these characteristics for fire management.

Recent studies have documented short-term fire effects in temperate grasslands (e.g. McDougall 1989; Lunt 1990a, 1994; Morgan 1996; Briggs and Muller 1997; Gilfedder and Kirkpatrick 1998a, b; Lunt and Morgan 1999c). Common outcomes from these studies include: a short-term decline in the biomass and cover of the dominant grasses (and all other species) as a result of burning (all studies); a short-

term increase in the abundance of opportunistic post-fire colonisers, most of which are annual exotic species which recruit from a large soil seed bank (Lunt 1990a, b; Briggs and Muller 1997; Gilfedder and Kirkpatrick 1998a); vigorous resprouting by all perennial species, with no obligate seed regenerators except annuals (Lunt 1990a; Morgan 1996); little or no post-fire seedling recruitment by many perennials (Lunt 1990a; Morgan 1998b); enhanced vigour and flower production of many perennial herbs in the spring after burning (Lunt 1994; Morgan 1996; Gilfedder and Kirkpatrick 1998b); and little change in plant composition after burning (Lunt 1990a; Morgan 1996, 1999; Briggs and Muller 1997). These trends are discussed in more detail below.

Seasonal phenology

Most grassland species exhibit strong seasonal growth patterns. In southern Victoria, above-ground vegetative growth commences from dormant buds in late autumn or winter after drought-breaking rains, then continues rapidly through spring when flowering is most pronounced, before seed is shed in December or January (Groves 1965; Morgan 1999). The above-ground parts of most plant species (including therophytes, geophytes and hemicryptophytes) then die back to dormant buds (or seeds for therophytes) to avoid the summer drought period (Morgan 1999). *Themeda triandra*, and other summer growing C4 grasses, are an obvious exception to this general pattern. *Themeda* commences rapid growth in late spring (October) before flowering and shedding seed in midsummer (January–February). It can remain green through most of summer, responding quickly to summer rainfall, before autumn frosts and low temperatures reduce growth rates in winter (Groves 1974; McDougall 1989).

Such strong growth rhythms presumably have great implications for burning in different fire seasons. Fires in summer and early autumn (before the autumn rainfall break) occur at a time when most grassland species are dormant (Morgan 1999). Hence, few plants are directly exposed to fire and, because grassland fires barely raise soil temperatures, dormant buds in the soil are unlikely to be affected by burning. Carbohydrate levels are often at a maximum in root storage organs at the time of summer burning. Presumably, since little labile carbohydrate is lost from plants in fire events, resprouting can be maximised in the post-fire conditions where light and (potentially) water availability are high. Of those species exposed to fire, many have their regenerative buds tightly held in tussock bases (e.g. *Themeda triandra*) and thus, buds are effectively insulated from direct fire damage. Buried seeds are unaffected by the passage of fire (Morgan 1999), presumably because soil heating is minimal during grassland fires. However, most seeds on the soil surface are incinerated (Morgan 1999).

Vegetative resprouters

Virtually all perennial grassland plants resprout after burning (Lunt 1990a; Morgan 1996, 1999), and 64% of perennial species in annually burnt grasslands in western Victoria were classed as obligate resprouters by Morgan (1999). A further 28% of species were classed as 'autoregenerating long-lived sprouters' (*sensu* Bell et al. 1984), i.e. species with good success at vegetative resprouting after fire coupled with limited successful seedling germination. The resprouting of native grassland plants in

the absence of substantial seedling regeneration is similar to the obligate vegetative sprouting syndrome described by Bell *et al.* (1984) for sandplain species in Western Australia.

Obligate seed regenerators (non-resprouting species which rely on seedling recruitment for post-fire regeneration) are very sensitive to frequent burning, as repeated fires may kill regenerating plants before they reach reproductive maturity (Gill 1981; Keith 1996). No perennial plants in temperate grasslands are known to possess this strategy, although it is possible that any species which did may have been eliminated by frequent burning in the past. The ability to resprout after burning, grazing or summer moisture stress appears to be shared by virtually all perennial grassland species in southeastern Australia.

Flower promotion

Many herbaceous species flower prolifically in the spring after burning, provided that adequate soil moisture is available (Lunt 1994; Morgan 1996). Flower abundance then declines in successive years as shoot competition from the dominant grasses increases. Exceptions to this generalisation include *Themeda triandra*, for which flowering was reduced by 70% after a late-summer fire (McDougall 1989), and the orchids *Diuris punctata* and *D. fragrantissima*, which appear to flower most prolifically in the second spring after autumn fires (Cropper 1993; Lunt 1994). For most perennial grassland species, the secondary juvenile period (i.e. the time required for resprouting plants to flower and set seed after burning: Johnson *et al.* 1994) would appear to be very short (i.e. <12 months) with most species capable of flowering in the first spring after an autumn fire (Lunt 1990a, 1994; Morgan 1996, 1999).

How fire promotes flowering remains poorly understood. Changes in daily temperature fluctuations, increased light penetration and soil temperatures, changes in physical and chemical characteristics of the soil, reduced competition, increased water availability, leaf removal and the production of ethylene have all been postulated as being important in other ecosystems (Hulbert 1988; Bond and van Wilgen 1996), and may also play a role in Australian temperate grasslands. For many grassland species, post-fire pulse flowering does not appear to be restricted to the period immediately after burning (e.g. 12 months), but appears to continue for as long as grass biomass levels remain low. This suggests that competition for light is a driving force influencing flower production in herbaceous species.

Soil seed banks

A conspicuous feature of the grassland flora of southeastern Australia is that many species shed readily germinable seed (Willis and Groves 1991; Lunt 1995; Morgan 1998d) and form small, transient soil seed banks (Gilfedder and Kirkpatrick 1993; Morgan 1995, 1998c; Lunt 1996, 1997d). For example, a seed burial study on the endangered daisy *Rutidosis leptorrhynchoides* found that all viable seeds germinated promptly in response to the autumn rainfall 'break' and no seeds survived in the soil for more than 16 weeks after burial (Morgan 1995).

In general, it would appear that only small-seeded, perennial forb species, such as *Hypericum gramineum, Juncus* and *Wahlenbergia* species form large, persistent soil seed banks (Lunt 1990b, 1997d; Willis *et al.* 1997; Morgan 1998c), as has been found in many soil seed bank studies in other ecosystems (Leck *et al.* 1989; Thompson *et al.* 1993). By contrast, annual species (most of which are exotic) typically form large soil seed banks and can account for more than 80% of seeds present in grassland seed banks (Lunt 1990b, 1997d; Gilfedder and Kirkpatrick 1993; Morgan 1998c). The absence of a soil seed bank means that the persistence of many species relies upon the maintenance of a resprouting 'bud and tuber bank' in the existing plant population. If existing plants die, then future recruitment cannot occur. This problem is accentuated in fragmented remnants as the chance of new propagules disseminating into isolated remnants is very low.

Seedling recruitment

Perhaps as a consequence of small soil seed banks, single grassland fires do not promote mass seed germination of most perennial species (Lunt 1990a;

Pyrke 1993; Foreman 1996; Morgan 1997b, 1999), in contrast to their effects on seedling recruitment in many heathland and forest communities (Wark *et al.* 1987). Seedling recruitment appears to occur infrequently for many native perennial species (Pyrke 1993; Morgan 1998d, 1999). In a 4-year study, Morgan (1997b) found seedling recruitment of native species to be rare in high quality, species-rich *Themeda* grasslands in western Victoria, even though over 70% of perennial native species were ultimately dependent on seedlings for regeneration. Given the widespread distribution of grassland species (McIntyre *et al.* 1993), a similar scenario may exist in other grassland regions of southeastern Australia. The 'storage effect' recruitment model proposed by Warner and Chesson (1985) appears to be relevant to productive *Themeda* grasslands. Under this model, infrequent but successful recruitment events are 'stored' into the adult population, allowing persistence during periods of extended, low-level recruitment.

The absence of a persistent soil seed bank for many species has significant implications for potential post-fire recruitment. If few seeds are produced in the year immediately before a fire, little seedling recruitment is possible in the year after burning. Since flower and seed production are strongly inhibited by a dense, closed grass layer, little seedling recruitment may be expected when a long-undisturbed, dense grass canopy is finally burnt. Instead, pulse flowering in the spring after burning may enhance opportunities for seedling recruitment in the second autumn–winter after burning (Lunt 1994). Thus, rather than directly triggering seedling recruitment, grassland fires may facilitate deferred post-fire recruitment, by providing open conditions for successful seed production in the following spring–summer, thereby permitting seedling recruitment in following years. Whilst this model would seem a logical consequence of transient soil seed banks, few empirical data are yet available, and further recruitment studies are needed.

Seedling survival is tightly coupled to light availability at ground level. In the absence of gaps in the canopy that allow high light levels to penetrate to the ground, seedling survival is negligible for many species (Morgan 1997a, b). Indeed, for some native species, gaps greater than 30 cm diameter are necessary to promote seedling survival (Morgan 1997a). Since grass biomass rapidly accumulates and gaps quickly disappear in productive grasslands (Figs 8.5, 8.6), only short periods (of 1–2 years) after burning may be suitable for abundant seed production and successful recruitment. Thus, fire may provide brief 'windows of opportunity' during which plants can successfully set seed and regenerate whilst grass biomass is low.

Fire regimes and fauna

The effects of grassland fire regimes on fauna have been poorly studied. Only one quantitative experimental study is available (Greenslade 1997), and most published information is based on casual observations of short-term responses of endangered reptiles and invertebrates to single fires. Despite this limited information base, frequent burning is widely perceived as having negative impacts on many fauna, particularly small species that are relatively immobile.

A number of reports on threatened grassland lizards, including the striped legless lizard (*Delma impar*) and grassland earless dragon (*Tympanocryptis lineata pinguicolla*) suggest that light grazing might be a more suitable management regime than frequent burning (Webster *et al.* 1992; Robertson and Cooper 1997; Robertson 1999).

Coulson (1990) recorded mortality of the threatened striped legless lizard (*Delma impar*) during a grassland fire, and suggested that predation was likely to increase as a result of the post-fire reduction in vegetation cover. Subsequent management recommendations for grasslands containing *D. impar* suggest that burning be conducted in late summer or autumn, when the dry clay soils are extensively cracked, to enable lizards to avoid incineration by hiding in deep cracks (Webster *et al.* 1992; Craigie 1995). Robertson and Cooper (1997) reported deaths of the endangered grassland earless dragon (*Tympanocryptis lineata pinguicolla*) due to incineration, although some individuals were able to escape

by retreating to burrows and perhaps remaining in them for long periods. Appropriate fire regimes have yet to be determined for this species.

The endangered golden sun-moth (*Synemon plana*) is most abundant in the Australian Capital Territory (ACT) in native grasslands that are grazed or mown rather than burnt. Frequent burning is considered likely to be detrimental to the species (ACT Government 1998). Sun-moth larvae feed on plant roots, especially *Austrodanthonia* species, and it has been suggested that root reserves would be reduced after burning as plants vigorously resprout, thereby reducing the food supply to larvae (ACT Government 1998). However, no information is available on post-fire changes in root reserves, and it is questionable whether root reserves are substantially reduced by fires.

In contrast to the above species, which usually occur in large, unburnt paddocks rather than small, frequently burnt grassland remnants, the endangered morabine grasshopper (*Keyacris scurra*) occurs in rail-line remnants in the ACT and NSW Southern Tablelands. These sites were historically burnt every 1–3 years. Little is known about the effects of fire on the species, but it is thought that burning in autumn (when eggs are hatching) or in spring (when adults are active) might present a hazard to these slow-moving flightless grasshoppers. On the other hand, post-fire reduction in grass competition is likely to aid the survival of food plants (Rowell and Crawford 1995).

Greenslade (1997) studied the responses of grassland invertebrates for 16 months after an autumn fire in an experimental small-plot study. She found that fire caused a short-term reduction in invertebrate richness and abundances compared to unburnt plots, but that these differences had disappeared after 16 months. The species composition of Collembola differed between burnt and unburnt plots at 16 months, which was considered to be due to differences in post-fire vegetation structure.

The challenge for managers of small grassland remnants that contain a diverse flora and endangered fauna is to maintain an open vegetation structure to maintain plant diversity whilst maintaining viable fauna populations. Where fire is used, far more information is required to balance short-term fire effects (e.g. immediate mortality and restricted foraging and/or elevated predation risk after burning), and the longer-term impacts on fauna populations. Short-term losses may only be sustainable if fauna benefit from post-fire conditions in the medium term, or can immigrate from nearby sites. However, the highly fragmented nature of grassland remnants strongly minimises opportunities for dispersal. Many grasslands containing threatened fauna may be easier to manage using intermittent, light grazing rather than frequent burning. Clearly, far more work is required to identify the responses of grassland fauna to long-term fire regimes.

Responses to fire regimes

Grassland plant species demonstrate considerable resilience to single fires. However, observations on the short-term impacts of single fires provide little insight into the effects of consecutive fires. As in other ecosystems, attributes of fire regimes including fire intensity, seasonality and frequency of occurrence are all likely to have significant effects on grassland composition and functioning. However, little is known about many of these factors.

The effects of grassland fire regimes have been investigated in two ways: (1) by regional comparisons of sites that have experienced different burning regimes for many decades (e.g. Stuwe and Parsons 1977; Lunt 1997b, c; Morgan 1998c) and (2) a small number of experimental studies (e.g. Robertson 1985; Foreman 1996; Henderson 1999).

Effects of historical fire regimes

In an early study, Stuwe and Parsons (1977) compared the plant composition of remnant *Themeda triandra* grasslands with different management histories in western Victoria. Remnants on rail easements which were ungrazed and frequently burnt (every 1–3 years in summer) were found to be significantly richer in native species than remnants in unburnt, lightly grazed paddocks or those on unburnt and ungrazed road reserves (these sites were probably grazed in the past). Inter-tussock forbs such as *Chrysocephalum apiculatum*, *Leptorhynchos squa-*

matus and *Plantago gaudichaudii* were best represented in frequently burnt sites, and no native species were favoured by fire exclusion. These differences were interpreted as responses to grazing and burning history, mediated through competition from the dominant grass, *Themeda triandra* (Stuwe and Parsons 1977). This study laid a firm foundation for the ongoing emphasis on frequent burning as a management tool in Victorian grasslands.

In a similar regional study, Lunt (1997b) compared the composition of frequently burnt, ungrazed, secondary grasslands in eastern Victoria (most of which were on rail easements), against unburnt and intermittently grazed, grassy forest and woodland remnants in the region. All remnants were assumed to be derived from the same grassy woodland ecosystem originally. Frequently burnt rail remnants possessed a different suite of plants to unburnt woodland remnants, and these differences reflected different management regimes rather than underlying soil features. Tall forbs were abundant in frequently burnt sites and relatively rare in grazed sites, whereas short species were abundant in intermittently grazed woodlands and rare in frequently burnt sites. Lunt (1997c) hypothesised that tall species were depleted in woodland sites by past stock grazing and that short species were depleted in frequently burnt sites as a result of competition from the dominant perennial grass, *Themeda triandra*, which dominated the burnt remnants. The second hypothesis is intriguing since it suggests that the depletion of many forbs in frequently burnt sites was a result of strong inter-specific competition *between* fire events rather than being a direct negative impact of frequent burning *per se*. If correct, this finding again suggests that the effects of grassland burning regimes are mediated through their effects on the dominant species. Burning regimes which promote dense swards of a competitively superior dominant (such as *Themeda triandra*) might be expected to deplete small species which are readily out-competed beneath a closed grass sward.

Are species-rich rail-line grasslands diverse because they have been frequently burnt, or because they have rarely been grazed? Unfortunately, the effects of historical patterns of grazing and fire man-agement are not easily separated in regional studies such as Stuwe and Parsons (1977) or Lunt (1997b). However, many of the species reported from fre-quently burnt grazing refugia by Stuwe and Parsons (1977) and McDougall and Kirkpatrick (1994) are also abundant in rarely burnt grazing refugia, such as travelling stock reserves and cemeteries, in *Eucalyptus albens* grassy woodlands in central NSW (Prober and Thiele 1995). This suggests that fire is not critically important to the perpetuation of these species.

Regional studies of fire history have provided invaluable information on the long-term effects of particular fire regimes (especially frequency and seasonality) on plant composition. In southern Victoria, many annually burnt grasslands are among the richest and most important remnants for plant conservation (Scarlett *et al.* 1992; McDougall and Kirkpatrick 1994). *Ipso facto*, such regimes are unlikely to be deleterious to species which are abundant in these sites. Surprisingly, rel-atively few grassland species appear to have been grossly depleted or eliminated by burning at such high frequencies. A notable exception may be members of the Fabaceae, including *Glycine*, *Psoralea*, *Desmodium*, *Swainsona* and *Lotus* species. Scarlett and Parsons (1982) suggested that these taxa were underrepresented on frequently burnt (1–3 years) rail and road remnants, and speculated that annual burning in late spring and early summer may have resulted in their demise.

Based on the 'intermediate disturbance hypothe-sis' (Grime 1979), one might theoretically expect diminished species diversity in sites subjected to extremely high fire frequencies (e.g. annual burning). In practice this does not seem to occur. Many annually burnt sites are very species-rich and very important for grassland conservation (McDougall and Kirkpatrick 1994). This unexpected response might reflect the fact that annual summer fires may not act as a 'disturbance' for many species – if disturbance is interpreted as 'partial or total destruction of the plant biomass' (Grime 1979, p. 7) – since at the time of summer burning, most plants are reduced to dormant rootstocks at or below ground level, and thereby escape destruction by fire.

Only one study is available which compares the effects of frequent burning against the long-term exclusion of fire and other disturbances (e.g. grazing): an unreplicated comparison of a long-unburnt area and an adjacent, frequently burnt area of productive *Themeda* grassland in western Melbourne (Lunt and Morgan 1999c). The long-unburnt area remained unburnt for 17 years from 1978 until 1995, whilst the frequently burnt zone was burnt six times, at 2–5 year intervals, during this period. In 1997 the rarely burnt zone was domi-nated by exotic species (49% exotic cover cf. 40% native cover), whereas the frequently burnt zone was dominated by native species (72% cover) with just 7% exotic cover. The most dramatic differences in species abundances between the two zones were for the exotic daisy *Hypochoeris radicata* which attained 33% mean cover in the rarely burnt zone compared to just 1% in the frequently burnt zone, and *Themeda triandra* which attained 22% in the rarely burnt zone compared to 63% in the fre-quently burnt zone. The density of live *Themeda* tus-socks in the rarely burnt area was only 30% of that in the frequently burnt zone (Lunt and Morgan 1999c). Thus, the long-term absence of burning appears to have killed many *Themeda* tussocks (as has been recorded elsewhere; Morgan and Lunt 1999) and promoted many perennial, exotic weeds.

Experimental fire studies

Apart from those studies which have described the outcomes of historical burning practices, most fire studies from temperate Australian grasslands have described short-term responses after single fire events. Few experimental manipulations of differ-ent fire regimes have been undertaken (Robertson 1985; Foreman 1996; Henderson 1999). Foreman (1996) conducted a replicated, small-plot experi-ment over 3 years to investigate the effects of burning, grazing and cultivation on a species-rich *Austrodanthonia* grassland in northern Victoria. Data were analysed by univariate statistics to describe changes in cover and density of functional plant groups (including annuals, perennials, forbs, grasses, exotics, natives). Burning reduced the cover of all functional groups compared to the control treatment, including annual and perennial native and exotic species. The notable exception was an increase in abundance of the exotic geophyte, *Romulea minutiflora*. For annual species, in particular, the changes induced by burning were of lesser magnitude than those caused by climatic variations during the experiment (i.e. drought). It should be noted that these results are not necessarily transferable to larger paddock scales, since all above-ground vegetation was consumed with an oxygen/acetylene torch and much of the vegetation would not have carried a fire normally.

Despite considerable conjecture, little is known of the effects of burning in different seasons. In the conservation management literature, spring burning has often been suggested as a possible mechanism to deplete annual exotic species in invaded grasslands by reducing their flowering and seed set (e.g. Stuwe 1986; McDougall 1989). Over several years, repeated spring burning may exhaust seed supplies and deplete these species from the community. By contrast, autumn fire regimes may help to promote exotic annual grasses by providing competition-free sites for establishment. Few data are available to support these hypotheses.

In an early study, Robertson (1985) documented the short-term effects (over 3 years) of annual and biennial burning in a long-grazed and rather species-poor *Eucalyptus camaldulensis*–*E. melliodora* open woodland with *Themeda triandra*-dominated understorey west of Melbourne. Annual burning consistently maintained lower biomass and slightly greater richness of herbaceous species than biennial burning or unburnt controls. By 2 years after burning, biennially burnt plots were similar in structure and floristics to unburnt areas. However, few significant differences were observed between spring and autumn burning, which had similar effects.

An ongoing experimental study near Melbourne may address a number of important questions about the effects of frequency and seasonality of burning on species composition, structure and nutrient cycling in productive, species-poor *Themeda* grasslands (Henderson 1999; Wijesuriya and Hocking 1999). A preliminary report by Henderson (1999) demonstrated substantial differences in the post-fire weed flora according to fire frequency, with annual autumn burning leading to a marked reduction in the abundance of the large-seeded, exotic annual grass *Briza maxima*, compared to triennial autumn burning. Furthermore, the total cover of exotic annual species was markedly lower in annual burnt plots than under triennial burning or annual or triennial slashing (Henderson 1999). Further long-term experimental studies are obviously required to identify the effects of different fire regimes.

Synthesis

From the above review, it is clear that further research is needed to broaden our understanding of most aspects of fire ecology in remnant native grasslands. Some of the many topics which deserve attention include: experimental studies of ecosystem responses to particular fire regimes (especially to compare the effects of different fire frequencies and seasons); seedling recruitment (when does it occur, under what conditions, how does fire affect it, how often is it necessary?); and the effects of different fire regimes on grassland fauna and soil nutrient fluxes.

Despite our limited understanding of grassland ecology and fire effects, we have attempted to develop a simple conceptual model of how productive *Themeda* grasslands might respond to different fire regimes, based on the studies reviewed above. Insufficient information is available to generalise to other grassland ecosystems. Our model *Themeda* grassland possesses the following characteristics.

It is dominated by a vigorous perennial grass which rapidly outcompetes associated species through biomass accumulation. The many, smaller, inter-tussock species are primarily herbaceous perennials, which suffer minimal mortality of genets and ramets due to burning, and possess small, transient soil seed banks. Seedling recruitment is not directly promoted by fire, but instead is relatively rare and may require a combination of open gap conditions and uncommon climatic events. Most plants die back to buds or tubers at or below ground level over summer. Climatic conditions permitting, plants flower and set seed abundantly when biomass levels are low. Most plants flower abundantly in the first spring after burning (climate permitting), and plant vigour and reproductive output decline as biomass levels accumulate. Ultimately, many plants die beneath a dense grass sward, and in the absence of a persistent soil seed bank, complete mortality of mature plants is irreversible, especially in small isolated remnants.

We do not wish to imply that all grassland species possess all of these characteristics (some certainly do not). However, apart from speculation about the conditions that might promote rare recruitment events, this character suite does appear to be shared by many grassland species in temperate Australia. It should also be noted that these characteristics are not necessarily shared by all grassy ecosystems, especially grassy woodlands and secondary grasslands with abundant shrubs. Frequent burning may deplete or eliminate woody species in these ecosystems.

What role might fire play in such a system? The major ecological force driving the model system is inter-specific competition, or more specifically, resource acquisition through biomass accumulation by the dominant grasses. Any activities which constrain the dominant grasses will facilitate coexistence of other subordinate species. This is Grime's (1979) archetypical 'competitive environment'. The primary role of burning would appear to be to reduce biomass accumulation and shoot competition from the dominant grasses. Fire is not critical to this process, since biomass can be removed or altered in structure in a number of ways, including burning, grazing and slashing. In relatively unproductive environments, accumulation rates may be so slow that no biomass reduction is necessary to maintain species richness and ecological processes. Furthermore, in this model ecosystem there appear to be few critical ecological processes that only fire can fulfil. Thus, burning may be more exchangeable as a management tool in grasslands than in other ecosystems such as heathlands or wet sclerophyll forests where critical ecosystem processes (such as

fire-stimulated seed germination) can only be practically accomplished by using fire.

In the model system, the processes which take place *between* fire events (which are driven by grass dominance) are likely to have more impact on ecosystem structure and composition than the fire events themselves or their immediate after-effects. Fire frequency, and in particular, the variable 'time-since-fire', is likely to exert the most profound influence on grassland composition and structure (rather than other attributes of the fire regime, such as fire season or intensity), simply because fire frequency exerts the greatest influence on accumulated biomass and structural dominance. As time extends after fire and biomass levels accumulate, the potential reproductive output and survival of inter-tussock species declines. The longer the interval between fires, the lower the potential for future re-establishment and recruitment.

It is clear that far more ecological research is required to enable burning regimes to be confidently tailored to individual grassland remnants. In the meantime, however, many productive grassland remnants need to be continually burnt to prevent further declines in plant diversity. We hope that the insights provided here may help to stimulate more applied research to be undertaken and to avert diversity losses until future results become available.

References

ACT Government (1998). Golden Sun Moth (*Synemon plana*): An endangered species. Environment ACT, Action Plan no. 7, Canberra.

Allen, M. R. (1998). Forest history projects for State Forests of New South Wales: case studies of three cypress pine forests on the Lachlan and Bogan River catchments, Forbes Forestry District on Back Yamma, Euglo South and Strahorn State Forests. State Forests of New South Wales, Pennant Hills.

Auld, T. D., and Bradstock, R. A. (1996). Soil temperatures after the passage of a fire: do they influence the germination of buried seeds? *Australian Journal of Ecology* **21**, 106–109.

Axelrod, D. I. (1985). Rise of the grassland biome, Central North America. *Botanical Review* **51**, 163–201.

Baker-Gabb, D. (1993). Managing native grasslands to maintain biodiversity and conserve the plains-wanderer. Royal Australasian Ornithological Union, Conservation Statement no. 8, Melbourne.

Bell, D. T., Hopkins, A. J. M., and Pate, J. S. (1984). Fire in the Kwongan. In *Kwongan: Plant Life of the Sandplain* (eds. J. S. Pate and J. S. Beard) pp. 178–204. (University of Western Australia: Nedlands.)

Benson, J. S. (1994). The native grasslands of the Monaro region: Southern Tablelands of NSW. *Cunninghamia* **3**, 609–650.

Benson, J. S., Ashby, E. M., and Porteners, M. F. (1997). The native grasslands of the Riverine Plain, New South Wales. *Cunninghamia* **5**, 1–48.

Bond, W. J., and van Wilgen, B. W. (1996). *Fire and Plants*. (Chapman and Hall: London.)

Bradstock, R. A., and Auld, T. D. (1995). Soil temperatures during experimental bushfires in relation to fire intensity: consequences for legume germination and fire management in south-eastern Australia. *Journal of Applied Ecology* **32**, 76–84.

Briggs, J. D., and Muller, W. J. (1997). Effects of fire and short term domestic stock grazing on the composition of a native, secondary grassland bordering the Australian Capital Territory. Report to Environment Australia and ACT Parks and Conservation Service, Canberra.

Costello, D. A. (1998). Ecological impacts of coast wattle (*Acacia sophorae*) on coastal grassland in south-eastern NSW. BSc thesis, Charles Sturt University, Albury.

Costin, A. B. (1954). *A Study of the Ecosystems of the Monaro Region of New South Wales with Special Reference to Soil Erosion*. (Soil Conservation Service of New South Wales: Sydney.)

Coulson, G. (1990). Conservation biology of the Striped Legless Lizard (*Delma impar*): an initial investigation. Arthur Rylah Institute for Environmental Research Technical Report no. 106, Department of Conservation and Environment Victoria, Melbourne.

Craigie, V. (1995). Derrimut Grassland Reserve: management issues. In *Management of Relict Lowland Grasslands, Proceedings of a Workshop and Public Seminar, 24–25 September 1993* (eds. S. Sharp and R. Rehwinkel) pp. 51–54. (ACT Parks and Conservation Service: Canberra.)

Cropper, S. (1993). *Management of Endangered Plants*. (CSIRO: Melbourne.)

Davies, R. J. P. (1997). *Weed Management in Temperate Native Grasslands and Box Grassy Woodlands in South Australia*. (Black Hill Flora Centre, Botanic Gardens of Adelaide: Athelstone, SA.)

Fensham, R. J. (1989). The pre-European vegetation of the Midlands, Tasmania: a floristic and historical analysis of vegetation patterns. *Journal of Biogeography* **16**, 29–45.

Fensham, R. J., and Fairfax, R. J. (1996). The disappearing grassy balds of the Bunya Mountains, South-eastern Queensland. *Australian Journal of Botany* **44**, 543–558.

Fensham, R. J., and Fairfax, R. J. (1997). The use of the land survey record to reconstruct pre-European vegetation patterns in the Darling Downs, Queensland, Australia. *Journal of Biogeography* **24**, 827–836.

Fensham, R. J., and Kirkpatrick, J. B. (1992). The eucalypt forest – grassland/grassy woodland boundary in central Tasmania. *Australian Journal of Botany* **40**, 123–138.

Foreman, P. W. (1996). Ecology of native grasslands on Victoria's Northern Riverine Plain. MSc thesis, La Trobe University, Melbourne.

Gilfedder, L., and Kirkpatrick, J. B. (1993). Germinable soil seed and competitive relationships between a rare native species and exotics in a semi-natural pasture in the Midlands, Tasmania. *Biological Conservation* **64**, 113–119.

Gilfedder, L., and Kirkpatrick, J. B. (1998a). Factors influencing the integrity of remnant bushland in subhumid Tasmania. *Biological Conservation* **84**, 89–96.

Gilfedder, L., and Kirkpatrick, J. B. (1998b). Distribution, disturbance tolerance and conservation of *Stackhousia gunnii* in Tasmania. *Australian Journal of Botany* **46**, 1–13.

Gill, A. M. (1981). Adaptive responses of Australian vascular plant species to fires. In *Fire and the Australian Biota* (eds. A. M. Gill, R. H. Groves and I. R. Noble) pp. 243–272. (Australian Academy of Science: Canberra.)

Greenslade, P. (1997). Short term effects of a prescribed burn on invertebrates in grassy woodland in south-eastern Australia. *Memoirs of the Museum of Victoria* **56**, 305–312.

Grime, J. P. (1979). *Plant Strategies and Vegetation Processes*. (Wiley: Chichester, UK.)

Groves, R. H. (1965). Growth of *Themeda australis* tussock

grassland at St. Albans, Victoria. *Australian Journal of Botany* 13, 291–302.

Groves, R. H. (1974). Growth of *Themeda australis* grassland in response to firing and mowing. *Field Station Records, Division of Plant Industry, CSIRO Australia* 13, 1–7.

Groves, R. H., and Williams, O. B. (1981). Natural grasslands. In *Australian Vegetation* (ed. R. H. Groves) pp. 293–316. (Cambridge University Press: Cambridge.)

Grubb, P. J. (1977). The maintenance of species-richness in plant communities: the importance of the regeneration niche. *Biological Reviews* 52, 107–145.

Henderson, M. (1999). How do urban grasslands respond to fire and slashing? Effects on gaps and weediness. In *Down to Grass Roots: Proceedings of a Conference on the Management of Grassy Ecosystems*, Victoria University St Albans, 9–10 July 1998 (eds. V. Craigie and C. Hocking) pp. 38–43. (Department of Conservation and Natural Resources and Parks Victoria: Melbourne.)

Howitt, A. W. (1890). The eucalypts of Gippsland: influence of settlement on the *Eucalyptus* forests. *Transactions of the Royal Society of Victoria* 1, 81–120.

Hulbert, L. C. (1988). Causes of fire effects in tallgrass prairie. *Ecology* 69, 46–58.

Johnson, K. A., Morrison, D. A., and Goldsack, G. (1994). Post-fire flowering patterns in *Blandfordia nobilis* (Liliaceae). *Australian Journal of Botany* 42, 49–60.

Keith, D. (1996). Fire-driven extinction of plant populations: a synthesis of theory and review of evidence from Australian vegetation. *Proceedings of the Linnean Society of New South Wales* 116, 37–78.

Leck, M. A., Parker, V. T., and Simpson, R. L. (1989). *Ecology of Soil Seed Banks*. (Academic Press: San Diego.)

Leigh, J. H., and Noble, J. C. (1972). *Riverine Plain of New South Wales: Its Pastoral and Irrigation Development*. (CSIRO Division of Plant Industry: Canberra.)

Lunt, I. D. (1990a). Impact of an autumn fire on a long-grazed *Themeda triandra* (Kangaroo Grass) grassland: implications for management of invaded, remnant vegetations. *Victorian Naturalist* 107, 45–51.

Lunt, I. D. (1990b). The soil seed bank of a long-grazed *Themeda triandra* grassland in Victoria. *Proceedings of the Royal Society of Victoria* 102, 53–57.

Lunt, I. D. (1994). Variation in flower production of nine grassland species with time since fire, and implications for grassland management and restoration. *Pacific Conservation Biology* 1, 359–366.

Lunt, I. D. (1995). Seed longevity of six native forbs in a closed *Themeda triandra* grassland. *Australian Journal of Botany* 43, 439–449.

Lunt, I. D. (1996). A transient soil seed bank for the Yam-daisy *Microseris scapigera*. *Victorian Naturalist* 113, 16–19.

Lunt, I. D. (1997a). The distribution and environmental relationships of native grasslands on the lowland Gippsland Plain, Victoria: an historical study. *Australian Geographical Studies* 35, 140–152.

Lunt, I. D. (1997b). Effects of long-term vegetation management on remnant grassy forests and anthropogenic native grasslands in south-eastern Australia. *Biological Conservation* 81, 287–297.

Lunt, I. D. (1997c). A multivariate growth-form analysis of grassland and forest forbs in south-eastern Australia. *Australian Journal of Botany* 45, 691–705.

Lunt, I. D. (1997d). Germinable soil seed banks of anthropogenic native grasslands and grassy forest remnants in temperate south-eastern Australia. *Plant Ecology* 131, 21–34.

Lunt, I. D. (1998). Two hundred years of land use and vegetation change in a remnant coastal woodland in southern Australia. *Australian Journal of Botany* 46, 629–647.

Lunt, I. D., and Morgan, J. W. (1999a). Lessons from grassland management of Laverton North and Derrimut Grassland Reserves: what can we learn for future grassland management? In *Down to Grass Roots: Proceedings of a Conference on the Management of Grassy Ecosystems*, Victoria University St Albans, 9–10 July 1998 (eds. V. Craigie and C. Hocking) pp. 69–74. (Department of Conservation and Natural Resources and Parks Victoria: Melbourne.)

Lunt, I. D., and Morgan, J. W. (1999b). Can competition from *Themeda triandra* inhibit invasion by the perennial exotic grass *Nassella neesiana* in native grasslands? *Plant Protection Quarterly* 15, 92–94.

Lunt, I. D., and Morgan, J. W. (1999c). Effect of fire frequency on plant composition at the Laverton North Grassland Reserve, Victoria. *Victorian Naturalist* 116, 84–90.

Lunt, I., Barlow, T., and Ross, J. (1998). *Plains Wandering: Exploring the Grassy Plains of South-Eastern Australia*. (Victorian National Parks Association and Trust for Nature: Melbourne.)

Mack, R. N. (1989). Temperate grasslands vulnerable to

plant invasions: characteristics and consequences. In *Biological Invasions: A Global Perspective* (eds. J. A. Drake, H. A. Mooney, F. diCastri, R. H. Groves, F. J. Kruger, M. Rejmanek and M. Williamson) pp. 155–179. (Wiley: Chichester, UK.)

McDougall, K. L. (1989). The re-establishment of *Themeda triandra* (kangaroo grass): implications for the restoration of grassland. Arthur Rylah Institute for Environmental Research Technical Report no. 89, Department of Conservation, Forests and Lands Victoria, Melbourne.

McDougall, K., and Kirkpatrick, J. B. (1994). *Conservation of Lowland Native Grasslands in South-Eastern Australia*. (World Wide Fund for Nature Australia: Sydney.)

McIntyre, S., Huang, Z., and Smith, A. P. (1993). Patterns of abundance in grassy vegetation of the New England Tablelands: identifying regional rarity in a threatened vegetation type. *Australian Journal of Botany* **41**, 49–64.

McIntyre, S., Lavorel, S., and Tremont, R. M. (1995). Plant life-history attributes: their relationship to disturbance response in herbaceous vegetation. *Journal of Ecology* **83**, 31–44.

Moore, C. W. E. (1953). The vegetation of the south-eastern Riverina, New South Wales. II. The disclimax communities. *Australian Journal of Botany* **1**, 548–567.

Moore, R. M. (1962). Effects of the sheep industry on Australian vegetation. In *The Simple Fleece: Studies in the Australian Wool Industry* (ed. A. Barnard) pp. 170–183. (Melbourne University Press: Melbourne.)

Morcom, L. A., and Westbrooke, M. E. (1998). The pre-settlement vegetation of the western and central Wimmera Plains of Victoria, Australia. *Australian Geographical Studies* **36**, 273–288.

Morgan, J. W. (1995). Ecological studies of the endangered *Rutidosis leptorrhynchoides*. II. Patterns of seedling emergence and survival in a native grassland. *Australian Journal of Botany* **43**, 13–24.

Morgan, J. W. (1996). Secondary juvenile period and community recovery following late-spring burning of a kangaroo grass *Themeda triandra* grassland. *Victorian Naturalist* **113**, 47–57.

Morgan, J. W. (1997a). The effect of grassland gap size on establishment, growth and flowering of the endangered *Rutidosis leptorrhynchoides* (Asteraceae). *Journal of Applied Ecology* **34**, 566–576.

Morgan, J. W. (1997b). Regeneration processes in an endangered, species-rich grassland community on the volcanic plains of western Victoria. PhD thesis, La Trobe University, Melbourne.

Morgan, J. W. (1998a). Small-scale plant dynamics in temperate *Themeda triandra* grasslands of southeastern Australia. *Journal of Vegetation Science* **9**, 347–360.

Morgan, J. W. (1998b). Patterns of invasion of an urban remnant of a species-rich grassland in southeastern Australia by non-native plant species. *Journal of Vegetation Science* **9**, 181–190.

Morgan, J. W. (1998c). Composition and seasonal flux of the soil seed bank of species-rich *Themeda triandra* grasslands in relation to burning history. *Journal of Vegetation Science* **9**, 145–156.

Morgan, J. W. (1998d). Comparative germination responses of 28 temperate grassland species. *Australian Journal of Botany* **46**, 209–219.

Morgan, J. W. (1998e). Importance of canopy gaps for recruitment of some forbs in *Themeda triandra*-dominated grasslands in south-eastern Australia. *Australian Journal of Botany* **46**, 609–627.

Morgan, J. W. (1999). Defining grassland fire events and the response of perennial plants to annual fire in temperate grasslands of south-eastern Australia. *Plant Ecology* **144**, 127–144.

Morgan, J. W., and Lunt, I. D. (1999). Effects of time-since-fire on the tussock dynamics of a dominant grass (*Themeda triandra*) in a temperate Australian grassland. *Biological Conservation* **88**, 379–386.

Osborne, W., Kukolic, K., and Jones, S. (1995). Management of threatened vertebrates in native grasslands: a case of clutching at straws? In *Management of Relict Lowland Grasslands, Proceedings of a Workshop and Public Seminar, 24–25 September 1993* (eds. S. Sharp and R. Rehwinkel) pp. 89–96. (ACT Parks and Conservation Service: Canberra.)

Prober, S. M., and Thiele, K. R. (1995). Conservation of the grassy white box woodlands: relative contributions of size and disturbance to floristic composition and diversity of remnants. *Australian Journal of Botany* **43**, 349–366.

Pyrke, A. (1993). The role of soil disturbance by small mammals in the establishment of rare plant species. PhD thesis, University of Tasmania, Hobart.

Reader, R. J. (1992). Herbivory as a confounding factor in an experiment measuring competition among plants. *Ecology* **73**, 373–376.

Rehwinkel, R. (1997). Grassy ecosystems of the South Eastern Highlands. Technical report: literature review, data audit, information gap analysis and research strategy, NSW National Parks and Wildlife Service, Queanbeyan.

Robertson, D. (1985). Interrelationships between kangaroos, fire and vegetation dynamics at Gellibrand Hill Park, Victoria. PhD thesis, University of Melbourne.

Robertson, P. (1999). Terrick Terrick National Park habitat assessment for five species of threatened vertebrates. Report to Parks Victoria, Melbourne.

Robertson, P., and Cooper, P. (1997). Recovery plan for the Grassland Earless Dragon (*Tympanocryptis lineata pinguicolla*). Report to Environment Australia, Canberra.

Rowell, A., and Crawford, I. (1995). A survey of the Morabine Grasshopper *Keyacris scurra* (Rehn) in the ACT. Report to ACT Parks and Conservation Service, Canberra.

Sauer, C. O. (1950). Grassland climax, fire and man. *Journal of Range Management* **8**, 117–121.

Scarlett, N. H. (1994). Soil crusts, germination and weeds: issues to consider. *Victorian Naturalist* **111**, 125–130.

Scarlett, N. H., and Parsons, R. F. (1982). Rare plants of the Victorian plains. In *Species at Risk: Research in Australia* (eds. R. H. Groves and W. D. L. Ride) pp. 89–105. (Australian Academy of Science: Canberra.)

Scarlett, N. H., and Parsons, R. F. (1990). Conservation biology of the southern Australian daisy *Rutidosis leptorrhynchoides*. In *Management of Small Populations* (eds. T. W. Clark and J. H. Seebeck) pp. 195–205. (Chicago Zoological Society: Chicago.)

Scarlett, N. H., and Parsons, R. F. (1993). Rare or threatened plants in Victoria. In *Flora of Victoria*, vol. 1, *Introduction* (eds. D. B. Foreman and N. G. Walsh.) pp. 227–255. (Inkata: Melbourne.)

Scarlett, N. H., Wallbrink, S. J., and McDougall, K. (1992). *Field Guide to Victoria's Native Grasslands*. (National Trust of Australia: Melbourne.)

Sharp, S. (1995). Floral diversity in lowland native grasslands in the ACT. In *Management of Relict Lowland Grasslands, Proceedings of a Workshop and Public Seminar*, 24–25 September 1993 (eds. S. Sharp and R. Rehwinkel) pp. 80–88. (ACT Parks and Conservation Service: Canberra.)

Stewart, O. C. (1956). Fire as the first great force employed by man. In *Man's Role in Changing the Face of the Earth* (ed. W. L. Thomas) pp. 115–133. (University of Chicago Press: Chicago.)

Stuwe, J. (1986). An assessment of the conservation status of native grasslands on the Western Plains, Victoria and sites of botanical significance. Arthur Rylah Institute for Environmental Research Technical Report Series no. 48, Department of Conservation, Forests and Lands, Melbourne.

Stuwe, J., and Parsons, R. F. (1977). *Themeda australis* grasslands on the Basalt Plains, Victoria: floristics and management effects. *Australian Journal of Ecology* **2**, 467–476.

Thompson, K., Band, S. R., and Hodgson, J. G. (1993). Seed size and shape predict persistence in soil. *Functional Ecology* **7**, 236–241.

Tremont, R. M. (1994). Life-history attributes of plants in grazed and ungrazed grasslands on the Northern Tablelands of New South Wales. *Australian Journal of Botany* **42**, 511–530.

Tremont, R. M., and McIntyre, S. (1994). Natural grassy vegetation and native forbs in temperate Australia: structure, dynamics and life histories. *Australian Journal of Botany* **42**, 641–658.

Vogl, R. J. (1974). Effects of fire on grasslands. In *Fire and Ecosystems* (eds. T. T. Kozlowski and C. E. Ahlgren) pp. 139–194. (Academic Press: New York.)

Wark, M. C., White, M. D., Robertson, D. J., and Marriott, P. H. (1987). Regeneration of heath and heath woodland in the north-eastern Otway Ranges following the wildfire of February 1983. *Proceedings of the Royal Society of Victoria* **99**, 51–88.

Warner, R. R., and Chesson, P. L. (1985). Coexistence mediated by recruitment fluctuations: a field guide to the storage effect. *American Naturalist* **125**, 769–787.

Watson, S. (1995). Seed ecology of five native forbs in a basalt plains grassland. Thesis, Burnley Horticultural College, Melbourne.

Webster, A., Fallu, R., and Preece, K. (1992). Striped Legless Lizard *Delma impar*. Action Statement no. 17,

Department of Conservation and Environment Victoria, Melbourne.

Wijesuriya, S., and Hocking, C. (1999). Why do weeds grow when you dig up native grasslands? The effects of physical disturbance on available nutrients, mineralisation and weed invasion in grassland soils. In *Down to Grass Roots: Proceedings of a Conference on the Management of Grassy Ecosystems*, Victoria University St Albans, 9–10 July 1998 (eds. V. Craigie and C. Hocking) pp. 31–37. (Department of Conservation and Natural Resources, and Parks Victoria: Melbourne.)

Williams, O. B. (1969). Studies in the ecology of the Riverine Plain. V. Plant density response of species in a *Danthonia caespitosa* grassland to 16 years of grazing by merino sheep. *Australian Journal of Botany* 17, 255–268.

Williams, O. B. (1970). Population dynamics of two perennial grasses in Australian semi-arid grassland. *Journal of Ecology* 58, 869–875.

Williams, O. B., and Roe, R. (1975). Management of arid grasslands for sheep: plant demography of six grasses in relation to climate and grazing. *Proceedings of the Ecological Society of Australia* 9, 142–156.

Williams, R. J., Gill, A. M., and Moore, P. H. R. (1998). Seasonal changes in fire behaviour in a tropical savanna in northern Australia. *International Journal of Wildland Fire* 8, 227–239.

Willis, A. J., and Groves, R. H. (1991). Temperature and light effects on the germination of seven native forbs. *Australian Journal of Botany* 39, 219–228.

Willis, A. J., Groves, R. H., and Ash, J. E. (1997). Seed ecology of *Hypericum gramineum*, an Australian forb. *Australian Journal of Botany* 45, 1009–1022.

Willis, J. H. (1964). Vegetation of the basalt plains in western Victoria. *Proceedings of the Royal Society of Victoria* 77, 397–418.

Withers, J. R., and Ashton, D. H. (1977). Studies on the status of unburnt *Eucalyptus* woodland at Ocean Grove, Victoria. I. The structure and regeneration. *Australian Journal of Botany* 25, 623–637.

Part IV

Ecosystems: shrublands

9

Fire regimes in Australian heathlands and their effects on plants and animals

DAVID A. KEITH, W. LACHIE McCAW AND ROBERT J. WHELAN

Abstract

Heathlands span the full latitudinal range of the Australian continent, though their occurrence may be restricted locally. Australian heathlands vary in regional climate and local soil characteristics, but have in common a scleromorphic physiognomy, dominance by a diverse range of shrubs and sedges, deficiency in soil nutrients, variability in soil moisture and proneness to recurring fires. Total fuel loads typically reach equilibrium within 1–2 decades after fire, although the dead fuel component continues increasing for a further decade, leading to greater flammability of older stands. Fire intensity varies up to 40 MW m^{-1} under extreme fuel and weather conditions, with forward rates of spread approaching 2 m s^{-1}. Fire return times vary from 1 to more than 50 years, but these extremes are not commonly observed and are detrimental to heathland biodiversity. Australian heathland floras exhibit a diverse range of fire related life-history characteristics that is unparalleled except in South African fynbos. Plant life-history syndromes widespread in Australian heathlands include vegetative mechanisms of fire survival, fire-cued seedling recruitment, serotinous seed banks, fire-stimulated soil seed banks, vegetative spread and pyrogenic flowering. Syndromes such as wide propagule dispersal and seedling establishment beyond the immediate post-fire period are very poorly represented except in tropical and alpine heathlands. Although life histories of component species have a profound influence on heathland vegetation dynamics, higher-order processes such as competition, predation and environmental interactions are also important. The responses of heathland fauna to fire regimes are quite poorly studied but appear to be highly variable. They are influenced by the interaction between animal characteristics (e.g. life history, mobility, behaviour and resource use) and fire and habitat characteristics (e.g. intensity and season of the particular fire and the fire history of the surrounding areas).

Introduction

In this chapter we describe fire regimes in Australian heathlands and review their role in shaping ecological patterns and processes across a range of spatial and temporal scales. Although often restricted locally, heathlands span the full latitudinal range of the Australian continent (12–42° S, Fig. 9.1). Within this range, the ecological characteristics of heathlands vary according to regional climatic differences, as well as more locally controlled environmental factors. In Table 9.1 we propose an ecological classification of Australian heathlands as a framework for describing and contrasting dynamic processes associated with recurring fires.

In the tropical north, heathlands on sandstone plateaux are subject to wet-summer tropical monsoons. Conversely, in southwestern Australia and the lower Murray district, heathlands on sand sheets and laterites experience wet-winter Mediterranean climates grading to the semi-arid zone. Other heathlands are scattered along the full length of the eastern seaboard from tropical climates in far northeast Queensland to less seasonal temperate climates further south which are, none the less, variable between years due to the El Niño/Southern Oscillation (ENSO). At their southern extremity in Tasmania, heathlands are exposed to a wet cool temperate Roaring Forties climate, while at their

Fig. 9.1. Distribution of Australian heathlands described in Table 9.1.

highest elevations in the southeast Australian alps and Tasmanian highlands, heathlands are prone to transient snow cover. Heathlands occur universally on low-nutrient acid soils usually derived from quartz-rich substrates. Soil chemistry is typified by phosphorus and nitrogen deficiency and high levels of immobilised iron and aluminium (Groves 1981). Heathlands often occur on landscape components with high levels of exposure to wind and solar radiation or where levels of soil moisture are periodically high (Haigh 1981). Recurring fires are ubiquitous across all types of Australian heathland, although their frequency, intensity, season of occurrence and spatial characteristics may vary widely (Gill and Groves 1981).

Heathlands are typically treeless, although sparsely scattered eucalypts may emerge from a diverse stratum of sclerophyllous small-leaved shrubs and graminoids, which provide a short but relatively continuous fuel. Plant diversity at small

Table 9.1. *Classification of Australian heathlands*

Heathland class	Climatic regime	Typical plant genera	Examples of typical and specialist fauna species	Distribution
Tropical heath[a]	Wet summer monsoon	Poorly known. *Acacia, Calytrix, Regelia, Grevillea, Hibbertia, Jacksonia, Triodia, Aristida*	Sandstone antechinus (*Pseudantechinus bilarni*), white-throated grass wren (*Amytornis woodwardi*), Leichhardt's grasshopper (*Petasida ephippigera*)	Arnhem escarpment, Mitchell Plateau? (Kimberley)
Sandplain kwongan[b]	Wet winter Mediterranean to semi-arid	Highly diverse, variable. *Banksia, Hakea, Conospermum, Verticordia, Melaleuca, Calothamnus, Acacia, Daviesia, Astroloma, Stylidium, Thysanotus, Antigozanthos, Xanthorrhoea*	Splendid fairy wren (*Malurus splendens*), rufous field wren (*Calamanthus campestris*), ash-grey mouse (*Pseudomys albocinereus*), honey possum (*Tarsipes rostratus*)	Southwestern Australia (Shark Bay to Swan Plain) west coast to wheat belt
Western mallee-heath[b]	Wet winter Mediterranean to semi-arid	Highly diverse, variable. Mallee *Eucalyptus, Banksia, Hakea, Petrophile, Dryandra, Boronia, Daviesia, Gompholobium, Astroloma, Leucopogon, Schoenus, Stylidium, Austrostipa*	Western ground parrot (*Pezoporus wallicus flaviventris*), western whipbird (*Psophodes nigrogularis*), honey possum (*Tarsipes rostratus*), western mouse (*Pseudomys occidentalis*)	South-western Australia (Cape Naturaliste to Cape Arid) south coast to wheat belt
Central lowland heath[c]	Wet winter Mediterranean to semi-arid	*Banksia, Allocasuarina, Leptospermum, Phyllota, Hibbertia, Lepidosperma*	Southern emu wren (*Stipiturus malachurus*)	Lower Murray River district of Victoria and South Australia
East coast dry heath[d]	Warm temperate (weakly seasonal) ENSO[1]	Diverse and variable. *Banksia, Persoonia, Hakea*, mallee *Eucalyptus, Allocasuarina, Leptospermum, Acacia, Pultenaea, Dillwynia, Epacris, Leucopogon, Xanthorrhoea, Lomandra,*	Eastern ground parrot (*Pezoporus wallicus*), southern emu wren (*Stipiturus malachurus*), eastern chestnut mouse (*Pseudomys gracillicaudatus*)	Deep coastal sand from northeast Queensland to southern Victoria, coastal plateaux from Sydney to Victoria, north and

Table 9.1. (cont.)

Heathland class	Climatic regime	Typical plant genera	Examples of typical and specialist fauna species	Distribution
		Caustis, Lepidosperma		east coast of Tasmania and associated islands
Eastern montane heath[e]	Cool temperate (weakly seasonal) ENSO	Allocasuarina, Banksia, Hakea, Persoonia, Brachyloma, Kunzea, Leptospermum, Gahnia, Schoenus		Great Dividing Range from Tenterfield to Bombala
Temperate wet heath[f]	Warm temperate (weakly seasonal) ENSO	Melaleuca, Callistemon, Leptospermum, Epacris, Sprengelia, Bauera, Banksia, Drosera, Xanthorrhoea, Gahnia, Lepidosperma, Gymnoschoenus, Schoenus, Empodisma, Lepyrodia, Leptocarpus, Baloskion	Broad-toothed rat (Mastacomys fuscus), freshwater crayfish (Eustachys spp.)	Coastal dunefields and flats from Queensland to Victoria. Headwater valleys on Great Dividing Range and coastal plateaux in southeastern Australia, Tasmanian buttongrass plains
Alpine heath[g]	Cold temperate, periodic snow	Epacris, Leucopogon, Richea, Phebalium, Prostanthera, Grevillea, Bossiaea, Oxylobium, Baeckea, Podocarpus, Olearia, Astelia, Carex, Oreobolis, Baloskion, Poa	Mountain pygmy possum (Burramys parvus), broad-toothed rat (Mastacomys fuscus)	Kosciusko range, Victorian alps, Tasmanian alps

Note:

[1] ENSO – El Niño/Southern Oscillation.

Sources: [a] Russell-Smith et al. (1998), J. C. Z. Woinarski (personal communication); [b] Bell et al. (1984); [c] Specht et al. (1958); [d] Myerscough et al. (1995), Benwell (1998) and Bradstock et al. (1997); [e] Keith and Benson (1988); [f] Keith and Myerscough (1993) and Keith (1995a); [g] Kirkpatrick and Dickinson (1984) and Williams and Ashton (1988).

Table 9.2. *Summary of survey by Kikkawa et al. (1979) of vertebrate species recorded in eight heathland regions of Australia, identifying the total numbers of species (only native species included) recorded in each region and those showing special association with heathland habitats of these regions (in parentheses)*

	Frogs	Reptiles	Birds	Mammals
Tropical heath				
Arnhem and Kimberley	16(1)	76(11)	200(6)	53(0)
Sandplain kwongan				
SW Western Australia	13(2)	70(16)	182(20)	38(6)
Central lowland heath				
South Australia	8(0)	37(0)	220(18[a])	31(4)
East coast dry heath and nearby temperate wet heath				
Cape York	12(0)	33(5)	196(3)	22(0)
SE Queensland and NE New South Wales	19(4)	58(2)	114(13)	39(0)
Sydney region	9(2)	43(3)	209(16)	33(2)
Victoria	22(3)	44(1)	219(13)	52(3)
Temperate wet heath				
Tasmania	9(0)	15(1)	91(13)	23(0)
Mean% specialists	9.8	9	7.9	5.1

[a] *Pezoporus wallicus* included even though now extinct in the SA heathlands.

spatial scales (<0.1 ha) varies among Australian heathlands from moderate to very high in global terms, the highest diversity occurring in southwestern Australia and the Sydney basin (George *et al.* 1979; Keith and Myerscough 1993). The vertebrate fauna is less diverse in heathlands than in some other habitats such as forest. However, a small but significant component of heathland specialists may represent up to 22% of recorded species in a taxonomic group in particular heathlands (e.g. frogs in northeast NSW, Table 9.2). The invertebrate faunas of heathlands are very poorly known, but are likely to show high levels of specialisation and endemism, given the roles of these species in pollination, herbivory, seed predation and nutrient cycling.

Fuels, fire behaviour and fire regimes

Australian heathlands, like those on other continents, are renowned for their flammability (Chandler *et al.* 1983). Features contributing to the fire-prone nature of heathlands and their ability to support intense fires include the presence of well-aerated fine fuel, the tendency for dead foliage to persist on some plant species, and the direct exposure of fuel particles to wind and solar radiation in the absence of tree canopies. Flammable terpenes and waxes present in the foliage of some shrubs may also promote combustion of live fuel components (Vines 1981). These factors result in heathlands remaining fire-prone throughout much of the year, unlike forest fuels dominated by leaf litter which may be too moist to burn during the wetter months, and grassy fuels which have a distinct fire season linked to the physiological process of annual curing (Cheney and Sullivan 1997). Understanding the factors that determine the occurrence and behaviour of fire in heathlands is a prerequisite for interpreting and managing fire regimes at the landscape level.

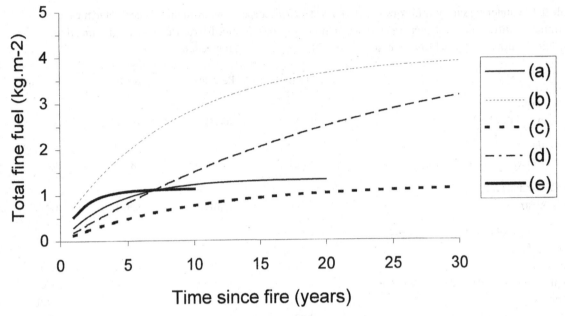

Fig. 9.2. Total fuel loading as a function of time since fire for selected Australian heathlands: (a) western mallee-heath at Stirling Range (McCaw (1997); (b) eastern lowland dry heath at Ku-ring-gai Chase (Conroy 1993); (c) temperate wet heath on low productivity sites in southwest Tasmania (Marsden-Smedley and Catchpole 1995a); (d) temperate wet heath on moderate productivity sites in southwest Tasmania (Marsden-Smedley and Catchpole 1995a); and (e) temperate wet heath at Cooloola, southeast Queensland (Sandercoe 1992). Curves have been drawn using equations provided by the original authors, and have not been extrapolated beyond the range of the original data.

Fuel structure and dynamics

The most important plant characteristics that affect fuel structure and load include height, bark texture, density and fineness of foliage and the degree of retention of dead foliage (Catchpole, this volume). Changes in fuel characteristics following fire have been examined in sandplain kwongan (Delfs et al. 1987) and western mallee-heath (McCaw 1997), central lowland heath (Specht 1966), east coast dry heath (Conroy 1993), and temperate wet heath (Sandercoe 1992, Marsden-Smedley and Catchpole 1995a) (Table 9.2). Despite wide variation in structure, floristic composition and environmental conditions amongst these communities, fuel load has generally been described as an exponential function of time since fire (T):

$$\text{FUEL LOAD} = a\,(1 - e^{-b.T})$$

where a is maximum load, and b is rate of accumulation. Similar models have been used to describe litter accumulation in eucalypt forests (Fox *et al.* 1979; Birk and Simpson 1980). Fuel loads in heathlands typically reach equilibrium within 10–15 years after fire, but may continue to accumulate for more than 30 years on productive sites (Fig. 9.2).

Equilibrium fuel loads are low (<1.5 kg m^{-2}) for heathlands on sites that are nutrient poor and prone to seasonal drought (Delfs *et al.* 1987; McCaw 1997) or waterlogging (Marsden-Smedley and Catchpole 1995a), but may exceed 3.0 kg m^{-2} on more productive sites (Specht 1969; Conroy 1993; Marsden-Smedley and Catchpole 1995a). Fine fuels comprise live and dead particles less than 6 mm diameter (McCaw 1991). The proportion of live fine fuel consumed during heathland fires varies according to the intensity of the fire, the location of the fuel in relation to the flaming zone, and the condition of the plants, among other things. Experimental fires in western mallee-heath rarely consumed live fuels >6 mm diameter, even when flames completely engulfed the standing vegetation (McCaw 1997). However, Whight and Bradstock (1999) reported that fires in east coast dry heath around Sydney consumed shoots up to 10 mm diameter, particularly where the vegetation had been unburnt for 15 years or more.

Trends in vegetation cover with time since fire generally follow a similar pattern to fine fuel load (Delfs et al. 1987; McCaw 1997), although cover is difficult to estimate reliably in the field (Marsden-Smedley and Catchpole 1995a; Catchpole et al. 1998).

The pattern of accumulation for particular components of the fuel may differ from that for total fuel load and cover. Central lowland heath and western mallee-heath in semi-arid climates may experience a flush of ephemeral grasses, for example Austrostipa spp., in the first 1 or 2 years following fire which temporarily provides sufficient fuel for a subsequent fire before shrubs return to dominance (Specht et al. 1958; McCaw 1997). In western mallee-heath and temperate wet heath, the load and proportion of dead fuel, including leaf litter, continues to increase for at least 20 years after fire thereby contributing to a general increase in the flammability of older vegetation, even though the total fine fuel load approaches an equilibrium level (Marsden-Smedley and Catchpole 1995a; McCaw 1997).

Fuel continuity has a major influence on fire behaviour in heathlands, particularly the conditions under which a fire will spread, and which layer of the fuel will dominate the spread process. Few heathlands provide a uniform continuous fuel, either in the horizontal or vertical dimension. Catchpole (1985) found considerable variation in shrub height and density at a localised scale in eastern montane heath in southern NSW, despite the apparent uniformity of the vegetation. Three separate vertical strata were recognisable, including a surface layer of dead litter, a middle layer of twigs and branches, and an upper layer of live foliage. Each of these strata may contribute differently to fire spread. In floristically richer heathlands, further heterogeneity in fuel characteristics may result from species-linked traits including leaf size and shape, and whether or not litter accumulates preferentially beneath particular shrub species. Litter cover and load in western mallee-heath have been found to be greatest beneath tall shrubs and mallees. Under marginal conditions for fire spread, flames will only be sustained in these areas of more continuous fuel (McCaw 1997). Heaths with abundant graminoids are notable for their continuous fuel structure (Marsden-Smedley and Catchpole 1995a).

A notable feature of most Australian heathlands is their ability to carry fire within a relatively short time following substantial rain, with as little as 24 hours required in the case of temperate wet heath in Tasmania (Marsden-Smedley and Catchpole 1995a). This phenomenon can in part be attributed to rapid drying of fine dead fuels which are exposed more directly to wind and solar radiation in heath than in forest where the rate of fuel drying is restricted by the presence of a tree canopy. Vegetation with a substantial component of elevated dead fuel, such as temperate wet heath, may sustain fires when the dead fuel is affected by surface moisture, at least up to a moisture content of 70% of oven dry weight (Marsden-Smedley and Catchpole 1995b).

In addition to rapid drying after rain or heavy dew, the exposure of dead fuels in heathlands means that fuel particles may attain low moisture contents under relatively mild weather conditions. For example, litter fuels on the ground in canopy gaps of western mallee-heath consistently dried to a minimum moisture content of 5% or less under typical summer conditions when air temperatures were in the range 25–30 °C (McCaw 1997). A moisture threshold of 7% represents a dangerous level of fuel dryness in eucalypt forest litter (Luke and McArthur 1978). Litter fuels were found to be consistently drier than elevated dead leaves in western mallee-heath, but this situation could be reversed in other heathlands with a deeper litter layer that retained more moisture, or where the litter was more completely shaded by shrubs.

Table 9.3. *Forward rate of spread (FROS) and fire-line intensity of selected wildfires in Australian heathlands in relation to fire weather conditions*

Heathland class	Fire location and date	Fuel age (years)	T[1] (°C)	RH[1] (%)	Wind speed[1,2] (km h^{-1})	FROS (m s^{-1})	Fire-line intensity (MW m^{-1})
Western mallee-heath[a]	Fitzgerald River (WA), December 1989	>20	35	9	30	2.08	35
	Fitzgerald River (WA), December 1997	>20	30	45	22	1.10	25
Central lowland heath[b]	Heywood (western Vic), February 1991	5	18	50	19	0.33	9
East coast dry heath[c]	Royal NP (NSW), January 1994	5	35	10	20	0.67	25
Temperate wet heath (Buttongrass, moorland)[d]	Birchs Inlet (Tas), October 1985	25	21	46	36	0.92	19
	Mulcahy Bay (Tas), November 1986	16	27	30	35	0.90	16

Notes:

[1] Values of air temperature (T), relative humidity (RH) and wind speed represent averages for the period of the fire run. [2] Wind speeds have been standardised to 2 m height in open conditions.

Sources: [a] McCaw *et al.* (1992); [b] Wouters (1993); [c] R.A. Bradstock (unpublished data); [d] Marsden-Smedley and Catchpole (1995b).

The effect of live fuel moisture content on fire behaviour has not been extensively studied in Australian heathlands, but in western mallee-heath variation in live moisture content from season to season, and year to year is quite limited (McCaw 1997). The issue of whether or not moisture content of live fuel declines substantially during prolonged drought warrants further investigation across the range of environments where heathlands occur.

Fire behaviour

As a general rule, forward rates of spread in heathland fires are higher than in open eucalypt forest but lower than in grasslands under equivalent weather conditions. There is a tendency for fine dead fuels to be drier in heathlands than in open-forests and wind speeds are substantially reduced by a forest canopy. Fuel particles in heathlands tend to be coarser and less evenly distributed than in most grasslands, both factors reducing the rate of forward spread (Catchpole, this volume). Correlations between fire spread in heathlands and predictions from models developed for Australian forests and grasslands have been found to be weak and inconsistent, and this has stimulated research to develop specific fire behaviour guides for important heathland fuel types (Marsden-Smedley and Catchpole 1995b; McCaw 1997; Catchpole *et al.* 1998). Functional relationships incorporated in these and other fire behaviour models have been reviewed by Catchpole (this volume). Prospects for the development of a generalised Australian heathland fire behaviour model are promising (Catchpole *et al.* 1998).

Heathland fires are capable of rapid rates of forward spread. Spread rates of around 2.1 m s^{-1}

and fire-line intensities of 35 MW m^{-1} have been documented for wildfires burning under severe weather conditions (Table 9.3). These values represent averages taken over periods of several hours and considerably higher spread rates and intensities would be expected during shorter pulses of peak fire behaviour.

In heathlands that have emergent tall shrubs, mallees or trees, flames may extend into, and spread through, the crown stratum. This is more likely when trees and tall shrubs have fibrous bark or long bark ribbons, abundant dead foliage is retained on the stem, and the crown foliage is fine and contains volatile compounds (Vines 1981). Critical thresholds of fire behaviour needed for crown fire initiation will vary between vegetation types; McCaw (1997) proposed that crowning would occur in mallee-heath when the forward rate of spread and fire-line intensity exceeded 0.4 m s^{-1} and 8.5 MW m^{-1}, respectively.

Fire regimes

Fire regimes experienced by heathlands vary across the Australian continent according to climate, terrain, land management objective and surrounding land use. The ability of heathlands to carry fire throughout much of the year makes it difficult to define a distinct fire season and to speculate on what fire regimes might occur in the absence of human influence in particular regions. In tropical heath, fires are largely restricted to the dry winter season. Early dry-season fires derive largely from human ignitions and are generally of lower intensity than those later in the season initiated by both human and lightning ignitions (Russell-Smith et al. 1998). In Mediterranean climate regions, fires may burn extensive areas of sandplain kwongan, western mallee-heath and central lowland heath at any time from early spring (August), throughout the long dry summer and into late autumn (May). Heathlands with a substantial component of elevated dead fuel may be capable of burning during dry spells even in the coolest and wettest months of June and July (W. L. McCaw, unpublished data). Lightning-caused ignition is most common in southwestern Australia during December and January, but has been recorded as early as October and as late as April. There may be a substantial delay between the timing of ignition and the occurrence of widespread fire activity, particularly in landscapes with a mosaic of heathland and woodland or forest where fires may smoulder for several months in logs and stumps, only to break out later during severe weather conditions.

Australia's eastern heathlands (east coast dry heath, eastern montane heath and temperate wet heath) experience a peak fire season that occurs in spring–early summer in the subtropics and shifts with increasing latitude to late summer–autumn in Tasmania. Rainfall reduces the probability of fire spread after midsummer in northern NSW and Queensland and before midsummer in the far south. Most of these heathlands have the capacity to carry fire throughout much of the year, although in western Tasmania the likelihood of re-ignition is reduced because rain-free periods of more than 14 days are rare (Marsden-Smedley and Catchpole 1995b). In the Sydney region the predominant season of prescribed burning is autumn–winter (April–July) (McLoughlin 1998), although peak activity of all fires is spring–midsummer (Conroy 1996). Fire occurrence is more seasonally defined in montane and alpine heathlands where the presence of snow and periods of cool humid weather prevent the ignition and spread of fires during winter months.

Heathlands may undergo a period of several years after fire when fuels are too sparse to support a spreading fire because fuel continuity is related to vegetation age. There is considerable variation in the minimum inter-fire periods documented for different Australian heathlands, with periods as short as 18 months in east coast dry heath near Sydney (Bradstock et al. 1997), 3 years in Tasmanian temperate wet heath (Marsden-Smedley and Catchpole 1995b), 5–8 years in western mallee-heath (McCaw et al. 1992; McCaw 1997), and more than 7 years in central lowland heath with an understorey of Melaleuca uncinata (Grant and Wouters 1993). Bradstock et al. (1998) defined the probability of a fire spreading in Sydney sandstone heathland according to the time since fire, representing fuel load and continuity, and the severity of burning conditions. They based this probability function on a combination of empirical observation and predictions from fire behaviour models. This type of analysis provides a useful framework for future observations and improving understanding of factors governing fire spread in young fuels.

Fire frequency is a critical component of fire regimes in heathlands because it is strongly linked to patterns of human activity, and because it can profoundly influence the composition and structure of the vegetation (see next section). Remnants of heathland and other native vegetation around urbanised areas are particularly vulnerable to degradation by this mechanism because of the high probability of ignition by arson and accidental fires. However, lightning fires may also burn areas at a frequency sufficient to threaten populations of slower maturing obligate seeders. For example, successive lightning-caused fires in 1983 and 1991 burnt more than 10 000 ha in Cape Arid National Park on the southern coast of Western Australia including heathlands containing shrubs with juvenile periods of 6–8 years.

Interactions between fuel load and moisture content, fire weather and landscape features will determine the patchiness of fire regimes. More topographically varied heathlands, such as those of the Arnhem escarpment, Sydney sandstone plateaux, Great Dividing Range, southwest Tasmania and the Stirling Range are likely to have greater fire patchiness than heathlands on extensive plains such as the sandplain kwongan, central lowland heath and east coast sandsheets.

Effects of fire regimes on plant populations and communities

Fire-related processes in plant populations

The life-history characteristics of plants have a profound influence over the response of their populations to fire (Gill 1975). Species that share life-history characteristics may be expected to have similar mechanisms of response to individual fire events (e.g. vegetative recovery, fire-stimulated ger-

mination, serotiny, propagule dispersal etc.). Different forms of life-history analysis are needed to describe the mechanisms of response and population outcomes of fire regimes (Noble and Slatyer 1980). Here, we defined six life-history attributes pertaining to survival, reproduction, dispersal and establishment characterisitics (Table 9.4) to describe mechanisms of plant population responses to fire regimes. Where data were available, we also examined the timing of critical life stages such as maturation and senescence. We examined variation within and between different types of Australian heathland (Table 9.1) by comparing the frequencies of species with different life-history characteristics. We interpreted life-history data from the literature according to the attributes defined in Table 9.4, and resolved some of the unknown attributes by contacting the authors and by drawing upon our own field experience in respective heathlands. The life-history data may be used to classify species into functional groups to examine the population outcomes of fire responses (e.g. Noble and Slatyer 1980), but such an analysis was beyond the scope of this review (see Whelan *et al.*, this volume for such an application).

Life-history attributes

A diverse range of plant life histories is represented in Australian heathlands, with variability in the representation of particular syndromes occurring at both local and regional scales (Table 9.5). Only South African heathlands may have a comparable diversity of fire related life-history types (Le Maitre and Midgley 1992), while those of the northern hemisphere appear to be less rich in species (Keith and Myerscough 1993), as well as life-history types (Zedler 1977; Hobbs *et al.* 1984). Nevertheless, some characteristics appear to be ubiquitous in vascular heathland floras worldwide, such as localised propagule dispersal, fire-related patterns of seedling recruitment, representation of myrmecochory, persistent soil seed banks and vegetative recruitment (Zedler 1977; Hobbs *et al.* 1984; Le Maitre and Midgley 1992).

Table 9.4. *Definition of life-history attributes pertaining to the survival, reproduction, dispersal and establishment characteristics of plant species*

Life-history attribute	Definition
Wide propagule dispersal	Plant propagules able to disperse frequently between habitat patches over scales of kilometres. Inferred from propagule morphology: buoyant on air or water; ingested and excreted by wide-ranging animals; adherent to fur or feathers of wide-ranging animals.
Vegetative recovery	Established plants survive when just subject to 100% leaf scorch and sprout new shoots from dormant or active recovery buds (sprouters).
Vegetative spread	Sprouters capable of expanding laterally over scales of tens of metres by way of vegetative organs such as rhizomes, stolons, suckers etc.
Seed bank type	Canopy: seeds retained within serotinous (or bradysporous) woody fruits. Seeds largely survive fire within fruits and are subsequently released. Persistent soil seed bank: seeds or fruits released from plants when mature and largely remain dormant until stimulated to germinate by fire. Transient seed bank: seeds or fruits released from plant when mature and germinate without dormancy when exposed to adequate moisture.
Pyrogenic reproduction	Exclusive: production of reproductive organs is stimulated only in response to fire. Facultative: production of reproductive organs is substantially enhanced in response to fire but occurs at reduced levels in absence of fire.
Pattern of recruitment	Post-fire: recruitment largely restricted to the immediate post-fire period when increased levels of resources are available. Few individuals recruited in fire intervals contribute to next generation. Fire interval: appreciable levels of recruitment and establishment occur in the interval between fires, beyond the period of increased resource availability.

Dispersal

Few species (<10%) in Australian heathland floras have long-distance dispersal mechanisms, except in alpine and tropical habitats (Table 9.5). Up to 50% of alpine heath floras comprise species whose propagules are buoyant in air (Asteraceae, Poaceae), adhere to mammal fur (Poaceae, Cyperaceae, Rosaceae) or are ingested by birds (fleshy-fruited Epacridaceae, Rubiaceae, Rosaceae). One-quarter of the tropical heath flora includes species with fleshy ingestible fruits. The few widely dispersed species in temperate heathlands include pteridophytes with air-borne spores and awned grasses transported on mammal fur. However, the majority of heath species in temper-ate regions have large heavy seeds, including some with wings, hairs or elaiosomes, or smaller seeds with no structures to assist dispersal. The few studies of seed dispersal in heath suggest that heavy winged seeds rarely move more than tens of metres from their source (Lamont 1985; Hammill *et al.* 1998), while ant-dispersed seeds rarely move more than 10–15 metres (Berg 1981).

The paucity of widely dispersed species in temperate heathlands contrasts with other habitats which include more fleshy-fruited species (mesic forests: French 1991), or species with air-buoyant diaspores (grassy woodlands and deserts). This means that vegetation dynamics in heathlands is

Table 9.5. *Number and frequency (%) of plant species with six life-history attributes in various Australian heathlands*

Location	Wide dispersal	Vegetative recovery	Vegetative spread	Seed bank type			Pyrogenic flowering		Post-fire recruitment
				Canopy	Persistent soil	Transient	Exclusive	Facultative	
Tropical heath									
Arnhem escarpment[a]	32(25)	60(46)	21(35)	3(2)	68(52)	3(2)	?(?)	1(2)	80–126(62–97)
Alpine heath									
Tasmanian alps[b]	39(48)	30(37)	18(60)	0(0)	20(25)	17(21)	1(3)	1(3)	3–62(4–77)
Victorian alps[c]	28(27)	87(83)	10(11)	2(2)	38(36)	15(14)	0(0)	3(3)	21–23(20–22)
Sandplain kwongan									
Badgingara[d]	7(5)	100(66)	14(14)	12(8)	75(50)	40(26)	8(8)	?(?)	102–151(68–100)
Jurien[e]	7(8)	57(62)	3(5)	11(12)	41(45)	7(8)	2(4)	10(18)	50–94(54–100)
Western mallee-heath									
Stirling Range[f]	3(2)	69(44)	3(4)	45(29)	77(49)	20(13)	1(1)	23(33)	139–156(87–98)
Central lowland heath									
Dark Island[g]	6(16)	22(59)	3(15)	4(11)	21(57)	1(3)	1(5)	?(?)	25–37(68–100)
Ngarkat[h]	6(8)	54(68)	7(13)	10(13)	47(59)	7(9)	4(7)	7(13)	61–79(76–99)
Ngarkat[e]	10(14)	43(58)	6(15)	7(9)	46(62)	7(9)	3(7)	6(14)	61–74(82–100)
Eastern montane heath									
Wadbilliga[i]	6(9)	47(70)	3(6)	9(13)	52(78)	2(3)	0(0)	5(11)	47–67(70–100)
Blue Mountains[e]	7(5)	81(60)	8(10)	19(14)	92(68)	9(7)	1(1)	15(19)	104–135(77–100)
Gibraltar Range[j]	5(6)	60(76)	10(17)	15(19)	38(48)	9(11)	4(7)	8(13)	56–79(71–100)
East coast dry heath									
Anglesea[k]	8(7)	84(76)	7(8)	7(6)	47(43)	39(35)	6(7)	27(32)	50–110(45–100)
O'Hares Creek[l]	6(4)	102(73)	17(17)	15(11)	72(52)	22(16)	9(9)	37(36)	86–139(62–100)
Myall Lakes[m]	5(6)	52(66)	13(25)	7(9)	53(67)	6(8)	0(0)	9(17)	60–79(76–100)
Broadwater[n]	3(3)	57(66)	11(19)	7(8)	67(78)	10(12)	2(4)	8(14)	64–88(74–100)
Brisbane Water: rock substrate[o]	1(2)	32(52)	6(19)	3(5)	41(66)	11(18)	4(13)	12(38)	43–62(69–100)

O'Hares Creek: rock substrate[p]	5(6)	47(55)	8(17)	13(15)	48(56)	9(10)	4(9)	13(28)	59–86(69–100)
Temperate wet heath									
Gibraltar Range[j]	4(7)	41(76)	10(24)	8(15)	21(39)	9(17)	4(10)	9(22)	31–54(57–100)
O'Hares Creek[l]	13(12)	81(76)	25(31)	15(14)	40(38)	20(19)	9(11)	27(33)	62–106(58–100)
Myall Lakes[m]	4(5)	67(76)	21(31)	13(15)	46(52)	9(10)	7(10)	14(21)	59–88(67–100)
Broadwater[n]	4(4)	67(70)	16(24)	13(14)	72(75)	9(9)	4(6)	12(18)	73–96(76–100)
Leura[q]	1(2)	49(78)	8(16)	10(16)	30(48)	5(8)	2(3)	16(25)	41–63(65–100)
Melaleuca[r]	9(9)	81(81)	21(27)	8(8)	34(34)	18(18)	5(5)	12(12)	47–97(47–97)
Gippsland High Plains[c]	6(11)	44(77)	11(25)	0(0)	10(18)	1(2)	0(0)	2(5)	2–5(4–12)

Sources: [a]Russell-Smith et al. (1998); [b]Kirkpatrick and Dickinson (1984); [c]Wahren et al. (1999); [d]Bell et al. (1984); [e]D. A. Keith (unpublished data); [f]McCaw (1997); [g]Specht et al. (1958); [h]Forward (1996); [i]B. Mackenzie, M. G. Tozer, D. A. Keith and R. A. Bradstock (unpublished data); [j]Williams (1995); [k]Wark et al. (1987); [l]Keith and Myerscough (1993 and unpublished data); [m]Myerscough et al. (1995); [n]Benwell (1998); [o]Bradstock et al. (1997); [p]Keith (1994 and unpublished data); [q]Holland et al. (1991); [r]Keith (1995a and unpublished data).

governed largely by disturbance and population processes at local spatial scales, much more so than in other plant communities. Once eliminated from local areas of heathland, species are thus likely to remain absent over very long time-scales before dispersal and re-establishment. Temporal patterns in Holocene charcoal and pollen of several locally dispersed Tasmanian alpine species are consistent with this model of fire-driven extinction and slow return (McPhail 1979). The naturally patchy distribution of heathlands and increasing fragmentation due to land use changes would exacerbate this effect.

Recruitment

Australian heathlands are dominated by species whose seedling recruitment is largely restricted to the immediate post-fire period and almost devoid of species whose recruitment is unrelated to fire (Table 9.5). The latter occur mainly in alpine heath, but may occur in long-unburnt heath as low-frequency vagrants from nearby rainforests. Patterns of seedling recruitment were unknown in 10%–50% of most heathland floras (Table 9.5), mainly because seedlings of geophytes and graminoids may be cryptic within established populations. For some woody species, low frequencies of recruitment have been observed in long-unburnt heathland (e.g. Specht *et al.* 1958; Gill and McMahon 1986; Witkowski *et al.* 1991). However, few of the individuals recruited during the intervals between fires may establish, grow to maturity and contribute to future generations before they are killed in subsequent fires (Specht *et al.* 1958; Tozer and Bradstock 1997).

Levels of seedling recruitment in the first post-fire year may exceed that in subsequent years by orders of magnitude (Fig. 9.3). Inter-fire seedling recruitment may be limited in heathlands by seed availability (Whelan *et al.* 1998), predation (Bradstock 1991) or competition (Tozer and Bradstock 1997). This pattern is consistent with observations in other heathlands around the world (e.g. Kruger 1983; Hobbs *et al.* 1984), but contrasts with stronger recruitment beyond the immediate post-fire period in other communities. Whelan *et al.* (1998), for example, found evidence of seedling establishment

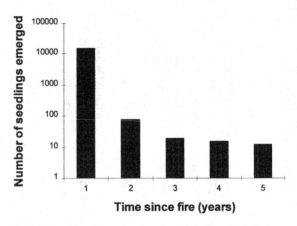

Fig. 9.3. Trend in seedling emergence with time since fire in 25 plant species of a temperate wet heathland near Sydney. Data from Keith (1991).

of *Banksia serrata* between fires in dry sclerophyll forests but not in nearby heathlands.

Many heathland species with unknown patterns of seedling recruitment are capable of vegetative reproduction. Unlike the recruitment of seedlings, vegetative recruitment may continue throughout the interval between fires, although the rate of clonal recruitment may be enhanced in the immediate post-fire period (Lamont and Runciman 1993). Some clonal species with horizontal rhizomes, suckers or other subterranean stems may extend their occupation of habitats appreciably by vegetative spread, but those reproducing by bulbs or tillers may not spread so widely. Rhizomes may become fragmented with time, leaving physiologically independent but genetically identical individuals spread over large areas (Sydes and Peakall 1998). Species that spread vegetatively were more frequent in temperate wet heathlands than east coast dry heathlands (Table 9.5).

Survival

Typically 60%–70% of each heathland flora comprises species capable of vegetative post-fire recovery (Table 9.5). Sprouting and non-sprouting forms of the same species may be found close by in different habitats (Keith 1991, 1995b; Benwell 1998). Tropical heathlands and some dry lowland heathlands in the east and west had relatively low proportions of sprouters (*c.* 45%–55%), while Tasmanian alpine heath was exceptional with only one-third of its species capable of vegetative recovery. Wet heathlands had a slightly higher proportion of sprouters compared with dry heathlands at the same locations, while dry heathlands on rocky substrates had the lowest proportions of sprouters (Table 9.5). Keith (1991) suggested that obligate seeders (non-sprouters) were not favoured in wet heathland habitats because survival and growth of seedlings were curtailed by competition from rapid vegetative regrowth of sprouting species, more so than in drier heathland habitats. Hence, in wetter habitats only the most rapidly growing non-sprouters reach maturity prior to canopy closure and propagate in abundance. In contrast, rocky habitats have the most open space, the least-developed soil and potentially the lowest levels of competition.

Heathland sprouters and non-sprouters have different characteristics of growth and nutritional storage. Seedlings of non-sprouters typically grow faster and mature earlier (Zammit and Westoby 1987; Pate *et al.* 1990, 1991; Hansen *et al.* 1991; Bell and Pate 1996), thus directing a high proportion of photosynthate towards establishment of a large seed bank before the next fire. Sprouters, conversely, develop a higher root:shoot biomass ratio and accumulate more starch in their root tissues (Pate *et al.* 1990, 1991; Bell and Pate 1996). The opportunity cost in terms of reduced growth rate and reproductive effort is offset by establishment of fire-resistant organs and starch reserves, which form the basis of renewed growth when leaves and shoots are killed by fire.

Levels of post-fire survival vary among and within sprouter species according to the size, location and degree of insulation of their recovery organs in relation to the heat output of individual fires (Gill 1981). In general, higher levels of heat exposure cause higher levels of mortality (e.g. Bradstock and Myerscough 1988). Variability in the size of recovery organs reflects variation in the age and growth rate of individuals within populations. Given similar levels of heat exposure, individuals with small recovery organs suffer greater fire-related mortality than larger individuals in a range

of woody heath species (Bradstock and Myerscough 1988; Auld 1990; Bradstock 1995; Keith 1996).

Buds and starch reserves may be depleted by high-frequency fires, even though they are replenished between fire events (Zammit 1988; Bowen and Pate 1993; Bell and Pate 1996). Some woody sprouters may be surprisingly susceptible to frequent fires of certain intensities or seasons because these may cause elevated levels of mortality and inhibit recruitment (e.g. Bradstock and Myerscough 1988; Zammit 1988; Noble 1989). If fire-resistant organs are slow to develop (e.g. Bradstock and Myerscough 1988; Keith 1996), fires at short intervals will repeatedly kill new recruits so that elevated losses of established plants cannot be offset. Keith and Tozer (1997) suggested this explanation for observed declines in populations of *Banksia oblongifolia*, subject to a 5-year fire interval, while those subject to a 14-year fire interval remained stable. In contrast, some herbaceous sprouters may be highly resilient to, or even favoured by, fire intervals as short as 3 years or less (Lamont and Runciman 1993; Tolhurst and Burgman 1994). Those capable of vegetative spread independent of fire, which are most abundant in temperate wet heath (Table 9.5), will also be resilient to infrequent fire regimes.

Most heathland sprouters have subterranean recovery organs which may avoid up to 95% of the heat generated in a fire above ground (Packham 1970). These organs include lignotubers, rhizomes and suckers, active buds, corms, tubers and bulbs (Bell *et al.* 1984). Less than 10% of heathland sprouters have aerial recovery buds, although such species are more prevalent in forest and woodland habitats. Several heath genera have dormant aerial epicormic buds whose survival is predicated by bark thickness in relation to heat exposure during fire (Gill and Ashton 1968; Bradstock 1985). Some emergent shrub species that lack epicormic regrowth may nevertheless survive low-intensity fires if some of their canopy is held above scorch height and their bark insulates vascular tissue sufficiently (Morrison 1995; Keith 1996). This mechanism is stage-dependent, since older emergent individuals with thick bark, a high canopy and vertically discontinuous fuel structure have a higher chance of survival than smaller individuals without these characteristics. Other aerial mechanisms of vegetative recovery represented rarely in heathland include active aerial buds insulated either by crowded leaf bases or, uniquely in alpine habitats, by cushion growth forms.

Reproduction

Species with canopy (serotinous) seed banks were most frequently represented in western mallee-heath, where several diverse proteaceous and myrtaceous genera account for more than one-quarter of all species (Table 9.5). Only one serotinous (or 'bradysporous': Whelan 1995) species was recorded in alpine heath and few in tropical heath, but they account for 5%–15% of the species in other heathlands. Several *Leptospermum* and *Melaleuca* species accounted for slightly higher frequencies in wet habitats. Serotinous seeds have a greater chance of survival during fire if they are protected within thick-walled fruits (Bradstock *et al.* 1994) or if their fruits are aggregated into compact clusters (Judd and Ashton 1991). Seed survival also depends on the moisture content of fruit walls, which may vary with fruit age (Ashton 1986). In *Banksia*, fire temperatures also influence the extent and rate of seed release, with some species requiring higher temperatures than others to stimulate early and complete release (Enright and Lamont 1989). Wet–dry cycles in post-fire weather may enhance rates of seed release in some serotinous species (Cowling and Lamont 1985a).

Rates of seed retention in the absence of fire vary substantially between serotinous heath species. Some, such as the Sydney mallee, *Eucalyptus luehmanniana*, release most seeds within 1–2 years of maturation (Davies and Myerscough 1991), while others such as *Banksia elegans* in southwestern Australia retain nearly all seeds indefinitely (Lamont 1988). Cowling and Lamont (1985b) found a climatically plastic fire response in kwongan Banksias. Rates of seed retention increased, while arborescence decreased, as the habitat became drier, hotter and presumably more fire-prone. Thus, serotiny was greatest where canopy-consuming fires were most probable and least where relatively few fires consume the plant canopies and their fruits.

Canopy seed banks are exhausted completely by a single fire and lose viability when maternal adult plants senesce (Gill and McMahon 1986). Obligate seeders with canopy seed banks that are unable to recruit seedlings in the absence of fire may decline to extinction as fire intervals approach either the age of seed bank establishment or the age of senescence (Keith 1996). Population extinctions reviewed by Gill and Bradstock (1995) demonstrate the sensitivity of such species to both high- and low-frequency fire regimes. The species-rich heath floras of southwestern Australia are especially susceptible to loss of diversity when exposed to these regimes because of their large complement of serotinous obligate seeders. In other heathlands, although less species-rich, this group of species often includes the structural dominants, whose abundance may influence other aspects of heathland dynamics (e.g. Keith and Bradstock 1994).

Trends in the frequency of persistent soil seed banks and transient seed banks were difficult to discern, given incomplete seed dormancy data. We thus assumed that a distinct pulse of seedling emergence in the first year after fire was evidence of a persistent soil seed bank (e.g. Specht *et al.* 1958; Auld and Tozer 1995; Benwell 1998). Despite a varying number of species that could not be ascribed to either seed bank type, the majority of non-serotinous species recorded in all of the heathlands appear to have persistent soil seed banks (Table 9.5). In some eastern heathlands more than three-quarters of all species had persistent soil seed banks. Many of these species are dispersed by ants, which are attracted by lipid-rich elaiosomes attached to the seeds (Berg 1981). Australian heathlands are renowned globally for a high diversity of myrmecochorous species, which may account for 20%–50% of each flora (Berg 1981).

Fire-related dormancy has been demonstrated in a wide range of heath species with persistent seed banks, although most also produce a fraction of non-dormant seeds (Auld and O'Connell 1991; Dixon *et al.* 1995; Keith 1996). Seed dormancy may be of several innate or enforced types (Whelan 1995). Although innate dormancy seems common among heath species (Auld and O'Connell 1991; Bell *et al.*

1993; Dixon *et al.* 1995), some species lacking innate dormancy may have enforced dormancy mechanisms. For example, *Kunzea ambigua* seeds lack dormancy in the laboratory, but have long-lived fire-related dormancy in the field (Auld *et al.* 2000). Dormancy patterns may vary within and between seed crops of the same species (Auld *et al.* 2000) and may change over time as seeds age (Morrison *et al.* 1992; Tozer 1996; Roche *et al.* 1997). The principal fire-related cues for germination appear to be heat shock and chemical stimuli derived from smoke (Dixon *et al.* 1995; Keith 1996).

Heat-stimulated seeds, such as many legumes, have a hard impervious testa that is cracked under elevated temperatures, allowing the seed to imbibe. Auld and O'Connell (1991) found that threshold temperatures for germination varied between 40 °C and 80 °C among a group of 35 eastern Australian Fabaceae mainly from heathland habitats. They also showed that temperatures above 120 °C were consistently lethal to seeds of all species examined. Studies of other species have yielded similar conclusions (e.g. Bell *et al.* 1993). The proportions of dormant seed banks that either are stimulated to germinate, remain dormant or are killed by heat shock therefore depend on the amount of heat transferred to the soil during fire and the depth of seed burial (Auld 1987a). Soil temperatures during fire are related to ground-fuel consumption (Bradstock and Auld 1995) and soil type and moisture content (Whelan 1995), while depth of burial depends on ant activity and soil movement (Auld 1986).

Mechanisms controlling smoke-stimulated germination are less well understood, but this kind of dormancy is apparently widespread in temperate Australian heathland floras (Dixon *et al.* 1995; Keith 1997; Roche *et al.* 1997; Morris 2000). The active chemical constituents of smoke, although currently unknown, are apparently ubiquitous in bushfire fuels (Dixon *et al.* 1995). The factors influencing their concentration and leaching through the soil profile are also poorly understood, but may lead to different patterns of dormancy release compared with heat shock (Keith 1997). While some species with smoke-mediated germination are

apparently not germinable by other means and others respond to heat shock and not to smoke (Dixon *et al.* 1995), these two cues may interact in the germination of other species. The combined effects of heat and smoke stimulated higher rates of germination in some species of *Epacris* (Keith 1997 and unpublished data) and *Grevillea* (Morris 2000) than separate application of either treatment.

Unlike canopy seed banks, persistent soil seed banks may not be exhausted by a single fire and therefore confer greater resilience to high-frequency fire. Nevertheless, species with persistent seed banks will decline under high fire frequencies, depending on how rapidly the residual seed bank is depleted by successive fires. A variable proportion of the seed bank of *Acacia suaveolens* may remain dormant, depending on soil temperatures during fire (Auld 1987a). High-intensity fires may effectively exhaust its seed bank, but only if seeds remaining ungerminated are buried too deep to emerge. Some non-leguminous soil seed banks from sandplain kwongan may be exhausted more readily by a single fire (Meney *et al.* 1994; Dixon *et al.* 1995). In contrast, a study by Bradstock *et al.* (1997) suggests that seed bank exhaustion is less common in east coast dry heath. There was no significant difference in the abundance of 25 out of 28 obligate seeders with persistent seed banks between sites that had burnt in three consecutive years and sites burnt only once in the same period. Only two of the species were unequivocally less abundant in frequently burnt heath, suggesting significant seed bank depletion, but not exhaustion.

Transient seed banks are characteristic of many taxa in lilioid families (Anthericaceae, Blandfordiaceae, Colchicaceae, Haemodoraceae, Orchidaceae and Xanthorrhoeaceae) and pteridophytes, as well as Droseraceae and some woody dicot taxa (e.g. *Angophora hispida*: Auld 1987b; and *Telopea speciosissima*: Bradstock 1995). Germination from transient seed banks occurs in the first year after seed release, or otherwise viability is lost. The availability of seeds therefore largely depends on the timing of fruiting (Auld 1994). Species with transient seed banks account for at least 5%–25% of heathland floras (Table 9.5). Only species shown experimentally to lack dormancy were scored as having transient seed banks in Table 9.5. A further 10%–30% of species could have either transient or persistent seed banks. Some of these latter taxa may have persistent soil seed banks controlled by mechanisms of enforced dormancy (Whelan 1995). Representation of transient seed banks could be higher in alpine and tropical heath because the nature of soil seed banks is largely unknown in these communities.

Many species with transient seed banks have fire-stimulated (pyrogenic) flowering (e.g. Auld 1987b, Lamont and Runciman 1993), a feature that heathland floras share with some other fire-prone habitats (Gill and Ingwersen 1976; Lamont and Downes 1979; Bond and van Wilgen 1996). Post-fire seedling recruitment must therefore be delayed until after seed release from the post-fire flowering event(s); and is usually later than in species with seeds released from canopy seedbanks or persistent soil seed banks. The magnitude of pyrogenic flowering responses may be sensitive to fire season (Gill and Ingwersen 1976; Le Maitre 1984; Lamont *et al.* 2000) and possibly other ecophysiological factors (Lamont *et al.* 2000). In Australian lowland heaths, up to 13% of sprouters may have an exclusive pyrogenic flowering response (Tables 9.4 and 9.5), and a further 10%–40% may have a facultative response, while such responses are rare in alpine or montane heathlands. The majority of pyrogenic flowering occurs in monocots, although several species are woody dicots (Gill 1981; Keith 1996).

A few heath species apparently lack effective means of arrival or persistence after fire, and hence may be eliminated by a single fire. These include the six Tasmanian alpine conifers and *Nothofagus gunnii* (Kirkpatrick and Dickinson 1984), which have locally dispersed, short-lived, non-dormant seeds and lack organs of vegetative recovery (though some may propagate vegetatively in the absence of fire). The annual grass of tropical heath, *Thaumastochloa major*, may also be susceptible to single fires. If burnt during the dry season, it regenerates effectively from a short-lived exhaustible soil seed bank. However, during its juvenile growth phase in the wet season, seed reserves have been exhausted by the seasonal germination event, leaving no means of recovery in the unlikely event of fire.

Timing of critical life stages

Data from seven heathlands show that most obligate seeders reach maturity by 3 years after fire (Fig. 9.4). However, a small proportion of these species takes at least 6 years to mature and possibly up to 10 years. The slowest heath flora to mature is that of an east coast dry heath on skeletal rocky soils at Brisbane Water, near Sydney (Fig. 9.4). Arguably, this has the poorest site quality of the seven heathlands examined. Rates of maturation are likely to depend on site quality and post-fire rainfall (Bradstock and O'Connell 1988), but this requires more detailed study in a wide range of species. Most obligate seeders are unlikely to produce many seeds in the first post-fire reproductive season and subsequent rates of seed bank accumulation may vary substantially between species. Keith *et al.* (this volume) suggest that at least three reproductive seasons are required in most obligate seeders before significant seed bank accumulation occurs. In sprouters, resprouting individuals may become reproductive quite rapidly, usually from 1 to 5 years after fire, but seedling maturation may take considerably longer than in obligate seeders (Keith 1991; Bell and Pate 1996).

Data on the longevity of standing plants are scarce, but suggest great variability among heath species. Fire-ephemeral herbs may live for 1–5 years (Specht *et al.* 1958; Pate *et al.* 1985; Keith 1991), while obligate seeding forbs and shrubs may live for 10–50 years (e.g. Auld 1987a; Bradstock and O'Connell 1988; Burgman and Lamont 1992). Resprouters tend to live longer than obligate seeders. Ages in excess of 500 years have been obtained for mallee eucalypts by radiocarbon dating (Head and Lacey 1988), while ages in excess of 300 years have been estimated for *Xanthorrhoea* and *Kingia* by counts of stem rings (Lamont and Downes 1979). The longevity of clonally reproducing individuals may be effectively infinite, unless interactions with competitors, predators or pathogens adversely affect survival. These examples may describe extremes; many resprouters are likely to senesce within 100 years.

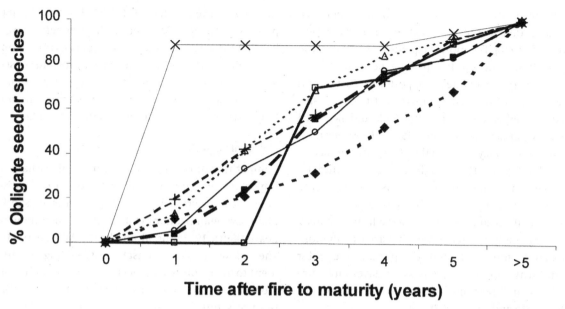

Fig. 9.4. Maturation times of obligate seeders in six Australian heathland floras: □ tropical heath on the Arnhem escarpment (Russell-Smith *et al.* 1998); × alpine heath in east Victorian highlands (Wahren *et al.* 1999); △ western mallee-heath on Stirling Ranges (McCaw 1997); + central lowland heath at Ngarkat Conservation Park (Forward 1996, and D. A. Keith unpublished data); ◆ east coast dry heath at Brisbane Water (Benson 1985); ■ east coast dry heath at O'Hares Creek (Keith 1991); and ○ temperate wet heath at O'Hares Creek (Keith 1991).

Data on the longevity of seeds is similarly limited. As stated previously, transient seed banks are exhausted within a year of release, while canopy seed banks are either soon exhausted or predisposed to incineration after death of parent plants (Lamont 1996). In these cases, populations with fire-related patterns of recruitment will become extinct when all standing plants senesce. In contrast, species with persistent seed banks may not become locally extinct in the absence of fire until decades after senescence of the standing population (Auld 1987a). Auld *et al.* (2000), assuming exponential rates of decay, found that half-lives of persistent soil seed banks of several east Australian temperate heath species varied from 0.4 to over 10 years. Le Maitre and Midgley (1992) noted that standing plants of obligate seeders with persistent soil seed banks are generally less long-lived than those with canopy seed banks that expire when adults senesce.

Fire-related processes in communities and ecosystems

Environmental interactions

The availability of soil moisture exerts a critical influence over the timing and levels of seedling emergence and survival (Cowling and Lamont 1987; Bradstock and O'Connell 1988; Lamont *et al.* 1991a; Whelan and York 1998). Recruitment may fail when fires are followed by protracted dry spells. Thus, the prevalence of fire-related seedling recruitment in heathland floras (Table 9.5) and the temporal variability of rainfall and temperatures in all heathland climates (Table 9.1) make post-fire weather one of the most important stochastic environmental factors affecting vegetation dynamics in Australian heathland. The Mediterranean and tropical heathlands have strongly seasonal rainfall (e.g. Taylor and Tulloch 1985), while in the southeast seasonal variation is small compared to variation between years related to the El Niño/Southern Oscillation (Nicholls 1990). The more frequently fires occur, the more often plant populations are exposed to the risk of poor recruitment due to dry post-fire weather, and the greater the risk of declines and extinctions, especially in obligate seeders (Bradstock and Bedward 1992; Burgman and Lamont 1992). These sorts of relationships may be crucial in the maintenance of heathland diversity through a period of climatic change (Cary, this volume). Landscape variation, upon which temporal variability is superimposed, may also affect soil moisture and seedling establishment. Spatially restricted sites that are rich in moisture and/or nutrients could be important refuges for some heath species exposed to unfavourable fire regimes and post-fire weather conditions (Keith 1991; Keith and Tozer 1997).

The habitat structure of certain alpine and wet heathlands is profoundly affected by fires that consume not only standing biomass, but also peaty substrates (Young 1982; Kirkpatrick and Dickinson 1984) and the subterranean vegetative recovery organs and seed banks that they contain (Gill 1996). The destruction of peat, loss of its capacity for moisture retention and cation exchange, and subsequent exposure of mobile gravels and sands is likely to influence the number and species of seedlings that are able to establish and grow in the post-fire environment. Long-lasting open space and unstable substrates seem likely to favour widely dispersed opportunists over species in which population persistence is predicated on the longevity of individuals and *in situ* recovery mechanisms. These opportunists account for a higher proportion of the alpine heath flora than any other heathland flora (Table 9.5). Kirkpatrick and Dickinson (1984) inferred that single fire events in Tasmanian alpine heath caused loss of peat, nutrients and mineral soil, and local extinctions of coniferous and deciduous shrub species. At several sites, a high proportion of the soil surface remained bare and conifers remained absent for at least 20 years after fire, while widely dispersed opportunistic species accounted for the majority of biomass. Once eliminated by such fires, alpine conifers may remain absent from a site for

thousands of years (McPhail 1979; Gill 1996). Similar processes apparently operate in temperate wet heathlands whose soils contain sequences of erosion and sedimentation indicative of peat fires and subsequent erosive rainfall events (Young 1982). Some European wet heathlands also show significant and long-lasting effects of peat fires (Maltby *et al.* 1990).

Biological interactions

Competition during the establishment phase has an important influence on the distribution of heathland species across landscape gradients in sandplain kwongan (Lamont *et al.* 1989), temperate wet heath (Keith 1991) and alpine heath (Wilson 1993). The widespread synchronisation of seedling recruitment to fire (Table 9.5) reflects partly the inability of seedlings to compete with established plants for resources, unless fire enhances resource supply and reduces competition from established plants. Post-fire seedling populations are subject to both thinning and reduced growth under high densities (Morris and Myerscough 1988). Competitive suppression of growth and reproduction is likely to reduce rates of seed bank accumulation and curtail the ability of such populations to persist under frequent fire regimes (Bond and van Wilgen 1996). Post-fire litter microsites support higher seedling densities, and hence stronger competition between seedlings than other parts of burnt landscapes. Lamont *et al.* (1989) observed that high-intensity fires were associated with greater levels of post-fire seedling competition than low-intensity fires because more seeds were released in response to higher canopy temperatures and litter microsites were possibly more restricted due to greater fuel consumption. However, low-intensity fires produced smaller recruitment rates, despite reduced competition between seedlings, due to the release of fewer seeds and competition for moisture from unburnt plants.

Whereas competition during recruitment in heathland appears to be driven by limited soil moisture (Lamont *et al.* 1989), light may be the limiting factor during the mature phase (Specht and Morgan 1981; Keith and Bradstock 1994). With increasing time since fire dominant overstorey shrubs overtop understorey plants, reducing their access to light and in turn, their fecundity and survival. The longer overstorey dominants monopolise light and soil resources during periods uninterrupted by fire, the greater the probability that understorey species will be eliminated from the site, leading to a decline in diversity of the community (Specht and Morgan 1981; Burrell 1981; Keith and Bradstock 1994). Species that are most susceptible to competitive elimination have low stature, intolerance of shade, short-lived seed banks and slow rates of growth and maturation (Keith and Bradstock 1994). Species with long-lived seed banks may be most resilient to competitive elimination, even though their standing plants may senesce prematurely, because their persistent seed banks provide the means for post-fire population recovery. The dominants are typically serotinous obligate seeders whose density is sensitive to fire frequency. Fire regimes will thus be a critical force in the maintenance of plant diversity in heathlands if fires repeatedly interrupt processes of competitive elimination (Keith and Bradstock 1994).

The dynamics of heath communities in response to long fire-free periods (>30 years) when overstorey dominants senesce is poorly known, partly because this scenario is rare, at least in recent decades. Inferences about heath vegetation dynamics in response to low fire frequencies are largely drawn from studies of chronosequences of sites of different post-fire ages (Specht et al. 1958; Brown and Podger 1982). However, it is possible to project likely changes from an understanding of life-history processes (Noble and Slatyer 1980; Gill and McMahon 1986; Keith and Bradstock 1994). Differential longevity, dormancy characteristics of propagules and establishment abilities of seedlings are likely to be critical to vegetation dynamics under low-frequency fire regimes (Keith and Bradstock 1994). In Victorian alpine heath, senescent dominant shrubs are replaced locally after an estimated 30–50 years by either snowgrass (Poa spp.) if a source of propagules is available, or shrubs depending on their life-history characteristics and the nature of gaps (Williams and Ashton 1988).

Predation of seeds and seedlings may be an important factor shaping post-fire plant commu-nities in Australian heathlands. Seed predation may occur at the pre-dispersal stage (e.g. Auld and O'Connell 1988; Witkowski et al. 1991) or at the post-dispersal stage (Andersen and Ashton 1985; Auld 1995). The principal pre-dispersal seed predators are larvae of beetles, wasps and moths (Andersen and New 1987), although in some plant species many seeds may be destroyed incidentally by avian preda-tors of these invertebrates (Lamont and van Leeuwin 1988; Vaughton 1998). The principal post-dispersal predators include ants (Andersen and Ashton 1985), other invertebrates and terrestrial mammals such as rodents (Auld and Denham 1999).

Lamont and van Leeuwin (1988), Andersen (1989), Vaughton (1998) and Auld and Denham (1999) reported examples where predators destroyed extremely high proportions (c. 90%) of the potential seed bank. If seed production is very high and safe sites for seedling establishment are relatively scarce, severe seed losses may ultimately have little impact on dynamics of the plant population (Andersen 1989). Nevertheless, in species with low fecundity, such high levels of predation may have a profound effect on rates of seed bank accumulation and hence the ability of the species to persist under high-fre-quency fire regimes (Auld 1995; Vaughton 1998; Auld and Denham 1999).

The effects of predation can be highly variable in space and time (Whelan and Main 1979; Auld and Denham 1999). Spatial variation may be explained by stochastic predator movement and behavioural patterns. Rates of predation may vary with time since fire (e.g. Lamont and Barker 1988), depending on the impact of fire on predator populations. Temporal variation in the production and release of seeds may reduce the impact of predation if the abundance of seeds is increased relative to the size of predator populations at critical times, an effect known as predator satiation (e.g. O'Dowd and Gill 1984; Auld and Myerscough 1986; Andersen 1988). Fire is the stimulus for the production and/or release of a superabundance of seeds in serotinous species (Lamont et al. 1991b), fire ephemerals (Pate et al. 1985) and species with pyrogenic flowering responses (Keith 1996), each of which are well repre-sented in various Australian heathlands (Table 9.5).

Thus, fire may be critical in the escape of seeds and seedlings from predators at the time most favourable for seedling emergence and establishment.

Root-borne fungal pathogens, particularly *Phytophthora cinnamomi*, are major causes of disease in temperate heathlands in winter maximum rainfall zones, including southwestern Australia, southern Victoria and Tasmania (Burdon 1991). The interactions between plants, fungal pathogens and fire are poorly understood, but potentially important to the long-term maintenance of heathland diversity. While standing plants of susceptible species are killed during the initial phase of a disease epidemic, soil seed banks may persist and might offer a means of disease evasion if pathogen activity declines with time since invasion (Keith 1995b). The occurrence and timing of fire, as a germination cue, may be crucial in the recovery of susceptible species with long-lived seed banks, but is unlikely to provide viable management options for those with canopy or transient seed banks.

Effects of fire regimes on fauna

Much of our knowledge of the responses of heathland animals to fires comes from very few studies at just a few sites. Work on mammals, birds and various invertebrate groups has previously been summarised by: Newsome *et al.* (1975); Kikkawa *et al.* (1979); Dwyer *et al.* (1979); Newsome and Catling (1979); Main (1981); Gill *et al.* (1981); Haigh (1981); Fox (1982); Friend (1993, 1995); and Woinarski (1999). General post-fire patterns of fauna abundance in heathland are typical of those observed after fires in virtually all other vegetation types. They can be summarised as follows (Whelan *et al.*, this volume):

- Fires can cause substantial mortality, varying in extent among species and among fires.
- Although some individuals die, many survive even the most intense fires. There is little information on the proportion of individuals surviving, nor on sex-, size- or age-specific survival of heathland fires.
- Some species that are recorded immediately after a fire disappear soon thereafter. This also

appears to vary among species and among fires.
- Predation levels are typically high after fire, and probably explain some of the post-fire disappearance and mortality of some species. The importance of predation in post-fire population dynamics is not clear.
- Some animals exhibit signs of starvation in the post-fire environment (weight loss, poor condition), and this may explain the disappearance of some species.
- Recolonising ability (mobility) influences the rate of reappearance of species that have been eliminated.
- Post-fire vegetation recovery influences the rate of reappearance in species that are dependent on vegetation cover and related habitat characteristics.
- Some species that were rare (or apparently absent) prior to a fire can become abundant soon after some fires but not others.

Below, we examine the features of animals, habitats and fires in heathlands that might explain these patterns and the variation in fire effects between and within species.

Animal characteristics

Variation in characteristics among animal species is likely to affect both immediate and longer-term responses to fire. For plants, classification schemes constructed using relevant life-history characteristics (e.g. Table 9.4) permit prediction of survival, recolonisation and recovery rates after individual fires, potential declines in the absence of fire and persistence under varying fire regimes. Although animals may not be so easily classified, here we attempt to apply the system presented by Whelan *et al.* (this volume). Thus, attributes such as microhabitat (canopy, shrubs, litter, soil, burrows), short-distance mobility, size, breeding season, specialisation of diet, susceptibility to predation, long-distance dispersal ability and intrinsic rate of population increase might all be expected to contribute to survival of fire and/or post-fire recolonisation and/or disappearance from long-unburnt habitat.

Table 9.6 compares two pairs of species that are

fire responses

Attribute	Example species			
	Pezoporus wallicus Ground parrot	*Dasyornis brachypterus* Eastern bristlebird	*Pseudomys gracilicaudatus* Eastern chestnut mouse	*Rattus lutreolus* Swamp rat
Microhabitat association	Ground layer[a]	Ground and shrub layers[b]	Ground layer	Ground layer
Ability to escape fire or find and use refuge	Good flier, observed to flee ahead of flames[a]	Poor flier, could only get to a very nearby refuge[b]	Could use burrows, rock outcrops	Could use burrows, rock outcrops and riparian vegetation
Breadth of habitat association	Broad (sparse to dense heaths and sedgelands)[a]	Narrow – needs particular vegetation structure[b]	Broader	Narrower – requires dense ground layer in moist habitat[g,h]
Breadth of diet	? Narrow, but these items common in habitat[a]	? Narrow (few data)[b]	Broader; seeds plus other vegetation components[e,i]	Narrow, mostly plant stems[e]
Susceptibility to competition	? Not affected	? Not affected	Susceptible[f]	Probably not affected[e,f]
Susceptibility to predation in open habitat	Possibly not susceptible[d]	Assume very susceptible[c]	Assume susceptible	Assume susceptible
Long-distance dispersability	Good	Poor[b]	Poor?[i]	Poor?[i]
Reproductive rate	High; 4 eggs, 50% fledging success	Low[b,c]; 1–2 eggs and 1 fledgling, once a year[c]	High	Lower
Prediction of fire response	Elimination by fire; rapid recolonisation even over long distances; potentially long span of occupancy post-fire.	Elimination by non-patchy fires; slow recovery, especially where refuges are distant; long span of occupancy post fire.	Survival of fire; decline immediately afterwards; rapid reappearance: rapid population growth, short occupancy of site.	Survival of fire; decline immediately after; slow reappearance; slow population growth; long span of occupancy post-fire.
Observed fire response	See Fig. 9.5a	See Fig. 9.5b	See Fig. 9.5c	See Fig. 9.5d

Sources: [a]Meredith *et al.* (1984); [b]Baker (1998); [c]Baker (1997); [d]Meredith (1983); [e]Luo and Fox(1995); [f]Higgs and Fox (1993); [g]Newsome and Catling (1979); [h]Newsome *et al.* (1975); [i]Fox (1982).

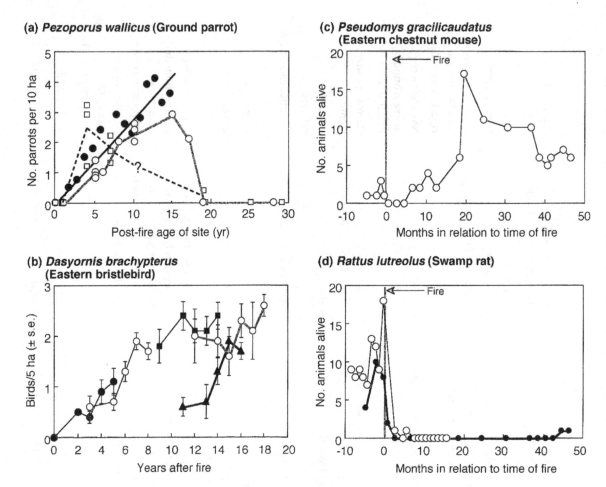

(a) *Pezoporus wallicus* (Ground parrot)

(b) *Dasyornis brachypterus* (Eastern bristlebird)

(c) *Pseudomys gracilicaudatus* (Eastern chestnut mouse)

(d) *Rattus lutreolus* (Swamp rat)

Fig. 9.5. Observed fire responses of four heathland vertebrate species, for comparison with predictions in Table 9.6. (a) Three different patterns for *Pezoporus wallicus* (ground parrot): open squares (and dashed line) are for shrub-heath and open circles (and grey line) for 'graminoid heath' in Victoria (data from Meredith 1983); solid circles (and black line) are for sedgeland in New South Wales (data from Baker and Whelan 1994). (b) *Dasyornis brachypterus* (Eastern bristlebird), Figure 3.3 from Baker (1998), with permission. (c) *Pseudomys gracilicaudatus* (Eastern chestnut mouse), data from Fox (1982). (d) *Rattus lutreolus* (swamp rat), data from Fox (1982) and Newsome *et al.* (1975).

typically found in east coast dry heath or temperate wet heath: birds (*Dasyornis brachypterus* and *Pezoporus wallicus*) and rodents (*Pseudomys gracilicaudatus* and *Rattus lutreolus*). These comparisons suggest that it may be feasible to define attributes of heathland animals in such a way as to predict fire responses. Importantly, this table illustrates that considerable autecological data are needed to apply such a classification and that, even for the well-known species in Table 9.6, predictions about fire response may be sensitive to uncertainty in a single attribute. We would predict slow recolonisation for a species with poor dispersal ability and a high reproductive rate, if we assumed it was eliminated by fire, whereas an assumption of survival within the burnt area would lead us to predict rapid reappearance and population growth. Figure 9.5 shows some observed responses for the taxa whose characteristics are described in Table 9.6. More work is needed before other taxa can be classified.

The attributes in Table 9.6 depend partly on extrinsic factors that are not characteristics of the organism, but characteristics of the habitat and fire (see below). The ability of heathland *Antechinus stuartii*, for example, to remain in an area immediately after a fire might be contingent on the nature of the habitat (e.g. heterogeneous topography provides refuge from predation and suitable nest sites cf. sandy soils with no unburnt vegetation; Whelan *et al.* 1996). The following sections focus more closely on attributes presented in Table 9.6 with relevant case studies.

Survival/mortality

Many animals are observed alive immediately after even the most intense of heathland fires. Mobility may give animals more opportunities to escape the direct impacts of fire, the most mobile species being the most likely to survive, but this is difficult to assess in the face of limited data. Records of recovered carcasses, which account for only some of the total fatalities in a fire, give limited insight into survival rates in populations. Even highly mobile honeyeaters and kangaroos were observed as carcasses after high-intensity wildfires burnt Nadgee heathlands (Fox 1978), and records of beach-washed bird carcasses included both highly mobile species associated with the heathlands (e.g. honeyeaters) and poorer fliers such as the southern emu wren (*Stipiturus malachurus*, Table 9.7). Despite these fatalities, some individuals of these species were also observed to survive fires.

Nevertheless, it would be expected that sedentary species closely associated with shrub canopies or ground vegetation are the most likely to die, whereas mobile species or even sedentary species that live in burrows or are associated with creeklines and other less fire-prone microhabitats (e.g. rock outcrops, lake margins) are more likely to survive. Fox (1978) reported that large numbers of lizards survived the 1972 Nadgee fire, presumably in burrows around the bases of shrubs (only to die soon afterwards, see below). Likewise, no wombat carcasses were observed after the fire and wombat burrows in heath at Jane Spiers beach were active: "The fire-stormed south end of Jane Spiers beach

Table 9.7. *Numbers of birds, representing 41 species, found dead on the beaches at Nadgee Nature Reserve after the 1972 wildfire*

Species	Number of carcasses
***Coturnix pectoralis* (stubble quail)**[a]	2
***Phaps elegans* (brush bronzewing)**[a]	4
Trichoglossus haematodus (rainbow lorikeet)	1
Alisterus scapularis (king parrot)	1
Platycercus elegans (crimson rosella)	7
***Pezoporus wallicus wallicus* (ground parrot)**[a]	5
Cacomantis pyrrhophanus (fan-tailed cuckoo)	1
Ninox novaseelandiae (boobook owl)	1
Tyto alba (barn owl)	1
Aegotheles cristatus (owlet nightjar)	2
Podargus strigoides (tawny frogmouth)	1
Turdus merula (blackbird)	5
Malurus cyaneus (superb blue wren)	3
***Stipiturus malachurus* (southern emu wren)**[a]	1
Acanthiza lineata (striated thornbill)	11
Acanthiza pusilla (brown thornbill)	25
Sericornis frontalis frontalis (white-browed scrub wren)	1
Petroica multicolor (scarlet robin)	1
Eopsaltria australis (southern yellow robin)	49
Myiagra rubecula (leaden flycatcher)	1
Pachycephala rufiventris (rufous whistler)	21
Colluricincla harmonica (grey shrike thrush)	1
Falcunculus frontatus frontatus (eastern shrike tit)	2
Psophodes olivaceus (eastern whipbird)	1
Neositta chrysoptera (orange-winged sitella)	2
Climacteris leucophaea (white-throated tree creeper)	10
Pardalotus punctatus (spotted pardalote)	1
***Meliphaga chryops* (yellow-faced honeyeater)**[a]	29
***Melithreptus brevirostris* (brown-headed honeyeater)**[a]	5
Melithreptus lunatus white-naped honeyeater	1
***Phiylidonyris novaehollandiae* (New Holland honeyeater)**[a]	266
***Acanthorhynchus tenuirostrtis* (eastern spinebill)**[a]	12
Manorina melanophrys (bell miner)	1
***Anthochaera chrysoptera* (little wattlebird)**[a]	120
Emblema bella (beautiful firetail)	3
Aegintha temporalis (red-browed finch)	3
Oriolus sagittatus (olive-backed oriole)	1
Strepera graculina (pied currawong)	2
Cracticus torquatus (grey butcherbird)	2
Ptilinorhynchus violaceus (satin bowerbird)	1

Note:
[a] Species in **bold** are those listed by Kikkawa *et al.* (1979) as having a special association with heathlands in southeast Australia.
Source: Fox (1978).

and hinterland however seemed lifeless except for the wombats which remained safe in their burrows having had 50 cm of sand insulating them from the searing heat above, heat which had melted aluminium signs (659 °C) and melted beer bottles flat (greater than 1200 °C)." (Fox 1978).

Several studies have identified unburnt patches of vegetation as critical for the survival of animals during and immediately after fire (Whelan 1995), mostly based on observations of animals alive in these refuges. Highly mobile animals may disperse to these refuges more readily but an important question is whether heathland animals actively seek out areas that are less likely to burn intensely, as a fire advances, or whether animals found in unburnt refugia after fire simply represent those that happened to be there when the fire passed.

A location does not have to be left unburnt to function as a refuge. For example, invertebrate fauna occupy the bases of burnt crowns of *Xanthorrhoea* and cycads (Whelan *et al.* 1980; Main 1981). Immediately after wildfire at Royal National Park in 1994, *Xanthorrhoea* crowns in burnt heath supported a higher diversity and abundance of invertebrates than in the few patches of unburnt vegetation (D. A. Keith and M. G. Tozer unpublished data). The abundance and diversity in burnt areas decreased over the ensuing year, becoming more similar to *Xanthorrhoea* crowns in the unburnt heath patches. It would be most interesting to see whether this pattern at a small scale is repeated with vertebrates in larger fragments of unburnt vegetation (refuges) within a burnt landscape. The identification of habitat components that might represent refuges for different animal taxa and the question of how important they are for maintenance of post-fire populations require much more study in heathlands.

Specialisation in diet and habitat

The close link with vegetation as the provider of food, shelter and nest sites produces patterns of animal species abundance in heathlands that have been interpreted as 'ecological succession' (Fig. 9.6) (Fox 1982; McFarland 1988). Even if animals are wide-ranging enough to disperse to burnt areas, the habitat may be inappropriate to attract or sustain a species.

The slow recolonisation of *Rattus lutreolus* in heaths has been attributed to its dependence on dense cover at ground level. *Rattus lutreolus* was eliminated by a fire in temperate wet heath at Myall Lakes and was first trapped 4 years afterwards, but apparently did not breed successfully until the 5th year (Fox 1982; Fig. 9.6a). The species invaded the burnt site before food availability and/or habitat were suitable for its persistence. In contrast, populations of *Pseudomys gracilicaudatus*, which has a less specialised diet and perhaps favours more open habitat, increased more immediately after fire (Fig. 9.6a).

McFarland (1988) inferred post-fire population changes in 15 heath birds from a chronosequence of sites surveyed immediately (year 0) and 2.5, 5.5, 6.5 and 10.5 years after fire in a subtropical east coast dry heath at Cooloola National Park. Among a variety of post-fire population patterns (Fig. 9.6b), he interpreted the peaks in abundance of white-cheeked honeyeaters to reflect the availability of food (nectar and insects) and nest sites. Mass flowering of *Xanthorrhoea fulva* provided a food resource in the first year after fire, but birds were not observed breeding in the area. The 10.5-year-old site presumably provided both nest sites and food resources. The early peak abundance in quail may be explained by their association with open habitats and grass seeds as food, both of which were facilitated by the fire. The peak in ground parrot numbers in the 5.5-yr site was associated with the maximum availability of sedges – their primary food source.

Pyrogenic flowering (Table 9.4) might provide an important food resource for many heathland animals. An *Antechinus stuartii* population, surviving in east coast dry heath at Royal National Park through the first winter after an intense 1994 wildfire, made extensive use of a large nectar resource in *Xanthorrhoea* species (Whelan *et al.* 1996). In Recher's (1981) study of similar heath north of Sydney, long-unburnt heath seemed unsuitable for avian heath specialists such as emu wrens and tawny-crowned honeyeaters. A fire in 1977 permitted these species to use the heath. Other species,

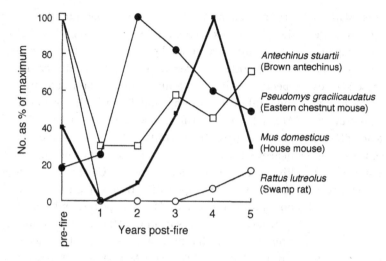

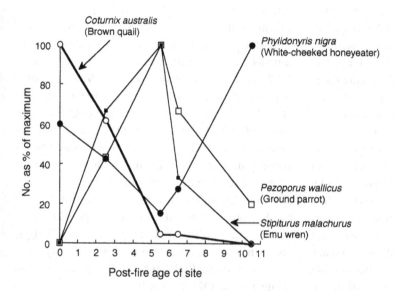

Figure 9.6. (a) Changes in abundances of four heathland mammal species at Myall Lakes, NSW. Numbers for each species are expressed as the percentage of the maximum abundance for that species recorded over the 5 years of annual censuses in this longitudinal study. Data from Fox (1982). (b) Abundances of four heathland bird species, expressed as a percentage of their maximum abundance across the sites. Data are from a 'synchronic' study (see Whelan 1995, Ch. 4) in which sites of differing post-fire age were compared. There was one site of each post-fire age except 2.5 years, which was represented by two sites. Data from McFarland (1988).

particularly those which used the nectar of *Banksia ericifolia* in winter, had reduced abundances during the plant's post-fire juvenile period.

There are few studies on long-unburnt heaths. However, several heath specialist birds in western mallee-heath appear to have a preference for tall, dense heath habitat associated with prolonged absence of fire (Cale and Burbidge 1993): western ground parrot (*Pezoporus wallicus flaviventris*), western bristlebird (*Dasyornis longirostris*) and the western whipbird (*Psophodes nigrogularis*). These three species increased in abundance after prolonged fire exclusion, along with the noisy scrub

bird (*Atrichornis clamosus*) – a species of low eucalypt forest and tall thickets on their margins (Smith 1985). In central lowland heath, Woinarski (1989) linked the time of peak abundance after fire of whipbirds, insectivores which forage in the litter of dense shrub vegetation, to vegetation density and availability of litter invertebrates.

Some animal species exhibit flexibility in their use of vegetation, as the resource changes in response to fire. In sandplain kwongan near Perth, many of the passerine bird species which were not eliminated by a fire showed considerable flexibility in nest locations after fire (Brooker and Rowley

1991). For example, splendid fairy wrens (*Malurus splendens*) altered their nest siting from obligate seeding shrubs before fire to the dense foliage of resprouters soon afterwards, returning to the foliage of the obligate seeders when these species increased in cover. Western thornbills (*Acanthiza inornata*) used holes in eucalypt trees and crevices beneath shedding bark as nest sites instead of *Hakea* foliage, which was unavailable after fire. The extent that high specificity for particular nest sites might explain the post-fire absence of other species warrants further study.

Susceptibility to predation

Fox (1978) highlighted the significance of post-fire predation in heathlands, reporting that there was substantial lizard mortality observed shortly after the 1972 Nadgee fire. There are many anecdotal reports of an influx of predators soon after a fire. Although many of these may scavenge carcasses, animals that survive the passage of a fire are undoubtedly at greater risk of predation soon afterwards. Though it seems likely that predation should be a major cause of population decline immediately after fire in heathlands, there are few direct studies of predation rates and their effects on prey population dynamics after fires (Sutherland 1998). Christensen's (1980) radio-tracking study of woylies after experimental fire in southwest Australian forest is a good model for the sort of studies that are needed in heathland, if we are to determine the role of predation in limiting post-fire populations of species that survive the passage of fire.

Bird research provides perhaps the clearest examples of the potential importance of post-fire predation in population dynamics of prey. Brooker and Brooker (1994) concluded that the main effect of fire on populations of splendid fairy wrens was the result of increased nest predation in the more exposed, open post-fire habitat.

Long-distance dispersability

Mobility may influence the time taken for animals to reappear in burnt areas. Among birds and insects that disappear from newly burnt sites, good fliers are generally the earliest to recolonise, while recov-

ery of sedentary species with low mobility is delayed. For example, Danks (1997) concluded that fires would have isolated populations of noisy scrub bird (*Atrichornis clamosus*), a poorly dispersing species, in sites containing suitable habitat (dense, long-unburnt heath around Two Peoples Bay near Albany). After fires around Mt Gardner, natural dispersal apparently permitted birds to recolonise nearby sites (the Lakes, about 4 km, and the Angove River area, about 6–8 km from Mt Gardner) as suitable habitat developed with time since fire. Rapid population growth at Mt Manypeaks, only after translocation of birds from the Mt Gardner population, indicated that birds had been unable to reach suitable habitat at that site. McFarland (1994) drew similar conclusions about limited dispersal of southern emu wrens (*Stipiturus malachurus*).

Eastern bristlebirds (*Dasyornis brachypterus*) in southeastern Australia are also poor fliers (Baker 1997, 1998). Consecutive fires in 1972 and 1980 appear to have driven the species close to extinction at Nadgee (Recher 1981), yet a large population persists 300 km north at Barren Grounds Nature Reserve, despite a major fire in 1968 followed by six smaller fires in the subsequent 30 years (Baker 1998). Closer proximity of suitable habitat patches at Barren Grounds may have permitted more rapid recolonisation than at Nadgee (despite 20 years without fire), where habitat patches are further apart. In contrast, populations of other bird species, capable of longer flight, recovered rapidly after the Nadgee fire in 1972 (Recher 1981).

Reproductive rate

Once re-established, species that most rapidly increase in abundance after fire should be those with the highest reproductive rates. Among heathland mammals, the rodents illustrate this clearly. *Pseudomys* and *Mus* have particularly rapid intrinsic rates of population growth and often appear in peak abundance soon after fires. There have been few demographic studies of post-fire animal populations, but a long-term study of splendid fairy wrens (*Malurus splendens*) in sandplain kwongan provides a good model for much-needed studies in other heathlands (Rowley and Brooker 1987; Brooker and

Rowley 1991; Russell and Rowley 1993; Woinarski 1999). Fires reduced natality and juvenile survival, thereby reducing population density, even though adult birds survived and territories were not vacated immediately after fire. Nesting in the year after fire was delayed by 3 to 5 weeks compared with birds in unburnt sites in the same year.

Interactions with habitat characteristics

Habitats vary among heathland sites in ways that affect responses of fauna to fires. For example, east coast dry heath and temperate wet heath may be juxtaposed, but differ in plant species composition and structure, rates of biomass recovery after fire and fauna composition. Heath on sand has different microtopography from that on sandstone. The latter substrate provides many more potential refugia and landscape discontinuities than the former.

There is considerable variation in fire responses within species at different sites or after different fires. Whelan et al. (1996) found no precipitous decline in Antechinus stuartii in heath of Royal National Park near Sydney in the year after the 1994 wildfire, whereas studies by Newsome et al. (1975) and Fox (1982) found local extinction and substantial population declines, respectively, after heath fires at Nadgee and Myall Lakes. Newsome et al. (1975) recorded local extinctions in some areas of Nadgee but only population declines in other areas. Habitat differences may explain some of these variations. Shelter provided by topographic heterogeneity of sandstone substrates in Royal National Park may have partly explained the maintenance of that population after fire. In contrast, the declines and extinctions at Nadgee and Myall Lakes occurred on more uniform sands, which lacked post-fire refuges and nest sites provided by rock outcrops (Whelan et al. 1996).

Post-fire dynamics of ground parrot (Pezoporus wallicus) populations also vary between sites. Meredith et al. (1984) studied a chronosequence of heath sites in southern Victoria and concluded that ground parrot numbers peaked more rapidly after fire in 'shrub heathland' (4 years) than in 'graminoid heathland' (15 years). McFarland (1988) found

an early peak (5.5 years) for the ground parrot in a chronosequence of heathlands in Queensland. In a longitudinal study at Barren Grounds Nature Reserve, Baker and Whelan (1994) found that ground parrot densities were still increasing 12 years after wildfire in 1983, and this trend has continued since then (R. J. Whelan unpublished data). Limited data for southwestern Australia (Cale and Burbidge 1993) indicate that the WA subspecies favours long-unburnt heathland.

One explanation for these patterns of variation is that fire responses are sensitive to interactions between habitat characteristics and attributes of the animals. Analogous interactions were discussed above for plants. Early peaks in parrot densities in some sites and sustained increases in others may be explained if the food and cover resources that are important for ground parrots respond differently to fire in different heath types (Gill 1996). In the shrub-dominated heaths, for example, increasing shrub density with time since fire may progressively exclude sedges and other plants which provide food for the parrots (Keith and Bradstock 1994). This may not happen for a very long time (if ever) in graminoid heaths. In contrast, the slow recovery of vegetation density after fire in low-rainfall southwestern Australia may explain the preference of ground parrots (Pezoporus wallicus flaviventris) for long-unburnt heath (Cale and Burbidge 1993).

Interactions with fire characteristics

The features of fire regimes (intensity, frequency, season), fire behaviour and spatial patterns are expected to affect the responses of animal populations. However, much of the current knowledge reviewed here only concerns processes and patterns following a single fire. The complexity of interactions between fire regimes, habitat and animal populations in heathland presently defies generalisation. The following are only a few examples of conflicting responses to fire regimes to illustrate this point.

Large numbers of bird carcasses have been observed after high-intensity fires in heathlands (Fox 1978), yet high survival rates after wildfires have also been reported (Brooker and Rowley 1991). Fox (1978) suggested that bird mortality may have been greater in areas of Nadgee burnt in a backfire than in areas burnt by the headfire, implying an inverse relationship between rate of spread and mortality.

Short-term survival of splendid fairy wrens was greater after a high-intensity summer fire than after cooler, winter and spring fires, perhaps because the summer fire was 5 weeks after the breeding season (Rowley and Brooker 1987). Catling and Newsome (1981) speculated that the summer season of the 1972 Nadgee fire might have permitted survival of heathland *Antechinus swainsonii* populations, as the fire came after the young had become independent. There is far too little work on the effects of fire season on population dynamics of animals.

Patches of unburnt vegetation left behind both the Nadgee fires (Fox 1978) and the 1994 Royal National Park fire (Whelan *et al.* 1996) illustrate the patchiness even of high-intensity and fast-moving heathland fires. Intuitively we expect that unburnt patches of vegetation will be important for the survival of animals both during and after a fire, yet there are few studies on this. Unburnt ribbons of creekline vegetation at Nadgee acted as refugia for small mammals during and after the big fires there (Fox 1978; Newsome and Catling 1979), and provided the sources of colonists as the heaths recovered as suitable habitat. However, the study by Whelan *et al.* (1996) at Royal National Park did not find a strong association of small mammals (mostly *Antechinus stuartii*) with patches of unburnt heath during the first winter after the 1994 wildfire.

Frequency of fire is perhaps the best-studied aspect of fire regime for heathland plants, but there are still too few animal studies to generalise. Frequent fires are likely to alter the nature of the habitat so as to make it less suitable for some species (Recher 1981; Brooker and Rowley 1991; Brooker and Brooker 1994; York 1999).

Heath-dependent species (Table 9.2) are those most likely to be affected by inappropriate fire regimes. For birds, studies of fire effects in heathlands are well represented in the literature relative to other habitat types. A quarter of the references in a recent literature review of over 350 papers dealing with fire and birds in Australia addressed heathlands (Woinarski 1999). Much of this research has focussed on only a few heathland-dependent species, most notably the ground parrot (*Pezoporus wallicus*). This situation clearly illustrates the many gaps in our knowledge of the responses of heathland animals to fire regimes, despite heathlands being one of the best-studied habitats. Few species have been studied, yet many are interesting and many are threatened. Inappropriate fire regimes are often identified as one of the possible threatening processes. However, causes of fire-driven changes in populations of heathland animals remain obscure and are certain to be highly complex.

Conclusions

Knowledge of fire regimes and their ecological effects in Australian heathlands has developed substantially in the last 20 years. About 90% of the references cited here post-date the publication of *Fire and the Australian Biota* in 1981. While the preceding heathland review focussed on description of patterns (Specht 1981), our emphasis is on a mechanistic understanding of processes that control fire regimes and their effects on biota. This reflects an expanded base of data on key fire and biotic parameters, as well as enhanced understanding of processes that influence change, spatial pattern and their variability in heathlands. Nevertheless there is a dearth of knowledge in several important areas, most notably the effects of fuels, landscape features and weather on fire size and patchiness and the influence of these spatial characteristics of fire on animal populations and plant predation. The effects of fire frequency on animal populations, particularly invertebrates, also warrants research attention and may fruitfully involve novel life-history approaches.

While our improved knowledge of fire regimes and their role in ecological processes provides a strong basis for fire management in heathlands into the 21st century, the conservation of heathlands presents a considerable challenge, especially given their juxtaposition in landscapes with human settlements, production industries and recreational uses. The high diversity and uniqueness of

Australian heathlands add imperative to the challenge. Adaptive management approaches that promote both temporal and spatial (i.e. mosaic) variability in fire regimes offer a means of resolving potential conflicts in the management of fire in remnant reserves surrounded by agricultural or urbanised land, as well as lands dedicated to multiple uses (Keith *et al.*, this volume).

Acknowledgements

We thank all researchers who contributed plant life-history data for adaptation in Table 9.6. Allan Burbidge, Peter Catling, Dick Schodde and John Woinarski suggested characteristic fauna for some heathland types listed in Table 9.1. Berin Mackenzie and Belinda Pellow assisted with manuscript preparation. Critical comments by Byron Lamont and Jann Williams did much to improve this paper. The Australian Fire Bibliography provided a valuable means for identifying relevant literature addressed in this review.

References

Andersen, A. N. (1988). Immediate and longer-term effects of fire on seed predation by ants in sclerophyllous vegetation in south-eastern Australia. *Australian Journal of Ecology* **13**, 285–293.

Andersen, A. N. (1989). How important is seed predation to recruitment in stable populations of long-lived perennials? *Oecologia* **81**, 310–315.

Andersen, A. N., and Ashton, D. H. (1985). Rates of seed removal by ants at heath and woodland sites in southeastern Australia. *Australian Journal of Ecology* **10**, 381–390.

Andersen, A. N., and New, T. R. (1987). Insect inhabitants of fruits of *Eucalyptus, Leptospermum* and *Casuarina* in southeastern Australia. *Australian Journal of Zoology* **35**, 327–336.

Ashton, D. H. (1986). Viability of seeds of *Eucalyptus obliqua* and *Leptospermum juniperinum* from capsules subjected to a crown fire. *Australian Forestry* **49**, 28–35.

Auld, T. D. (1986). Population dynamics of the shrub *Acacia suaveolens* (Sm.) Willd.: dispersal and dynamics of the soil seedbank. *Australian Journal of Ecology* **11**, 235–254.

Auld, T. D. (1987a). Population dynamics of the shrub *Acacia suaveolens* (Sm.) Willd: survivorship throughout the life cycle, a synthesis. *Australian Journal of Ecology* **12**, 139–151.

Auld, T. D. (1987b). Post-fire demography in the resprouting shrub *Angophora hispida* (Sm.) Blaxell: flowering, seed production, dispersal, seedling establishment and survival. *Proceedings of the Linnean Society of New South Wales* **109**, 259–269.

Auld, T. D. (1990). The survival of juvenile plants of the resprouting shrub *Angophora hispida* (Myrtaceae) after a simulated low-intensity fire. *Australian Journal of Botany* **38**, 255–260.

Auld, T. D. (1994). The role of soil seedbanks in maintaining plant communities in fire-prone habitats. In *Proceedings of the 2nd International Conference on Forest Fire Research* (ed D. X. Viegas.) pp. 1069–1078. (University of Coimbra: Coimbra.)

Auld, T. D. (1995). Burning grevilleas, ants, rats and wallabies. *CALMScience Supplement* **4**, 159–164.

Auld, T. D., and Denham, A. J. (1999). The role of ants and mammals in dispersal and post-dispersal seed predation of the shrubs *Grevillea* (Proteaceae). *Plant Ecology* **144**, 201–213.

Auld, T. D., and O'Connell, M. A. (1988). Changes in predispersal seed predation levels after fire for two Australian legumes, *Acacia elongata* and *Sphaerolobium vimineum*. *Oikos* **54**, 55–59.

Auld, T. D., and O'Connell, M. A. (1991). Predicting patterns of post-fire germination in 35 eastern Australian Fabaceae. *Australian Journal of Ecology* **16**, 53–70.

Auld, T. D., and Myerscough, P. J. (1986). Population dynamics of the shrub *Acacia suaveolens* (Sm.) Willd.: seed production and predispersal seed predation. *Australian Journal of Ecology* **11**, 219–234.

Auld, T. D., and Tozer, M. G. (1995). Patterns in emergence of *Acacia* and *Grevillea* seedlings after fire. *Proceedings of the Linnean Society of New South Wales* **115**, 5–15.

Auld. T. D., Keith, D. A., and Bradstock, R. A. (2000). Patterns in longevity in soil seedbanks in the Sydney region of southeastern Australia. *Australian Journal of Botany* **48**, 539–548.

Baker, J. R. (1997). The decline, response to fire, status and management of the Eastern Bristlebird. *Pacific Conservation Biology* **3**, 235–243.

Baker, J. R. (1998). Ecotones and fire and the conservation of the eastern bristlebird. PhD thesis, University of Wollongong.

Baker, J. R., and Whelan, R. J. (1994). Ground parrots and fire at Barren Grounds, New South Wales: a long-term study and an assessment of management implications. *Emu* **94**, 300–304.

Bell, D. T., Hopkins, A. J. M., and Pate, J. S. (1984). Fire in the Kwongan. In *Kwongan: Plant Life of the Sandplain* (eds. J. S. Pate and J. S. Beard) pp. 178–204. (University of Western Australia Press: Nedlands.)

Bell, D. T., Plummer, J. A., and Taylor, S. K. (1993). Seed germination ecology in southwestern Western Australia. *The Botanical Review* **59**, 24–73.

Bell, T. L., and Pate, J. S. (1996). Growth and fire response of selected Epacridaceae of south-western Australia. *Australian Journal of Botany* **44**, 509–526.

Benson, D. H. (1985). Maturation periods for fire sensitive shrub species in Hawkesbury sandstone vegetation. *Cunninghamia* **1**, 339–349.

Benwell, A. S. (1998). Post-fire seedling recruitment in coastal heathland in relation to regeneration strategy and habitat. *Australian Journal of Botany* **46**, 75–101.

Berg, R. Y. (1981). The role of ants in seed dispersal in Australian lowland heathland. In *Heathlands and Related Shrublands* (ed. R. L. Specht) pp. 51–59. (Elsevier: Amsterdam.)

Birk, E. M., and Simpson, R. W. (1980). Steady state and the continuous input model of litter accumulation and decomposition in Australian eucalypt forests. *Ecology* **61**, 481–485.

Bond, W. J., and van Wilgen, B. W. (1996). *Fire and plants.* (Chapman and Hall: London.)

Bowen, B. J., and Pate, J. S. (1993). The significance of root starch in post-fire shoot recovery of the resprouter *Stirlingia latifolia* R.Br. (Proteaceae). *Annals of Botany* **72**, 7–16.

Bradstock, R. A. (1985). Plant population dynamics under varying fire regimes. PhD thesis, University of Sydney.

Bradstock, R. A. (1990). Demography of woody plants in relation to fire: *Banksia serrata* L.f. and *Isopogon anemonifolius* (Salisb.) Knight. *Australian Journal of Ecology* **15**, 117–132.

Bradstock, R. A. (1991). The role of fire in establishment of seedlings of serotinous species from the Sydney region. *Australian Journal of Botany* **39**, 347–356.

Bradstock, R. A. (1995). Demography of woody plants in relation to fire: *Telopea speciosissima. Proceedings of the Linnean Society of New South Wales* **115**, 25–33.

Bradstock, R. A., and Auld, T. D. (1995). Soil temperatures during experimental bushfires in relation to fire intensity: consequences for legume germination and fire management in south-eastern Australia. *Journal of Applied Ecology* **32**, 76–84.

Bradstock, R. A., and Bedward, M. (1992). Simulation of the effect of season of fire on post-fire seedling emergence of two *Banksia* species based on long-term rainfall records. *Australian Journal of Botany* **40**, 75–88.

Bradstock, R. A., and Myerscough, P. J. (1988). The survival and population response to frequent fires of two woody resprouters *Banksia serrata* and *Isopogon anemonifolius. Australian Journal of Botany* **36**, 415–431.

Bradstock, R. A., and O'Connell, M. A. (1988). Demography of woody plants in relation to fire: *Banksia ericifolia* L.f. and *Petrophile pulchella* (Schrad) R.Br. *Australian Journal of Ecology* **13**, 505–518.

Bradstock, R. A., Gill, A. M., Hastings, S. M., and Moore, P. H. R. (1994). Survival of serotinous seedbanks during bushfires: comparative studies of *Hakea* species from southeastern Australia. *Australian Journal of Ecology* **19**, 276–282.

Bradstock, R. A., Tozer, M. G., and Keith, D. A. (1997). Effects of high frequency fire on floristic composition and abundance in a fire-prone heathland near Sydney. *Australian Journal of Botany* **45**, 641–655.

Bradstock, R. A., Bedward, M., Kenny, B. J., and Scott, J. (1998). Spatially explicit simulation of the effect of prescribed burning on fire regimes and plant extinctions in shrublands typical of south-eastern Australia. *Biological Conservation* **86**, 83–95.

Brooker, L. C., and Brooker, M. G. (1994). A model for the effects of fire and fragmentation on the population viability of the Splendid Fairy-wren. *Pacific Conservation Biology* **1**, 334–358.

Brooker, M. G., and Rowley, I. (1991). Impact of wildfire on the nesting behaviour of birds in heathland. *Wildlife Research* **18**, 249–263.

Brown, M. J., and Podger, F. D. (1982). Floristics and fire regimes of a vegetation sequence from sedgeland-heath to rainforest at Bathurst Harbour, Tasmania. *Australian Journal of Botany* **30**, 659–676.

Burdon, J. J. (1991). Fungal pathogens as selective forces in plant populations and communities. *Australian Journal of Ecology* **16**, 423–432.

Burgman, M. A., and Lamont, B. B. (1992). A stochastic model for the viability of *Banksia cuneata* populations: environmental demographic and genetic effects. *Journal of Applied Ecology* **29**, 719–727.

Burrell, J. P. (1981). Invasion of coastal heaths of Victoria by *Leptospermum laevigatum* (J. Gaertn.) F.Muell. *Australian Journal of Botany* **29**, 747–764.

Cale, P. G., and Burbidge, A. H. (1993). *Research Plan for the Western Ground Parrot, Western Whipbird and Western Bristlebird*, Endangered Species Program Project no. 228. (Australian National Parks and Wildlife Service: Canberra.)

Cary, G. J. (2001). Importance of a changing climate for fire regimes in Australia. In *Flammable Australia: The Fire Regimes and Biodiversity of a Continent* (eds. R. A. Bradstock, J. E. Williams and A. M. Gill) pp. 26–46. (Cambridge University Press: Cambridge.)

Catchpole, W. R. (1985). Predicting biomass for *Casuarina nana* for input into the Rothermel fire model. Department of Mathematics, University of New South Wales, Report no. 24/85, Duntroon.

Catchpole, W. R. (2001). Fire properties and burn patterns in heterogeneous landscapes. In *Flammable Australia: The Fire Regimes and Biodiversity of a Continent* (eds. R. A. Bradstock, J. E. Williams and A. M. Gill) pp. 49–76. (Cambridge University Press: Cambridge.)

Catchpole, W. R. , Bradstock, R. A., Choate, J., Fogarty, L., Gellie, N., McCarthy, G., McCaw, W. L., Marsden-Smedley, J., and Pearce, G. (1998). Cooperative development of equations for heathland fire behaviour. In *Proceedings of the 3rd International Conference on Forest Fire Research/14th Fire and Forest Meteorology Conference* (ed. D. X. Viegas) pp. 631–645. (University of Coimbra: Coimbra.)

Catling, P. C., and Newsome, A. E. (1981). Responses of the Australian vertebrate fauna to fire: an evolutionary approach. In *Fire and the Australian Biota* (eds. A. M. Gill, R. H. Groves and I. R. Noble) pp. 273–310. (Australian Academy of Science: Canberra.)

Chandler, C., Cheney, N. P., Thomas, P., Trabaud, L., and Williams D. (1983). *Fire in Forestry*, vol. 1, *Forest Fire Behaviour and Effects*. (Wiley-Interscience: New York.)

Cheney, N. P., and Sullivan, A. (1997). *Grassfires: Fuel, Weather and Fire Behaviour*. (CSIRO: Melbourne.)

Christensen, P. E. S. (1980). The biology of *Bettongia penicillata* (Gray 1837) and *Macropus eugenii* (Demarest 1917) in relation to fire. Forests Department of Western Australia, Bulletin no. 91, Perth.

Conroy, B. (1993). Fuel management strategies for the Sydney region. In *The Burning Question: Fire Management in NSW* (ed. J. Ross) pp. 73–81. (University of New England: Armidale, NSW.)

Conroy, R. W. (1996). To burn or not to burn? A description of the history, nature and management of bushfires within Ku-ring-gai Chase National Park. *Proceedings of the Linnean Society of New South Wales* **116**, 80–96.

Cowling, R. M., and Lamont, B. B. (1985a). Seed release in *Banksia*: the role of wet–dry cycles. *Australian Journal of Ecology* **10**, 169–171.

Cowling, R. M., and Lamont, B. B. (1985b). Variation in serotiny of three *Banksia* species along a climatic gradient. *Australian Journal of Ecology* **10**, 345–350.

Cowling, R. M., and Lamont, B. B. (1987). Post fire recruitment of four co-occurring *Banksia* species. *Journal of Applied Ecology* **24**, 645–658.

Danks, A. (1997). Conservation of the Noisy Scrub-bird: a review of 35 years of research and management. *Pacific Conservation Biology* **3**, 341–349.

Davies, S. J., and Myerscough, P. J. (1991). Post-fire demography of the wet mallee *Eucalyptus luehmanniana* F.Muell. (Myrtaceae). *Australian Journal of Botany* **39**, 459–466.

Delfs, J. C., Pate, J. S., and Bell, D. T. (1987). Northern sandplain kwongan: community biomass and selected species response to fire. *Journal of the Royal Society of Western Australia* **69**, 133–138.

Dixon, K. W., Roche, S., and Pate, J. S. (1995). The promotive effect of smoke derived from burnt native vegetation on seed germination of Western Australian plants. *Oecologia* **101**, 185–192.

Dwyer, P. D., Kikkawa, J., and Ingram, C. J. (1979). Habitat relations of vertebrates in subtropical heathlands of coastal southeastern Queensland. In *Ecosystems of the world*, vol. 9A, *Heathlands and Related Shrublands: Descriptive Studies* (ed. R. L. Specht) pp. 281–299. (Elsevier: Amsterdam.)

Enright, N. J., and Lamont, B. B. (1989). Fire temperatures and follicle-opening requirements in 10 *Banksia* species. *Australian Journal of Ecology* **14**, 107–113.

Forward, L. R. (1996). Managing fire for biodiversity in

Ngarkat Conservation Park: development of methodology. Department of Environment and Natural Resources, Adelaide.

Fox, A. M. (1978). The '72 fire of Nadgee Nature Reserve. *Parks and Wildlife* 2, 5–24.

Fox, B. J. (1982). Fire and mammal secondary succession in an Australian coastal heath. *Ecology* 63, 1332–1341.

Fox, B. J., Fox, M. D., and McKay, G. M. (1979). Litter accumulation after fire in a eucalypt forest. *Australian Journal of Botany* 27, 157–165.

French, K. (1991). Characteristics and abundance of vertebrate-dispersed fruits in temperate wet sclerophyll forest in southeastern Australia. *Australian Journal of Ecology* 16, 1–14.

Friend, G. R. (1993). The impact of fire on small vertebrates in mallee woodlands and heathlands of temperate Australia: a review. *Biological Conservation* 65, 99–114.

Friend, G. R. (1995). Fire and invertebrates: a review of research methodology and the predictability of post-fire response patterns. *CALMScience Supplement* 4, 165–174.

George, A. S., Hopkins, A. J. M., and Marchant, N. G. (1979). The heathlands of Western Australia. In *Ecosystems of the World*, vol. 9A, *Heathlands and Related Shrublands: Descriptive Studies* (ed. R. L. Specht) pp. 211–230. (Elsevier: Amsterdam.)

Gill, A. M. (1975). Fire and the Australian flora: a review. *Australian Forestry* 38, 4–25.

Gill, A. M. (1981). Adaptive responses of Australian vascular plant species to fire. In *Fire and the Australian Biota* (eds. A. M. Gill, R. H. Groves and I. R. Noble) pp. 243–272. (Australian Academy of Science: Canberra.)

Gill, A. M. (1996). How fires affect biodiversity. In *Fire and Biodiversity: The Effects and Effectiveness of Fire Management*, Biodiversity Series Paper no. 8 (ed. J. R. Merrick) pp. 47–55. (Department of Environment, Sport and Territories: Canberra.)

Gill, A. M., and Ashton, D. H. (1968). The role of bark type in relative tolerance to fire of three central Victorian eucalypts. *Australian Journal of Botany* 16, 491–498.

Gill, A. M., and Bradstock, R. A. (1995). Extinction of biota by fires. In *Conserving Biodiversity: Threats and Solutions* (eds. R. A. Bradstock, T. D. Auld, D. A. Keith, R. T. Kingsford, D. Lunney and D. P. Sivertsen) pp. 309–322. (Surrey Beatty and Sons: Chipping Norton, NSW.)

Gill, A. M., and Groves, R. H. (1981). Fire regimes in heathlands and their plant-ecological effects. In *Ecosystems of the World*, vol. 9B, *Heathlands and Related Shrublands: Analytical Studies* (ed. R. L. Specht) pp. 61–84. (Elsevier: Amsterdam.)

Gill, A. M., and Ingwersen, F. (1976). Growth of *Xanthorrhoea australis* R.Br. in relation to fire. *Journal of Applied Ecology* 13, 195–203.

Gill, A. M., and McMahon, A. (1986). A post-fire chronosequence of cone, follicle and seed production in *Banksia ornata*. *Australian Journal of Botany* 34, 425–433.

Gill, A. M., Groves, R. H., and Noble, I. R. (eds.) (1981). *Fire and the Australian biota*. (Australian Academy of Science: Canberra.)

Grant, S., and Wouters, M. (1993). The effect of fuel reduction burning on the suppression of four wildfires in Western Victoria. Fire Management Branch, Victorian Department of Conservation and Natural Resources, Research Report no. 41, Melbourne.

Groves, R. H. (1981). Heathland soils and their fertility status. In *Ecosystems of the World*, vol. 9B, *Heathlands and Related Shrublands: Analytical Studies* (ed. R. L. Specht) pp. 143–150. (Elsevier: Amsterdam.)

Haigh, C. (1981). *Heaths in New South Wales*. (NSW National Parks and Wildlife Service: Sydney.)

Hammill, K. A., Bradstock, R. A., and Allaway, W. G. (1998). Post-fire seed dispersal and species re-establishment in proteaceous heath. *Australian Journal of Botany* 46, 407–419.

Hansen, A., Pate, J. S., and Hansen, A. P. (1991). Growth and performance of a seeder and a resprouter species of *Bossiaea* as a function of plant age after fire. *Annals of Botany* 67, 497–509.

Head, M. J., and Lacey, C. J. (1988). Radiocarbon age determinations from lignotubers. *Australian Journal of Botany* 36, 93–100.

Higgs, P., and Fox, B. J. (1993). Interspecific competition, a mechanism for rodent succession after fire in wet heathland. *Australian Journal of Ecology* 18, 193–201.

Hobbs, R. J., Mallik, A. U., and Gimingham, C. H. (1984). Studies on fire in Scottish heathland communities. III. Vital attributes of the species. *Journal of Ecology* 72, 963–976.

Holland, W. N., Benson, D. H., and McRae, R. H. D. (1991). Spatial and temporal variation in a perched headwater valley in the Blue Mountains: geology, geomorphology, vegetation, soils and hydrology. *Proceedings of the Linnean Society of New South Wales* 113, 271–295.

Judd, T. S., and Ashton, D. H. (1991). Fruit clustering in the Myrtaceae: seed survival in capsules subjected to experimental heating. *Australian Journal of Botany* **39**, 241–245.

Keith, D. A. (1991). Coexistence and species diversity in upland swamp vegetation: the roles of an environmental gradient and recurring fires. PhD thesis, University of Sydney.

Keith, D. A. (1994). Floristics, structure and diversity of natural vegetation in the O'Hares Creek catchment, south of Sydney. *Cunninghamia* **3**, 543–594.

Keith, D. A. (1995a). How similar are geographically separated stands of the same vegetation formation? A moorland case study from Tasmania and mainland Australia. *Proceedings of the Linnean Society of New South Wales* **115**, 61–75.

Keith, D. A. (1995b). Surviving fire and fungal pathogens: are there life-history solutions for threatened species of *Epacris*? In Landscape Fires '95 (ed. A. Blanks) (Tasmanian Parks and Wildlife Service: Hobart.)

Keith, D. A. (1996). Fire-driven extinction of plant populations: a synthesis of theory and review of evidence from Australian vegetation. *Proceedings of the Linnean Society of New South Wales* **116**, 37–78.

Keith, D. A. (1997). Combined effects of heat shock, smoke and darkness on germination of *Epacris stuartii* Stapf, an endangered fire-prone Australian shrub. *Oecologia* **112**, 340–344.

Keith, D. A., and Benson, D. H. (1988). The natural vegetation of the Katoomba 1:100,000 map sheet. *Cunninghamia* **2**, 107–143.

Keith, D. A., and Bradstock, R. A. (1994). Fire and competition in Australian heath: a conceptual model and field investigations. *Journal of Vegetation Science* **5**, 347–354.

Keith, D. A., and Myerscough, P. J. (1993). Floristics and soil relations of upland swamp vegetation near Sydney. *Australian Journal of Ecology* **18**, 325–344.

Keith, D. A., and Tozer, M. G. (1997). Experimental design and resource requirements for monitoring flora in relation to fire. In *Bushfire '97: Proceedings of the Australian Bushfire Conference* (eds. B. J. McKaige, R. J. Williams and W. M. Waggitt) pp. 274–279. (CSIRO: Darwin.)

Keith, D. A., Williams, J. E., and Woinarski, J. C. Z. (2001). Fire management and biodiversity conservation: key approaches and principles. In *Flammable Australia: The Fire Regimes and Biodiversity of a Continent* (eds. R. A. Bradstock, J. E. Williams and A. M. Gill) pp. 401–425. (Cambridge University Press: Cambridge.)

Kikkawa, J., Ingram, G. J., and Dwyer, P. D. (1979). The vertebrate fauna of Australian heathlands: an evolutionary perspective. In *Ecosystems of the World*, vol. 9A, *Heathlands and Related Shrublands: Descriptive Studies* (ed. R. L. Specht) pp. 231–279. (Elsevier: Amsterdam.)

Kirkpatrick, J. B., and Dickinson, K. J. M. (1984). The impact of fire on Tasmanian alpine vegetation and soils. *Australian Journal of Botany* **32**, 613–629.

Kruger, F. J. (1983). Plant community diversity and dynamics in relation to fire. In *Mediterranean-Type Ecosystems: The Role of Nutrients*, Ecological Studies no. 48 (eds. F. J. Kruger, D. T. Mitchell and J. U. M. Jarvis) pp. 446–472. (Springer-Verlag: Berlin.)

Lamont, B. B. (1985). Dispersal of the winged fruits of *Nuytsia floribunda* (Loranthaceae). *Australian Journal of Ecology* **10**, 187–194.

Lamont, B. B. (1988). Sexual versus vegetative reproduction in *Banksia elegans*. *Botanical Gazette* **149**, 370–375.

Lamont, B. (1996). Conservation biology of banksia in southwestern Australia. In *Gondwanan heritage: past, present and future of the Western Australian biota* (eds. S. D. Hopper, J. A. Chappill, M. S. Harvey and A. S. George) pp. 292–298. (Surrey Beatty: Chipping Norton, NSW.)

Lamont, B. B., and Barker, M. J. (1988). Seedbank dynamics of a serotinous, fire-sensitive *Banksia* species. *Australian Journal of Botany* **36**, 193–203.

Lamont, B. B., and Downes, S. (1979). The longevity, flowering and fire history of the grasstrees *Xanthorrhoea preissii* and *Kingia australis*. *Journal of Applied Ecology* **16**, 893–899.

Lamont, B. B., and Runciman, H. V. (1993). Fire may stimulate flowering, branching, seed production and seedling establishment in two kangaroo paws (Haemodoraceae). *Journal of Applied Ecology* **30**, 256–264.

Lamont, B. B., and van Leeuwin, S. J. (1988). Seed production and mortality in a rare *Banksia*. *Journal of Applied Ecology* **25**, 551–559.

Lamont, B. B., Enright, N. J., and Bergl, S. M. (1989). Coexistence and competitive exclusion of *Banksia hookeriana* in the presence of congeneric seedlings along a topographic gradient. *Oikos* **24**, 39–42.

Lamont, B. B., Connell, S. W., and Bergl, S. M. (1991a). Seedbank and population dynamics of *Banksia cuneata*: the role of time, fire, and moisture. *Botanical Gazette* **152**, 114–122.

Lamont, B. B., Le Maitre, D. C., Cowling, R. M., and Enright, N. J. (1991b). Canopy seed storage in woody plants. *Botanical Review* **57**, 277–317.

Lamont, B. B., Swanborough, P. W., and Ward, D. (2000). Plant size and season of burn affect flowering and fruiting of the grasstree *Xanthorrhoea preissii*. *Austral Ecology* **25**, 268–272.

Le Maitre, D. C. (1984). A short note on seed predation in *Watsonia pyramidata* (Andr.) Stapf in relation to season of burn. *Journal of South African Botany* **50**, 407–415.

Le Maitre, D. C., and Midgley, J. J. (1992). Plant reproductive ecology. In *Fynbos: Nutrients, Fire and Diversity* (ed. R. M. Cowling) pp. 135–174. (Oxford University Press: Oxford.)

Luke, R. H., and McArthur, A. G (1978). *Bushfires in Australia*. (Australian Government Publishing Service: Canberra.)

Luo, J., and Fox, B. J. (1995). Competitive effects of *Rattus lutreolus* presence on food resource use by *Pseudomys gracilicaudatus*. *Australian Journal of Ecology* **20**, 556–564.

Main, A. R. (1981). Fire tolerance of heathland animals. In *Ecosystems of the World*, vol. 9B, *Heathlands and Related Shrublands: Analytical Studies* (ed. R. L. Specht) pp. 85–90. (Elsevier: Amsterdam.)

Maltby, E., Legg, C. J., and Proctor, M. C. F. (1990). The ecology of severe moorland fire on the North York Moors: effects of the 1976 fires, and subsequent surface and vegetation development. *Journal of Ecology* **78**, 490–518.

Marsden-Smedley, J. B., and Catchpole, W. R. (1995a). Fire behaviour modelling in Tasmanian buttongrass moorlands. I. Fuel characteristics. *International Journal of Wildland Fire* **5**, 203–214.

Marsden-Smedley, J. B., and Catchpole, W. R. (1995b). Fire behaviour modelling in Tasmanian buttongrass moorlands. II. Fire behaviour. *International Journal of Wildland Fire* **5**, 215–228.

McCaw, W. L. (1991). Measurement of fuel quantity and structure for bushfire research and management. In *Proceedings of a Conference on Bushfire Modelling and Fire Danger Rating Systems* (eds. N. P. Cheney and A. M. Gill) pp. 147–156. (CSIRO: Canberra.)

McCaw, W. L. (1997). Predicting fire spread in Western Australian mallee-heath shrubland. PhD thesis, University of New South Wales.

McCaw, W. L., Maher, T., and Gillen, K. (1992). Wildfires in the Fitzgerald River National Park, Western Australia, December 1989. Department of Conservation and Land Management, Technical Report no. 26, Perth.

McFarland, D. C. (1988). The composition, microhabitat use and response to fire of the avifauna of subtropical heathlands in Cooloola National Park, Queensland. *Emu* **88**, 249–257.

McFarland, D. C. (1994). Notes on the Brush Bronzewing (*Phaps elegans*) and Southern Emu Wren (*Stipiturus malachurus*) in Cooloola National Park. *Sunbird* **24**, 14–17.

McLoughlin, L. C. (1998). Season of burning in the Sydney region: the historical records compared with recent prescribed burning. *Australian Journal of Ecology* **23**, 393–404.

McPhail, M. K. (1979). Vegetation and climates in southern Tasmania since the last glaciation. *Quarternary Research* **11**, 306–341.

Meney, K. A., Neilssen, G. M., and Dixon, K. W. (1994). Seedbank patterns in Restionaceae and Epacridaceae after wildfire in kwongan in southwestern Australia. *Journal of Vegetation Science* **5**, 5–12.

Meredith, C. W. (1983). RAOU Conservation Statement No. 1: The Ground Parrot. *RAOU Newsletter* **55**, 6–11.

Meredith, C. W., Gilmore, A. M., and Isles, A. C. (1984). The ground parrot (*Pezoporus wallicus* Kerr) in south-eastern Australia: a fire-adapted species? *Australian Journal of Ecology* **9**, 367–380.

Morris, E. C. (2000). Germination response of seven east Australian *Grevillea* species (Proteaceae) to smoke, heat exposure and scarification. *Australian Journal of Botany* **48**, 179–189.

Morris, E. C., and Myerscough, P. J. (1988). Survivorship, growth and self-thinning in *Banksia ericifolia*. *Australian Journal of Ecology* **13**, 181–189.

Morrison, D. A. (1995). Some effects of low-intensity fires on co-occurring small trees in the Sydney region. *Proceedings of the Linnean Society of New South Wales* **115**, 109–119.

Morrison, D. A., Auld, T. D., Rish, S., Porter, C., and McClay, K. (1992). Patterns of testa-imposed seed dormancy in native Australian legumes. *Annals of Botany* **70**, 157–163.

Myerscough, P. J., Clarke, P. J, and Skelton, N. J. (1995). Plant coexistence in coastal heaths: floristic patterns and species attributes. *Australian Journal of Ecology* **20**, 482–493.

Newsome, A. E., and Catling, P. C. (1979). Habitat preferences of mammals inhabiting heathlands of warm temperate coastal, montane and alpine regions of southeastern Australia. In *Ecosystems of the World*, vol. 9A, *Heathlands and Related Shrublands: Descriptive Studies* (ed. R. L. Specht) pp. 301–316. (Elsevier: Amsterdam.)

Newsome, A. E., McIlroy, J., and Catling, P. C. (1975). The effects of an extensive wildfire on populations of twenty ground vertebrates in south-east Australia. *Proceedings of the Ecological Society of Australia* **9**, 107–123.

Nicholls, N. (1990). Predicting the El Niño – Southern Oscillation. *Search* **21**, 165–167.

Noble, I. R., and Slatyer, R. O. (1980). The use of vital attributes to predict successional changes in plant communities subject recurrent disturbances. *Vegetatio* **43**, 5–21.

Noble, J. C. (1989). Fire regimes and their influence on herbage and mallee coppice dynamics. In *Mediterranean Landscapes in Australia: Mallee Ecosystems and their Management* (eds. J. C. Noble and R. A. Bradstock) pp. 168–180. (CSIRO: Melbourne.)

O'Dowd, D. J., and Gill, A. M. (1984). Predator satiation and site alteration following fire: mass reproduction of alpine ash in south-eastern Australia. *Ecology* **65**, 1052–1066.

Packham, D. R. (1970). Heat transfer above a small ground fire. *Australian Forest Research* **5**, 19–24.

Pate, J. S., Casson, N. E., Rullo, J., and Kuo, J. (1985). Biology of the fire ephemerals of the sandplains of the kwongan of south-western Australia. *Australian Journal of Plant Physiology* **12**, 641–655.

Pate, J. S., Froend, R. H., Bowen, B. S., Hansen, A., and Kuo, J. (1990). Seedling growth and storage characteristics of seeder and resprouter species of mediterranean-type ecosystems of S.W. Australia. *Annals of Botany* **65**, 585–601.

Pate, J. S., Meney, K. A., and Dixon, K. W. (1991). Contrasting growth and morphological characteristics of fire-sensitive (obligate seeders) and fire-resistant (resprouter) species of Restionaceae (S. Hemisphere Restiads) from south-western Australia. *Australian Journal of Botany* **39**, 505–525.

Recher, H. (1981). Bird communities of heath and their management and conservation requirements. In *Heaths in New South Wales* (ed. C. Haigh) pp. 27–40. (New South Wales National Parks and Wildlife Service: Sydney.)

Roche, S., Dixon, K. W., and Pate, J. S. (1997). Seed aging and smoke: partner cues in the amelioration of seed dormancy in selected Australian native species. *Australian Journal of Botany* **45**, 783–815.

Rowley, I., and Brooker, M. (1987). The response of a small insectivorous bird to fire in heathlands. In *Nature Conservation: The Role of Remnants of Native Vegetation* (eds. D. A. Saunders, G. W. Arnold, A. A. Burbidge and A. J. M Hopkins.) pp. 211–218. (Surrey Beatty: Chipping Norton, NSW.)

Russell, E. M., and Rowley, I. (1993). Demography of the cooperatively breeding Splendid Fairy Wren, *Malurus splendens* (Maluridae). *Australian Journal of Zoology* **41**, 475–505.

Russell-Smith, J., Ryan, P. G., Klessa, D., Waight, G., and Harwood, R. (1998). Fire regimes, fire-sensitive vegetation and fire management of the sandstone Arnhem Plateau, monsoonal northern Australia. *Journal of Applied Ecology* **35**, 829–846.

Sandercoe, C. (1992). Fire management of Cooloola National Park: fuel dynamics of the western catchment. In *Fire Research in Rural Queensland* (ed. B. R. Roberts) pp. 367–384. (University of Southern Queensland: Toowomba.)

Smith, D. T. (1985) Fire effects on populations of the noisy scrub-bird (*Atrichornis clamosus*), western bristle-bird (*Dasyornis longirostris*) and western whip-bird (*Psophodes nigrogularis*). In *Fire Ecology and Management of Western Australian Ecosystems* (ed. J. R. Ford) pp. 95–102. (Western Australia Institute of Technology: Perth.)

Specht, R. L. (1966). The growth and distribution of mallee–broombush (*Eucalyptus incrassata–Melaleuca uncinata*) and heath vegetation near Dark Island Soak, Ninety Mile Plain, South Australia. *Australian Journal of Botany* **14**, 361–371.

Specht, R. L. (1969). A comparison of the sclerophyllous vegetation characteristics of the mediterranean type climates in France, California and southern Australia. II. Dry matter, energy and nutrient accumulation. *Australian Journal of Botany* **17**, 293–308.

Specht, R. L. (1981). Responses to fire of heathlands and

related shrublands. In *Fire and the Australian Biota* (eds. A. M. Gill, R. H. Groves and I. R. Noble) pp. 395–415. (Australian Academy of Science: Canberra.)

Specht, R. L., and Morgan, D. G. (1981). The balance between the foliage projective covers of overstorey and understorey strata in Australian vegetation. *Australian Journal of Ecology* **6**, 193–202.

Specht, R. L., Rayson, P., and Jackman, M. E. (1958). Dark Island heath (Ninety Mile Plain, South Australia). VI. Pyric succession changes in composition, coverage, dry weight and mineral nutrient status. *Australian Journal of Botany* **6**, 59–88.

Sutherland, E. F. (1998). Fire, resource limitation, and small mammal populations in coastal eucalypt forest. PhD thesis, University of Sydney.

Sydes, M. A., and Peakall, R. (1998). Extensive clonality in the endangered shrub *Haloragodendron lucassii* (Haloragaceae) revealed by allozymes and RAPDs. *Molecular Ecology* **7**, 87–93.

Taylor, J. A., and Tulloch, D. (1985). Rainfall in the wet–dry tropics: extreme events at Darwin and similarities between years during the period 1870–1983. *Australian Journal of Ecology* **10**, 281–295.

Tolhurst, K. G., and Burgman, M. A. (1994). Simulation of bracken cover in forested areas in Victoria in response to season, overstorey and fire conditions. *Australian Journal of Ecology* **19**, 306–318.

Tozer, M. G. (1996). The invasive potential of *Acacia saligna* (Labill.) H. L. Wendl. in vegetation communities occurring on soils derived from Hawkesbury sandstone. MSc thesis, University of Sydney.

Tozer, M. G., and Bradstock, R. A. (1997). Factors infuencing the establishment of seedlings of the mallee, *Eucalyptus luehmanniana* (Myrtaceae). *Australian Journal of Botany* **45**, 997–1008.

Vaughton, G. (1998). Soil seedbank dynamics in the rare obligate seeding shrub, *Grevillea barkleyana* (Proteaceae). *Australian Journal of Ecology* **23**, 375–384.

Vines, R. G. (1981). Physics and chemistry of rural fires. In *Fire and the Australian Biota* (eds. A. M. Gill, R. H. Groves and I. R. Noble) pp. 129–150. (Australian Academy of Science: Canberra.)

Wahren, C.-H. A., Papst, W. A., and Williams, R. J. (1999). Post-fire regeneration in Victorian alpine and subalpine vegetation. In *Bushfire '99: Proceedings of the Australian Bushfire Conference* (eds. B. Lord, A. M. Gill and R. A. Bradstock) pp. 425–431. (Charles Sturt University: Albury, NSW.)

Wark, M. C., White, M. D., Robertson, D. J., and Marriott, P. F. (1987). Regeneration of heath and heath woodland in the north-eastern Otway Ranges following the wildfire of February 1983. *Proceedings of the Royal Society of Victoria* **99**, 51–88.

Whelan, R. J. (1995). *The Ecology of Fire*. (Cambridge University Press: Cambridge.)

Whelan, R. J., and Baker, J. R. (1999). Fire in Australia: coping with variation in ecological effects of fire. In *Protecting the Environment, Land, Life and Property: Proceedings of the Bushfire Management Conference* (eds. F. Sutton, J. Keats, J. Dowling and C. Doig) pp. 71–79. (New South West Nature Conservation Council: Sydney.)

Whelan, R. J., and Main, A. R. (1979). Insect grazing and post-fire plant succession in south-west Australian woodland. *Australian Journal of Ecology* **4**, 387–389.

Whelan, R. J., and York, J. (1998). Post-fire germination of *Hakea sericea* and *Petrophile sessilis* after spring burning. *Australian Journal of Botany* **46**, 367–376.

Whelan, R. J., Langedyk, W., and Pashby, A. (1980). The effects of a wildfire on arthropod populations in jarrah–*Banksia* woodland. *Western Australian Naturalist* **14**, 214–220

Whelan, R. J., Ward, S., Hogbin, P., and Wasley, J. (1996). Responses of heathland *Antechinus stuartii* to the Royal National Park wildfire in 1994. *Proceedings of the Linnean Society New South Wales* **116**, 97–108.

Whelan, R. J., De Jong, N. H., and Von der Burg, S. (1998). Variation in bradyspory and seedling recruitment without fire among populations of *Banksia serrata* (Proteaceae). *Australian Journal of Ecology* **23**, 121–128.

Whelan, R. J., Rodgerson, L., Dickman, C. R., and Sutherland, E. F. (2001). Critical life cycles of plants and animals: developing a process-based understanding of population changes in fire-prone landscapes. In *Flammable Australia: The Fire Regimes and Biodiversity of a Continent* (eds. R. A. Bradstock, J. E. Williams and A. M. Gill) pp. 94–124. (Cambridge University Press: Cambridge.)

Whight, S., and Bradstock, R. A. (1999). Indices of fire characteristics in sandstone heath near Sydney, Australia. *International Journal of Wildland Fire* **9**, 145–153.

Williams, P. (1995). Floristic patterns within and between

sedge-heath swamps of Gibraltar Range National Park, New South Wales. BSc thesis, University of New England, Armidale.

Williams, R. J., and Ashton, D. H. (1988). Cyclical patterns of regeneration in subalpine heathland communities on the Bogong High Plains, Victoria. *Australian Journal of Botany* **36**, 605–619.

Wilson, S. D. (1993). Competition and resource availability in heath and grassland in the Snowy Mountains of Australia. *Journal of Ecology* **81**, 445–451.

Witkowski, E. T. F., Lamont, B. B., and Connell, S. J. (1991). Seedbank dynamics of three co-occurring Banksias in south coastal Western Australia: the role of plant age, cockatoos, senescence and interfire establishment. *Australian Journal of Botany* **39**, 385–397.

Woinarski, J. C. Z. (1989). The vertebrate fauna of broombush *Melaleuca uncinata* vegetation in north-western Victoria, with reference to the effects of broombush harvesting. *Australian Wildlife Research* **16**, 217–238.

Woinarski, J. C. Z. (1999). Fire and Australian birds: a review. In *Australia's Biodiversity: Responses to Fire: Plants, Birds and Invertebrates*, Biodiversity Technical Paper no. 1 (eds. A. M. Gill, J. C. Z. Woinarski and A. York) pp. 55–111. (Environment Australia: Canberra.)

Wouters, M. A. (1993). Wildfire behaviour in heath and other elevated fuels: a case study of the 1991 Heywood fire. Research Report no. 36, Victorian Department of Conservation and Environment, Melbourne.

York, A. (1999). Long-term effects of repeated prescribed burning on forest invertebrates: management implications for the conservation of biodiversity. In *Australia's Biodiversity: Responses to Fire: Plants, Birds and Invertebrates*, Biodiversity Technical Paper no. 1 (eds. A. M. Gill, J. C. Z. Woinarski and A. York) pp. 181–266. (Environment Australia: Canberra.)

Young, A. R. M. (1982). Upland swamps (dells) on the Woronora Plateau, New South Wales. PhD thesis, University of Wollongong.

Zammit, C. (1988). Dynamics of resprouting in the lignotuberous shrub *Banksia oblongifolia*. *Australian Journal of Ecology* **13**, 311–320.

Zammit, C., and Westoby, M. (1987). Seedling recruitment strategies in obligate-seeding and resprouting *Banksia* shrubs. *Ecology* **68**, 1982–1992.

Zedler, P. (1977). Life history attributes of plants and the fire cycle: a case study in chaparral dominated by *Cupressus forbesii*. In *Proceedings of a Symposium on the Environmental Consequences of Fire and Fuel Management in Mediterranean Ecosystems* (eds. H. A. Mooney and C. E. Conrad) pp. 451–458. U.S. Department of Agriculture Forest Service General Technical Report WO-3.

10

Fire regimes and biodiversity in semi-arid mallee ecosystems

ROSS A. BRADSTOCK AND JANET S. COHN

Abstract
The mallee shrublands of semi-arid, southern Australia are significant for their biodiversity. Mallee (dominated by *Eucalyptus* spp. with a multi-stemmed habit) inhabits dunefields, sandplains and flats. Soils affect the density of the overstorey and understorey composition, which in turn affects flammability. Coarse-scale fire patterns and fire regimes are probably influenced by the mix of community types. Fire behaviour is governed by the discontinuous nature of fuels, with cover of eucalypt litter often insufficient to sustain spreading fires. Perennial hummock grasses (e.g. *Triodia scariosa*) or ephemerals provide fuel continuity in some cases. Thus mallee exhibits regular and stochastic patterns of fuel accumulation and flammability. Fire sizes of up to *c.* 100 000 ha can result from ignitions by lightning and severe weather associated with summer storms and current fire regimes are dominated by such fires on a decadal basis. A variety of functional groupings of plants occurs within mallee with distinct fire regime requirements, ranging from dominant obligate seeder *Callitris* spp. and resprouting *Eucalyptus* spp. to ephemeral herbs and grasses. Differing fire regimes will affect floristic composition and structure. Patterns of animal response to fires (e.g. reptiles and birds) are affected by the plant community floristics and structure. Notable are endangered birds (e.g. the malleefowl, *Leipoa ocellata*) that are most abundant in long-unburnt mallee. Quantitative studies of fire regimes are required to resolve management dilemmas posed by the necessity to conserve a range of biota with disparate fire regime requirements. The fragmented nature of mallee vegetation requires particular consideration.

Introduction

The word 'mallee' has many meanings. It evokes broad regions, types of vegetation, groups of *Eucalyptus* species and a growth habit of woody plants. These senses of the word have a conjunction in the semi-arid lands of southern Australia where shrubby vegetation ('mallee scrub') occurs. Such shrublands are dominated by *Eucalyptus* species (mallees) with a characteristic multi-stemmed habit and a Mediterranean climate (Parsons 1994). While much of the mallee regions has been greatly altered by pastoral and agricultural activities, significant areas of mallee shrublands remain more or less intact. This reflects their relatively poor status in terms of agricultural production but is fortuitous given their inherent value for biodiversity conservation. Large areas of mallee shrublands are now either reserved specifically for biodiversity conservation or else managed sympathetically. Fire is a prominent feature of the semi-arid lands in general and the mallee shrublands in particular. Management of fire is crucial to sustainable management of mallee lands, whether the objective is agricultural production or conservation (Noble 1984). Accordingly, the objective of this chapter is to describe the nature of fire regimes and their interactions with biodiversity.

Mallee communities

Within the semi-arid rainfall zone (200–500 mm annual rainfall), mallee shrublands occur on lateritic sandplains in southern Western Australia

Fig. 10.1. The extent of semi-arid, mallee vegetation (shaded areas) in southern Australia. From Noble (1984).

(WA), and in southeastern Australia in the aeolian landscapes of southern South Australia (SA), northwestern Victoria and western and central New South Wales (NSW) (Fig. 10.1) (Hill 1989). Differing geomorphological processes have shaped these main mallee regions, with much of the central and southeastern mallee characterized by dune and swale systems (Hill 1989). By contrast, the landforms in WA on which mallee communities now exist are older. Local relief is slight, but major soil differences occur across the landforms (Hill 1990). A common feature of many mallee shrublands is a surface layer of sand. Variations in the thickness of the layer affect structural and floristic composition of mallee shrublands. In turn such variations affect the nature of fuel, its flammability and consequent fire regimes.

In southwestern WA, the floristic diversity of mallee *Eucalyptus* species is higher than in the southeast and may relate to a mosaic of relict soils and climatic changes which resulted in isolation and subsequent speciation (Hill 1990). Hill (1990) postulated that the origin of most species within mallee vegetation is in southwestern WA. This is also true for many other plant groups (Parsons 1994).

Despite floristic differences in mallee between southwestern and southeastern Australia, similarities in structure exist. Although the mallee eucalypt overstorey varies little between whipstick (2–5 m tall) and bull mallee (5–10 m), the understorey is more variable (Beadle 1981). Beadle (1981) identified a number of understorey types in mallee communities across Australia, which relate chiefly to soil, i.e. hummock grasses (*Triodia* spp.), *Melaleuca* spp. and/or *Casuarina* spp., and semi-succulent subshrubs (*Maireana* spp., *Atriplex* spp.). These are briefly described.

On sandy soils in lower rainfall areas, a herbaceous layer of hummock grasses (*Triodia* spp.) often dominates the understorey (Beadle 1981; Parsons 1994). *Triodia scariosa* is the most widespread and common hummock grass in semi-arid mallee and tends to be in higher densities on dune slopes and crests than in swales, where soils have a heavier texture (e.g. Noble and Vines 1993). Typically, in NSW, this community also contains shrubs (Fabaceae, Euphorbiaceae, Asteraceae, Myrtaceae) and *Callitris verrucosa* (Bradstock 1989; Noble 1989a). In southern WA, another hummock grass, *Plectrachne rigidissima*, is present (Beadle 1981).

A dense thicket-like understorey of *Melaleuca* spp. and/or *Casuarina* spp. (Beadle 1981) commonly excludes a herbaceous understorey and mallee *Eucalyptus* spp. are either absent or sparsely distributed (Bradstock 1989). Blackburn and Wright (1989) note that *Melaleuca uncinata*, a widespread species, is often found on solodised solonetz and solodic soils on sandplains and swales of dunefields. Whilst these soils have a variable surface texture of loose sand to compact sandy loam, the subsoils are dense, coarsely structured, sandy clays or clays (Blackburn and Wright 1989).

In contrast, there are communities that lack both hummock grasses and woody shrubs. In NSW and Victoria, this type of understorey occurs on soils of heavier texture. Mallee eucalypts may be up to 10 m tall (Noble and Mulham 1980), and as rainfall declines, an understorey of semi-succulent subshrubs (*Maireana* spp., *Atriplex* spp., *Zygophyllum* spp.) may appear (Cheal and Parkes 1989; Sparrow 1989). During years of high rainfall, this community may support annuals, including more or less continuous swards of *Stipa* spp. (Land Conservation Council 1987).

A review by Menkhorst and Bennett (1990) found that the variation in the floristics and structure of mallee communities, especially the understorey, has led to a diverse and distinctive array of vertebrate species associated with any given geographic area. Most of these species, however, except for passerine birds, are not mallee specialists, relying also on the adjacent communities for their existence. Compared with the adjacent woodlands and heathlands, mallee has higher reptile species diversity, similar mammal diversity and lower bird and amphibian diversity.

Fuels and their arrangement in mallee

Where fuels are highly discontinuous, such as in mallee, Gill *et al.* (1995) advocated a multi-faceted approach to the problem of prediction of fire spread. The threshold for the propagation of fire is of pivotal interest to the study of fires and fire regimes. Where fuel discontinuity is high, the arrangement of fuel and ambient conditions may critically affect the threshold for propagation. Once conditions are sufficient for fire propagation, the average fuel load may be an important influence on rate of spread and intensity.

The principal surface fuel, eucalypt litter, is usually clustered around individual plants and estimates of load vary between 0.5 and 1.5 kg m^{-2} (e.g. Rawson 1982; Noble 1984; Bradstock and Gill 1993). Suspended material, such as bark strands, dead leaves and twigs, typically accumulates among the stem bases in the centre of mallee eucalypt plants, up to 1 metre above ground. While the mass of such material may be low, it can play a prominent role in promoting fires in live eucalypt crowns and in spotting (Noble 1984). Shrub litter loads are generally low (e.g. 0.15 kg m^{-2}) and due to the fine-leaved nature of many common species, litter beds are tightly packed and often partially buried by sand (Bradstock and Gill 1993). Grasses and herbs constitute the other primary source of fuel. The most notable perennial component of surface fuel, aside

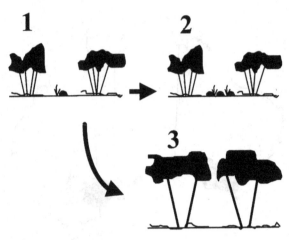

Fig. 10.2. Variations in the spatial arrangement of fuel in mallee. In (1) cover of eucalypt litter and other elements in the gaps between eucalypts (e.g. shrubs and hummock grasses) is insufficient for the propagation of fire. In (2) other elements in the gaps are sufficiently dense for fire propagation to occur. In (3) cover of eucalypt litter is sufficient for propagation. Pathways between states (arrows) represent changes with increasing time since fire (see text).

from the eucalypt litter, are the hummocks of *Triodia. Triodia* hummock mass ranges from 1.5 to 3.0 kg m^{-2} and hummock development typically peaks within 15–30 years after fire (Bradstock and Gill 1993; Noble and Vines 1993). Ephemeral herbs and grasses vary substantially in biomass according to rainfall (Noble 1989b). Maximum values of 0.8 kg m^{-2} have been reported after a sequence of average or above-average years, soon after fire (Noble 1989b).

Thresholds for fire spread will be related to characteristics of eucalypt litter and other fuel elements. Three basic fuel states may be identified (Fig. 10.2) where: (1) all fuel elements are arranged so that fire propagation is not possible; (2) eucalypt litter arrangement is inadequate, but the additive contribution of other fuels renders propagation possible; and (3) eucalypt litter arrangement alone is adequate. Particular types of mallee communities may conform to one of these particular states or alternately shift between states as circumstances change.

Estimates of eucalypt cover of *c.* 20%–50% (e.g. Cohn 1995; Forward and Robinson 1996) are common on sandy soils in dune landscapes. Noble

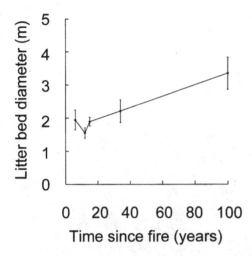

Fig. 10.3. Changes in the mean diameter of litter beds beneath mallee *Eucalyptus* species in relation to time since fire. Samples taken from a chronosequence of sites in Round Hill Nature Reserve, central-western NSW.

and Vines (1993) give an example of typical variation in cover, ranging from 50% on dunes and sandplains to 30% in swales. Crown cover can be reasonably equated to the cover of litter beds beneath mallee eucalypts (Bradstock and Gill 1993; authors' unpublished data). Such data, however, do not account for changes in cover as a function of time since last fire. A study of temporal changes in litter bed dimensions surrounding individual eucalypts indicated a significant increase in diameter (Fig. 10.3) beneath long-unburnt mallees (*c.* 100 years) compared with relatively recently burnt plants (10–20 years after fire).

Thus average density and cover of eucalypts are commonly insufficient for propagation of fires for many decades after fire. Growth and changes in crown and litter dimensions after the prolonged absence of fire may result in a doubling of the effective area of litter (Fig. 10.3). The transition from state (2) to (3) (Fig. 10.2) may be possible only when lengthy periods without fire occur (*c.* 100 years). In some mallee communities, such as on heavier textured soils, eucalypt density may be insufficient for fire spread irrespective of time after fire (Cheal *et al.* 1979). Other fuel elements commonly provide spatial continuity in the fuel bed (state 2). The

various grasses and herbs have a prominent role as supplementary fuel elements. In terms of flammability, a distinction between communities can be made on the basis of the presence of *Triodia* spp.

Bradstock and Gill (1993) showed that the dimensions of flames were dependent on the height and diameter of individual *Triodia* hummocks and concluded that the size and spacing of hummocks is important for fire propagation. Flame propagation in beds of *Triodia* hummocks may be dependent on direct flame contact. Flames from a burning hummock must bridge the gap to an adjacent hummock and the duration of contact must be sufficient to initiate ignition. Green (1983) predicted that wind speed may affect the critical density of discrete fuel elements needed for propagation of a spreading fire. At high wind speeds, the critical density of *Triodia* hummocks will be less than at low wind speeds. Griffin and Allan (1984) and Burrows *et al.* (1991) produced fire-spread models for arid hummock grasslands that relate rate of spread to overall cover, between-hummock gap size and wind speed. The range of cover sufficient for fire spread (30%–50%), observed by Griffin and Allan (1984) is probably applicable to the patches of *Triodia* that occupy gaps among eucalypts in mallee. Observations of *T. scariosa* in central NSW at the time of a major summer fire in 1985 give insight into fire propagation (authors' unpublished data). Ten years had elapsed since the previous fire. The typical cover on dune slopes and swales was respectively, 20%–30% and <10%. Fire spread was extensive on dune slopes, but many unburnt patches were evident on dune crests and swales. In more moderate weather, fire propagation, at a similar time after fire, was usually restricted to dune slopes (authors' unpublished data).

Mallee vegetation with *Triodia* often contains a prominent shrub component (Fig. 10.4a). Bradstock and Gill (1993) concluded that it is unlikely that individuals or small clusters of these shrubs contribute independently to fire spread. Low litter-loadings and crown characteristics, such as bulk density, are insufficient to sustain flaming without an additional source of heat. Shrubs are most likely to play a synergistic role in conjunction with

(a)

(b)

Fig. 10.4. (a) Propagation of a fire in a mallee community in central NSW, illustrating the role of the shrub layer in promoting the passage of flames into the crowns of overstorey eucalypts. (b) A mallee community on heavier textured soils exhibiting a relatively high ground-cover dominated by ephemeral grasses (*Stipa* spp.). This cover developed following several years of above-average rainfall, hence providing a high level of surface fuel continuity when cured. (Photographs by R. A. Bradstock.)

primary surface fuel elements (*Triodia* hummocks and eucalypt litter). They observed that shrubs contribute to the development of fires in the crowns of overstorey eucalypts (e.g. Fig. 10.4a). In old *Callitris verrucosa* stands (*c.* 100 years), litter loads remain low, containing tightly packed leaves often intermixed with soil and the crown–base height is sufficient to prevent active transfer of flames into the foliage (authors' unpublished data). The promotion of crown fires in eucalypts by shrubs may enhance fire spread through spotting and radiant heat transfer from ensuing long flames.

In communities without *Triodia*, flammability is a property of ephemeral herbs, forbs and other grasses (Fig. 10.4b). Stochastic variations in the cover and biomass of ephemerals are governed by rainfall. Noble (1989b) demonstrated that the flush of ephemerals is most pronounced in the 3 years following fire. In long-unburnt mallee the response of ephemerals to rainfall was weak. In adjacent, non-mallee communities, massive herbage responses to sequences of drought followed by above-average rainfall have been described (Leigh *et al.* 1989). Biomass may be tightly related to cumulative rainfall effects in such events (Robertson 1988). Such exceptional responses await formal description in mallee (Gill 1997). None the less, the switch from a non-flammable to a flammable state (states (1) to (2) as above) in mallee, as a function of drought followed by high rainfall, has been well described (e.g. Noble *et al.* 1980).

When is mallee flammable? Some claim that dune mallee is unable to burn for 10–20 years after fire (e.g. Cheal *et al.* 1979; Noble 1984). Evidence suggests that communities containing *Triodia* may burn extensively at about 10 years after fire particularly in severe weather (e.g. fires in central NSW in 1974 and 1984: Pickard 1987). In open mallee without *Triodia*, where large-scale fires followed by above-average rainfall have resulted in vigorous herbage growth, areas have burnt twice in 3 years (Condon and Alchin 1977). Noble (1984) noted that whilst open mallee in swales carrying dense herbage burnt 3 years after last fire, adjacent dunes with a *Triodia* understorey did not burn, due to inadequate surface fuel cover. Variations in the density

of *Triodia* populations may also affect flammability. Obligate seeder populations of *Triodia* are subject to substantial variations in density over time due to the influences of rainfall and grazing on post-fire seedling establishment and survival (Noble and Vines 1993; Cohn and Bradstock 2000), thereby weakening regular temporal patterns of flammability. Resprouting by *Triodia* may also lead to variable patterns of flammability if rainfall affects the rate of regrowth (Griffin and Allan 1984).

Bradstock (1990) noted the possibility of feedback effects between prominent components of the vegetation and flammability. Thus, changes to the density of key obligate seeder species such as *Triodia* and *Callitris* as a function of particular fire regimes have the potential to alter the subsequent flammability of the community. In mesic communities dominated by eucalypts, fuel accumulation patterns and consequent flammability may be less susceptible to feedback of this kind, than in mallee. Replacement of litter from woody plants by herbaceous material, under certain fire regimes, may be more rapid and regular than in semi-arid systems.

Fire behaviour and incidence

Fire-weather over much of the mallee regions is shaped by hot, dry summers, though as Noble (1984) noted, significant summer rainfall can occur in many parts of the region due to storms. Summer storms resulting from cold fronts, or commonly, the southward drift of tropical depressions (Cullenward 1989) can produce instability and lightning. When such storms are dry, multiple ignitions can ensue in large landscapes (Cheal *et al.* 1979; Noble *et al.* 1980; Rawson 1982). Erratic wind changes associated with such storms, plus high temperatures and low humidity, may result in rapid development of fires ignited from lightning. Shifts in wind direction (e.g. with the passage of cold-fronts) may be accompanied by an increase in wind speed, with lengthy flanks of fires converted to fast-running headfires (Condon 1980; Rawson 1982).

Gill and Moore (1990) showed that days of Extreme or Very High Fire Danger Rating (McArthur's Forest Fire Danger Index, FFDI) were relatively common in the southeastern mallee region (i.e. Mildura, Victoria). Maxima and average daily values of 15.00 hr FFDI were consistently higher in the summer months than at stations in more mesic or coastal localities (Gill and Moore 1990). On average their analysis indicates 6 Extreme and 44 Very High days per annum (FFDI >50 and 30–50 respectively). The potential for severe fire behaviour in mallee regions is therefore substantial in any year.

There are few quantitative reports of fire behaviour in mallee vegetation. Prolific spotting, presumably due to burning brands of bark or leaves from eucalypts, is often described (e.g. up to 2 km distance: Rawson 1982). Although the contribution of spotting to fire behaviour has not been quantified, adverse effects on fire suppression activities are likely. Rawson (1982) documents spread rates in summer wildfires of between 0.8 and 1.5 m s^{-1} in mallee with *Triodia* as a surface layer (about 15 years post-fire). Such estimates corresponded to wind speeds in the 2.2–5.5 m s^{-1} range (FFDI *c.* 30). Rawson (1982) reported similar spread rates for a prescribed burn in spring with similar windspeeds (4.5–5.0 m s^{-1}) but under more humid conditions and lower FFDI (12–17). Noble *et al.* (1980) reported average spread rates ranging from 0.4 to 1.5 m s^{-1} corresponding with 'light' winds and 'typical' summer weather conditions. These observations were made in long-unburnt mallee with presumed high litter and herbage loads. By contrast, Noble (1989b) and Bradstock *et al.* (1992) reported spread rates as low as 0.08 m s^{-1} in prescribed fires in autumn, winter and spring.

While spread rates can be extremely high in mallee shrublands relative to other eucalypt-dominated communities (more mesic woodlands and forests), overall loadings of fuel are lower in mallee. Higher spread rates may be due to the more open canopy structure of the mallee (Rawson 1982). Thus surface wind speeds may be higher in mallee than in more closed communities for any given ambient wind (lower wind reduction factor). The situation is similar to that found in eucalypt-dominated tropical savannas, where spread rates of fires are higher but fuel loads lower than in southern forests, due to the open canopy structure (Williams and Bradstock

2000; Williams *et al.*, this volume). Conversely, the discontinuous nature of fuel in mallee means that a threshold wind speed must be exceeded before fires are likely to spread (McCaw 1997), whereas in more mesic forests and woodlands with continuous surface fuels, the spread of fire without wind may be possible. Disjunct patterns of burning in mallee may be expected as a result. Maximum intensities in the vicinity of 20–30 000 Kw m^{-1} may be expected in *Triodia* mallee given average fuel loads and upper levels of rates of spread (see above). These predictions accord with observations of McCaw (1997) for fires in mallee-heath.

The principal source of contemporary fires in mallee is lightning (e.g. Cheal *et al.* 1979; Noble 1984; McCaw 1996), in contrast to many fire-prone ecosystems in Australia, where human ignitions currently dominate (Williams and Bradstock 2000). In landscapes bounded by cereal cropping, such as in northwestern Victoria, escaped stubble fires and accidental ignitions or arson accounted for a substantial percentage of ignitions (Cheal *et al.* 1979; Land Conservation Council 1987), but lightning-ignited fires account for the bulk of area burnt. This may reflect a higher density of human population as well as agricultural practices in these regions. Where land use is dominated by rangeland pastoralism and/or conservation, ignitions appear to be almost solely attributable to lightning. In the Western Division of NSW this reflects the prevailing management of pastoralists who attempt to exclude fire in order to maximise herbage for stock (Noble 1984).

Noble (1984) and Noble and Vines (1993) hypothesised that size and timing of fires in mallee lands are linked. Thus in average years, ignitions from lightning were unlikely to produce substantial fires due to fuel discontinuity at the landscape scale. Noble and Vines (1993) suggested that the periodic occurrence of large fires in the mallee lands was a function of above-average antecedent rainfall. As discussed above, such rainfall will promote prolific herbage growth in areas with heavier textured soils (in both mallee shrublands and other adjacent woodlands) and will result in high levels of fuel continuity at both fine and coarse spatial scales. Resultant fires from lightning ignitions will be large due to high rates of spread if accompanied by strong winds.

Noble and Vines's (1993) hypothesis assumes that the incidence of significant fires in mallee systems is essentially fuel-limited. That is, fire spread is limited by an absence of fuel continuity, particularly the ephemeral component, and ignitions are ubiquitous. The supposed linkage between significant fires and 'wet' years is of considerable importance to the management of fire regimes in mallee. It implies an immutable fire regime that could not be substantially altered by changes in ignition rate and consequent feedback effects through the fuel system. Some alternatives to their hypothesis are that the incidence and consequent size of fire are constrained by: (1) ignition sources (or lack thereof), (2) regular patterns of accumulation of fuel derived from perennial plants such as eucalypt litter and (3) a positive association between wet years and subsequent dry-lightning activity. There are few available data to devise a test that would critically discriminate between these hypotheses. Krusel et al. (1993) found that incidence of fire in northwestern Victoria was weakly related to measures such as Forest and Grass Fire Danger Index, but positively related to ambient temperature. They interpreted this as evidence in support of Noble and Vines's (1993) fuel limitation hypothesis.

Noble and Vines (1993) postulated an approximate major cycle of major rainfall events for the Western Division of NSW in the order of 1–2 decades. Evidence from mallee areas in SA (Forward and Robinson 1996), the Western Division of NSW (Pickard 1987) and northwestern Victoria (Cheal et al. 1979; Rawson 1982) unequivocally illustrates the association of large fires with preceding years of above-average rainfall. However, not all rainfall events of this kind inevitably produce large fires. For example, major fires in 1957 were confined to central western NSW with no large fires in the southwest (Pickard 1987).

The occurrence of significant fires is unlikely to be limited in any particular fire season by lack of appropriate fire-weather. Rather, the incidence of large fires may be limited either by ignition sources or by fuel factors at a landscape level. Following above-average wet periods some areas may miss out on ignitions or else lightning may be accompanied by rain. The nature and frequency of lightning storms and consequent ignitions may vary substantially from year to year. Condon (1975), Alchin (1978) and Condon and Alchin (1977, 1979) described how the frequency of dry lightning storms in the summer of 1974/75 in the Western Division of NSW was substantially higher than in subsequent years. They attributed decreased fire activity in those years to this factor.

Size of mallee fires and the nature of fire regimes

The fire regime in many mallee areas is dominated by very large fires on a decadal basis with few or no fires in intervening periods (Morelli and Forward 1996; Willson 1999; NSW National Parks and Wildlife Service unpublished records). The coincidence of unplanned fires at a relatively short interval (e.g. 10 years or less) occasionally occurs, sometimes over a significant area (e.g. >10 000 ha: Condon and Alchin 1977). Notably, the postulated decadal cycle of wet weather (Noble and Vines 1993) is roughly coincident with the peak in flammability associated with the regular accumulation of eucalypt litter (unaffected by antecedent rain) and maximum development of Triodia hummocks (see above). Confounding of these elements makes it difficult to disentangle cause and effect. A single large fire following a sequence of wet years may be enough to synchronise the regular component of fuel accumulation with the stochastic component (herbage). This sort of cycle of fuel development may fit well with a scenario of non-ubiquitous ignition (low annual chance of ignition). Ignition risk may be low annually but reasonably high on a decadal time-scale in large landscapes. Under this scenario both fuel and ignition are limiting factors which operate on coincident time-scales. If a subsequent wet period was not followed by ignitions, the opportunity would exist to decouple the regular and irregular components of fuel.

While Noble and Vines (1993) contend that ignitions in 'normal' years are essentially immediately self-extinguishing (and thus are unnoticed), there is ample evidence of self-sustaining prescribed or experimental burning in such years, in differing localities (Cheal *et al.* 1979; Noble 1984; Bradstock *et al.* 1992). Rawson (1982) commented on the need for prescribed burning to be conducted in weather conditions more extreme (i.e. late spring or even early summer) than that recommended for forests in order to produce effective burns of reasonable size. Morelli and Forward (1996) described how most fire activity in the northern Murray mallee of SA, including large fires, was a result of either deliberate or accidental human ignitions.

Cheal *et al.* (1979) reported that large fires in northwestern Victoria were associated with both above-average and below-average antecedent spring rainfall. In part this may reflect more diverse causes of ignitions in these areas, with human ignitions resulting in a large proportion of fires and total area burnt (Cheal *et al.* 1979). The largest fires in the study period occurred in 1974/75 and were associated with above-average spring and annual rainfall. Rawson (1982) confirmed the contribution of massive herbage growth to these fires. None the less, lightning-ignited fires in January 1981 in both the Sunset and Big Desert areas which exceeded the size of fires in the mid 1970s (>100000 ha) in each case (Rawson 1982; Land Conservation Council 1987) followed a year of average rainfall at Mildura and Ouyen.

The data of Cheal *et al.* (1979) suggest that in southern parts of the region (e.g. below the Murray River), significant fires occur in dry conditions or years of average rainfall. A similar tendency (Forward 1996) can be seen for Ngarkat National Park (SA) which is contiguous with the mallee and mallee-heath systems of the Big Desert. One possibility is that southern parts of the region are more prone to lightning ignition associated with cold fronts than areas further north that are less strongly affected by such fronts. The dominance of very old age classes in mallee areas north of the Murray in western NSW and SA (Morelli and Forward 1996; Willson 1999) may reflect such a trend.

Individual wildfires, ignited from severe dry lightning storms, appear to have a maximum size of about 100000 to 120000 ha, whether associated with wet years or not. This may reflect the duration of weather favourable to rapid fire spread (strong winds). Sequences of days with strong wind are probably finite. This size limit could also be affected by suppression operations, particularly once severe weather has abated. McCaw (1996) reported mallee and mallee-heath fires of this size in southeastern WA. This area is largely uninhabited and unresourced in terms of fire suppression (McCaw 1996). Given the size of fragments and conservation reserves in the mallee regions an upper limit to fire size of *c.* 100000 ha is important. A single ignition has the potential to burn many large remnant landscapes and reserves to a substantial degree.

The current decadal cycle of fire found in many mallee landscapes is a possible outcome of the interaction between the regular pattern of accumulation of eucalypt litter and *Triodia* and the chance of ignition by lightning. Regional variations in fire regimes may relate to trends in chance of ignition. Landscapes with higher proportions of open mallee prone to 'grassing-up' in wet years may have a different range of potential fire regimes than those with a lesser component of open communities. Such effects are yet to be fully explored.

Responses of plant species to fire regimes

The periodic incidence of fire in mallee communities has led many to regard resident flora and fauna as being 'adapted to fire' (e.g. Gardner 1957). This tendency continues despite the development of concepts that relate the life histories of organisms to disturbance regimes in general and fire regimes in particular (Gill 1975; Bond and van Wilgen 1996). The dichotomous categorisation of species and communities as either 'fire-adapted' or 'fire-sensitive', while seemingly innocuous, masks the subtlety and complexity of responses to fire regimes, reduces predictive power and may lead to erroneous management of biodiversity.

A range of functional groups (after Noble and Slatyer 1980) can be distinguished in the mallee on the basis of attributes such as mode of recovery from disturbance, maturation rates, seed bank characteristics and life spans. These include perennial obligate seeders (serotinous and soil-stored seed banks), perennial resprouters (woody and herbaceous) and ephemerals. The proportion of perennial, obligate seeders is similar to many other sclerophyllous communities (e.g. Williams and Bradstock 2000). For example Morelli and Forward (1996) reported fire responses of 66 species in eastern SA, of which 30 species were obligate seeders.

Callitris and *Hakea* spp. are the only obligate seeders with serotinous seed banks (on plant storage of seeds in woody cones) in the mallee (Morelli and Forward 1996). In *C. verrucosa* smaller plants are generally killed by fire, larger individuals may survive across a range of fire conditions and those older than 50 years have characteristics sufficient to survive intense fires (authors' unpublished data). Seeds of *C. verrucosa* are released in the first year after fire and the extent of release is proportional to the amount of the plant killed by fire (Bradstock 1989). The highest germination of seeds is attained at low temperatures, typical of winter in southeastern Australia (Adams 1985; Bradstock 1989). Seedling establishment may be strongly affected by grazing pressure, particularly from rabbits (Zimmer 1944; Cheal et al. 1979). The primary juvenile period (age when first sexually mature) is between 11 and 13 years in central NSW but may vary across the range of the species (Bradstock 1989). If fires are excluded from the mallee for more than 100 years, *C. verrucosa* may overtop the mallee and dominate (Cheal et al. 1979; Bradstock 1989; McCaw 1996). The life span of *C. verrucosa* is unknown, though the sizes of plants suggest potential longevity well in excess of 100 years (authors' unpublished data).

The majority of perennial obligate seeders are shrubs with a soil-stored seed bank (e.g. *Acacia*, *Senna*, *Beyeria* spp.) and *Triodia scariosa*. All these species generally suffer high levels of mortality, irrespective of fire intensity. The size of the annual seed crop of *Acacia* species and *T. scariosa* may relate to rainfall (Bradstock 1989). Seeds are dispersed by ants and stored in the soil, though some may be lost in this process and taken by predators such as mallee-

fowl (Andersen 1982; Bradstock 1989; Harlen and Priddel 1996). Although up to 50% of *Acacia* seeds in any crop may be non-dormant, the remainder respond to heat, either from high summer temperatures or fire (Bradstock 1989). Similar responses were reported for *T. scariosa* though germination rates were substantially lower. Noble (1989a) found that seedling recruitment of *T. scariosa* was significantly higher after spring fires than autumn and winter fires, though Bradstock (1989) reported the opposite effect for *Acacia* spp. The juvenile period of a wide range of species of these woody shrubs and *T. scariosa* ranges from 2 to 5 years (Cheal et al. 1979; Bradstock 1989; Morelli and Forward 1996). During the juvenile period, survival of germinants of species such as *T. scariosa* and various acacias is strongly influenced by rainfall and grazing by vertebrate herbivores, especially rabbits and kangaroos (Cohn and Bradstock 2000). Observations in central NSW indicate a life span of *T. scariosa* and *Acacia* spp. between 30 and 40 years (Bradstock 1989). Data on other genera are lacking but given similarities in size and juvenile periods among woody taxa, a similar range may be assumed.

The chenopods are usually obligate seeders with no effective post-fire seed bank, short-range seed dispersal and low flammability, which may be attributed to high concentrations of NaCl in the leaves (Sharma et al. 1972) and to semi-succulence. Some species seem to exhibit intra-specific variation in fire response (Gill 1997). Seeds of *Atriplex vesicaria* and *Maireana sedifolia* do not have protection of embryos by thick endocarp walls to protect them from lethal temperatures during fires (Hodgkinson and Griffin 1982).

Common facultative resprouters include *Dodonaea viscosa*, *Senna artemisioides*, *Eremophila mitchellii*, *Eucalyptus* spp., *Myoporum* spp., selected *Acacia* spp., *Melaleuca uncinata*, and some populations of *T. scariosa* (Hodgkinson and Griffin 1982; Noble and Vines 1993; Parsons 1994). Apart from their resprouting capability in response to fire, *Eucalyptus* spp. and *M. uncinata* release seeds from woody capsules in the canopy, whilst other species rely on soil seed banks. Ecotypic differences within the species may relate to variations in factors such as soil texture (Bradstock 1989).

Noble (1989a) found that the timing and the frequency of fires affected the survival of selected adult *Eucalyptus* spp. Mallee populations regenerated with little mortality from one fire every 5 to 10 years, irrespective of season. A more frequent occurrence resulted in significantly higher mortality after autumn burns than spring burns. Since growth of mallee eucalypts usually occurs in the warmer months, burning in autumn may have caused physiological stress by forcing out-of-phase growth (Noble 1989a). The secondary juvenile period (time to sexual maturity after resprouting) ranges from 3 to 8 years (Gardner 1957; Parsons 1968; Noble 1982). This may be influenced by post-fire environmental conditions such as rainfall.

Recruitment of seedlings of mallee *Eucalyptus* spp. can vary depending on the season of the fire. For example (Noble 1989a) found it significantly higher after spring fires than either autumn or winter fires. Establishment is affected by rabbit grazing, competition from mature eucalypts and lignotuber growth and rainfall (Wellington 1989; Parsons 1994). Noble (1982) argued that seedling establishment need only occur once every few centuries, since eucalypts may live for around 200 years (Wellington 1989). Other woody species such as *Melaleuca uncinata* follow broadly similar patterns of survivorship, regrowth maturation and seedling establishment (Bradstock 1989).

In recently burnt mallee, species richness is higher than in long-unburnt mallee due to the occurrence of fire-induced ephemerals (Hopkins 1985; Land Conservation Council 1987; Noble 1989b). In NSW, Noble (1989a) found that forbs were only conspicuous in the first post-fire season following warm-season fires and full development of populations relied on adequate rainfall in autumn. The nature of seed banks, germination cues and seed dispersal of ephemerals awaits detailed study.

Cheal *et al.* (1979) noted that there is apparently no functional group of plants that appears only after a lengthy absence of fire. An exception to this are lower plants such as mosses and lichens that form extensive crusts in gaps with bare soil. Increases in cover of non-vascular plants (cryptogams) and changes in species composition were recorded with increasing time since fire (Eldridge and Bradstock 1994). Cheal *et al.* (1979) speculated that the cryptogam layer inhibits establishment of vascular plants in long-unburnt mallee. Disruption of the surface crust may facilitate post-fire germination of vascular plants. In the absence of fire, evidence for the successful establishment of seedlings of many perennial species is slight (Cheal *et al.* 1979; Holland 1986). Whether establishment is limited by release of seeds from dormant seed banks, competition with established plants or factors such as grazing is unclear. Conditions after a fire lead to a temporary reduction in competition, and predator and grazer satiation (Wellington 1989).

Vegetation dynamics

While knowledge of some vital attributes (Noble and Slatyer 1980) is sketchy or lacking (e.g. longevity of soil storages of seeds of ephemeral herbs and grasses) there is sufficient to predict some general pathways of change in relation to alternative fire regimes (Bradstock 1989). A hypothesis is that major functional groups in mallee communities are in tension (Bradstock 1989, 1990). Competition for resources, principally water, may lead to dominance on the one hand and decline and exclusion on the other. Promotion of a particular fire regime by one group through its role as fuel (see above) may result in exclusion of another. Such mechanisms may be interlinked, creating potentially complex dynamics.

Potential interactions exist between the overstorey and understorey in mallee communities. When long unburnt, *Callitris verrucosa* may overtop mallee eucalypts. Bradstock (1989) and Cheal *et al.* (1979) have speculated that this may cause eventual elimination of eucalypts. Alternatively, this may happen if *C. verrucosa* were to outlive eucalypts (Noble and Slatyer 1977). The low-fuel characteristics of old *C. verrucosa* stands may reinforce such effects by lowering the probability of burning. Another interaction is that between *C. verrucosa* and the perennial understorey composed of *Triodia scariosa* and various shrub species. Burning cycles approximating the juvenile period of *C. verrucosa* (e.g. 10–15 years) may cause its local elimination, maintaining high densities of the other understorey species. Slightly longer cycles of burning (20–25 years) should lead to coexistence of all elements, though high densities of *C. verrucosa* may elevate mortality in *Acacia* spp. and *T. scariosa* (authors' unpublished data). A third interaction is that between ephemerals and perennials, particularly the obligate seeder perennials and the special case of chenopods with no effective post-fire seed bank. Successive flushes of ephemerals following closely spaced wet episodes create the potential for elimination of perennials through a short fire interval.

Bradstock (1989) related differing states observed in dune-mallee to the interval between successive fires, based on these interactions. A number of pathways of vegetation change were identified, resulting in various highly contrasting states. Effects of intensity and sizes of fires were considered as supplementary factors. Bradstock (1989) emphasised that each pathway of change was likely to have a different probability, this being dependent on the likelihood of occurrence of particular fire regimes. Thus maintenance of the system in a flammable state with a dense sward of *T. scariosa* and a layer of shrubs was considered more likely than other pathways.

Responses of animal species to fire regimes

Vertebrates

For vertebrates, mallee vegetation provides a range of resources such as food, shelter and nest sites (Menkhorst and Bennett 1990). Suites of vertebrate species characterize mallee communities defined by structurally different understories, i.e. shrub, hummock grass and chenopod (Menkhorst and Bennett 1990). As with other communities, studies on the fire ecology of animals in mallee have mainly been concerned with seral effects (time since fire) on populations and communities. Relative to other groups, the seral responses of reptiles and birds have received most attention. Effects of fire regimes have been inferred in some cases, particularly where responses of populations have been closely linked to the dynamics of the vegetation. Detailed studies of animals and fire regimes are a priority.

Reptiles and amphibians

Friend (1993) considered reptiles and amphibians in mallee systems more resilient to fire than mammals, reflecting an ability to cope with dry conditions, strong seasonal activity patterns, their burrowing habit and a preference for open, relatively non-inflammable microhabitats. He speculated that for both groups, the intensity and the timing of fire are important.

For reptiles, the importance of *Triodia* spp. is in providing a microclimatic refuge from extremes in temperature and humidity (Cogger 1984). Several studies have found lower reptile diversity and abundance in mallee long unburnt than more recently burnt (Cogger 1984). Some believe this is related to the dynamics of *Triodia* (Cheal *et al.* 1979; Cogger 1984; Schlesinger *et al.* 1997). Caughley (1985) found the peak abundances of three groups of reptiles exploiting different niches, i.e. burrowers that forage on open ground, species associated with *Triodia* hummocks and litter foragers, occurred at respectively 3, 6 and 25 years after fire. Schlesinger *et al.* (1997) found that the abundance of nocturnal reptiles was significantly lower 18 years after fire than in more recently burnt sites, since they require open ground in which to forage (Cogger 1992).

Birds

Woinarski and Recher (1997) and Woinarski (1999) reviewed post-fire successional patterns in the occurrence of bird species in mallee vegetation. These patterns vary in response to the landscape position (crest vs. swale, soil type) and the spatial context of fires (extent, patchiness, patch isolation etc.).

In recently burnt mallee (<1 year), widespread opportunists are most common. Those that become more common in 1 to 10 years after fire are predominantly species that use mallee as well as other communities, such as heath or woodland. At an intermediate stage after fire (10–30 years), as the vegetation becomes taller and denser, bird species endemic to the mallee are favoured. For example, red-lored whistler (*Pachycephala rufogularis*) forages in dense ground cover with a very open overstorey and southern scrub-robin (*Drymodes brunneopygia*) is found in mallee–broombush. Mallee emu-wren (*Stipiturus mallee*), rufous-crowned emu-wren (*Stipiturus ruficeps mallee*), and striated grasswren (*Amytornis striatus*) nest or shelter in *Triodia* hummock grasses and the latter two prefer large hummocks. Species preferring older, taller mallee with an open understorey (>30 years) include endangered species such as the malleefowl (*Leipoa ocellata*), which is reliant on litter for nest building, the black-eared miner (*Manoria flavifula melanotis*) which forages in the decorticating bark of very old mallee eucalypts and hollow-nesting birds such as the striated pardalote (*Pardalotus striatus*) and the regent parrot (*Polytelis anthopeplus*).

Mallee dominated by *Callitris verrucosa* has fewer species of birds than other mallee communities and this is presumably related to the absence of shrubs. There is evidence that after extensive fires, unburnt patches of vegetation such as this are used at least by malleefowl, for colonisation or emigration.

Mammals

A review by Friend (1993) found few data on the effects of fire on small mammals in semi-arid mallee communities, throughout Australia. Research in other shrubby communities indicates that differing rates of recovery among mammal species after fire are a response to changes in structure and floristics of the vegetation. In mallee heath, Cockburn (1981) found that whilst the optimum habitat for *Pseudomys apodemoides* was during the early seral stages after fire, when floristic diversity was highest, ensuring a more reliable source of seeds, Mitchell's hopping-mouse (*Notomys mitchelli*) preferred vegetation with greater vertical diversity (*Leptosperum* and *Melaleuca* spp.). Cockburn (1983) believed a decline in the breeding success of *P. apodemoides* would occur as the vegetation matured and plant diversity declined with increasing time since fire. Bennett *et al.* (1989) found that mallee ningaui (*Ningaui yvonneae*) was most abundant where closely spaced hummock grasses occurred.

Large mammals, such as kangaroos (*Macropus* spp.), have the mobility to escape fires. Conflicting data on the relative presence of kangaroos in mallee after fire (Cheal *et al.* 1979; Caughley *et al.* 1985; Noble 1986) may be dependent on the availability of herbage, which can be abundant and widespread after good rains, thus discouraging convergence grazing of the new germinants (Caughley *et al.* 1985).

Invertebrates

Although there has been limited research on the effects of fire on invertebrates in semi-arid shrublands, a review by Friend (1996) suggests it is unwise to draw from research in more temperate climates. In general, fires in woodland and shrubland communities have less effect on invertebrate populations than those in open forest, possibly as a result of adaptations to aridity (Friend 1996). With exception, however, the Araneae group (spiders) consistently stood out as a potential environmental indicator, which is not surprising given its position at the top of the invertebrate food chain.

The few and varied studies on fire and invertebrates in mallee communities make generalizations of trends difficult. Immediately following a wildfire, Andersen and Yen (1985) found that the doubling of ant species on the ground, and the halving of their overall abundance was in response to habitat simplification and the demise of a dominant species. Woinarski (1989) described an increase in abundance of ants soon after fire, with a peak in total invertebrate density at 26 years after fire in mallee–broombush. Strehlow (1993) found that a high-intensity fire had a significant short-term effect on spider abundance and richness, with a greater decline in web than ground or burrow dwellers. Yen (1989) compared sites burnt 3 and 21 years previously and found a higher mean number of predators on older foliage, reflecting the increased availability of microhabitats for web building by spiders and more cryptic microhabitats for shelter. Schlesinger *et al.* (1997) found that although the number of terrestrial beetle species was higher in recently burnt mallee (3–5 years), the overall abundance of beetles did not vary between different fire histories. This may, in part, reflect differences in the activity of animals in relation to structural and other habitat changes.

Fire regimes, coexistence and management

Are current fire regimes in mallee landscapes adverse to biodiversity conservation? We contend that there is little evidence of adverse regimes in the large expanses of remaining mallee communities. Short intervals between fires sometimes occur over significant areas and while these may result, locally, in adverse changes to some plant species, there is little evidence of large-scale effects. Losses of long-unburnt mallee during wildfires, and thus habitat for certain animal species, are an ongoing concern. None the less significant areas of long-unburnt mallee remain in most landscapes within the mallee regions. There has been no systematic analysis of contemporary trends in fire regimes. Such insights would provide a clearer basis for debate over appropriate management. Fragmentation of mallee systems may lead to a reduction of probability of fire in small remnants. Conversely, changes in land use and human usage may affect ignition rates in ways that are difficult to predict. Given that human ignition rates are generally low, the scope for an increase in human ignitions coincident with severe fire weather is high. Any resultant change in mallee fire regimes is likely to be significant.

Discussion about the nature of fire regimes in mallee and interactions with biodiversity is typically focussed on intervention (e.g. Heislers et al. 1982; Noble et al. 1990), encompassing technological suppression of unplanned fires, creation of trails and fuel breaks, prescribed burning and other methods of manipulation of fuel. Arguments about the need for intervention are shaped by 'historical', 'mosaic' (or seral) and 'demographic' (or life-history) paradigms (Bradstock et al. 1995; Gill et al., this volume).

The notion that past fire regimes can be re-created and that conservation of biodiversity will automatically ensue as an emergent property is unrealistic and impractical. The past is poorly known and methods of unravelling the nature of past regimes at a level of scale relevant to management are fraught with difficulty. The context of mallee landscapes is now unique, as a result of clearing, European land uses and human population density. Functional knowledge of the interplay between biodiversity and fire regimes within a contemporary context is required. Resolution of the differing perspectives offered by the 'mosaic' and 'life-history' paradigms is also required. Respectively these reflect animal and plant-based knowledge of fire ecology (see above). There is some predictive knowledge of responses of plant species and communities in relation to fire regimes, but there is little comparable knowledge for animals. Approaches to fire management of animals have therefore been concerned with provision of a range of age classes. Such an approach is 'event' (e.g. major wildfires) rather than 'regime' oriented. The presence of endangered species of birds with an apparent requirement for long-unburnt stands of mallee gives strong impetus for 'event' management and corresponding decision-support systems (e.g. Richards et al. 1999).

Management cannot focus entirely on 'events' for two reasons. Firstly, seral trends in animal abundance usually reflect the influence of vegetation as habitat. Since fire regimes affect mallee vegetation, the true effects of any particular intervention to manage age classes in a landscape ('mosaic' management) for the benefit of animal species cannot be understood without concomitant appraisal of effects on fire regimes. A focus on 'events' without attention to 'regimes' may be seriously misleading in terms of ultimate, indirect effects on animal populations through their habitat. Secondly, the critical proportion of a particular seral state in a landscape is so far undefined for any animal species. Every fire results in some sort of mosaic, but the appropriate degree of spatial heterogeneity for animal species has received virtually no attention. Thus 'event' oriented systems for managing age classes have relied on arbitrary assumptions in this regard (e.g. Richards et al. 1999).

Approaches that allow resolution of both seral and fire regime requirements of species are required. The problem of coexistence of both plant and animal groups with seemingly varying requirements is not unique to mallee. An emerging view is that spatial and temporal variation in fire regimes is necessary for the coexistence of species with disparate requirements (Gill and McCarthy 1998). Landscape level

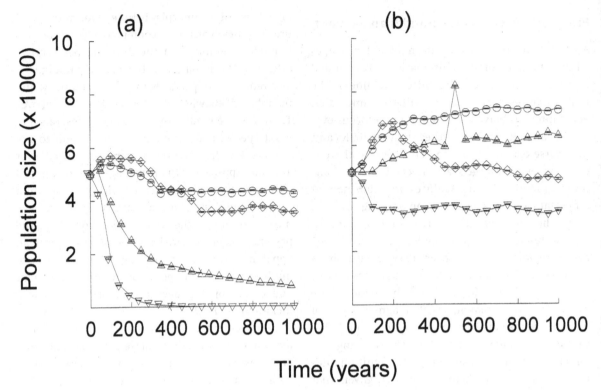

Fig. 10.5. Spatially explicit simulations (gridded model) of the dynamics of *Callitris verrucosa* populations (mean percentage of landscape occupied), in (a) sandplain and (b) dune landscapes, in relation to different levels of prescribed burning (◇ 1%, ○ 5%, △ 10%, ▽ 20% of the landscape per annum). Probability of wildfire = 1 in 20 years. Results are for five replicates, with standard errors obscured by symbols.

approaches to the investigation of fire regimes (McCarthy and Cary, this volume) offer such a solution. Simple models indicate that variation in fire interval and age classes in landscapes are related to the chance of burning (McCarthy and Cary, this volume).

Previous work with grid-based, spatial models has established thresholds of extinction for plant species in relation to the frequency and sizes of fires (e.g. Bradstock *et al.* 1996). Such a model was used to simulate the dynamics of the obligate seeder *Callitris verrucosa*, in a flat, *Triodia*–mallee landscape (based on an increase in flammability with time since fire, peaking to an asymptote). Results indicated that declining populations with a high risk of extinction (Fig. 10.5a) occur when about half the intervals between fire in a representative cell (point

scale) are unfavourable (i.e. ≤ the juvenile period of 10–15 years). Such a scenario was most likely at high levels of annual prescribed burning (20% of the landscape per annum: Fig. 10.5a).

An analogy can be drawn between the juvenile period in plants and the occupancy of seral age classes by animals; e.g. a species such as the malleefowl may not inhabit areas <20 years after fire (Benshemesh 1990). Ostensibly, such a species would then be more susceptible to high levels of ignition than *C. verrucosa*. Given that adult birds occupy home ranges of about 4 km² and that juveniles may disperse and forage across a range of several km (Booth 1987; Benshemesh 1990; Priddel and Wheeler 1995), the general dispersal capability of malleefowl greatly exceeds that of *C. verrucosa* (seed dispersal range may be less than 50 m:

authors' unpublished data). Such dispersal capability will engender greater resistance to habitat fragmentation through adverse fire intervals (<20 years) than for a species such as *C. verrucosa*. Thus selection of a prescribed burning strategy to minimise the risk of extinction of *C. verrucosa* in a landscape should, prima facie, be an adequate strategy for the conservation of a malleefowl population. A strategy of low to moderate prescribed burning (1%–5% of the landscape per annum; Fig. 10.5a) maximises the proportion of old stands of *C. verrucosa*, whilst restricting the chance of a single fire burning the whole landscape (authors' unpublished data).

Substantially different fire regimes emerge from a model landscape composed of dunes with differential flammability on crests, slopes and swales (Fig. 10.5b). Sizes of fires in severe weather were smaller (authors' unpublished data), due to discontinuities in fuel resulting in longer average intervals between fire, and *C. verrucosa* populations persist under all scenarios of prescribed burning (Fig. 10.5b). Such models (Fig. 10.5a, b) indicate the role that intrinsic variability of landscape features may play in determining the nature of fire regimes and the subsequent response of species. Topographic diversity and consequent local variations in flammability may buffer populations of species requiring long-unburnt conditions and/or long intervals between fire, from potentially adverse rates of ignition. These results emphasise the importance of local conditions. Management strategies that apply in one landscape may not apply elsewhere.

While there are many potential approaches to intervention, a middle path has evolved. Division of large tracts of mallee into substantial blocks, through trail construction, has occurred in many areas. Trails facilitate suppression of wildfires and the creation of strategic buffers through manipulation of fuel (e.g. prescribed burning or chaining: Cullenward 1989). Such an approach seems reasonable in large landscapes, offering some influence on large unplanned fires and consequent fire regimes, while being realistic in terms of resource requirements. While debate continues about the degree to which this can be pursued, the ideal level and comparisons with other strategies await systematic research. Constraints through wilderness declarations in large mallee areas and the increasing use of aircraft for suppression and strategically targeted prescribed burning may lessen the necessity for planned trail and fuel-break construction and the corresponding use of heavy machinery during suppression. Trails may also increase rates of human ignition, through provision of access.

In small mallee remnants, options are more constrained. The achievable level of variation in fire regimes is likely to be considerably less than in large landscapes. Mechanical disturbances, through trail and fuel-break construction, may carry high risks of associated weed invasion (Wapshere 1989) in small remnants. In all cases, the occurrence of wet years may render fuel break and prescribed burning areas more, rather than less, liable to carry a subsequent wildfire (Gill 1997). Practical opportunities for response to 'outbreaks' of high levels of ephemeral fuel may be extremely limited (curing may occur over a time-scale of days and weeks). Actions aimed at manipulation of such fuels may be limited in area and need careful planning to be effective.

Planning systems that define an acceptable range of fire regimes for major plant communities and specific groups of plant and animal species are now being applied to some mallee conservation reserves (Willson 1999; Keith *et al.*, this volume). A hallmark of this adaptive approach is that the manager is compelled, on an ongoing basis, to evaluate the state of fire regimes in a reserve against ecological criteria. Thus no set prescription is followed but rather an ever-changing set of prescriptions is derived to cope with circumstances at the time. The onus is to continually scrutinise, interpret and respond to the nature of fire regimes and to use simple summaries of ecological knowledge as a basis for response (Bradstock 1999). Such a system allows for developments and alterations to the knowledge base through research. It encourages monitoring for assessment of performance and scrutiny of potential problems. Willson (1999) has described a prudent range of actions (e.g. strategic prescribed burning in association with fuel breaks, experimental management of long-unburnt stands) stemming from this approach in a medium-sized mallee reserve in western NSW.

In many respects mallee landscapes offer an ideal laboratory for refinement of fire management strategies for biodiversity conservation. The scope for variation in fire regimes is large due to the size of many remaining tracts and the relatively low rate of human ignitions. The presence of important species with seemingly disparate requirements and the potential complexity of interactions between biodiversity and fire regimes add spice to the mix. While there is much to learn about fire in the mallee, there is also much to be learned from it.

Acknowledgements

We are grateful to Malcolm Gill, Jim Noble, Bob Parsons, Jann Williams and Andrew Willson for many valuable comments and suggestions regarding the manuscript.

References

Adams, R. B. (1985). Aspects of the distribution and ecology of two species of *Callitris* Vent. in Victoria. PhD thesis, LaTrobe University, Melbourne.

Alchin, B. M. (1978). A report on bushfires in the Western Division of New South Wales in the summer of 1977–78. Western Lands Commission, Sydney.

Andersen, A. N. (1982). Seed removal by ants in the mallee of north-western Victoria. In *Ant–Plant Interactions in Australia* (ed. R. C. Buckley) pp. 31–43. (Dr W. Junk: The Hague.)

Andersen, A. N., and Yen, A. L. (1985). Immediate effects of fire on ants in the semi-arid mallee region of north-western Victoria. *Australian Journal of Ecology* 10, 25–30.

Beadle, N. C. W. (1981). *The Vegetation of Australia*. (Cambridge University Press: Cambridge.)

Bennett, A. F., Lumsden, L. F., and Menkhorst, P. W. (1989). Mammals of the mallee, southeastern Australia. In *Mediterranean Landscapes in Australia: Mallee Ecosystems and their Management* (eds. J. C. Noble and R. A. Bradstock) pp. 191–220. (CSIRO: Melbourne.)

Benshemesh, J. (1990). Management of malleefowl with regard to fire. In *The Mallee Lands: A Conservation Perspective* (eds. J. C. Noble, P. J. Joss and G. K. Jones) pp. 206–211. (CSIRO: Melbourne.)

Blackburn, G., and Wright, M. J. (1989). Soils. In *Mediterranean Landscapes in Australia: Mallee Ecosystems and their Management* (eds. J. C. Noble and R. A. Bradstock) pp. 35–53. (CSIRO: Melbourne.)

Bond, W. J., and van Wilgen, B. W. (1996). *Fire and Plants*. (Chapman and Hall: London.)

Booth, D. T. (1987). Home range and hatching success of Malleefowl, *Leipoa ocellata* Gould (Megapodiidae), in Murray Mallee near Renmark, South Australia. *Australian Wildlife Research* 14, 95–104.

Bradstock, R. A. (1989). Dynamics of a perennial understorey. In *Mediterranean Landscapes in Australia: Mallee Ecosystems and their Management* (eds. J. C. Noble and R. A. Bradstock) pp. 141–154. (CSIRO: Melbourne.)

Bradstock, R. A. (1990). Relationships between fire regimes, plant species and fuels in mallee communities. In *The Mallee Lands: A Conservation Perspective* (eds. J. C. Noble, P. J. Joss and G. K. Jones) pp. 218–225. (CSIRO: Melbourne.)

Bradstock, R. A. (1999). 'Thresholds' for biodiversity: the National Parks and Wildlife Service approach to planning of fire management for conservation. In *Bushfire Management Conference Proceedings: Protecting the Environment, Land, Life and Property* (eds. F. Sutton, J. Keats, J. Dowling and C. Doig) pp. 11–18. (Nature Conservation Council of New South Wales: Sydney.)

Bradstock, R. A., and Gill, A. M. (1993). Fire in semi-arid mallee shrublands: size of flames from discrete fuel arrays and their role in the spread of fire. *International Journal of Wildland Fire* 3, 3–12.

Bradstock, R. A., Auld, T. D., Ellis, M. E., and Cohn, J. S. (1992). Soil temperatures during bushfires in semi-arid mallee shrublands. *Australian Journal of Ecology* 17, 433–440.

Bradstock, R. A., Keith, D. A., and Auld, T. D. (1995). Fire and conservation: imperatives and constraints on managing for diversity. In *Conserving Biodiversity: Threats and Solutions* (eds. R. A. Bradstock, T. D. Auld, D. A. Keith, R. Kingsford, D. Lunney and D. Sivertsen) pp. 323–333. (Surrey Beatty Chipping Norton, NSW.)

Bradstock, R. A., Bedward, M., Scott, J., and Keith, D. A. (1996). Simulation of the effect of spatial and temporal variation in fire regimes on the population viability of a *Banksia* species. *Conservation Biology* 10, 776–784.

Burrows, N. D., Ward, B. and Robinson, A. (1991). Fire behaviour in spinifex fuels on the Gibson Desert Nature Reserve, Western Australia. *Journal of Arid Environments* **20**, 189–204.

Caughley, J. (1985). Effect of fire on the reptile fauna of mallee. In *The Biology of Australasian Frogs and Reptiles* (eds. G. Grigg, R. Shine and H. Ehrman) pp. 31–34. (Surrey Beatty: Chipping Norton, NSW)

Caughley, G., Brown, B., and Noble, J. (1985). Movement of kangaroos after a fire in mallee woodland. *Australian Wildlife Research* **12**, 349–353.

Cheal, D. C., and Parkes, D. M. (1989). Mallee vegetation in Victoria. In *Mediterranean Landscapes in Australia: Mallee Ecosystems and their Management* (eds. J. C. Noble and R. A. Bradstock) pp. 125–140. (CSIRO: Melbourne.)

Cheal, P. D., Day, J. C., and Meredith, C. W. (1979). Fire in the national parks of north-western Victoria. Report by National Parks and Wildlife Service, Melbourne.

Cheney, N. P. (1996). The effectiveness of fuel reduction burning for fire management. In *Fire and Biodiversity: The Effects and Effectiveness of Fire Management* (ed. J. R. Merrick) pp. 9–16. Commonwealth Department of Environment Sport and Territories Biodiversity Series, Biodiversity Unit, Paper no. 8, Canberra.

Cockburn, A. (1981). Diet and habitat preference of the silky desert mouse *Pseudomys apodemoides* (Rodentia). *Australian Wildlife Research* **8**, 475–497.

Cockburn, A. (1983). Silky mouse *Pseudomys apodemoides*. In *Australian Museum Complete Book of Australian Mammals* (ed. R. Strahan) p. 412. (Angus and Robertson: Sydney.)

Cogger, H. G. (1984). Reptiles in the Australian arid zone. In *Arid Australia* (eds. H. G. Cogger and E. Cameron) pp. 235–252. (Australian Museum: Sydney.)

Cogger, H. G. (1992). *Reptiles and Amphibians of Australia* (Reed Books: Sydney.)

Cohn, J. S. (1995). The vegetation of Nombinnie and Round Hill Nature Reserves, central-western New South Wales. *Cunninghamia* **4**, 81–101.

Cohn, J. S., and Bradstock, R. A. (2000). Factors affecting post-fire seedling establishment of selected mallee understorey species. *Australian Journal of Botany* **48**, 59–70.

Condon, R. W. (1975). Report on bushfires in the Western Division of New South Wales, November 1974 to March 1975. Report by Western Lands Commission, Sydney.

Condon, R. W. (1980). Fire behaviour and fire fighting in relation to landscape types in the Western Division of New South Wales. Report by Western Lands Commission, Sydney.

Condon, R. W., and Alchin, B. M. (1977). A report on bushfires in the Western Division of New South Wales in the summer of 1976–77. Report by Western Lands Commission, Sydney.

Condon, R. W., and Alchin, B. M. (1979). A report on bushfires in the Western Division of New South Wales in the 1978–79 fire period. Report by Western Lands Commission, Sydney.

Cullenward, G. B. (1989). Fire in the mallee rangelands: a resource document for the development of fire management plans for the Mallee Bushfire Prevention Scheme. Report by Western Lands Commission, Sydney.

Eldridge, D. J., and Bradstock, R. A. (1994). The effect of time since fire on the cover and composition of cryptogamic soil crusts on a eucalypt shrubland soil. *Cunninghamia* **3**, 521–527.

Forward, L. R. (1996). Managing fire for biodiversity. Development of methodology. Unpublished Report by Department of Environment and Natural Resources, Adelaide.

Forward, L. R., and Robinson, A. C. (eds.) (1996). A biological survey of the South Olary Plains, South Australia. Unpublished report by Department of Environment and Natural Resources, Adelaide.

Friend, G. R. (1993). Impact of fire on small vertebrates in mallee woodlands and heathlands of temperate Australia: a review. *Biological Conservation* **65**, 99–114.

Friend, G. R. (1996). Fire and invertebrates: a review of research methodology and predictability of post-fire response patterns. *CALMScience Supplement* **4**, 165–173.

Gardner, C. A. (1957). The fire factor in relation to the vegetation of Western Australia. *Western Australian Naturalist* **5**, 166–173.

Gill, A. M. (1975). Fire and the Australian flora: a review. *Australian Forestry* **38**, 4–25.

Gill, A. M. (1997). A land of drought and fire and 'flooding rains': towards an understanding of fire regimes, applied and 'natural', in mallee lands. Report by CSIRO Plant Industry, Centre for Plant Biodiversity Research, Canberra.

Gill, A. M., and McCarthy, M. C. (1998). Intervals between prescribed fires in Australia: what intrinsic variation should apply? *Biological Conservation* **85**, 161–169.

Gill, A. M., and Moore, P. H. R. (1990). Fire intensities in *Eucalyptus* forests of southeastern Australia. *Proceedings of the International Conference on Forest Fire Research*, Coimbra, Portugal, 1990, pp. B24/1–12.

Gill, A. M., Burrows, N. D., and Bradstock, R. A. (1995). Fire modelling and fire weather in an Australian desert. *CALMScience Supplement* 4, 29–34.

Gill, A. M., Bradstock, R. A., and Williams, J. E. (2001). Fire regimes and biodiversity: legacy and vision. In *Flammable Australia: The Fire Regimes and Biodiversity of a Continent* (eds. R. A. Bradstock, J. E. Williams and A. M. Gill) pp. 429–446. (Cambridge University Press: Cambridge.)

Green, D. G. (1983). Shapes of simulated fires in discrete fuels. *Ecological Modelling* 20, 21–32.

Griffin, G. F., and Allan, G. E. (1984). Fire behaviour. In *Anticipating the Inevitable: A Patch-Burn Strategy for Fire Management at Uluru (Ayers Rock–Mt Olga) National Park* (ed. E. C. Saxon) pp. 55–68. (CSIRO: Melbourne.)

Harlen, R., and Priddel, D. (1996). Potential food sources available to Malleefowl *Leipoa ocellata* in marginal mallee lands during drought. *Australian Journal of Ecology* 21, 418–428.

Heislers, A., Lynch, P., and Walters, B. (eds.) (1982). Fire Ecology in Semi-Arid Lands, unpublished proceedings of a Workshop, Mildura, May 1981. (CSIRO Division of Land Resources Management Communications Group: Deniliquin, NSW.)

Hill, K. D. (1989). Mallee eucalypt communities: their classification and biogeography. In *Mediterranean Landscapes in Australia: Mallee Ecosystems and their Management* (eds. J. C. Noble and R. A. Bradstock) pp. 93–108. (CSIRO: Melbourne.)

Hill, K. D. (1990). Biogeography of the mallee eucalypts. In *The Mallee Lands: A Conservation Perspective* (eds. J. C. Noble, P. J. Joss and G. K. Jones) pp. 16–20. (CSIRO: Melbourne.)

Hodgkinson, K. C., and Griffin, G. F. (1982). Adaptation of shrub species to fires in the arid zone. In *Evolution of the Flora and Fauna of Arid Australia* (eds. W. R. Barker and P. J. M. Greenslade) pp. 145–152. (Peacock Publications: Adelaide.)

Holland, P. G. (1986). Mallee vegetation: steady state or successional? *Australian Geographer* 17, 113–120.

Hopkins, A. J. M. (1985). Fire in the woodlands and associated formations of the semi-arid region of south-western Australia. In *Fire Ecology and Management in Western Australian Ecosystems* (ed. J. Ford) pp. 83–90. (Western Australian Institute of Technology: Perth.)

Keith, D.A., Williams, J.E., and Woinarski, J.C.Z. (2001). Fire management and biodiversity conservation: key approaches and principles. In *Flammable Australia: The Fire Regimes and Biodiversity of a Continent* (eds. R. A. Bradstock, J. E. Williams and A. M. Gill) pp. 401–425. (Cambridge University Press: Cambridge.)

Krusel, N., Packham, D., and Tapper, N. (1993). Wildfire activity in the mallee shrubland of Victoria. *International Journal of Wildland Fire* 3, 217–227.

Land Conservation Council (1987). Report on the Mallee Area Review. Government of Victoria, Melbourne.

Leigh, J. H., Wood, D. H., Holgate, A., and Stanger, M. G. (1989). Effects of rabbit and kangaroo grazing on two semi-arid grassland communities in central-western New South Wales. *Australian Journal of Botany* 37, 375–396.

McCarthy, M. A., and Cary, G. J. (2001). Fire regimes in landscapes: models and realities. In *Flammable Australia: The Fire Regimes and Biodiversity of a Continent* (eds. R. A. Bradstock, J. E. Williams and A. M. Gill) pp. 77–93. (Cambridge University Press: Cambridge.)

McCaw, L. (1996). *Callitris preissii* on Bald Island, Western Australia: preliminary observations on distribution, stand structure and tree age. Report by Department of Conservation and Land Management, Perth.

McCaw, L. (1997). Predicting fire spread in western Australian mallee-heath shrubland. PhD thesis, University of New South Wales, Sydney.

Menkhorst, P. W., and Bennett, A. F. (1990). Vertebrate fauna of mallee vegetation in southern Australia. In *The Mallee Lands: A Conservation Perspective* (eds. J. C. Noble, P. J. Joss and G. K. Jones) pp. 39–53. (CSIRO: Melbourne.)

Morelli, J., and Forward, L. R. (1996). Mallee fire ecology. In *A Biological Survey of the South Olary Plains, South Australia* (eds. L. R. Forward and A. C. Robinson) pp. 263–272. Report by Department of Environment and Natural Resources, South Australia.

Noble, I. R., and Slatyer, R. O. (1977). The effect of disturbance on plant succession. *Proceedings of the Ecological Society of Australia* 10, 135–145.

Noble, I. R., and Slatyer, R. O. (1980). The use of vital attributes to predict successional changes in plant communities subject to recurrent disturbances. *Vegetatio* **43**, 5–21.

Noble, J. C. (1982). The significance of fire in the biology and evolutionary ecology of mallee *Eucalyptus* populations. In *Evolution of the Flora and Fauna of Arid Australia* (eds. W. R. Barker and P. J. M. Greenslade) pp. 153–160. (Peacock Publications: Adelaide.)

Noble, J. C. (1984). Mallee. In *Management of Australia's Rangelands* (eds. G. N. Harrington, A. D. Wilson and M. D. Young) pp. 223–240. (CSIRO: Melbourne.)

Noble, J. C. (1986). Prescribed fire in mallee rangelands and the potential role of aerial ignition. *Australian Rangeland Journal* **8**, 118–135.

Noble, J. C. (1989a). Fire regimes and their influence on herbage and mallee coppice dynamics. In *Mediterranean Landscapes in Australia: Mallee Ecosystems and their Management* (eds. J. C. Noble and R. A. Bradstock) pp. 168–180. (CSIRO: Melbourne.)

Noble, J.C. (1989b). Fire studies in mallee (*Eucalyptus* spp.) communities of western New South Wales: the effects of fires applied in different seasons on herbage productivity and their implications for management. *Australian Journal of Ecology* **14**, 169–187.

Noble, J. C., and Mulham, W. E. (1980). The natural vegetation of aeolian landscapes in semi-arid south-eastern Australia. In *Aeolian Landscapes in Semi-Arid Zone of South Eastern Australia* (eds. R. R. Storrier and M. E. Stannard) pp. 125–139. (Australian Society of Soil Science: Wagga Wagga.)

Noble, J. C., and Vines, R. G. (1993). Fire studies in mallee (*Eucalyptus* spp.) communities of western New South Wales: grass fuel dynamics and associated weather patterns. *Rangeland Journal* **15**, 270–297.

Noble, J. C., Smith, A. W., and Leslie, H. W. (1980). Fire in the mallee shrublands of western New South Wales. *Australian Rangeland Journal* **2**, 104–114.

Noble, J. C., Joss, P. J., and Jones, G. K. (eds.) (1990). *The Mallee Lands: A Conservation Perspective*. (CSIRO: Melbourne.)

Parsons, R. F. (1968). An introduction to the regeneration of mallee eucalypts. *Proceedings of the Royal Society of Victoria* **81**, 59–68.

Parsons, R. F. (1994). Eucalypt scrubs and shrublands. In *Australian Vegetation* (ed. R. H. Groves) pp. 291–319. (Cambridge University Press: Cambridge.)

Pickard, J. (1987). Mallee management strategy: a discussion paper. Report by New South Wales Western Lands Commission, Sydney.

Priddel, D., and Wheeler, R. (1995). The biology and management of the Malleefowl (*Leipoa ocellata*) in New South Wales. New South Wales National Parks and Wildlife Service, Species Management Report no. 19, Hurstville.

Rawson, R. (1982). Fire behaviour. In *Fire Ecology in Semi-Arid Lands* (eds. A. Heislers, P. Lynch and B. Walters) Proceedings of a Workshop. Mildura, May 1981. (CSIRO Division of Land Resources Management Communications Group: Deniliquin, NSW.)

Richards, S. A., Possingham, H. P., and Tizard, J. (1999). Optimal fire management for maintaining habitat diversity. *Ecological Applications* **9**, 880–892.

Robertson, G. (1988). Effect of rainfall on biomass, growth and dieback of pastures in an arid grazing system. *Australian Journal of Ecology* **13**, 519–528.

Schlesinger, C. A., Noble, J. C., and Weir, T. (1997). Fire studies in mallee (*Eucalyptus* spp.) communities of western New South Wales: reptile and beetle populations in sites of differing fire history. *Rangeland Journal* **19**, 190–205.

Sharma, M. L., Tunny, J., and Tongway, D. J. (1972). Seasonal changes in sodium and chloride concentrations of saltbush (*Atriplex* spp.) leaves as related to soil and plant potential. *Australian Journal of Agricultural Research* **23**, 1007–1019.

Sparrow, A. (1989). Mallee vegetation in South Australia. In *Mediterranean Landscapes in Australia: Mallee Ecosystems and their Management* (eds. J. C. Noble and R. A. Bradstock) pp. 109–124. (CSIRO: Melbourne.)

Strehlow, K. H. (1993). Impact of fires on spider communities inhabiting semi-arid shrublands in Western Australia's wheatbelt. BSc thesis, Murdoch University, Perth.

Wapshere, A. J. (1989). Biological control of weeds. In *Mediterranean Landscapes in Australia: Mallee Ecosystems and their Management* (eds. J. C. Noble and R. A. Bradstock) pp. 443–463. (CSIRO: Melbourne.)

Wellington, A. B. (1989). Seedling regeneration and the population dynamics of eucalypts. In *Mediterranean Landscapes in Australia: Mallee Ecosystems and their Management* (eds. J. C. Noble and R. A. Bradstock) pp. 155–167. (CSIRO: Melbourne.)

Williams, R. J., and Bradstock, R. A. (2000). Fire regimes

and the management of biodiversity in temperate and tropical *Eucalyptus* forest landscapes in Australia. In *Fire and Forest Ecology: Innovative Silviculture and Vegetation Management, Tall Timbers Fire Ecology Conference Proceedings no. 21* (eds. W. K. Moser and C. F. Moser) pp. 139–150. (Tall Timbers Research Station: Tallahassee, FL.)

Williams, R. J., Griffiths, A. D. and Allan, G. (2001). Fire regimes and biodiversity in the savannas of northern Australia. In *Flammable Australia: The Fire Regimes and Biodiversity of a Continent* (eds. R. A. Bradstock, J. E. Williams and A. M. Gill) pp. 281–304. (Cambridge University Press: Cambridge.)

Willson, A. (1999). Tarawi Nature Reserve: draft fire management plan. Report by New South Wales National Parks and Wildlife Service, Sydney.

Woinarski, J. C. Z. (1989). The vertebrate fauna of broombush *Melaleuca uncinata* vegetation in north-western Victoria, with reference to effects of broombush harvesting. *Australian Wildlife Research* **16**, 217–238.

Woinarski, J. C. Z. (1999). Fire and Australian birds: a review. In *Australia's Biodiversity: Responses to Fire*, pp. 55–112. Department of Environment and Heritage, Biodiversity Technical Paper no. 1, Canberra.

Woinarski, J. C. Z., and Recher, H. F. (1997). Impact and response: a review of the effects of fire on the Australian avifauna. *Pacific Conservation Biology* **3**, 183–205.

Yen, A. L. (1989). Overstorey invertebrates in the Big Desert, Victoria. In *Mediterranean Landscapes in Australia: Mallee Ecosystems and their Management* (eds. J. C. Noble and R. A. Bradstock) pp. 285–299. (CSIRO: Melbourne.)

Zimmer, W. J. (1944). Notes on the regeneration of Murray pine (*Callitris* spp.). *Transactions of the Royal Society of South Australia* **68**, 183–190.

11

Fire regimes in *Acacia* wooded landscapes: effects on functional processes and biological diversity

KEN C. HODGKINSON

Abstract

The *Acacia* communities of semi-arid and arid Australia are subject to recurrent fire. Fire occurrence is strongly dependent on herbaceous plants that become dominant after irregular high-rainfall periods and therefore fire frequency is low (e.g. 30–50 years). The exception is where *Acacia* species occur over hummock grasses; steady growth of the grasses predisposes the community to fire once litter exceeds a critical level. Under pastoral management of these lands the frequency and extent of fires have generally declined. However, ignition by humans in parts of Western Australia and near settlements has locally increased fire frequency. Fire as a consumer of living and dead plant material exposes landscapes to erosive forces and if heavy grazing occurs the effects may be compounded. The landscape function model of Ludwig *et al.* (1997) is used to explore how the fire (and grazing) regime may affect landscape function attributes and linked biodiversity values. Implications of these relationships on prospective assessment and monitoring procedures are discussed.

Introduction

All *Acacia* wooded landscapes are fire-prone. Fire has travelled through these lands for millennia. The fire regimes that evolved when landscapes began burning in the Mid Tertiary, and which changed over the millennia, would have significantly shaped the traits of the flora and fauna that survived to today. Human interference with fire regimes is a very recent phenomenon. The *Acacia* ecological systems and the organisms that comprise them have evolved with, and are a product of, various stresses, perturbations and disturbance regimes of which the fire regime is significant. Present populations of flora and fauna species are well able to cope with a range of fire regimes.

However, the addition of widespread grazing and changed fire regimes probably threatens survival of increasing numbers of species through dysfunctional landscapes. *Acacia* wooded landscapes are vast but remote from the experience of most urban Australians. For the few people that live and work in 'outback' areas, fire management is an important issue. For example, Aboriginal people gathering food items from the wild want to apply fire regimes that encourage and conserve the plants and animals important to their culture and way of life. Pastoralists also prescribe burn to shape the vegetation for easier herbivore management and more profitable businesses.

Animal production is the main land use but pastoralists, although knowing the utility of prescribed fire for reshaping vegetation composition and structure for the benefit of livestock production, struggle for various reasons to apply it in a timely and effective manner. Staff of governments, community-based groups and individual land managers are increasingly interested in assessment and monitoring the status of these lands for tactical decision-making about fire and grazing and for long-term audit purposes.

Fire ecology and management should be addressed at the landscape level – fires occur at this scale. Recently, Ludwig *et al.* (1997) described a conceptual framework for the function of semi-arid *Acacia*-dominated woodland landscapes. This framework was developed primarily to assess the impact

Fig. 11.1. Distribution of the two largest *Acacia*-dominated plant communities in Australia and the limit for the *Acacia* communities. After Australian Surveying and Land Information Group (1990) and Johnson and Burrows (1994).

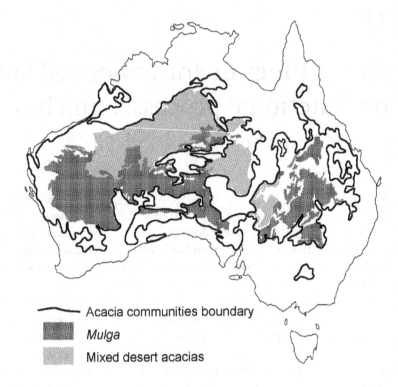

— Acacia communities boundary

Mulga

Mixed desert acacias

Acacia communities

In arid and semi-arid Australia, shrubs and trees of the genus *Acacia* dominate the woody vegetation. Different combinations of *Acacia* and other plant species are associated with particular soils and landscapes and the repeated patterns are the basis for many distinctive plant communities and unique faunal assemblages. At the continental scale these *Acacia* communities form a wide band of shrubland, woodland or open-forest, stretching from a section of the Western Australian coast to the woodlands and rain forests of the Great Divide along the eastern coast of Australia (Fig. 11.1). To the north, east and south of the central arid zone, the genus *Eucalyptus* becomes dominant (Williams and Woinarski 1997).

Across southern Australia, *Acacia*-dominated com-

munities occur where the long-term annual rainfall is below 250 mm and where it occurs regularly in the winter season. Across northern Australia these communities occur where the annual rainfall, which is monsoonal, falls below 350 mm. In eastern Australia, the *Acacia* communities occur in more mesic environments where they often form open-forests and woodlands dominated by single species. For example, open-forests of brigalow (*A. harpophylla*) occur where long-term annual rainfall averages 500–750 mm but forest pockets will occur where annual rain is as low as 350 mm and as high as 900 mm. *Acacia* species also occur ubiquitously in the understorey of *Eucalyptus* open-forests and woodlands.

Beadle (1981) and Johnson and Burrows (1994) have comprehensively described the flora of *Acacia* dominated plant communities and their associated soil characteristics. The Australian Survey and Land Information Group (1990), from which Fig. 11.1 was derived, mapped distributions of the 16 major communities. There are two prominent communities: mulga (*A. aneura*) shrublands with tussock grasses and *Acacia* spp. over hummock grasses (Johnson and Burrows 1994). The latter community occurs to the north of the mulga shrublands in western and central Australia. The other 14 identifiable *Acacia* communities are much smaller in extent.

The communities of *Acacia* occur on a very wide range of soil and landscape types (Johnson and Burrows 1994). Different parent materials and geomorphic processes are associated with each of the dominant *Acacia* species and their associated suites of plant and animal species. The soils include shallow lateritic red and yellow earths, coarse-textured skeletal soils developed on laterised sandstone (*A. shirleyi* or *A. catenulata* dominate these), shallow laterite podsolic soils, deep clays (*A. harpophylla* occurs on these), texture contrast soils, red earths (*A. aneura* or *A. cambagei* dominate these) and yellow earths. All these types of soil, except for the clays, are infertile, often highly so, with low levels of nitrogen, phosphorus and carbon (Beeston and Webb 1977).

The communities of mulga shrubland extend from near the Western Australian coast, across the southern edge of the central deserts and into western New South Wales and southwest Queensland. They comprise a near-continuous belt estimated to cover about 1.5 million km² or 20% of the Australian continent. The *Acacia* spp. over hummock grasses communities occur in 22% of Australia (see Allan and Southgate, this volume). There is a separate but substantial occurrence in the north of the central arid area in the Northern Territory (see Fig. 11.1).

The geographic distribution of the main species, *A. aneura*, appears to be strongly influenced by rainfall; it is found on areas receiving 200–500 mm of annual rainfall. However within this rainfall envelope it does not occur where summer or winter drought are regular. Nix and Austin (1973) suggest that, for this and certain other arid-zone plant species to persist, recharge of soil water must be possible at any time of the year. *Acacia aneura* most commonly grows on plains and sandplains that receive additional water by run-off from adjacent hills, low ranges or just from upslope. The species also occurs on a variety of soils and land forms: dissected residual soils on ridges, desert sand hills, solonised brown soils, red earths and texture contrast soils of the flats and plains. It is however most commonly found on red earth soils. These are light-textured with hard coherent subsoil. Typically all soils on which this species occurs have very low levels of plant-available phosphorus.

Throughout its range, mulga is associated with different suites of sparsely occurring tree species, shrub species, tussock grasses and annual and perennial forbs. The communities vary widely in density and composition. Trees, if they are present, may include *Eucalyptus populnea*, *E. intertexta*, *E. melanophloia*, *E. kingsmilli*, *E. oleosa*, *Brachychiton* spp., *Canthium* spp. and others. Shrubs in addition to *Acacia* spp. include species of *Atalaya*, *Atriplex*, *Cassia*, *Dodonea*, *Eremophila*, *Hakea*, *Maireana*, *Myoporum* and *Senna*. Grasses are ubiquitous and tussock in form. Common genera and species include *Aristida* (*A. contorta*, *A. jerichoensis*), *Eragrostis* (*E. eriopoda*, *E. sedifolia*), *Enneapogon* (*E. avenaceous*), *Enteropogon*, *Eriachne*, *Monachather*, *Stipa* and *Thyridolepis*. There is an abundance of forbs, both annual and perennial species, in these lands.

The associations of forbs and grasses under mulga vary greatly, especially in the north. Perry and Lazarides (1962) identified 18 types of ground storey communities under mulga in the northern Territory and Speck (1963) identified 75 in the Wiluna-Meekatharra region of Western Australia. Johnson and Burrows (1994) give more detail on the array of plant species in the mulga communities.

Mulga communities contain rich faunal assemblages. For example, there are 18 species of birds resident throughout its longitudinal distribution as 'core' species, 28 as 'peripheral' species and 35 as 'casual' species (Cody 1994). Equally, there are many marsupials, reptiles, springtails, grasshoppers and innumerable species of ants and other insects and some frogs and small mammals (see Landsberg *et al.* 1999 for inventories of these groups at four widely spaced mulga shrubland communities).

Although these communities have generally not been cleared, a significant number of the species naturally occurring within the *Acacia* communities have been placed on national lists of rare or threatened plant and animal species (Short and Smith 1994; Briggs and Leigh 1996). One faunal example is the bilby (*Macrotis lagotis*), once extensive in *Acacia* communities and now confined to a few isolated locations in spinifex-dominated hummock grasslands (Smyth and Philpott 1968; Johnson 1988; Allan and Southgate, this volume).

At regional and local scales the *Acacia* communities are rarely continuous and are often interspersed with a variety of other plant communities associated with particular soil types, floodplains and localised run-on systems. For instance chenopod shrublands, eucalypt woodlands along creeklines, hummock grasslands on dunes, saltmarshes around salt lakes, communities on rocky outcrops etc. occur within the broad-scale area mapped as mulga communities.

In this chapter the focus is on the mulga communities but knowledge is drawn from the less extensive *Acacia* communities where this is available. The spinifex grasslands (*Triodia* and *Plectrachne* species), including a prominent component of *Acacia* shrub spp., is dealt with elsewhere (Allan and Southgate, this volume).

Land-use history and fire management

Aboriginal people have lived in *Acacia* lands for at least 60000 years (Flood 1999) and deliberately burnt most, but not all, of the vegetation for a number of purposes linked to their survival, sense of custodianship and enjoyment. The diaries of explorers and other travellers, and conversations with Aboriginal people in central Australia led Kimber (1983) to conclude that 'large Aboriginal fires were not accidental, random or otherwise uncontrolled'. Furthermore he found evidence for small-patch burning, in the March–August period and more extensive burning in general in the October–January period. There were 32 highly directed uses of fire in central Australia showing that fire is still a major tool of Aboriginal people. Reid *et al.* (1993) also give a contemporary account of the use of fire by Aboriginal people around Uluru–Kata–Tjuta, Northern Territory.

Today, Aboriginal people manage a significant proportion of *Acacia* lands, especially in South Australia, Western Australia and the Northern Territory. Many pursue pastoral activities. Some

jointly manage National Parks and increasingly Aboriginal knowledge is becoming a critical driver of fire management. Most live semi-traditionally in permanent settlements. Presumably, most if not all of these communities continue to use fire as a tool and many strive to follow traditional practices.

The dominant land use in most *Acacia* communities is pastoralism. This is increasingly being challenged (e.g. Morton *et al.* 1995; Stafford Smith *et al.* 2000) in some regions on sustainability grounds (economic, social and environmental). Cattle are grazed in the north; sheep dominate in the south and in an area of Queensland. The majority of the pastoral properties are leased from the Crown. The value of the pastoral products in 1991/92 was A$0.8 billion but has declined since. Mining in these lands was valued at A$10 billion and tourism was valued at A$3 billion (Stafford Smith 1994).

Within the vast area classified as *Acacia* communities there is diversity in the biology of individual species, in the ecology of the plant and animal communities, in land use and in the management of vegetation. Significant knowledge of this diversity has been captured in two recent symposia on mulga communities (Sattler 1986; Page and Beutel 1995).

Today, Aboriginals, pastoralists, managers of conservation lands, miners and others continue, deliberately, and sometimes inadvertently, to shape the botanical composition of *Acacia* vegetation (and presumably the assemblages of fauna) in many places by changing the naturally occurring fire regimes (Leigh and Noble 1981). One example is in the higher-rainfall portions of *Acacia* communities in eastern Australia where pastoralists have deliberately uncoupled fire occurrence from woody plant recruitment events by fire suppression and through an increase in grazing pressure (domestic, feral and native herbivores). Woody plant biomass has increased and perennial grass biomass decreased in the last 120 years (Hodgkinson and Harrington 1985). Another example is in western and central Australia where uncontrolled fires, lit by pastoralists and others, appear to constrict populations of many fire-sensitive, or 'seeder', plant species such as mulga (*A. aneura*) in some areas (Griffin and Friedel 1984; Start 1986) but not others (Bowman *et al.* 1994).

Compared with Australia's mesic lands dominated by *Eucalyptus* species, the *Acacia* communities appear to have retained their plant diversity.

The *Acacia* vegetation has been least influenced by direct human activity when compared with the rest of Australia, because the land is sparsely populated and too dry for extensive cropping; the exception is the extensively cleared brigalow (*A. harpophylla*) community in the northeast. Management of fire regimes is only a local issue in *Acacia* communities where the people who live on the land and who locally desire change in the abundance of certain elements of the biota have the authority and opportunity to prescribe burn. It is problematic whether fires additional to those that occur naturally from lightning strikes are required for nature conservation over much of these lands. However in more intensive pastoral areas, prevailing large herbivore grazing restricts accumulation of grass fuel and significantly reduces the chance of wildfire.

Fire regimes

The distribution and amounts of rainfall, vegetation patterns in the landscapes and the grazing of domestic livestock shape the fire regimes in *Acacia* landscapes. To these must be added the human factor, in both starting and extinguishing fires. Lightning starts most fires but increasingly city- and town-based people, without prescribed-fire intentions, inadvertently or deliberately start fires, especially while travelling public roads.

In general the frequency of fires in the *Acacia* plant communities is very low. Luke and McArthur (1978) estimate the frequency of wildfire in *Acacia* lands to be one in 30–50 years. Independent calculations based on fuel accumulation rates and other factors support these estimates and indicate the potential fire frequency is in the order of 10–50 years, depending on location (Walker 1981). This low frequency is because periods of prolonged high rainfall needed for substantial grass growth are infrequent and the high fuel levels are of short duration. Rainfall, and the subsequent plant growth, is the driver of fires in these lands not accumulation of litter. Morrisey (1984) reported extensive wildfire in

the mulga lands of Western Australia during 1942, 1963 and 1975 following La Niña periods. In the decade 1970–80 there were many fires in central Australia and Griffin *et al.* (1983) related these to antecedent rainfall; both the number of fires and the area burnt being a linear function of the total rain over the preceding 3-year period. Mean fire size was 381 km² and was attributed to poor control in their early stages. Griffin *et al.* (1983) noted 'while the mulga country was afforded a high protection priority by land managers and fire fighters, control was often difficult because of limited access and high intensity canopy fires during extreme weather conditions.' The grazing of livestock will substantially reduce the chance of wild or prescribed fire after plant growth from prolonged and high rainfall.

There is no season of the year when a fire can not occur but the chance of fire is greatest in the warm to hot summer months, September to March (Walker 1981). Prevailing weather conditions and the spatial arrangement of the fuel densities determine the intensity and patchiness of fires. A feature of all *Acacia* communities, but most evident in mulga communities, is the distinctive grove/inter-grove patterning (Slatyer 1961; Mabbutt and Fanning 1987; Tongway and Ludwig 1990) as shown in Fig. 11.2. Such patterning strongly affects the passage and characteristics of a travelling fire.

The patterning is linked to the redistribution by rainfall and wind of scarce resources from 'source' to 'sink' in these landscapes. This redistribution strongly shapes the botanical composition and structure of vegetation, and creates extreme variation in fuel quantity and type over short distances. This has important implications for fire characteristics. Firstly, the highly variable distribution of fuel predetermines high spatial variability in fire intensity. In the grassy inter-grove areas, combustible vegetation may reach 800–1800 kg ha⁻¹ whereas in the mulga groves or bands, fuel load may be as high as 7000 kg ha⁻¹ (Friedel 1981). Here most of the fuel is mulga leaf litter but high levels of cured grass fuel may also be present. Secondly, there is a greater chance of areas of low fuel load not burning, especially during evening, night and dawn when humid-

Fig. 11.2. Aerial view of the mulga community at Lake Mere CSIRO long-term study site. Here the mulga occurs as a shrubland in the run-on zone of the landscape and as short bands of groves across the slope of the run-off zone as seen in the foreground. (Photograph by A.W. Knight, CSIRO.)

ity (and hence fuel moisture) rises and temperatures and/or wind decline (Hodgkinson *et al.* 1984). This uneven distribution of fuel may collapse fires during the night period and is the cause of the patchy nature of fire in these plant communities. Thirdly, the transient nature of the fine grass fuel, especially in the inter-grove areas, predisposes vegetation to fire only occasionally following prolonged periods of above-average rainfall.

Together these fuel characteristics, shaped strongly by the spatial and temporal patterning of these ecosystems, dictate the fire regimes within these shrublands and woodlands. In general terms, fire frequency is low, the fire intensity highly variable in space and time, the fire season wide and the patchiness of burning high. This temporal and spatial variability in fire intensity has significance for vegetation dynamics, the local survival of some species and possibly the conservation of biodiversity.

Landscape functional approach to fire ecology

The *Acacia* landscapes

Landscapes in *Acacia* communities comprise resource-rich units or 'fertile patches' separated by more open, resource-poor zones (Ludwig and Tongway 1995). Patches occur at a range of scales: individual plants, clumps of trees, shrubs and grasses and zones of vegetation within landscapes. The patches at the two larger scales can be detected and assessed in functional terms by remote sensing (Knight 1995, 1998).

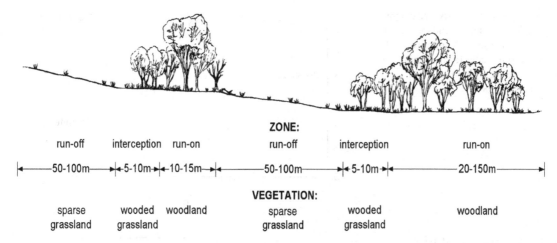

Fig. 11.3. Cross-sectional, stylised diagram of a typical patterned mulga tall shrubland showing the banded *Acacia aneura* groves on and at the bottom of slopes, the hydrologic features of each zone and the vegetation types. Source: Anderson and Hodgkinson (1997).

At the largest scale of 'fertile patches' are groves of shrubs and trees that occur at the bottom of slopes or across slopes when they form an island or band (Fig. 11.3). These are often repeated up a slope. The grass plants within the band, particularly in the upper and grass-dense interception zone, slow the water shed from grassland upslope and physically trap litter items and soil particles washed there during storms. The grass roots channel water into the soil so that infiltration is double that of soil made bare by heavy grazing (Greene *et al.* 1994). Levels of mineralisable nitrogen and organic carbon are much greater in soil within the mulga groves than in open inter-grove areas (Tongway and Ludwig 1990). Grass-mediated capture of the resource flows maintains the bands of woody vegetation across slopes. Where grass populations have collapsed due to over-grazing, mulga shrubs experience extended periods of water stress and, during prolonged drought, they die prematurely (Anderson and Hodgkinson 1997).

At the intermediate scale are log mounds (Fig. 11.4). At their functional centre is a log that traps saltating soil particles and provides a concentrated source of carbon for termites and other soil organisms (Tongway *et al.* 1989). Consequently, soil within the raised mounds differs from surrounding areas

and is more favourable for the growth of perennial forbs and grasses. The mound is nutrient-rich and water infiltrates much more rapidly into the soil, and such mounds may be regarded as 'environmental resource patches'. About 100 mounds per hectare have been measured by Noble *et al.* (1989). They are refugia for a wide range of animals and plants during times of prolonged environmental stress. These and larger-scale 'patches' may be critical for the long-term survival of some species such as small marsupials (Morton 1990).

The trigger–transfer–reserve–pulse framework

In the framework developed by Ludwig *et al.* (1997), the functional elements of landscapes are integrated. The framework recognises the importance of redistribution of scarce resources as well as biotic and abiotic feedback processes that stabilise overall system behaviour.

In this framework (Fig. 11.5), the master input or **trigger** is rainfall. Some rainfall events, because of their high amount and sustained distribution, trigger dramatic recruitment, growth and reproduction events in populations of flora and fauna and biomass in general. If these events are diminished or do not occur, the landscape has become

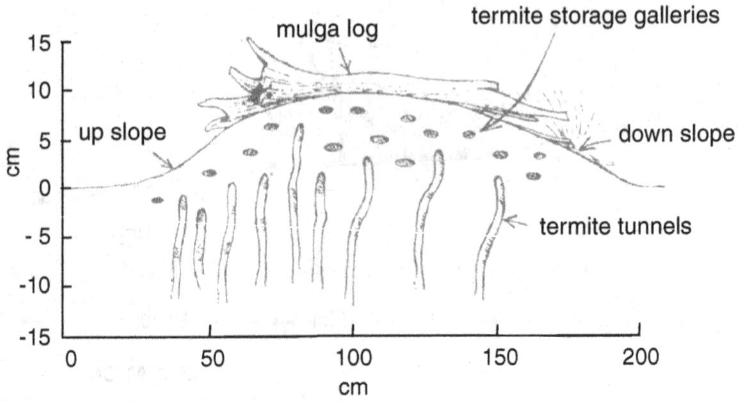

Fig. 11.4. Cross-sectional, stylised diagram of a typical mulga log mound showing the general location and orientation of mulga logs and termite galleries and tunnels (the sizes have been exaggerated). From Tongway *et al.* (1989).

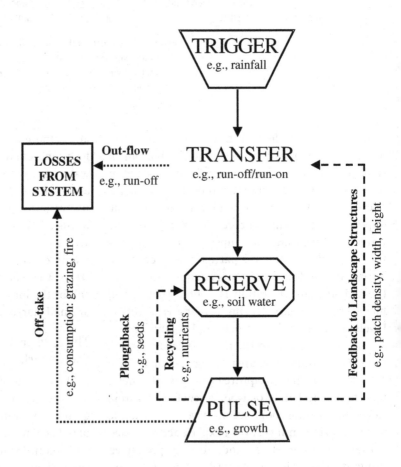

Fig. 11.5. The trigger–transfer–reserve–pulse framework for arid and semi-arid landscapes. Source: Ludwig *et al.* (1997).

dysfunctional. Fire, or any other process threatening the survival of local populations of plant and animal species, can not be the trigger because by itself, fire does not cause a growth pulse.

The term **transfer** is used to describe the horizontal redistribution of materials across a landscape. Water, wind and some biota are agents of the trans-

port. Water is the major agent and will erode and then usually deposit soil, seed and litter downslope. A change from low-energy run-off and transport to highly energetic erosion processes (Pickup 1985) occurs when landscapes lose function.

Transported materials are trapped and deposited within 'safe places' – the **reserve** element of the landscape. In functional landscapes there are many reserves or patches; there is little leakage of resources into watercourses. Reserves include nutrients, seeds, dead wood (standing, lying or decomposed), organisms and fine litter. These reserves are critical for the revival of biological activity when triggered by rainfall and need to be renewed. Dysfunctional landscapes are 'leaky' and with their poor reserves, these landscapes may not be able to grow sufficient and connected plant biomass to carry a fire.

In many cases, plants, animals and soil microorganisms are quiescent for much of the time. To respond to rainfall and produce a **pulse** of growth and probably reproduction, these organisms require certain critical or threshold amounts of available resources built up in landscape reserves. The level of resource availability determines the size of the pulse when rainfall triggers plant growth (Hodgkinson and Freudenberger 1997). It is only the large irregular and infrequent pulses of plant growth that predispose *Acacia* landscapes to fire.

Within the framework there are two distinct and vital feedback loops leading out of the **pulse** (see Fig. 11.5). The feedback loops comprise (1) *ploughback* and *recycling* to the **reserve** and (2) *feedbacks* that modify resource regulation structures (patches) in the landscape, affecting future transfer events and processes. *Ploughback* involves plant structures and products that contribute to the reserve's capacity to support continued plant growth. In contrast, *recycling* products are those materials 'shed' by the plant and turned-over to the reserve pool of carbon and nutrients.

For a plant to persist at a location, it must produce sufficient carbohydrates and proteins for growth and maintenance of perennial tissues and seed production. Ephemeral plants use a high proportion of a production pulse for constructing seeds, whereas perennials use only a small fraction of a pulse for seeds. For example, plants in central Australia vary greatly in their reliance on seed for persistence – many produce seeds of low germinability but potentially long dormancy in order to 'spread the risk' of germinating under conditions unsuitable for growth and reproduction (Jurado

and Westoby 1992). A regular input of seed into the **reserve** is required to compensate for continuous losses in the soil seed pool by predation (Hughes and Westoby 1990) and death (Silcock and Smith 1990).

Plant production pulses also contribute to the return of the nutrients to the soil reserves. This *recycling* occurs through the numerous and diverse groups of soil organisms such as termites and micro-arthropods which are ubiquitous in these lands (Whitford *et al.* 1992; Noble *et al.* 1996). Fire as an abiotic process rapidly cycles plant material (fuel) into ash, much of which may be deposited locally, or which is lost on the winds.

Recycling of nutrients back into soil pools (reserves) by processes other than fire is slower than *ploughback* of carbon and nutrients into plant tissues. There is a time-lag for plant tissues to die, be shed, and then be broken down by most abiotic and biotic processes. Some plant pulses are probably limited by available nutrients because most of the long-term soil pool is tied up in plant tissues (Charley and Cowling 1968).

The tight recycling of nutrients is essential for the persistence of these landscapes. Most of the soils are very infertile. What little fertility exists is bound up in organic matter cycled and recycled through plants. Tight linkages between plants, soil organisms and the soil itself are essential to the maintenance of fertility and the capacity of the system to respond to fleeting and unpredictable periods of soil water.

Landscapes lose function when an excessive amount of system inputs, such as water from rainfall **triggers**, is lost as out-flows (Fig. 11.5). In other words, compared with what is available, rainwater and nutrients are not efficiently captured and stored within landscape reserves (patches). This means a smaller production **pulse** (or no pulse at all if reserves of water and nutrients remain below critical thresholds). A smaller production pulse results in smaller *ploughbacks* (seeds, carbohydrates etc.) into reserves (seed pools, storage roots etc.). Also, since less organic matter is produced, less litter is available for *recycling*, and there are smaller feedbacks to patches themselves. Structure and integrity of patches are not maintained which means even less capture of water during the next rainfall event. The system becomes increasingly 'leaky'.

Fire: a consumptive process for ploughback and feedback

Fire and the invertebrate grazers, termites (Watson and Gay 1970; Watson 1982), are the major consumers of standing dead biomass throughout these *Acacia* landscapes. In contrast, the large vertebrate grazers (kangaroos, cattle, sheep etc.) select and depend on the green leaf of herbaceous plants and some woody plants. This domination by fire and grazers is due to the fact that cellulose is the principal component of the post-pulse plant material. Cellulose can only be burnt or digested by specialised groups of protozoa, bacteria and fungi. Fire as a consuming agent of plant material, principally dead, differs substantially from large herbivore and termite grazing in that it is a non-selective process that releases an enormous, but short-lived, quantity of energy in the form of heat.

Fire periodically consumes the pulses of plant production in *Acacia* communities. For populations of organisms to survive periodic fire, species have evolved over many millennia a suite of adaptive traits (Gill 1975). For the shrub flora these include sprouting, fire-enhanced flowering, early reproductive maturity, heavy seed production, protected seed embryos, fire-induced seed fall and fire-promoted germination (Hodgkinson and Griffin 1982). In the case of non-sessile organisms most escape the damage of heat by moving away from the firefront or sheltering under rocks or hollows of standing or fallen trees. Many surviving organisms eventually benefit for a short time from the post-fire nutrient flush and reduced competition. Fire has probably been an episodic consumer of pulses ever since the Australian continent began to dry out about 60 million years ago. For landscape function, fire is a relatively infrequent but rapid and intense consumer of above-ground plant material, both alive and dead.

The feedback effects of fire on landscape function have not been measured in a systematic way. However, a few predictions can be made. Fire influences landscape function via two pathways. Fire influences the recycling of plant materials, such as seed (Harrington and Driver 1995), into the reserve, and fire affects the density and size of patches that, in turn, influence the surface transfer of water and nutrients. Fire rapidly recycles plant material into the reserve through the nutrient-laden ash. Fire rapidly recycles lignified cellulose in wood that may otherwise take decades to break down. However, fire naturally reduces much plant material to CO_2 representing, in turn, a net loss of above-ground organic carbon to the reserve. Airborne ash will return to soil reserves, but not necessarily locally (that is, ash is an off-take or loss from the local landscape).

Fire also rapidly reduces patch density and size and the degree of reduction is scale-dependent. At broader scales fire would usually temporarily increase patchiness by leaving burnt and unburnt vegetation but at finer scales of individual grass and tree clumps the patchiness would be decreased. Surface obstructions provided by perennial grass tussocks are temporarily reduced, although this obstruction is rapidly re-established when there is sufficient soil moisture for tussocks to tiller and grow in the absence of heavy grazing. Surface obstructions such as logs and fine fuel may be reduced to gas and ash. Fire may increase surface obstruction by reducing tree and shrub densities thereby allowing growth of herbs and grasses to form smaller-scale surface obstructions. However, when fire becomes intense nearly all plant material is consumed, even at the tops of trees and shrubs, leaving potentially open woodland in its wake. The consumption of logs and standing dead trees by fire would affect population size and even survival of some organisms, especially birds.

The net result of fire on landscape function can be a dramatically enhanced pulse of plant growth when triggered by subsequent rainfall, particularly amongst the annual forbs and/or perennial grasses. Following an experimental fire in mulga woodland (Hodgkinson 1990), pulse size and duration were much greater in burnt treatments than in unburnt controls. Fire also temporarily shifted the species composition of the plant growth pulse. In the autumn-burnt treatment, annual forbs dominated the pulse in the next and several subsequent years, whereas annual forbs were a lower proportion of the biomass in unburnt sites. A summer fire in central Australia significantly decreased the grass component also while increasing the proportion of forbs (Griffin and Friedel 1984). Many Aboriginal people know the population responses of particular species to fire and this knowledge may be used to temporarily shift the composition of plant communities to some desired state for increased abundance of particular species important as food plants.

In addition to producing a pulse of nutrients from the consumption of litter and other dead and alive plant material, the heat generated by a travelling fire kills varying proportions of organisms, the most noticeable of which are the shrubs. Less noticeable are the cryptogams (Greene et al. 1990), and perennial herbaceous plants such as creepers and small subshrubs.

The mortality of shrubs is both species- and height-dependent (Hodgkinson 1998). Species differ widely (4%–94%) in sprouting success; a high percentage of established seedlings of all species is killed by fire but survival increases with height reaching a maximum at 25–60 cm (depending on the species). Sprouting success is not related either to bark thickness or lack of active meristems. Relationships between shrub height and sprouting success for species with contrasting strategies are shown in Fig. 11.6. 'Sprouters' (e.g. Eremophila sturtii), the species relying more on resprouting than recruitment for population persistence, maintain maximum sprouting success with height growth and gain sprouting ability along stems once they reach and exceed 1 m in height. In contrast, species largely relying on recruitment from seed to maintain populations (obligate seeders or 'nonsprouters'; e.g. Acacia aneura, Dodonaea viscosa and Senna artemisiodes) are either not able to sprout after seedling establishment or steadily lose ability to maintain sprouts with growth beyond 60 cm and do not develop axillary buds along stems at any height.

After fire damage, most tree species are 'sprouters', but in some genera, e.g. Callitris, the species are 'nonsprouters'. Similarly, most perennial grass species survive fire damage and tiller again (Hodgkinson 1986). However, at least one species, Eragrostis erio-

poda, suffers high mortality when burnt under high water stress; the tiller bases of this species are above the ground surface as rhizomes and are consumed by fire when dry (K. C. Hodgkinson, unpublished data).

The death of perennial plants following fire damage creates new spaces within landscapes. Seed of many shrub species have increased germination capacity after fire, including species of Acacia, Senna and Dodonaea (Hodgkinson and Oxley 1990). Fire enhanced the germination of two of seven principal shrub species in woodland containing Acacia aneura and other Acacia species (Hodgkinson 1991). The pattern of shrub recruitment across the landscape was strongly influenced by the fire-line intensity and spring, summer and autumn fires enhanced germination more than winter fires. The amount and timing of post-fire rainfall influenced and governed the proportion of shrub seedlings that established.

In summary, fire is an important driving force in Acacia communities (Stafford Smith and Morton 1990), because it maintains plant and animal species diversity by killing adult plants and making space for recruits. The plant species, especially perennials, that win the race to reoccupy vacated resources affect the botanical composition and hence the structure of the biodiversity often for decades. Recycling of the nutrients locked up in wood has much shorter effects on the production and probably the diversity of annual and perennial plants. The effect of fire on fauna is less well understood but survivors would have either an improved source of food, e.g. birds of prey, or would temporarily switch to other food sources and/or move to patches of unburnt vegetation.

Fire–grazing interactions

Grazing, particularly by large herbivores (domestic, feral and native), is a strong vector for change in these lands, but unlike fire, is persistent. On its own or in combination with fire, grazing is a major and widespread threatening process for many species in Acacia communities. The increased number of large herbivores in pastoral landscapes is the result of the development of artificial waters

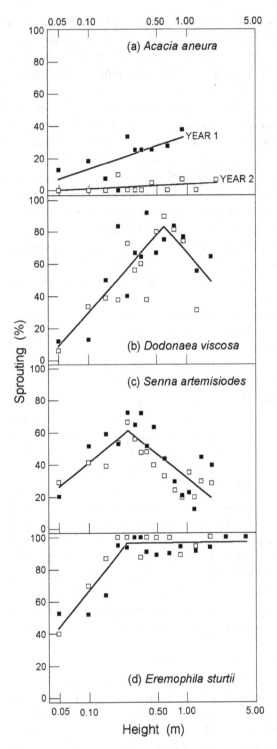

Fig. 11.6. Sprouting success of shrub species in height classes after low-intensity winter fires: (a) *Acacia aneura* (1978) and (1979), (b) *Dodonaea viscosa*, (c) *Senna artemisiodes* and (d) *Eremophila sturtii*. Source: Figure 1, Hodgkinson (1998).

in the form of surface dams and artesian and sub-artesian bores. Few areas of pastoral Australia are today further than 10 km away from water, the limit for cattle to graze (Landsberg and Gillieson 1996). This allows large herbivores – principally sheep and cattle but also kangaroo, goats and other feral herbivores – to graze nearly all the lands, and for predators, such as cats, to more easily prey on birds and other fauna that come to rely on permanent waters. The perceived impact on biological diversity of this combined grazing of the herbivores in rangelands (Total Grazing Pressure) is of growing concern to conservation-minded members of pastoral communities, scientists (e.g. Burbidge and McKenzie 1989; Morton 1990) and the general public.

Change in abundance of plant and animal species with distance from permanent artificial water points in mulga plant communities has been studied by Landsberg *et al.* (1999). Between 19% and 46% of species in six different taxonomic groups appeared to decrease in abundance close to water and between 5% and 38% appeared to increase. The remaining species within each taxonomic group did not change in abundance away from water. A significant proportion, 6% to 13% on average but as high as 21%, of species within a taxonomic group was only found at sites far from water; sites were grazed only lightly and intermittently after sustained rainfall. A similar decreasing trend for the number of plant and grasshopper species away from water points has been reported in a *Eucalyptus* woodland and an *Astrebla* grassland in northern Australia and is associated with decline in landscape function values (Ludwig *et al.* 1999).

Data indicate that in pastoral regions where mulga communities prevail, the local abundance of native species is extremely variable in space and the gradients may be quite steep. Species intolerant to grazing may have disappeared from vast areas where the distance to permanent water is less than 10 km. Other species have dramatically increased at artificial water points. These changes have occurred because of the settled nature of pastoralism and the artificial water points established by pastoral businesses. These large and now predictable effects of addition of artificial water points to these landscapes must be considered when thinking about periodic fires. Studies have yet to examine the interaction between grazing and fire for biological diversity. However it is reasonable to assume that in combination, especially during early drought, these two disturbances may have a dramatic and longlasting influence on biological diversity. For example, high grazing pressure will often occur after a local fire because of the attractiveness to herbivores of new and highly nutritious forage. If drought conditions develop and intensify, the combined stresses of grazing disturbance before and after fire will collapse populations of palatable, and then unpalatable grasses (Hodgkinson and Cook 1995), and increase the chance of shrub seedling survival (Harrington 1991).

Links between landscape function and biological diversity

Organisms depend on the continued functioning of the ecosystem(s), of which they are a part, to persist as populations. The relationships between measures of landscape function and the biological diversity that exists in a landscape, or collection of landscapes, remain to be established for the *Acacia* wooded landscapes. The search for these relationships must be a priority for ecologists because the cost and inadequacies of point-based sampling of all elements in the biota demand landscape or higher-level methods. The monitoring of all components of biological diversity at state and national scales is presently prohibitive. The way forward is to develop surrogate measures of biological diversity based on assessment of landscape function to supplement a few benchmark sites.

Fire and other threatening processes, particularly when in combination, may change the function of landscapes so that scarce resources of water and nutrients are poorly held in the landscape. The onset of dysfunction reduces the chance of suitable edaphic and habitat conditions for recruitment and survival of many plant species, and persistence of particular populations may be severely threatened. Change in plant community composition from landscape dysfunction and from the direct effects

of plant consumption by fire and grazers will have flow-on effects for fauna populations, including species of grasshoppers, ants, termites, reptiles, birds and small mammals (Morton *et al.* 1996). Relationships between plant community structure and faunal responses are a priority for ecological research and these could be derived from grazing-gradient studies described earlier.

Fire management for biological diversity

It is widely accepted that the Aboriginal people, who have been living in these lands for at least 60 000 years, shaped by fire the landscapes that were subsequently colonised by Europeans. The effect of their landscape burning on the maintenance of the rich and unique biodiversity in *Acacia* landscapes is complex and contentious (Bowman 1998). Current and future managers of landscapes will need to construct appropriate fire and grazing regimes on the basis of the best scientific knowledge available and the goals for vegetation structure and flora and fauna species composition desired by the wider local community.

Whether the recent declines and extinctions of small marsupials throughout the *Acacia* shrublands can be linked to changed fire regimes is problematic. A multitude of causes including modification of habitat via changed fire regimes and the impact of grazing by domestic animals or introduced feral rabbits on vegetation have been proposed. These include the diversion of environmental resources to humans and introduced species, a reduction in vegetative cover by exotic herbivores, changed fire regimes and destruction by grazing of refugia considered to be essential for populations to survive prolonged drought periods (Morton 1990). Burbidge and McKenzie (1989) argue that since European settlement the habitat for terrestrial vertebrates has become less suitable by a reduction and greater variability in the availability of forage sought by vertebrates. Recently this emphasis has been shifted by Short and Turner (1994) who showed that whilst burning (and oil exploration) had an influence on the habitat heterogeneity of Barrow Island, the effect on three mammals was not significant. Short (1998) later assembled compelling evidence that predation by foxes brought about the demise of most if not all of the rat-kangaroo species across much of their mainland range.

Debate (see Start 1986; Bowman *et al.* 1994) on whether perceived change in fire regimes is detrimental to biodiversity and landscape function along the interface between mulga shrublands and hummock grass communities needs resolution. It could be argued that the boundary has never been a fixed one and that in places mulga shrublands contract and expand depending on rainfall sequences and the penetration of fires started in the hummock grass communities. Aboriginal burning may have conceivably mostly prevented the penetration of fire into mulga shrubland communities but it is not possible to find evidence for this. The way forward is to adaptively manage fire to achieve the desired plant community and landscape function for the land use practised. Conservationists, pastoralists and others will have different value judgements about the state of landscape function or vegetation depending on the purpose of the landuse (see Figure 5.1 in Ludwig *et al.* 1997).

There is no case established on scientific evidence for prescribed fires in general to meet some biodiversity objective in *Acacia* landscapes of pastoral or non-pastoral land. Equally, wildfire when started by lightning and possibly by human activity should be allowed to travel unhindered until human life and infrastructure such as houses, fences etc., are threatened. The infrequent occurrence of wildfire should be sufficient stimulus to keep populations of plant species viable and certain parts of landscapes from becoming biologically moribund.

However, there are many landscapes in these lands where fire management is required to meet some production or conservation goal. Fire is advocated for reduction of shrub density in pastoral lands where the levels reach 'woody weed' status and inhibit effective management of domestic animals and the control of goat and kangaroo numbers (Hodgkinson and Harrington 1985; Hodgkinson 1993). Exclusion of fire may also be needed in specific locations to prevent local extinction of rare species such as species of *Acacia*, e.g. *A.*

ammobia, *A. undoolyana* and *A. sherlyii*. In these cases, fire exclusion by prescribed burning around the specific locations is required (Reid *et al.* 1993; Duguid 1999). Populations of all plant species are locally vulnerable when there is insufficient time for new individuals to reach reproductive maturity before another fire occurs. This threshold in fire frequency is species-dependent (Hodgkinson 1998) and should be considered in fire management planning. In some areas where hummock grasses predominate, a patch-burn strategy may be necessary to maximise the abundance of faunal species of particular interest, although more research is required to consolidate the basis for this view (Allan and Southgate, this volume).

The way forward: testing and improving the framework

The landscape function framework presented here was largely derived from, and tested on, grazed rangelands. The framework for fire effects in *Acacia* communities discussed here is based mainly on research from small-plot experiments. Research is required to evaluate the framework at large scales for explaining the temporal and spatial changes that occur following fire(s), with and without grazing, in the *Acacia* communities.

A way forward would be to conduct prescribed landscape fires at widely spaced locations and at operational scales that include as much of the landscape and plant community diversity as is practicable. Before and several times after fires, landscape function and floral and faunal assemblages should be assessed at set locations in the critical zones of the landscapes. At the time of the fire(s), measurements to calculate fire indices (Byram 1959) throughout the fire areas would be needed. Post-fire assessment would need to be after substantial rains.

Datasets collected from these locations would be analysed for change in landscape function indices and abundance and composition of species. The changes would be compared with predictions from the conceptual framework and if necessary the framework would be adjusted. These national sites would be ideal for detailed studies by postgraduate students.

This testing would best be conducted in partnership with managers and landowners. They would use the data collected by scientists to monitor the resource and adapt their future management to minimise adverse impacts of fire (and grazing) in their landscapes.

Assessment and monitoring for conservation management in *Acacia* wooded landscapes

There is substantial and growing public support for the conservation of Australia's unique flora and fauna and for sustainable and biodiversity-conserving pastoral systems. Two responses to this public concern are emerging. Firstly, managers of some pastoral properties and some state governments are calling for assessment and monitoring procedures for use at paddock and property levels to meet the requirements for 'clean-and-green' product certification. Secondly, state and federal governments are searching for accurate and cost-effective procedures to monitor the status of biological diversity in *Acacia* systems (and other ecosystems in Australia) through the National Land and Water Resources Audit. The information coming from the Audit will be used in meeting Australia's obligations under international treaties and agreements such as the International Conventions on Desertification and Biodiversity.

The cost of setting up and assessing sites over such a vast area as the *Acacia* communities is enormous. A. Hopkins and N. McKenzie (personal communication) in 1994 estimated that a network of benchmark and 'threatening process' sites in the rangelands would cost A\$35 million to establish and A\$8–9 million each year to monitor. It would be tempting to cobble together the emerging but different systems being developed for other purposes at property, region and state levels. However, differences in methodology would make national auditing an extremely difficult if not impossible task.

The landscape function framework is a logical approach to assessment and monitoring in these lands. However a much better understanding of the nature of relationships between elements of biological diversity and landscape function would need to be established for the major land system types. If these relationships prove robust then landscape function could be used as an indicator of biological diversity to supplement any benchmark sites to be intensively sampled. Clearly the animal–animal predation and habitat factors need to be more strongly developed within the trigger–transfer–reserve–pulse framework for landscape function, as outlined by Ludwig *et al.* (1997). It may be tempting to dismiss this framework because it conflicts with the mind-sets of some ecologists and many members of the public who are focussed on charismatic and much-publicised large animals. However, the conceptual linkage between landscape function and biological diversity, is a strong reason to explore further how it can be used at each scale of interest.

The second and related problem is the lack of knowledge on where to locate benchmark sites for sampling biodiversity. This is no easy matter given the myriad of steep gradients that exist for landscape function and flora and fauna species compositions and relative abundances over short distances within *Acacia* lands used for pastoralism. The challenge is to locate within landscapes the zones where the effects of threatening processes (fire, grazing, animal–animal predation etc.) are likely to have most influence and where early warning signs of adverse change can be first detected. Furthermore, location of sampling points needs to be scientifically acceptable from a statistical viewpoint for monitoring at the appropriate scales.

The third problem is to establish the limitations of remote sensing for monitoring change in landscape function. Remote sensing is a technique with much promise (Pickup 1989) for cost-effective monitoring of the vast interior of semi-arid and arid Australia (Milham *et al.* 1996). Using spatially explicit procedures for Landsat Thematic Mapper images, Knight (1998) was able satisfactorily to detect and interpret change in cover, plant composition, productive potential, stability and resilience of mulga landscapes. In summary, the technology and conceptual frameworks exist to use remote sensing technology to monitor change in landscape function attributes in *Acacia* landscapes. The major impediments to uptake are lack of broad-scale and independent testing and the cost of setting up teams of skilled technicians to apply the technology.

Acacia-dominated communities form about 50% of Australia's lands and fire, pest and grazing management in these lands is critical to the survival of viable populations of the native flora and fauna species. Fire is particularly important as a tool for shaping the vegetation and associated fauna for designed landscapes. The landscape function framework presented here is a logical and unifying framework for redesigning and extending the existing network of monitoring sites across these lands. The framework can be used for adaptive management purposes and national auditing. It is a significant advance on traditional assessment of range condition. The assessments are objective and can be interpreted for a variety of purposes depending on personal judgement about the value of the landscape for a given purpose, including fire management.

Acknowledgements

I appreciate the helpful comments provided by Alan Newsome, Julian Reid, Jann Williams and two anonymous referees on earlier versions of this chapter – their insights greatly improved the chapter. Discussions with Graham Griffin, Graham Harrington and Jim Noble over the past 25 years about the ecology of fire in rangelands significantly influenced my writing of this chapter and I gratefully acknowledge these interactions.

References

Allan, G., and Southgate, R. (2001). Fire regimes in the spinifex landscapes of Australia. In *Flammable Australia: The Fire Regimes and Biodiversity of a Continent* (eds. R. A. Bradstock, J. E. Williams and A. M. Gill) pp. 145–176. (Cambridge University Press: Cambridge.)

Anderson, V. J., and Hodgkinson, K. C. (1997). Grass-mediated capture of resource flows and the maintenance of banded mulga in a semi-arid woodland. *Australian Journal of Botany* 55, 331–342.

Australian Surveying and Land Information Group (1990). *Atlas of Australian Resources.* vol. 6, *Vegetation* (AUSLIG: Canberra.)

Beadle, N. C. W. (1981). *The Vegetation of Australia.* (Cambridge University Press: Cambridge.)

Beeston, G. R., and Webb, A. A. (1977). The ecology and control of *Eremophila mitchellii.* Botany Branch, Queensland Department of Primary Industry Technical Bulletin no. 2, Department of Primary Industries, Brisbane.

Bowman, D. M. J. S. (1998). Tansley Review no. 101. The impact of Aboriginal landscape burning on the Australian biota. *New Phytologist* 140, 385–410.

Bowman, D. M. J. S., Latz, P. K., and Panton, W. J. (1994). Pattern and change in an *Acacia aneura* shrubland and *Triodia* hummock grassland mosaic on rolling hills in central Australia. *Australian Journal of Botany* 43, 25–37.

Burbidge, A. A., and McKenzie, N. L. (1989). Patterns in the modern decline of Western Australia's vertebrate fauna: causes and conservation implications. *Biological Conservation* 50, 143–198.

Briggs, J. D., and Leigh, J. H. (1996). *Rare or Threatened Australian Plants,* revised edn, Special Publication no. 14. (Australian National Parks and Wildlife Service: Canberra.)

Byram, G. M. (1959). Combustion of forest fuels. In *Forest Fire: Control and Use* (ed. K. P. Davis) pp. 61–80. (McGraw Hill: New York.)

Charley, J. L., and Cowling, S. W. (1968). Changes in soil nutrient status resulting from over grazing and their consequences in plant communities of semi-arid areas. *Proceedings of the Ecological Society of Australia* 3, 28–38.

Cody, M. L. (1994). Mulga bird communities. I. Species composition and predictability across Australia. *Australian Journal of Ecology* 16, 206–219.

Duguid, A. (1999). Protecting *Acacia undoolyana* from wildfires: an example of off park conservation from central Australia. In *Proceedings of the Australian Bushfire Conference,* Albury, Australia, 7–9 July, 1999, pp. 127–131. (Charles Sturt University: Albury.)

Flood, J. (1999). *Archaeology of the Dreamtime: The Story of Prehistoric Australia and its People,* revised edn. (HarperCollins: Pymble.)

Friedel, M. H. (1981). Studies of central Australian semidesert rangelands. I. Range condition and the biomass dynamics of the herbage layer and litter. *Australian Journal of Botany* **29**, 219–231.

Gill, A. M. (1975). Fire and the Australian flora: a review. *Australian Forestry* **38**, 4–25.

Greene, R. S. B., Chartres, C. J., and Hodgkinson, K. C. (1990). The effects of fire on the soil in a degraded semi-arid woodland. I. Cryptogam cover and physical and micromorphological properties. *Australian Journal of Soil Research* **28**, 755–777.

Greene, R. S. B., Kinnell, P. I. A., and Wood, J. T. (1994). Role of plant cover and stock trampling on runoff and erosion from semi-arid wooded rangelands. *Australian Journal of Soil Science* **32**, 953–973.

Griffin, G. F., and Friedel, M. H. (1984). Effects of fire on central Australian rangelands. I. Fire and fuel characteristics and responses of herbage and nutrients. *Australian Journal of Ecology* **9**, 381–393.

Griffin, G. F., Price, N. F., and Portlock, H. F. (1983). Wildfires in the central Australian rangelands, 1970–1980. *Journal of Environmental Management* **17**, 311–323.

Harrington, G. N. (1991). Effects of soil moisture on shrub seedling survival in a semi-arid grassland. *Ecology* **72**, 1138–1149.

Harrington, G. N., and Driver, M. A. (1995). The effect of fire and ants on the seed-bank of a shrub in a semi-arid grassland. *Australian Journal of Ecology* **20**, 538–547.

Hodgkinson, K. C. (1986). Responses of rangeland plants to fire in water limited environments. In *Rangelands: A Resource Under Siege, Proceedings of the 2nd International Rangeland Congress*, Adelaide, 1984 (eds. P. J. Joss, P. W. Lynch and O. B. Williams) pp. 437–441. (Australian Academy of Science: Canberra.)

Hodgkinson, K. C. (1990). The ecological basis of prescribed burning for shrub control in the semi-arid pastoral zone of Australia. In *Proceedings of the International Symposium on Grassland Vegetation*, Huhhot, Republic of China, August 1987 (ed. Yang Hanxi) pp. 557–564. (The Peoples' Science Press: Beijing.)

Hodgkinson, K. C. (1991). Shrub recruitment response to intensity and season of fire in a semi-arid woodland. *Journal of Applied Ecology* **28**, 60–70.

Hodgkinson, K. C. (1993). Prescribed fire for shrub control in sheep rangelands. In *Pests of Pastures: Weed, Invertebrate and Disease Pests of Australian Sheep Pastures*.

(ed. E. S. Delfosse) pp. 219–225. (CSIRO Information Services: Melbourne.)

Hodgkinson, K. C. (1998). Sprouting success of shrubs after fire: height-dependent relationships for different strategies. *Oecologia* **115**, 64–72.

Hodgkinson, K. C., and Cook, J. D. (1995). The ecology of perennial grass collapse under grazing. In *Ecological Research and Management in the Mulgalands*, Conference Proceedings, 5–6 July 1994 (eds. M. J. Page and T. S. Beutal) pp. 203–207. (University of Queensland Gatton College: Lawes.)

Hodgkinson, K. C., and Freudenberger, D. O. (1997). Production pulses and flow-ons in rangeland landscapes. In *Landscape Ecology, Function and Management: Principles from Australia's Rangelands* (eds. J. A. Ludwig, D. J. Tongway, D. O. Freudenberger, J. C. Noble and K. C. Hodgkinson) pp. 23–34. (CSIRO: Melbourne.)

Hodgkinson, K. C., and Griffin, G. F. (1982). Adaptation of shrub species to fires in the arid zone. In *Evolution of the Flora and Fauna of Arid Australia* (eds. W. R. Barker and P. J. M. Greenslade) pp. 145–152. (Peacock Publications: Adelaide.)

Hodgkinson, K. C., and Harrington, G. N. (1985). The case for prescribed burning to control shrubs in eastern semi-arid woodlands. *Australian Rangeland Journal* **7**, 64–74.

Hodgkinson, K. C., and Oxley, R. E. (1990). Influence of fire and edaphic factors on germination of the arid zone shrubs *Acacia aneura*, *Cassia nemophila* and *Dodonaea viscosa*. *Australian Journal of Botany* **38**, 269–279.

Hodgkinson, K. C., Harrington, G. N., Griffin, G. F., Noble, J. C., and Young, M. D. (1984). Management of vegetation with fire. In *Management of Australia's Rangelands* (eds. G. N. Harrington, A. D. Wilson and M. D. Young) pp. 141–156. (CSIRO: Melbourne.)

Hughes, L., and Westoby, M. (1990). Removal rates of seeds adapted for dispersal by ants. *Ecology* **71**, 138–148.

Johnson, K. A. (1988). Bilby *Macrotis lagotis*. In *The Australian Museum Complete Book of Australian Mammals*, 2nd edn (ed. R. Strahan) pp. 107–108. (Angus and Robertson: Sydney.)

Johnson, R. W., and Burrows, W. H. (1994). *Acacia* open-forests, woodlands and shrublands. In *Australian Vegetation*, 2nd edn (ed. R. H. Groves) pp. 257–290. (Cambridge University Press: Cambridge.)

Jurado, E., and Westoby, M. (1992). Germination biology of selected central Australian plants. *Australian Journal of Ecology* **17**, 341–348.

Kimber, R. G. (1983). Black lightning: Aborigines and fire in central Australia and the western desert. *Archaeology in Oceania* **18**, 38–45.

Knight, A. W. (1995). REMA: a neutral model to reveal patterns and processes of cover change in wooded rangelands. *Remote Sensing of Environment* **52**, 1–14.

Knight, A. W. (1998). Prediction and assessment of forage production and ecological function in wooded rangelands using remote sensing. PhD Thesis, University of New South Wales, Australian Defence Force Academy, Canberra.

Landsberg, J. J., and Gillieson, D. (1996). Looking beyond the piospheres to locate biodiversity reference areas in Australia's rangelands. In *Rangelands in a Sustainable Biosphere, Proceedings of the 5th International Rangeland Congress*, Salt Lake City, Utah, 23–28 July 1995 (ed. N. West) pp. 304–305. (Society of Range Management: Denver, CO.)

Landsberg, J. J., James, C. D., Morton, S. R., Hobbs, T. J., Stol, J., Drew, A., and Tongway, H. (1999). The effects of artificial sources of water on rangeland biodiversity. Department of Environment and Heritage, Biodiversity Technical Paper no. 3, Canberra.

Leigh, J. H., and Noble, J. C. (1981). The role of fire in the management of rangelands in Australia. In *Fire and the Australian Biota* (eds. A. M. Gill, R. H. Groves and I. R. Noble) pp. 471–495. (Australian Academy of Science: Canberra.)

Ludwig, J. A., and Tongway, D. J. (1995). Spatial organisation of landscapes and its function in semi-arid woodlands, Australia. *Landscape Ecology* **10**, 51–63.

Ludwig, J. A., Tongway, D. J., Freudenberger, D. O., Noble, J. C., and Hodgkinson, K. C. (eds.) (1997). *Landscape Ecology, Function and Management: Principles From Australia's Rangelands*. (CSIRO: Melbourne.)

Ludwig, J. A., Eager, R. W., Williams, R. J., and Lowe, L. M. (1999). Declines in vegetation patches, plant diversity, and grasshopper diversity near cattle watering-points in the Victoria River District, Northern Australia. *Rangeland Journal* **21**, 135–149.

Luke, R. H., and McArthur, A. G. (1978). *Bushfires in Australia*. (Australian Government Publishing Service: Canberra.)

Mabbutt, J. A., and Fanning, P. C. (1987). Vegetation banding in arid Western Australia. *Journal of Arid Environments* **12**, 41–59.

Milham, N., Pickup, G., and Bastin, G. (1996). A remote sensing and range monitoring assessment: a cost-benefit analysis of the grazing gradient approach. *Australian Journal of Environmental Management* **3**, 58–73.

Morrisey, J. G. (1984). Arid mulga woodlands. In *Management of Australia's Rangelands* (eds. G. N. Harrington, A. D. Wilson and M. D. Young) pp. 285–298. (CSIRO: Melbourne.)

Morton, S. R. (1990). The impact of European settlement on the vertebrate animals of arid Australia: a conceptual model. *Proceedings of the Ecological Society of Australia* **16**, 201–213.

Morton, S. R., Stafford Smith, D. M., Friedel, M. H., Griffin, G. F., and Pickup, G. (1995). The stewardship of arid Australia: ecology and landscape management. *Journal of Environmental Management* **43**, 195–217.

Morton, S. R., James, C. D., and Landsberg, J. (1996). Plant community processes and their roles in maintaining fauna. In *Rangelands in a Sustainable Biosphere, Proceedings of the 5th International Rangeland Congress*, vol. 2 (ed. N. E. West) pp. 38–42. (Society for Range Management: Denver, CO.)

Nix, H. A., and Austin, M. C. (1973). Mulga: a bioclimatic analysis. *Tropical Grasslands* **7**, 9–22.

Noble, J. C., Diggle, P. J., and Whitford, W. G. (1989). The spatial distribution of the termite pavements and hummock feeding sites in a semi-arid woodland in eastern Australia. *Oecologica/Oecologia Generalis* **10**, 355–376.

Noble, J. C., Whitford, W. G., and Kaliszweski, M. (1996). Soil and litter microarthropod populations from two contrasting ecosystems in semi-arid eastern Australia. *Journal of Arid Environments* **32**, 329–346.

Page, M. J., and Beutel, T. S. (eds.) (1995). *Ecological Research and Management in the Mulgalands*, Conference Proceedings. (University of Queensland: Gatton.)

Perry, R. A., and Lazarides, M. (1962). Vegetation of the Alice Springs area. CSIRO, Australia, Land Research Series no. 6, pp. 208–36, Canberra.

Pickup, G. (1985). The erosion cell: a geomorphic approach to landscape classification in range assessment. *Australian Rangeland Journal* **7**, 114–121.

Pickup, G. (1989). New land degradation survey

techniques for arid Australia: problems and prospects. *Australian Rangeland Journal* **11**, 74–82.

Reid, J. R. W., Kerle, J. A., and Morton, S. R. (eds.) (1993). *Uluru Fauna: The Distribution and Abundance of Vertebrate Fauna of Uluru (Ayers Rock–Mt Olga) National Park, N.T.* Kowari, vol. 4. (Australian National Parks and Wildlife Service: Canberra.)

Sattler, P. S. (ed.) (1986). *The Mulga Lands.* (Royal Society of Queensland: Brisbane.)

Short, J. (1998). The extinction of rat-kangaroos (Marsupialia: Potoroideae) in New South Wales, Australia. *Biological Conservation* **86**, 365–377.

Short, J., and Smith, A. (1994). Mammal decline and recovery in Australia. *Journal of Mammalogy* **75**, 288–297.

Short, J., and Turner, B. (1994) A test of the vegetation mosaic hypothesis: a hypothesis to explain the decline and extinction of Australian mammals. *Conservation Biology* **8**, 439–449.

Silcock, R. G., and Smith, F. T. (1990). Viable seed retention under field conditions by western Queensland pasture species. *Tropical Grasslands* **24**, 65–74.

Slatyer, R. O. (1961). Methodology of a water balance study conducted on a desert woodland (*Acacia aneura* F. Muell.) community in central Australia. *UNESCO Arid Zone Research* **16**, 15–26.

Smyth, D. R., and Philpott, C. M. (1968). Field notes on rabbit bandicoots (*Macrotis lagotis*) Reid (Marsupialia), from central Western Australia. *Transactions of the Royal Society of South Australia* **92**, 3–17.

Speck, N. H. (1963). Vegetation of the Wiluna-Meekatharra area. CSIRO, Australia, Land Research Series no. 7, pp. 143–61, Canberra.

Stafford Smith, D. M. (1994). Sustainable production systems and natural resource management in the rangelands. In *Outlook 94: Natural Resources*, pp. 148–159. (ABARE: Canberra.)

Stafford Smith, D. M., and Morton, S. R. (1990). A framework for the ecology of arid Australia. *Journal of Arid Environments* **18**, 255–278.

Stafford Smith, D. M., Morton, S. R., and Ash, A. J. (2000). Towards sustainable pastoralism in Australia's rangelands. *Australian Journal of Environmental Management* **7**, 190–203.

Start, A. N. (1986). Status and management of mulga in the Pilbara region of Western Australia. In *The Mulga Lands* (ed. P. S. Sattler) pp. 136–139. (Royal Society of Queensland: Brisbane.)

Tongway, D. J., and Hodgkinson, K. C. (1992). The effects of fire on the soil in a degraded semi-arid woodland. III. Nutrient pool sizes, biological activity and herbage response. *Australian Journal of Soil Research* **30**, 17–26.

Tongway, D. J., and Ludwig, J. A. (1990). Vegetation and soil patterning in semi-arid mulga lands of Eastern Australia. *Australian Journal of Ecology* **15**, 23–34.

Tongway, D. J., Ludwig, J. A., and Whitford, W. G. (1989). Mulga log mounds: fertile patches in the semi-arid woodlands of eastern Australia. *Australian Journal of Ecology* **14**, 263–268.

Walker, J. (1981). Fuel dynamics in Australian vegetation. In *Fire and the Australian Biota* (eds. A. M. Gill, R. H. Groves and I. R. Noble) pp. 101–149. (Australian Academy of Science: Canberra.)

Watson, J. A. L. (1982). Distribution, biology and speciation in the Australian harvester termites, *Drepanotermes* (Isoptera: Termitinae). In *Evolution of the Flora and Fauna of Arid Australia* (eds. W. R. Barker and P. J. M. Greenslade) pp. 263–265. (Peacock Publications: Adelaide.)

Watson, J. A. L., and Gay, F. J. (1970). The role of grass-eating termites in the degradation of a mulga ecosystem. *Search* **1**, 43.

Whitford, W. G., Ludwig, J. A., and Noble, J. C. (1992). The importance of subterranean termites in semi-arid ecosystems in south-eastern Australia. *Journal of Arid Environments* **22**, 87–91.

Williams, J. E., and Woinarski, J. C. Z. (eds.) (1997). *Eucalypt Ecology: Individuals to Ecosystems.* (Cambridge University Press: Cambridge.)

Part V

Ecosystems: woodlands

12

Fire regimes and biodiversity in the savannas of northern Australia

RICHARD J. WILLIAMS, ANTHONY D. GRIFFITHS AND GRANT E. ALLAN

Abstract

Tropical savannas dominate the landscape of the wet–dry tropics of northern Australia. The savannas are both fire-prone, and relatively intact ecologically, hence the management of biodiversity requires an understanding of the interactions between fire regimes and flora and fauna. Fire is extensive and frequent in Australia's savannas, and most fires are lit by humans. There is regional variation in both the extent and frequency of fire, both being greater in northwestern Australia (northern Northern Territory and Western Australia) than in northeastern Australia (Cape York). Fire frequency is higher in mesic savannas than in semi-arid savannas. Fire regimes are the product of consistent patterns, both inter-annually and inter-seasonally, in the occurrence of moderate–severe fire-weather, and the production and curing of fuels. Early dry-season fires are less intense and extensive than late dry-season fires. Across the savannas early dry-season prescribed burning is used extensively in an attempt to reduce the extent and impact of late dry season fires. The ecological effects of fire vary with intensity and frequency, depending on life form. For some components of the biota – e.g. much of the ground stratum vegetation, invertebrates, herpetofauna – interseasonal conditions in moisture appear to be a more forceful determinant of community composition and structure than fire. Tree mortality is higher following late dry-season fires than early dry-season fires. While there are species and communities in the savannas that can persist in the landscape in the face of frequent fires, the abundance of some plant life forms (e.g. obligate seeding shrubs), and some animals (some medium-sized mammals), and seedling recruitment in some dominant eucalypts may be reduced by frequent fires. Further research is required to integrate point-based studies of fire impacts with a spatial understanding of ignitions, fire behaviour and ecosystem dynamics.

Introduction

Tropical savannas are a substantial biome globally. Savannas consist of a discontinuous stratum of trees over a more or less continuous layer of grasses, and occur in tropical areas where the climate is distinctly seasonal – a summer wet season, followed by a dry season. The primary determinants of the composition and structure of savannas are variations in moisture and soil nutrients, with secondary determinants being disturbances due to fire and herbivory (Brouliere 1983; Tothill and Mott 1985; Skarpe 1992; Scholes and Walker 1995). Thus, an understanding of fire and its effects on savanna landscapes is integral to their ecology and management.

Australia's tropical savannas occupy approximately 2 million square kilometres, north of about 20° S in northern Western Australia (WA), the northern half of the Northern Territory (NT) and inland of the Great Dividing Range in north Queensland (Mott *et al.* 1985). The savannas of northern Australia are biologically diverse, with relatively little tree-clearing for agriculture and forestry compared with southern Australia (Ridpath 1985; Braithwaite and Werner 1987). Despite their apparent structural simplicity, the savannas have high species richness of both plants and animals (Taylor and Dunlop

1985; Woinarski and Braithwaite 1990; Bowman *et al.* 1993).

The distinct seasonality of rainfall is a primary determinant of both the biota and the fire regimes of the region. The combination of regular wet-season rainfall, and an extended, warm, dry season produces fine grassy fuels, which cure every year. There are also high rates of ignition due, in the past, to lightning and, at present, to both planned and unplanned ignitions across virtually all land tenures. Fire is therefore both frequent and extensive (Braithwaite and Estbergs 1985; Gill *et al.* 1990; Russell-Smith *et al.* 1997a). The management of fire regimes in relation to biodiversity is an issue common to all savanna landscapes in northern Australia (Craig 1997, 1999; Williams *et al.* 1997a; Crowley and Garnett 1998; Saint and Russell-Smith 1998).

Land uses in the savannas include pastoralism, nature conservation, tourism, defence and mining, and much of the savanna is Aboriginal freehold land (Ash 1996). Until the early 1980s, the focus of fire research in Australia's savannas was largely agricultural (e.g. Norman 1969; Mott *et al.* 1981), and such work continues to be important (Ash 1996; Craig 1997, 1999; Dyer *et al.* 1997). However, since the early 1980s, fire research has broadened to include studies of fire behaviour, the use of fire by Aborigines and fire as a tool for the management of biodiversity, both on- and off-reserve (Haynes 1985; Bowman *et al.* 1988; Fensham 1997; Andersen *et al.* 1998; Bowman 1999). More research has been undertaken in the savannas of Top End of the NT, than in the Kimberley or Cape York. Research has also tended to focus on the open-forests and woodlands of the lowland plains, with less research on rocky country, floodplains and the semi-arid savannas generally.

Given that savannas are extensive, are relatively intact, contain important biodiversity and are generally predisposed to extensive occurrence of fires, the interactions between fire and biodiversity in savannas are of considerable global importance. In this chapter we review some of these interactions, comment on regional differences in patterns, and highlight some major management issues.

The tropical savannas of northern Australia

Climate

Annual rainfall in the savannas varies from about 400 mm to more than 2000 mm, almost all of which falls in summer (McCown 1981; Taylor and Tulloch 1985; McDonald and McAlpine 1991; Cook 1998). Temperatures are generally high, with 15.00 hr maxima generally >25 °C year-round. Broadly, we may recognize mesic savannas, where mean annual rainfall exceeds about 900 mm, and semi-arid savannas, where mean annual rainfall is less than 900 mm (Williams *et al.* 1996).

The occurrence of the wet and dry seasons is predictable from year to year, but there is considerable inter-annual and regional variability in the duration and the timing of both the onset and the cessation of the wet season (Table 12.1) (McCown 1981; Taylor and Tulloch 1985; Mollah and Cook 1996). The Kimberley, the Top End of the NT and Cape York all differ in these features of the climate, and thus patterns of fire-weather. The onset of the dry season is rapid, with sharp declines in both atmospheric and surface soil moisture from late March onwards (Gill *et al.* 1996; Duff *et al.* 1997; Cook 1998).

Aboriginal people further subdivide the wet and dry seasons (Braithwaite and Estbergs 1988; Russell-Smith *et al.* 1997b). There is, for example, the early–cool and late–hot dry season, the early storms and peak monsoon part of the wet season, and, over most of northern Australia, the 'build-up' period between dry and wet seasons.

Savanna landscapes

In some parts of the world, the savannas have been derived from closed forest due to frequent, recent (late Pleistocene) burning, e.g. some parts of the higher-rainfall zones of the African seasonal tropics (Menaut *et al.* 1996; Cook and Mordelet 1997). In northern Australia, however, the wet–dry tropical climate, with its intense dry season, has existed since at least the end of the Tertiary. Australian savannas have not been derived from more extensive rainforests as a consequence of extensive and frequent Aboriginal burning (Bowman 1999;

Table 12.1. *Timing of the onset and finish of the wet season ('green season' of McCown 1981) in western and eastern regions of the Kimberley (Western Australia), Top End of the Northern Territory and Cape York Peninsula (Queensland), and for semi-arid savannas in general*

	Kimberley		Top End		Cape York		Semi-arid
	West	East	West	East	West	East	
Start	late Nov to mid Dec	early–mid Nov	mid Oct to early Nov	early–mid Nov	early–mid Nov	mid Nov	mid Nov
Finish	mid–late May	late May	mid June	mid June to late July	mid May to mid June	early–late August	mid April to early May

Sources: McCown (1981); Crowley and Garnett (2000).

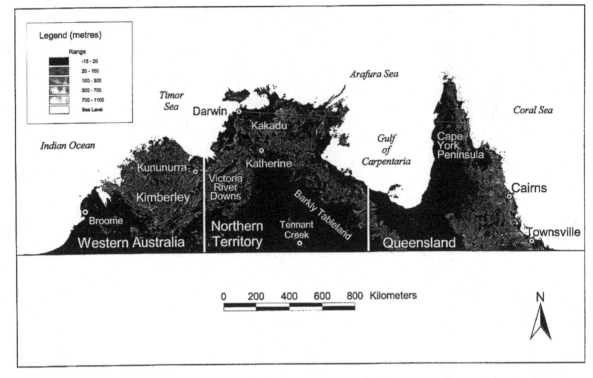

Fig. 12.1. Map of northern Australia, indicating relative relief, and major areas where fire/biodiversity research in the savannas has been undertaken. Relief calculated from 9s Digital Elevation Model. Source: Mr G. Connors, Northern Territory Parks and Wildlife Commission.

Russell-Smith and Stanton, this volume). Australian savannas are therefore old, and highly evolved eco-systems, which is reflected in their high biological diversity (Bowman *et al.* 1993).

Landforms over most of the savanna biome are less than 500 m a.s.l. Local relief is generally less than 100 m (Fig. 12.1) and has been throughout the Tertiary (Williams 1991). In both the mesic and semi-arid savannas there are three broad landform classes. Firstly, and most extensive in all climatic settings, there are the flat lowland plains, which are essentially Tertiary erosional and depositional sur-faces where the substrate is typically strongly later-ised sedimentary or metamorphic material of various ages. The main soils are sands (rudosols and tenosols) and loams (kandosols). Secondly, there are the rocky escarpments, slopes and plateaux of the 'stone country', the bedrock of which in many areas is pre-Cambrian. Extensive areas of such country

occur, for example in eastern Arnhem Land and parts of the Victoria River District in the NT, the Kimberley in WA and northeastern Cape York in Queensland. Soils are typically shallow rudisols, although deep sandy soils may occur on the outwash slopes at the base of slopes and plateaux (Isbell 1983; Williams 1991). Thirdly, there are the 'black-soil plains', where the dominant soils are cracking clays. In the mesic savannas these are the floodplains of the major river systems (Russell-Smith 1985; Woodroffe *et al.* 1989). In the semi-arid savannas, the clay plains may be Quaternary depos-its, or they may be the product of *in situ* weathering of extensive areas of pre-Tertiary fine sediments and basalts (Isbell and Hubble 1983).

Most vegetation types are dominated by *Eucalyptus* (*sensu lat.*). Broad-leaf, pan-tropical trees and shrubs (e.g. *Terminalia*) may occur in the upper–mid storeys (Wilson *et al.* 1990, 1996; Williams *et al.* 1996) and may form the tree stratum of some savannas. Structure varies from open forest (*sensu* Specht 1981) in the mesic coastal and subcoastal regions, to woodlands and low open woodlands of the more arid interior. Treeless grasslands may occur on heavier soils, or in sites of impeded drainage, in both the mesic and semi-arid savannas (Wilson *et al.* 1990; Williams *et al.* 1996).

Mesic savannas (monsoon tall-grass savannas: Mott *et al.* 1985) occur in the higher-rainfall, coastal and near-coastal regions of northwestern and northeastern Australia (see Fig. 12.2a). The dominant eucalypts are 10–20 m tall. In the NT and north Kimberley, the understorey is typically dominated by species of the tall, annual grass *Sorghum* in conjunction with perennial and other annual grasses. *Sorghum* is uncommon on Cape York where other annual grasses such as *Schizachyrium* spp. dominate the annual grass flora (Crowley and Garnett 2000). On the floodplains, floodplain margins and other localised areas of impeded drainage, the dominant vegetation types are open-forests and woodlands dominated by *Melaleuca* spp., grasslands and sedgelands, and some rainforest (Neldner and Clarkson 1995; Wilson *et al.* 1996; Crowley and Garnett 1998). On the stony escarpments, plateaux and slopes, heaths with hummock grasses – *Triodia* spp. – occur (Wilson *et al.* 1990; Russell-Smith *et al.* 1997a, 1998).

The semi-arid savannas (mid-grass savannas: Mott *et al.* 1985) are woodlands and open woodlands on the sands and loams (Fig. 12.2b) with grasslands on the clay soils (Wilson *et al.* 1990). The dominant trees are 10–15 m tall. Perennial grasses, rather than tall annual grasses, dominate the ground stratum; common taxa are *Themeda*, *Heteropogon*, *Chrysopogon* and *Sehima*. On the clay soils the dominant grasses are *Astrebla*, *Dichanthium* and *Iseileama* (Fig. 12.2c). Whereas the woodlands are extensive and more or less continuous, grasslands of the clay soil plains have a much more fragmented distribution across the savanna region (Orr 1975). On the escarpments (Fig 12.2d) and other areas of poor, shallow soils in the lower rainfall areas, tree cover is sparse, and hummock grasses – *Triodia* spp. – occur rather than tussock grasses (Wilson *et al.* 1990).

Diversity and community composition – both plant and animal – are strongly related to annual rainfall, season and soil clay content. Tree cover, diversity of plants and frequency of some vertebrates decline with declining annual rainfall; values are generally lower on the clay soils than on the sands and loams (Woinarski 1992a,b; Williams *et al.* 1996; Woinarski *et al.* 1999a). Temporal variation in mammal composition is strongly related to long-term variations in the spatial distribution of permanent water (Braithwaite and Muller 1997). For invertebrates, there are distinct wet and dry season faunas (Orgeas and Andersen 2001). There are several foci of high diversity which have national and international conservation significance, for example Kakadu and the Gulf of Carpentaria (NT), the Mitchell Plateau and the Bungle Bungles (WA) and the Iron Range (Cape York).

Fire regimes in tropical savannas

Fire is extensive and frequent in Australia's tropical savannas. The fire regimes are the product, both currently and historically, of high rates of ignition, a relatively flat landscape, and a wet–dry climate that consistently produces moderate–severe fire–weather and dry fuels.

Fire-weather and ignition

Fire-weather in tropical savannas is determined by the annual arrival and departure of the monsoon (Tapper *et al.* 1993). Gill *et al.* (1996) examined the seasonal changes in McArthur's Forest Fire Danger Index (FFDI: Luke and McArthur 1978; Noble *et al.* 1980) for a 12-year period in Jabiru, NT. During the peak monsoon period of January to early March, when the majority of the rain falls, the average daily FFDI is below 5 (Fig. 12.3). During this period the vegetation is essentially non-flammable. Despite the rapid onset of the dry season, average daily FFDI in the early dry season (May–June) remains below 20. Given the variation in the cessation of the wet season across northern Australia, conditions of the 'early' dry season (June in the Top End of the NT) may not occur until August on Cape York (Crowley 1995).

Fig. 12.3. Fire weather patterns. Seasonal 15.00 hr Forest Fire Danger Index (FFDI) for Jabiru, 250 km east of Darwin, based on 12 years of data. AbsMAX = absolute maximum FFDI for month; AvMAX = average maximum FFDI for month; AvDay = average daily FFDI for month. From Gill *et al.* (1996).

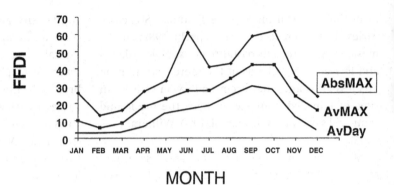

Average daily FFDI increases at Jabiru to around 25 in the September/October period. Average maximum FFDI during these months is about 40. The most extreme value was 60, well below peak levels of 100 that can occur on extreme days in southeastern Australia (Gill and Moore 1990; Williams and Bradstock 2000). FFDI declines again with the onset of the wet season, but days when FFDI is in the range of 20–30 can occur in November/December and fire is possible during this period as a consequence of lightning and prescribed burning (Stocker and Mott 1981; Williams and Lane 1999).

At present, ignition in the savannas is overwhelmingly due to human activity. Less than 1% of ignitions and area burnt is due to lightning. Locally, however, lightning fires may be extensive, e.g. at Murgenella, NT, where about 25% of a 1000-km² study area was burnt by lightning-induced fires over a 14-year period (Bowman 1988). In pre-human times lightning during the build-up and early wet-season periods would have been the ignition source of savanna fires (Braithwaite 1991). Lightning is very common in the Top End of the NT, but is less common on Cape York (Crowley 1995).

Fuels

Fine fuels are primarily grass and tree leaf litter. Both the fuel load and the proportion of leaf and twig components in the woodlands increase as the dry season progresses, due to leaf fall in both deciduous and evergreen trees (Williams *et al.* 1997b). Ladder fuels – e.g. bark ribbons, and interconnected layers of shrubs that link into the tree canopy – are

not present (Gill *et al.* 1990). Fuel moisture declines with the progression of the dry season (Bowman and Wilson 1988; Williams *et al.* 1998). The tall annual grasses commence curing during the late wet season (April) whereas perennials do not usually begin to cure until the early dry season (June).

Annual inputs of fine fuel are about 2–5 t ha^{-1}, in both lowland woodlands and grassland, and maximum fuel loads generally remain within this domain with annual burning, in both mesic and semi-arid savannas (Mott and Andrew 1985; Cook 1994; Dyer *et al.* 1997; Williams *et al.* 1998). Equilibrium fuel loads in mesic savannas are about 10 t ha^{-1} (Fig. 12.4), and can be achieved in 2–3 years without fire (Cook *et al.* 1995). For a perennial-grass-dominated savanna near Katherine (annual rainfall 950 mm) Mott and Andrew (1985) reported grass fuel levels of 2–4 t ha^{-1} in biennially burnt systems, and 6 t ha^{-1} in pastures protected from fire for 4 years. In Rockhampton (annual rainfall 890 mm), Walker (1981) indicated levels of 3 t ha^{-1} in annually burnt savannas, and 6–7 t ha^{-1} 3 years post-fire. Extreme values would appear to be the 15–25 t ha^{-1} (of which about 50% is fine fuel) in tall eucalypt forest savannas bordering rainforest in the less seasonal Atherton Tableland region in subcoastal northeastern Queensland (Unwin *et al.* 1985). The fuel dynamics of the vegetation of the stone country differ from those of the lowland woodlands (Russell-Smith *et al.* 1997a; 1998). Here, the vegetation is sparse woodland and heath, and the dominant grass spinifex (*Triodia; Plectrachne* spp.). Equilibrium fuel loads are 10–20 t ha^{-1}, and these levels are attained over 5–10 rather than 2–3 years.

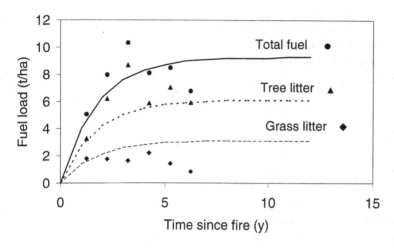

Fig. 12.4. Fuel accumulation curves for grass litter, tree litter and total fuels from *Eucalyptus miniata – Sorghum* open-forest savanna with annual grass understorey at Kapalga, NT; annual rainfall *c*. 1400 mm. Source: G. Cook, CSIRO Sustainable Ecosytems, Darwin, unpublished data.

Fire behaviour

Byram fire intensity is generally 500–10 000 kW m^{-1}, and fires in excess of 20 000 kW m^{-1} are rare. There is a distinct seasonality to fire behaviour as a function of both fire-weather and fuel dynamics. Average fire intensity for early dry-season fires at Kapalga (annual rainfall 1400 mm) was 2200 kW m^{-1} compared with 7700 kW m^{-1} for late dry-season fires (Williams *et al.* 1998). At nearby Munmarlary, the comparative figures were about 1000 kW m^{-1} cf. 4000 kW m^{-1} (J. Russell-Smith unpublished data). Figures of 500–2000 kW m^{-1} were reported for dry-season fires in the north Kimberley (White 1998) and in the Atherton region (Unwin *et al.* 1985). Fuel consumption tends to be complete for fires greater than 2000 kW m^{-1}. Below this intensity some of the fuel bed remains, creating a heterogeneous mix of burnt and unburnt patches across the landscape (Williams *et al.* 1998). In Kakadu, early dry-season fires are smaller than late dry-season fires, e.g. 50–100 ha compared with several thousand ha (Russell-Smith *et al.* 1997a).

Several features of the fuels and fires in the savannas differ from those in temperate eucalypt forests and woodlands of southern Australia (Table 12.2.) (Gill *et al.* 1987; Williams and Bradstock 2000; Gill and Catling, this volume). In the savannas, equilibrium fuel loads are lower (10 t ha^{-1} versus >30 t ha^{-1}), as is the time to equilibrium (2–3 years versus 10–30 years). Rate of spread in the predominantly grassy fuel of the savannas is independent of fuel

load (Cheney *et al.* 1993; Cheney and Sullivan 1997). Peak fire intensity appears to be 20 000 kW m^{-1}, and crown fires with spot fires up to several kilometres ahead of the main front, have never been observed, even in fires >10 000 kW m^{-1} (Stocker and Mott 1981; Williams *et al.* 1998). Flame height and leaf scorch height for a given fire intensity appear to be substantially lower in the savannas than in the temperate forests and woodlands (Williams *et al.* 1998).

Areal extent, frequency and seasonality of fire at landscape scales

Fire is extensive in Australia's savannas, although there is substantial variation in the extent of fire according to state or territory and between the mesic and the semi-arid savannas (Fig. 12. 5; Tables 12.2 and 12.3). In 1997, approximately 26% of the savannas in the NT were burnt; the comparative figures were 16% in WA and 3% in Queensland. Within each state, the percentage of country burnt in 1997 increased as a function of long-term mean annual rainfall, and more country was burnt in the late dry season than in the early dry season (Table 12.3). In some regions, e.g. the mesic savannas of the Darwin–Alligator Rivers region of the NT, as much as 50%–70% of the landscape may be burnt annually (Braithwaite and Estbergs 1985; Press 1988; Russell-Smith *et al.* 1997a; Gill *et al.* 2000). In contrast, the area burnt and frequency of fire is substantially less on Cape York (Crowley and Garnett 2000).

Given the extent of fire, fire frequency in the

Table 12.2. *Fire features in wet–dry tropical and temperate areas of Australia; a comparison of the broad fire-weather, fuel and fire behaviour characteristics of the landscapes of the mesic savanna forests and woodlands of the Darwin region, and the open and tall open forests of the Sydney region of southeastern Australia*

	Darwin	Sydney
Peak fire-weather period	Sept/Oct	Nov
Maximum FFDI	60	100
Equilibrium fuel load	$10\,t\,ha^{-1}$	$30\,t\,ha^{-1}$
Major fuels	Grass	Leaf and twig
Ladder fuels	No	Yes
Maximum fire intensity	$20\,000\,kW\,m^{-1}$	$50\,000–100\,000\,kW\,m^{-1}$
Landscape relief	10–100 m	50–500 m
Crown fires	No	Yes

Source: Williams and Bradstock (2000).

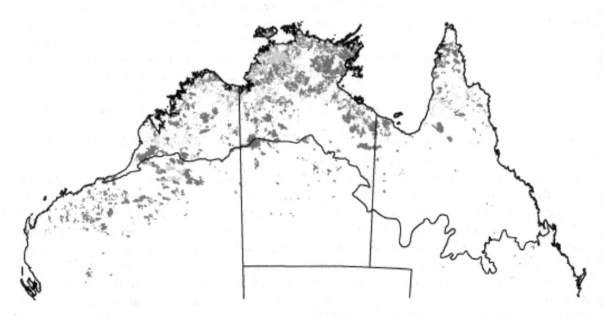

Fig. 12.5. Map of northern Australia derived from AVHRR data showing extent of fire in 1997. Light grey indicates areas burnt prior to July 15; dark grey indicates areas burnt after July 15.

savannas is high. AVHRR data for the NT north of 18° S over the period 1993–1998 indicate that 50% of the area north of 15° S (i.e. the mesic savannas) was burnt at least 3 years in 6 (Table 12.4) (see Fig. 7.4, Allan and Southgate, this volume). There was a substantially lower fire frequency in the semi-arid

savannas (between 15° and 18° S) with nearly 80% of the area being either unburnt or burnt once over the same period. O'Neill *et al.* (1993) reported a similar proportion of unburnt country over a 3-year period for a 1000-km² study area in a semi-arid savanna in the East Kimberley. This difference in

Table 12.3. *Extent (percentage of landscape) and timing (early, pre-July and late, post-July) of fire in northern Australia*

| State | Annual rainfall | | | | | |
| | 400–600mm | | 600–1000 mm | | 1000–2000 mm | |
	Early	Late	Early	Late	Early	Late
Western Australia	2.9	11.5	4.3	10.4	8.6	14.6
Northern Territory	0.2	7.7	6.4	19.7	15.2	26.0
Queensland	0	0.5	0.5	2.9	3.7	8.3

Note:

Data expressed as a function of state (Western Australia, Northern Territory, Queensland) in the savanna biome of northern Australia. Each state/territory has been further sub-divided into three broad rainfall zones according to mean annual rainfall: 400–600 mm, 600–1000 mm and 1000–2000 mm. The figures refer to the area burnt as a percentage of the area of each rainfall zone within each state.

Table 12.4. *Fire frequency in the savannas of the Northern Territory north of 18° S*

Number of times burnt:	0	1	2	3	4	5	6
North of 15° S	12	20	22	19	15	9	4
18° S to 15° S	47	32	15	5	2	<1	0

Note:

Data are the percentage of the landscape burnt either 0, 1, 2, 3, 4, 5, or 6 times over the 1993–98 period, for the region north of 15° S (232 300 km²) and between 18° S and 15° S (291 000 km²).

period for the NT (Table 12.5) (see Fig. 7.4 in chapter by Allan and Southgate, this volume), the majority of burnt areas resulted from fires in the latter half of the dry season (after mid-July) rather than from the early half. Of the eight operational districts of the Northern Territory Bushfires Council within the savannas, only two (Arafura West and Vernon) had a higher proportion of country affected by early dry-season fires than by late dry-season fires, due to the prescriptive use of early dry-season fire. Extensive late dry-season fires also appear to be a feature of the west Kimberley (Fig. 12.5), but they are not frequent on Cape York (Stanton 1992; Crowley and Garnett 2000).

extent between the mesic savannas and semi-arid savannas may be partly related to climate, but is also related to land use. Pastoralism is the main land use in the semi-arid savanna belt. Cattle consume grass, and therefore reduce fuel loads, and pastoralists in the semi-arid savannas do not use fire extensively (Dyer *et al.* 1997) because of potential threats to pasture reserves.

Fires in the savannas can occur from March to December. During the 1997 dry season for northern Australia (Fig. 12.5; Table 12.4) and the 1993–98

Impacts of fire regimes in the savannas

Fire ecology and critical life-history processes

Fire impacts in the savannas, as in all ecosystems, must be seen in the context of the timing, frequency and intensity of fire – fire regimes – in relation to critical life-history processes (Whelan 1995). For plants, we must consider such features as the relative sensitivities of life forms and age, the ability to

Table 12.5. *Extent (percentage of landscape) and timing (pre-July; post July) of fires as a function of region in the Northern Territory*

Region[a]	1993		1994		1995		1996		1997		1998	
	Pre	Post	Pre	Post	Pre	Post	Pre	Post	Pre	Post	Pre	Post
Top End NT	7	9	9	20	9	17	12	21	9	20	9	9
Melville Is	14	7	12	16	2	18	21	31	18	38	6	10
Arafura (W)	28	21	31	21	34	29	52	21	31	23	38	15
Kakadu NP	22	14	30	11	27	11	38	21	21	18	26	13
Arafura (E)	4	15	5	29	2	23	4	54	3	35	10	2
Vernon	39	11	38	5	32	18	48	13	30	10	24	13
Katherine	10	34	15	37	13	45	22	24	13	37	18	10
VRD	3	3	6	17	6	11	6	9	6	12	10	3
Gulf	2	4	4	22	4	16	5	17	5	17	8	7

Note:

[a] Regions defined by Bushfires Council of the NT: Top End of the Northern Territory; Melville Island; Arafura (West); Kakadu National Park; Arafura (East); Katherine; Victoria River District; Gulf of Carpentaria. Top End (not a BFC region) refers to the northern section of the NT, north of about Katherine.

Table 12.6. *Fire and life history: major life-history characters of savanna flora and fauna in relation to fire regimes in the savannas*

Flora	Fauna
Resprouters – dominant; few obligate seeding shrubs in lowlands, locally abundant in sandstone country	Dispersal – both mobile (e.g. most birds) and sedentary species (e.g. lizards, mammals) occur
Phenology – herbs wet season peak, dry season senescence; trees mainly dry season, especially late dry season	Reproduction – primarily wet season; some dry season
Serotiny – largely absent	Dormancy – not universal and relatively limited; some reduced activity in dry season (e.g. frogs, lizards)
Succession – not a general feature; rapid responses to fire in ground stratum and tree canopy	Habitat – most species habitat generalists, some specialists; core refugia important

resprout and phenology. For animals, mobility, dispersal and habitat requirements are important determinants of responses to fire regimes (Table 12.6).

Most of the constituent perennial plants, both woody and herbaceous, have the capacity to resprout vegetatively following fire. All of the eucalypts are resprouters (rather than obligate seeders: Lacey 1974; Gill 1997). Re-establishment of the canopy of the dominant eucalypts is usually complete by the end of the first wet season following fire (Wilson and Bowman 1987; R. J. Williams unpublished data). The grasses also regenerate within 1 year to more or less pre-fire levels of abundance, in both the mesic and semi-arid savannas (Mott and Andrew 1985; Dyer *et al.* 1997). Such rapid post-fire responses in the vegetation, coupled with a high fire frequency, mean that the gradual post-fire vegetation succession, whether classically facilitatory, or via initial floristic composition mechanisms (Noble and Slatyer 1981) does not generally occur in the savannas (Woinarski and Braithwaite 1990). Whilst the annual, obligate-seeding herbs may be a notable component of the ground stratum, the abundance of woody obligate seeders is generally low – less than 1% of the trees and shrubs of the lowland woodlands in the Alligators Rivers/Kakadu region (Brennan 1996b). However, there are some vegetation types where woody obligate seeders are locally abundant – e.g. 50% of the flora of heaths of the sandstone slopes and plateaux (Russell-Smith *et al.* 1998).

The phenology of the vegetation is an important determinant of fire impact. The key pheno-phases in the ground stratum, both vegetative and reproductive, occur during the wet season. Most of the ground stratum is then senescent – physiologically dormant – in the dry season (Brennan 1996a). In contrast, key pheno-phases in many tree species occur in the dry season, when most fires occur, with a peak in both leaf flushing and budding or flowering in the late dry season, when the most intense fires occur (Gillison 1994; Williams *et al.* 1997b; 1999a). Adult eucalypts and other canopy subdominants are capable of flowering both in the season of fire, and in the season following fire (Williams 1997). Most trees are non-serotinous (Dunlop and Webb 1991; Setterfield and Williams 1996; Williams *et al.* 1999a).

Fire regimes impact on fauna in direct and indirect ways. The direct impacts of fire include behavioural responses (attraction to or movement away from flames) or death of individuals or populations. The mobility of birds (Braithwaite 1985), reptiles (Braithwaite 1987; Griffiths and Christian 1996) and some mammals affords protection from most fires. Indirect impacts of fire on fauna include changes in habitat structure, food resources and reproductive

opportunities. We argue that these indirect impacts are more important to the maintenance of faunal abundance and diversity than are direct impacts such as mortality.

Fire, by altering the spatial and temporal patterning of resources, thus creates a mosaic of different environments within landscapes. This has been shown to be important in maintaining vertebrate diversity in southern eucalypt forests (e.g. Newsome and Catling 1983; Catling 1991) and is likely to have similar functional significance in savannas, given their extensiveness and structural simplicity (Belsky 1995; Braithwaite 1995b). Alternatively, fire may destroy core refugial areas that may be necessary for post-fire dispersal of individuals into burnt landscapes (Gill and Bradstock 1995). Fire regime may also interact with the seasonality of migration and the capacity of animals to disperse. Mobile species and sedentary species are likely to respond differently to variation in fire regimes. Those species with limited dispersal or mobility are likely to be disadvantaged by extensive, intense fires. Finally, as a consequence of the general lack of vegetation succession, at least in the lowland woodlands, the vertebrate fauna does not generally display post-fire, successional sequences of species, as it may in temperate forests, woodlands and heaths (e.g. Fox 1982; Caughley 1985).

Fire regime impacts on savanna vegetation

Given the strong seasonality in fire behaviour in the savannas, what are the comparative impacts of early and late dry-season fires? The more intense, late dry-season fires, both as individual fires and repeat fires, can have substantial impacts on both the savannas and non-savanna elements of the landscape. Monsoon rainforest patches, which are generally surrounded by savanna vegetation, may contract locally in the face of repeat late dry-season fires, despite containing a number of dominant trees that are resprouters (Bowman 1991; Panton 1993; Russell-Smith and Stanton, this volume). Individuals of native cypress (*Callitris intratropica*), one of the very few obligate seeder trees, may die following complete canopy scorch. Stands of this species have contracted in area in the past 50–100 years, due to an increased incidence of late dry-season fires (Bowman and Panton 1993). Indeed, these authors argue that the decline in the vigour of stands of *Callitris* in the savannas of the NT is a strong indication of the broad-scale shift towards a fire regime characterised by more intense late dry-season fires in the last century. Lancewood (*Acacia shirleyi*), extensive areas of which occur in the semi-arid savanna belt, is also killed following 100% crown scorch and is showing signs of stand degradation due to repeated late dry-season fires (Woinarski and Fisher 1995). In some of the heathland and shrublands of the stone country, which contain a high proportion of woody obligate seeder species with 2–4-year juvenile periods, significant reductions in species diversity may occur under a regime of repeated, extensive late dry-season fires (Russell-Smith *et al.* 1998).

Within the eucalypt-dominated savannas, there are also considerable impacts of late dry-season fires on the trees. Tree mortality increases linearly with fire intensity (Williams 1995; Williams *et al.* 1999b). There is clear differential species susceptibility to the more intense fires. Stem mortality in the deciduous, broad-leaved trees (e.g. *Terminalia*), and the blood-wood group of eucalypts (e.g. *Eucalyptus porrecta*) is higher at fire intensities of 7000–20 000 kW m^{-1} than is mortality in the dominant eudesmid group of eucalypts (e.g. *E. miniata*: Lonsdale and Braithwaite 1991; Williams *et al.* 1999b). Late dry-season fires can also reduce the floral and fruit reserves across all functional groups of tree by more than 50% for 2–5 years (Setterfield 1997; Williams 1997). The impact of such a regime on structure and phenology not only reduces diversity and potentially inhibits regeneration of the dominant trees, but it may reduce habitat suitability for fauna.

Late dry-season fires, via the incremental effect on tree biomass, may also reduce the nutrient capital of the savanna (Cook 1994; Hurst *et al.* 1996). Late dry-season fires, which result in reduced ground cover, may also increase surface run-off and erosion at the time of the first rains, which may impact upon hydrology and stream biology (Douglas 1999; Townsend and Douglas 2000). Late dry-season fires can result in increased transport of suspended solids into small streams at the commencement of the wet season. This does not, however, appear to result in increases in the export of nutrients to streams. There were differences at Kapalga between late and unburnt regimes in the richness of aquatic invertebrates, and in the abundance and biomass of aquatic macrophytes. However, the causes of these differences in streams between fire regimes, and any relationships between changed ground cover, sediment and nutrient fluxes, and stream biota, are not understood (Douglas 1999).

What are the impacts of early dry-season fires? Evidence is accumulating that the landscape-scale impacts of such a regime are either relatively benign or indeed advantage some elements of the biota. The Munmarlary and Kapalga fire experiments (Bowman *et al.* 1988; Andersen *et al.* 1998) have indicated relatively few impacts on: understorey composition and diversity (Bowman *et al.* 1988; R. J. Williams unpublished data), the density and survival of woody sprouts (R. J. Williams unpublished data), tree basal area and mortality (R. J. Williams *et al.* 1999b; R. J. Williams unpublished data) and water quality (Townsend and Douglas 2000). There appear to be variable impacts on tree reproductive phenology. The dominant trees, which flower in the mid dry season (Williams 1997), continue to flower and set fruit in the weeks and months following early dry-season fires, and flowering and fruiting is not reduced in the year(s) following annual early dry-season fires (Williams 1997). However, seed output may be reduced in the dominant eucalypts following early dry-season fire (Setterfield 1997), and the recruitment of seedlings may be inhibited by annual early dry-season fire.

Does fire frequency matter? The impacts of repeat late dry-season fires have been highlighted above, but what about frequent low-intensity, early dry-season fires? Such fires have been hypothesised to represent a bottleneck to recruitment (Hoare *et al.* 1980; Braithwaite and Werner 1985; Werner 1986), by regularly destroying the tree seedling and small sapling bank, and/or preventing the recruitment of saplings into the tree canopy. However, on the basis of stand population structure, Fensham and Bowman (1992) discounted this hypothesis in the tall humid savannas on deeper soils on Melville

Island. On poorer soils, e.g. those with low nutrients, or seasonally inundated, frequent fire may suppress woody species, and prevent the development of a tree stratum (Wilson and Bowman 1993). R. J. Williams (unpublished data) found no reduction in the size of the seedling and short sapling bank due to annual fire (early or late), compared with unburnt treatment, in all of the major tree species at Kapalga. However, Setterfield's (1997) data suggest that some years without fire may be necessary for seedling recruitment of the dominant trees in mesic savannas.

With respect to the grass layer, R. J. Williams (unpublished data) and Bowman *et al.* (1988) detected no overall differences in the composition of the ground stratum of mesic savannas over 5–15-year periods in annually burnt savannas, compared with unburnt savannas. At Kapalga, the main changes in composition were driven by variation in annual rainfall. Annual burning may decrease the yield of some perennial grass pastures in the semi-arid savannas, but biennial burning had little impact on these variables (Mott and Andrew 1985). In the semi-arid savannas in the Victoria River District of the NT, Dyer *et al.* (1997) have indicated the importance of soil type and land condition as determinants of fire response in pastures. Annual–biennial fire had little impact on composition and productivity on tussock grasslands of the black soil plains. In contrast, the impact on savanna woodlands on red soils depended on condition. Red earths in poor condition – low cover of perennial grasses, and a high cover of annuals and bare ground – suffered subsequent declines in condition after fire, whereas red earths in good condition showed few effects of fire.

Although most of the fires in northern Australia occur during the dry season, prescribed fire may be used during the early wet season. It is used primarily in the mesic savannas, to reduce the fuel load of annual species of *Sorghum*. These grasses do not set seed until the end of the wet season, and have a transient seed bank (Andrew and Mott 1983). Thus, burning during the wet season at the time of seedling emergence may reduce their abundance (Stocker and Sturtz 1966). *Sorghum* cover is reduced more by fires in late December (after the seed bank has emerged fully, but prior to seed set) than those of early December, or late January (subsequent to the onset of the monsoon: Williams and Lane 1999). The impacts of wet-season burning on the other elements of the ground stratum appeared to be restricted primarily to the abundance of annual grasses other than *Sorghum*, and annual herbs, both of which may increase in abundance following wet-season burning (Lane and Williams 1997).

Fire regime impacts on fauna

Birds

Some bird species such as the black kite (*Milvus milvans*), tree martin (*Hirundo nigricans*) and pied butcherbird (*Cracticus nigrogularis*) are attracted to the combustion zone of fires (Braithwaite and Estbergs 1987). In contrast, some ground-nesting species that breed during the dry season may have nests and young destroyed by dry-season fires. These include partridge pigeon (*Geophaps smithii*) and masked finch (*Peophila personata*) (Woinarski and Recher 1997).

Bird diversity in general is positively correlated with structural diversity in the savannas. The diversity of birds may be lower in frequently burnt savannas than in unburnt savannas, at a decadal scale, because of the increase in structural diversity over time in unburnt savanna (Woinarski 1990). Time since last fire, as a determinant of vegetation structure, was also important in determining the composition of bird communities in a semi-arid woodland (Woinarski *et al.* 1999b).

The bird fauna in northern lowland woodlands is very mobile in response to the highly seasonal nature of food resources. Fires may affect the availability of food resources for many bird species. Insectivores and raptors may increase in abundance for up to 6 months following individual fires (Braithwaite and Estbergs 1987; Woinarski 1990; Crowley and Garnett 1999; Woinarski *et al.* 1999b). Granivores gain improved access to seed after early dry-season fires following the removal of dry grass, but seed reserves may be depleted more rapidly in such burnt areas during the early wet season than in unburnt areas (Crowley and Garnett 2000). Late dry-season fires can reduce seed resources for species such as the Gouldian finch (*Erythura gouldiae*) and partridge pigeon, by destroying the seeds lying on the ground (Woinarski 1990). Frugivores in woodland may also be disadvantaged by repeat late dry-season fires, due to declines in the availability of fleshy fruits (Woinarski 1990; Williams 1997). The availability of nectar is extremely seasonal, with a superabundance during the dry season followed by a prolonged period of scarcity in the wet season (Franklin 1997; Woinarski *et al.* 1999c). Declines in the abundance of flowers of *Eucalyptus* due to late dry-season fires (Setterfield 1997; Williams 1997, Williams *et al.* 1999a) may cause local reduction in nectar availability.

The magpie goose (*Anseranas semipalmata*) is a distinctive, high-profile bird of the floodplains in northern Australia. Geese forage extensively across a mosaic of grassland habitats, feeding on native sedges and grasses. Late dry-season fire increases the abundance of an annual wild rice *Oryza rufipogon*, an important food source for juvenile and adult geese. This species also tends to nest in areas burnt late in the previous dry season (Whitehead and McGuffog 1997). Therefore, complete exclusion of fire on the floodplains may reduce the availability of local nesting sites for this species. Fire exclusion, combined with unseasonably high rainfall, may have also contributed to a decline in the amount of grassland on Cape York, which may have contributed to the contraction of the range of the seed-eating golden-shouldered parrot (*Psephotus chrysopterygius*) (Garnett and Crowley 1994; Keith *et al.*, this volume).

The semi-arid stone county is home to two endemic species of grasswrens, the white-throated grasswren (*Amytornis woodwardi*) and the carpentarian grasswren (*Amytornis dorotheae*). Both species prefer unburnt habitat (McKean and Martin 1989; Woinarski 1992c). Extensive and frequent burning of this habitat reduces the availability of long-unburnt, mature stands of spinifex. Most populations exist on isolated ranges and massifs, and frequent fires may threaten these populations.

Mammals

The terrestrial mammals of the savannas are mainly small marsupials and rodents (<5 kg body mass). The dispersability of these species is low and the life cycle is relatively short for many species (1–3 years). Habitat structure at the local level (1 ha – 1 km^2) is a primary determinant of the distribution and abundance of small mammals as it influences the availability of food and shelter. For example, savannas with dense mid-level tree foliage, abundant logs and leaf litter have the highest abundance and diversity of small terrestrial mammals (Friend and Taylor 1985). Similarly, at Kapalga the abundance of most mammal species was highest in the unburnt treatment, compared with the three annually burnt treatments (annual early, late and progressive dry-season fires). However, overall species richness was similar among burnt and unburnt treatments over 7 years (R. W. Braithwaite and co-workers unpublished data).

Individual species displayed a variety of responses to fire at Kapalga. For example the abundance of the northern bandicoot (*Isoodon macrourus*) was higher in unburnt habitats than in habitats burnt annually, whether late or early. This species appears to require fire-free intervals of 2–3 years for its persistence. In contrast, the northern quoll (*Dasyurus hallucatus*) appeared to benefit from annual early dry-season fire, declining in both the annual late dry-season fire regime and the unburnt treatment (R. W. Braithwaite and co-workers unpublished data).

Reptiles and frogs

Reptiles exhibit a wide range of responses to different fire regimes. As with the other vertebrates, habitat structure has a significant influence on the abundance and composition of reptile communities. Many skink species that require dense litter (e.g. *Glaphyromorphus douglasi*) are generally less abundant in annually burnt areas. The dragons (Agamidae) exhibit a general preference for open habitat in tropical savannas, and as a consequence members of this family increase in abundance in recently burnt woodland (Braithwaite 1987; Trainor and Woinarski 1994; Woinarski *et al.* 1999b). The frillneck lizard (*Chlamydosaurus kingii*), a conspicuous member of the dragon family, responds very positively to fire. It shows a strong preference for burnt habitat (including that burnt by late dry-season fires, which may cause 30% mortality) because post-fire habitat provides better accessibility to their invertebrate prey (Griffiths and Christian 1996). Generally, reptiles are one of the most resilient faunal groups with respect to fire in the savannas, given their preference for more open, structurally simple environments.

Most frog species in the Australian savannas are relatively inactive during the dry season. At Kapalga Corbett *et al.* (1997) detected increases in numbers of arboreal frogs in the annual late dry-season fire regime. Terrestrial frogs decreased in abundance in the late and unburnt regimes but the mechanisms for this are unclear. Reduced ground cover following fire in riparian habitats may also reduce frog abundance.

Invertebrates

The diversity of invertebrates in Australian savannas is high (Andersen and Lonsdale 1990). At Kapalga the overall abundance of the majority of ordinal-level taxa of insects and other arthropods were relatively unaffected by fire (Andersen and Müller 2000). There were some fire regime effects, however. The total abundance of ground-dwelling beetles in the early wet season decreased after repeat late dry-season fires, but only when rainfall in the wet season prior to sampling was high (Blanche *et al.* 2001). In contrast, the abundance of grass-layer beetles declined in the unburnt regime, with little difference between annual early and annual late dry-season fire regimes (Orgeas and Andersen 2001). Fire regimes in the forests and woodlands at Kapalga had relatively little effect on grasshopper abundance and diversity. However, the abundance of Leichhardt's grasshopper (*Petasida ephippigera*), which occurs in the 'stone country' of Kakadu and the East Kimberley, is closely associated with the small perennial herb *Pityrodia* that may be adversely affected by late dry-season fires (Lowe 1995).

Ant communities are probably the most fire-responsive of the invertebrate groups. At both Kapalga and Munmarlary, fire exclusion resulted in marked reductions in ant abundance and diversity. Annually burnt savannas were dominated by arid-adapted species such as *Iridomyrmex* and opportunistic *Rhytidoponera* genera whilst long-unburnt savanna contained greater numbers of cryptic species (Andersen 1991; A. N. Andersen and co-workers unpublished data). These differences resulted primarily from differences in habitat structure, with frequent fire producing more open habitats. However, competitive interactions due to the presence of *Iridomyrmex* may also affect species composition.

Management considerations

One of the key questions which influences management of fire in the savannas is: how have the fire regimes changed over time? Prior to the arrival of Aboriginal people in the savannas, at least 40 000 years ago (Mulvaney and Kamminga 1999), fire was most likely to have occurred during the dry season/wet season transition period, the only period with both dry, combustible fuel, and lightning. Aboriginal people modified the timing of fire, by burning earlier in the dry season. Haynes (1985) and Braithwaite (1991) indicated that burning commenced very early in the dry season (March), and continued for a 9–10-month period until the monsoon arrived, with a peak of activity in June/July. There is some regional variation in this pattern of burning within Arnhem Land (Russell-Smith *et al.* 1997b), and on Cape York, Aborigines appeared to use late dry-season fire more frequently than in the Top End of the NT (Crowley and Garnett 2000). Nevertheless, the current fire regimes, with the frequent occurrence of spatially extensive late dry-season fires, contrast markedly with the historical Aboriginal regimes.

The prevention of repeated, intense, extensive late dry-season fires and hence their unwanted effects on the savanna biota is one of the management objectives in the savannas (Russell-Smith 1995; Saint and Russell-Smith 1998). A key tool to achieve this is prescriptive, fuel-reduction burning in the early dry season. There is good evidence that such prescriptive burning works. For example, median fire size declined from hundreds of hectares to tens of hectares in Kakadu National Park, as a consequence of the prescriptive use of early dry-season fires between 1980 and 1994 (Russell-Smith *et al.* 1997a; Gill *et al.* 2000).

Not all late dry-season fires are unwanted or detrimental, however. They are used in pastoral country to control tree growth (so-called 'woody weed control': Dyer *et al.* 1997). On Cape York Peninsula, they are an explicit part of conservation management, and are used to prevent grassland habitats of the endangered golden-shouldered parrot from being invaded by melaleucas (Crowley 1995; Crowley and Garnett 1998, 1999).

Early dry-season fuel-reduction burning in savannas will continue to be used as a general management tool, as this is the most effective means of minimising the area of more intense, late dry-season fires and for many elements of the biota the effects of these fires appears to be relatively benign. Such fires need to be of relatively low intensity (<1000 kW m^{-1}) if potential impacts on tree diversity, stem survival and seed outputs are to be minimised. Alternatively, low-intensity wet-season burning can reduce fuels substantially by reducing the abundance of annual *Sorghum* (Williams and Lane 1999).

However, annual prescribed-fire may have deleterious effects on elements of the biota, such as small mammals, litter-dwelling lizards, ground-nesting birds, and key ecological processes, such as tree seedling recruitment. The optimal configuration and size of unburnt patches needed to sustain such taxa and processes remains to be determined, but such fire–habitat interactions need to be a recognised part of the decision-making processes by which prescribed fire is deployed.

Should we try to exclude fire from the savannas over longer periods of time and over greater spatial scales? Decadal-scale absence of fire can certainly result in a more diverse tree stratum (Bowman *et al.* 1988; Fensham 1990; Bowman and Panton 1995; Braithwaite 1995a, b). However, herb diversity may decline in the longer-term absence of fire (Fensham 1990). Excluding fire from the savannas is essentially an unrealistic option, at least in the mesic savannas, because fuel loads can effectively double to maximal loads (10 t ha^{-1}) within 2–3 years of fire exclusion (Fig. 12.4) (Gill *et al.* 1990; Cook *et al.* 1995). This can predispose savannas to high-intensity fires, spatially extensive fires if ignited in the late dry season. However, some unburnt country needs to be maintained for several years to allow effective recruitment of some plant and animal groups.

The control of exotic grassy weed species is another critical management issue. These species are spreading in the savannas, particularly the mesic savannas. Perennials such as mission grass (*Pennisetum polystachion*) and gamba grass, (*Andropogon gayanus*) can dramatically increase fuel loads (to 30 t ha^{-1}: Panton 1993; Barrow 1995). Both are capable of invading native savannas, both on and off conservation reserves. They cure later than native species thereby making intense, late dry-season fires more likely.

The management of fire in the extensive grasslands on clay soils in the semi-arid savannas is a major issue, for both conservation and pastoral production reasons. These systems are nutrient rich, but relatively poor with respect to both floral and faunal diversity, and a relatively small area is committed to conservation reserves, with numerous species not represented in conservation reserves (Woinarski 1992a). These grasslands are structurally simple, and may be further simplified by fire regimes – either by frequent, extensive late-dry season fires or by fire exclusion. Fire exclusion is the dominant management regime of pastoralists, who manage most of this biome. Attitudes to fire, however, are changing (Dyer *et al.* 1997) and the judicious use of prescribed fire may be the key to maximising habitat heterogeneity and conserving biodiversity in these systems (Fisher 1999).

The savannas are largely intact, and most of the vertebrate species in the mesic savannas occur in at least one conservation reserve (Woinarski 1992a), and relatively few species are endangered. Given that the range of some mammals and granivorous birds has contracted across northern Australia, especially in the semi-arid zone (Braithwaite and Griffiths 1994, 1996; Franklin 1997), then the role of fire in creating and preserving critical habitats for these faunal elements in the landscape requires further research. It is axiomatic that no single fire regime is appropriate in northern Australia for the effective conservation of all components of biodiversity, and indeed for the management of other landscape values. The basic problem facing fire managers and researchers in the savannas in the coming decades is the determination, for biodiversity conservation, of the optimal mix of fire regimes and characteristics of individual fires that may underpin these. As with elsewhere in the continent, further research is required to integrate point-based studies of fire impacts with a spatial understanding of ignitions, fire behaviour and ecosystem dynamics.

Acknowledgements

We thank Alan Andersen, Ross Bradstock, David Bowman, Garry Cook, Andrew Craig, Gabriel Crowley, Rod Fensham, Stephen Garnett, Malcolm Gill, Jeremy Russell-Smith and John Woinarski for comments on the draft manuscript. Greg Connors (NT Parks and Wildlife Commission) compiled Fig. 12.1. Garry Cook provided the data for Fig. 12.4. Help in preparing the manuscript was given by Eric Hollies, Wendy Waggitt and John Walker.

References

Allan, G., and Southgate, R. (2002). Fire regimes in the spinifex landscapes of Australia. In *Flammable Australia: the Fire Regimes and Biodiversity of a Continent* (eds. R. A. Bradstock, J. E. Williams and A. M. Gill) pp. 145–176. (Cambridge University Press: Cambridge.)

Andersen, A. N. (1991). Responses of ground-feeding ant communities to three experimental fire regimes in a savanna forest of tropical Australia. *Biotropica* 23, 575–585.

Andersen, A. N., and Lonsdale, W. M. (1990). Herbivory by insects in Australian tropical savannas: a review. *Journal of Biogeography* 17, 433–444.

Andersen, A. N., and Müller, W. J. (2000). Arthropod responses to experimental fire regimes in an Australian tropical savannah: ordinal level responses. *Austral Ecology* 25, 199–209.

Andersen, A. N., Braithwaite, R. W., Cook, G. D., Corbett, L. K., Williams, R. J., Douglas, M. M., Setterfield, S. A., and Muller, W. J. (1998). Fire research for conservation management in tropical savannas: introducing the Kapalga fire experiment. *Australian Journal of Ecology* 23, 95–110.

Andrew, M. H., and Mott, J. J. (1983). Annuals with transient seed banks: the population biology of indigenous *Sorghum* species of tropical north-western Australia. *Australian Journal of Ecology* 8, 265–276.

Ash, A. (Ed.) (1996). *The Future of Tropical Savannas: An Australian Perspective.* (CSIRO: Melbourne.)

Barrow, P. (1995). Ecology and management of Gamba grass (*Andropogon gayanus* Kunth). Report to Australian Nature Conservation Agency, Canberra.

Belsky, A. J. (1995). Spatial and temporal landscape patterns in arid and semi-arid African Savannas. In *Mosaic Landscapes and Ecological Processes* (eds. L. Hansson, L. Fahrig and G. Merriam) pp. 31–56. (Chapman and Hall: London).

Blanche, K. R., Andersen, A. N., and Ludwig, J. A. (2001). Rainfall-contingent detection of fire impacts: responses of beetles to experimental fire regimes. *Ecological Applications* 11, 86–96.

Bowman, D. M. J. S. (1988). Stability amid turmoil? Towards an ecology of north Australian eucalypt forests. *Proceedings of the Ecological Society of Australia* 15, 149–158.

Bowman, D. M. J. S. (1991). Recovery of some northern Australian monsoon forest species following fire. *Proceedings of the Royal Society of Queensland* 101, 21–25.

Bowman, D. M. J. S. (1999). Tansley Review no. 101. The impact of Aboriginal landscape burning on the Australian biota. *New Phytologist* 140, 385–410.

Bowman, D. M. J. S., and Panton, W. J. (1993). Decline of *Callitris intratropica* R.T. Baker and H.G. Smith in the Northern Territory: implications for pre- and post-European colonization fire regimes. *Journal of Biogeography* 20, 373–381.

Bowman, D. M. J. S., and Panton, W. J. (1995). Munmarlary revisited: responses of *Eucalyptus tetrodonta* savanna protected from fire for 20 years. *Australian Journal of Ecology* 20, 526–531.

Bowman, D. M. J. S., and Wilson, B. A. (1988). Fuel characteristics of coastal monsoon forests, Northern Territory, Australia. *Journal of Biogeography* **15**, 807–817.

Bowman, D. M. J. S., Wilson, B. A., and Hooper, R. J. (1988). Response of *Eucalyptus* forest and woodland to four fire regimes at Munmarlary, Northern Territory, Australia. *Journal of Ecology* **76**, 215–232.

Bowman, D. M. J. S., Woinarski, J. C. Z., and Menkhorst, K. A. (1993). Environmental correlates of tree species diversity in Stage III of Kakadu National Park, northern Australia. *Australian Journal of Botany* **41**, 649–660.

Braithwaite, R. W. (1985). Fire and fauna. In *Kakadu Fauna Survey: Final Report to Australian National Parks and Wildlife Service*, vol. 3 (ed. R. W Braithwaite) pp. 634–650. (CSIRO: Canberra).

Braithwaite, R. W. (1987). Effects of fire regimes on lizards in the wet–dry tropics of Australia. *Journal of Tropical Ecology* **3**, 265–275.

Braithwaite, R. W. (1991). Aboriginal fire regimes of monsoonal Australia in the 19th century. *Search* **22**, 247–249.

Braithwaite, R. W. (1995a). Fire intensity and the maintenance of habitat heterogeneity in a tropical savanna. *CALMScience Supplement* **4**, 189–196.

Braithwaite, R. W. (1995b). Biodiversity and fire in the savanna landscape. In *Biodiversity and Savanna Ecosystem Processes: A Global Perspective*, Ecological Studies no. 121, (eds. O. T. Solbrig, E. Medina and J. F. Silva) pp. 121–140. (Springer-Verlag: Berlin.)

Braithwaite, R. W., and Estbergs, J. A. (1985). Fire pattern and woody vegetation trends in the Alligator Rivers region of northern Australia. In *Ecology and Management of the World's Tropical Savannas* (eds. J. C. Tothill and J. J. Mott) pp. 359–364. (Australian Academy of Science: Canberra.)

Braithwaite, R. W., and Estbergs, J. A. (1987). Fire-birds of the Top End. *Australian Natural History* **22**, 299–302.

Braithwaite, R. W., and Estbergs, J. A. (1988). Tuning in to the six seasons of the wet–dry tropics. *Australian Natural History* **22**, 445–449.

Braithwaite, R. W., and Griffiths, A. D. (1994). Demographic variation and range contraction in the Northern Quoll, *Dasyurus hallucatus* (Marsupialia: Dasyuridae). *Wildlife Research* **21**, 203–217.

Braithwaite, R. W., and Griffiths, A. D. (1996). The paradox of *Rattus tunneyi*: the endangerment of a native pest. *Wildlife Research* **23**, 1–21.

Braithwaite, R. W., and Muller, W. J. (1997). Rainfall, groundwater and refuges: predicting extinctions of Australian tropical mammal species. *Australian Journal of Ecology* **22**, 57–67.

Braithwaite, R. W., and Werner, P. A. (1985). The biological value of Kakadu National Park. *Search* **18**, 296–301.

Brennan, K. (1996a). Flowering and fruiting phenology of native plants in the Alligator Rivers Region with particular reference to the Ranger uranium mine lease area. Supervising Scientist Report no. 107, Canberra.

Brennan, K. (1996b). An annotated checklist of the vascular plants of the Alligator Rivers Region, Northern Territory, Australia. Supervising Scientist Report no. 109, Canberra.

Brouliere, F. (ed.) (1983). *Ecosystems of the World*, vol. 13, *Tropical Savannas*. (Elsevier: Amsterdam.)

Catling, P. C. (1991). Ecological effects of prescribed burning practices on the mammals of southeastern Australia. In *Conservation of Australia's Forest Fauna*. (ed. D. Lunney) pp. 354–363. (Royal Zoological Society of New South Wales: Mossman.)

Caughley, J. (1985). Effects of fire on the reptile fauna of mallee. In *The Biology of Australian Frogs and Reptiles* (eds. G. Grigg, R. Shine and H. Ehmann) pp. 31–34. (Royal Zoological Society of New South Wales: Sydney.)

Cheney, N. P., and Sullivan, A. (1997). Grassfires: Fuel, Weather and Fire Behaviour. (CSIRO: Melbourne.)

Cheney, N. P., Gould, J. S., and Catchpole, W. R. (1993). The influence of fuel, weather and fire shape variables on fire-spread in grasslands. *International Journal of Wildland Fire* **3**, 31–44.

Cook, G. D. (1994). The fate of nutrients during fires in a tropical savanna. *Australian Journal of Ecology* **19**, 359–365.

Cook, G. D. (1998). Rainfall reckoning in rangelands: the case of north Australian savannas. *Range Management Newsletter* **98**, 10–12.

Cook, G. D., and Mordelet, P. (1997). A tale of two savannas: the effects of fire on vegetation patterns in West Africa and the Northern Territory. In *Bushfire97: Proceedings of the Australasian Bushfire Conference* (eds. B. J. McKaige, R. J. Williams and W. M. Waggitt) pp. 45–50. (CSIRO: Darwin.)

Cook, G. D., Hurst, D. F., and Griffith, D. W. T. (1995).

Atmospheric trace gas emissions from tropical Australian savanna fires. *CALMScience Supplement* **4**, 123–128.

Corbett, L. K., Hertog, A., and Gill, A. M. (1997). The effect of fire on terrestrial vertebrates in Kakadu National Park. In *Bushfire97: Proceedings of the Australasian Bushfire Conference* (eds. B. J. McKaige, R. J. Williams and W. M. Waggitt) p. 66. (CSIRO: Darwin.)

Craig, A. (1997). A review of information on the effects of fire in relation to the management of rangelands in the Kimberley high-rainfall zone. *Tropical Grasslands* **31**, 161–187.

Craig, A. (1999). Fire management of rangelands in the Kimberley low-rainfall zone: a review. *Rangeland Journal* **21**, 39–70.

Crowley, G. M. (1995) Fire on Cape York Peninsula: Cape York Land Use Strategy. Office of the Co-ordinator General of Queensland, Brisbane, and Department of Environment, Sport and Territories, Canberra.

Crowley, G. M., and Garnett, S. T. (1998). Vegetation change in the grasslands and grassy woodlands of east-central Cape York Peninsula. *Australia Pacific Conservation Biology* **4**, 132–148.

Crowley, G. M., and Garnett, S. T. (1999). Seeds of the annual grasses *Schizachyrium* spp. as a food resource for tropical granivorous birds. *Australian Journal of Ecology* **24**, 208–220.

Crowley, G. M. and Garnett, S. T. (2000). Changing fire management in the pastoral lands of Cape York Peninsula: 1623 to 1996. *Australian Geographical Studies* **38**, 10–26.

Douglas, M. M. (1999) Tropical savanna streams and their catchments: seasonal dynamics and response to catchment disturbance. PhD thesis, Monash University, Melbourne.

Duff, G. A, Myers, B. A., Williams, R. J., Eamus, D., Fordyce, I., and O'Grady, A. (1997). Seasonal patterns in canopy cover and microclimate in a tropical savanna near Darwin, northern Australia. *Australian Journal of Botany* **45**, 211–224.

Dunlop, C. R., and Webb, L. J. (1991). Flora and vegetation. In *Monsoonal Australia: Landscape Ecology and Man in the Northern Lowlands* (eds. C. D. Haynes, M. G. Ridpath and M. A. J. Williams) pp. 41–61. (Balkema: Rotterdam.)

Dyer, R., Cobiac, C., Cafe, L., and Stockwell, T. (1997). Developing sustainable pasture management practices for the semi-arid tropics of the Northern Territory. Northern Territory Department of Primary Industries and Fisheries, Meat Research Corporation Final Report NTA 022, Katherine.

Fensham, R. J. (1990). Interactive effects of fire frequency and site factors in tropical *Eucalyptus* forests. *Australian Journal of Ecology* **16**, 363–374.

Fensham, R. J. (1997). Aboriginal fire regimes in Queensland, Australia: analysis of the explorers' record. *Journal of Biogeography* **24**, 11–22.

Fensham, R. J., and Bowman, D. M. J. S. (1992). Stand structure and the influence of overwood on regeneration in tropical eucalypt forest on Melville Island. *Australian Journal of Botany* **40**, 335–352.

Fisher, A. (1999). Wildlife of the Mitchell grasslands of northern Australia. Parks and Wildlife Commission of the Northern Territory, Australian Heritage Commission Final Report, Darwin.

Fox, B. J. (1982). Fire and mammalian secondary succession in an Australian coastal heath. *Ecology* **63**, 1332–1341.

Franklin, D. C. (1997). The foraging behaviour of avian nectarivores in a monsoonal Australian woodland over a six month period. *Corella* **21**, 48–54.

Friend, G. R., and Taylor, J. A. (1985). Habitat preferences of small mammals in tropical open-forests of northern Australia. *Australian Journal of Ecology* **10**, 173–185.

Garnett, S. T., and Crowley, G. M. (1994). The ecology and conservation of the golden-shouldered parrot. Cape York Peninsula Land Use Strategy.

Gill, A. M. (1997). Eucalypts and fires: interdependent or independent? In *Eucalypt Ecology: Individuals to Ecosystems* (eds. J. E Williams and J. C. Z. Woinarski) pp. 151–167. (Cambridge University Press: Cambridge.)

Gill, A. M, and Bradstock, R. A. (1995). Extinction of biota by fires. In *Conserving Biodiversity: Threats and Solutions*. (eds. R. A. Bradstock, T. D. Auld, D. A. Keith, R. T. Kingsford, D. Lunney and D. P. Siverston) pp. 309–322. (Surrey Beatty: Sydney.)

Gill, A. M., and Catling, P.C. (2001). Fire regimes and biodiversity of forested landscapes of southern Australia. In *Flammable Australia: The Fire Regimes and Biodiversity of a Continent* (eds. R. A. Bradstock, J. E. Williams and A. M. Gill) pp. 351–369. (Cambridge University Press: Cambridge.)

Gill, A. M., and Moore, P. H. R. (1990). Fire intensities in eucalypt forests of south-eastern Australia. *Proceedings of the International Conference on Forest Fire Research,* Coimbra, Portugal, 1990, **B. 24**, 1–12.

Gill, A. M., Christian, K. R., Moore, P. H. R., and Forrester, R. J. (1987). Bushfire incidence, fire hazard and fuel reduction burning. *Australian Journal of Ecology* **12**, 299–306.

Gill, A. M., Hoare, J. R. L., and Cheney, N. P. (1990). Fires and their effects in the wet–dry tropics of Australia. In *Fire in the Tropical Biota: Ecosystem Processes and Global Challenges* (ed. J. G. Goldammer) pp. 159–178. (Springer-Verlag: Berlin.)

Gill, A. M., Moore, P. H. R., and Williams, R. J. (1996). Fire weather in the wet–dry tropics: Kakadu National Park, Australia. *Australian Journal of Ecology* **21**, 302–308.

Gill, A. M., Ryan, P. G., Moore, P. H. R., and Gibson, M. (2000). Fire regimes of World Heritage Kakadu National Park, Australia. *Austral Ecology* **25**, 616–625.

Gillison, A. N. (1994). Woodlands. In *Australian Vegetation.* 2nd Edn. (ed. R. H. Groves) pp. 227–255. (Cambridge University Press: Cambridge.)

Griffiths, A. D., and Christian, K. A. (1996). The effects of fire on the frill-neck lizard (*Chlamydosaurus kingii*) in northern Australia. *Australian Journal of Ecology* **21**, 386–398.

Haynes, C. D. (1985). The pattern and ecology of munwag: traditional Aboriginal fire regimes in north-central Arnhem Land. *Proceedings of the Ecological Society of Australia* **13**, 203–214.

Hoare, J. R. L., Hooper, R. J., Cheney, N. P., and Jacobsen, K. L. S. (1980). A report on the effect of fire in tall open-forest and woodland with particular reference to fire management in Kakadu National Park in the Northern Territory. Australian National Parks and Wildlife Service, Canberra.

Hurst, D. F., Griffith, D. W. T., and Cook, G. D. (1996). Trace gas emissions from biomass burning in Australia. In *Biomass Burning and Global Change* (ed. J. S. Levine) pp. 787–792. (MIT Press: Cambridge, MA.)

Isbell, R. F. (1983). Kimberley–Arnhem–Cape York (III). In *Soils: An Australian Viewpoint* (ed. CSIRO Division of Soils) pp. 189–199. (CSIRO: Melbourne.)

Isbell, R. F., and Hubble, R. F. (1983). North-eastern plains. In *Soils: An Australian Viewpoint* (Ed. CSIRO Division of Soils) pp. 201–209. (CSIRO: Melbourne.)

Keith, D. A., Williams, J. E., and Woinarski, J. C. Z. (2001).

Fire management and biodiversity conservation: key approaches and principles. In *Flammable Australia: The Fire Regimes and Biodiversity of a Continent.* (eds R. A. Bradstock, J. E. Williams and A. M. Gill) pp. 401–425. (Cambridge University Press: Cambridge.)

Lacey, C. J. (1974). Rhizomes and tropical eucalypts and their role in the recovery from fire damage. *Australian Journal of Botany* **22**, 29–38.

Lane, A. M., and Williams, R. J. (1997). The effect of wet season burning on understorey vegetation in a humid tropical savanna in northern Australia. In *Bushfire97: Proceedings of the Australasian Bushfire Conference* (eds. B. J. McKaige, R. J. Williams and W. M. Waggitt) pp. 25–30. (CSIRO: Darwin.)

Lonsdale, W. M., and Braithwaite, R. W. (1991). Assessing the effects of fire on vegetation in tropical savannas. *Australian Journal of Ecology* **16**, 363–374.

Lowe, L. (1995). Preliminary investigations of the biology and management of Leichhardt's grasshopper, *Petasida ephippigera* White. *Journal of Orthoptera Research* **4**, 219–221.

Luke, R. H., and McArthur, A. G. (1978). *Bushfires in Australia.* (Australian Government Publishing Service: Canberra.)

McCown, R. L. (1981). The climatic potential for beef cattle production in tropical Australia. III. Variation in the commencement, cessation and duration of the green season. *Agricultural Systems* **7**, 163–178.

McDonald, N. S., and McAlpine, J. (1991). Floods and drought: the northern climate. In *Monsoonal Australia. Landscape, Ecology and Man in the Northern Lowlands* (eds. C. D. Haynes, M. G. Ridpath and M. A. J. Williams) pp. 19–29. (Balkema: Rotterdam.)

McKean, J. L., and Martin, K. C. (1989). Distribution and status of the Carpentarian Grass-wren *Amytornis dorothea. Northern Territory Naturalist* **11**, 12–19.

Menaut, J. C., Lepage, M., and Abbadie, L. (1996). Savannas, woodlands and dry forests in Africa. In *Seasonally Dry Tropical Forests* (eds. S. H. Bullock, H. A. Mooney and E. Medina) pp. 64–92. (Cambridge University Press: Cambridge.)

Mollah, W. S., and Cook, I. M. (1996). Rainfall variability and agriculture in the semi-arid tropics: the Northern Territory, Australia. *Agricultural and Forest Meteorology* **79**, 39–60.

Mott, J. J., and Andrew, M. H. (1985). The effect of fire on the population dynamics of native grasses in tropical savannas of north-west Australia. *Proceedings of the Ecological Society of Australia* **13**, 231–239.

Mott, J. J., Tothill, J. C., and Weston, E. (1981). The native woodlands and grasslands of northern Australia as a grazing resource for low cost animal production. *Journal of the Australian Institute of Agricultural Science* **47**, 132–141.

Mott, J. J., Williams, J., Andrew, M. A., and Gillison, A. N. (1985). Australian savanna ecosystems. In *Ecology and Management of the World's Savannas* (eds. J. C. Tothill and J. J. Mott) pp. 56–82. (Australian Academy of Science: Canberra.)

Mulvaney, J., and Kamminga, J. (1999). *Prehistory of Australia.* (Allen and Unwin: Sydney.)

Neldner, V. J. and Clarkson, J. R. (1995). Cape York Peninsula Vegetation Mapping, Cape York Land Use Strategy. Office of the Co-Ordinator General of Queensland, Brisbane, Department of Environment, Sport and Territories, Canberra and Department of Environment and Heritage, Brisbane.

Newsome, A. E., and Catling, P. C. (1983). Animal demography in relation to fire and shortage of food and some indicative models. *Ecological Studies* **43**, 490–505.

Noble, I. R., and Slatyer, R. O. (1981). Concepts and models of succession in vascular plant communities subject to recurrent fire. In *Fire and the Australian Biota* (eds. A. M. Gill, R. H. Groves and I. R. Noble) pp. 312–335. (Australian Academy of Science: Canberra.)

Noble, I. R., Bary, G. A. V, and Gill, A. M. (1980). McArthur's fire danger meters expressed as equations. *Australian Journal of Ecology* **5**, 201–203.

Norman, M. J. T. (1969). The effect of burning and seasonal rainfall on native pastures at Katherine, N. T. *Australian Journal of Experimental Agriculture and Animal Husbandry* **9**, 295–298.

Orgeas, J., and Andersen, A. N. (2001). Fire and biodiversity: responses of grass-layer beetles to experimental fire regimes in an Australian tropical savanna. *Journal of Applied Ecology* **38**, 49–62.

O'Neill, A. L., Head, L. M., and Marthick, J. K. (1993). Integrating remote sensing and spatial analysis techniques to compare Aboriginal and pastoral fire patterns in the East Kimberley, Australia. *Applied Geography* **13**, 67–85.

Orr, D. M. (1975). A review of *Astrebla* (Mitchell grass) pastures in Australia. *Tropical Grasslands* **9**, 21–36.

Panton, W. J. (1993). Changes in post World War II distributions and status of monsoon rainforests in the Darwin area. *Australian Geographer* **24**, 50–59.

Press, A. J. (1988). Comparisons of the extent of fire in different land management systems in the Top End of the Northern Territory. *Proceedings of the Ecological Society of Australia* **15**, 167–175.

Ridpath, M. G. (1985). Ecology of the wet-dry tropics: how different? *Proceedings of the Ecological Society of Australia* **13**, 3–20.

Russell-Smith, J. (1985). A record of change: studies of Holocene vegetation history in the South Alligator River region, Northern Territory. *Proceedings of the Ecological Society of Australia* **13**, 191–202.

Russell-Smith, J. (1995). Fire management. In *Kakadu: Natural and Cultural Heritage and Management* (eds. A. J. Press, D. A. M. Lea, A. L. Webb and A. D. Graham.) pp. 217–223. (Australian Nature Conservation Agency/North Australian Research Unit: Darwin.)

Russell-Smith, J. and Stanton, P. (2001). Fire regimes and fire management of rainforest communities across northern Australia. In *Flammable Australia: The Fire Regimes and Biodiversity of a Continent* (eds. R. A. Bradstock, J. E. Williams and A. M. Gill) pp. 329–350. (Cambridge University Press: Cambridge.)

Russell-Smith, J., Ryan, P. G., and Durieu, R. (1997a). A LANDSAT MSS-derived fire history of Kakadu National Park, monsoonal northern Australia, 1980–94: seasonal extent, frequency and patchiness. *Journal of Applied Ecology* **34**, 748–766.

Russell-Smith, J., Lucas, D., Gapindi, M., Gunbunuka, B., Kaparigia, N., Namingum, G., Lucas, K., Giulani, P., and Chaloupka, G. (1997b). Aboriginal resource utilization and fire management practice in western Arnhem Land, monsoonal northern Australia: notes for pre-history, lessons for the future. *Human Ecology* **25**, 159–195.

Russell-Smith, J., Ryan, P. G., Kleesa, D., Waight, G., and Harwood, R. (1998). Fire regimes, fire sensitive vegetation, and fire management of the sandstone Arnhem Plateau, monsoonal northern Australia. *Journal of Applied Ecology* **35**, 829–846.

Saint, P., and Russell-Smith, J. (eds.) (1998). *Malgarra: Burning the Bush*, 4th North Australia Fire Management Workshop, Kalumburu, North Kimberley, Western Australia. (Cooperative Research Centre for Tropical Savannas: Darwin.)

Scholes, R. J., and Walker, B. H. (1995). *An African Savanna: Synthesis of Nylsvley Study*. (Cambridge University Press: Cambridge.)

Setterfield, S. A. (1997). The impact of experimental fire regimes on seed production in two tropical eucalypt species in northern Australia. *Australian Journal of Ecology* **22**, 279–287.

Setterfield, S. A., and Williams, R. J. (1996). Patterns of flowering and seed production in *Eucalyptus miniata* and *Eucalyptus tetrodonta* in a tropical savanna woodland, northern Australia. *Australian Journal of Botany* **44**, 107–122.

Skarpe, C. (1992). Dynamics of savanna ecosystems. *Journal of Vegetation Science* **3**, 293–300.

Specht, R. L. (1981). Projective foliage cover and standing biomass. *In Vegetation Classification in Australia* (eds A. N. Gillison and D. J. Anderson.) pp. 10–21. (CSIRO: Canberra.)

Stanton, J. P. (1992). J. P. Thomson Oration. The neglected lands: recent changes in the ecosystem of Cape York Peninsula and the challenge of their management. *Queensland Geographical Journal* **7**, 1–18.

Stocker, G. C., and Mott, J. J. (1981). Fire in the tropical forests and woodlands of northern Australia. *In Fire and the Australian Biota* (eds. A. M. Gill, R. H Groves and I. R. Noble) pp. 423–439. (Australian Academy of Science: Canberra.)

Stocker, G. C., and Sturtz, J. D. (1966). The use of fire to establish Townsville lucerne in the Northern Territory. *Australian Journal of Experimental Agriculture and Animal Husbandry* **6**, 277–279.

Tapper, N. J., Garden, G., Gill, J., and Fernon, J. (1993). The climatology and meteorology of high fire danger in the Northern Territory. *Rangeland Journal* **15**, 339–351.

Taylor, J. A., and Dunlop, C. R. (1985). Plant communities of the wet–dry tropics of Australia: the Alligator Rivers region, Northern Territory. *Proceedings of the Ecological Society of Australia* **13**, 83–127.

Taylor, J. A., and Tulloch, D. (1985). Rainfall in the wet–dry tropics: extreme events at Darwin and similarities between years in the period 1870–1983. *Australian Journal of Ecology* **10**, 281–295.

Tothill, J. C., and Mott, J. J. (eds.) (1985). *Ecology and Management of the World's Tropical Savannas*. (Australian Academy of Science: Canberra.)

Townsend, S. A. and Douglas, M. M. (2000). The effect of three fire regimes on stream water quality, water yield and export co-efficients in a tropical savanna (northern Australia). *Journal of Hydrology* **229**, 118–137.

Trainor, C. R., and Woinarski, J. C. Z. (1994). Responses of lizards to three experimental fires in the savanna forests of Kakadu National Park. *Wildlife Research* **21**, 131–148.

Unwin, G. L., Stocker, G. C., and Sanderson, K. D. (1985). Fire and the forest ecotone in the Herberton highland, north Queensland. *Proceedings of the Ecological Society of Australia* **13**, 215–224.

Walker, J. (1981). Fuel dynamics in Australian vegetation. *In Fire and the Australian Biota* (eds. A. M. Gill, R. H Groves and I. R. Noble) pp. 101–127. (Australian Academy of Science: Canberra.)

Werner, P. A. (1986). Population dynamics and productivity of selected forest trees in Kakadu National Park. CSIRO, Report to Australian National Parks and Wildlife Service, Canberra.

Whelan, R. J. (1995). *The Ecology of Fire*. (Cambridge University Press: Cambridge.)

White, K. (1998). Fire behaviour and on ground evidence of fire intensity for remote sensing evaluation. In *Malgarra: Burning the Bush*, 4th North Australia Fire Management Workshop, Kalumburu, North Kimberley, Western Australia (eds. J. Russell-Smith and P. Saint) pp. 20–25. (Cooperative Research Centre for Tropical Savannas: Darwin.)

Whitehead, P. J., and McGuffog, T. (1997). Fire and vegetation pattern in a tropical floodplain grassland: a description from the Mary River and its implications for wetland management. In *Bushfire97: Proceedings of the Australasian Bushfire Conference* (eds. B. J. McKaige, R. J. Williams and W. M. Waggitt) pp. 115–120. (CSIRO: Darwin.)

Williams, M. A. J. (1991). Evolution of the landscape. In *Monsoonal Australia. Landscape, Ecology and Man in the Northern Lowlands* (eds. C. D. Haynes, M. G. Ridpath and M. A. J. Williams) pp. 5–17. (Balkema: Rotterdam.)

Williams, R. J. (1995). Tree mortality in relation to fire intensity in a tropical savanna of the Kakadu region, Northern Territory, Australia. *CALMScience Supplement* **4**, 77–82.

Williams, R. J. (1997). Fire and floral phenology in a humid tropical savanna at Kapalga, Kakadu National Park. In *Bushfire97: Proceedings of the Australasian Bushfire Conference* (eds. B. J. McKaige, R. J. Williams and W. M. Waggitt) pp. 54–59. (CSIRO: Darwin.)

Williams, R. J., and Bradstock, R. A. (2000). Fire regimes and the management of biodiversity in temperate and tropical *Eucalyptus* forest landscapes in Australia. In *Fire and Forest Ecology: Innovative Silviculture and Vegetation Management*, Tall Timbers Fire Ecology Conference Proceedings, no. 21 (eds. W. K. Moser and C. F. Moser) pp. 139–150. (Tall Timbers Research Station: Tallahassee, FL.)

Williams, R. J., and Lane, A. (1999). Wet season burning as a fuel management tool in wet–dry tropical savannas: applications at Ranger Mine, Northern Territory, Australia. In *People and Rangelands*, Proceedings of the 6th International Rangelands Congress (eds. D. Eldridge and D. Freudenberger) pp. 972–977. (International Rangelands Congress: Townsville, Qld.)

Williams, R. J., Duff, G. A., Bowman, D. M. J. S., and Cook, G. D. (1996). Variation in the composition and structure of tropical savannas as a function of rainfall and soil texture along a large-scale climatic gradient in the Northern Territory, Australia. *Journal of Biogeography* 23, 747–756.

Williams, R. J., Cook, G. D., Ludwig, J. L., and Tongway, D. (1997a). Torch, trees, teeth and tussocks: disturbance in the tropical savannas of the Northern Territory (Australia). In *Frontiers in Ecology: Building the Links* (eds. N. Klomp and I. Lunt) pp. 55–66. (Elsevier: Oxford.)

Williams, R. J., Myers, B. A., Muller, W. A., Duff, G. A., and Eamus, D. (1997b). Leaf phenology of woody species in a northern Australian tropical savanna. *Ecology* 78, 2542–2558.

Williams, R. J., Gill, A. M., and Moore, P. H. R. (1998). Seasonal changes in fire behaviour in a tropical savanna in northern Australia. *International Journal of Wildland Fire* 8, 227–239.

Williams, R. J., Myers, B. A., Duff, G. A., and Eamus, D. (1999a). Reproductive phenology of woody species in a north Australian tropical savanna. *Biotropica* 31, 626–636.

Williams, R. J., Cook, G. D., Gill, A. M., and Moore, P. H. R. (1999b). Fire regime, fire intensity and tree survival in a tropical savanna in northern Australia. *Australian Journal of Ecology* 24, 50–59.

Wilson, B. A., and Bowman, D. M. J. S. (1987). Fire, storm, flood and drought: the vegetation ecology of Howard's Peninsula, Northern Territory, Australia. *Australian Journal of Ecology* 12, 165–174.

Wilson, B. A., and Bowman, D. M. J. S. (1993). Factors influencing tree growth in tropical savanna: studies of an abrupt *Eucalyptus* boundary at Yapilika, Melville Island, northern Australia. *Journal of Tropical Ecology* 10, 103–120.

Wilson, B. A., Brocklehurst, P. S., Clark, M. J., and Dickinson, K. J. M. (1990). Vegetation Survey of the Northern Territory, Australia. Conservation Commission of the Northern Territory, Technical Report no. 49, Darwin.

Wilson, B. A., Russell-Smith, J., and Williams, R. J. (1996). Terrestrial vegetation. In *Landscape and Vegetation Ecology of the Kakadu Region, Northern Australia* (eds. C. M. Finlayson and I. von Oertzen) pp. 57–79. (Kluwer, Dordrecht.)

Woinarski, J. C. Z. (1990). Effects of fire on the bird communities of tropical woodlands and open-forests in northern Australia. *Australian Journal of Ecology* 15, 1–22.

Woinarski, J. C. Z. (1992a). Biogeography and conservation of reptiles, mammals and birds across north-western Australia: an inventory and base for planning an ecological reserve system. *Wildlife Research* 19, 665–705.

Woinarski, J. C. Z. (1992b). A survey of the wildlife and vegetation of Purnululu (Bungle Bungle) National Park and adjacent area. Department of Conservation and Land Management, Research Bulletin no. 6, Perth.

Woinarski, J. C. Z. (1992c). The conservation status of the White-throated Grasswren *Amytornis woodwardii*, an example of problems in status designation. *Northern Territory Naturalist* 13, 1–5.

Woinarski, J. C. Z., and Braithwaite, R. W. (1990). Conservation foci for Australian birds and mammals. *Search* 21, 65–69.

Woinarski, J. C. Z., and Fisher, A. (1995). Wildlife of lancewood (*Acacia shirleyii*) thickets and woodlands in northern Australia. I. Variation in vertebrate species composition across the environmental range occupied by lancewood vegetation in the Northern Territory. *Wildlife Research* 22, 379–411.

Woinarski, J. C. Z., and Recher, H. F. (1997). Impact and response: a review of the effects of fire on the Australian avifauna. *Pacific Conservation Biology* 3, 183–205.

Woinarski, J. C. Z., Fisher, A., and Milne, D. (1999a). Distribution patterns of vertebrates in relation to an extensive rainfall gradient and soil variation in the tropical savannas of the Northern Territory, Australia. *Journal of Tropical Ecology* 15, 381–398.

Woinarski, J. C. Z., Brock, C., Fisher, A., and Oliver, B. (1999b). Response of birds and reptiles to fire regimes on pastoral land in the Victoria River District, Northern Territory. *Rangeland Journal* 21, 24–38.

Woinarski, J. C. Z., Franklin, D., and Connors, G. (1999c). Landscape mapping of nectar resources as a base for conservation management. *Australian Biologist* 12, 97–105.

Woodroffe, C. D., Chappell, J., Thom, B. G., and Wallensky, E. (1989). Depositional model of a macrotidal estuary and floodplain, South Alligator River, NT, Australia. *Sedimentology* 36, 737–756.

13

Fire regimes and their effects in Australian temperate woodlands

RICHARD HOBBS

Abstract

Temperate woodlands were once extensive across southern Australia, but have now been largely cleared or modified for agriculture and grazing. Relatively little is known about the fire ecology of woodlands, given their extent and variety. The development of some open woodlands may have been related to Aboriginal burning practices, and frequent fire was a tool used in the development of woodland areas by Europeans for agriculture. There is evidence from some woodland types that fire, or other disturbance, is required for regeneration of tree and understorey plant species, and that fire frequency, season and intensity all affect the post-fire vegetation development. Interactions between fire, grazing and weed invasion are also important. Relatively little is known of the relationship between either individual fires or fire regimes on woodland fauna. Consideration of fire management in the highly fragmented or modified woodland systems that remain today has to take account of the importance of weed invasion and a variety of other factors affecting plant recruitment. Relatively few woodland areas are likely to receive the degree of management needed, and instead most are virtually unmanaged or open to inadvertent burning. Fire, if used effectively, can promote woodland regeneration and diversity, but can otherwise contribute to the eventual decline of temperate woodlands in Australia.

Introduction

In this chapter I focus on the woodlands of temperate southern Australia including Tasmania. They were once widespread in what are now Australia's agricultural heartlands (Fig. 13.1), and also occurred in humid-zone coastal areas with low nutrients. In the southeast of the continent they formed a relatively continuous vegetation on the inland side of the Great Dividing Range occurring from approximately 27° S in southern Queensland to the lower southeast of South Australia with a narrow strip running north and south of Adelaide (Moore 1970). Throughout this zone woodlands occurred in a mosaic with smaller areas of treeless grasslands on valley bottoms, and dry sclerophyll forests on stony oligotrophic hills. Similarly, in Tasmania eucalypt woodlands occurred interspersed with grasslands and sclerophyll forests throughout the northeast and in the Midlands as far south as 42° S (Moore 1970). Eucalypt woodlands were also widespread in southwestern Australia and often formed a mosaic with heathlands, mallee and salt lakes (Beard 1990). Although not strictly included as temperate woodlands, sub-alpine woodlands are also discussed here, where relevant studies are available.

Temperate woodlands are amongst the most poorly conserved and threatened ecosystems in Australia having borne the brunt of agricultural development and land degradation for well over 150 years. In recent decades they have been the scene for some of the most dramatic and spectacular examples of landscape and ecosystem collapse with thousands of hectares of woodland being affected by tree dieback and secondary salinity. Considering the magnitude of the problems which face many of the remaining eucalypt woodlands in temperate Australia, the scientific literature is not large and is spread across disparate sources (Yates and Hobbs 1997; Hobbs and Yates 2000a). In this chapter I review the available information relating to fire in

Fig. 13.1. Areas extensively cleared of native vegetation for agriculture which roughly corresponds to the distribution of woodlands as mapped by AUSLIG (1990).

26°

woodland communities. I use examples from all woodland types, but the information is biased towards the shrubby woodlands of coastal eastern Australia and southwestern Australia. The restricted availability of information on fire in the grassy woodlands of inland eastern Australia is a major hindrance in providing a comprehensive overview of fire in Australian temperate woodlands. Further information on herbaceous components of grassy woodland systems can be found in Lunt and Morgan (this volume).

Woodland communities: structure and composition

The term 'woodland' is used generally to describe ecosystems that contain widely spaced trees with their crowns not touching (projected foliage cover <30%). The height and density of trees in woodlands varies with biophysical and anthropogenic factors. Thus Australian woodland communities vary structurally from low open woodland (low trees up to 10 m high forming up to 10% projective foliage cover) to tall woodland (tall trees greater than 30 m high forming between 10% and 30% projective foliage cover) (Specht 1970). The understorey of woodlands also varies with biophysical and anthropogenic factors. Thus Australian woodlands may have a

patchy understorey of low trees and shrubs (shrubby woodlands) or a more continuous ground layer of grasses, herbs and graminoids (grassy woodlands) (Table 13.1) (Moore 1970; Beadle 1981; AUSLIG 1990; Beard 1990).

Prior to European settlement, Australian woodlands occurred across a wide range of environments varying in climate, soil type, soil parent material and hydrology (Moore 1970; Beadle 1981; Beard 1990). In temperate mainland Australia, woodlands were generally dominated by *Eucalyptus* spp., although large areas of woodlands dominated by species of *Callitris*, *Casuarina*, *Allocasuarina* and *Banksia* also occurred. Climate in this region varies from Mediterranean in southwestern and south Australia to summer rainfall dominant in northern New South Wales and southern Queensland where temperate and subtropical eucalypt woodlands merge (Moore 1970). Temperate eucalypt woodlands were most prevalent in sub-humid regions with a mean annual rainfall of between 200 and 800 mm forming a transitional zone between the higher-rainfall forested margins of the continent and the shrublands and grasslands of the arid interior (Beadle 1981; AUSLIG 1990). They also occurred, but to a lesser extent, in higher-rainfall zones in coastal areas on nutrient poor sandy soils and in sub-alpine regions above 1500 m and at lower-altitude sites

Table 13.1. *Specht's classification of Australian woodlands based on the height and projected foliage cover of the tallest stratum and AUSLIG's extension of this to include the growth form of the next tallest stratum with cover of more than 10%*

Structural form (Specht 1970)	AUSLIG (1990)
Woodland (trees 10–30m high; 10%–30% foliage cover)	Woodland with low trees and tall shrubs
	Woodland with low shrubs
	Woodland with hummock grasses
	Woodland with tussock grasses
Open woodland (trees 10–30m high; <10% foliage cover)	Open woodland with low trees and tall shrubs
	Open woodland with hummock grasses
	Open woodland with tussock grasses
	Open woodland with other herbaceous plants
Low woodland (trees <10m high; 10%–30% foliage cover)	Low woodland with tall shrubs
	Low woodland with low shrubs
	Low woodland with hummock grasses
	Low woodland with tussock grasses
	Low woodland with other herbaceous plants
Low open woodland (trees <10m high; <10% foliage cover)	Low open woodland with tall shrubs
	Low open woodland with low shrubs
	Low open woodland with hummock grasses
	Low open woodland with tussock grasses
	Low open woodland with other herbaceous plants
	Low open woodland with no significant lower stratum

receiving cold air drainage (Beadle 1981; Brown *et al.* 1998). Eucalypt woodlands also occurred in Tasmania under similar conditions to temperate southern Australia, but in sub-alpine regions occurred at altitudes of between 915 and 1200 m (Beadle 1981). In areas of higher rainfall, woodlands are often extensions of adjacent forest onto drier sites.

The floristic composition of the pre-European temperate eucalypt woodlands is little known and because of widespread clearing and modification the original understorey composition of woodlands in some regions remains uncertain (Prober 1996). However, the available evidence suggests that floristic composition of woodlands was influenced by climate, soil type, topography, hydrology, biotic interactions such as grazing, and large-scale disturbances such as fire, windstorms, floods and droughts (Costin 1954; Moore 1970; Beadle 1981;

Kirkpatrick *et al.* 1988; Beard 1990; Yates *et al.* 1994b; Prober 1996). In woodlands with a well-developed shrub understorey, particularly in low-nutrient coastal areas and in southwestern Australia, members of the Cupressaceae, Casuarinaceae, Epacridaceae, Fabaceae, Mimosaceae, Myrtaceae, Myoporaceae and Proteaceae were common. The shrub layer was typically sclerophyllous, but semi-succulent Chenopodiaceae became dominant on alkaline soils in lower-rainfall areas of Western Australia and on some floodplains of the Murray and Murrumbidgee River systems (Beadle 1981; AUSLIG 1990; Beard 1990). In woodlands with a well-developed herbaceous ground layer and sparse shrub understorey, particularly in inland eastern Australia, the dominant family in terms of cover was the Poaceae; members of the Asteraceae, Fabaceae, Liliaceae, Orchidaceae, Juncaceae and Cyperaceae were common but less conspicuous components of the flora (Costin 1954; Trémont and McIntyre 1994).

Beyond the broad classification of woodlands given in Table 13.1, no consistent treatment of woodlands across Australia is available. More detailed discussions of woodland communities are given on a state-by-state basis in Hobbs and Yates (2000a).

Until recently, there have been few comprehensive accounts of woodland fauna, excepting that forest and woodland birds were discussed by Keast *et al.* (1985). For eastern Australia, Lunt and Bennett (2000) have suggested that a distinctive woodland fauna cannot be identified due to overlaps with adjacent vegetation types. Nevertheless, this account, together with other regional accounts (e.g. Hobbs *et al.* 1993; Saunders and Ingram 1995; Bennett *et al.* 1998) provide an indication of the fauna likely to be associated with woodland vegetation. Lunt and Bennett (2000) discuss sets of species with differing distributions in eastern Australia, with particular reference to Victoria. Firstly, a suite of species extends (or formerly extended) in a broad zone across the inland side of the Great Dividing Range: e.g. the yellow-footed antechinus (*Antechinus flavipes*), squirrel glider (*Petaurus norfolcensis*), brush-tailed phascogale (*Phascogale tapoatafa*), diamond firetail (*Stagonopleura guttata*), bush stone-curlew, superb parrot (*Polytelis swainsonii*), turquoise parrot (*Neophema pulchella*), fuscous honeyeater (*Lichenostomus fuscus*), black-chinned honeyeater (*Melithreptus gularis*), woodland blind snake (*Ramphotyphlops proximus*), Dwyer's snake (*Unechis dwyeri*), olive legless lizard (*Delma inornata*) and Sloane's froglet (*Ranidella sloanei*). A second group of woodland species are those that occur (or formerly occurred) more widely in inland Australia, but in Victoria are primarily associated with dry woodlands in the north. These include species such as the red-tailed phascogale (*Phascogale calura*), pig-footed bandicoot (*Chaeropus ecaudatus*), bridled nail-tail wallaby (*Onychogalea fraenata*), apostlebird (*Struthidae cinerea*), cockatiel (*Nymphicus hollandicus*), red-capped robin (*Petroica goodenovii*), curl snake (*Suta suta*), hooded scaly-foot (*Pygopus nigriceps*) and Boulenger's skink (*Morethia boulengeri*).

Tree hollows in mature trees provide important nesting sites for a suite of obligate hollow-using species. Parrots and cockatoos (families Psittacidae and Cacatuidae) use tree hollows for nesting and a number also forage in the ground layer for seed-bearing grasses and forbs, culms and bulbs; e.g. the sulphur-crested cockatoo (*Cacatua galerita*), long-billed corella (*Cacatua tenuirostris*), galah (*Cacatua roseicapilla*), eastern rosella (*Platycercus eximius*) and red-rumped parrot (*Psephotus haematonotus*). In drier areas, woodlands provide nesting areas for species such as the mallee ringneck (*Barnardius barnardi*), blue bonnet (*Northiella haematogaster*), mulga parrot (*Psephotus varius*) and Major Mitchell's cockatoo (*Cacatua leadbeateri*). The open ground layer of temperate woodlands also provides foraging habitat for other seed-eating species [e.g. diamond firetail, peaceful dove (*Geopelia striata*), budgerigar (*Melopsittacus undulatum*), crested pigeon (*Ocyphaps lophotes*)], as well as numerous ground-foraging birds that feed on invertebrates [e.g. red-capped robin (*Petroic goodenovii*), hooded robin (*Melanodryas cuculata*), white-winged chough (*Corcorax melanoramphos*), apostlebird, restless flycatcher (*Myiagra inquieta*), yellow-rumped thornbill (*Acanthiza chrysorrhoa*), Australian magpie (*Gymnorina tibicen*)]. Temperate woodlands also support diverse assemblages of reptiles, including families such as geckoes (Gekkonidae), legless lizards (Pygopodidae) and blind snakes (Ramphotyphlopidae).

Forest birds such as the golden whistler (*Pachycephala pectoralis*), crimson rosella (*Platycercus elegans*), white-naped honeyeater (*Melithreptus lunatus*), pied currawong (*Strepera graculina*) and flame robin (*Petroica phoenicea*) are seasonal migrants, as are species from northern Australia, such as rufous whistler (*Pachycephala rufiventris*), Horsfield's bronze cuckoo (*Chrysococcyx basalis*), tree martin (*Hirundo nigricans*) and sacred kingfisher (*Todiramphus sanctus*).

Woodland fauna has been greatly affected by the widespread loss of habitat and fragmentation, as discussed later. The greatest impacts were on the medium-sized ground-dwelling mammals, many of which became extinct, except in limited areas of Western Australia which retain elements of a distinctive woodland mammal fauna including species such as the woylie (*Bettongia penicillata*) and numbat

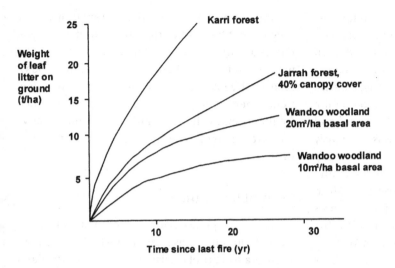

Fig. 13.2. Leaf litter fuel accumulation rates for three communities in Western Australia: karri (*Eucalyptus diversicolor*) forest, jarrah (*E. marginata*) forest, and woodland dominated by wandoo (*E. wandoo*) at two densities. Redrawn from Burrows (1985).

(*Myrmecobius fasciatus*). Yates *et al.* (2000a) also indicate that many of Western Australia's threatened animal species occur in woodlands. Of the species listed as Declared Threatened Fauna, 32% of mammal species, 19% of bird species and 9% of reptile species were found to occur in woodland habitats. Saunders and Ingram (1995) present data suggesting that a large proportion of woodland bird species has declined in abundance or range during the 20th century.

Historical fire regimes

The open woodlands which cover much of temperate Australia are thought to have developed, or spread, as a result of climatic drying and increased incidence of fire during the latter part of the Pleistocene. Scattered palynological records indicate marked shifts in both vegetation and fire activity from the start of the last interglacial period, about 128 000 years ago (Singh *et al.* 1981; Singh 1982; but see also Williams and Gill 1995 and Bowman 1998). These records, for instance from Lake George, also provide evidence for human alteration of fire regimes from about 38 000 years ago. There seems little doubt that Aboriginal burning had a significant influence on the vegetation in the zones currently occupied by woodlands, as elsewhere (Pyne 1991; Bowman 1998). Beyond this generalisation, however, it is difficult to provide a clear picture of the likely frequency or pattern of Aboriginal burning in woodland communities. Only sporadic data are available on the subject, for instance derived from the analysis of *Xanthorrhoea* stems (Gill and Ingwersen 1976; Lamont and Downes 1979), and anecdotal material compiled for particular areas (Hallam 1975; Clark and McLoughlin 1986; Ryan *et al.* 1995). While both these sources give information for certain sites, it is difficult to extrapolate this information to wider areas. Hence, for woodlands, as for other vegetation types, determining historical fire regimes is a difficult proposition (Benson and Redpath 1997; Gill and McCarthy 1998; Lunt 1998a).

Despite this, some generalisations have been made concerning fire in the temperate woodlands in southwestern Australia. In particular, it is likely that parts of these woodlands experienced fire less frequently than the vegetation types in higher-rainfall areas, particularly the forests. This is likely to be a simple result of lower rates of litter accumulation in woodlands compared with forests (Fig. 13.2) (Walker 1981; Burrows *et al.* 1987). Litter accumulation is significantly slower in the woodland communities, firstly because the canopy is more open and secondly because there is significant litter harvest by termites (Park *et al.* 1993). This pattern may reflect historical regimes, but on the other hand, some woodland types, and areas of woodland frequented by Aboriginal populations, may have been burnt more frequently.

Following European settlement, human impacts increased greatly in intensity in the woodland areas. The introduction of livestock resulted in increased levels of grazing and changes in soil properties, while clearing of woodlands for agriculture and tree harvesting for timber and firewood significantly reduced the areas of woodlands present and altered the vegetation structure. Fire was frequently used as a tool in vegetation clearance (Johnson and Purdie 1981), and it seems likely that the remaining uncleared vegetation frequently was burnt inadvertently during this process. Lamont and Downes's (1979) study indicated an increase in fire frequency following European settlement in Western Australia, changing from a frequency of once every 40–50 years to once every 6–8 years. In addition, Burrows et al. (1987) presented data indicating that land-clearing fires in adjacent areas were responsible for most of the fires occurring in Dryandra State Forest in Western Australia, particularly during the period 1950–1960. While significantly less clearing now occurs in the agricultural area in Western Australia, fires spreading from stubble burns on agricultural land are still a major cause of fires in woodland areas (author's unpublished observations).

The data presented by Burrows et al. (1987) illustrate a major difficulty with any attempt to generalise about historical fire regimes – in recent historical times, the influences on fire regime have changed dramatically and often. The extent of burning in the landscape has waxed and waned, not only in relation to vegetation clearance, but also in relation to fire management policies. Fire prevention and prescribed burning policies are introduced and adapted through time, meaning that any one area is likely to have been subjected to a variety of fire management policies and fire treatments. Alternatively, large areas of woodland have been significantly undermanaged, or received no active management at all – fire in these areas has thus occurred sporadically and its occurrence and impacts on the vegetation have generally been unrecorded. The fire history of any one location is thus generally unknown. Benson and Redpath (1997) have recently discussed the dangers of trying to base current management regimes on purported historical observations. They suggest that 'In light of the paucity of data on pre-European burning and the fragmented nature of the natural southern Australian landscape today, modern fire management should be based on scientific understanding of species and their habitats (Williams and Gill 1995), rather than on selective interpretations of some of the early explorers' observations.'

Current fire regimes

Current fire regimes in temperate woodlands are as difficult to summarise as the historical regimes. There is a paucity of information for most woodland types, and in particular the woodlands of inland eastern Australia, where a herbaceous understorey is prevalent. Fire is not a common component of many of these systems today, and may to some extent have been replaced by livestock grazing as the major force influencing system dynamics. Hence there are few studies of fire in many eastern Australian woodland communities. Slightly more information is available for woodlands with shrubby understories, either from more coastal areas of eastern Australia, or from Western Australia. These are reported here, and the reader is referred to Lunt and Morgan (this volume) for a more detailed account of fire in grassland communities, which contain many of the herbaceous elements also found in eastern woodland communities.

Over much of the extent of temperate woodlands, current regimes are determined either deliberately or inadvertently by humans. Burning in woodland communities is extremely rare in eastern Australia (Lunt 1995, 1998a). Fire management in woodland areas may be undertaken for fuel reduction purposes, with the aim of protecting life and property in urban and rural areas, or more rarely to promote regeneration of particular species, or for the maintenance of habitat or floristic diversity. More typically in eastern Australia, fuel management is achieved by stock grazing rather than burning. In coastal Western Australia recommendations for fuel reduction generally involve burning at fairly high frequencies – for instance Burrows and

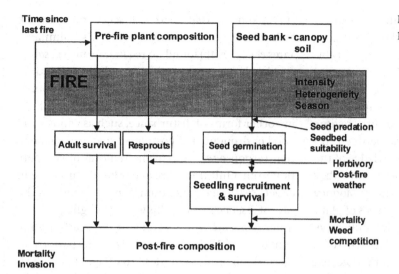

Time since last fire

Mortality Invasion

Fig. 13.3. Conceptual model of woodland plant dynamics in relation to fire.

McCaw (1989) recommend burning *Banksia* woodland with shrubby understories every 3–4 years for fuel reduction purposes. Burning to encourage regeneration is carried out on the basis that fire provides a suitable seedbed for germination and seedling establishment, particularly through the release of nutrients in the ash bed (Burrows *et al.* 1990) and may temporarily overcome lack of recruitment caused by seed predation by ants (Andersen 1987, 1988; Yates *et al.* 1994a). Finally, the use of fire for habitat management and maintenance of diversity assumes that there will be some optimum frequency which will maintain floristic diversity by encouraging the establishment and growth of species that may be dependent on fire or other disturbance for regeneration, but at the same time allowing species that are slow-growing or do not reproduce until a certain age to remain in the system. Clearly, this is a complex task, made more difficult by the lack of detailed information on the population dynamics of many woodland species and by the species diversity of many woodland vegetation types. It may be impossible to establish a fire regime which encompasses all requirements, and hence a pluralistic approach is likely to be required – i.e. high-frequency fires in some areas, with lower frequencies in others, or a mixed fire regime with mosaics burned at different intervals.

While this level of complexity may be possible in some settings, it is unlikely to be achieved in many woodland systems. In urban areas, it may be difficult to prevent frequent fire through arson and accident or to carry out prescribed burns because of risk to adjacent property, while in many rural areas, there are not sufficient conservation resources to devote to detailed fire management prescriptions. Hence woodlands remain either unburnt or are burned by accident, for instance when stubble burns in adjacent paddocks get out of control. Studies quoted below indicate that both individual fires and changes in fire regime have the potential to significantly alter the structure and composition of woodland systems. Unfortunately, often too little thought appears to be devoted to the management of fire in woodland systems, perhaps because of the large number of other factors currently influencing them, as described below.

Fire and woodland dynamics

Relatively few studies have been conducted on the fire-related dynamics of temperate woodland communities in Australia. Figure 13.3 illustrates a conceptual model of the response of the plant community to fire and the factors affecting it. Early studies of post-fire vegetation development concluded that the process of vegetation redevelopment would be strongly influenced by the species composition at the time of the fire and the nature of the fire (Purdie and Slatyer 1976). Post-fire vegetation dynamics have been treated more generally by, for instance, Noble and Slatyer (1980). Their general approach was one of identifying the vital life-history attributes of the species making up the community and relating these to the fire regime experienced, particularly the interval between successive fires. This in turn requires knowledge of the behaviour of individual species and how they respond to and survive fire (Gill 1981a, b; Mooney and Hobbs 1986). In Fig. 13.3 the main types of plant response to fire are illustrated: namely, adult survival through the possession of protective structures or coverings, vegetative resprouting from protected buds or meristems, and germination of seed either from seed stored in the plant canopy and released after fire or from seed stored in the soil. Plant population processes in temperate woodlands in eastern Australia have recently been reviewed by Clarke (2000) and Windsor (2000), and both these authors highlight a general lack of information on the role of fire in these processes. Much of the information that is available relates more to woodlands with shrubby understories than to the grassy woodlands of inland areas. More information on herbaceous species response can be found in Lunt and Morgan (this volume).

Plant regeneration following fire

Numerous studies indicate that many plant species require fire, or some other form of disturbance, to regenerate from seed. Fire may act to allow seed germination through breaking dormancy in some way, allowing seed release from canopy stores, or by altering the environment to provide better conditions for germination, by providing a temporary improvement in seedbed substrate and/or making scarce resources more available.

Seed germination and seedling survival are frequently either rare or non-existent in many unburned woodland communities. For instance, Barker (1988) indicated that past recruitment of *Eucalyptus pauciflora* in sub-alpine woodland probably resulted from fire events, while Burrows *et al.*

(1990) showed that *E. wandoo* regenerated in ash beds following fire, but rarely otherwise. Similarly, Yates *et al.* (1994b) found no regeneration of *E. salmonophloia* in mature woodland stands, but abundant seedlings in recently burnt areas. They also found, however, that seedlings were also abundant following other forms of disturbance, such as windstorm, flood or drought, and postulated that the regeneration response was due to the removal of competition from adult trees. Fire and other disturbance in effect created a regeneration 'gap', although, in the case of *E. salmonophloia*, this was not a light gap but a water gap. Patchy fire leading to the death of individual trees leads to a characteristic pattern of recruitment around the dead tree (Fig. 13.4). *Callitris* spp. are also obligate seeders which regenerate mainly following disturbance, particularly fire (Noble and Slatyer 1977; Clayton-Greene and Ashton 1990; Harris and Kirkpatrick 1991).

As well as removing adult trees (where these are killed by fire, rather than surviving and resprouting), fire may act to encourage seedling establishment by providing a favourable seedbed. The ash-bed effect has been discussed by numerous authors (e.g. Hatch 1960; Chambers and Attiwill 1994), and represents a temporary provision of a sterile, nutrient-rich substrate which allows enhanced seedling growth. Burrows *et al.* (1990) indicated that ash beds were important for *E. wandoo* regeneration, and Yates *et al.* (1996) reported increased germination of *E. salmonophloia* on ashbeds, but only where supplementary watering was given. This indicates the importance not only of substrate availability after fire, but also of the timing and amount of rainfall in the season following the fire, a point noted again later.

A further factor influencing seedling recruitment following fire is the influence of the fire in triggering a massive release of canopy-stored seed. Some serotinous species require fire or long periods of drying to open their fruits, while others such as *Callitris* release seed in summer (Noble and Slatyer 1977) and many *Eucalyptus* species release small quantities of seed more or less continuously (Yates *et al.* 1994a). However, most eucalypt seed that is released is rapidly consumed by ants, and it is only

Fig. 13.4. Regeneration of *Eucalyptus salmonophloia* in Western Australia following a patchy fire which killed individual adult trees. Note the dead tree in the centre of the regeneration clump (Photograph by R. J. Hobbs.)

when there is a massive seed release triggered by fire that 'predator satiation' occurs and reasonable quantities of seed remain to germinate (Andersen 1987, 1988; Yates 1995).

While fire has been implicated as a factor encouraging regeneration of woodland species on many occasions, its importance is not always clear. Gill (1997) suggests that it is possible for some eucalypts to regenerate in the absence of fire, and hence care is needed in generalising about the role of fire. Windsor (1998) found no effect of fire on the regeneration of *E. melliodora* in the Central Tablelands of eastern Australia, while other authors have reported enhanced seedling establishment following fire (e.g. Curtis 1990). It should be noted, however, that Windsor's (1998) study was conducted on individual paddock trees in highly modified systems. It is now very difficult to assess the likely role of fire due to the massive modification that has occurred over most of the area previously covered by the woodland systems. In many woodland areas, livestock grazing is a far more prevalent influence, and hence has received considerably more research attention.

Vegetation structure and composition

Fire can have a significant impact on vegetation structure and composition, removing much of the plant biomass and initiating a process of recolonisation by species that were present before the fire, species present as seeds in the soil, and/or species invading from elsewhere. It is also likely that fire can result in a more-or-less permanent structural change in vegetation. For instance, Hopkins and Robinson (1981) describe the conversion of a woodland dominated by *Eucalyptus cylindriflora, E. diptera* and *E. eremophila* to a mallee heath as a result of a single fire event which removed adult tree-form eucalypts and resulted in mallee-form regeneration. They suggest that mallee-heaths may often be the result of such conversions, but point out that the change in community structure is not coupled with an overall change in floristics.

Studies elsewhere have also indicated that the absence of fire over long periods of time can also result in shifts in vegetation structure and composition. For instance, Withers and Ashton (1977), Withers (1978a, b) and Lunt (1998b) indicated that long-unburned stands of *Eucalyptus ovata/E. viminalis/E. leucoxylon* woodland were being replaced by scrub dominated by *Casuarina littoralis*. This change appears to be facilitated by the presence of a dense understorey grass sward which prevents establishment of eucalypts but not of *Casuarina*. Recruitment of eucalypts and *Casuarina* then later becomes impossible beneath the dense *Casuarina* canopy that develops. Hazard and Parsons (1977) similarly indicated that *Leucopogon parviflorus* would replace *L. laevigatum* in the absence of fire. Muir (1985) made the suggestion that the absence of fire may be important for the retention of some vegetation types and structures, and that we need to take care when assuming that fire is always a necessary component of the system.

Fire frequency

While there is little information available on fire regimes in woodlands, some information can be presented on the likely impacts of changing fire frequencies on plant composition in some woodland communities, on the basis of both population and community studies. Some studies have examined woodland community composition in stands that had been burned at different times in the past, while others have estimated the age individual species need to reach before they flower and set seed.

Most studies of this nature come from the coastal humid woodlands with shrubby understories. The only example I have found from the interior woodlands considers *Callitris* woodland. *Callitris columellaris* is a slow-growing conifer which occurs over most of the inland woodland area of eastern Australia and parts of Western Australia, where it may be interspersed with eucalypt species or may occur in pure stands (Noble and Slayter 1977; Boland *et al.* 1984; Clayton-Greene and Ashton 1990). Noble and Slatyer (1977) considered the dynamics of communities containing *Callitris* in relation to its life-history characteristics, using the vital attributes approach. *Callitris* is killed by fire, and regenerates from canopy-stored seed. It can reach reproductive age at 6 years old. Where *Callitris* is mixed with eucalypt species capable of vegetative regeneration following fire, a number of pathways are possible after fire, depending on the time between fires. If fire is frequent (i.e. less than 6–8 years between fires), *Callitris* will be lost from the plant community since the adults will be killed and have not had time to set seed. If fires are less frequent (i.e. every few decades), both *Callitris* and eucalypts can re-establish. However, if the fire interval is considerably longer (>100yrs?), eucalypts may be lost from the community, and *Callitris* will dominate. If a fire occurs after that length of time, there may be a period of dominance by herbaceous species, followed by a re-development of a *Callitris* canopy. Noble and Slatyer (1977) put forward this scheme to explain the presence of woodland stands with different mixes of eucalypts and *Callitris* in areas of New South Wales, but did not follow this up with detailed field observations. They also pointed out that some observed differences may be due to edaphic and climatic differences, rather than fire history, a point which is further suggested by the work of Clayton-Greene and Ashton (1990).

While herbaceous species form an important part of some phases of the plant community discussed here, and are also widely important in the grassy eucalypt woodlands, relatively little is known of the response of individual species to fire frequencies. Some clues may be obtained from studies of other disturbance types in herbaceous communities (e.g. McIntyre and Lavorel 1994a, b; McIntyre *et al.* 1995): this is discussed more fully in Lunt and Morgan (this volume).

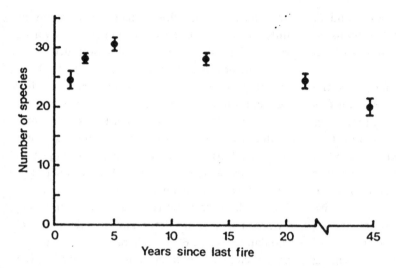

Fig. 13.5. Number of vascular plant species (mean ± s.e.) in ten 5 × 5 m quadrats in stands of *Banksia* woodland of different ages since last fire. From Hobbs and Atkins (1990).

Other studies in shrubby woodlands provide some insights into fire frequency effects. Fox and Fox (1986) studied the effect of fire frequency on understorey structure and composition in a *Eucalyptus pilularis* woodland. Comparing a site which had been burnt 14, 8 and 2 years previously with one that had been burnt only 14 and 2 yrs previously, they found significantly more species and a higher shrub cover and density in the more frequently burnt site. The species present in the less frequently burnt site but absent from the more frequently-burnt site were all obligate seed regenerating species, while those found only in the more frequently burnt site were mostly resprouting species. Fox and Fox (1986) suggested, therefore, that increasing fire frequency would lead to the loss of obligate seeder species, but conversely that reducing frequency could result in the loss of resprouting species.

In a study of a *Banksia* woodland in Western Australia, Hobbs and Atkins (1990) examined stands ranging from 1 to >44 years since the last fire. Species richness was greatest in the 5-year-old stand (Fig. 13.5), and many shrub species were most abundant 2–5 years after fire. Both of the dominant *Banksia* species, *B. attenuata* and *B. menziesii*, had mixed size structures including seedlings in stands of all ages, indicating that neither is dependent on fire for recruitment. Cowling *et al.* (1990) examined both these species, which also occur in sandplain

vegetation further north, and estimated that *B. attenuata* only started producing seed 4 years after a fire, and *B. menziesii* only after 6 years. However, for both species it took several more years before seed production rates increased. They suggested, therefore, that high fire frequencies could have an impact on seed production and availability. In the woodland situation, however, low-intensity fires are likely to impact only the understorey, leaving the dominant trees untouched.

Other studies of seed production by the woodland understorey species *Banksia ericifolia* and *Petrophile pulchella* (Bradstock and O'Connell 1988) indicated that seeds of these species were first available at 5 and 6 years post-fire, respectively, although up to 13 years after fire may be required to provide sufficient seed for population replacement if seedling establishment rates were low. Bradstock and O'Connell (1988) found that the age at which a stand was burned, together with the proportion of seeds which emerge as seedlings, accounted for the biggest changes in numbers and density of these species between generations.

Fire characteristics and post-fire conditions

As in previous sections, there is little information on this from grassy woodlands. Here I report studies from shrubby woodlands in coastal areas and in Western Australia. The fire characteristics discussed primarily here are season of burning, intensity and heterogeneity. Whelan and Tait (1995) have commented on the relative lack of information on the importance of fire season on the outcome of fire. A few studies have provided some insight into this. For instance, Hobbs and Atkins (1990) found that autumn and spring burns in *Banksia* woodland in southwestern Australia displayed different patterns of regeneration, with seedling regeneration occurring only in the autumn burn area. Vegetative regrowth was more rapid and post-fire species numbers were higher in the spring burn area.

Clark (1988) compared the outcomes of spring and autumn fires on understorey species on Hawkesbury sandstone, and also found that plots burned in spring had faster regrowth and greater species diversity for the first 6 year after fire. On the other hand, for individual species, approximately equivalent numbers of species did better (in terms of population numbers) in either the spring- or the autumn-burned plots. Clark (1988) pointed out that some species were influenced by the fire treatment directly (in terms of differences in intensity) but that others were more influenced by post-fire rainfall or temperature. In a simulation study of the effect of season of fire on post-fire seedling emergence of *Banksia ericifolia* and *B. serrata* Bradstock and Bedward (1992) suggested that inter-year variation in rainfall and its influence on seedling emergence precluded any prediction of the effects of fire season. They suggested that the timing of fire relative to wet and dry years may be of equal importance to fire season in its effect on the *Banksia* populations, a suggestion echoed by Whelan and Tait (1995).

While some of the studies discussed above implicitly assumed that fires in different seasons were also different in intensity, few studies of fire intensity or temperature have been undertaken in woodland systems. Hobbs and Atkins (1988) measured fire temperatures in two woodland types and illustrated that fire in open woodland communities is liable to be very heterogeneous, based on the distribution of fuel in the form of understorey shrubs or herbaceous plants. Even where a thick, uniform litter

layer occurred and fire temperatures were very uniform at the soil surface, temperatures varied from 50 °C to 800 °C, 2 cm below the soil surface.

The heterogeneity in fire intensity is likely to have important influences on the survival and/or regeneration of woodland species. For instance, Hobbs and Atkins (1991) recorded significantly greater numbers of seedlings of *Eucalyptus loxophleba* in lightly burned areas than in areas subject to intense fire. In another study Atkins and Hobbs (1995) illustrated the impact of varying fire temperatures on the germination of seeds of three understorey *Acacia* species in salmon gum (*E. salmonophloia*) woodland (Fig. 13.6). In laboratory experiments, one species, *A. hemeiteles* germinated well in the unheated control treatment and at 50 °C and 100 °C, but germination was significantly reduced at 150 °C. *Acacia microbotrya*, on the other hand, germinated significantly better than the unheated control in the 50 °C and 100 °C treatments, and *A. colletiodes* germinated well only in the 150 °C treatment. Hence the germination of each species of *Acacia* is enhanced by different treatments. If these treatments are taken as representative of possible heat treatments experienced by seeds in the soil during fires, this indicates that spatial variation in fire characteristics, and also variation in intensity among different fires, is likely to be important in maintaining these different species in the understorey. These data could also be interpreted as indicating that the different functional responses of the suite of *Acacia* species present in the woodland system confer resilience to the system since at least one species is likely to germinate following a fire of a particular intensity (Chapin *et al.* 1997).

Grazing regimes

Grazing of woodland communities by introduced stock is known to significantly alter the structure and composition of the woodland community (Landsberg 1993; Scougall *et al.* 1993; Grice and Barchia 1995; Petit *et al.* 1995). Such changes undoubtedly alter the distribution and amount of fuel available, and hence the likely fire behaviour in areas that have been grazed. A further impact of

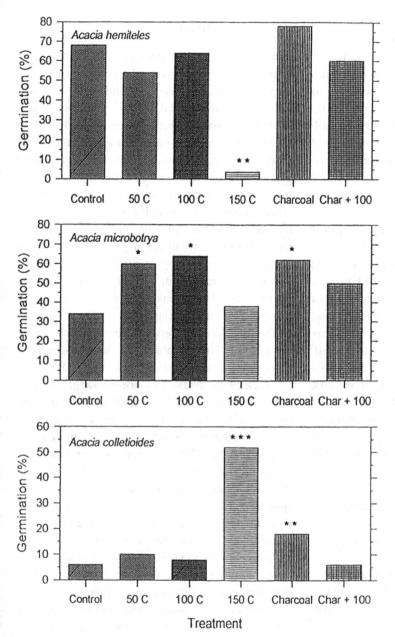

Fig. 13.6. Germination (%) of three *Acacia* species that are common components of the understorey in *Eucalyptus salmonophloia* woodland in the central wheatbelt of Western Australia, following various treatments. Seeds were scarified and subjected to different temperatures in an oven for 10 min, placed on filter paper in petri dishes and watered regularly for 38 days. In addition, charcoal fragments were added to some dishes. Controls received no heat or charcoal. Five dishes were assigned to each treatment, with ten seeds per dish. Asterisks indicate significant differences from control, as determined by t test on square root transformed data: * $p < 0.05$, ** $p < 0.01$, *** $p < 0.001$. From Atkins and Hobbs (1995).

stock grazing in some woodland systems is the alteration of soil surface and subsurface properties, in particular the removal of the surface crust and litter layer, and the compaction of the soil (Wylie and Landsberg 1987; Scougall and Majer 1991). Yates *et al.* (2000b) provided data indicating that such compaction significantly reduced water infiltration rates and altered the subsurface temperature regime. The buffering effect of the litter and surface crust was effectively lost, and hence soil temperatures at depth fluctuated much more than in ungrazed situations. Reduced soil moisture retention and increased heat transmission through the soil are both likely to alter the impact of fire on the soil and stored seed, although this has not been examined to date.

In eastern Australia, Cluff and Semple (1994) found that eucalypt regeneration was greater where grazing was absent or where cattle rather than sheep were grazed. It has also been suggested that grazing in recently burned areas, both by vertebrate and invertebrate grazers, can have important effects on the post-fire vegetation development. Whelan and Main (1979) illustrated how grazing of seedlings by grasshoppers (*Goniaea* spp.) affected post-fire regeneration of canopy and understorey species in a *Banksia* woodland in Western Australia. Of particular interest is their observation that grazing impacts were greatest where the areas burned were relatively small. Leigh and Holgate (1979) and Leigh *et al.* (1991) found that grazing profoundly affected species composition, density and height, particularly of certain species after fire, and suggested that grazing by native animals may exert a more profound effect on plant survival than fire itself. Almost three to four times more seedlings established on ungrazed plots relative to grazed plots. This has also been suggested by studies on Rottnest Island in Western Australia, where post-fire regeneration of the woodland tree species *Callitris preissii* and *Melaleuca lanceolata* in the 1940s and 1950s was prevented by grazing by increasing populations of the quokka (*Setonix brachyuris*) (Pen and Green 1983; Rippey and Rowland 1995; Hobbs 1998). This resulted in a marked reduction in the area of woodland on the island, and a replacement with low scrub dominated by *Acanthocarpus preissii*.

In some instances, grazing and fire regimes have been regarded as substitutable. For instance, Kirkpatrick (1986) suggested that frequent fire may be necessary to maintain the open conditions required to maintain a set of understorey species in grassy woodland in Tasmania, in the absence of grazing by native mammals that are no longer present in the urban setting. I have already suggested the proposition that stock grazing may have replaced fire as the major driving force over much of the grassy woodland area of eastern Australia.

Weed competition

Invasion by non-native species is a problem in many woodland systems. Many weed species are capable of invading woodland areas, predominantly grass and forb species, both annual and perennial, particularly of Mediterranean or South African origin. Particular species of note are perennial grasses such as *Ehrharta calycina*, *Hyparrhenia hirta* and *Eragrostis* spp., annual grasses such as *Avena*, *Bromus* and *Phalaris* spp., and forbs such as *Hypochaeris* spp. and *Arctotheca calendula*. Weed invasion in woodlands is thought to be important in terms of fire management for several reasons. Firstly, the incidence of fire may enhance weed invasion. Weeds in turn may alter fuel distributions and lead to more frequent or more intense fires. Finally, the presence of weeds may prevent the regeneration of native perennial species.

The relationship between fire and weeds has been explored in a number of instances. Reflecting a more widespread phenomenon (D'Antonio and Vitousek 1992), invasion by non-native grasses into woodland systems has been implicated in the development of a 'grass–fire' cycle, whereby the invasion of a non-native grass provides a more complete fuel cover and hence alters fire frequency and intensity (Baird 1977; Wycherley 1984). In turn, fire enhances the establishment and spread of the invasive grass species, often at the expense of native species. This situation is particularly evident in urban bushland areas, where effective control of the invasive species requires active management with, for instance, herbicide, coupled with measures to attempt to reduce the incidence of deliberate and accidental fire.

Direct evidence for the impact of weed species on the regeneration of native species comes from studies by Hobbs and Atkins (1991), who found a strong negative correlation between the survival of seedlings of *Eucalyptus loxophleba* and cover of non-native annuals in the year following a fire in *E. loxophleba*/*Acacia acuminata* woodland. Other studies report similar impacts on tree regeneration of rapid weed growth following fire (e.g. Curtis 1990; Windsor 1998). This impact of weed invasion of woodland systems is particularly important in extensively fragmented areas, as discussed below.

Fire may not always lead to increased weed growth and cover. For instance Hester and Hobbs (1992) found no evidence for increased abundances of non-native annuals in a *E. loxophleba/A. acuminata* woodland after fire. Data from studies in *Banksia* woodland (Hobbs and Atkins 1990) also suggest that the degree of weed invasion may be related to the season of burning, with autumn fire areas having higher weed abundances than spring fire areas. Hobbs and Atkins (1990) also found that weed species were present only in relatively recently burned areas and had largely disappeared by 5 years after fire. It seems likely that weeds present a major problem mostly in highly disturbed areas, and that fire simply acts as another factor enhancing weed establishment and growth (Hobbs and Atkins 1991).

Fire regimes and fauna habitat

Relatively little information is available on the effects of fire on fauna habitat in woodland systems. Friend (1993) has reviewed the information available on the impact of fire on small vertebrates in mallee woodlands and heathlands, but there has been relatively little written on woodlands *per se*. In Western Australia, Christensen (1980) and Hopkins (1985) outlined the importance of thickets of *Gastrolobium microcarpum* shrubs within *Eucalyptus wandoo/E. accedens* woodland as refuges for woylies (*Bettongia penicillata*) and tammar wallabies (*Macropus eugenii*). *Gastrolobium* is an obligate seeder species, and thickets favoured as habitat were generally greater than 15 years old, and could be up to 50 years old. Similarly, Hopkins (1985) reports that sites most frequently used by tammars at another location consisted of thickets of *Allocasuarina huegelii* which had probably been burned 30 years previously. These studies indicate an interesting interaction between fire frequency and habitat availability. Fire is required to stimulate regeneration of the thicket species, but too frequent fire in any area could potentially prevent the formation of suitable habitat. Studies by Bamford (1985, 1992, 1995) examined the faunal communities in stands of *Banksia* woodland burned at different times in the past, ranging between 0 and 23 years. He found that the reptile

assemblage varied mainly in the first 2 years after fire, due to the differential survival of adult and immature individuals immediately after fire. Only one species, the agamid *Tympanocryptis adelaidensis*, was most abundant 3–6 years after the fire, possibly because of changes in habitat structure. Bamford (1992) found that, of three species of frog found in the study areas, *Helioporus eyrei* showed no apparent population changes with time since fire, while *Limnodynastes dorsalis* and *Myobatrachus gouldii* were more abundant in long-unburned areas. Bamford (1995) and Friend (1993) concur in their assessment that small-mammal faunas in woodlands in general exhibit resilience to individual fires, although their resilience to particular fire regimes remains unclear.

Fire regimes can be, in general, expected to impact fauna populations to the extent that they modify habitat characteristics. Hence, where fire regimes significantly alter habitat structure and/or composition, for instance by causing a switch in species dominance or by altering understorey composition, this is likely to decrease habitat suitability for some species and increase it for others. As an example, fire-induced mortality of adult trees may increase the number of hollows available for hollow-nesting species, but decrease the resources available for foliovores or bark-feeding species.

Fire management in fragmented landscapes

Following European settlement the relatively fertile woodland soils quickly attracted attention and large areas were cleared for cropland and towns, or were grazed and converted to exotic pasture (Burvill 1979; Fensham 1989; Lunt 1991; Prober and Thiele 1993, 1995). The conversion of temperate eucalypt woodlands to agricultural land represents one of the most significant vegetation changes in Australian history. In total approximately 500 000 km^2 of woodlands have been cleared (AUSLIG 1990). In many areas woodlands are now restricted to remnants of varying size, quality and isolation (Kirkpatrick *et al.* 1988; Hobbs *et al.* 1993; Prober and Thiele 1993, 1995; McIntyre 1994; McIntyre and Lavorel 1994a, b).

Changes in the area occupied by woodlands from

Table 13.2. *The areas of woodlands cleared and remaining in Australia; the area lost has been predominantly in temperate southern Australia*

Structural form	Area remaining ($\times 000\,km^2$)	Area lost ($\times 000\,km^2$)
Woodland	538	461
Open woodland	141	34
Low woodland	432	141
Low open woodland	1453	21

Source: AUSLIG (1990).

before European settlement to the present have been estimated by AUSLIG (1990) and are summarised in Table 13.2. For example in the wheat belts of southeastern and southwestern Australia, eucalypt woodlands have been almost completely eliminated from the landscape with as much as 95% of the native vegetation removed in some districts (AUSLIG 1990; Hobbs *et al.* 1993; Robinson and Traill 1996). In many cases the woodlands that remain are under continued threat from further clearing, rising saline water tables and increased inundation, livestock grazing, nutrient enrichment, soil structural decline, the invasion of exotic species, firewood harvesting, disease, outbreaks of phytophagous insects and loss of interactions which sustain the ecosystem (Yates and Hobbs 1997). In addition, many of the coastal woodlands are under similar threats due to urban development. When considering fire regimes in woodland systems, it is important to do so in the context of the vastly altered landscapes and communities that currently exist.

In particular, the interactions between fire, grazing and weed invasion discussed above are particularly important in fragmented and altered systems. Especially in terms of tree regeneration, the impacts of these factors cannot be underestimated. Where in a relatively unmodified system, fire may encourage regeneration from seed, this may be prevented in the fragmented system because of degradation of the seedbed by stock grazing and because of competition from weeds. If, as appears likely from the studies discussed above, regenera-

tion is also affected by the weather conditions in the season after the fire, effective regeneration of some species may therefore depend on the coincidence of a fire or other disturbance and good rainfalls in the subsequent months. This suggests that woodland regeneration may have been markedly episodic in the past. The danger is that now, when a suitable episode occurs, effective regeneration may be prevented by the impacts of grazing and weeds.

Another important consideration relating to fire management in fragmented systems is that the limited studies available suggest that local plant extinction is possible under either very frequent or infrequent fire regimes. In a continuous habitat, these local extinctions would be accompanied by persistence or recolonisation elsewhere in the landscape. In a fragmented system, on the other hand, such broader landscape processes are disrupted (Hobbs 1987), and local extinctions may not be followed either by recolonisation of that particular site, or perpetuation of the population elsewhere in the landscape. Fragmented woodland systems may thus be very prone to fire-regime-driven plant extinctions, either as a result of very frequent burning such as is the case in some urban remnants, or because of lack of fire over long periods as is the case over much of the grassy woodlands of eastern Australia. The same considerations apply to fauna. Whereas in continuous habitat, a fire regime in one area may locally alter habitat quality and resource availability, suitable habitat would generally have been available elsewhere in the landscape. In fragmented systems, alternative habitat areas either may no longer exist or may be out of reach.

Where, then, does this leave fire management in woodland systems? Clearly, fire may be an effective tool in woodland management, but it must be used with care, and must be accompanied by whatever other management treatments are required to ensure persistence and regeneration. Setting clear management aims is of primary importance, and then the use of fire needs to be assessed in the context of its utility in achieving these aims. Hobbs and Yates (2000b) have recently summarised the need for adequate assessment of woodland condition and regeneration potential, leading to the selection of appropriate management responses. In some cases, other alternatives to fire may be more appropriate. In some urban areas, the task is to reduce the incidence and impact of accidental and deliberately lit fires. In others, a problem is the inability to burn urban remnants because of potential damage to adjacent property. In urban and rural areas, finding simple, low-cost management treatments is the primary need. The degree of success in finding these will in large part determine the future of Australia's remaining temperate woodlands.

References

Andersen, A. N. (1987). Effects of seed predation by ants on seedling densities at a woodland site in SE Australia. *Oikos* **48**, 171–174.

Andersen, A. N. (1988). Immediate and longer-term effects of fire on seed predation by ants in sclerophyllous vegetation in south-eastern Australia. *Australian Journal of Ecology* **13**, 285–293.

Atkins, L., and Hobbs, R. J. (1995). Measurement and effects of fire heterogeneity in south-west Australian wheatbelt vegetation. *CALMScience Supplement* **4**, 67–76.

AUSLIG (1990). *Atlas of Australian Resources: Vegetation.* (Australian Government Printing Service: Canberra.)

Baird, A. M. (1977). Regeneration after fire in King's Park, Perth, Western Australia. *Journal of the Royal Society of Western Australia* **60**, 1–22.

Bamford, M. (1985). The fire-related dynamics of small vertebrates in *Banksia* woodland: a summary of research in progress. In *Fire Ecology and Management in Western Australian Ecosystems* (ed. J. W. Ford) pp. 107–110. (WAIT Environmental Studies Group: Bentley, WA.)

Bamford, M. J. (1992). The impact of fire and increasing time after fire upon *Heleioporus eyrei, Limnodynastes dorsalis* and *Myobatrachus gouldii* (Anura : Leptodactylidae) in *Banksia* woodland near Perth, Western Australia. *Wildlife Research* **19**, 169–178.

Bamford, M. J. (1995). Response of reptiles to fire and increasing time after fire in *Banksia* woodland. *CALMScience Supplement* **4**, 175–186.

Barker, S. (1988). Population structure of snow gum (*Eucalyptus pauciflora* Sieb. ex Spreng.) subalpine woodland in Kosciusko National Park. *Australian Journal of Botany* **36**, 483–501.

Beadle, N. C. (1981). *The Vegetation of Australia.* (Cambridge University Press: Cambridge.)

Beard, J. S. (1990). *Plant Life in Western Australia.* (Kangaroo Press: Kenthurst, NSW.)

Bennett, A., Brown, G., Lumsden, L., Hespe, D., Krasna, S., and Silins, J. (1998). Fragments for the future: wildlife in the Victorian Riverina (the Northern Plains). Department of Natural Resources and the Environment, Melbourne.

Benson, J. S., and Redpath, P. A. (1997). The nature of pre-European native vegetation in south-eastern Australia: a critique of Ryan, D. G., Ryan, J. S., and Starr, B. J. (1995) *The Australian Landscape: Observations of Explorers and Early Settlers. Cunninghamia* **5**, 285–328.

Boland, D. J., Brooker, M. I. H., Chippendale, G. M., Hall, G. M., Hyland, B. P. M., Johnston, R. D., Kleinig, D. A., and Turner, J. D. (1984). *Forest Trees of Australia.* (Nelson–CSIRO: Melbourne.)

Bowman, D. M. J. S. (1998). The impact of Aboriginal landscape burning on the Australian biota. *New Phytologist* **140**, 385–410.

Bradstock, R. A., and Bedward, M. (1992). Simulation of the effect of season of fire on post-fire seedling emergence of two *Banksia* species based on long-term rainfall records. *Australian Journal of Botany* **40**, 75–88.

Bradstock, R. A., and O'Connell, M. A. (1988). Demography of woody plants in relation to fire: *Banksia ericifolia* L.f. and *Petrophile pulchella* (Schrad) R.Br. *Australian Journal of Ecology* **13**, 505–518.

Brown, K. L., Gadd, L. S., Norton, T. W., Williams, J. E., and Klomp, N. I. (1998). The effect of fire on fauna in the Australian Alps National Parks: a database. A report to the Australian Alps Liaison Committee, The Johnstone Centre, Charles Sturt University, Albury.

Burrows, N. D. (1985) Planning fire regimes for nature conservation forests in south western Australia. In *Fire Ecology and Management in Western Australian Ecosystems* (ed. J. W. Ford) pp. 129–138. (WAIT Environmental Studies Group: Bentley, WA.)

Burrows, N. D., and McCaw, W. L. (1989). Fuel characteristics and bushfire control in *Banksia* low woodlands in Western Australia. *Journal of Environmental Management* **31**, 229–236.

Burrows, N. D., McCaw, W. L., and Maisey, K. G. (1987). Planning for fire management in Dryandra Forest. In *Nature Conservation: The Role of Remnants of Native*

Vegetation (eds. D. A. Saunders, G. W. Arnold, A. A. Burbidge and A. J. M. Hopkins) pp. 305–312. (Surrey Beatty: Chipping Norton, NSW.)

Burrows, N. D., Gardiner, G., Ward, B., and Robinson, A. (1990). Regeneration of *Eucalyptus wandoo* following fire. *Australian Forestry* 53, 248–258.

Burvill, G. H. (1979). The first sixty years, 1829–1889. In *Agriculture in Western Australia 1829–1979*. (ed. G. H. Burvill) pp. 4–17. (University of Western Australia Press: Nedlands.)

Chambers, D. P., and Attiwill, P. M. (1994). The ash-bed effect in *Eucalyptus regnans* forest: chemical, physical and microbiological changes in soil after heating or partial sterilisation. *Australian Journal of Botany* 42, 739–749.

Chapin, F. S., Walker, B. H., Hobbs, R. J., Hooper, D. U., Lawton, J. H., Sala, O. E., and Tilman, D. (1997). Biotic control over the functioning of ecosystems. *Science* 277, 500–504.

Christensen, P. E. (1980). Biology of *Bettongia penicillata* Gray 1837 and *Macropus eugenii* Desmarest 1804 in relation to fire. Western Australian Forest Department Bulletin no. 91, Perth.

Clark, S. S. (1988). Effects of hazard-reduction burning on populations of understorey plant species on Hawkesbury sandstone. *Australian Journal of Ecology* 13, 473–484.

Clark, S. S., and McLoughlin, L. C. (1986). Historical and biological evidence for fire regimes in the Sydney region prior to the arrival of Europeans: implications for future bushland management. *Australian Geographer* 17, 101–112.

Clarke, P. J. (2000) Plant population processes in temperate woodlands of eastern Australia: premises for management. In *Temperate Eucalypt Woodlands in Australia: Biology, Conservation, Management and Restoration* (eds. R. J. Hobbs and C. J. Yates) pp. 248–270. (Surrey Beatty: Chipping Norton, NSW.)

Clayton-Greene, K. A., and Ashton, D. H. (1990). The dynamics of *Callitris columellaris/Eucalyptus albans* communities along the Snowy River and its tributaries in south-eastern Australia. *Australian Journal of Botany* 38, 403–432.

Cluff, D., and Semple, B. (1994). Natural regeneration: in 'Mother Nature's' own time. *Australian Journal of Soil and Water Conservation* 7, 28–33.

Costin, A. B. (1954). *A Study of the Ecosystems of the Monaro Region of New South Wales*. (Government Printer: Sydney.)

Cowling, R. M., Lamont, B. B., and Enright, N. J. (1990). Fire and management of south-western Australian banksias. *Proceedings of the Ecological Society of Australia* 16, 177–183.

Curtis, D. (1990). *Natural Regeneration of Eucalypts in the New England Region: Sowing the Seeds*. (Greening Australia: Adelaide).

D'Antonio, C. M., and Vitousek, P. M. (1992). Biological invasions by exotic grasses, the grass/fire cycle, and global change. *Annual Review of Ecology and Systematics* 23, 63–87.

Fensham, R. J. (1989). The pre-European vegetation of the Midlands, Tasmania: a floristic and historical analysis of vegetation patterns. *Journal of Biogeography* 16, 29–45.

Fox, M. D., and Fox, B. J. (1986). The effect of fire frequency on the structure and floristic composition of a woodland understorey. *Australian Journal of Ecology* 11, 77–85.

Friend, G. R. (1993). Impact of fire on small vertebrates in mallee woodlands and heathlands of temperate Australia: a review. *Biological Conservation* 65, 99–114.

Gill, A. M. (1981a). Adaptive responses of Australian vascular plant species to fire. In *Fire and the Australian Biota*. (eds. A. M. Gill, R. H. Groves and I. R. Noble) pp. 243–271. (Australian Academy of Science: Canberra.)

Gill, A. M. (1981b) Coping with fire. In *The Biology of Australian Plants*. (eds. J. S. Pates and A. R. McComb) pp. 65–87. (University of Western Australia Press: Nedlands.)

Gill, A. M. (1997) Eucalypts and fires: interdependent or independent? In *Eucalypt Ecology: Individuals to Ecosystems*. (eds. J. E. Williams and J. C. Z. Woinarski) pp. 151–167. (Cambridge University Press: Cambridge.)

Gill, A. M., and Ingwersen, F. (1976). Growth of *Xanthorrhoea australis* R.Br. in relation to fire. *Journal of Applied Ecology* 13, 195–203.

Gill, A. M., and McCarthy, M. A. (1998). Intervals between prescribed fires in Australia: what intrinsic variation should apply? *Biological Conservation* 85, 161–169.

Grice, A. C., and Barchia, I. (1995). Changes in grass density in Australian semi-arid woodlands. *Rangelands Journal* 17, 26–36.

Hallam, S. J. (1975). *Fire and Hearth: A Study of Aboriginal Usage and European Usurpation in South-Western Australia.* (Institute of Aboriginal Studies: Canberra.)

Harris, S., and Kirkpatrick, J. B. (1991). The distributions, dynamics and ecological differentiation of *Callitris* species in Tasmania. *Australian Journal of Botany* 39, 187–202.

Hatch, A. B. (1960). Ash bed effects in Western Australian forest soils. *Western Australian Forests Department Bulletin* 64, 3–19.

Hazard, J., and Parsons, R. F. (1977). Size-class analysis of coastal scrub woodland, Western Port, southern Australia. *Australian Journal of Ecology* 2, 187–197.

Hester, A. J., and Hobbs, R. J. (1992). Influence of fire and soil nutrients on native and non-native annuals at remnant vegetation edges in the Western Australian wheatbelt. *Journal of Vegetation Science* 3, 101–108.

Hobbs, R. J. (1987). Disturbance regimes in remnants of natural vegetation. In *Nature Conservation: The Role of Remnants of Native Vegetation.* (eds. D. A. Saunders, G. W. Arnold, A. A. Burbidge and A. J. M. Hopkins) pp. 233–40. (Surrey Beatty: Chipping Norton, NSW.)

Hobbs, R. J. (1998). Impacts of land use on biodiversity in southwestern Australia. In *Landscape Degradation in Mediterranean-Type Ecosystems* (eds. P. W. Rundel, G. Montenegro and F. Jaksic) pp. 81–106. (Springer-Verlag: New York.)

Hobbs, R. J., and Atkins, L. (1988). Variability of experimental fires in SW Western Australia. *Australian Journal of Ecology* 13, 295–329.

Hobbs, R. J., and Atkins, L. (1990). Fire-related dynamics of a *Banksia* woodland in south-western Australia. *Australian Journal of Botany* 38, 97–110.

Hobbs, R. J., and Atkins, L. (1991). Interactions between annuals and woody perennials in a Western Australian wheatbelt reserve. *Journal of Vegetation Science* 2, 643–654.

Hobbs, R. J., and Yates, C. J. (eds.) (2000a). *Temperate Eucalypt Woodlands in Australia: Biology, Conservation, Management and Restoration.* (Surrey Beatty: Chipping Norton, NSW.)

Hobbs, R. J., and Yates, C. J. (2000b). Priorities for action and management guidelines. In *Temperate Eucalypt Woodlands in Australia: Biology, Conservation, Management and Restoration.* (eds. R. J. Hobbs and C. J. Yates) pp. 400–414. (Surrey Beatty: Chipping Norton, NSW.)

Hobbs, R. J., Saunders, D. A., Lobry de Bruyn, L. A., and

Main, A. R. (1993). Changes in biota. In *Reintegrating Fragmented Landscapes: Towards Sustainable Production and Nature Conservation* (eds. R. J. Hobbs and D. A. Saunders) pp. 65–106. (Springer-Verlag: New York.)

Hopkins, A. J. M. (1985). Fire in the woodlands and associated formations of the semi-arid region of south-western Australia. In *Fire Ecology and Management in Western Australian Ecosystems* (ed. J. W. Ford) pp. 83–90. (WAIT Environmental Studies Group: Bentley, WA.)

Hopkins, A. J. M., and Robinson, C. J. (1981). Fire induced structural change in a Western Australian woodland. *Australian Journal of Ecology* 6, 177–188.

Johnson, R. W., and Purdie, R. W. (1981). The role of fire in the establishment and management of agricultural systems. In *Fire and the Australian Biota* (eds. A. M. Gill, R. H. Groves and I. R. Noble) pp. 497–528. (Australian Academy of Science: Canberra.)

Keast, A., Recher, H. F., Ford, H., and Saunders, D. (eds.) (1985). *Birds of Eucalypt Forests and Woodlands: Ecology, Conservation, Management.* (Surrey Beatty: Chipping Norton, NSW.)

Kirkpatrick, J. B. (1986). The viability of bush in cities: ten years of change in an urban grassy woodland. *Australian Journal of Botany* 34, 691–708.

Kirkpatrick, J., Gilfedder, L., and Fensham, R. (1988). *City Parks and Cemeteries: Tasmania's Remnant Grasslands and Grassy Woodlands.* (Tasmanian Conservation Trust: Hobart.)

Lamont, B. B., and Downes, S. (1979). The longevity, flowering and fire history of the grasstrees *Xanthorrhoea preissii* and *Kingia australis. Journal of Applied Ecology* 16, 893–899.

Landsberg, J. (1993). Rural dieback and insect damage in remnants of native woodlands. *Victorian Naturalist* 110, 37.

Leigh, J. H., and Holgate, M. D. (1979). The response of the understorey of forests and woodlands of the Southern Tablelands to grazing and burning. *Australian Journal of Ecology* 4, 25–45.

Leigh, J. H., Wood, D. H., Slee, A. V., and Holgate, M. D. (1991). The effects of burning and grazing on productivity, forage quality, mortality and flowering of eight subalpine herbs in Kosciusko National Park. *Australian Journal of Botany* 39, 97–118.

Lunt, I. D. (1991). Management of remnant lowland

grasslands and grassy woodlands for nature conservation: a review. *Victorian Naturalist* **108**, 57–66.

Lunt, I. D. (1995). European management of remnant grassy forests and woodlands in south-eastern Australia: past, present and future? *Victorian Naturalist* **112**, 239–249.

Lunt, I. D. (1998a). *Allocasuarina* (Casuarinaceae) invasion of unburnt coastal woodland at Ocean Grove, Victoria: structural changes 1971–1996. *Australian Journal of Botany* **46**, 649–656.

Lunt, I. D. (1998b). Two hundred years of land use and vegetation change in a remnant coastal woodland in southern Australia. *Australian Journal of Botany* **46**, 629–647.

Lunt, I. D., and Bennett, A. F. (2000). Temperate woodlands in Victoria: distribution, composition and conservation. In *Temperate Eucalypt Woodlands in Australia: Biology, Conservation, Management and Restoration* (eds. R. J. Hobbs and C. J. Yates) pp. 17–31. (Surrey Beatty: Chipping Norton, NSW.)

Lunt, I. D., and Morgan, J. W. (2001). The role of fire regimes in temperate lowland grasslands of south-eastern Australia. In *Flammable Australia: The Fire Regimes and Biodiversity of a Continent* (eds. R. A. Bradstock, J. E. Williams and A. M. Gill) pp. 177–196. (Cambridge University Press: Cambridge.)

McIntyre, S. (1994). Integrating agricultural land-use and management for conservation of a native grassland flora in a variegated landscape. *Pacific Conservation Biology* **1**, 236–244.

McIntyre, S., and Lavorel, S. (1994a). How environmental and disturbance factors influence species composition in temperate Australian grasslands. *Journal of Vegetation Science* **5**, 373–384.

McIntyre, S., and Lavorel, S. (1994b). Predicting richness of native, rare, and exotic plants in response to habitat and disturbance variables across a variegated landscape. *Conservation Biology* **8**, 521–531.

McIntyre, S., Lavorel, S., and Trémont, M. (1995). Plant life-history attributes: their relationship to disturbance response in herbaceous vegetation. *Journal of Ecology* **83**, 45–54.

Mooney, H. A., and Hobbs, R. J. (1986). Resilience at the level of the individual plant. In *Resilience in Mediterranean Ecosystems* (eds. B. Dell, A. J. M. Hopkins and B. B. Lamont) pp. 65–82. (Dr W. Junk: The Hague.)

Moore, R. M. (1970). South-eastern temperate woodlands and grasslands. In *Australian Grasslands* (ed. R. M. Moore) pp. 169–190. (Australian National University Press: Canberra.)

Muir, B. G. (1985). Fire exclusion: a baseline for change? In *Fire Ecology and Management in Western Australian Ecosystems* (ed. J. W. Ford) pp. 119–128. (WAIT Environmental Studies Group: Bentley, WA.)

Noble, I. R., and Slatyer, R. O. (1977). Post-fire succession of plants in Mediterranean ecosystems. In *Symposium on the Environmental Consequences of Fire and Fuel Management in Mediterranean Ecosystems* (eds. H. A. Mooney and C. E. Conrad) pp. 27–35. (U.S. Department of Agriculture Forest Service: Palo Alto, CA.)

Noble, I. R., and Slatyer, R. O. (1980). The use of vital attributes to predict successional changes in plant communities subject to recurrent disturbances. *Vegetatio* **43**, 5–21.

Park, H. C., Majer, J. D., Hobbs, R. J., and Bae, T. U. (1993). Harvesting rate of the termite *Drepanotermes tamminensis* (Hill) within native woodland and shrubland of the Western Australian wheatbelt. *Ecological Research* **8**, 269–275.

Pen, L. J., and Green, J. W. (1983). Botanical exploration and vegetation changes on Rottnest Island. *Journal of the Royal Society of Western Australia* **66**, 20–24.

Petit, N. E., Froend, R. H., and Ladd, P. G. (1995). Grazing in remnant woodland vegetation: changes in species composition and life form groups. *Journal of Vegetation Science* **6**, 121–130.

Prober, S. M. (1996). Conservation of grassy white box woodlands: rangewide floristic variation and implications for reserve design. *Australian Journal of Botany* **44**, 57–77.

Prober, S. M., and Thiele, K. R. (1993). The ecology and genetics of remnant grassy white box woodlands in relation to their conservation. *Victorian Naturalist* **110**, 30–36.

Prober, S. M., and Theile, K. R. (1995). Conservation of the grassy white box woodlands: relative contributions of size and disturbance to floristic composition and diversity of remnants. *Australian Journal of Botany* **43**, 349–366.

Purdie, R. W., and Slatyer, R. O. (1976). Vegetation succession after fire in sclerophyll woodland communities in south-eastern Australia. *Australian Journal of Ecology* **1**, 223–236.

Pyne, S. J. (1991). *Burning Bush: A Fire History of Australia.* (Henry Holt: New York.)

Rippey, E., and Rowland, B. (1995). *Plants of the Perth Coast and Islands.* (University of Western Australia Press: Nedlands.)

Robinson, D., and Traill, B. J. (1996). Conserving woodland birds in the wheat and sheep belts of southern Australia. *Wingspan (Supplement)* **6**, 1–16.

Ryan, D. G., Ryan, J. E., and Starr, B. J. (1995). *The Australian Landscape: Observations of Explorers and Early Settlers.* (Murrumbidgee Catchment Management Committee: Wagga Wagga, NSW.)

Saunders, D. A., and Ingram, J. (1995). *Birds of Southwestern Australia: An Atlas of Changes in the Distribution and Abundance of the Wheatbelt Fauna.* (Surrey Beatty: Chipping Norton, NSW.)

Scougall, A., and Majer, J. D. (1991). The impact of fencing on soil, flora and soil fauna in remnants of open Jam-York Gum woodlands in the Western Australian wheatbelt. In *Soil Science and the Environment*, pp. 11–30. (Australian Society of Soil Science: Albany, WA.)

Scougall, S. A., Majer, J. D., and Hobbs, R. J. (1993). Edge effects in grazed and ungrazed Western Australian wheatbelt remnants in relation to ecosystem reconstruction. In *Nature Conservation*, vol. 3, *Reconstruction of Fragmented Ecosystems, Global and Regional Perspectives* (eds. D. A. Saunders, R. J. Hobbs and P. R. Ehrlich) pp. 163–178. (Surrey Beatty: Chipping Norton, NSW.)

Singh, G. (1982) Environmental upheaval: the vegetation of Australasia during the Quaternary. In *A History of Australasian Vegetation* (ed. J. M. B. Smith) pp. 90–108. (McGraw-Hill: Sydney.)

Singh, G., Kershaw, A. P., and Clark, R. (1981). Quaternary vegetation and fire history in Australia. In *Fire and the Australian Biota* (eds. A. M. Gill, R. Groves and I. R. Noble) pp. 23–54. (Australian Academy of Science: Canberra.)

Specht, R. L. (1970). Vegetation. In *The Australian Environment* (ed. G. W. Leeper) pp. 44–67. (CSIRO and Melbourne University Press: Melbourne.)

Trémont, R. M., and McIntyre, S. (1994). Natural grassy vegetation and native forbs in temperate Australia: structure, dynamics and life histories. *Australian Journal of Botany* **42**, 641–658.

Walker, J. (1981). Fuel dynamics in Australian vegetation. In *Fire and the Australian Biota* (eds. A. M. Gill, R. H. Groves and I. R. Noble) pp. 101–127. (Australian Academy of Science: Canberra.)

Whelan, R. J., and Main, A. R. (1979). Insect grazing and post-fire plant succession in south-west Australian woodland. *Australian Journal of Ecology* **4**, 387–398.

Whelan, R. J., and Tait, I. (1995). Responses of plant populations to fire: fire season as an under-studied element of fire regime. *CALMScience Supplement* **4**, 147–150.

Williams, J. E., and Gill, A. M. (1995). *The Impact of Fire Regimes on Native Forests in Eastern New South Wales.* (New South Wales National Parks and Wildlife Service: Hurtsville, NSW.)

Windsor, D. M. (1998). A landscape approach to optimise recruitment of woodland species in an intensive agricultural environment in the Central Tablelands of NSW. PhD thesis, Charles Sturt University, Bathurst, NSW.

Windsor, D. M. (2000). A review of factors affecting regeneration of box woodlands in the Central Tablelands of New South Wales. In *Temperate Eucalypt Woodlands in Australia: Biology, Conservation, Management and Restoration* (eds. R. J. Hobbs and C. J. Yates) pp. 271–285. (Surrey Beatty: Chipping Norton, NSW.)

Withers, J. R. (1978a). Studies on the status of unburnt *Eucalyptus* woodland at Ocean Grove, Victoria. II. The differential seedling establishment of *Eucalyptus ovata* Labill. and *Casuarina littoralis* Salisb. *Australian Journal of Botany* **26**, 465–483.

Withers, J. R. (1978b). Studies on the status of unburnt *Eucalyptus* woodland at Ocean Grove, Victoria. III. Comparative water relations of the major tree species. *Australian Journal of Botany* **26**, 819–835.

Withers, J. R., and Ashton, D. H. (1977). Studies on the status of unburnt Eucalyptus woodland at Ocean Grove, Victoria. I. Structure and regeneration. *Australian Journal of Botany* **25**, 623–637.

Wycherley, P. (1984) People, fire and weeds: can the vicious spiral be broken? In *The Management of Small Bush Areas in the Perth Metropolitan Region* (ed. S. A. Moore) pp. 11–17. (Department of Fisheries and Wildlife: Perth.)

Wylie, F. R., and Landsberg, J. (1987) The impact of tree decline on remnant woodlots on farms. In *Nature Conservation: the Role of Remnants of Native Vegetation* (eds.

D. A. Saunders, G. W. Arnold, A. A. Burbidge and A. J. M. Hopkins) pp. 331–332. (Surrey Beatty: Chipping Norton, NSW.)

Yates, C. J. (1995). Factors limiting the recruitment of *Eucalyptus salmonophloia* in remnant woodlands. II. Post-dispersal seed predation and soil seed reserves. *Australian Journal of Botany* **43**, 145–155.

Yates, C. J., and Hobbs, R. J. (1997). Temperate eucalypt woodlands: a review of their status, processes threatening their persistence and techniques for restoration. *Australian Journal of Botany* **45**, 949–973.

Yates, C. J., Hobbs, R. J., and Bell, R. W. (1994a). Factors limiting the recruitment of *Eucalyptus salmonophloia* in remnant woodlands. I. Pattern of flowering, seed production and seed fall. *Australian Journal of Botany* **42**, 531–542.

Yates, C. J., Hobbs, R. J., and Bell, R. W. (1994b). Landscape-scale disturbances and regeneration of semi-arid woodlands of south-western Australia. *Pacific Conservation Biology* **1**, 214–221.

Yates, C. J., Hobbs, R. J., and Bell, R. W. (1996). Factors limiting the recruitment of *Eucalyptus salmonophloia* in remnant woodlands. III. Conditions necessary for seed germination. *Australian Journal of Botany* **44**, 283–296.

Yates, C. J., Hobbs, R. J., and True, D. T. (2000a). The distribution and status of eucalypt woodlands in Western Australia. In *Temperate Eucalypt Woodlands in Australia: Biology, Conservation, Management and Restoration* (eds. R. J. Hobbs and C. J. Yates) pp. 86–106. (Surrey Beatty: Chipping Norton, NSW.)

Yates, C. J., Norton, D. A., and Hobbs, R. J. (2000b). Grazing effects on soil and microclimate in fragmented woodlands in south western Australia: implications for restoration. *Australian Journal of Ecology* **25**, 36–47.

Part VI

Ecosystems: forests

14

Fire regimes and fire management of rainforest communities across northern Australia

JEREMY RUSSELL-SMITH AND PETER STANTON

Abstract

This chapter reviews information concerning the effects of fire regimes on, and conservative fire management of, the diverse array of rainforest structural and floristic assemblages which generally lie north of 20° S, from south of Townsville in north Queensland to around Broome in Western Australia. The palaeoecological record indicates that major decline of rainforests across northern Australia, and their replacement by savanna formations, occurred in the Late Tertiary under increasingly seasonally arid conditions. Late Quaternary palaeoecological data, particularly from northeastern Queensland, illustrate landscape-scale fluctuations in the relative distributions of rainforest and savanna vegetation types; rainforest vegetation being more extensive at times of higher rainfall (such as the present) under interglacial, high sea-level conditions. Palynological and fossil charcoal studies, undertaken at a range of extant rainforest sites in the wet tropics belt from Ingham to Cooktown, demonstrate that savanna woodland vegetation was regionally more extensive, and frequently burnt, from the height of the last Ice Age up until the Early Holocene; presumably at least some of this burning was undertaken by people. Today, rainforests are continuing to expand at the expense of wet sclerophyll forest and woodland savanna communities in many non-agricultural landscapes of northeastern Queensland, under conditions of low fire intensity and frequency. Conversely, elsewhere across the more seasonal, monsoonal landscapes of northern Australia, frequent, relatively intense late dry-season fires are causing significant damage to the margins of typically small monsoon rainforest patches. While many rainforest tree and shrub species possess the capacity to resprout from at least basal shoots after aerial stems have been damaged by fire or other disturbance (e.g. cyclones, logging), effective rehabilitation of disturbed, fire-prone rainforest margins (including exclusion of flammable grasses/weeds) requires the regrowth of closed sub-canopy/canopy conditions. This may take as little as 3–5 years on moist, fertile sites to decades on seasonally xeric substrates. Conservative fire management of rainforests and other relatively 'fire-sensitive' assemblages across northern Australia is still a long way from being realised given societal attitudes and practical considerations.

Introduction

In the two decades following the 1978 conference which gave birth to *Fire and the Australian Biota* (Gill *et al.* 1981), much has been learned about the rainforests of northern Australia, including the significance of fire. For much of the region relatively detailed mapping of rainforest distribution has been undertaken, and regional flora and fauna inventories are now available. A nation-wide rainforest floristic framework has been developed (Webb *et al.* 1984), and this scheme has been applied, more or less successfully, to a number of comprehensive regional surveys. Considerable research has been undertaken on regeneration pathways in north Australian rainforest vegetation, the dynamics of patch boundaries, fuel loads, and the responses of plant species to fire regimes. Valuable, if limited, descriptions of traditional/contemporary Aboriginal applications of fire for management of rainforest resources have become available. Various comprehensive regional rainforest conservation assessments have been undertaken, notably various papers included in the landmark *Australian National Rainforests Study* (Werren and Kershaw 1987–1991).

Powerful insights into the longer-term vegetation history of northern rainforests have also been developed. Building on the earlier palynological studies of Late Pleistocene vegetation history from the Atherton Tableland, in perhumid northeast Queensland (Kershaw 1970, 1971, 1974, 1975, 1976, 1978), more recent research has substantially extended that record (e.g. Kershaw and Sluiter 1982; Kershaw 1983, 1985, 1986, 1994). Relatively fine-scale data are even available for Early Holocene fire regimes (Chen 1986). Expanded regional perspectives of long-term vegetation change and dynamics in northeast Queensland have been provided through palynological investigation of a 1.5 Ma deep sea core (Kershaw *et al.* 1993), and notably through Late Pleistocene–Holocene soil fossil charcoal studies (Hopkins *et al.* 1990a, 1993, 1996). The reporting of mega-rich, rainforest fossil fauna assemblages from now semi-arid Riversleigh, northwestern Queensland, has afforded an exceptional view of the role of developing Late Tertiary seasonality and aridity on the collapse of formerly more widespread complex rainforest communities (Archer *et al.* 1989, 1991, 1994). With all too few exceptions, however, relevant fossil assemblages for other parts of monsoonal northern Australia continue to be elusive.

The above synopsis illustrates the generally broad body of research information which now is available to assist with developing appropriate fire management regimes for north Australian rainforest assemblages. It is fair to say that available knowledge, if used, is sufficient to either conserve or promote boundary assemblages (and thereby maintain the integrity of patches), or rapidly degrade or destroy at least small rainforest patches. However, over much of northern Australia conservative fire management of rainforest and associated ecotonal communities (e.g. tall wet sclerophyll forests of northeast Queensland) remains a problematic, little-applied, ideal.

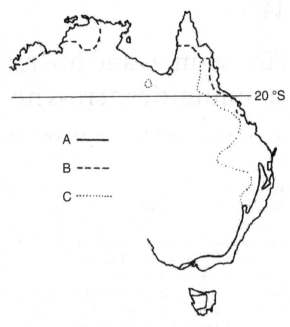

Fig. 14.1. Generalised distribution of rainforest across Australia, showing ecofloristic regions as defined by Webb *et al.* (1984). North Australian rainforests are contained within (B) northern forests and (C) mid east-coast and subcoastal forests. (From Webb, L. J., Tracey, J. G., and Williams, W.T. (1984) A floristic framework of Australian rainforest, *Australian Journal of Ecology* 9, 169–198, Figure 3a; reproduced with permission.)

In this paper our focus is on the effects of fire regimes on, and conservative fire management of, the diverse array of rainforest structural and floristic assemblages which generally lie north of 20° S, from south of Townsville in north Queensland to around Broome in Western Australia (WA) (Fig. 14.1). We include as well the narrow band of tall *Eucalyptus*-dominated wet sclerophyll forest (tall open forest) associated particularly with the western margin of major rainforest blocs in northeast Queensland, but omit discussion of inland scrubs dominated by *Acacia* (see Hodgkinson, this volume); many of the latter include significant rainforest species components. We first describe the broad distribution of rainforest types, and then consider relevant implications from available long-term palaeogeographic and vegetation records. The remainder of the chapter addresses current knowledge concerning the ecological effects of fire regimes on north Australian rainforests and associated vegetation types, particularly in relation to the growth rates of component species under different climatic and edaphic conditions. Management implications are considered in the final section. Further details and references concerning major rainforest ecology and palaeovegetation themes explored here are given in Adam (1992), Hill (1994), White (1994) and Bowman (1999).

North Australian rainforest

Rainforest assemblages across northern Australia describe a range of structural and floristic types. Following the structural terminology of Webb (1959, 1968, 1978) and Webb and Tracey (1981a, b), these include structurally complex vine-forests developed on relatively nutrient rich, moist but well-drained soils developed *in situ* on basalt and colluvium/alluvium, with mostly mesophyll-sized (Complex Mesophyll Vine-forest – CMVF) or, at higher elevations, notophyll-sized (CNVF) canopy leaves, through structurally simple evergreen types on oligotrophic, moist soils (e.g. SNEVF), to seasonally semi-deciduous and deciduous types, often of low canopy height (3–20+ m), under strongly seasonal, monsoonal rainfall conditions (e.g. Semi-Deciduous Notophyll Vine-forest – SDNVF; Deciduous Vine Thicket – DVT). While above authors note that rainforest and sclerophyll vegetation (i.e. dominated by *Acacia*, *Eucalyptus*, *Melaleuca* etc.) types often form successional admixtures, their structural schema strictly excludes such assemblages save allowance for sclerophyll components as emergents in the final stages of development to rainforest.

A floristic framework of Australian rainforests was presented by Webb *et al.* (1984), based on numerical classification of an Australia-wide dataset comprising 561 sites × 1316 tree species. North Australian rainforests are contained within two of three defined 'ecofloristic regions', B – the northern forests, and C – the mid east-coast and subcoastal forests (Fig. 14. 1). Ecofloristic region B comprises three subgroups (or 'ecofloristic provinces') as follows: B1 – disjunct tracts, but more typically small patches, of semi-deciduous and evergreen vine-forests across higher-rainfall regions of monsoonal northern Australia; B2 – 'the diverse and relatively large rainforest massif of the tropical humid lowlands and uplands between Ingham and Cooktown', northeast Queensland; and B3 – a drier version of B1, comprising patches of deciduous vine thicket, with a similar scattered extent across monsoonal northern Australia.

Ecofloristic region C comprises two subgroups: C1 – disjunct patches along the east coast under humid–subhumid (annual rainfall typically between 1000 and 1500 mm), warm subtropical climates, optimally with *Araucaria* emergents; and C2 – scattered patches of semi-evergreen or deciduous vine thicket, typically under more inland, lower-rainfall conditions, minus *Araucaria*. While this floristic framework has found useful application in other north Australian studies, classification of rainforests based on structural terminology (see above) has presented difficulties at least for drier types given considerable overlap/ambiguity with terms such as 'semi-deciduous', 'deciduous' and 'semi-evergreen' (Russell-Smith 1991; Fensham 1995).

The term 'rainforest' as used here thus describes a variety of types, ranging from those occurring under: (1) more-or-less perhumid megathermal or, at increasingly higher elevations, mesothermal and microthermal bioclimates – often referred to as tropical, subtropical and temperate rainforest, respectively; and (2) those occurring under mega- to mesothermal, strongly seasonal rainfall conditions – referred to variously as monsoon forest, monsoon vine-forest, monsoon rainforest or dry rainforest. Of note, however, is the potentially ambiguous use of the term 'monsoon forest' (*sensu* Schimper 1903) given that, in Malesian studies, this term denotes a range of woody vegetation types, including those dominated by *Eucalyptus*, under monsoonal rainfall regimes (e.g. van Steenis 1957; Whitmore 1984 a, b). Differing definitions also are associated with the use of the term 'dry rainforest' (cf. Gillison 1987; Fensham 1995). In this paper the terms rainforest or monsoon rainforest are used throughout.

Available (but incomplete) data indicate that north Australian rainforests cover collectively a

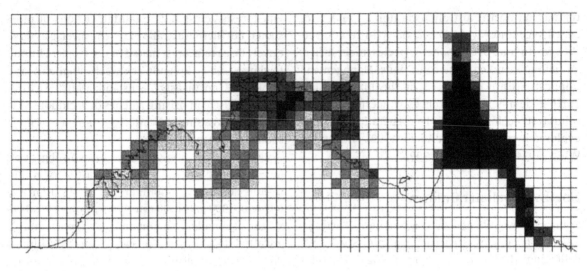

Legend

☐ No Rainforest
▨ 0 to 1 km²
▨ 1 - 10 km²
▧ 10 - 100 km²
■ >100 km²

Fig. 14.2. Distribution of rainforest in northern Australia per 30′ × 30′ cells. Distribution of rainforest in Western Australia after Kimber *et al.* (1991), Northern Territory after J. Russell-Smith and D. Lucas (unpublished data), Wet Tropics region of northeast Queensland after Tracey and Webb (1975) and Cape York Peninsula after Neldner and Clarkson (1995). Note missing map data for much of subcoastal (but see Fensham 1996) and western Queensland.

planar area somewhat greater than 18 000 km², or *c.* 0.01% of continental Australia north of 20° S (Fig. 14.2). For north Queensland this total comprises an estimate of 7910 km² for the major rainforest massif between Townsville and Cooktown (Tracey and Webb 1975), 6961 km² for Cape York Peninsula north of 16° S (Neldner and Clarkson 1995) excluding the Wet Tropics area already mapped by Tracey and Webb (1975), and 829 km² for subcoastal rainforest patches in northeast Queensland (Fensham 1996). Data for other north Queensland areas are not available. Based on comprehensive rainforest mapping undertaken in Western Australia (Kimber *et al.* 1991) and the Northern Territory (J. Russell-Smith and D. Lucas unpublished data), the combined total area

for northwest and northern Australia amounts to 2750 km², comprising over 1500 patches in WA and 15000 patches in the Northern Territory (NT) (Russell-Smith *et al.* 1992). Of that combined total area 41% comprises rainforest dominated by *Allosyncarpia ternata* (Myrtaceae), a tree species restricted to the rugged western Arnhem Land region of the NT (Russell-Smith *et al.* 1993).

Decline of the northern rainforests

Late Tertiary

Major decline of a former more extensive cover of rainforest across Australia occurred from the Mid-to Late Miocene (*c.* 15–10 Ma ago), associated with developing rainfall seasonality and aridity (Truswell 1990, 1993; Kershaw *et al.* 1994; Martin 1994). That decline may have commenced earlier in northwestern Australia given the absence of strict rainforest pollen types from an Early–Mid Miocene deep sea core regional fossil assemblage (McMinn and Martin 1992). Non-rainforest vegetation (comprising leaf fossils of probable Cupressaceae, Proteaceae and *Melaleuca*), of uncertain but likely Mid–Late Tertiary provenance, has also been reported from Melville Island, NT (Pole and Bowman 1996).

Certainly, by the (possibly Early) Pliocene (*c.* 5 Ma), the former species-rich rainforests of Riversleigh in now semi-arid northwestern Queensland had given way to open sclerophyll vegetation, possibly with a grassy understorey (Archer *et al.* 1989, 1991). In central Australia, while rainforest vegetation persisted in at least some locations into the Pliocene (Truswell and Harris 1982), records of sclerophyll vegetation are available from as early as the Late Miocene–Pliocene (Martin 1990). Dramatic increase in fossil charcoal deposition is recorded first from one Late Miocene wet sclerophyll site, from central western New South Wales (Martin 1991). The Pliocene saw also the early radiation of grazing marsupials (Archer *et al.* 1994), although fossil pollen evidence suggests that the widespread development of savanna formations, including those in northwestern Australia (McMinn and Martin 1992), may not have occurred until the Late Pliocene (Martin 1994).

By the start of the Quaternary (*c.* 2 Ma – present) fire was thus already an element fashioning northern Australia vegetation (Martin 1994). Available but limited data indicate that savannas were already extensive in some areas of northern Australia by the end of the Tertiary (Archer *et al.* 1989, 1991; McMinn and Martin 1992). Contrary to widely espoused, current opinion (e.g. Stocker and Mott 1981; Kershaw 1985; Flannery 1994), therefore, 'Aborigines could not have caused the widespread decline of rainforest by burning . . . The decline in rainforest was caused by climatic change.' (Martin 1994, p. 127)

The Quaternary

The Quaternary describes the modern epoch, an era of marked, cyclic climatic change associated with, at one extreme of each cycle, growth of polar ice sheets and concomitant lowering of sea levels by as much as 150 m, and alternating relatively warm, humid, high sea-level interglacials, at periodicities of *c.* 100 ka (e.g. Chappell and Shackleton 1986; Hope 1994). The major development of dunefields in central Australia over the past 500 ka implies amplification of the Late Tertiary trend for developing aridity (Wasson 1984, 1986; White 1994). Relevant palaeorecords for northern Australia relate mostly to perhumid northeast Queensland.

Fossil pollen data from Lynch's Crater on the Atherton Tableland (Kershaw 1974, 1976, 1978, 1985, 1986) provide a relatively detailed history of the flux of rainforest and sclerophyll vegetation over a period attributed to the past *c.* 200 ka, encompassing the last two glacial–interglacial cycles. Data indicate that mesic evergreen rainforest occupied the catchment during warmer, wetter interglacial periods, with mostly drier types of rainforest replacing them in lower-rainfall, cooler phases. However, within the radiocarbon-dated part of the sequence (*c.* 40 ka BP), drier rainforest pollen types (e.g. *Araucaria*) were replaced gradually by sclerophyll types (*Casuarina*, *Eucalyptus*), with the transition to sclerophyll vegetation near-complete by *c.* 26 ka BP. At Lynch's Crater these sclerophyll types remain dominant until c. 10 ka BP, around the start of the Holocene or Recent era, when evergreen rainforest vegetation again resumes dominance. The Early–Mid Holocene expansion of evergreen rainforest, and concomitant replacement of sclerophyll vegetation, is documented also for four other fossil pollen sites on the Atherton Tableland (Kershaw 1970, 1971, 1975; Chen 1986, 1988; Walker and Chen 1987).

Charcoal particles occur throughout the Lynch's Crater sedimentary sequence, but are concentrated particularly in upper sections dated from *c.* 38 ka BP (e.g. Singh *et al.* 1981; Kershaw 1985, 1986). The ostensibly gradual replacement of drier rainforest types by sclerophyll vegetation over a *c.* 12 ka period, and maintenance of sclerophyll vegetation thereafter until the Holocene, is attributed by above authors to frequent fires lit by Aboriginal people. While the marked increase in charcoal production clearly indicates localised fires in the Lynch's Crater catchment, associated fire regimes are less certain. Previously Kershaw noted that sediment charcoal concentrations could be exaggerated since it 'coincides with a change to a drier swamp surface which could carry fire' under diminishing rainfall (Kershaw 1985, p. 186), and/or presumably increased rainfall seasonality, conditions. Elsewhere it has been suggested (e.g. Singh et al. 1981; Clark 1983) that frequent, low-intensity fires produce relatively little charcoal, whereas high charcoal concentrations and associated large fluctuations in sediments (as in the Lynch's Crater record) may be produced by intermittent, high-intensity fires.

Fine-resolution analysis of an Early Holocene sedimentary core from Lake Barrine on the Atherton Tableland provides a relatively detailed assessment of fire regimes, from before rainforest reinvasion until rainforest establishment (Chen 1986). Data indicate that, before the appearance of any rainforest pollen, fires in grassy woodland occurred 'certainly greater than once in 50 years, probably once in ten years or even higher'. From the time of first appearance of rainforest pollen at 9.3 ka BP, until significant pollen concentrations of rainforest pioneer species appear c. 6.8 ka BP, fires apparently occurred at a mean of 220–240 year intervals. Thereafter, fires were absent from the site associated with the development of rainforest.

At a more regional scale, but encompassing the upland catchments of rivers draining the Atherton Tablelands, fossil pollen studies from a sea core on the continental slope, c. 60 km off the north Queensland coast, afford a record attributed to the past 1.5 Ma (Kershaw et al. 1993). Rainforest pollen types are represented notably by gymnosperms throughout, with marked decline of Araucariaceae in the upper part of the sequence, from c. 150–100 ka BP. Sclerophyll taxa, including grass and Casuarina pollen types, occur throughout the core, as does charcoal with peaks both at the bottom and top of the sequence. The authors attribute the decline in

Araucariaceae, and the most recent peaks in charcoal, to burning by Aboriginal people.

The former wider regional extent of sclerophyll vegetation in the period prior to recent rainforest expansion in the Holocene is demonstrated graphically by studies undertaken on the distribution of fossil charcoal in soils underlying contemporary major rainforest massifs throughout the Ingham to Cooktown area (Hopkins et al. 1990a, 1993, 1996). Eucalyptus charcoal embedded in soils underlying rainforest in 17 upland sites, and from the lowland Daintree area, provide radiocarbon dates with ages ranging from 27 to 1.3 ka BP, with most in the period 13–8 ka BP. These data thus support and significantly expand the Late Pleistocene palaeovegetation perspective provided by palynological studies from the Atherton Tableland.

Relevant palaeorecords for the remainder of monsoonal northern Australia are scarce. Nevertheless, on the basis of studies from the Carpentaria Basin (Torgersen et al. 1988), the extension of mobile dunefields into the Kimberley and Gulf of Carpentaria under significantly stronger southeasterly winds (Jennings 1975; Wasson 1984, 1986), and marine cores from the Timor Sea (van der Kaars 1989), it is evident that, as for much of Australia in the Late Pleistocene, regional climates were semi-arid to arid (Hope 1994). Climate models presented by Nix and Kalma (1972) and Webster and Streten (1978) suggest that rainfall across northern Australia during this period of lowered sea-level may have been reduced by as much as half present values. Climate amelioration from the Early Holocene is evident from shallow lake sediments on Groote Eylandt (Shulmeister 1992), with possible higher rainfall conditions than present in the Mid Holocene (Nix and Kalma 1972; Jennings 1975; Shulmeister 1992).

Direct evidence concerning the distribution of rainforest across monsoonal Australia in the Late Pleistocene and Holocene is scarce. Certainly, during periods of major aridity mesic rainforest species and forest types would have been reduced to small patches associated with perennial springs and seeps. Rainforest types developed on seasonally dry substrates would have been particularly vulnerable to fire. In northwest Arnhem Land evergreen rainforest dominated by the hardy myrtaceous sclerophyll *Allosyncarpia ternata* presumably persisted as extensive tracts in fire-protected gorges and open sand sheets much as it does in the present day, given very limited vagility in this species (Russell-Smith *et al.* 1993). The distribution of former orange-footed scrub-fowl (*Megapodius reinwardt*) and/or scrub-turkey (*Alectura lathami*) incubation mounds in sclerophyll vegetation indicates greater, if localised rainforest extent in parts of Cape York Peninsula (Lavarack and Godwin 1987), and coastal areas of the Northern Territory (Bowman *et al.* 1994). Such mounds have been radiocarbon-dated as being as much as 8 ka old (Stocker 1971a). Conversely, much extant rainforest across monsoonal northern Australia is of Recent origin, given its occurrence on Holocene landforms such as coastal riverine floodplains and dunes (Russell-Smith and Dunlop 1987; Bowman *et al.* 1991; McKenzie *et al.* 1991; Russell-Smith and Lee 1992; Fensham 1993), and young basalt lava flows (Fensham 1995).

Summary

Major contraction of rainforest across northern Australia occurred in the Late Tertiary associated with developing rainfall seasonality and aridity. That trend has been amplified in the latter part of the Quaternary, with mesic rainforest remnants in northeastern Queensland expanding in relatively wetter, warmer periods, and contracting at other times. Similar patterns, albeit on a much reduced scale, are likely to have operated elsewhere across northern Australia. While fire doubtless helped fashion regional savanna landscapes throughout the Quaternary, available data from northeast Queensland suggest that fire impact on rainforests was particularly prevalent just prior to the Holocene under relatively low-rainfall, probably amplified seasonal moisture conditions. Burning by Aboriginal people possibly was involved in the replacement of araucarian rainforest with sclerophyll vegetation in the Late Pleistocene, and doubtless maintained sclerophyll vegetation and/or slowed rainforest expansion in the Holocene. As considered further below, these palaeoecological data provide a temporal context, albeit at scales of hundreds and thousands of years, relevant for development and implementation of conservative, dynamic management regimes.

Fire regimes and north Australian rainforests

Fires affecting the margins of north Australian rainforests typically are borne in grassy understoreys. Fuel loads are generally of the order of 2–10 t ha^{-1}, depending on component species, site conditions and number of years since last burnt (Walker 1981; R. J. Williams *et al.*, this volume). In some circumstances grassy fuel loads on rainforest margins may approach or even exceed 30 t ha^{-1} (dry weight), for example: highly productive floodplain grasslands [e.g. *Imperata*, or introduced para grass (*Urochloa mutica*)]; introduced dry-land perennials such as *Andropogon gayanus* and *Pennisetum polystachyon* (Panton 1993); and long-unburnt spinifex in higher-rainfall areas (Russell-Smith *et al.* 1998). In the Wet Tropics region, fuel loads of introduced grass species may attain dry weight equilibria in as little as 2 years following fire [e.g. molasses grass (*Melinus minutiflora*) *c.* 18 t ha^{-1}; guinea grass (*Panicum maximum*) *c.* 12 t ha^{-1}: Wallmer, in Ritchie 1996]. High fuel loads may develop also at patch margins given invasion of flammable weedy shrubs, for example *Lantana camara* in drier scrubs in northeast Queensland (Fensham *et al.* 1994).

Intense savanna fires (>5000 kW m^{-1}) may develop from fully cured grassy fuel loads of as little as 4 t ha^{-1} (Williams *et al.* 1998), under late dry-season climatic conditions (e.g. low humidity, high temperature, high afternoon wind speeds: Gill *et al.* 1996). Low-intensity fires tend to be readily extinguished on shrubby marginal vegetation (Unwin *et al.* 1985), or on moist, compacted litter under rainforest canopies (Bowman and Wilson 1988). Over much of monsoonal northern Australia the major dry-season fire period occurs between May and December, and grass productivity is such that fires can be borne on an annual – biennial basis (Stocker and Mott 1981; Walker 1981; R. J. Williams *et al.*, this volume). In wetter areas, particularly northeastern

Queensland, the fire season may be truncated to dry spells between September and November.

Rainforests of northeast Queensland

A characteristic feature of rainforest in the wetter parts of northeast Queensland is the often abrupt boundary with eucalypt-dominated savanna. In the region between Ingham and Cooktown, wet sclerophyll forests (WSF, or tall open forest), dominated by tall (>40 m) eucalypts, typically form a discontinuous narrow fringe, mostly much less than 300 m wide (P. Stanton, personal observation); cf. 'usually less than 4 km wide' (Harrington and Sanderson 1994. p. 319) between rainforest and the surrounding expanse of open forest and woodland savanna. WSF and lower-statured eucalypt forests and woodlands also occur in pockets and on exposed ridges within larger rainforest tracts. Fire is essential for maintaining sclerophyll dominance of the canopy layer and, in the absence of fire, such marginal sclerophyll communities are rapidly infiltrated by rainforest species over a 20–30 year period (Ashton 1981; Webb and Tracey 1981a, b; Unwin et al. 1985; Ash 1988; Stocker and Unwin 1989; Unwin 1989; Hopkins et al. 1993; Harrington and Sanderson 1994). With longer-term absence of fire the transition from sclerophyll emergents in WSF or former open forest/woodland vegetation to rainforest may be complete in 100+ years (Unwin 1989; Harrington and Sanderson 1994). Fire impacts are likely to be significantly increased following severe frosts (Duff and Stocker 1989) and cyclonic events (Webb 1958; Unwin et al. 1988a). Succinct diagrammatic models describing interactions between fire regime and rainforest/sclerophyll vegetation dynamics are presented in Hopkins et al. (1993) (Fig. 14.3) and Harrington (1995) (Fig. 14.4).

Few empirical studies have been undertaken on the effects of fire on species responses and vegetation dynamics of north Australian moist evergreen rainforests. Stocker (1981) demonstrated that a wide range of tree species regenerated from stem basal resprouts (coppice), suckers, and from seed, following logging and burning of a NVF rainforest plot. Of 82 tree species observed regenerating after the fire treatment, 74 coppiced from stumps, 10 produced root suckers, and 34 regenerated from seed. Seven of 20 canopy tree species recorded before logging did

not regenerate. In a study of the effects of a low- to moderate-intensity experimental fire on an abrupt NVF–sclerophyll boundary on the western edge of the Atherton Tableland, Unwin (1983, Table 9.3) found that of seedlings/saplings of 30 rainforest woody species actually burnt (mostly represented by few individuals), all but 6 species regenerated from basal coppice or stem resprouts. Studies of soil seed banks in a range of regional rainforest structural types have shown that these are dominated by secondary (or pioneer) trees and shrubs and, where close to cultural disturbance (e.g. roads, pastures), especially by herbaceous species including exotics (Hopkins and Graham 1983, 1984a, b; Graham and Hopkins 1990; Hopkins et al. 1990b).

A detailed study of vegetation dynamics across two abrupt NVF rainforest–tall WSF (dominated by *Eucalyptus grandis*)–medium canopy height eucalypt open-forest transitions is reported by Unwin (1989). At one site the outer rainforest margin expanded 12 m in the absence of any burning over the 10-year study period. At the other site there was clear evidence from the occurrence of overmature *E. grandis* remnants up to 150 m inside the developing rainforest margin that rainforest expansion had already been under way for some considerable period; *E. grandis*, like other eucalypts, is unable to establish under rainforest canopy, low light conditions (Doley 1979; Duff 1987). At this second site Unwin (1983, 1989) suggested that *E. grandis* was itself expanding into the surrounding eucalypt open-forest ahead of the developing rainforest fringe.[1] Unwin (1983)

[1] One reviewer familiar with this site made the following observation: 'I have inspected this site. The measurements on *E. grandis* were confined to a plot. Outside the plot there are mature *E. grandis* close to the measured regeneration, i.e. the regeneration was within the existing limits of *E. grandis*. This observation has been used to suggest that WSF will advance in front of the rainforest. In my experience it doesn't.' On rare occasions, however, seedlings of *E. grandis* have been observed to establish in burnt areas in surrounding eucalypt forest (G. Duff personal communication); whether they survive to maturity on poor sites is unknown.

The preceding comments illustrate the paucity of information available concerning regeneration requirements of dominant WSF species in this region. For example, Ashton (1981, Table 1) notes simply the presence of lignotubers in *E. cloeziana, E. acmenoides, E. resinifera*, and their absence in *E. grandis, E. torelliana*; and Florence (1996) comments on the relative light tolerance of aforementioned species – *E. grandis* being very intolerant (i.e. does not regenerate in small gaps, and relies on fire). Clearly more research on the regeneration requirements of WSF dominants is needed.

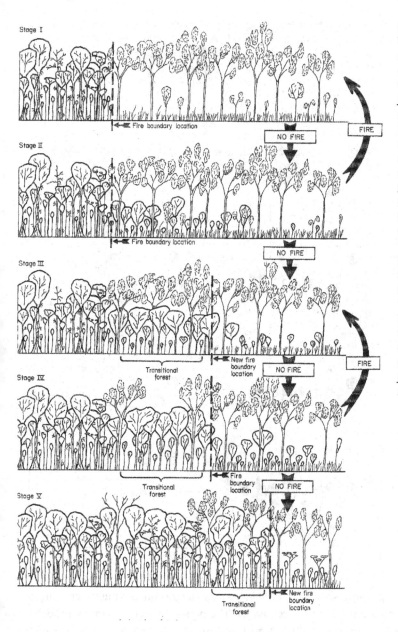

Fig. 14.3. Development of 'transitional' or Wet Sclerophyll Forest margins in northeastern Queensland under influence of fire. From Hopkins *et al.* (1993). The caption to the original figure ran: *The formation of 'transitional' forests as the location of the boundary between rainforest and Eucalyptus forest alters in relation to fire. In the developing 'transitional' forest, rainforest colonizers form a closed canopy under the eucalypts and eliminate the herbaceous groundstorey. Normal intensity fires occurring at Stages II or IV penetrate the developing transitional forest, and the boundary reverts to its earlier positions in Stages I and III. Once the 'transitional' forest is established, the effective location of the fire boundary alters (between Stages II and III, and IV and V). Normal intensity fires do not penetrate the established 'transitional' forest (Stages III and V), the boundary remains unaltered, and the transitional forest continues to develop. Catastrophic fires may penetrate the rainforest in extreme drought periods, particularly following natural physical destruction by storms or cyclones.*

demonstrated that *E. grandis* does not regenerate in moist, grassy open-forest situations in the absence of fire (or presumably other disturbance also).

Ash (1988) undertook a qualitative assessment of the environmental correlates (particularly climate, substrate type, topography, slope) of rainforest boundaries in northeast Queensland, in two large study regions encompassing respectively the Atherton Tablelands and adjoining coastal lowlands, and the inland McBride basalt province to the west of the Atherton Tablelands. Based on recon-

struction of the pre-European rainforest distribution from available 1:100 000 rainforest mapping derived from 1:25 000–1:50 000 aerial photos (Tracey and Webb 1975), historical records etc., and 1:250 000 geological mapping, Ash found that 38% of boundaries at the Atherton study site coincided with major geological boundaries, and at the more inland site rainforest boundaries generally conformed to edges of lava flows. He proposed a model similar to that of Webb (1968) whereby substrate fertility status, or favourability, for rainforest could be

Fig. 14.4. Development of 'transitional' or Wet Sclerophyll Forest margins in northeastern Queensland under influence of fire. From Harrington (1995).

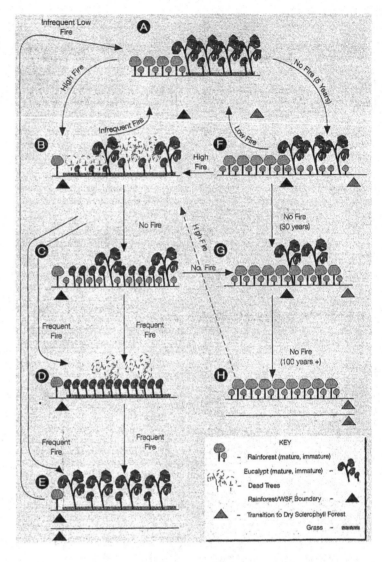

ranked as follows: (5) recent stony basalt >(4) weathered basalt >(3) alluvium >(2) metamorphic sedimentary rocks >(1) granites. Higher nutrient status substrates would thus compensate for otherwise poorer, more fire-prone sites (lower rainfall/substrate moisture conditions; upper slope positions). While noting that some generalisations typically held true (e.g. 'no boundaries were observed in which rainforest was present on the less fertile substrate with pyrophytic vegetation occupying the more fertile substrate'; and 'once slopes exceeded about 20° there was no substrate sufficiently fertile to maintain a rainforest upslope of pyrophytic vegetation'), actual locations of rainforest boundaries

were insufficiently predictable in the absence of detailed site knowledge concerning vegetation and fire history. The paper provides a general model for rainforest expansion as an irregular advance across landscapes: firstly, 'into favourable habitats and then spreading on the flanks to occupy less favourable rainforest sites'; and secondly, over unfavourable habitats (e.g. ridgetops) and substrates through short- or longer-distance seed dispersal.

A number of authors have noted that rainforests are expanding at the expense of sclerophyll vegetation in various parts of northeast Queensland at the present time. This has been attributed principally to reduced fire frequency since Aboriginal occupancy, combined with widespread cattle grazing (Unwin *et al.* 1988b; Stocker and Unwin 1989; Unwin 1989; Harrington and Sanderson 1994). For example, based on a detailed comparison of vegetation change associated with WSF communities in three widely separated areas of northeast Queensland using 1940s and early 1990s aerial photography, Harrington and Sanderson (1994) found that there had been major invasion of former grassy WSF types by developing rainforest understoreys (69% and 57% in two WSF types, respectively), and significant conversion of sclerophyll open-forest to WSF. Extensive recent rainforest expansion has been reported also from south of Cooktown (Unwin *et al.* 1988b), and coastal ranges generally between Cairns and Tully (Hopkins *et al.* 1996).

Further, in the course of current fine-scale mapping of rainforest vegetation throughout the Wet Tropics region, one of the authors (P.S.) notes that rainforest invasion is now well advanced in most sclerophyll communities of the 1800+ mm coastal/sub-coastal rainfall belt, and over large areas has reached the stage where such communities are essentially closed forests with remnant sclerophyll emergents. Aerial photo comparisons show that the process of rainforest invasion has been extremely rapid, with some areas having developed from open grassy woodland to closed forest within 30 years. Such collective observations are at variance, however, with those of Ash (1988) who, in his study of rainforest boundary characteristics (see above), considered 95% of boundaries as being apparently 'stable' (i.e. transitions <20 m wide), and only 5% of mapped boundaries as being 'successional' (i.e. admixtures of rainforest and sclerophyll vegetation 200+ m wide); this interpretation clearly does not apply to the broad region at the present.

Intense fires will burn even in areas of WSF under advanced rainforest shrub invasion, especially where flammable *Eucalyptus grandis* bark streamers are ignited under windy, dry conditions (G. Harrington personal communication). With the formation of a rainforest subcanopy, however, intense defoliating/stem-killing fires will carry only under extreme fire weather conditions, or extreme fire regimes, or after devastating climatic events (e.g.

cyclones, frosts). One of us (P.S.) observed a fire burning under such conditions (gale-force westerly winds, mostly low humidities, high temperatures), up steep mountainsides in December 1995, near Gordonvale. Despite these conditions, the fire penetrated only a narrow margin of forest dominated by *Eucalyptus grandis* over a relatively young rainforest understorey less than 10 m in height; it is axiomatic that maintaining open WSF habitats requires regular burning.

Monsoon rainforests

As for above moist rainforests, the margins between monsoon rainforests and surrounding sclerophyll vegetation types typically are very narrow or abrupt. For example, in a study of 144 rainforest–savanna boundaries in the Northern Territory and Western Australia, Bowman (1992a) found that 18% of rainforests studied on seasonally dry sites ($n=97$) had narrow ecotones (mean width 30 m, range 10–70 m), contrasting with 49% of rainforests at moist sites ($n=47$; mean width 40 m, range 10–120 m). That study illustrates also the more general point that the locations of monsoon rainforest margins and ecotones frequently do not conform to changes in edaphic or substrate conditions; rather, such differences as are observed (e.g. nutrient accumulation in rainforest soil organic matter) are the product of vegetation development itself. General recognition that much monsoon rainforest is developed on harsh, seasonally xeric substrates (Russell-Smith 1986; Unwin and Kriedemann 1990; Fordyce *et al.* 1997a), and on a wide range of nutrient-rich and -poor substrates (Kikkawa *et al.* 1981; Kenneally *et al.* 1991; Russell-Smith 1991; Fensham 1995), has led many regional studies to conclude that fire regimes play a key role in the localised distribution and boundary characteristics of monsoon rainforest (e.g. Stocker 1966; Webb 1968; Kikkawa *et al.* 1981; Langcamp *et al.* 1981; Stocker and Mott 1981; Beard *et al.* 1984; Clayton-Greene and Beard 1985; Russell-Smith and Dunlop 1987; Bowman *et al.* 1990; Bowman 1991a, 1992b; Russell-Smith 1991; Kershaw 1992; Fensham 1995).

Following fire, many monsoon rainforest woody plants coppice or reproduce clonally, and this capacity appears to be particularly well developed on seasonally dry sites. For example, in a study of sapling distribution in 33 monsoon rainforest patches, 29% and 43% of saplings were derived from clonal species in two seasonally dry forest types respectively, whereas 4% and 21% occurred in two moist forest types (Russell-Smith 1996). Bowman (1991b) monitored the response after burning in the late dry season of 32 tree species in seasonally dry rainforest types. Fire-affected stems of all but two species (*Aidia racemosa*, *Ficus racemosa*) resprouted, although three others (*Hibiscus tiliaceus*, *Myristica inspida*, *Terminalia microcarpa*) suffered mortalities of 30% or more. For *Allosyncarpia* forest on seasonally dry sites in western Arnhem Land, data given in Russell-Smith *et al.* (1998) indicate that of 50 sampled rainforest tree and shrub species, all (including *Aidia racemosa*) but three shrub species (*Alyxia ruscifolia*, *Psychotria daphnoides*, *Trema tomentosa*) exhibited either strong coppice or clonal regeneration capabilities. *Allosyncarpia ternata* itself is recognised as a strong resprouter, although stems of all sizes are vulnerable to intense fires. In one patch of *Allosyncarpia* forest subjected to two late dry-season fires in four years, Bowman (1994) recorded an overall stem mortality of 36% ($n = 289$), with 75% stem mortality in the 1–5 cm DBH size class, 14% for all stems >5 cm DBH, and 17% for stems >50–100 cm DBH. From observational and experimental studies, Fordyce *et al.* (1997b) showed that all 18-month *Allosyncarpia* seedlings died following burning, whereas over half of a 3-year-old seedling cohort would probably recover from lignotuber resprouts.

Fire in monsoon rainforest patches associated with peaty soils at moist sites can be catastrophic given that humic layers, once ignited, can consume roots and soil seed banks alike. Observations on the effects of fire on an endangered, endemic palm restricted to moist rainforest sites on clay loams in the Darwin region, *Ptychosperma bleeseri*, indicate that one of two major extant populations of this species lost more than 75% of adult and 50% of juvenile plants in the period 1992–95, as a result of two wildfires in that period (Barrow *et al.* 1993; Liddle *et al.* 1996). In moist rainforests in the Northern Territory resprouting ability following severe fires is

apparently common in canopy species, and lesser numbers of mainly pioneer and secondary species (e.g. *Alphitonia*, *Elaeocarpus*, *Ficus*, *Nauclea orientalis*, *Terminalia microcarpa*, *Timonius timon*) possess significant soil dormant seed banks (Russell-Smith and Lucas 1994). Those authors estimated that some 20% of the Northern Territory rainforest flora ($n = c.$ 640: Liddle *et al.* 1994) possesses dormant diaspores. In time, it is possible that soil seed banks, particularly of drier margins, may become dominated by exotic weeds as shown for the Forty Mile Scrub, south west of the Atherton Tablelands (Hopkins *et al.* 1990b; Fensham *et al.* 1994).

Assessments of contemporary fire impacts on monsoon rainforests in northern and northwestern Australia indicate significant widespread death of rainforest trees, particularly on patch margin situations (McKenzie and Belbin 1991; Russell-Smith and Bowman 1992). In some situations substantial portions, even entire patches, have been razed in the recent past (McKenzie and Belbin 1991; Panton 1993; Lucas *et al.* 1997). Such impacts have been attributed mostly to recurring intense fires under late dry-season conditions, often in combination with feral animal impact (particularly cattle and Asian water-buffalo). Additionally, intense fires in cyclonic debris are likely to have contributed to significant damage to rainforest around Darwin following Cyclone Tracy (Panton 1993; Bowman and Panton 1994). Low to moderately intense fires (say < 2000 kW m^{-1}), even if frequent, are unlikely to cause significant damage given the regenerative capabilities of component, especially marginal/ecotonal species. For example, Bowman (1992b) observed only slight effects on a small patch of SDNVF on the Karslake Peninsula, Melville Island, from annual, mid dry-season, low–moderate-intensity burning undertaken for the purposes of hunting macropods, over a 22-year period.

In the absence or only occasional occurrence of low–moderate-intensity fire, the expansionary response of marginal monsoon rainforest vegetation on seasonally dry substrates is characteristically slow. For example, Bowman and Fensham (1991) found only moderate ecotonal development (range 0–80 m from the patch margin) in one sector of a patch of SDNVF near Weipa, Cape York Peninsula, after 15 years of fire exclusion. Although that site enjoys high annual rainfall (c. 2200 mm), it is highly seasonal with 90% falling between November and April, and the soil is a deep, well-drained red earth. Observations of *Allosyncarpia*-dominated rainforest margins on deep well-draining sands in western Arnhem Land (annual rainfall c. 1300 mm, 90% between November and April), indicate likewise that in situations unburnt for at least 15 years, ecotonal development is negligible. Given lack of evident clonal capacity in *Allosyncarpia* (Russell-Smith 1986), and very restricted vagility and light intolerance of its germinating seedlings (e.g. Bowman 1991a), extension of *Allosyncarpia* patch margins thus occurs by incremental stages, with the rate of development of each stage being dependent on the time taken for plants established at the margin to attain reproductive maturity; this is likely to take decades (Russell-Smith and Dunlop 1987).

Above observations illustrate the generally slow rate of monsoon rainforest seedling establishment and development in savanna habitats on seasonally dry substrates. For example, based on ecological studies from relatively long unburnt plots at Munmarlary, in present-day Kakadu National Park (15 years: Bowman *et al.* 1988), and from the Solar Village, near Darwin (10 years; Fensham 1990), these authors reported no, and low numbers, of rainforest seedlings, after 15 and 10 years fire exclusion, respectively. Ten or so years later, substantial numbers of rainforest shrubs now occur on the woodland (but not open forest) plots at Munmarlary (J. Russell-Smith *et al.*, unpublished data), and large numbers of rainforest species and individuals have established at unburnt rocky sites at the Solar Village (J. Brock, personal communication). Even though absence of nearby rainforest seed sources at both sites doubtless has contributed to slow rates of rainforest species immigration, under contemporary fire regimes 'a succession model describing the broad-scale transformation of eucalypt open-forest to closed monsoon forest is untenable' (Fensham 1990, p. 261). Experimental studies suggest that mycorrhizas, soil fertility and soil moisture are all important factors in regulating monsoon rainforest

seedling establishment and growth (Stocker 1971b; Bowman and Panton 1993).

Conversely, relatively rapid but localised rainforest expansion has been observed on moist sites under low fire pressure. These include significant expansion in: some regions of Cape York Peninsula (Crowley 1995); over the past 50 years of floodplain forests associated with the Reynolds River, south of Darwin (R.E. Petherick personal communication); and nearby rainforests associated with seepage zones from the Tolmer Tableland, Litchfield National Park (D. Bowman and D. Milne unpublished data). Recent expansion of SDNVF rainforest on seasonally moist substrates is also observable in some coastal, higher-rainfall areas where these are expanding into former eucalypt woodland, for example: northwest Melville Island and eastern Elcho Island (J. Russell-Smith, personal observation); and Gunn Point, northeast of Darwin (Bowman and Dunlop 1986).

Synthesis: site conditions, growth rates, fire regimes

Rainforest vegetation types across northern Australia are adapted to a wide range of site conditions, from mesic to xeric, from eutrophic to oligotrophic. Growth rates of constituent rainforest species thus vary from the rapid, associated with year-round available soil moisture, to the relatively very slow, on freely draining substrates under seasonal, low-rainfall conditions. In savanna environments any condition that slows growth rates of woody plants results in vegetation being more fireprone (Kellman 1984). Many north Australian rainforest trees and shrubs (possibly the great majority) possess the capacity to regenerate from rhizomes and/or coppice after death of aerial shoots/stems, following at least occasional intense fires. Data for one monsoon rainforest species (*Allosyncarpia ternata*) indicate that some 3 years (growing seasons) are required following germination before lignotubers are sufficiently well developed to effect successful recovery after fire; presumably this takes longer on harsh sites. Relatively few species, and these mostly pioneers, possess the capacity to regenerate

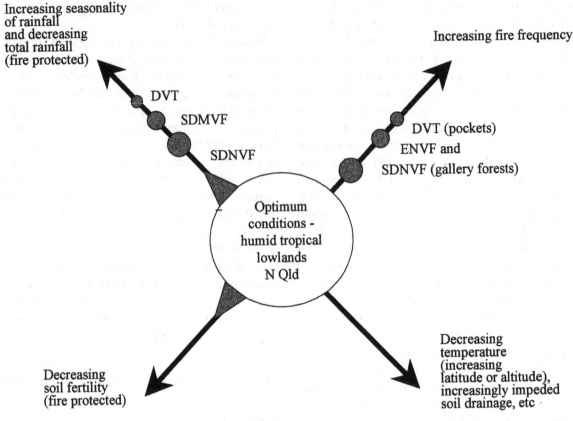

Increasing seasonality of rainfall and decreasing total rainfall (fire protected)

DVT
SDMVF
SDNVF

Increasing fire frequency

DVT (pockets)
ENVF and
SDNVF (gallery forests)

Optimum conditions - humid tropical lowlands N Qld

Decreasing soil fertility (fire protected)

Decreasing temperature (increasing latitude or altitude), increasingly impeded soil drainage, etc ·

Fig. 14.5. Diagrammatic representation of environmental gradients contributing to the biotic and structural diversity of northern Australian rainforest; hatched portions of the gradients represent monsoon rainforest assemblages. From Kikkawa *et al.* (1981).

from dormant soil seed banks, either additionally or solely.

Low- to moderate-intensity savanna fires cause little significant damage to north Australian rainforest margins, even if of frequent/annual occurrence. Indeed, on moist, relatively productive sites where such fires may be patchy, rainforest expansion/development may still occur albeit at reduced rates. Regimes dominated by intense (total scorch, stem death) frequent fires are catastrophic to rainforest margins. When combined with frequent fires of otherwise low to moderate intensities, even occasional intense fires are highly destructive given that they facilitate invasion by fire-carrying grasses and herbs. Hence, the integrity/maintenance of rainforest margins is dependent foremost on the periodic-

ity of intense fires. Following an intense fire, regeneration of closed marginal rainforest sub-canopies may be effected on moist, productive sites in the absence of recurring fires in 3–5 years. On harsh, xeric sites such regeneration takes decades. A generalised framework concerning the role of fire in the ecology of northern Australian rainforests (Kikkawa *et al.* 1981), is given in Fig. 14.5.

Management

It is evident that fire was, and in areas still is, used by Aboriginal people to conservatively manage rainforest resources in different regions of northern Australia, perhaps especially in seasonal environments where rainforest patches (resources) are small and scattered (e.g. Jones 1975, 1980; Mangglamara *et al.* 1991; Lucas *et al.* 1997; Russell-Smith *et al.* 1997; Bridgewater *et al.* 1998; Yibarbuk *et al.* 2001). Such management practice was/is associated both with spiritual/ceremonial responsibilities, as well as resource management requirements. In humid northeastern Queensland it is evident that recent Aboriginal burning practices maintained open walking tracks linking uplands with coastal areas, and open canopied sclerophyll woodland pockets within larger rainforest tracts (e.g. diary of Christie Palmerston, 25 October 1886, as reported in Savage 1989, p. 219). Hill *et al.* (1999) provide a useful description of the cultural usage of fire by Kuku-Yalanji people in the Wet Tropics region. Collectively, the above studies point to the intrinsic ecologic value of traditional knowledge systems, and the need to learn from and accommodate such values in contemporary management contexts.

The pressing challenge for conservative fire management of the monsoon rainforest archipelago over much of northern and northwestern Australia concerns reducing the extent and destructiveness of typically intense, frequent, late dry-season fires (McKenzie and Belbin 1991; Russell-Smith and Bowman 1992; Russell-Smith *et al.* 1992). Fuel management of individual rainforest patch margins may be effected through a number of practices, for example: back-burning from rainforest patch margins as soon as fuels cure; control of vigorous, flammable weeds; and reduction of annual *Sorghum* fuels through wet-season burning (see R.J. Williams *et al.*, this volume). While effective management models exist, for example intensive management as still undertaken on clan estates in parts of Arnhem Land (Jones 1980; Yibarbuk *et al.* 2001), under present-day patterns of human settlement and economy conservative fire management must be addressed at wider landscape scales and broader, inclusive social contexts. In this regard a number of collaborative regional programmes, involving landholders, representative organisations and responsible agencies, have commenced recently in Arnhem Land and the principally pastoral Sturt Plateau and Victoria River District of the Northern Territory, the Kimberley region of Western Australia, and northeast Queensland.

By contrast, in moist areas of northeast Queensland serious consideration needs to be given to regular burning of threatened WSF or grassland habitats for conservative management purposes (Harrington and Sanderson 1994; Harrington 1995). This applies particularly to the case of conserving WSF sclerophyll habitat and dependent fauna [e.g. yellow-bellied glider (*Petaurus australis*); eastern yellow robin (*Eopsaltria australis magnirostris*) (Chapman and Harrington 1998); northern bettong (*Bettongia tropica*) (Laurance 1997)], fringing the wet tropics region between Townsville and Cooktown. As well, in the Iron Range–McIlwraith Range region of eastern Cape York Peninsula, reduced fire frequency in the present day is leading to rainforest expansion onto open grassland habitats associated with a number of lowland river systems, and into small woodland pockets which provide nesting habitat for birds such as Palm Cockatoos, *Probosciger aterrimus* (P. Stanton and D. Storch personal observations). Further, little is known of the dependency of understorey plant species, and faunal groups such as insects, frogs, and reptiles, on WSF habitat conditions (K. McDonald personal communication).

A critical issue thus is to maintain/develop habitat diversity to ensure no loss of species or regional populations; it is not simply a matter of conserving rainforest *per se*. Management practice must be dynamic (Walker 1990), at scales ranging from years and decades for practical management purposes (e.g. weed control, fuel reduction), to hundreds and thousands of years to accommodate for longer-term regional changes associated with climatic fluctuations. Management on such timeframes clearly has its tensions, especially when considering politicised frameworks involving regulatory agencies, agricultural industry (especially cane farming) and local communities (Ritchie and Collins 1994), and indigenous rights and responsibilities (Hill *et al.* 1999). To date, however, there has been a singular reluctance on the part of responsible agencies and research institutions to address fire management and associated monitoring requirements of WSF habitats. We concur with Harrington and Sanderson (1994, p. 326) that these matters require urgent attention, and that 'conservation managers will need fire management strategies which are based on careful research and widespread consultation'. To continue to ignore such management responsibilities in northeast Queensland effectively 'hands the decision to the next yahoo with a cigarette' (Harrington 1995: p. 11).

Acknowledgements

Kris Abbott (Parks and Wildlife Commission of the Northern Territory) is thanked for her excellent library assistance. Lisa Roeger helped prepare the figures. The manuscript benefited from comments provided by Rod Fensham, Graham Harrington, Rosemary Hill, Keith McDonald, Garry Werren, Dick Williams and John Woinarski.

References

Adam, P. (1992). *Australian Rainforests*, Oxford Monographs on Biogeogeography no. 6. (Clarendon Press: Oxford.)

Archer, M., Godthelp, H., Hand, S. J., and Megirian, D. (1989). Fossil mammals of Riversleigh, northwestern Queensland: preliminary overview of biostratigraphy, correlation and environmental change. *Australian Zoologist* 25, 29–65.

Archer, M., Hand, S. J., and Godthelp, H. (1991). *Riversleigh*. (Reed Books: Sydney.)

Archer, M., Hand, S. J., and Godthelp, H. (1994). Patterns in the history of Australia's mammals and inferences about palaeohabitats. In *History of Australian Vegetation: Cretaceous to Recent* (ed. R. S. Hill) pp. 81–103. (Cambridge University Press: Cambridge.)

Ash, J. (1988). The location and stability of rainforest boundaries in north-eastern Queensland, Australia. *Journal of Biogeography* 15, 619–630.

Ashton, D. H. (1981). Tall open-forests. In *Australian Vegetation* (ed. R. H. Groves) pp. 121–151. (Cambridge University Press: Cambridge.)

Barrow, P., Duff, G. A., Liddle, D., and Russell-Smith, J. (1993). Threats to monsoon rainforest habitat in northern Australia: the case of *Ptychosperma bleeseri* Burrett (Arecaceae). *Australian Journal of Ecology* 18, 463–471.

Beard, J. S., Clayton-Greene, K. A., and Kenneally, K. F. (1984). Notes on the vegetation of the Bougainville Peninsula, Osborn and Institut Islands, North Kimberley District, Western Australia. *Vegetatio* 57, 3–13.

Bowman, D. M. J. S. (1991a). Environmental determinants of *Allosyncarpia ternata* forests that are endemic to western Arnhem Land, northern Australia. *Australian Journal of Botany* 39, 575–589.

Bowman, D. M. J. S. (1991b). Recovery of some northern Australian monsoon forest tree species following fire. *Proceedings of the Royal Society of Queensland* 101, 21–25.

Bowman, D. M. J. S. (1992a). Monsoon forests in north-western Australia. II. Forest-savanna transitions. *Australian Journal of Botany* 40, 89–102.

Bowman, D. M. J. S. (1992b). Evidence for gradual retreat of dry monsoon forests under a regime of Aboriginal burning, Karslake Peninsula, Melville Island, northern Australia. *Proceedings of the Royal Society of Queensland* 102, 25–30.

Bowman, D. M. J. S. (1994). Preliminary observations on the mortality of *Allosyncarpia ternata* stems on the Arnhem Land plateau, northern Australia. *Australian Forestry* 57, 62–64.

Bowman, D. M. J. S. (1999). *Australian Rainforests: Islands of Green in the Land of Fire*. (Cambridge University Press: Cambridge.)

Bowman, D. M. J. S., and Dunlop, C. R. (1986). Vegetation pattern and environmental correlates in coastal forests of the Australian monsoon tropics. *Vegetatio* 65, 99–104.

Bowman, D. M. J. S., and Fensham, R. J. (1991). Response of a monsoon forest-savanna boundary to fire protection, Weipa, northern Australia. *Australian Journal of Ecology* 16, 111–118.

Bowman, D. M. J. S., and Panton, W. J. (1993). Factors that control monsoon-rainforest seedling establishment and growth in north Australian *Eucalyptus* savanna. *Journal of Ecology* 81, 297–304.

Bowman, D. M. J. S., and Panton, W. J. (1994). Fire and cyclone damage to woody vegetation on the north coast of the Northern Territory, Australia. *Australian Geographer* 25, 32–35.

Bowman, D. M. J. S., and Wilson, B. A. (1988). Fuel characteristics of coastal monsoon forests, Northern Territory, Australia. *Journal of Biogeography* 15, 807–817.

Bowman, D. M. J. S., Wilson, B. A., and Hooper, R. J. (1988).

Response of *Eucalyptus* forest and woodland to four fire regimes at Munmarlary, Northern Territory, Australia. *Journal of Ecology* **76**, 215–232.

Bowman, D. M. J. S., Wilson, B. A., and Fensham, R. J. (1990) Sandstone vegetation pattern in the Jim Jim Falls region, Northern Territory, Australia. *Australian Journal of Ecology* **15**, 163–174.

Bowman, D. M. J. S., Panton, W. J., and McDonough, L. (1991). Dynamics of forest clumps on chenier plains, Cobourg Peninsula, Northern Territory. *Australian Journal of Botany* **38**, 593–601.

Bowman, D. M. J. S., Woinarski, J. C. Z., and Russell-Smith, J. (1994). Environmental relationships of Orange-footed Scrubfowl *Megapodius reinwardt* nests in the Northern Territory. *Emu* **94**, 181–185.

Bridgewater, P. B., Russell-Smith, J., and Cresswell, I. D. (1998). Vegetation science in a cultural landscape: the case of Kakadu National Park. *Phytocoenologia* **28**, 1–17.

Chapman, A., and Harrington, G. N. (1998). Responses by birds to fire regime and vegetation at the wet sclerophyll/tropical rainforest boundary. *Pacific Conservation Biology* **3**, 213–220.

Chappell, J. M. A., and Shackleton, N. J. (1986). Oxygen isotopes and sea level. *Nature* **334**, 137–140.

Chen, Y. (1986). Early Holocene vegetation dynamics of Lake Barrine Basin, northeast Queensland, Australia. PhD thesis, Australian National University, Canberra.

Chen, Y. (1988). Early Holocene population expansion of some rainforest trees at Lake Barrine Basin, Queensland. *Australian Journal of Ecology* **13**, 225–233.

Clark, R. L. (1983). Pollen and charcoal evidence for the effects of Aboriginal burning on the vegetation of Australia. *Archaeology in Oceania* **18**, 32–37.

Clayton-Greene, K. A., and Beard, J. S. (1985). The fire factor in vine thicket and woodland vegetation of the Admiralty Gulf region, north-west Kimberley, Western Australia. *Proceedings of the Ecological Society of Australia* **13**, 225–230.

Crowley, G. M. (1995). Fire on Cape York Peninsula. *Cape York Peninsula Land Use Strategy*. (Office of the Co-ordinator General of Queensland, Brisbane, and Department of Environment, Sport and Territories, Canberra.)

Doley, D. (1979). Effects of shade on xylem development in seedlings of *Eucalyptus grandis* Hill ex Maiden. *New Phytologist* **82**, 545–555.

Duff, G. A. (1987). Physiological ecology and vegetation dynamics of north Queensland upland rainforest-open-forest transitions. PhD thesis, James Cook University, Townsville, Qld.

Duff, G. A., and Stocker, G. C. (1989). The effects of frosts on rainforest/open-forest ecotones in the highlands of north Queensland. *Proceedings of the Royal Society of Queensland* **100**, 49–54.

Fensham, R. J. (1990). Interactive effects of fire frequency and site factors in tropical *Eucalyptus* forests. *Australian Journal of Ecology* **15**, 255–266.

Fensham, R. J. (1993). The environmental relations of vegetation pattern on chenier ridges on Bathurst Island, Northern Territory. *Australian Journal of Botany* **41**, 275–291.

Fensham, R. J. (1995). Floristics and environmental relations of inland dry rainforest in north Queensland, Australia. *Journal of Biogeography* **22**, 1047–1063.

Fensham, R. J. (1996). Land clearance and conservation of inland dry rainforest in north Queensland, Australia. *Biological Conservation* **75**, 289–298.

Fensham, R. J., Fairfax, R. J., and Cannell, R. J. (1994). The invasion of *Lantana camara* L. in Forty Mile Scrub National Park, north Queensland. *Australian Journal of Ecology* **19**, 297–305.

Flannery, T. M. (1994). *The Future Eaters*. (Reed Books: Sydney.)

Florence, R.G. (1996). *Ecology and Silviculture of Eucalypt Forests*. (CSIRO: Melbourne.)

Fordyce, I. R., Duff, G. A., and Eamus, D. (1997a). The water relations of *Allosyncarpia ternata* (Myrtaceae) at contrasting sites in the monsoonal tropics of northern Australia. *Australian Journal of Botany* **45**, 259–274.

Fordyce, I. R., Eamus, D., Duff, G. A., and Williams, R. J. (1997b). The role of seedling age and size in the recovery of *Allosyncarpia ternata* following fire. *Australian Journal of Ecology* **22**, 262–269.

Gill, A. M., Groves, R. H., and Noble, I. R. (eds.) (1981). *Fire and the Australian Biota*. (Australian Academy of Science: Canberra.)

Gill, A. M., Moore, P. H. R., and Williams, R. J. (1996). Fire weather in the wet-dry tropics of the World Heritage Kakadu National Park, Australia. *Australian Journal of Ecology* **21**, 302–308.

Gillison, A. N. (1987). The 'dry' rainforests of *Terra Australis*. In *Australian National Rainforests Study*, vol. 1, *The*

Rainforest Legacy (eds. G. L. Werren and A. P. Kershaw) pp. 305–321. (Australian Government Publishing Service: Canberra.)

Graham, A. W., and Hopkins, M. S. (1990). Soil seed banks of adjacent unlogged rainforest types in north Queensland. *Australian Journal of Botany* 38, 261–268.

Harrington, G. N. (1995). Should we play God with the rainforest? *Wildlife Australia* 1995, 8–11.

Harrington, G. N., and Sanderson, K. D. (1994). Recent contraction of wet sclerophyll forest in the wet tropics of Queensland due to invasion by rainforest. *Pacific Conservation Biology* 1, 319–327.

Hill, R., Baird, A., and Buchanan, D. (1999). Aborigines and fire in the Wet Tropics of Queensland, Australia: ecosystem management across cultures. *Society and Natural Resources* 12, 205–223.

Hill, R. S. (1994). *History of Australian Vegetation: Cretaceous to Recent*. (Cambridge University Press: Cambridge.)

Hodgkinson, K. C. (2001). Fire regimes in *Acacia* wooded landscapes: effects on functional processes and biological diversity. In *Flammable Australia: The Fire Regimes and Biodiversity of a Continent* (eds. R. A. Bradstock, J. E. Williams and A. M. Gill) pp. 259–277. (Cambridge University Press: Cambridge.)

Hope, G. S. (1994). The Quaternary. In *History of Australian Vegetation: Cretaceous to Recent* (ed. R. S. Hill) pp. 368–389. (Cambridge University Press: Cambridge.)

Hopkins, M. S., and Graham, A. W. (1983). The species composition of soil seed banks beneath lowland tropical rainforests in north Queensland, Australia. *Biotropica* 15, 90–99.

Hopkins, M. S., and Graham, A. W. (1984a). The role of soil seed banks in canopy gaps in Australian tropical lowland rainforest: preliminary field experiments. *Malaysian Forester* 47, 146–158.

Hopkins, M. S., and Graham, A. W. (1984b). Viable soil seed banks in disturbed lowland tropical rainforest sites in North Queensland, Australia. *Australian Journal of Ecology* 9, 71–79.

Hopkins, M. S., Graham, A. W., Hewett, R., Ash, J., and Head, J. (1990a). Evidence of late Pleistocene fires and eucalypt forest from a north Queensland humid tropical rainforest site. *Australian Journal of Ecology* 15, 345–347.

Hopkins, M. S., Tracey, J. G., and Graham, A. W. (1990b). The size and composition of soil seed-banks in remnant patches of three structural rainforest types in North Queensland. *Australian Journal of Ecology* 15, 43–50.

Hopkins, M. S., Ash, J., Graham, A. W., Head, J., and Hewett, R. K. (1993). Charcoal evidence of the spatial extent of the *Eucalyptus* woodland expansions and rainforest contractions in north Queensland during the late Pleistocene. *Journal of Biogeography* 20, 357–372.

Hopkins, M. S., Head, J., Ash, J. E., Hewett, R. K., and Graham, A. W. (1996). Evidence of a Holocene and continuing recent expansion of lowland rainforest in humid, tropical North Queensland. *Journal of Biogeography* 23, 737–745.

Jennings, J. N. (1975). Desert dunes and estuarine fill in the Fitzroy estuary (North-western Australia). *Catena* 2, 215–262.

Jones, R. (1975). The Neolithic, Palaeolithic and the hunting gardeners: man and land in the antipodes. In *Quaternary Studies* (eds. R. P. Suggate and M. M. Cresswell) pp. 21–34. (The Royal Society of New Zealand: Wellington.)

Jones, R. (1980) Hunters in the Australian coastal savanna. In *Ecology in Savanna Environments* (ed. D. Harris) pp. 107–146. (Academic Press: London.)

Kellman, M. (1984). Synergistic relationships between fire and low soil fertility in Neotropical savannas: a hypothesis. *Biotropica* 16, 158–160.

Kenneally, K. F., Keighery, G. J., and Hyland, B. P. M. (1991). Floristics and phytogeography of Kimberley rainforests, Western Australia. In *Kimberley Rainforests Australia* (eds. N. L. McKenzie, R. B. Johnston and P. G. Kendrick) pp. 93–131. (Surrey Beatty: Sydney.)

Kershaw, A. P. (1970). A pollen diagram from Lake Euramoo, north-eastern Queensland, Australia. *New Phytologist* 69, 785–805.

Kershaw, A. P. (1971). A pollen diagram from Quincan Crater, northeast Queensland, Australia. *New Phytologist* 70, 669–681.

Kershaw, A. P. (1974). A long continuous pollen sequence from north-eastern Australia. *Nature* 251, 222–223.

Kershaw, A. P. (1975). Stratigraphy and pollen analysis of Bromfield Swamp, north-eastern Queensland, Australia. *New Phytologist* 75, 173–190.

Kershaw, A. P. (1976). A late Pleistocene and Holocene pollen diagram from Lynch's Crater, north-eastern Queensland, Australia. *New Phytologist* 77, 469–498.

Kershaw, A. P. (1978). Record of last interglacial–glacial cycle from north-eastern Queensland. *Nature* **272**, 159–161.

Kershaw, A. P. (1983). A Holocene pollen diagram from Lynch's Crater, north-eastern Queensland, Australia. *New Phytologist* **94**, 669–682.

Kershaw, A. P. (1985). An extended late Quaternary vegetation record from northeastern Queensland and its implications for the seasonal tropics of Australia. *Proceedings of the Ecological Society of Australia* **13**, 179–189.

Kershaw, A. P. (1986). Climatic change and Aboriginal burning in north-east Australia during the last two glacial/interglacial cycles. *Nature* **322**, 47–49.

Kershaw, A. P. (1992). The development of rainforest/savanna boundaries in tropical Australia. In *The Nature and Dynamics of the Forest–Savanna Boundary* (eds. P. A. Furley, J. Proctor and J. A. Ratter) pp. 255–271. (Chapman and Hall: London.)

Kershaw, A. P. (1994). Pleistocene vegetation of the humid tropics of north-east Queensland, Australia. *Palaeogeography, Palaeoclimatology, Palaeoecology* **109**, 399–412.

Kershaw, A. P., and Sluiter, I. R. (1982). Late Cenozoic pollen spectra from the Atherton Tableland, north-eastern Australia. *Australian Journal of Botany* **30**, 279–295.

Kershaw, A. P., McKenzie, G. M., and McMinn, A. (1993). A Quaternary vegetation history of northeastern Queensland from pollen analysis of ODP Site 820. *Proceedings of the Ocean Drilling Program, Scientific Results* **133**, 107–114.

Kershaw, A. P., Martin, H. A., and McEwen Mason, J. R. C. (1994). The Neogene: a period of transition. In *History of Australian Vegetation: Cretaceous to Recent* (ed. R. S. Hill) pp. 299–327. (Cambridge University Press: Cambridge.)

Kikkawa, J. (1990). Specialisation in the tropical rainforests. In *Australian Tropical Rainforests: Science, Values, Meaning* (eds. L. J. Webb and J. Kikkawa) pp. 67–73. (CSIRO Publications: Melbourne.)

Kikkawa, J., Webb, L. J., Dale, M. B., Monteith, G. B., Tracey, J. G., and Williams, W. T. (1981). Gradients and boundaries of monsoon forests in Australia. *Proceedings of the Ecological Society of Australia* **11**, 39–52.

Kimber, P. C., Forster, J. E., and Behn, G. A. (1991). Mapping of Kimberley rainforests: an overview. In *Kimberley Rainforests Australia* (eds. N. L. McKenzie, R. B. Johnston and P. G. Kendrick) pp. 37–40. (Surrey Beatty: Sydney.)

Langcamp, M. J., Ashton, D. H., and Dalling, M. J. (1981). Ecological gradients in forest communities on Groote Eylandt, Northern Territory, Australia. *Vegetatio* **48**, 27–46.

Laurance, W. F. (1997). A distributional survey and habitat model for the endangered Northern Bettong *Bettongia tropica* in tropical Queensland. *Biological Conservation* **82**, 47–60.

Lavarack, P. S., and Godwin, M. (1987). Rainforests of northern Cape York Peninsula. In *Australian National Rainforests Study*, vol. 1, *The Rainforest Legacy* (eds. G. L. Werren and A. P. Kershaw) pp. 201–222. (Australian Government Publishing Service: Canberra.)

Liddle, D. T., Russell-Smith, J., Brock, J., Leach, G. J., and Connors, G. T. (1994). *Flora of Australia*, Supplementary Series no. 3, *Atlas of Vascular Rainforest Plants in the Northern Territory, Australia*. (Australian Government Publishing Service: Canberra.)

Liddle, D. T., Taylor, S. M., and Larcombe, D. R. (1996). Population changes from 1990 to 1995 and management of the endangered rainforest palm *Ptychosperma bleeseri* Burret (Arecaceae). In *Back from the Brink: Refining the Threatened Species Recovery Process* (eds. S. Stephens and S. Maxwell) pp. 110–113. (Surrey Beatty: Sydney.)

Lucas, D., Gapindi, M., and Russell-Smith, J. (1997). Cultural perspectives of the South Alligator River floodplain: continuity and change. In *Tracking Knowledge in Northern Australian Landscapes* (eds. D. B. Rose and A. Clarke) pp. 120–140. (North Australia Research Unit, Australian National University: Darwin.)

Mangglamara, G., Burbidge, A., and Fuller, P. J. (1991). Wunambul words for rainforest and other Kimberley plants and animals. In *Kimberley Rainforests Australia* (eds. N. L. McKenzie, R. B. Johnston and P. G. Kendrick) pp. 413–421. (Surrey Beatty: Sydney.)

Martin, H. A. (1990). The palynology of the Namba Formation in Wooltana-1 bore, Callabona Basin (Lake Frome), South Australia and its relevance to Miocene grasslands in central Australia. *Alcheringa* **14**, 247–255.

Martin, H. A. (1991). Tertiary stratigraphic palynology and palaeoclimate of the inland river systems in New South Wales. In *The Cainozoic of Australia: A Re-Appraisal of the Evidence* (eds. M. A. J. Williams, P. De Deckker and

A. P. Kershaw) pp. 181–194. (Geological Society of Australia: Sydney.)

Martin, H. A. (1994). Australian Tertiary phytogeography: evidence from palynology. In *History of Australian Vegetation: Cretaceous to Recent* (ed. R. S. Hill) pp. 103–142. (Cambridge University Press: Cambridge.)

McKenzie, N. L., and Belbin, L. (1991). Kimberley rainforest communities: reserve recommendations and management considerations. In *Kimberley Rainforests Australia* (eds. N. L. McKenzie, R. B. Johnston and P. G. Kendrick) pp. 453–468. (Surrey Beatty: Sydney.)

McKenzie, N. L., Belbin, L., Keighery, G. J., and Kenneally, K. F. (1991). Kimberley rainforest communities: patterns of species composition and Holocene biogeography. In *Kimberley Rainforests Australia* (eds. N. L. McKenzie, R. B. Johnston and P. G. Kendrick) pp. 423–451. (Surrey Beatty: Sydney.)

McMinn, A., and Martin, H. A. (1992). Late Cainozoic pollen history from Site 765, Eastern Indian Ocean. *Proceedings of the Ocean Drilling Program Scientific Results*, **123**, 421–427.

Neldner, V. J., and Clarkson, J. R. (1995) Vegetation survey and mapping of *Cape York Peninsula. Cape York Peninsula Land Use Strategy.* (Office of the Co-ordinator General of Queensland, Brisbane. Department of Environment, Sport and Territories, Canberra, and Queensland Department of Environment and Heritage, Brisbane.)

Nix, H. A., and Kalma, J. D. (1972). Climate as a dominant control in the biogeography of northern Australia and New Guinea. In *Bridge and Barrier: The Natural and Cultural History of Torres Strait* (ed. D. Walker) pp. 61–92. (Australian National University: Canberra.)

Panton, W. J. (1993). Changes in post World War II distribution and status of monsoon rainforests in the Darwin area. *Australian Geographer* **24**, 50–59.

Pole, M. S., and Bowman, D. M. J. S. (1996). Tertiary plant fossils from Australia's Top End. *Australian Systematic Botany* **9**, 113–126.

Ritchie, T. W. (1996). Overview of recent fire research on the margins of the Wet Tropics of Queensland World Heritage area in the Cairns–Atherton region. In *Proceedings of the Fire Ecology Workshop*, Darwin, Northern Territory, August 1994 (ed. S. van Cuylenburg) pp. 28–44. (Parks and Wildlife Commission of the Northern Territory: Darwin.)

Ritchie, T. W., and Collins, S. (1994). Public perceptions of fire in the World Heritage Area and adjacent hillslopes of cairns and solutions to fire problems. In *Proceedings of a Workshop on Fire Management on Conservation Reserves in Tropical Australia*, Malanda, Queensland, 26–30 July 1993 (eds. K. R., McDonald, and D. Batt) pp. 123–150. (Queensland Department of Environment and Heritage: Brisbane.)

Russell-Smith, J. (1986). The forest in motion: exploratory studies in western Arnhem Land, northern Australia. PhD thesis, Australian National University, Canberra.

Russell-Smith, J. (1991). Classification, species richness, and environmental relations of monsoon rain forest in northern Australia. *Journal of Vegetation Science* **2**, 259–278.

Russell-Smith, J. (1996). Regeneration of monsoon rainforest in northern Australia: the sapling bank. *Journal of Vegetation Science* **7**, 889–900.

Russell-Smith, J., and Bowman, D. M. J. S. (1992). Conservation of monsoon rainforest isolates in the Northern Territory, Australia. *Biological Conservation* **59**, 51–63.

Russell-Smith, J., and Dunlop, C. R. (1987). The status of monsoon vine forests in the Northern Territory: a perspective. In *Australian National Rainforests Study*, vol. 1, *The Rainforest Legacy* (eds. G. L. Werren and A. P. Kershaw) pp. 227–288. (Australian Government Publishing Service: Canberra.)

Russell-Smith, J., and Lee, A. H. (1992). Plant populations and monsoon rain forest in the Northern Territory, Australia. *Biotropica* **24**, 471–487.

Russell-Smith, J., and Lucas, D. E. (1994). Regeneration of monsoon rain forest in northern Australia: the dormant seed bank. *Journal of Vegetation Science* **5**, 161–168.

Russell-Smith, J., McKenzie, N. L., and Woinarski, J. C. Z. (1992). Conserving vulnerable habitat in northern and north-western Australia: the rainforest archipelago. In *Conservation and Development Issues in Northern Australia* (eds. I. Moffatt and A. Webb) pp. 64–69. (North Australia Research Unit, Australian National University: Darwin.)

Russell-Smith, J., Lucas, D. E., Brock, J., and Bowman, D. M. J. S. (1993). *Allosyncarpia*-dominated rain forest in monsoonal northern Australia. *Journal of Vegetation Science* **4**, 67–82.

Russell-Smith, J., Lucas, D., Gapindi, M., Gunbunuka, B.,

Kapirigi, N., Namingum, G., Lucas, K., Giuliana, P., and Chaloupka, G. (1997). Aboriginal resource utilization and fire management practice in western Arnhem Land, monsoonal northern Australia: notes for prehistory, lessons for the future. *Human Ecology* **25**, 159–195.

Russell-Smith, J., Ryan, P. G., Klessa, D., Waight, G., and Harwood, R. (1998). Fire regimes, fire-sensitive vegetation, and fire management of the sandstone Arnhem Plateau, monsoonal northern Australia. *Journal of Applied Ecology* **36**, 829–846.

Savage, P. (1989). Christie Palmerston, explorer. James Cook University of North Queensland, Records of North Queensland History no. 2, Townsville, Qld.

Schimper, A. F. W. (1903). *Plant Geography upon a Physiological Basis.* (Groom and Balfour: Oxford.)

Singh, G., Kershaw, A. P., and Clark, R. (1981). Quaternary vegetation and fire history in Australia. In *Fire and the Australian Biota* (eds. A. M. Gill, R. H. Groves and I. R. Noble) pp. 23–54. (Australian Academy of Science: Canberra.)

Shulmeister, J. (1992). A Holocene pollen record from lowland tropical Australia. *The Holocene* **2**, 107–116.

Stocker, G. C. (1966). Effects of fire on vegetation in the Northern Territory. *Australian Forestry* **30**, 223–230.

Stocker, G. C. (1971a). The age of charcoal from old jungle fowl nests and vegetation change on Melville Island. *Search* **2**, 28–30.

Stocker, G. C. (1971b). Fertility differences between the surface soils of monsoon and *Eucalyptus* forest in the Northern Territory. *Australian Forest Research* **4**, 31–38.

Stocker, G. C. (1981). Regeneration of a North Queensland rain forest following felling and burning. *Biotropica* **13**, 86–92.

Stocker, G. C., and Mott, J. J. (1981). Fire in the tropical forests and woodlands of northern Australia. In *Fire and the Australian Biota.* (eds. A. M. Gill, R. H. Groves and I. R. Noble) pp. 425–439. (Australian Academy of Science: Canberra.)

Stocker, G. C., and Unwin, G. L. (1989). The rainforests of northeastern Australia: their environment, evolutionary history and dynamics. In *Tropical Rain Forest Systems* (eds. H. Lieth and M. J. A. Werger) pp. 241–259. (Elsevier: Amsterdam.)

Torgersen, T., Luly, J., De Deckker, P., Jones, M. R., Searle, D. E., Chivas, A. R., and Ullman, W. J. (1988). Late Quaternary environments of the Carpentaria Basin, Australia. *Palaeogeography, Palaeoclimatology, Palaeoecology* **67**, 245–261.

Tracey, J. G. and Webb, L. J. (1975). *Vegetation of the Humid Tropical Region of North Queensland: Maps and Key.* (CSIRO Division of Plant Industry: Indooroopilly, Brisbane.)

Truswell, E. M. (1990). Australian rainforests: the 100 million year record. In *Australian Tropical Rainforests: Science, Value, Meaning* (eds. L. J. Webb and J. Kikkawa) pp. 7–22. (CSIRO: Melbourne.)

Truswell, E. M. (1993). Vegetation changes in the Australian Tertiary in response to climatic and phytogeographic forcing factors. *Australian Systematic Botany* **6**, 533–557.

Truswell, E. M., and Harris, W. K. (1982). The Cainozoic palaeobotanical record in arid Australia: fossil evidence for the origins of an arid-adapted flora. In *Evolution of the Flora and Fauna of Arid Australia* (eds. W. R. Barker and P. J. M. Greenslade) pp. 67–76. (Peacock Publications: Adelaide.)

Unwin, G. L. (1983). Dynamics of the rainforest–eucalypt forest boundary in the Herberton Highland, north Queensland. MSc thesis, James Cook University, Townsville, Qld.

Unwin, G. L. (1989). Structure and composition of the abrupt rainforest boundary in the Herberton Highland, north Queensland. *Australian Journal of Botany* **37**, 413–428.

Unwin, G. L., and Kriedemann, P. E. (1990). Drought tolerance and rainforest tree growth on a north Queensland rainfall gradient. *Forest Ecology and Management* **30**, 113–123.

Unwin, G. L., Stocker, G. C., and Sanderson, K. D. (1985). Fire and the forest ecotone in the Herberton Highland, north Queensland. *Proceedings of the Ecological Society of Australia* **13**, 215–224.

Unwin, G. L., Applegate, G. B., Stocker, G. C., and Nicholson, D. I. (1988a). Initial effects of Tropical Cyclone 'Winifred' on forests in north Queensland. *Proceedings of the Ecological Society of Australia* **15**, 283–296.

Unwin, G. L., Stocker, G. C., and Sanderson, K. D. (1988b). Forest successions following European settlement and mining in the Rossville area, north Queensland. *Proceedings of the Ecological Society of Australia* **15**, 303–305.

van der Kaars, W. A. (1989). Aspects of Late Quaternary palynology of eastern Indonesian deep sea cores. *Netherlands Journal of Sea Research* **24**, 495–500.

van Steenis, C. G. G. J. (1957). Outline of vegetation types in Indonesia and some adjacent islands. *Proceedings of the Pacific Science Congress* **8**, 61–97.

Walker, D. (1990). Directions and rates of tropical rainforest processes. In *Australian Tropical Rainforests: Science, Value, Meaning* (eds. L. J. Webb and J. Kikkawa) pp. 23–32. (CSIRO: Melbourne.)

Walker, D., and Chen, Y. (1987). Palynological light on tropical rainforest dynamics. *Quaternary Science Reviews* **6**, 77–92.

Walker, J. (1981). Fuel dynamics in Australian vegetation. In *Fire and the Australian Biota* (eds. A. M. Gill, R. H. Groves and I. R. Noble) pp. 101–127. (Australian Academy of Science: Canberra.)

Wasson, R. J. (1984). Late Quaternary palaeoenvironments in the desert dunefields of Australia. In *Late Cainozoic Palaeoclimates of the Southern Hemisphere* (ed. J. C. Vogel) pp. 419–432. (Balkema: Rotterdam.)

Wasson, R. J. (1986). Geomorphology and Quaternary history of the Australian continental dunefields. *Geographical Review of Japan* **59**, 55–67.

Webb, L. J. (1958). Cyclones as an ecological factor in tropical lowland rainforest, north Queensland. *Australian Journal of Botany* **6**, 220–228.

Webb, L. J. (1959). A physiognomic classification of Australian rainforests. *Journal of Ecology* **47**, 551–570.

Webb, L. J. (1968). Environmental relationships of the structural types of Australian rainforest. *Ecology* **49**, 296–311.

Webb, L. J. (1978). A general classification of Australian rainforests. *Australian Plants* **9**, 349–363.

Webb, L. J., and Tracey, J. G. (1981a). Australian rainforests: patterns and change. In *Ecological Biogeography in Australia* (ed. A. Keast) pp. 605–694. (Dr W. Junk: The Hague.)

Webb, L. J., and Tracey, J. G. (1981b). The rainforests of northern Australia. In *Australian Vegetation* (ed. R. H. Groves) pp. 67–101. (Cambridge University Press: Cambridge.)

Webb, L. J., Tracey, J. G., and Williams, W. T. (1984). A floristic framework of Australian rainforests. *Australian Journal of Ecology* **9**, 169–198.

Webster, P. J., and Streten, N. A. (1978). Late Quaternary ice age climates of tropical Australasia: interpretations and reconstructions. *Quaternary Research* **10**, 279–309.

Werren, G., and Kershaw, A. P. (eds.) (1987–1991). Australian *National Rainforests Study*, vols. 1–3, *The Rainforest Legacy*. (Australian Government Publishing Service: Canberra.)

White, M. E. (1994). *After the Greening: the Browning of Australia*. (Kangaroo Press: Sydney.)

Whitmore, T. C. (1984a). *Tropical Rain Forests of the Far-East*, 2nd Edn. (Clarendon Press: Oxford.)

Whitmore, T. C. (1984b). A vegetation map of Malesia at 1:5 million. *Journal of Biogeography* **11**, 461–471.

Williams, R. J., Gill, A. M., and Moore, P. H. R. (1998). Seasonal changes in fire behaviour in a tropical savanna in northern Australia. *International Journal of Wildland Fire* **8**, 227–239.

Williams, R. J., Griffiths, A. D., and Allan, G. (2001). Fire regimes and biodiversity in the savannas of northern Australia. In *Flammable Australia: The Fire Regimes and Biodiversity of a Continent* (eds. R. A. Bradstock, J. E. Williams and A. M. Gill) pp. 281–304. (Cambridge University Press: Cambridge.)

Yibarbuk, D., Whitehead, P. J., Russell-Smith, J., Jackson, D., Godjuwa, C., Fisher, A., Cooke, P., Choquenot, D., and Bowman, D. M. J. S. (2001). Fire ecology and aboriginal land management in central Arnhem Land, northern Australia: a tradition of ecosystem management. *Journal of Biogeography*, **28**.

15

Fire regimes and biodiversity of forested landscapes of southern Australia

A. MALCOLM GILL AND PETER C. CATLING

Abstract

The forests of southern Australia fall into three main regions: southeastern mainland Australia; Tasmania; and southwestern Australia. The forests consist of 'open-forests', 'tall open-forests' and 'closed forests' (rainforests). Open-forests are relatively common and support many species of *Eucalyptus* (*sensu lato*). Tall open-forests, also dominated by eucalypts, form the world's tallest hardwood forests when fully grown. The relatively rare closed forests are dominated by genera other than *Eucalyptus*. Southeastern Australian open-forests are richer in eucalypt species and vertebrate animals than open-forests in other regions. In southeastern Australia closed forests are often found in valleys, the most mesic sites, while tall open-forests and open-forests are found on increasingly drier sites, respectively. There is no closed forest in southwestern Australia. Open-forests are the most frequently burnt because of the rapid accumulation of fuels in relatively dry habitats; closed forests may accumulate similar quantities of fuels but are less likely to burn because of moist conditions there. Mean fire intervals may encompass two orders of magnitude in these forests. The quasi-equilibrium fuel loads reached mean that fire intensities can be extremely high in all forests but the relatively protected locations and deep shade of closed forests suggests that intensities in general will be lower there. Models and observations of fire behaviour and crown-scorch patterns indicate that there is a wide range of intensity within and between fires. Peat fires are rare but do occur. Habitats of vertebrate animals may be described in terms of tree-hollow characteristics, the mix of understorey and overstorey densities, vertical continuity of foliage, ground surface features and the presence of logs and other forest 'debris' on the ground. Fire regimes affect many of these habitat components. In general, forests with denser understoreys support more mammals, birds and insects, but not necessarily all taxa of these organisms known from the area. Hollow formation may be assisted by particular fire regimes, curtailed by others, but these regimes have not yet been quantified. Plant species may increase or decrease in population according to fire regimes, and in extreme circumstances, to extinction. Further work is needed to determine: the mean probability of a fire at a point, its variance and its relationship to the time since the last fire for the variety of situations found in forests; the mean and variance of fire characteristics, and seasons, for points in the landscape through time; the full range of forest biota – especially invertebrates and non-vascular plants; and the indicator species that define limits for the survival of the entire biota under a variety of fire regimes in any one area of the forested estate.

Introduction

The forests of southern Australia have significant commercial, aesthetic and conservation value. They are located near large population centres and have been highly utilised since white settlement began in 1788. They have been at the centre of controversy for decades. The focus of this chapter is not the contemporary use of these forests but, rather, the biodiversity and fire regimes of southern Australian forests prior to European settlement.

The definitions of 'forest' used here follow Specht

Fig. 15.1. Map of Australia with a sketch of southern forest regions as used in this chapter. The mean annual rainfall is usually higher on the coastal side of the 600 mm isohyet.

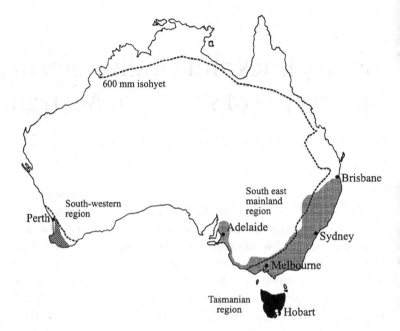

(1970). 'Open-forests' have a cover of 30%–70% and are 10–30 m tall when mature; 'tall open-forests' have the same cover range but are taller than 30 m when mature; and 'closed forests' are more than 10 m tall but have a cover greater than 70%. 'Forested landscapes' are landscapes clothed predominantly by forests but may also include shrublands, grasslands and woodlands.

The main forest regions of southern Australia (after Carnahan 1976) are discrete rather than continuous (Fig. 15.1). The southeastern mainland region, the largest, is separated from the Tasmanian region by sea (Bass Strait) and from the southwestern Australian region by arid treeless country (the Nullarbor Plain). The southeastern region is particularly rich in vertebrate fauna; the Tasmanian region is notable for its areas of coniferous closed forest; and the southwestern region has a diverse understorey flora but no closed forest. A common feature is that all these areas are prone to fires, some of which may attain unrivalled high intensities (Gill and Moore 1990).

In this chapter we sketch a picture of the forests of southern Australia and their biota, review the literature on the factors predisposing these areas to, and outline the evidence for, forest-fire regimes just

prior to the start of white settlement in Australia just over 200 years ago. We outline the principles relating fire regimes to their effects on flora and fauna and examine changes in the forest biota according to the time since the last fire and the between-fire interval.

Forested landscapes of southern Australia

Forested terrain in much of the southeastern region and in Tasmania is rugged but in the southwestern region the terrain is relatively subdued. Closed forests are found in wetter locations as a general rule, particularly in deep valleys and at higher altitudes. Australia's southern forests generally occur where the rainfall is greater than 600 mm per year (Fig. 15.1) (Christensen *et al.* 1981). In drier (generally inland), colder (generally at higher altitudes) and nutritionally more impoverished sites, forests give way to woodlands with shrubby or grassy understoreys. Saline and freshwater swamps may abut forests.

Open-forests are dominated by *Eucalyptus* (*sensu lato*), the closed forests by a number of non-eucalypt species such as *Nothofagus cunninghamii*, *Athrotaxis selaginoides* and *Phyllocladus aspleniifolius* (Jarman *et* al. 1994), *Doryphora sassafras, Eucryphia moorei* and *Ceratopetalum apetalum* (Helman 1987). Wherever forests are found there are catenary or regional gradients in the composition and structure of the vegetation.

In the southeast region, local variation in the eucalypt component of open-forests can be quite marked. On the south coast of New South Wales, a ridge may support open-forests of *Eucalyptus maculata* (syn. *Corymbia maculata*), *E. paniculata* and *E. globoidea*, the slopes *E. pilularis* and *E. maculata* and the valley a closed forest. There are numerous variants on this pattern with different radiation environments, geological substrates, proximity to the sea, latitude and altitude (Austin 1978). With altitude, a sequence of forest communities on sediments in a transect from the coast to the mountains may begin with those featuring *E. paniculata, E. muellerana, E. longifolia, E. gummifera* and *E. maculata* (up to 200 m altitude) then pass through *E. agglomerata, E. muellerana, E. cypellocarpa* (from 200 to 550 m altitude) to *E. sieberi, E. fastigata* and *E. cypellocarpa* (from 550 m to 100 m) thence to *E. radiata, E. viminalis* and *E. fastigata* (above 1000 m) (Austin 1978). Along this transect numerous other eucalypt species will be encountered. Many examples have been illustrated for Tasmania by Jackson (1965).

Tall open-forests are found in all three regions (Ashton 1981a; Ashton and Attiwill 1994). Of these, mature mountain ash (*Eucalyptus regnans*) of southeastern Australia (including Tasmania) dominate the world's tallest hardwood forests. Tall open-forests may have tall-shrub or tree understoreys. Common in the southern part of southeastern Australia are the genera *Acacia, Olearia, Bedfordia* and *Pomaderris* (Ashton 1981a) while in the northern part are *Syncarpia, Acacia, Rapanea* and *Elaeocarpus* for example (Ashton 1981a). Common in the southwestern region are *Acacia, Bossiaea* and *Trymalium* (Ashton 1981a; Bradshaw and Lush 1981). Various tree and ground ferns, palms, cycads, lianes and herbs are also to be found in tall open-forests (Ashton 1981a). Rainforest taxa may occur in the understoreys of eastern Australian tall open-forests (Ashton 1981a; Ashton and Attiwill 1994).

Open-forests are more widespread than either closed forests or tall open-forests. In open-forests, a greater diversity of eucalypt species is found both within stands (Gill 1994) and more generally. In south-eastern Australia, in zones 'O' and 'P' of Gill *et al.* (1985), there were 220 and 73 species of eucalypts respectively while in the smaller southwestern forest zone 'L', 56 species occurred. Common shrubs in these forests include species of *Acacia, Daviesia, Hakea, Leptospermum* and *Pultenaea* while herb genera include *Dianella, Lepidosperma* and *Lomandra*; the commonest fern is likely to be *Pteridium*, the bracken fern (Gill 1994). While numbers of vascular plant species for forest areas are difficult to obtain, the estimate for a 784 000-ha area of southeastern New South Wales was 1100 species and subspecies (Drielsma 1994). For a larger area of jarrah forest (*E. marginata*) in the southwestern region, the number of species estimated was 'at least 784' (Bell and Heddle 1989).

Closed forests in southeastern Australia generally have few species of vascular plants (Read and Hill 1981) and vertebrate animals, unlike the species-rich stereotype of the tropical rainforest. Non-vascular plants may be common as epiphytes (Ashton and McRae 1970). In Tasmanian closed forests there may be as few as 5 species of mammals and 6 of birds whereas open-forests in southern Australia may support 15–27 species of mammals and 50 to 154 species of birds (Catling and Newsome 1981). Bird faunas in Tasmania and in southwestern region forests are generally impoverished in comparison to those of the southeastern Australian mainland (Keast 1981). Recently, Woinarski *et al.* (1997) summarised data for vertebrate taxa (excluding birds) in southern Australian forests. In Tasmanian 'dry forest' (i.e. 'open-forest') there were 9 species of frogs, 14 species of reptiles and 27 species of non-aquatic mammals (including 8 species of bats) but Tasmanian 'wet forests' (presumably 'tall open-forests') were poorer in all groups. Southeastern region forests had 21 species of frogs, 41 species of reptiles and 52 species of non-aquatic mammals (including 21 species of bats) but the southwestern Australian jarrah forest, with a similar number of reptile species, had fewer frogs and mammals (including bats). Arboreal marsupials – such as the

354 A. MALCOLM GILL AND PETER C. CATLING

koala and marsupial gliders and various possums – are a feature of some forests in the southeastern region.

Invertebrates in the canopy and ground strata of open-forest are diverse. Majer *et al.* (1997) summarised the evidence. Over 400 species of invertebrates were found in crowns of jarrah in the southwestern region, most of the species being rare. Lepidoptera, Hemiptera and Araneae were the most common groups on jarrah. Even richer faunas have been recorded in the canopies of eucalypt forests of the southeastern region. Sap-feeding insects in the form of psyllids were the most abundant canopy guild there. In contrast to the canopy faunas, the invertebrate ground fauna was dominated by Acarina (mites and ticks) and Collembola (springtails) but ants were also common.

Factors predisposing southern forests to fires of different intensity

Fires occur in all southern Australian forests – including closed forests (e.g. Victoria: Chesterfield *et al.* 1991; Tasmania: Barker 1991). Fires occur when there is a conjunction of an ignition source and dry fuels. Senescence of the vegetation creates fuels which desiccate and moisten according to the weather. The amount of fuel determines the *amount* of heat that can be released while the *rate* of heat release is affected by fuel properties, weather, wind direction and slope of the land (Catchpole, this volume). In this section we consider fuel loadings and fire-weather as precursors to fires and fire intensities in southern Australian forests.

Fuel loadings increase in a way characteristic of the time since the last fire eventually reaching a quasi-equilibrium quantity (Walker 1981). Eucalypts supply the bulk of the fuel in both open-forest types. The tree crowns shed a great deal of woody material as well as leaves (Pook *et al.* 1997). Because the leaves decompose quicker than wood the proportion of wood in the fuel is even higher than that in the litter fall. Fuel accumulation curves may rise to a quasi-equilibrium amount in a relatively smooth fashion on a yearly scale but, within a seasonal cycle, fuel loadings may vary substantially (Mercer

et al. 1995) due to seasonally varying litter fall (usually in summer: Pook *et al.* 1997) and decomposition (likely to be in winter or spring in most areas). Litter fuel loadings at the time of the quasi steady state often fall within the range of 11 to 24 t ha^{-1} (Walker 1981). Closed forests may reach similar amounts (Barker 1992). In tall open-forest in Western Australia litter fuel loads reach quasi steady state at a larger amount, *c.* 35 t ha^{-1} (O'Connell 1987). Live shrubby fuels less than 4 mm in diameter in open-forests of jarrah in the southwestern region contribute up to *c.* 4 t ha^{-1} while rough dead bark on tree trunks may add up to another 10 t ha^{-1} (Burrows 1994). The extent to which these fuels, and indeed the tree crowns, are consumed by fires, and thus part of the fuel array, will be influenced by the height of the flames from the litter and understorey.

The rate of heat release at a point on the fire edge is given by the 'fire intensity', the product of the heat yield of the fuel, the amount of fuel per unit area ('fuel loading') and the rate of spread (ROS) of the fire (Byram 1959). A crude scale of intensities would be: 'low', less than 350 kW m^{-1}; 'high', 350–3500 kW m^{-1}; 'very high', 3500–35 000 kW m^{-1}; and, 'extreme', greater than 35 000 kW m^{-1}. Gill and Moore (1994) argued that the maximum intensity of fire likely to be encountered in Australian forests was about 100 000 kW m^{-1}, but the estimate was made by extrapolation.

The heat-yield component of the intensity calculation may vary relatively little, but the other components vary substantially (fuel loading) or enormously (rate of spread). Rates of spread of the fires are a major component of the intensity equation. Rates of spread of fires moving up a slope may double with each 10 degrees of slope (McArthur 1967) so the steepness of the terrain is most important. Where fires are burning downhill, or against the wind, rates of spread are considered to be the same as those expected without wind on level ground (Viegas 1994). Fuel dryness will also have a strong influence on ROS, the moisture content being a function of temperature and relative humidity and soil moisture (Pook and Gill 1993). ROS of a fire burning with the wind is directly related to Forest Fire Danger Index (FFDI) (McArthur 1967; Noble *et al.* 1980).

Fire-weather, using FFDI, has been examined for meteorological stations in areas marginal to, if not within, temperate forests in the Sydney area (Gill and Moore 1996, 1998), in Melbourne and Adelaide (Gill and Moore 1990) and in East Sale and Hobart (Beer *et al.* 1988). Most of the year the values of FFDI are low but in the fire season (spring, summer, autumn, or combinations of these) values may rise to 'extreme' levels (of 50 to 100) for a day or two. Values are usually low overnight. Average values vary from year to year.

Fire-intensity variation is an ecologically important topic. Within a fire the variation can be seen as variation in the death of leaves (scorch) in crowns of trees of the same height. Intense fires can cause immediate defoliation through combustion but delayed defoliation follows complete leaf scorch of tree crowns. Low-intensity fires may scorch leaves of the lower crown only.

The Forests Department of Western Australia (now part of the Department of Conservation and Land Management) reported areas of open-forests burnt by unplanned fires having intensities causing leaf death up to, or greater than, 9 m height (see Gill and Moore 1997). As most of these fires occurred in summer, the threshold intensity for this separation of classes is likely to be about 300 kW m^{-1} (Burrows 1994), a 'low' intensity. A compilation of these data (Gill and Moore 1997) revealed that an average of 44% of the areas burnt by unplanned fires experienced low-intensity fires. In the southeastern region, the large fire documented by Dexter *et al.* (1977) had 64% of its area 'lightly burnt'. In an *E. regnans* forest fire 70% of the area was affected to the extent that less than 75% of the leaves of the trees were scorched (Squire *et al.* 1991). Models for lightning-caused fires in the forested landscapes of the Australian Capital Territory suggested a predominance of low intensities (Cary and Banks 2001). A crude estimate from available measurements, then, is that an average of about 50% of the areas burnt by unplanned forest fires would be of relatively low intensity. Ideally, data would be available for frequency distributions of intensities for individual

fires and series of fires at a point, like the modelled data.

Forest fire regimes two centuries ago

A 'fire regime' is the representation of a sequence of fires having different types (peat or above ground), intensities, seasons of occurrence and between-fire intervals (Gill 1975). The fire-regime concept is a point-based concept. In any of the fire-regime components, temporal variation can be expected. There is no suitable theory yet for the discussion of variation in seasonal occurrence in a statistical way. Intensity variation has been discussed above, while variation in 'type of fire' may be considered as a two-way split into above-ground ('surface') and below-ground fires. The latter are ignited by the former – so have a conditional probability – and must be relatively rare because not all surface fires ignite peat and peat takes long periods of time to accumulate. Variation in the intervals between fires could be non-random or random. Random patterns are assumed to occur in 'natural' systems (i.e. where any human influence is not contrived to be regular). Random patterns may take many forms (Johnson and Gutsell 1994; McCarthy *et al.* 2001; Clark *et al.* this volume; McCarthy and Cary, this volume).

Most discussion of past fire regimes revolves around fire intervals because other aspects of fire regimes are difficult or impossible to discern from the record (Clark *et al.*, this volume; Kershaw *et al.*, this volume). Random patterns of fire intervals arise as a consequence of the relationship between the probability of burning at a point (PBP) and the time elapsed since the last fire (Johnson and Gutsell 1994). In southern Australian forests the PBP function, based on *a priori* reasoning, would be shaped like a fuel accumulation curve (Lang 1997; McCarthy *et al.* 2001). When closed forest is invading a eucalypt forest a dip in the curve at relatively long periods after fire may be expected as a result of decreased wind speeds and increased moisture in the forest (McCarthy *et al.* 2001). If grassy fuels were the same a year after fire then a constant value of PBP would apply. Variance around PBP functions arises from variations in numbers and sizes of individual fires and in weather from year to year. PBP functions mathematically imply the proportions of land having had different times since fire and between-fire intervals (Johnson and Gutsell 1994).

There are many records of forest burning by Aboriginal people (Hallam 1975; Benson and Redpath 1997) but the determination of fire regimes from them is fraught with difficulty (e.g. McBryde and Nicholson 1978; Williams and Gill 1995; Benson and Redpath 1997). A popular model for burning by Aboriginal people, based on explorers' evidence, is for frequent (low-intensity) fires keeping treed vegetation types open but, in fact, dense understoreys were reported by explorers too (Benson and Redpath 1997). We note that, just as prescribed fires today do not prevent intense fires occurring from time to time, it is likely that intense fires, ignited by lightning or people, were a part of the forest environment 200 years ago. Indeed, it appears that intense fires were necessary for the perpetuation of some forests like those of mountain ash (Ashton 1981b).

In open-forests of the southwestern region, tree rings and fire scars have been examined in order to determine changes in fire intervals (Burrows *et al.* 1995). Unfortunately, 'the resilience of jarrah to injury by fire and the limitations of ring counting as an ageing technique, prevented an accurate reconstruction of the fire frequency'. Even so, Burrows *et al.* (1995) were able to estimate the average interval between tree-scarring fires as about 80 years. In a new technique, the annual growth-bulges and fire-induced stains on the stems of the suffrutescent monocotyledon *Xanthorrhoea* spp. – or 'grasstrees' – have provided fire-interval data (Ward and Van Didden 1997). The data show negative effects of rainfall, sample height and time on the average number of fires per decade from about 1750 until the present. In the pre-closer-settlement period to 1859 there appear to have been just under three fires per decade on average – a mean interval of about 3.4 years. Further information from this dataset will arise when the 'crude preliminary test of the hypotheses' mediated through multiple regression (Ward and Van Didden 1997) is followed by more sophisticated analyses. The degree to which the grasstree data represent all jarrah landscapes has yet to be determined.

In the Tasmanian region, Duncan and Brown (1995) reported estimates of intervals between fires in forests occupying dissected landscapes. For open-forest ('dry sclerophyll forest') they noted an interval of 10–20 years between fires but longer intervals occurred in gully vegetation. For tall open-forest ('tall wet sclerophyll forest'), an interval of 30–50 years was given. In parts of Tasmania – where the 'summer rainfall is in excess of 2 inches per month' – Jackson's (1968) conceptual scheme shows a frequency distribution of fire interval for tall open-forest ('mixed forest') with a mean of 200 years while the mean for closed forest was 300 years.

In tall open-forest of mountain-ash region in the southeastern mainland zone (Fig. 15.1), McCarthy *et al.* (1999) estimated the mean fire interval by studying the stand age distribution, tree life history and the time course of the presence of tree hollows. They found a mean fire interval for stand-killing fires of about 100 years. By reference to data on the fire-intensity distribution in forests (above) they suggested that the actual mean interval was nearer 50 years [and therefore similar to that estimated by Duncan and Brown (1995) in Tasmania]. Thus, exposure of the understorey to fires was about double that of the forest canopy.

To summarise, we can say that most fires were likely to have been surface fires although peat fires also occurred (Cremer 1962; Churchill 1968). Mean intervals between fires likely increased in the drier to wetter sequence from open-forest to tall open-forests to closed forests. The grasstree data may indicate the minimum mean interval between fires in open-forests – say 3 years – while the mean in the closed forests of Tasmania may have been two orders of magnitude longer.

Changes in the biota with the time since the last fire: immediate effects

Plant mortality varies with surface fire intensity. Some species have individuals that are readily killed by fires whereas others are remarkably resilient to even high fire intensities. If a comparison between species or even individuals is to be made, however, it should be made with a biologically equivalent degree of exposure to fire. A mistletoe high in a tree may have quite a different response to a plant of the same species situated closer to the flames in the same fire. Using a threshold condition (e.g. complete scorch, to the point where topmost foliage is just killed) is useful for making inter-specific comparisons (Gill 1981). Two classes of plants emerge from such comparisons: the 'seeders', which have mature plants killed by such exposures and regenerate from seed after fire, and the 'sprouters' which have mature plants which resprout after this degree of damage. Noble and Slatyer (1980, 1981) described a system for modelling the effects of fires which is dependent on 'vital attributes' like these but open-ended with respect to intensity and therefore more generally applicable. Mechanisms affecting the mortality and recovery of trees exposed to fires have been reviewed by Gill (1995).

Many of the trees of tall open-forests are seeders. *Eucalyptus regnans*, aforementioned, is a classic. *Eucalyptus grandis*, *E. delegatensis* and *E. pilularis* in eastern Australia and *E. diversicolor* in Western Australia are other examples (Ashton 1981b). This is not to say that populations of these species will always be killed by fire although even-aged regeneration is common. Rather, fires may burn in mature forests of these types without killing the dominants while multi-aged stands may occur also (Ashton 1981b; Bowman and Kirkpatrick 1984; McCarthy and Lindenmayer 1998). Open-forest dominants are predominantly resprouters. In closed forests there are resprouters like *Nothofagus cunninghamii* (Ashton 1981b) and seeders like *Phyllocladus aspleniifolius* (Read and Hill 1981). Peat fires could kill all plants present whether seeders or sprouters (Cremer 1962; Wark 1997). Seeders in the form of short shrubs or herbs will be killed by any fire, whether a surface, or a peat, fire.

Ashton (1981b) tabulated the proportion of seeders in the vascular flora of three tall open-forests, two in the southeastern region and one in the southwest. All had about 50%–60% seeders but the proportion of these that were shrubby varied. In three closed forests in eastern Victoria, McMahon

(1987) found that just over a third of the native vascular plant species were seeders. In an open-forest in Canberra, the equivalent figure was 27% (Purdie and Slatyer 1976). Open-forests in southwestern Victoria had 38% seeders (Wark 1997) while in the jarrah forest the figure was 25%–30% (Burrows and Friend 1998). There is no predictive or explanatory model for the expected proportion of seeders.

Of great significance to understanding the species dynamics of forest is knowing the proportion of 'seeders with canopy-stored seed' (cf. those seeders with seed in the soil); these species seem more prone to extinction than others (Gill and Bradstock 1995) especially if their juvenile periods are long. Juvenile periods vary widely from say 5 years for a species in a heath-type understorey (after Bradstock and O'Connell 1988) to say 15–20 years for *E. regnans* (after Ashton 1981b; McCarthy et al. 1999) and decades longer for *Athrotaxis selaginoides*, a Tasmanian endemic (see Gill 1996, and below). Such species are relatively uncommon in the understoreys of open-forests (cf. the dominant seeders of tall open-forests). Seeders of shrubby-understorey genera like *Casuarina*, *Hakea* and *Banksia* – all with woody fruits – fall into this category as do some native conifers.

With fires of intensities at least sufficient to kill, directly or indirectly, pedicels of the capsules of eucalypts and the woody fruits of other species with canopy-stored seed, accelerated seed dispersal will result. Seed fall may be rapid and heavy (O'Dowd and Gill 1984). Seed-harvesting ants foraging for eucalypt seed after a fire may be satiated by the prolific supply of seed to the extent that many seeds survive predation and generate a new crop of trees (O'Dowd and Gill 1984). Seed survival like this is likely to be essential for species persistence if populations of seeders are killed by a fire.

Shrubby thickets create substantial cover, an important variable affecting the abundance and species composition of ground-dwelling mammals in forests (Newsome and Catling 1979; Catling 1991; Catling and Burt 1995). Precursors to thickets are 'banks' of protected buds (in lignotubers, roots and rhizomes of particular species) and seeds (in the soil or in woody capsules on living plants). The 'bank' will be obvious for seeders with canopy-stored seed but hidden away as dormant seed in the soil for other species. The forest understorey can be mostly free of shrubs before a fire but be followed by a thicket. A high-intensity fire is the most likely trigger for thicket formation (Catling 1991) because:

(1) canopy damage (due to high-intensity fire) may lead to compensatory growth in the understorey (Specht 1983);
(2) fire-stimulated germination of seed is a function of fire intensity, fuel loading, moisture content of the litter (Christensen and Kimber 1975) and soil moisture;
(3) seed dispersal from serotinous seeders may be augmented by crown-scorching fires (e.g. O'Dowd and Gill 1984) although some seed death may arise where crown consumption occurs (Ashton 1986; Bradstock et al. 1994);
(4) a mineral-soil seedbed (free of organic matter and competing plants) is ideal for the establishment of seeds from seeders with canopy-stored seed (e.g. O'Dowd and Gill 1984); and
(5) stimulation of root suckering will occur when the stems of species with suckering ability are killed (Gill 1981).

That vertebrate animals will die in a fire is not a foregone conclusion (see also Whelan et al., this volume). Sometimes some do; sometimes many do. Among birds, death rates in severe fires in an East Gippsland area of closed forest, open-forest and heath was about 40% (Lloyn 1997) while in heaths and woodlands in southeastern New South Wales the figure was 54% (Fox 1978). Arboreal animals such as koalas, unprotected in tree canopies, will be killed or injured to the extent that they are exposed to fire, a function of location height and fire intensity.

Among ground-dwelling mammals, the numbers of dead animals to be seen after fire may be only a small proportion of the total present before the fire (see below). Post-fire predation seems to be the most important cause of death (Christensen 1980; Newsome et al. 1983; Kinnear et al. 1984), especially when fidelity to home range is strong, e.g. in the case of the woylie (Bettongia penicillata) in the southwest

(Christensen 1980). If there is no cover then predation will be high. The corollary is that in the absence of particular predators – like the introduced fox and the native dingo – the need for cover may be alleviated for some animals, like the tammar (Macropus eugenii) (Christensen 1980). Johnson (1995) found no post-fire predation of the eastern bettong, (Bettongia gaimardia) in Tasmania, the island state without foxes or dingoes. When predators are present, post-fire survival may be maximised where the fire has only partially burnt home ranges. As the home ranges of animals in any one area vary widely (usually in relation to their body weight: Swihart et al. 1988), the scale of patchiness appropriate to the maximum persistence of all ground-dwelling animals is that commensurate with the smallest home range. In reality, the populations of animals of different species would be expected to fluctuate within a region according to the scales of fire patterning associated with each event.

Among invertebrates, death rates for animals living only in litter may be expected to be high if the litter is totally consumed while foliage-feeding invertebrates in the canopy may be lost when canopies are defoliated by crown fire (Majer et al. 1997).

Cover, protecting animals from adverse weather and predators, whether it is in the form of hollow logs, tree hollows, shrubs or grass tussocks, may be affected by fire. In the same way, food items may be lost (e.g. invertebrates in litter, foliage, nectar) or, for carnivores, made temporarily more available because of lack of cover for the prey. Diet switching may occur because of the changed availability of food (Newsome et al. 1983).

Variables such as tree and shrub crown density, and the abundance of rocks and hollow logs on the forest floor, may be formed into a 'habitat complexity score' which correlates with the species richness and abundance of small mammals (Newsome and Catling 1979). The score changes as the period after fire increases. Some larger animals may be more common where there is a low habitat complexity score, as is the case soon after fire (Catling 1991).

Tree hollows have particular significance for many species of Australian animals. For example, over 40% of Australian mammals may depend on hollows (Ambrose 1982 in Gibbons and Lindenmayer 1997). Their size varies widely – from large [e.g. the mountain brushtail possum (*Trichosurus caninus*), weighing up to 4 kg] to tiny [the feathertail glider (*Acrobates pygmaeus*), weighing only *c.* 15 g] (Lindenmayer 1996). The same may be said of birds; large owls and tiny pardalotes occupy hollows (Gibbons and Lindenmayer 1997). It is obvious that large animals need large hollows and that large hollows are more likely to be found in large trees. Indeed, trees need to have reached certain minimum dimensions (and ages) before hollows will form while hollows in general predominate in the older and larger trees (Fig. 15.2) (Gibbons and Lindenmayer 1997). For hollow-dwelling species to persist at a site, fire must not destroy suitable hollows. If the fire is of a stand-killing type, hollows may persist in stags (Lindenmayer 1996) but there could be long periods at the site without hollows after the stags have fallen and before the new stand has individuals big enough to develop them. For example, stags in *E. regnans* forests may last decades but hollows suited to Leadbeater's possum (*Gymnobelideus leadbeateri*) only appear in live trees if over about 190 years old (Lindenmayer 1996). Once hollows have formed, animals have to migrate from other suitable sites if colonisation is to take place. Hollows are more likely in fire-scarred trees (Gibbons and Lindenmayer 1997) so certain fire regimes may be expected to promote hollow formation (Fig. 15. 2). On the other hand, if resprouters are maintained by particular fire regimes as a coppice then stems may never attain the minimum size necessary for hollow formation (Fig. 15. 2).

The regenerating vegetation on a burnt patch – the 'green pick' – is an attraction to herbivores. Wombats and macropods (kangaroos and wallabies) may find new shoots of grasses and shrubs particularly palatable. With complete concentration of animals on the burnt patch, the stocking rate of animals is magnified above the norm, provided that only part of the home range has been burnt. Leigh and Holgate (1979) have demonstrated experimentally that the mortality of resprouting plants due to the burning–grazing interaction on small burnt patches is greater than that of either burning or grazing alone. After unplanned

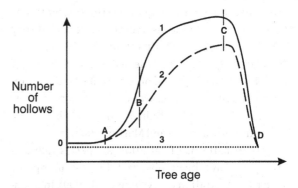

Fig. 15.2. A hypothetical scheme for the effects of fire regimes on the formation of hollows in trees. The extreme fire regimes depicted either promote the maximum number of hollows ('1', uppermost line) or prevent their occurrence by maintaining a coppice in which stems never reach the minimum size ('A') for hollow formation ('3', lowest line). Unburnt trees may form hollows ('2', middle line) but not to the extent that those with the optimum fire regime does. At 'C', the trees die but the hollows persist in the dead trees.

fires the regeneration of grasses may be immediate in certain parts of the landscape (e.g. on deeper soils) and precede that of shrubs. While this has not been formally studied it suggests that there may be spatial and temporal effects on plant populations due to the burning–grazing interaction that extend beyond those related to the overlap of home ranges with a burnt area.

Changes in the biota with the time since the last fire: longer-term effects

In this section, various forest conditions associated with years after fire are described. From previous sections it may be obvious that the starting conditions for time-since-fire sequences vary. Variables affecting initial conditions include the nature of the fire pattern, fire properties, and immediate post-fire conditions (including weather and the extent of predation or herbivory) (e.g. Ashton and Martin 1996a, b) as well as the condition of the ecosystem at the time of the fire. Because scientists can observe at any time differences between stands burnt at different times in the past in terms of population sizes, structure, species richness etc., such observations are made relatively frequently (Gill 1999). When such observations are arranged into time sequences, the assumption is that the initial conditions were the same or that they were unimportant. Other methods of studying time-since-fire sequences include experiment (Tolhurst and Flinn 1992), repeated observation of plots set up after unplanned fire (e.g. Wark 1997) or observations of plots, previously studied, that are burnt by unplanned fire (Ashton and Martin 1996a, b).

Several patterns of variation in species richness after fire have been observed for vascular-plant species in Australian ecosystems (Gill 1999). In open-forest the most common one was for richness to be more or less constant after a short recovery period, or show a slow decline in numbers, for the first decade or so. Examples were reported by Christensen and Kimber (1975) and Bell and Koch (1980) in the southwest, Purdie and Slatyer (1976) in the Australian Capital Territory and Venning (1978) for a South Australian 'woodland'. This pattern may be repeated in wetter eucalypt forests too (Ashton and Martin 1996a). In southwestern tall open-forests the number of species can increase quite substantially soon after fire (Christensen and Kimber 1975). Tall open-forests can also be colonised by closed-forest species causing, over time, a turnover of species (Ashton 1981a) and a net decline in richness (Gill 1999). Burnt closed forests could recycle or be displaced by eucalypt forest (Noble and Slatyer 1981). The small increase in richness after fire that is seen in some places is caused by ephemeral species.

In the southeast region (central Victoria) after a detailed study of plots examined for over a decade, Ashton and Martin (1996a) found a shift in dominance of both canopy and understorey species after fire in a 74-year-old stand of *Eucalyptus regnans*, *E. cypellocarpa* and *E. obliqua*. Three native ephemeral species from a total of over 60 vascular species present appeared for at least a few years after fire. Pre-fire mosses took 6–8 years to return to the site. After 50–100 years some species of vascular plants may be present only as seeds in the soil (Ashton and Martin 1996b). These results may be compared with those for forests burnt about the same time and studied for a similar period but in southwestern Victoria (Wark 1997). In 'gully complexes' dominated by *E. cypellocarpa* and *E. obliqua*, vascular-plant species richness of quadrats dropped by over 40% during the decade after crown fire. In nearby drier open-forests of *E. tricarpa*, vascular-plant species richness declined by about 30%. The small change in the central Victorian case of Ashton and Martin (1996a, b) compared with the large change in the southwestern case may be influenced by the regional *forest* context of the first compared with the general *heath*-, and *heath–woodland*, context (Wark 1997) of the second.

Generalisations as to the patterns of plant species richness to be found after fire in forests must be seen as hypotheses only at present (Gill 1999). The methods used and the time-scale of observation may affect the results. There is a need for many more case histories having regard to species characteristics and homology of sites. There are likely to be many variations in patterns according to species' characteristics (see Noble and Slatyer 1981), soil fertility, regional floral contexts and site histories. Non-vascular plant patterns need further study, the few data available suggesting a rise in richness with time after intense fire (Wark 1997).

Species richness is not the only item of post-fire studies. Thus, the numbers of individual plants of a species can be tracked as a function of time since fire also (Tolhurst and Oswin 1992; Ashton and Martin 1996b). The patterns in density with time will vary according to patterns of germination (and processes like suckering) and mortality and these, in turn, vary according to the characteristics of the species and the characteristics of fires (see above) beginning the sequence. Thus a species able to disseminate seed and establish in intervals between fires (e.g. some *Callitris* species) will show a different pattern to a species with hard seed stimulated to germinate by fires with appropriate properties (like many *Acacia* spp.) but unlikely to regenerate at other times. Shrub species with regenerative properties stimulated by fires will show a marked increase in density soon after fire followed by a progressive decline with time due to thinning. Height and cover may show a similar pattern with time but shrubberies may collapse eventually to leave only a short, sparse understorey. Examples of this are to be found in open-forest with *Pultenaea muelleri* (Gill 1964 and subsequent observations) and in tall open-forests of karri (*E. diversicolor*) with *Bossiaea* understorey (L. McCaw, personal communication).

The development of a 'habitat complexity score' provided a way of describing habitats of ground-dwelling mammals independent of plant species (Newsome and Catling 1979). In a similar way the habitat for birds can be described on the basis of the complexity of forest structure. Because the complexity of forest structure correlates with bird abundance (Recher 1985) and the abundance is correlated with bird species richness in eucalypt forests (Recher 1985; R. Lloyn, personal communication), the abundance and species richness of birds will change as structural complexity changes with time after fire. The extent of change will vary according to the immediate effects of fire and the subsequent rates of recovery of food sources and structure (Christensen *et al.* 1985). Lloyn (1997) found that, after an initial dip in the population, numbers increased for the following 3 years of the study. Similarly, Christensen *et al.* (1985) found increased numbers of bird species compared with an 'unburnt' area for 3 years after a fire. Following an intense fire in tall open-forest in the Australian Capital Territory species composition stayed constant but abundance changed (Catling and Newsome 1981). A single figure for richness, let alone figures depicting a changing richness, may obscure the presence of variations in species composition through time.

Changes occur in the abundance and/or species richness of invertebrates after fire just as those for plants and vertebrates do. York (1996) found a decrease in the richness of ant communities in open-forest on the north coast of New South Wales for approximately 8 years, then a levelling off in species numbers at the 14-year mark. It was noted though that the figures for species richness tended to obscure the fact that species replacement was occurring too (York 1996). The full complement of ant species was only to be found across sites with a variety of post-fire ages. Using a wider taxonomic range of invertebrates – bugs, beetles, spiders and flies – but a narrower range of fire histories in the same forest (frequently burnt versus unburnt), York

(1999) found changed community compositions to be the usual case.

The explanation of the changes in abundance and richness of animal species, including invertebrates, after fire resides in the changes of habitat created by the recovering plant community together with the life-history attributes of the species. Friend (1993) examined the characteristics of vertebrate species showing different abundance as mallee woodlands and heathlands recovered after fire, results that also apply to open-forests (G. Friend, personal communication). Species of small mammal that are abundant after fires (like *Pseudomys* spp.) live in burrows, are generalist feeders and opportunistic breeders; species more abundant later (like *Antechinus* species) live above ground in hollow logs among dense understoreys with thick litter and are monoestrus (Friend 1993). Native *Rattus* numbers, often peaking relatively late in 'succession' also have 'somewhat broader shelter and food requirements, and less rigid reproductive strategies than the two *Antechinus* species' (Friend 1993). Because this series of changes is driven by habitat, shifts in significant habitat variables may allow 'early successional' species like *Pseudomys* to become relatively abundant much later in the 'succession' than expected from short-term results (Fox 1996). Other examples would follow short-term or long-term shifts in habitat complexity, for ground-dwelling mammals, or structural complexity, for birds. It is the details of the habitats, the characteristics of the animals and how the habitat variables change with time – and with fire regime (Catling 1991; York 1999) – that largely determine the composition of faunal communities.

Changes in the biota according to the between-fire interval

Short intervals between fires may eliminate some species. Stand-killing fires within the juvenile period (time to flowering) of seeder species will eliminate them unless seed dispersal into the area is effective after the short inter-fire interval. Local extinction is to be expected in parts of the ranges of such species. Some, but not all, species of *Eucalyptus*,

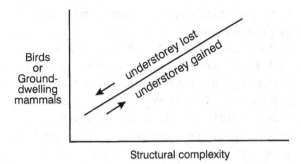

Fig. 15.3. Theoretical depiction of the changes in the abundance of birds and ground-dwelling mammals as a function of structural complexity. Structural complexity of the vegetation is a function of intervals between fires. As complexity changes, changes in species composition would be expected.

Hakea and *Banksia* could be affected in this way. There are few specific instances (Gill and Bradstock 1995). The lengths of juvenile periods of some seeder species have already been mentioned. Sprouter species are less prone to extinction (Gill and Bradstock 1995) because they can survive one or more fires. That does not mean they are not at all vulnerable because fires at intervals shorter than the time for seedling or sapling survival, repeated often, will mean that the adult population can be whittled away and all recruitment fail (Bradstock and Myerscough 1988). High-intensity fires may cause mortality in plant populations of some sprouter species.

Oft repeated low-intensity fires may simplify understorey structure and, by so doing, reduce the abundance and species richness of ground-dwelling mammals (Catling 1991), birds (after Recher 1985) and insects (York 1999). Figure 15.3 illustrates this for vertebrates. Note that species changes are likely with changes in complexity.

Fires at intervals too long to allow for the persistence of propagules in the ecosystem could cause local extinctions. The most obvious case of this is when *E. regnans* exceeds its longevity; the understorey of these forests is too dense to allow any regeneration between fires (Ashton 1981a). For species with long-dormant seed, like *Acacia dealbata*,

plant populations may die out but the species persists as seed in the soil, possibly for centuries (Gilbert 1959); germination is ineffective then until a fire with suitable characteristics allows germination and seedling establishment.

The studies of Ashton on the tall open-forest (*E. regnans*) at Wallaby Creek in central Victoria, extending over a 50-year period, provide an example of the sorts of plant-species dominance that may be expected in such forests with repeated fires at different mean intervals. *Cassinea aculeata* and bracken may dominate the understorey in the early years after fire but persist longest in physiologically dry sites (relatively low-rainfall sites or those sites with shallow soils). After four decades or so, the understorey may change to one with dominance of *Pomaderris apetala*, *Olearia argophylla* and *Bedfordia arborescens*, a condition that may persist for a century in the absence of a nearby source of closed-forest propagules (Ashton 1981b). Then *Pomaderris* may senesce and die out while tree ferns and ground ferns like *Polystichum proliferum* increase in abundance (Ashton and Attiwill 1994). If closed forests are nearby, the understorey may become a closed forest that in the absence of further fire in the centuries ahead will replace the eucalypt forest (Ashton and Attiwill 1994). Crown fires at intervals less than 15–20 years will eliminate *E. regnans*; fires at 5–10 years may remove *Pomaderris* while, in Tasmania, humus fires may destroy the entire community (Ashton and Attiwill 1994). Removal of the eucalypt may leave forests of other species like *A. dealbata*. There are many community compositions possible even for a group of just four potentially co-occurring taxa – *Acacia* and *Eucalyptus*, typical of tall open-forest, and *Nothofagus* and *Atherosperma*, typical of closed forest – given a variety of fire intervals (Noble and Slatyer 1981). A multi-dimensional matrix of factors is at work determining the composition and structure of these forest communities.

The effects of randomness of fire intervals have rarely been examined in Australia (but see McCarthy *et al.* 1999 for an exception). Randomness in a tall open-forest with a mean fire interval of 100 years may mean that there are patches of forest that have not burnt for 500 years. There may be areas in which fire intervals have been very short and in which species have been eliminated. Vulnerability to extinction will be great where intervals have been short and fires large (reducing the chances of effective dispersal and establishment). Under these circumstances, a stochastic fire interval may be less likely to cause extinction than a fixed interval, even though the mean was the same. If the region is large enough and dispersal processes effective then the expectation is of a pattern of occupancy and extinction (McCarthy and Lindenmayer 1998), population boom and bust and greater and lesser herbivory according to the fire history of each point and the nature of the plants and animals in the landscape.

Discussion and conclusions

The southern forests of Australia vary widely in composition and structure. In the period before the arrival of European settlers, the forests were occupied by Aboriginal people who set fires in them. Their activity did not preclude fires ignited by lightning. Early European explorers reported that the understoreys of eucalypt communities were open but thickets also occurred (Benson and Redpath 1997). In tall open-forests, and perhaps closed forests, high-intensity fires probably occurred along with less intense fires; peat fires occurred also. From the grasstree evidence (Ward and Van Didden 1997) it would appear that very frequent fires were usual in sectors of open-forest of jarrah, at least. If the mean between-fire interval in open-forest was 3 years (Ward and Van Didden 1997) and intervals were randomly distributed in time, then some quite long intervals between fires would have occurred and, in those areas, thickets may have formed. In those thickets species dependent on the longer fire intervals may have been found. Present-day observations of vertebrate fauna and insects would suggest that in areas with an open understorey would contain sub-maximum numbers of birds and ground-dwelling mammals. Frequent low-intensity fires would not provide the maximum number of hollows for hollow-dwellers on the basis of present evidence.

Mean fire intervals were longer in tall open-forests than in open-forests. A range of forest under-storeys dependent on the variation about the mean of the fire interval would have been expected. The closed forests would have had a mean fire interval longer than the other forests based on their present locations in gullies surrounded by open-forests or tall open-forests. These forests were quite simple in composition and confined to eastern Australia in the recent past.

Using the present as a guide to the past is useful but certain caveats must be kept in mind. Contemporary studies are influenced by fragmentation of forests, the influences of changed patterns of native animal abundance and distribution (including extinctions), and the presence of exotic plants, animals and diseases. Without these influences our understanding may be altered. With the reduction of numbers of the introduced fox, for example, the habitat needs of the woylie and tammar might be different (Christensen 1980).

Our knowledge of fire regimes and their effects is constantly growing but new insights create new challenges (see Gill et al., this volume). For example, there has been no long-term fire experiment in tall open-forest nor in closed forest. Between-fire intervals in experiments so far have been short. All experiments have used fixed between-fire intervals. There is a need to develop experiments in which the effects of stochastic regimes are examined.

Modelling has great potential for the discernment of past fire regimes and their effects. As further changes are imposed on the Australian landscape by development, the use of the life histories and habitats of species selected on the basis of their indicator value for fire regime components could provide the valuable insight into the fire regimes of the past. As the biological (e.g. exotics), economic and ecological context (e.g. fragmentation) of our forested landscapes changes, the value of our knowledge of the past fire regimes and their effects on biodiversity for management of today's forested landscapes will diminish.

Acknowledgements

We would like to thank Drs Neil Burrows and Gordon Friend for reading and commenting on a draft of the manuscript. Peter H. R. Moore provided valuable technical support.

References

Ambrose, G. J. (1982). An ecological and behavioural study of vertebrates using hollows in eucalypt branches. PhD thesis, La Trobe University, Melbourne.

Ashton, D. H. (1981a). Tall open-forests. In *Australian Vegetation* (ed. R. H. Groves) pp. 121–151. (Cambridge University Press: Cambridge.)

Ashton, D. H. (1981b). Fire in tall open-forests (wet sclerophyll forests). In *Fire and the Australian Biota* (eds. A. M. Gill, R. H. Groves and I. R. Noble) pp. 339–366. (Australian Academy of Science: Canberra.)

Ashton, D. H. (1986). Viability of seeds of *Eucalyptus obliqua* and *Leptospermum juniperinum* from capsules subject to a crown fire. *Australian Forestry* 49, 28–35.

Ashton, D. H., and Attiwill, P. M. (1994). Tall open-forests. In *Australian Vegetation*, 2nd Edn (ed. R. H. Groves) pp. 157–196. (Cambridge University Press: Cambridge.)

Ashton, D. H., and Martin, D. G. (1996a). Changes in a spar-stage ecotonal forest of *Eucalyptus regnans*, *Eucalyptus obliqua* and *Eucalyptus cypellocarpa* following wildfire on the Hume Range in November 1982. *Australian Forestry* 59, 32–41.

Ashton, D. H., and Martin, D. G. (1996b). Regeneration in a pole-stage forest of *Eucalyptus regnans* subjected to different fire intensities in 1982. *Australian Journal of Botany* 44, 393–410.

Ashton, D. H., and McRae, R. F. (1970). The distribution of epiphytes on beech (*Nothofagus cunninghamii*) trees at Mt. Donna Buang, Victoria. *Victorian Naturalist* 87, 253–261.

Austin, M. P. (1978). Vegetation. In *Land Use on the South Coast of New South Wales: A Study in Methods of Acquiring and Using Information to Analyse Regional Land-Use Options* (eds. M. P. Austin and K. D. Cocks) pp. 44–67. (CSIRO: Melbourne.)

Barker, M. J. (1991). The effect of fire on West Coast lowland rainforest. Tasmanian National Rainforest Conservation Program Report no. 7, Forestry Commission, Hobart.

Barker, P. C. J. (1992). Fire and the stability of a Victorian cool temperate rainforest. In *Victoria's Rainforests: Perspectives on Definition, Classification and Management*

(eds. P. Gell and D. Mercer) pp. 127–131. (Monash University: Melbourne.)

Beer, T., Gill, A. M., and Moore, P. H. R. (1988). Australian bushfire danger under changing climatic regimes. In *Greenhouse: Planning for Climatic Change* (ed. G. Pearman) pp. 421–427. (CSIRO: Melbourne.)

Bell, D. T., and Heddle, E. M. (1989). Floristic, morphologic and vegetational diversity. In *The Jarrah Forest* (eds. B. Dell, J. J. Havel and N. Malajczuk) pp. 53–66. (Kluwer: Dordrecht.)

Bell, D. T., and Koch, J. M. (1980). Post-fire succession in the northern jarrah forest of Western Australia. *Australian Journal of Ecology* 5, 9–14.

Benson, J. S., and Redpath, P. A. (1997). The nature of pre-European native vegetation in south-eastern Australia: a critique of Ryan, D. G., Ryan, J. R., and Starr, B. J. (1995) *The Australian Landscape: Observations of Explorers and Early Settlers. Cunninghamia* 5, 285–328.

Bowman, D. M. J. S., and Kirkpatrick, J. B. (1984). Geographic variation in the demographic structure of stands of *Eucalyptus delegatensis* R.T. Baker on dolerite in Tasmania. *Journal of Biogeography* 11, 427–437.

Bradshaw, F. J., and Lush, A. R. (1981). *Conservation of the Karri Forest.* (Forests Department of Western Australia: Perth.)

Bradstock, R. A., and O'Connell, M. A. (1988). Demography of woody plants in relation to fire: *Banksia ericifolia* L.f. and *Petrophile pulchella* (Schrad) R.Br. *Australian Journal of Ecology* 13, 505–518.

Bradstock, R. A., and Myerscough, P. J. (1988). The survival and population response to frequent fires of two woody resprouters *Banksia serrata* and *Isopogon anemonifolius. Australian Journal of Botany* 36, 415–431.

Bradstock, R. A., Gill, A. M., Hastings, S. M., and Moore, P. H. R. (1994). Survival of serotinous seedbanks during bushfires: comparative studies of *Hakea* species from southeastern Australia. *Australian Journal of Ecology* 19, 276–282.

Burrows, N. D. (1994). Experimental development of a fire management model for jarrah (*Eucalyptus marginata* Donn ex Sm) forest. PhD thesis, Australian National University, Canberra.

Burrows, N. D., and Friend, G. (1998). Biological indicators of appropriate fire regimes in southwest Australian ecosystems. *Tall Timbers Fire Ecology Conference Proceedings* 20, 413–421.

Burrows, N. D., Ward, B., and Robinson, A. D. (1995). Jarrah forest fire history from stem analysis and anthropological evidence. *Australian Forestry* 58, 7–16.

Byram, G. M. (1959). Combustion of forest fuels. In *Forest Fire: Control and Use* (ed. K. P. Davis) pp. 61–89. (McGraw-Hill: New York.)

Carnahan, J. A. (1976). Natural vegetation. Commentary on the map 'Natural Vegetation' in *Atlas of Australian Resources*, 2nd series. (Commonwealth Department of Natural Resources: Canberra.)

Cary, G., and Banks, J. C. G. (2001). Fire regime sensitivity to global climate change: an Australian perspective. In *Biomass Burning and its Inter-Relationships with the Climate System* (eds. J. L. Innes, M. Beniston and M. M. Verstraete) pp. 233–246. (Kluwer Academic Publishers: Dordrecht and Boston.)

Catchpole, W. (2001). Fire properties and burn patterns in heterogeneous landscapes. In *Flammable Australia: The Fire Regimes and Biodiversity of a Continent* (eds. R. A. Bradstock, J. E. Williams and A. M. Gill) pp. 49–76. (Cambridge University Press: Cambridge.)

Catling, P. C. (1991). Ecological effects of prescribed burning practices on the mammals of southeastern Australia. In *Conservation of Australia's Forest Fauna* (ed. D. Lunney) pp. 353–363. (Royal Zoological Society: Sydney.)

Catling, P. C., and Burt, R. J. (1995). Studies of the ground-dwelling mammals of eucalypt forests in southeastern New South Wales: the effect of habitat variables on distribution and abundance. *Wildlife Research* 22, 271–288.

Catling, P. C., and Newsome, A. E. (1981). Responses of the Australian vertebrate fauna to fire: an evolutionary approach. In *Fire and the Australian Biota* (eds. A. M. Gill, R. H. Groves and I. R. Noble) pp. 273–310. (Australian Academy of Science: Canberra.)

Chesterfield, E. A., Taylor, S. J., and Molnar, C. D. (1991). Recovery after wildfire: warm temperate rainforest at Jones Creek, East Gippsland, Victoria. *Australian Forestry* 54, 157–173.

Christensen, P. E. S. (1980). The biology of *Bettongia penicillata* Gray, 1837, and *Macropus eugenii* (Desmarest, 1817) in relation to fire. Forests Department of Western Australia Bulletin no. 91, Perth.

Christensen, P. E. S., and Kimber, P. C. (1975). Effect of prescribed burning on the flora and fauna of south-west Australian forests. *Proceedings of the Ecological Society of Australia* 9, 85–106.

Christensen, P. E. S., Recher, H., and Hoare, J. (1981). Responses of open-forests (dry sclerophyll forests) to fire regimes. In *Fire and the Australian Biota* (eds. A. M. Gill, R. H. Groves and I. R. Noble) pp. 367–393. (Australian Academy of Science: Canberra.)

Christensen, P. E. S., Wardell-Johnson, G., and Kimber, P. C. (1985). Birds and fire in southwestern forests. In *Birds of Eucalypt Forests and Woodlands: Ecology, Conservation, Management* (eds. A. Keast, H. F. Recher, H. Ford and D. Saunders) pp. 291–299. (Surrey Beatty: Sydney.)

Churchill, D. M. (1968). The distribution and prehistory of *Eucalyptus diversicolor* F. Muell., *E. marginata* Donn ex Sm., and *E. calophylla* R.Br. in relation to rainfall. *Australian Journal of Botany* 16, 125–151.

Clark, J. S., Gill, A. M., and Kershaw A. P. (2001). Spatial variability in fire regimes: its effects on recent and past vegetation. In *Flammable Australia: The Fire Regimes and Biodiversity of a Continent* (eds. R. A. Bradstock, J. E. Williams and A. M. Gill) pp. 125–141. (Cambridge University Press: Cambridge.)

Cremer, K. W. (1962). The effect of fire on eucalypts reserved for seeding. *Australian Forestry* 26, 129–154.

Dexter, B. D., Heislers, A., and Sloan, T. (1977). The Mount Buffalo fire. Forests Commission of Victoria, Bulletin no. 26, Melbourne.

Drielsma, J. H. (1994). *Proposed Forestry Operations in Eden Management Area*, vol. A, *Environmental Impact Statement Main Report*. (State Forests of New South Wales: Sydney.)

Duncan, F., and Brown, M. J. (1995). Edaphics and fire: an interpretive ecology of lowland forest vegetation on granite in northeast Tasmania. *Proceedings of the Linnean Society of New South Wales* 115, 45–60.

Fox, A. (1978). The '72 fire of Nadgee Nature Reserve. *Parks and Wildlife* 2, 5–24.

Fox, B. J. (1996). Long-term studies of small-mammal communities from disturbed habitats in eastern Australia. In *Long-Term Studies of Vertebrate Communities* (eds. M. L. Cody and J. A. Smallwood) pp. 567–501. (Academic Press: San Diego.)

Friend, G. R. (1993). Impact of fire on small vertebrates in mallee woodlands and heathlands of temperate Australia: a review. *Biological Conservation* 65, 99–114.

Gibbons, P., and Lindenmayer, D. B. (1997). *Forest Issues 2: Conserving hollow-dependent fauna in timber production forests*, New South Wales National Parks and Wildlife Service Environmental Heritage Monograph Series no. 3. (New South Wales National Parks and Wildlife Service: Sydney.)

Gilbert, J. M. (1959). Forest succession in the Florentine Valley, Tasmania. *Papers and Proceedings of the Royal Society of Tasmania* 93, 129–151.

Gill, A. M. (1964). Soil-vegetation relationships near Kinglake West, Victoria. MSc thesis, University of Melbourne.

Gill, A. M. (1975). Fire and the Australian flora: a review. *Australian Forestry* 38, 4–25.

Gill, A. M. (1981). Adaptive responses of Australian vascular plant species to fires. In *Fire and the Australian Biota* (eds. A. M. Gill, R. H. Groves and I. R. Noble) pp. 243–271. (Australian Academy of Science: Canberra.)

Gill, A. M. (1994). Patterns and processes in open-forests of *Eucalyptus* in southern Australia. In *Australian Vegetation*, 2nd edn. (ed. R. H. Groves) pp. 197–226. (Cambridge University Press: Cambridge.)

Gill, A. M. (1995). Stems and fires. In *Plant Stems: Physiology and Functional Morphology* (ed. B. Gartner) pp. 323–342. (Academic Press: San Diego.)

Gill, A. M. (1996). How fires affect biodiversity. In *Fires and Biodiversity: The Effects and Effectiveness of Fire Management*, (Biodiversity Series Paper no. 8, 47–55. Australian Department of Environment, Sports and Territories: Canberra.)

Gill, A. M. (1999). Biodiversity and bushfires: an Australia-wide perspective on plant-species changes after a Fire event. In *Australia's Biodiversity: Responses to Fire*, Biodiversity Technical Paper no. 1, pp. 9–53. (Environment Australia: Canberra.)

Gill, A. M., and Bradstock, R. A. (1995). Extinctions of biota by fires. In *Conserving Biodiversity: Threats and Solutions* (eds. R. A. Bradstock, T. D. Auld, D. A. Keith, R. Kingsford, D. Lunney and D. Sivertsen) pp. 309–322. (Surrey Beatty: Sydney.)

Gill, A. M., and Moore, P. H. R. (1990). Fire intensities in *Eucalyptus* forests of southeastern Australia. *Proceedings of the 1st International Conference on Forest Fire Research*, Coimbra, Portugal, Paper B24.

Gill, A. M., and Moore, P. H. R. (1994). Some ecological research perspectives on the disastrous Sydney fires of January 1994. *Proceedings of the 2nd International Conference on Forest Fire Research*, Coimbra, Portugal, volume 1, pp. 63–72.

Gill, A. M., and Moore, P. H. R. (1996). Regional and historic fire weather patterns pertinent to the January 1994 Sydney bushfires. *Proceedings of the Linnean Society of New South Wales* 116, 27–36.

Gill, A. M., and Moore, P. H. R. (1997). *Contemporary Fire Regimes in the Forests of Southwestern Australia*, Report for Environment Australia. (CSIRO Plant Industry: Canberra.)

Gill, A. M., and Moore, P. H. R. (1998). Big versus small fires: the bushfires of greater Sydney, January 1994. In *Large Forest Fires* (ed. J. M. Moreno) pp. 49–68. (Backhuys: Leiden.)

Gill, A. M., Belbin, L., and Chippendale, G. M. (1985). *Phytogeography of* Eucalyptus *in Australia*, Bureau of Flora and Fauna, Australian Flora and Fauna Series no. 3. (Australian Government Publishing Service: Canberra.)

Gill, A. M., Bradstock, R. A., and Williams, J. E. (2001). Fire regimes and biodiversity: legacy and vision. In *Flammable Australia: The Fire Regimes and Biodiversity of a Continent* (eds. R. A. Bradstock, J. E. Williams and A. M. Gill) pp. 429–446. (Cambridge University Press: Cambridge.)

Hallam, S. J. (1975). *Fire and Hearth: A Study of Aboriginal Usage and European Usurpation in South-Western Australia*. (Australian Institute of Aboriginal Studies: Canberra.)

Helman, C. (1987). Rainforest in New South Wales. In *Australian National Rainforests Study*, vol. 1, *The Rainforest Legacy*, pp. 47–70. (Australian Government Publishing Service: Canberra.)

Jackson, W. D. (1965). Vegetation. In *Atlas of Tasmania*. (ed. J. L. Davies) pp. 30–35. (Tasmanian Lands and Survey Department: Hobart.)

Jackson, W. D. (1968). Fire, air, water and earth: an elemental ecology of Tasmania. *Proceedings of the Ecological Society of Australia* 3, 9–16.

Jarman, S. J., Kantvilas, G., and Brown, M. J. (1994). Phytosociological studies in Tasmanian cool temperate rainforest. *Phytocoenologia* 22, 355–390.

Johnson, C. N. (1995). Interactions between fire, mycophagous mammals, and dispersal of ectomycorrhizal fungi in *Eucalyptus* forests. *Oecologia* 104, 467–475.

Johnson, E. A., and Gutsell, S. L. (1994). Fire frequency

models, methods and interpretations. *Advances in Ecological Research* **25**, 239–287.

Keast, A. (1981). The evolutionary biogeography of Australian birds. In *Ecological Biogeography of Australia* (ed. A. Keast) pp. 1586–1635. (Dr W. Junk: The Hague.)

Kershaw, A. P., Clark, J. S., Gill A. M., and D'Costa, D. M. (2001). A history of fire in Australia. In *Flammable Australia: The Fire Regimes and Biodiversity of a Continent* (eds. R. A. Bradstock, J. E. Williams and A. M. Gill) pp. 3–25. (Cambridge University Press: Cambridge.)

Kinnear, J., Onus, M., and Bromilow, B. (1984). Foxes, feral cats and rock wallabies. *Swans* **14**, 3–8.

Lang, S. (1997). Burning the bush: a spatio-temporal analysis of jarrah forest fire regimes. BSc. thesis, Australian National University, Canberra.

Leigh, J. H., and Holgate, M. D. (1979). The responses of the understorey of forests and woodlands of the Southern Tablelands to grazing and burning. *Australian Journal of Ecology* **4**, 25–45.

Lindenmayer, D. (1996). *Wildlife and Woodchips*. (University of New South Wales: Sydney.)

Lloyn, R. H. (1997). Effects of an extensive wildfire on birds in far eastern Victoria. *Pacific Conservation Biology* **3**, 221–234.

Majer, J. D., Recher, H. F., Wellington, A. B., Woinarski, J. C. Z., and Yen, A. L. (1997). Invertebrates of eucalypt formations. In *Eucalypt Ecology: Individuals to Ecosystems* (eds. J. E. Williams and J. C. Z. Woinarski) pp. 278–302. (Cambridge University Press: Cambridge.)

McArthur, A. G. (1967). Fire behaviour in eucalypt forests. Commonwealth of Australia Forestry and Timber Bureau, Leaflet no. 107, Canberra.

McBryde, I., and Nicholson, P. (1978). Aboriginal man and the land in south-western Australia. *Studies in Western Australian History* **3**, 38–42.

McCarthy, M. A., and Lindenmayer, D. B. (1998). Multi-aged mountain ash forest, wildlife conservation and timber harvesting. *Forest Ecology and Management* **104**, 43–56.

McCarthy, M. A., and Cary, G. J. (2001). Fire regimes in landscapes: models and realities. In *Flammable Australia: The Fire Regimes and Biodiversity of a Continent* (eds. R. A. Bradstock, J. E. Williams and A. M. Gill) pp. 77–93. (Cambridge University Press: Cambridge.)

McCarthy, M. A., Gill, A. M., and Lindenmayer, D. B. (1999). Fire regimes in mountain ash forest: evidence from forest age structure, extinction models and wildlife

habitat. *Forest Ecology and Management* **124**, 193–203.

McCarthy, M. A., Gill, A. M., and Bradstock, R. A. (2001). Theoretical fire-interval distributions. *International Journal of Wildland Fire* **10**, 73–77.

McMahon, A. R. G. (1987). The effects of the 1982–83 bushfires on sites of significance. Victorian Department of Conservation Forests and Lands, Environmental Studies Publication Series no. 411, Melbourne.

Mercer, G. N., Gill, A. M., and Weber, R. O. (1995). A flexible, non deterministic, litter accumulation model. In *Bushfire '95: Presented Papers*, (Forestry Tasmania, Parks and Wildlife Service Tasmania, Tasmania Fire Service: Hobart.)

Newsome, A. E., and Catling, P. C. (1979). Habitat preferences of mammals inhabiting heathlands of warm temperate coastal, montane and alpine regions of southeastern Australia. In *Ecosystems of the World*, vol. 9A, *Heathlands and Related Shrublands: Descriptive Studies* (ed. R. L. Specht) pp. 301–316. (Elsevier: Amsterdam.)

Newsome, A. E., Catling, P. C., and Corbett, L. K. (1983). The feeding ecology of the dingo. II. Dietary and numerical relationships with fluctuating prey populations in south-eastern Australia. *Australian Journal of Ecology* **8**, 345–366.

Noble, I. R., and Slatyer, R. O. (1980). The use of vital attributes to predict successional changes in plant communities subject to recurrent disturbances. *Vegetatio* **43**, 5–21.

Noble, I. R., and Slatyer, R. O. (1981). Concepts and models of succession in vascular plant communities subject to recurrent fire. In *Fire and the Australian Biota* (eds. A. M. Gill, R. H. Groves and I. R. Noble) pp. 311–335. (Australian Academy of Science: Canberra.)

Noble, I. R., Bary, G. A. V., and Gill, A. M. (1980). McArthur's fire danger meters expressed as equations. *Australian Journal of Ecology* **5**, 201–203.

O'Connell, A. M. (1987). Litter dynamics in karri (*Eucalyptus diversicolor*) forests of south-western Australia. *Journal of Ecology* **75**, 781–796.

O'Dowd, D. J., and Gill, A. M. (1984). Predator satiation and site alteration following fire: mass reproduction of alpine ash (*Eucalyptus delegatensis*) in southeastern Australia. *Ecology* **65**, 1052–1066.

Pook, E. W., and Gill, A. M. (1993). Variation of live and dead fine fuel moisture in *Pinus radiata* plantations of the Australian Capital Territory. *International Journal of Wildland Fire* 3, 155–168.

Pook, E. W., Gill, A. M., and Moore, P. H. R. (1997). Long-term variation of litter fall, canopy leaf area and flowering in *Eucalyptus maculata* forests on the South Coast of New South Wales. *Australian Journal of Botany* 45, 737–755.

Purdie, R.W., and Slatyer, R. O. (1976). Vegetation succession after fire in sclerophyll woodland communities in south-eastern Australia. *Australian Journal of Ecology* 1, 223–236.

Read, J., and Hill, R. S. (1981). Rainforest. In *The Vegetation of Tasmania* (ed. W. D. Jackson) pp. 73–84. (Botany Department, University of Tasmania: Hobart.)

Recher, H. F. (1985). Synthesis: a model of forest and woodland bird communities. In *Birds of Eucalypt Forests and Woodlands: Ecology, Conservation, Management* (eds. A. Keast, H. F. Recher, H. Ford and D. Saunders) pp. 129–135. (Surrey Beatty: Sydney.)

Specht, R. L. (1970). Vegetation. In *The Australian Environment* (ed. G. W. Leeper) pp. 44–67. (CSIRO and Melbourne University Press: Melbourne.)

Specht, R. L. (1983). Foliage projective covers of overstorey and understorey strata of mature vegetation in Australia. *Australian Journal of Ecology* 8, 433–439.

Squire, R. O., Campbell, R. G., Wareing, K. J., and Featherston, G. R. (1991). The Mountain Ash forests of Victoria: ecology, silviculture and management for wood production. In *Forest Management in Australia* (eds. F. H. McKinnell, E. R. Hopkins and J. E. D. Fox) pp. 38–57. (Surrey Beatty: Sydney.)

Swihart, R. K., Slade, N. A., and Bergstrom, B. J. (1988). Relating body size to the rate of home range use in mammals. *Ecology* 69, 393–399.

Tolhurst, K. G., and Flinn, D. (eds.) (1992). *Ecological Impacts of Fuel Reduction Burning in Dry Sclerophyll Forest: First Progress Report*, Research Report no. 349. (Forest Research Section, Victorian Department of Conservation and Environment: Melbourne.)

Tolhurst, K. G., and Oswin, D. A. (1992). Effects of spring and autumn low intensity fire on understorey vegetation in open eucalypt forest in west-central Victoria. In *Ecological Impacts of Fuel Reduction Burning in Dry Sclerophyll Forest: First Progress Report*, Research Report no. 349 (eds. K. G. Tolhurst and D. Flinn) pp. 3.1–3.60. (Forest Research Section, Victorian Department of Conservation and Environment: Melbourne.)

Venning, J. (1978). Post-fire responses of a *Eucalyptus baxteri* woodland near Penola in South Australia. *Australian Forestry* 41, 192–206.

Viegas, D. X. (1994). Letter to the editor. Some thoughts on the wind and slope effects on fire propagation. *International Journal of Wildland Fire* 4, 63–64.

Walker, J. (1981). Fuel dynamics in Australian vegetation. In *Fire and the Australian Biota* (eds. A. M. Gill, R. H. Groves and I. R. Noble) pp. 101–127. (Australian Academy of Science: Canberra.)

Ward, D., and Van Didden, G. (1997). Reconstruction of the fire history of the Jarrah forest of south-western Australia. A report to Environment Australia under the Regional Forest Agreement, Perth.

Wark, M. (1997). Regeneration of some forest and gully communities in the Angahook–Lorne State Park (north-eastern Otway Ranges) 1–10 years after the wildfire of February 1983. *Proceedings of the Royal Society of Victoria* 109, 7–36.

Whelan, R. J., Rodgerson, L., Dickman, C. R., and Sutherland, E. F. (2001). Critical life cycles of plants and animals: developing a process-based understanding of population changes in fire-prone landscapes. In *Flammable Australia: The Fire Regimes and Biodiversity of a Continent* (eds. R. A. Bradstock, J. E. Williams and A. M. Gill) pp. 94–124. (Cambridge University Press: Cambridge.)

Williams, J. E., and Gill, A. M. (1995). *Forest Issues 1: The impact of fire regimes on native forests in eastern New South Wales*, New South Wales National Parks and Wildlife Service Environmental Heritage Monograph Series no. 2. (New South Wales National Parks and Wildlife Service: Sydney.)

Woinarski, J. C. Z., Recher, H. F., and Majer, J. D. (1997). Vertebrates of eucalypt formations. In *Eucalypt Ecology: Individuals to Ecosystems* (eds. J. E. Williams and J. C. Z. Woinarski) pp. 303–341. (Cambridge University Press: Cambridge.)

York, A. (1996). Long-term effects of fuel-reduction burning on invertebrates in a dry sclerophyll forest. In *Fires and Biodiversity. The Effects and Effectiveness of Fire Management*, Biodiversity Series Paper no. 8, pp. 163–181. (Australian Department of Environment, Sports and Territories: Canberra.)

York, A. (1999). Long-term effects of repeated prescribed burning on forest invertebrates: management implications for the conservation of biodiversity. In *Australia's Biodiversity: Responses to Fire*, Biodiversity Technical Paper no. 1, pp. 181–266. (Environment Australia: Canberra.)

Part VII

Applications

16

Fire regimes in semi-arid and tropical pastoral lands: managing biological diversity and ecosystem function

JAMES C. NOBLE AND ANTHONY C. GRICE

Abstract

After briefly describing the nature of the dominant pastoral industries in the temperate semi-arid and tropical grazing lands of Australia, this chapter focusses on changes in fire regimes following upon European settlement and the impacts these changes have had on resident biota. In this context, the interactions between fire, rainfall, grazing and browsing are seen as pivotal in mediating fundamental ecological processes, and subsequent habitat status, in pastoral landscapes. While considerable effort has been expended on elucidating the impacts of different fire regimes on individual species in these ecosystems, very little research at a faunal community level has been undertaken since the topic was last reviewed two decades ago. Few management options are available to pastoralists other than those directed towards regulating the consumption of vegetation by either grazing or fire. Prescribed fire is now recognised as one of the few cost-effective tools available for maintaining biological diversity at an appropriate management, i.e. paddock, scale. None the less, considerable challenges still remain in developing reliable procedures for defining optimal fire/vegetation mosaics required to maintain or increase diversity of resident biota. Finally, after briefly discussing the functional significance of biodiversity in pastoral ecosystems, the chapter concludes by discussing theoretical considerations relating to ecosystem resilience and multiple stable states, and the roles of contrasting rainfall, fire and grazing regimes in mediating transitions from one state to another.

Introduction

Pastoralism is a term used in Australia to describe a land use 'relating to, or occupied in, the care of flocks or herds' (Ramson 1988). In semi-arid regions particularly, pastoral lands are closely allied with the rangelands used primarily for grazing purposes under extensive systems of management. Early European settlers viewed pastoralism as a pioneer or 'seral' phase during 'agricultural succession' supposedly leading inexorably to a 'climax' in land development characterised by stable populations of yeoman farmers (Powell 1988). As settlement advanced into the semi-arid hinterland, the inevitable droughts ultimately put pay to the popular fallacy that 'rainfall followed the plough'.

Eventually, pastoralism was recognised as providing the only plausible means of profitably exploiting much of the arid interior. As experience in the late 19th century subsequently demonstrated however, even this enterprise was fraught with considerable environmental and economic hazards (Noble 1997a). The first European settlers relied on nomadic systems whereby sheep were tended by shepherds who moved flocks from one natural waterhole to another while yarding them at night for protection against attacks by native dogs (*Canis familiaris dingo*). With the advent of fencing wire and artificial waters, livestock movement became increasingly confined and the country was exposed to impacts from increasing grazing pressures to which it was totally unaccustomed (Noble and Tongway 1986b, Noble *et al.* 1998b).

These pressures were not just the result of

Fig. 16.1. 'The rabbit pest.' Two sides of a netting fence; Mr Rodier is in the sulky on the Tambua side. Cobar/Bourke, October 1905. The scene clearly illustrates the severe grazing pressures being experienced on some properties, and their obvious impact on fuel availability. (Photograph by W. Leaney. Taken from the series *At Work and Play*, no. 02766, held by the State Library of New South Wales, Sydney.)

increasing numbers of livestock following a run of favourable seasons in the 1880s, but also burgeoning populations of native herbivores, especially kangaroos (*Macropus* spp.), that responded to more reliable water supplies. Meggy (1885, p. 149) recorded how 'A great deal of damage was done, too, last year, by the kangaroos which abounded in the district, 12,000 being killed on the Yanda run alone, and nearly 80,000 in the remainder of the district.' Feral species such as the goat (*Capra hircus*) and the European rabbit (*Oryctolagus cuniculus*) later contributed significantly to the total grazing pressure.

With the advent of severe drought conditions in the 1890s, serious environmental deterioration in the semi-arid pastoral lands of eastern Australia was bound to follow. The Minutes of Evidence of the 1900 Royal Commission enquiring into the condition of the crown tenants of the Western Division of New South Wales abound with graphic descriptions of environmental deterioration following excessive grazing pressures (Fig. 16.1) imposed during the last two decades of the 19th century (Noble 1997a). The loss of surface vegetative cover from excessive grazing pressures resulted in dust storms of awesome proportions as dense swirling dust rapidly converted daylight into Stygian gloom (Noble and Tongway 1986a). A major component of this wind erosion was the erosion of nutrients, the majority of which were stored in the surface soil and therefore highly vulnerable to depletion. Once this topsoil was removed, potential plant productivity was significantly diminished (Charley and Cowling 1968; Cowling 1977). Seed banks and 'safe sites' for seedling establishment also declined as landscape surfaces became increasingly homogenised following accelerated erosion (Anderson *et al.* 1996).

Climate, soil, herbivores and fires are recognised as the four major determinants of vegetation structure in pastoral lands (Walker 1985). At a finer scale, other factors such as minimum temperature, plant available moisture (PAM) and plant available nutrients (PAN) dictate the kinds of plants found within a rangeland community (Walker 1993). The widespread deterioration in surface-soil condition described earlier resulted in a major positional shift within the response surface delineated by PAM–PAN axes. While strategies can now be devised for restoring soil fertility in degraded landscapes (Tongway and Ludwig 1996), their successful implementation at a landscape scale is heavily constrained by time, economics and community acceptance (Noble *et al.* 1997).

While soil factors have a fundamental influence on vegetation composition and subsequent herbage fuel production, it is the interactions between grazing and fire regimes, as well as rainfall and topography, that provide the complex spatial heterogeneity characteristic of rangeland vegetation. Of the three major variables used to characterise a fire regime (Gill 1977), season and frequency are by far the most significant in a pastoral context. It is their interactions though, especially during infrequent multiple fire sequences, that are most likely to significantly enhance the final biological impact of different fire regimes (Hodgkinson 1986; Noble *et al.* 1986).

While rangelands also occur in smaller, non-arable regions of high rainfall, we focus primarily in this paper on fire and resource management in semi-arid and tropical pastoral systems (Fig. 16.2). Income from pastoralism in the semi-arid zone today is derived from a variety of enterprises including those concentrating solely on production from domestic livestock as well as others that involve a mix of activities including periodic harvesting of feral and native animals. These pursuits are inextricably linked to the 70% of continental Australia that is too arid for either cropping or sown pastures (Perry 1967, 1977). Around 60% of the total area of these rangelands is devoted to pastoral activities. Aboriginal homelands comprise 15% of the remaining land, 4% is reserved for conservation while 21%

is technically unoccupied although much is currently under claim by Aboriginal people (State of the Environment Advisory Council 1996).

Wool production has traditionally been the dominant land use in the temperate semi-arid zone. With the drastic decline in the price of wool, increasing interest is now being shown in meat production based on new breeds of sheep adapted to these semi-arid regions. Unlike the ubiquitous Merino, some breeds shed their coat resulting in a significant reduction in animal husbandry costs normally associated with shearing, crutching, blowfly (e.g. *Lucilia euprina*) control etc.

Until recently, there has been little interest in the use of prescribed fire as a management tool in temperate semi-arid rangelands. Major wildfire seasons only occur episodically following a sequence of above-average rainfall seasons that produce abundant herbage fuels (Noble *et al.* 1980; Leigh and Noble 1981).

Pastoral enterprises in the tropical regions of Australia are dominated by those based on beef production, primarily from *Bos indicus* breeds tolerant both of the climate and the cattle tick (*Boophilus microplus*). The use of prescribed fire to enhance animal productivity is well established in many northern pastoral lands (Tothill 1976; Ash *et al.* 1982; Andrew 1986; Winter 1987).

Managing fire regimes in pastoral landscapes

When compared with other agricultural systems, pastoral ecosystems are distinguished not only by their ecological complexity, but also by the severe management constraints imposed by this complexity when contemplating management intervention such as prescribed fire (Noble 1986a). These constraints relate primarily to the large areas involved, especially in the semi-arid zone where paddocks may range in size from 4000 up to 8000 hectares. In addition, pastoralists need to have some confidence in their ability to define what particular fire regime is required to achieve a given management objective.

A number of management objectives justifying

Fig. 16.2. Map showing the locations and climatic conditions of the semi-arid (delineated by the heavy border) and tropical pastoral zones of Australia. After UNESCO (1979).

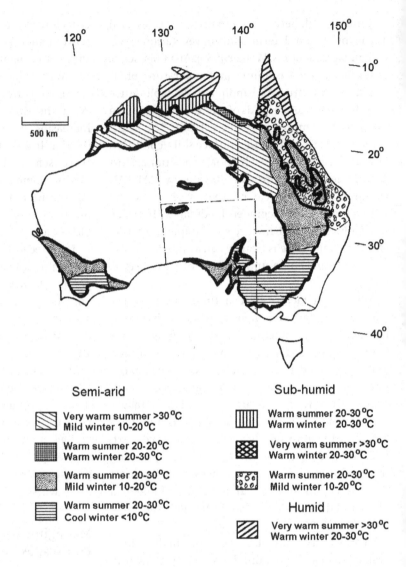

Semi-arid

Very warm summer >30 °C
Mild winter 10-20 °C

Warm summer 20-20 °C
Warm winter 20-30 °C

Warm summer 20-30 °C
Mild winter 10-20 °C

Warm summer 20-30 °C
Cool winter <10 °C

Sub-humid

Warm summer 20-30 °C
Warm winter 20-30 °C

Very warm summer >30 °C
Warm winter 20-30 °C

Warm summer 20-30 °C
Mild winter 10-20 °C

Humid

Very warm summer >30 °C
Warm winter 20-30 °C

the use of prescribed fire in pastoral systems have been cited elsewhere including the removal of rank feed; extending the useful growing and grazing seasons; control of woody species; and hazard or fuel reduction (Leigh and Noble 1981; Noble 1997b). Growing public awareness of biodiversity issues has led to the recognition of the need to conserve all rangeland resources, including endemic biota (Morton *et al.* 1995a, b, c, 1996; Noble *et al.* 1998b). More recently, the role of periodic prescribed fire in enhancing biodiversity has received increasing attention. Particular fire regimes, both on and off reserves, are required to ensure the long-term conservation of endangered biota such as, for example,

the golden-shouldered parrot (*Psephorus chrysoptery-gius*) on Cape York Peninsula (Garnett and Crowley 1997).

The importance of the planning process in defining, in ecological terms, clear management objectives for prescribed fire, has been discussed elsewhere (Noble 1997b). The operational requirements involved in undertaking prescribed fire have also been comprehensively reviewed (e.g. Griffin 1984b; Hodgkinson *et al.* 1984; Hodgkinson and Harrington 1985; Noble 1986b, 1997b; Dyer *et al.* 1997). There are obvious logistical and economic advantages accruing when technologies such as aerial ignition are employed to overcome problems of scale, especially when discontinuous, patchy fuel limits the spread of fire (Noble 1986b; Noble et al. 1997).

A single prescribed fire is unlikely to achieve lasting benefits. This is especially so where the management objective is to reduce the density of woody species. Integrated shrub control strategies involving combinations of control options such as, for example, chaining followed by prescribed fire, together with management of total grazing pressure (Robson 1995), can be used to exploit potential synergy. Other integrated strategies may involve the use of low-concentration chemical defoliation of young (i.e. 1–2-year-old) shrub seedlings and coppices regenerating after prescribed fire (Noble et al. 1991, 1992, 1997). Again, aerial application of defoliant has obvious benefits in terms of cost-effectiveness. Only burnt patches within a paddock require secondary treatment and these are readily identified from the air.

The density and potential productivity of such fuel-producing areas fundamentally reflects landscape function, defined in this context as the collective action of many ecosystem structures and processes that together regulate the redistribution, capture and storage of soil water and nutrients (Ludwig et al. 1997). Increasing emphasis is now being applied to sustain, and enhance, landscape processes and productivity from fertile patches (Noble et al. 1997). While the management of such 'keystone' areas (Noble and Brown 1997) is clearly important, so too is management of the less productive sites since their role as run-off or transfer zones can be crucial in maintaining redistribution processes.

Spatial variation in herbage, and hence fuel, production in these landscapes is strongly influenced by redistribution of rainfall from 'source' areas at the top of a catenary sequence into lower 'sinks' or fertile patches which intercept overland flow (Tongway and Ludwig 1997). This process not only redistributes rainfall, but also surface soil and litter so that a disproportionate amount of a paddock's total herbage production is produced from a relatively small area (Noble et al. 1998a). In terms of both biodiversity and pastoral productivity, high-production 'islands' become scattered throughout

a 'sea' of low-production, low-quality vegetation. Where the vegetation is in 'good condition', light grazing pressures may even enhance this redistribution function through a moderate reduction in soil-surface cover. While fire effects are important at a landscape scale, significant post-fire differences in nutrient availability may also occur at a smaller microsite, e.g. soil hummock, scale (Noble et al. 1996a).

In the two decades since the topic of fire in Australian rangelands was last reviewed (Leigh and Noble 1981), considerable research has been undertaken into various aspects of fire ecology and fire management in contrasting rangeland communities including mallee (Eucalyptus spp.) (Bradstock 1989, 1990; Noble 1989a, b; Bradstock et al. 1992; Noble and Vines 1993; Noble et al. 1996a; Bradstock and Cohn, this volume), mulga (Acacia aneura) (Griffin and Friedel 1984a, b; Harrington 1986, 1991; Hodgkinson and Oxley 1990; Hodgkinson 1991, 1992, 1998, this volume; Harrington and Driver 1995), mitchell grass (Astrebla spp.) (Scanlan 1980, 1983; Orr and Holmes 1984) and tropical (Orr et al. 1991; Orr and Paton 1993; Grice and Brown 1996; Dyer et al. 1997; Grice 1997, 1998; Grice and Slatter 1997) ecosystems.

The pastoral importance of hummock grasslands dominated by species of Triodia varies from one region to another. However, fire is universally important in all these ecosystems as a major environmental variable (Griffin 1984a). Fire has been successfully used as a management tool to promote pastoral productivity in such diverse regions as the Pilbara region in northwestern Western Australia (Suijdendorp 1981) and the semi-arid mallee region in southwestern New South Wales (Noble et al.1980) where Triodia-dominant communities are widespread. The biomass produced by ephemeral forbs in burnt Triodia areas can provide extremely valuable forage for livestock, especially during dry periods. In a biodiversity context, the ecological impacts of fire on the spinifex (Triodia spp.) communities of Uluṟu–Kata Tjuṯa National Park in central Australia have also been studied by several workers (e.g. Griffin 1984b, 1992; Masters 1993; Reid et al. 1993a; Allan and Southgate, this volume).

Biodiversity assumed national importance in 1993 once Australia became a signatory to the 1992 International Convention on Biological Diversity. This was followed by the National Strategy for the Conservation of Australia's Biological Diversity that was signed by state and territory governments (Saunders and Walker 1998). The Australian Government later established a Biological Advisory Committee to develop a national strategy for conserving biological diversity. The net impacts of various forms of land use on regional biodiversity have been reviewed by Walker and Nix (1993) while Holling *et al.* (1995) have identified key knowledge gaps. This chapter does not attempt to emulate these comprehensive reviews but instead aims to use them as a framework on which to address those ecological questions relating to fire and biodiversity interactions in Australian rangelands. Socio-economic aspects of biodiversity, while not discussed in detail in this chapter, are none the less recognised as central to the management of complex, adaptive pastoral ecosystems.

The interrelationships between fire and biodiversity in the rangelands are best comprehended in terms of three major forces. The first relates to the nature of fire itself and its influence in shaping the vegetation. Secondly, the greater proportion, by far, of the nation's pastoral lands occur in the semi-arid and arid regions where the native biota has evolved through Recent geological time to cope successfully with these climatic constraints. Finally, the dominant land use – pastoralism – has few tools available for managing much of the native biota apart from regulating consumption of vegetation, and thereby modifying habitat, by grazing and/or fire (Freudenberger and Noble 1997).

Altered fire regimes and their impacts on vegetation in pastoral lands

Little is known today about the precise ways in which the Aboriginal inhabitants used fire across most temperate, semi-arid pastoral landscapes. Despite this, it is generally assumed that the use of fire probably differed little from other semi-arid regions such as central Australia where much of the

traditional use of fire has remained (Latz and Griffin 1978; Griffin and Allan 1986). Yet there is still incomplete understanding of every nuance of Aboriginal use of fire throughout much of the pastoral zone (Craig 1997, 1999; Langton 1998). Fire was used for a multitude of purposes with patches burnt almost on a daily basis (Kimber 1983). Even in traditional lands, contemporary Aboriginal firing practices have clearly been subjected to major European influences. Altered habitation and grazing, as well as roads and motor vehicles, have all strongly influenced fire regimes and patterns of burnt landscapes, especially during the latter half of the 20th century.

Early historical records suggest there has been a significant reduction in fire frequency in the temperate semi-arid zone following European settlement around 130 years ago (Heathcote 1977). This reduction has been attributed not only to the removal of grasses which are the primary source of fuel but also by a change in management philosophy from one of fire encouragement (Aboriginal) to one of fire suppression (European) (Wilson 1990). During the 1900 Royal Commission, John Thomas Quinn, a landholder who had resided in the Canonbar/Nyngan district for the previous 38 years, claimed 'After that country became stocked, bush fires were less frequent. Previous to that every summer large bush fires swept through all that country, and that tended to keep down the scrub and undergrowth' (Royal Commission 1901, p. 730). Quinn's observation that summer fires occurred annually is of interest given the semi-arid nature of the country but may require some qualification. The relative importance of human (i.e. Aboriginal) versus abiotic (i.e. lightning) ignition sources responsible for more frequent pre-European fire regimes remains unknown.

Post-fire mosaics created by pre-European fires in the eastern semi-arid woodlands, for example, were dependent on the presence of perennial grass fuels. Sandplains and other patches of light-textured soils were generally dominated by relatively unpalatable C4 tussock grasses such as woollybutt (*Eragrostis eriopoda*) that could be burnt at most times of the year, especially during dry seasonal conditions. During

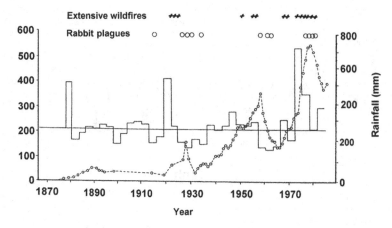

Fig. 16.3. Annual rainfall for Alice Springs and total cattle numbers for the central Australian pastoral district. Mean annual rainfall is 263 mm (for the period). Annual rainfall is grouped into 3-year averages to highlight peaks and troughs (droughts). Cattle numbers generally trend upwards but peak several years after high-rainfall periods then drop dramatically; rabbit plagues coincide with peaks in cattle numbers. Extensive wildfires follow periods of high rainfall more immediately. After Friedel *et al.* (1990).

favourable rainfall seasons, these potential grass fuels were augmented by palatable, perennial C3 species such as bandicoot grass (*Monachather paradoxa*) and mulga mitchell grass (*Thyridolepis mitchelliana*), as well as the weakly perennial speargrass (*Stipa* spp.).

It has since been postulated that increasing herbivore populations following European settlement consumed much of the palatable grasses that would otherwise have senesced to provide fuel (Harrington *et al.* 1984; Hodgkinson and Harrington 1985). Grazing studies in the temperate semi-arid woodlands have provided indisputable evidence that perennial C3 grasses such as bandicoot grass and mulga Mitchell grass are preferred by domestic livestock as fodder species and have suffered much higher rates of grazing-induced mortality (Grice and Barchia 1992; Hodgkinson *et al.* 1995). Not only has increasing grazing pressure resulted in a significant reduction in total biomass of these palatable grasses (Freudenberger *et al.* 1999) but there has also been a significant reduction in rainfall use efficiency (kg dry matter per ha per mm rainfall) (Noble *et al.* 1998a).

Consequently, instead of frequent small-scale fires, post-settlement fire regimes have been characterised by episodic, large-scale wildfires such as those experienced during the 1973/74 and 1984/85 summers in western New South Wales. A dearth of areas where perennial fuels had been recently burnt resulted in huge wildfires with extended fire perimeters creating major difficulties for fire suppression crews (Noble *et al.* 1980; Friedel *et al.* 1990). Extensive wildfires in Australia have generally been closely linked to antecedent seasonal conditions (Leigh and Noble 1981; Griffin and Friedel 1985; Noble *et al.* 1986). Griffin *et al.* (1983) found, for example, that the total number of wildfires in central Australia (see Fig. 16.3) was closely correlated with 3 years of accumulated antecedent rainfall (although these relationships were based on a limited data set extending over a 10-year period only).

The frequency of fuel-producing seasons in different pastoral landscapes today is a function of the amount, timing and regularity of rainfall, together with composition of the vegetation. The mitchell grasslands (*Astrebla* spp.) in northern Australia, for example, are primarily influenced by summer rainfall (Orr and Holmes 1984), and burn relatively frequently. Rainfall regimes in southern semi-arid rangelands, in contrast, tend to be dominated by winter rains (Harrington *et al.* 1984) and are usually only capable of supporting extensive wildfires at long and irregular intervals (Noble *et al.* 1986; Noble and Vines 1993).

In semi-arid mallee shrublands of western New South Wales, wildfires have tended to occur when substantial cool-season (April–September) and warm-season (October–March) rains are in-phase (Fig. 16.4). Areas of short-lived speargrass (*Stipa* spp.) fuels link those containing perennial fuels such as porcupine grass (*Triodia scariosa*), shrub litter and elevated canopy fuels. Once fuel continuity has been established, wildfires of major proportions

Fig. 16.4. Three-year moving averages of cool-season (April–September) and warm-season (October–March) rainfall over the past century for three mallee districts in western New South Wales illustrating the tendency for major wildfire seasons (arrowed) to occur when above-average rainfall occurs in both seasons.

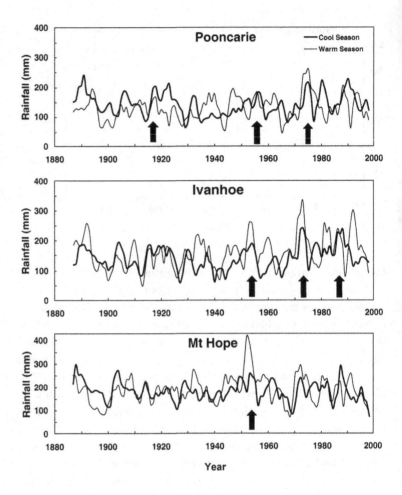

may burn for weeks across vast areas of sparsely populated rangelands (Noble *et al.* 1980). These large wildfires usually result from multiple lightning ignitions. In the absence of 'dry' electrical storms, areas carrying adequate fuel loads may still remain unburnt (Harrington 1991). Despite abundant fuel loads during the early 1970s in the Mount Hope district of New South Wales following above-average cool- and warm-season rainfall, no major wildfires occurred.

Probably the most obvious, and certainly earliest, manifestation of environmental decline following European-induced changes in fire regime has been the proliferation of native tree and shrub species in what had originally been open, grassy woodlands. A rapid transformation of community structure occurred throughout much of the semi-arid woodlands of western New South Wales and Queensland with the advent of sedentary pastoralism. As early as 1881, just 20 years after European settlement, the *Votes and Proceedings of the New South Wales Assembly* expressed concern about the widespread expansion of white cypress pine (*Callitris glaucophylla*) (Noble 1997a). Problems with other scrub species followed and by 1900, Robert William Peacock, then Manager of Coolabah Experiment Farm, was describing how 'Upon the red country the pine-scrub, box-seedlings [i.e. poplar box, *Eucalyptus populnea*], and budda [*Eremophila mitchellii*] have taken complete possession, to the thorough exclusion of even the worst grasses' (Peacock 1900, p. 655).

There is little doubt that pastoralists in the temperate semi-arid rangelands suffered severely as herbage stocks, and especially the more palatable and nutritious perennial grasses, disappeared during the severe drought and economic depression of the 1890s and early 1900s. While 'noxious scrub' was blamed for much of the decline in productivity, there was an inability, or reluctance, on the part of the Commissioners to accept the notion that the changed composition of the vegetation was a symptom, rather than the cause, of the problem. To a large extent this reluctance still persists despite strong evidence that scrub proliferation is closely correlated with a reduction in fire frequency.

Elsewhere, the same, and related, species of *Callitris* occur in much lower densities and are not regarded as 'woody weeds'. Episodic, high-intensity wildfires that now burn large areas have been cited as the principal reason for the contraction of such species as *Callitris glaucophylla* in central Australia (Bowman and Latz 1993) and *C. intratropica* in northern Australia (Bowman and Panton 1993). These workers claim that under a traditional, low-intensity burning regime employed by Aborigines, these species would have been able to survive in unburnt refugia.

While fire regimes have also changed markedly following pastoral settlement in tropical rangelands, subsequent changes in vegetation structure vary considerably (Dyer *et al.* 1997). Tothill and Gillies (1992) nominated several pasture systems regarded as being under threat by woody regrowth although changes in fire regime were probably influential in only three of these, viz. southern black speargrass (*Heteropogon contortus*), *Aristida–Bothriochloa* and mulga (*Acacia aneura*) communities. In the Kimberley region of Western Australia there is concern that the high incidence of extensive fires during the middle and late dry seasons is threatening a reduction in both pastoral and environmental values. Adverse impacts include, amongst others, losses in forage availability, damage to fencing, increased erosion risk, reduced reproductive success of mid-storey trees, adverse changes to vegetation structure and animal habitat, and a reduction in seed sources leading to a decline in granivorous birds (Craig 1997).

Interactions between fire regime, rainfall, grazing and browsing

Pastoral industries in Australia depend on the exploitation of both native and some naturalised plant species such as medics (*Medicago* spp.) in temperate areas, buffel grass (*Cenchrus ciliaris*) in semi-arid subtropical areas, and stylos (*Stylosanthes* spp.) in tropical regions. Consequently, fire research over the past two decades has focussed mainly on plant community response to fire, particularly in terms of forage production. With increasing awareness of the desirability of maintaining, and where appropriate restoring, biodiversity in pastoral ecosystems, habitat values are now assuming greater importance in fire management.

In this context, interactions between fire, rainfall and grazing and/or browsing are often far more important than the effects of fires alone. For example, in the tropical grasslands dominated by black speargrass (*Heteropogon contortus*), a combination of annual spring burning followed by deferred grazing can minimise invasion by undesirable *Aristida* species (Paton and Rickert 1989; Orr *et al.* 1991; Orr and Paton 1993). Major landscape changes have occurred only in the last 40 years, especially in northeastern Australia, as cattle numbers have increased. The addition of supplementary protein in the form of urea-based licks has enabled low-quality, senescent grass residues, previously consumed by fire, to be utilised instead as fodder (Tothill 1971; Tothill and Gillies 1992). This has subsequently led to a reduction in surface cover followed by accelerated erosion.

These environmental problems have been further compounded by another menace – invasion by exotic shrubs. Increasing densities of mesquite (*Prosopis* spp.), prickly acacia (*Acacia nilotica*), rubbervine (*Cryptostegia grandiflora*) and Chinee apple or Indian jujube (*Ziziphus mauritiana*) (Grice 1997, 1998; Grice and Brown 1996; Brown and Carter 1998), now constitute a major threat to pastoralism, and biodiversity, in these northern rangelands. The threat is particularly relevant in highly productive mitchell grass (*Astrebla* spp.) communities where the grassy understorey, and consequently fuel, can be completely eliminated by shrub competition, especially close to open bore drains (Burrows *et al.* 1990; Dyer *et al.* 1997). Rubbervine stands not only reduce herbage productivity and impair efficient livestock management, but also threaten long-term conservation values, especially in riparian ecosystems.

While the amount, and timing, of rainfall are critical in determining the availability of short-lived fuel species in semi-arid rangelands (Noble and Vines 1993), the rainfall regime following fire dictates the composition and abundance of succeeding herbage communities (e.g. Noble 1989a). This, in turn, influences the degree of grazing pressure generated by a mix of domestic, feral and native herbivores, and occasionally invertebrates as well. Apart from the innovative study undertaken over 30 years ago by Suijdendorp (1969) demonstrating the advantages of grazing deferral following fire in the soft spinifex association of Western Australia, no further studies have been undertaken on this important topic.

Plant communities commonly include both obligate seed regenerators that are killed by fire, and facultative sprouters that are not. Truly obligate sprouters probably do not exist despite the fact that seedlings of some long-lived species are rarely seen (Noble *et al.* 1986). While virtually all woody plants in semi-arid woodlands are vulnerable to fire during the seedling stage (Hodgkinson and Griffin 1982), many regenerate freely after fire by coppicing, regardless of post-fire seasonal conditions (Hodgkinson 1998). Despite this apparent tolerance of many sprouting species to repeated fire, experiments using artificial fuel have clearly demonstrated that most are vulnerable to multiple decapitation, but only when decapitation is imposed at critical times. Mallee eucalypts, long regarded as archetypal fire sprouters, have been found to be susceptible to repeated autumn fires with survival increasing as the interval between successive fires increased (Fig. 16.5a). Only 18% survived two consecutive autumn fires whereas the same proportion survived eight consecutive spring fires (Noble 1997a). Similarly, heavy mortality of budda (*Eremophila mitchellii*) and turpentine (*E. sturtii*) was recorded after just two consecutive

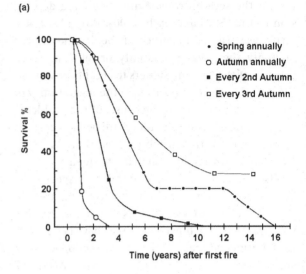

(a)

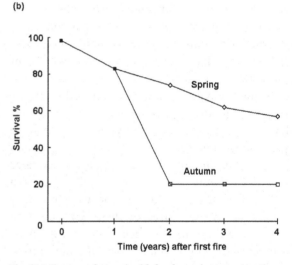

(b)

Fig. 16.5. Response by semi-arid shrub species to contrasting experimental fire regimes: (a) survival of mallee eucalypts at Pooncarie, New South Wales (after Noble 1997a); and (b) survival of budda at Coolabah, New South Wales (after Hodgkinson 1986).

autumn fires (Fig. 16.5b). Unlike mallee, though, survival increased significantly once the interval between successive fires increased to 2 years (Hodgkinson 1986). Furthermore, like mallee, *Eremophila* survival was significantly higher under an annual spring fire regime.

While fire has important direct effects upon actively growing plants in pastoral landscapes, dormant seeds in the soil seed bank are not necessarily sheltered from these effects. Fire effects on seed buried at shallow depths are generally regarded as non-lethal and in some cases, heat scarification of such seed promotes germination. This applies particularly to species whose seed banks contain a large proportion of 'hard' seeds, particularly leguminous species of the Mimosaceae (e.g. *Acacia* spp.) and Caesalpiniaceae (e.g. *Cassia* (*syn. Senna*) spp.). Seeds of non-leguminous species such as the hopbushes (*Dodonaea* spp.) and native fuchsias (*Eremophila* spp.) may also respond to heat treatment (Turnbull 1972; Hodgkinson and Oxley 1990).

The degree of heat scarification depends upon the amount of seed present at various depths, the amount and depth of fine fuel present, and the residence time of the fire. In semi-arid mallee, the highest temperatures were recorded under mallee litter (140 °C at 2 cm) followed by *Triodia* and *Acacia* litter (60–70 °C at 2 cm) (Bradstock *et al.* 1992). Elevated soil temperatures generated by fire may be only part of the story. Studies in three sclerophyllous woodland sites found that following a summer fire, soil temperatures in the surface 0.5 cm could reach 60 °C, sufficiently high to break the dormancy of leguminous seeds in the soil seed bank (Auld and Bradstock 1996). Soil temperatures following a winter fire, or in unburnt country during the summer however, did not exceed 40 °C. The importance of smoke in stimulating seed germination of semi-arid and tropical pastoral species has yet to be determined on a comprehensive scale.

Fire residence time (i.e. the time taken for a fire to pass a particular point), especially where substantial litter fuel occurs, is probably more important biologically than fire intensity. Traditionally calculated on the basis of heat released in kilowatts per metre of firefront (Byram 1959), fire-line intensity is a useful tool for predicting fire behaviour but because it integrates several variables, significant biological relationships have been difficult to establish. In semi-arid pastoral lands, serotinous plants such as the mallee eucalypts require high temperatures for sufficient seed to be released to more than assuage the immediate demands of seed-harvesting ants (Greenslade and Greenslade 1989; Wellington 1989). Whether mass release of seed is a direct response to fire-line intensity, or instead is related to other behavioural characters such as flame height and vertical profiles of sensible heat (Bradstock and Gill 1993), remains to be resolved.

Like plant communities, the effects of fire on faunal communities may be either direct, e.g. lethal effects on either adults or juveniles, or indirect when fire adversely affects habitat. Although wildfires in tussock grasslands rarely kill mobile species such as the red kangaroo (*Macropus rufus*), the impacts of such severe fires on rare terrestrial avifauna such as the plains-wanderer (*Pedionomus torquatus*), also resident in temperate semi-arid grasslands (Baker-Gabb *et al.* 1990), are currently unknown.

High-intensity fires commonly occur in hummock grasslands dominated by species of *Triodia*, and birds that nest within these hummocks, especially rare and endangered species such as the striated grasswren (*Amytornis striatus*), are obviously vulnerable to the direct effects of fire (Brickhill 1980; Pedler 1997). Although Reid *et al.* (1993c) made several pertinent remarks regarding fire in their *Uluru Fauna* report, there has been no rigorous analysis of the relative importance of the three major fire regime variables to faunal communities. One major study in this context was conducted by Woinarski *et al.* (1999) who examined the impact of experimental fires applied at different frequencies and seasons on reptile and bird populations in the Victoria River District of the Northern Territory. They found that 12 of the 30 species recorded from at least four plots were significantly associated with time since last fire. Most of these responses were correlated with the extremes, some preferring recently burnt areas whereas others were found mainly in long-unburnt areas.

Several autecological studies of various endangered species including the malleefowl (*Leipoa ocellata*) (Priddel 1989; Benshemesh 1990); the striated grass-wren (*Amytornis striatus*) (Brickhill 1980; Pedler 1997) and the western hare-wallaby (*Lagorchestes hirsutus*) (Bolton and Latz 1978) have led to general relationships with fire being inferred from observations taken following wildfire. Little additional

Fig. 16.6. Cover relationships over time between different functional elements in a semi-arid mallee ecosystem following prescribed fire applied in contrasting seasons on 'Birdwood' Station, Pooncarie, New South Wales.
(a) Control – no fire; (b) autumn fire; (c) winter fire; and (d) spring fire. After Noble (1989a).

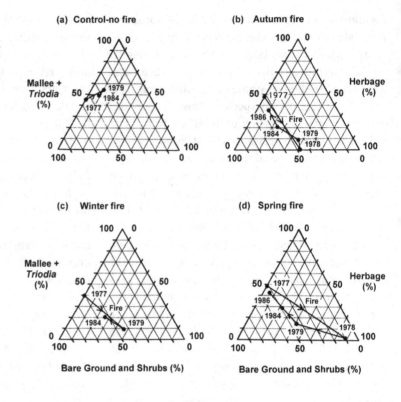

information has been forthcoming at a community level since the topic was last reviewed in any depth for pastoral landscapes (e.g. Catling and Newsome 1981; Christensen and Burrows 1986).

The reptiles however, are one faunal group that has received some attention, particularly in relation to the effects of fire in semi-arid mallee communities where the understorey is dominated by porcupine grass (*Triodia scariosa*). *Triodia*-dependent species such as *Ctenophorus fordi*, *Ctenotus schomburgkii* and *C. atlas* tended to be most common in vegetation that had not been burnt for at least 6 years. Several other species though, including *Egernia inornata*, *Lucasium damaeum*, *Rhynchoedura ornata* and *Ctenophorus pictus*, were more abundant in recently burnt areas (Caughley 1985; Schlesinger *et al.* 1997). While diurnal lizards showed no significant differences between areas with different fire histories, there were significantly fewer nocturnal reptiles, especially geckos, in the site burnt 18 years earlier compared with more recently burnt sites. Similarly, Reid *et al.* (1993c) found some species of reptiles were more commonly caught in burnt spinifex com-

pared with mature spinifex communities. While carabids dominated contemporaneous captures of terrestrial beetles in the mallee studies, unlike the geckos there was no significant relationship between the numbers of these beetles and time since last fire (Schlesinger *et al.* 1997).

Biodiversity, landscape function and fire regime

Because of the unpredictability of rainfall, there is rarely any ordered succession of plant communities following fire disturbance in pastoral landscapes (Fig. 16.6). Compositional changes observed across many different vegetation communities (Remmert 1991; Griffin 1992) generally conform to the following pattern:

(1) early occupation and dominance by short-lived forbs and grasses depending on post-fire rainfall (Wilson *et al.* 1988; Noble 1989a);

(2) re-establishment of perennial species through basal regeneration and/or seedling recruitment;

(3) progressive dominance by perennial species, both woody and herbaceous; and

(4) slowing down of rate of compositional and structural change once perennials have matured.

Faunal changes might be expected to follow a similar variety of post-fire successional patterns. The few studies of such taxa as reptiles (e.g. Fyfe 1980; Caughley 1985; Schlesinger et al. 1997; Woinarski et al. 1999), birds (e.g. Benshemesh 1990; Woinarski 1990; Woinarski and Recher 1997; Woinarski et al. 1999) and some invertebrates (e.g. Andersen and Yen 1985; Noble et al. 1996b; Yen 1989), do not enable any generalisations to be made.

The passage of fire through a landscape may produce a multitude of effects. These may dramatically alter the dynamics of the biota, hydrology, and nutrient status of the affected landscape (Gill et al. 1981). While fire may be lethal for some elements of the biota, it may also provide new opportunities for others. Which of these persist and prosper will depend not only on conditions before and during the fire, but also rainfall patterns after fire. Following a prescribed fire in a semi-arid Triodia–mallee community in western New South Wales, kangaroo numbers on burnt and unburnt areas did not vary significantly because good winter rains following the fire ensured widespread forage supplies (Caughley et al. 1985). In contrast, kangaroos congregated in large numbers (c. 5–8 kangaroos per hectare) as seasonal conditions became increasingly drier in a Stipa–mallee community burnt twice in the preceding 7 years (Noble 1989b).

Ultimately, landscape pattern is a function of vegetation composition and past fire regime. The creation of mosaics of burnt and unburnt patches following low- to moderate-intensity fire is particularly important to fauna since spatially complex habitats generally support richer faunal communities (Pianka 1979). The apposition of relatively dense, unburnt vegetation providing shelter with more open, regenerating vegetation, provides a greater variety, and quantity, of food (Morton 1990). With a decline in moderate-intensity fires and a corresponding increase in larger, less frequent, high-intensity wildfires, there has been a significant reduction of fine-scale mosaics preferred by some animal species.

The need for periodic fire to maintain optimal habitat heterogeneity is therefore seen as of paramount importance to ensure survival of several endangered species. These include the western hare-wallaby (Lagochestes hirsutus) (Bolton and Latz 1978; Latz 1995) and bilby (Macrotis lagotis) in the Tanami Desert of the Northern Territory and the golden-shouldered parrot (Psephotus chrysopterygius) on Cape York Peninsula (Garnett and Crowley 1997). Appropriate fire research has also been identified as being of high priority for enhancing the survival prospects of smaller rare and endangered species such as the desert skink (Egernia kintorei) (McAlpin 1997), the mulgara (Dasycercus cristicauda) (Masters 1997), and the striated grasswren (Amytornis striatas) (Pedler 1997).

Clearly conservation of the nation's diverse biological resources cannot be achieved simply by establishing more conservation reserves managed by government agencies. Many pastoralists are not only aware of these issues, but are also prepared to negotiate major adjustments to property rights to accommodate broader community interests while achieving desired property development. Some pastoralists in the southwestern mallee region of New South Wales have relinquished control over major areas of land in order to provide corridors of uncleared country linking riverine nesting habitat for the regent parrot (Polytelis anthopeplus) with feeding habitat in adjoining mallee scrub (Priddel 1989). In return, clearing concessions for nearby land designated as being of lower conservation value within that part of the region have subsequently been negotiated (Freudenberger et al. 1997).

Such negotiated agreements between landholder and government however, do not eliminate the need to maintain, or where appropriate improve, landscape integrity in reserves on both public and private land. Here landscape integrity applies specifically to the effectiveness of key ecosystem processes such as rainfall redistribution and nutrient cycling (Ludwig et al. 1997). Implicit in such management is the need to impose appropriate fire management strategies although responsibilities in

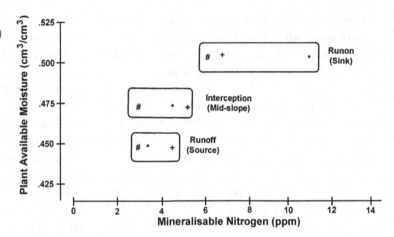

Fig. 16.7. Influence of plant available moisture (PAM) and available nitrogen (AN) in determining species composition of communities in three geomorphic zones (run-off, interception and run-on zones) characteristic of 'hard' red mulga (*Acacia aneura*) landscapes in western New South Wales.
Key: + 'Glenvue' Station, Cobar; * 'Booroomugga' Station, Cobar; and # 'Glenora' Station, Bourke. (J. C. Noble, unpublished data.)

this context have yet to be clearly defined. Because most of these fundamental processes are, to varying degrees, mediated by biotic elements, the functional role of biodiversity is coming under increasing scientific scrutiny (Chapin *et al.* 1997). Given the need to maintain pastoral landscapes in perpetuity by developing sustainable land-use systems, there is an increasing need to establish the functional role, and relative importance in terms of process regulation, of various groups of biota found within individual pastoral ecosystems.

Despite the controversy and misunderstanding often associated with community perceptions of biodiversity, Walker and Nix (1993) claim there are essentially three different kinds of values, or reasons, justifying its conservation. The first relates to direct economic values including products based on food and fibre; tourism; as well as genes for genetic engineering. The second group involves 'ecosystem services' whereby biological diversity plays a fundamental role in maintaining such basic ecosystem functions as nutrient cycling and water redistribution. Finally, the third group incorporates most of the aesthetic factors characterising 'quality of life' values – analogous to cultural values like fine art and music. It is the second group though that is currently attracting the attention of many rangeland ecologists concerned with maintaining biologically mediated processes fundamental to ecosystem functioning.

As mentioned earlier, the response to fire and

herbivory will depend primarily on the relative positions of individual communities within the plane defined by PAM (plant available moisture) and PAN (plant available nutrients) or AN (available nitrogen) axes. Vegetation communities growing along topographic gradients such as, for example, 'hard' red earth catenae in semi-arid mulga (*Acacia aneura*) landscapes, clearly separate over the PAM–AN response surface (Fig. 16.7). Scattered herbs and unpalatable perennial grasses, e.g. woollybutt, characterise communities found in the run-off zone. Mulga groves and palatable perennial grasses, e.g. mulga mitchell grass and bandicoot grass or mulga oats generally predominate in the midslope interception zone. Finally, poplar box, together with a mixed understorey of perennial and annual herbs, distinguish plant communities found in the run-on zone (Tongway and Ludwig 1990).

This mosaic or patch structure in some semi-arid rangelands strongly influences the relationships between woody plants and their understorey so that woody plants often dominate vegetation in fertile patches, eventually leading to a decline in pastoral productivity (Tongway and Ludwig 1997). Similar changes are often evident in artificial fertile patches engineered by water-spreading and ponding technologies used to restore vegetative cover in 'scalded' areas (Noble *et al.* 1997). As soil fertility increases over time in these run-on areas, so does the density of woody plants (Batianoff and Burrows 1973), highlighting the need for periodic prescribed fire to reg-

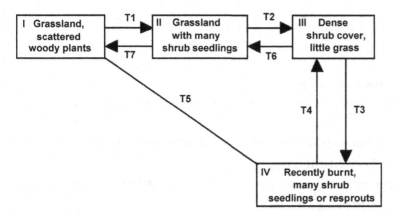

Fig. 16.8. A state-and-transition framework for semi-arid woodlands and grasslands of western New South Wales and southwestern Queensland based on four states and seven transitions. Individual states and transitions are catalogued in Table 16.1. After Westoby *et al.* (1989).

ulate such changes in community structure. Because of the spatial variability inherent in these plant communities however, there may be difficulties in applying fire as a control option over an entire paddock.

Ecosystem resilience to varying fire regimes

Changes between various components of these complex vegetation mosaics represent discontinuous, and sometimes irreversible, switches between the various equilibria of the systems (Perrings and Walker 1997). Perrings and Walker (1997) claim that semi-arid rangelands have several features in common including:

- the sensitivity of system dynamics to 'rare' or 'extreme' events such as fire;
- the importance of interactions between grazing and/or browsing and extreme events in causing transitions between states; and
- the existence of multiple, locally stable states.

Such multiple stable states can be portrayed as 'state-and-transition' models or frameworks (Westoby *et al.* 1989). As shown in Fig. 16.8 and Table 16.1, various 'states' of a semi-arid woodland ecosystem are characterised by different proportions of shrub and perennial grass components. 'Transitions' between contiguous 'states' can occur either through changes in rainfall or wildfire regimes or alternatively, through specific management intervention such as prescribed fire (see also

Noble 1997a). The capacity of an ecosystem to withstand shocks that may cause it to flip from one state to another before returning to its original state defines its resilience (Holling 1973).

Historical evidence suggests that interactions between grazing, fire and browsing of regenerating vegetation produced complex mosaics in a wide range of Australian ecosystems prior to European settlement. These included the white cypress pine (*Callitris glaucophylla*) communities of the Pilliga Scrub in New South Wales (Rolls 1981) as well as eucalypt woodlands in both the central tablelands of New South Wales (Nicholson (1993) and the Gippsland region in Victoria (Barr and Cary 1992). Here post-fire browsing of regenerating woody plants by medium-sized marsupials, particularly the bettongs (*Bettongia* spp.), was claimed to be instrumental in promoting vegetation mosaics of open, semi-open and closed plant communities. Other examples can be found elsewhere. For example, in maritime communities found on Rottnest Island off Western Australia, exclosure studies demonstrated that another mesomarsupial, the quokka (*Settonix brachyurus*), prevented tree seedlings establishing after fire (Main 1992).

Similar medium-sized marsupials were widespread throughout the semi-arid and arid zones prior to the arrival of Europeans although their numbers fluctuated considerably according to seasonal conditions (Finlayson 1958). Prior to pastoral settlement, light grazing by native herbivores such as large kangaroos (*Macropus* spp.) allowed sufficient fuel to accumulate which, in turn, allowed more

Table 16.1. *Catalogues for individual states and transitions portrayed in Fig. 16.8*

Catalogue of states

State I	Grassland, scattered woody plants.
State II	Grassland with many shrub seedlings. Transient: converts to State I or State III within a few years.
State III	Dense cover of shrubs with a soil seed bank and/or high potential for vegetative regeneration after fire. Little or no grass.
State IV	Recently burnt, many resprouting shrubs and/or shrub seedlings. Transient: converts to State I or State III within a few years.

Catalogue of transitions

Transition 1	Two or more very good rainfall years to produce many shrub seedlings. Frequency 2–5 per century. Substantial fuel of ephemerals and perennial grasses will also be produced. Fuel persistence depends on grazing and on floristics of the understorey at the particular location.
Transition 2	Inevitable over time (10–20 years) as shrub seedlings grow and establish a seed bank in the absence of Transition 7.
Transition 3	Fire following sufficient rain to provide the fuel of ephemerals. Depending on location and on completeness of shrub dominance, sufficient rain could fall 2–30 times per century. Can be blocked by grazing after the rain to remove the fuel, or by fire suppression.
Transition 4	Inevitable over time as shrub regeneration grows to maturity, in the absence of Transition 5. Transition would take about 5 years if shrubs are sprouters, 10–15 years if a shrub seed bank needed to be re-established.
Transition 5	Fire following rains adequate to produce ephemeral fuel before Transition 4 has re-established the regeneration capacity of the shrub population. Probably less rain required than for Transition 3 due to reduced shrub competition. Blocked by grazing heavy enough to remove the fuel, in which case, Transition 4 occurs instead.
Transition 6	Fire, following exceptional rains to provide the fuel of ephemerals. This transition found for fire-sensitive shrubs, compare to Transition 3 for resprouting shrubs.
Transition 7	Fire or competition from grasses kills shrub seedlings. Blocked by grazing or fire suppression.

Source: Westoby *et al.* (1989).

frequent fire. The mounting evidence suggests therefore that the significance of interactions between grazing, fire and browsing in regulating vegetation dynamics in semi-arid ecosystems may have been understated. In the event of minimal post-fire rain falling, it has been postulated (Noble 1995) that the green biomass produced at ground level by regenerating shrub coppices may have been heavily utilised by medium-sized marsupials including the once-widespread burrowing bettong (*Bettongia lesueur*) and brush-tailed bettong (*B. penicillata*).

The foliage of many contemporary 'woody weeds' species, particularly species of *Eremophila*, is highly aromatic and generally unpalatable to domestic livestock. Recent dietary studies however of the bridled nailtail wallaby (*Onychogalea fraenata*), another medium-sized marsupial, have shown that it is dependent on foliage of budda or false sandalwood (*Eremophila mitchellii*) as a source of green forage during the winter dry season (Ellis *et al.* 1992). It is postulated that shrub seedlings surviving in unburnt patches were also vulnerable to mesomarsupial browsing thereby maintaining vegetation mosaics, or multiple stable states, throughout much of the semi-arid woodlands (Noble 1996a, b, 1997a).

Many graziers in the temperate semi-arid zone claim that rabbits provided some degree of biological control of native 'woody weeds', particularly prior to the arrival of the myxomatosis virus in the early 1950s. Not only was the foliage of small shrubs browsed directly, larger mature shrubs could also be killed by stripping off the basal bark and sapwood (Noble 1997a). The ability of rabbits to successfully invade these semi-arid ecosystems, and to persist through extended drought periods, was due largely to the ready availability of refuge warrens constructed by burrowing bettongs (Myers and Parker 1965). Any rabbit browsing of woody plants was probably quite localised since rabbits rarely graze more than 300–400 m away from their warrens (Leigh *et al.* 1989) whereas burrowing bettongs can forage over much longer distances ranging from 5 to 7 km (Flannery *et al.* 1990; Hoser 1991).

Why changes in fire regimes in the semi-arid and arid zones have impacted so severely on mammals, especially on the medium-sized species, remains an enigma (Morton 1990). Whether mesomarsupials were keystone species, or ecosystem 'drivers', whose regional extinction resulted in a cascade effect with critical ecosystem processes becoming increasingly degraded as perennial grasses and fire frequency decreased while shrub density increased, remains hypothetical (Noble 1997a). In this context however, Walker (1992) has suggested that, ecologically speaking, not all species are necessarily created equal with 'drivers' and 'passengers' occupying opposite ends of a continuum. While the removal of 'passengers' may not necessarily lead to adverse changes in an ecosystem, it does not automatically follow that their loss is unimportant to a specific functional group since ultimately there is a decline in resilience if such species continue to be lost. The redundancy hypothesis therefore makes two explicit points: (1) species redundancy in ecosystems is an important property that contributes to ecosystem resilience; and (2) the highest priority needs to be placed on single species that are the sole representatives of their functional groups (Ehrlich and Walker 1998).

Plant species redundancy assumes particular importance in the context of ecosystem resilience following fire disturbance when the early post-fire flora is likely to be dominated by ephemeral herbs. Given the presence of a substantial viable seed bank in the soil, together with adequate post-fire rainfall, such herbage communities in burnt mallee country, for example, are often dominated by a few ephemeral species such as toothed raspwort (*Haloragis odontocarpa*). This particular species is highly favoured by pastoralists because it provides abundant, nutritious forage (Noble *et al.* 1980). Yet it is insignificant in unburnt communities thus appearing to be ecologically redundant. Species importance ranking may also differ substantially according to the criterion applied. Species like *Calotis hispidula* and *Harmsiodoxa blennodioides*, for example, may be highly rated as early cover after fire in mallee communities but contribute little to total biomass (Noble 1989a).

Despite the fact that many ephemerals spend by far the greater portion of their life histories as dormant populations, any apparent redundancy is nullified by their often 'miraculous' appearance after occasional fire. Many species previously thought extinct have been 'rediscovered' in recently burnt areas (Gill and Bradstock 1995). The native pea *Swainsona colutoides* grew in abundance in the first year following a prescribed fire in semi-arid mallee, the first time the species had been recorded in New South Wales (Noble 1989a). 'Minor and unimportant species may emerge as keystone species in a changed environment. We therefore cannot and should not diminish the importance of apparently ecologically redundant species' (Walker 1995, p. 751).

Scientific confirmation that future sustainability of pastoral enterprises is critically dependent on maintaining biological diversity within these ecosystems has become the 'holy grail' for many ecologists. The functional significance of biodiversity in pastoral systems has been highlighted in recent times by detailed studies on the functioning of semi-arid mulga (*Acacia aneura*) landscapes on 'Lake Mere' Station, northwest of Louth (144° 54' E, 30° 16' S) in northwestern New South Wales (Ludwig *et al.* 1997). Empirical studies included detailed descriptions of the abundance and distributions of various surface-soil features such as termite pavements and log hummocks (Noble *et al.* 1989; Tongway *et al.* 1989) and the role of termites in the decomposition of different forms of senescent plant material (Noble 1993b).

Comprehensive studies have also been undertaken into the distribution of termitaria in tropical pastoral ecosystems (Spain *et al.* 1983, 1986), as well as their influence on adjacent vegetation (Spain and McIvor 1988). At 'Lake Mere', relict warrens and nesting sites of the burrowing bettong (*Bettongia lesueur*) and malleefowl (*Leipoa ocellata*) respectively were also described (Noble 1993a). Because of their important role as litter 'decomposers' during nutrient cycling, additional studies focussed on the abundance and composition of micro-arthropod communities in contrasting micro-habitats (Noble *et al.* 1996b).

While such studies have provided useful scientific information in their own right, it is the interactions between keystone species and exogenous disturbances such as fire that are likely to be of greater importance (Noble *et al.* 1999). Experimental field testing of many hypotheses relating to semi-arid mulga systems, however, is likely to pose considerable difficulties in the short term given that some biota such as the bettongs and the malleefowl are currently regionally extinct. Furthermore, reintroductions of such endangered fauna are doomed to failure in the absence of effective strategies for controlling exotic predators such as the European fox (*Vulpes vulpes*) and feral cat (*Felix catus*).

Conclusions

No comprehensive investigation has been undertaken into the effects of fire in mediating long-term shifts in the relative proportions of functional species groups across all resident biota. Little information is available detailing changes occurring in the proportions of such groups, both dormant and non-dormant, for all plant and animal taxa following the imposition of contrasting fire regimes. Studies similar to that undertaken by Landsberg *et al.* (1997) examining the impacts of artificial water distribution on regional biodiversity are sorely needed. The theory underlying their classification of taxa into functional groups has strong Clementsian undertones, closely resembling the system developed by Dyksterhuis (1949) for determining range condition based on the relative abundance of 'decreasers', 'increasers' and 'invaders'. Whether this approach might also be useful in determining different categories of 'biodiversity condition' remains to be seen.

A process-oriented approach to rangeland management must ultimately depend on an ability to detect, and manage, changes in interactions between different patches within a given landscape mosaic, as much as the changes within individual patches (Remmert 1991; Stafford Smith and Pickup 1993). It is also axiomatic that such management depends heavily on a monitoring system capable of identifying, and quantifying, any changes in potential productivity and biodiversity values (Brown 1994). Many management research projects are being undertaken under the currently fashionable paradigm of complex adaptive systems (D. Cocks personal communication). Active adaptive management or 'learning by doing' (*sensu* Holling 1978, 1995; Walters and Holling 1990) implies a systems approach to analysing various ways in which natural systems (e.g. rangeland ecosystems) interact with human systems (e.g. social, economic and political).

For active adaptive management to be effective, it is essential that landholders collaborate fully in the development of management models. An important attribute of systems thinking in a group setting is its ability to create new knowledge and understanding beyond that held by each individual member. Such active learning requires, firstly, an evolving model of the resource dynamics, plus its response to management actions; and secondly, large-scale experiments undertaken at the management scale to refine the evolving model (Walker 1998). A natural consequence of the long-term feedback of such rangeland management actions, and the predisposition of action to restrain, or enhance, the scope for future actions, is to create a decision-making environment requiring a dynamic and iterative approach to management (Scifres and Hamilton 1993). Whatever approach is selected, it should be consistent with well-considered, and clearly articulated, management goals and objectives. Ultimately, regional- and landscape-scale management models should also be compatible with larger-scale models capable of predicting the impacts of alternative fire management options on greenhouse gas emissions (Moore *et al.* 1997), and conversely, the responses by rangeland ecosystems to future changes in global climate (Walker 1994; Walker and Steffen 1997).

Acknowledgements

Gil Pfitzner and Steve Marsden constructed the figures used in this paper and their expert assistance is gratefully acknowledged. Ross Bradstock, Peter Catling, John Ludwig, Steve Morton, John Pickard, Wal Whalley and Jann Williams provided valuable comments on earlier drafts of this paper. The photograph shown in Fig. 16.1 was reproduced under the Bicentennial Copying Project with the permission of the State Library of New South Wales.

References

Allan, G., and Southgate, R. (2001). Fire regimes in the Spinifex Landscapes of Australia. In *Flammable Australia: The Fire regimes and Biodiversity of a Continent* (eds. R. A. Bradstock, J. E. Williams and A. M. Gill) pp. 145–176. (Cambridge University Press: Cambridge.)

Andersen, A. N. (1991). Responses of ground-foraging ant communities to three experimental fire regimes in a savanna forest of tropical Australia. *Biotropica* **23**, 575–585.

Andersen, A. N., and Braithwaite, R. W. (1992). Burning for conservation of the Top End's savannas. In *Conservation and Development Issues in Northern Australia* (eds. I. Mofatt and A. Webb) pp. 117–122. (Northern Australia Research Unit: Darwin.)

Andersen, A. N, and Yen, A. L. (1985). Immediate effects of fire on ants in the semi-arid mallee region of north-western Victoria. *Australian Journal of Ecology* **10**, 25–30.

Andersen, A. N., Braithwaite, R. W., Cook, G. D., Corbett, L. K., Williams, R. J., Douglas, M. M., Gill, A. M., Setterfield, S. A., and Muller, W. J. (1998). Fire research for conservation management in tropical savannas: introducing the Kapalga fire experiment. *Australian Journal of Ecology* **23**, 95–110.

Anderson, V. J., Hodgkinson, K. C., and Grice, A. C. (1996). The influence of recent grazing pressure and landscape position on grass recruitment in a semi-arid woodland of eastern Australia. *Rangeland Journal* **18**, 3–9.

Andrew, M. H. (1986). Use of fire for spelling monsoon tallgrass pasture grazed by cattle. *Tropical Grasslands* **20**, 69–78.

Ash, A. J., Prinsen, J. H., Myles, D. J., and Hendricksen, R. E. (1982). Short-term effects of burning native pasture in spring on herbage and animal production in south-eastern Queensland. *Proceedings of the Australian Society of Animal Production* **14**, 377–380.

Auld, T. D., and Bradstock, R. A. (1996). Soil temperatures after the passage of a fire; do they influence the germination of buried seeds? *Australian Journal of Ecology* **21**, 106–109.

Baker-Gabb, D. J., Benshemesh, J. S., and Maher, P. N. (1990). A revision of the distribution, status and management of the plains-wanderer *Pedionomus torquata*. *Emu* **90**, 161–168.

Barr, N., and Cary, J. (1992). *Greening a Brown Land: The Australian Search for Sustainable Land Use*. (Macmillan Education Australia: South Melbourne.)

Batianoff, G. N., and Burrows, W. H. (1973). Studies in the dynamics and control of woody weeds in semi-arid Queensland. 2. *Cassia nemophila* and *C. artemisioides*. *Queensland Journal of Agriculture and Animal Science* **30**, 65–71.

Benshemesh, J. (1990). Management of malleefowl with

regard to fire. In *The Mallee Lands: A Conservation Perspective* (eds. J. C. Noble, P. J. Joss and G. K. Jones) pp. 206–211. (CSIRO: Melbourne.)

Bolton, B. L., and Latz, P. K. (1978). The western hare-wallaby, *Lagorchestes hirsutus* (Gould) (Macropodidae), in the Tanami Desert. *Australian Wildlife Research* **5**, 285–293.

Bowman, D. M. J. S., and Latz, P. K. (1993). Ecology of *Callitris glaucophylla* (Cupressaceae) on the MacDonnell Ranges, Central Australia. *Australian Journal of Botany* **41**, 217–225.

Bowman, D. M. J. S., and Panton, W. J. (1993). Decline of *Callitris intratropica* R. T. Baker and H. G. Smith in the Northern Territory: implications for pre- and post-European colonization fire regimes. *Journal of Biogeography* **20**, 373–378.

Bradstock, R. A. (1989). Dynamics of a perennial understorey. In *Mediterranean Landscapes in Australia: Mallee Ecosystems and their Management* (eds. J. C. Noble and R. A. Bradstock) pp. 141–154. (CSIRO: Melbourne.)

Bradstock, R. A. (1990). Relationships between fire regimes, plant species and fuels in mallee communities. In *The Mallee Lands: A Conservation Perspective* (eds. J. C. Noble, P. J. Joss and G. K. Jones) pp. 218–223. (CSIRO: Melbourne.)

Bradstock, R. A., and Cohn, J. S. (2001). Fires regimes and biodiversity in semi-arid mallee ecosystems. In *Flammable Australia: The Fire Regimes and Biodiversity of a Continent* (eds. R. A. Bradstock, J. E. Williams and A. M. Gill) pp. 238–258. (Cambridge University Press: Cambridge.)

Bradstock, R. A., and Gill, A. M. (1993). Fire in semi-arid, mallee shrublands: size of flames from discrete fuel arrays and their role in the spread of fire. *International Journal of Wildland Fire* **3**, 3–12.

Bradstock, R. A., Auld, T. D., Ellis, M. E., and Cohn, J. S. (1992). Soil temperatures during bushfires in semi-arid, mallee shrublands. *Australian Journal of Ecology* **17**, 433–440.

Braithwaite, R. W. (1987). Effects of fire regimes on lizards in the wet–dry tropics of Australia. *Journal of Tropical Ecology* **3**, 347–354.

Braithwaite, R. W. (1995). Fire intensity and the maintenance of habitat heterogeneity in a tropical savanna. *CALMScience Supplement* **4**, 189–196.

Brickhill, J. (1980). Striated grass wren. In *Endangered Animals of New South Wales* (ed. C Haigh) p. 68. (National Parks and Wildlife Service: Sydney.)

Brown, J. R. (1994). State and transition models for rangelands. 2. Ecology as a basis for rangeland management: performance criteria for testing models. *Tropical Grasslands* **28**, 206–213.

Brown, J. R., and Carter, J. (1998). Spatial and temporal patterns of exotic shrub invasion in an Australian tropical grassland. *Landscape Ecology* **13**, 93–102.

Burrows, W. H., Carter, J. O., Scanlan, J. C., and Anderson, E. R. (1990). Management of savannas for livestock production in north-east Australia: contrasts across the tree-grass continuum. *Journal of Biogeography* **17**, 503–512.

Byram, G. M. (1959). Combustion of forest fuels. In *Forest Fire: Control and Use* (ed. K. P. Davies) pp. 61–89. (McGraw-Hill: New York.)

Catling, P. C., and Newsome, A. E. (1981). Responses of the Australian vertebrate fauna to fire: an evolutionary approach. In *Fire and the Australian Biota* (eds. A. M. Gill, R. H. Groves and I. R. Noble) pp. 273–310. (Australian Academy of Science: Canberra.)

Caughley, G., Brown, B., and Noble, J. (1985). Movement of kangaroos after a fire in mallee woodland. *Australian Wildlife Research* **12**, 349–353.

Caughley, J. (1985). Effect of fire on the reptile fauna of mallee. In *Biology of Australasian Frogs and Reptiles* (eds. G. C. Grigg, R. Shine and H. Ehmann) pp. 31–34. (Royal Zoological Society of New South Wales and Surrey Beatty: Sydney.)

Chapin, F. S. III, Walker, B. H., Hobbs, R. J., Hooper, D. U., Lawton, J. H., Sala, O. E., and Tilman, D. (1997). Biotic control over the functioning of ecosystems. *Science* **277**, 500–504.

Charley, J. L., and Cowling, S. W. (1968). Changes in soil nutrient status resulting from overgrazing and their consequences in plant communities of semi-arid areas. *Proceedings of the Ecological Society of Australia* **3**, 28–38.

Christensen, P. E., and Burrows, N. D. (1986). Fire: an old tool with a new use. In *Ecology of Biological Invasions* (eds. R. H. Groves and J. J. Burdon) pp. 97–105. (Cambridge University Press: Cambridge.)

Cogger, H. G. (1989). Herpetofauna. In *Mediterranean Landscapes in Australia: Mallee Ecosystems and their Management* (eds. J. C. Noble and R. A. Bradstock) pp. 250–265. (CSIRO: Melbourne.)

Cowling, S. W. (1977). Effects of herbivores on nutrient cycling and distribution in rangeland ecosystems. In *The Impact of Herbivores on Arid and Semi-Arid Rangelands*, Proceedings of the 2nd US/Australia Rangeland Panel, Adelaide, 1972, pp. 277-298. (Australian Rangeland Society: Perth.)

Craig, A. B. (1997). A review of information on the effects of fire in relation to the management of rangelands in the Kimberley high-rainfall zone. *Tropical Grasslands* 31, 161-187.

Craig, A. B. (1999). Fire management of rangelands in the Kimberley low-rainfall zone: a review. *Rangeland Journal* 21, 39-70.

Dyer, R., Craig, A., and Grice, A. C. (1997). Fire in northern pastoral lands. In *Fire in Northern Australian Pastoral Lands* (eds. A. C. Grice and S. M. Slatter) pp. 24-40. (Tropical Grassland Society of Australia: St Lucia, Qld.)

Dyksterhuis, E. J. (1949). Condition and management of rangeland based on quantitative ecology. *Journal of Range Management* 2, 104-115.

Ehrlich, P., and Walker, B. (1998). Rivets and redundancy. *BioScience* 48, 387.

Ellis, B. A., Tierney, P. J., and Dawson, T. J. (1992). The diet of the bridled nailtail wallaby (*Onychogalea fraenata*). I. Site and seasonal influences and dietary overlap with the black-striped wallaby (*Macropus dorsalis*) and domestic cattle.*Wildlife Research* 19, 65-78.

Fensham, R. J., and Bowman, D. M. J. S. (1992). Stand structure and the influence of overwood on regeneration in tropical eucalypt forest on Melville Island. *Australian Journal of Botany* 40, 335-352.

Finlayson, H. H. (1958). On central Australian mammals (with notice of related species from adjacent tracts). III. The Potoroinae. *Records of the South Australian Museum* 13, 235-307.

Flannery, T., Kendall, P., and Wynn-Moylan, K. (1990). Burrowing bettong (*Bettongia lesueur*). In *Australia's Vanishing Mammals: Endangered and Extinct Native Species*, pp. 98-101. (RD Press: Surrey Hills.)

Freudenberger, D., and Noble, J. (1997). Consumption, regulation and off-take: a landscape perspective on pastoralism. In *Landscape Ecology, Function and Management: Principles from Australia's Rangelands* (eds. J. Ludwig, D. Tongway, D. Freudenberger, J. Noble and K. Hodgkinson) pp. 35-48. (CSIRO: Melbourne.)

Freudenberger, D., Noble, J. ,and Morton, S. (1997). *A Comprehensive, Adequate and Representative Reserve System for the Southern Mallee of New South Wales: Principles and Benchmarks.* (CSIRO Wildlife and Ecology: Canberra.)

Freudenberger, D., Wilson, A., and Palmer, R. (1999). The effects of perennial grasses, stocking rate and rainfall on sheep production. *Rangeland Journal* 21, 199-219.

Friedel, M. H., Foran, B. D., and Stafford Smith, D. M. (1990). Where the creeks run dry or ten feet high: pastoral management in arid Australia. *Proceedings of the Ecological Society of Australia* 16, 185-194.

Fyfe, G. (1980). The effects of fire on lizard communities in central Australia. *Herpetofauna* 12, 1-9.

Garnett, S., and Crowley, G. (1997). The golden-shouldered parrot of Cape York Peninsula: the importance of cups of tea to effective conservation. In *Conservation Outside Nature Reserves* (eds. P. Hale and D. Lamb) pp. 201-205. (Centre for Conservation Biology, University of Queensland: Brisbane.)

Gill, A. M. (1977). Management of fire-prone vegetation for plant species conservation in Australia. *Search* 8, 20-26.

Gill, A. M., and Bradstock, R. A. (1995). Extinction of biota by fires. In *Conserving Biodiversity: Threats and Solutions* (eds. R. A. Bradstock, T. D. Auld, D. A. Keith, R. T. Kingsford, D. Lunney and D. P. Sivertsen) pp. 309-322. (Surrey Beatty: Chipping Norton, NSW.)

Gill, A. M., Groves, R. H., and Noble, I. R. (eds.) (1981). *Fire and the Australian Biota.* (Australian Academy of Science: Canberra.)

Greenslade, P. J. M., and Greenslade, P. (1989). Ground layer invertebrate fauna. In *Mediterranean Landscapes in Australia: Mallee Ecosystems and their Management* (eds. J. C. Noble and R. A. Bradstock) pp. 266-284. (CSIRO: Melbourne.)

Grice, A. C. (1997). Post-fire regrowth and survival of the invasive tropical shrubs *Cryptostegia grandiflora* and *Ziziphus mauritiana. Australian Journal of Ecology* 22, 49-55.

Grice, A. C. (1998). Ecology in the management of invasive rangeland shrubs: a case study of Indian jujube (*Ziziphus mauritiana*). *Weed Science* 46, 467-474.

Grice, A. C., and Barchia, I. (1992). Does grazing reduce survival of indigenous perennial grasses of the semi-arid woodlands of western New South Wales? *Australian Journal of Ecology* 17, 195-205.

Grice, A. C., and Brown, J. R. (1996). The population ecology of the invasive tropical shrubs *Cryptostegia grandiflora* and *Ziziphus mauritiana* in relation to fire. In *Frontiers of Population Ecology* (eds. R. B. Floyd, A. W. Sheppard and P. J. De Barro) pp. 589–97. (CSIRO: Melbourne.)

Grice, A. C., and Slatter, S. M. (eds.) (1997). *Fire in the Management of Northern Australian Pastoral Lands.* (Tropical Grassland Society of Australia: St Lucia, Qld.)

Griffin, G. F. (1984a). Hummock grasslands. In *Management of Australia's Rangelands* (eds. G. N. Harrington, A. D. Wilson and M. D. Young) pp. 271–284. (CSIRO: Melbourne.)

Griffin, G. F. (1984b). Strategic timing of patch-burns. In *Anticipating the Inevitable: A Patch-Burn Strategy for Fire Management at Uluru (Ayers Rock – Mt Olga) National Park* (ed. E. C. Saxon) pp. 79–87. (CSIRO: Melbourne.)

Griffin, G. F. (1992). Will it burn – should it burn? Management of the spinifex grasslands of inland Australia. In *Desertified Grasslands: Their Biology and Management* (ed. G. P. Chapman) pp. 63–76. (Academic Press: New York.)

Griffin, G. F., and Allan, G. E. (1986). Fire and the management of Aboriginal owned lands in central Australia. In *Science and Technology for Aboriginal Development* (eds. B. Foran and B. Walker) paper 2.6. (CSIRO: Melbourne.)

Griffin, G. F., and Friedel, M. H. (1984a). Effects of fire on central Australian rangelands. I. Fire and fuel characteristics and changes in herbage and nutrients. *Australian Journal of Ecology* 9, 381–393.

Griffin, G. F., and Friedel, M. H. (1984b). Effects of fire on central Australian rangelands. II. Changes in tree and shrub populations. *Australian Journal of Ecology* 9, 395–403.

Griffin, G. F., and Friedel, M. H. (1985). Discontinuous change in central Australia: some implications of major ecological events for land management. *Journal of Arid Environments* 9, 63–80.

Griffin, G. F., Price, N. F., and Portlock, H. F. (1983). Wildfires in the central Australian rangelands, 1970–1980. *Journal of Environmental Management* 17, 311–323.

Harrington, G. N. (1986). Critical factors in shrub dynamics in eastern mulga lands. In *The Mulga Lands* (ed. P. S. Sattler) pp. 90–92. (Royal Society of Queensland: Brisbane.)

Harrington, G. N. (1991). Effects of soil moisture on shrub seedling survival in a semi-arid grassland. *Ecology* 72, 1138–1149.

Harrington, G. N., and Driver, M. A. (1995). The effect of fire and ants on the seed bank of a shrub in a semi-arid grassland. *Australian Journal of Ecology* 20, 538–547.

Harrington, G. N., Mills, D. M. D., Pressland, A. J., and Hodgkinson, K. C. (1984). Semi-arid woodlands. In *Management of Australia's Rangelands* (eds. G. N. Harrington, A. D. Wilson and M. D. Young) pp. 189–207. (CSIRO: Melbourne.)

Heathcote, R. L. (1977). Pastoral Australia. In *Space and Society* (ed. D. N. Jeans) pp. 259–300. (Sydney University Press: Sydney.)

Hodgkinson, K. C. (1986). Responses of rangeland plants to fire in water limited environments. In *Rangelands: A Resource Under Siege*, Proceedings of the 2nd International Rangeland Congress, Adelaide (eds. P. J. Joss, P. W. Lynch and O. B. Williams) pp. 437–441. (Australian Academy of Science: Canberra/Cambridge University Press: Sydney.)

Hodgkinson, K. C. (1991). Shrub recruitment response to intensity and season of fire in a semi-arid woodland. *Journal of Applied Ecology* 28, 60–70.

Hodgkinson, K. C. (1992). Water relations and growth of shrubs before and after fire in a semi-arid woodland. *Oecologia* 90, 467–473.

Hodgkinson, K. C. (1998). Sprouting success of shrubs after fire: height-dependent relationships for different strategies. *Oecologia* 115, 64–72.

Hodgkinson, K. C. (2001). Fire regimes in *Acacia* wooded landscapes: effects on functional processes and biological diversity. In *Flammable Australia: The Fire Regimes and Biodiversity of a Continent* (eds. R. A. Bradstock, J. E. Williams and A. M. Gill) pp. 259–277. (Cambridge University Press: Cambridge.)

Hodgkinson, K. C., and Griffin, G. F. (1982) Adaptation of shrub species to fire in the arid zone. In *Evolution of the Flora and Fauna of Arid Australia* (eds. W. R. Barker and P. J. M. Greenslade) pp. 145–152. (Peacock Publications: Adelaide.)

Hodgkinson, K. C., and Harrington, G. N. (1985). The case for prescribed burning to control shrubs in eastern semi-arid woodlands. *Australian Rangeland Journal* 7, 64–74.

Hodgkinson, K. C., and Oxley, R. E. (1990). Influence of fire and edaphic factors on germination of the arid zone shrubs *Acacia aneura*, *Cassia nemophila* and *Dodonaea viscosa*. *Australian Journal of Botany* **38**, 269–279.

Hodgkinson, K. C., Harrington, G. N., Griffin, G. F., Noble, J. C., and Young, M. D. (1984). Management of vegetation with fire. In *Management of Australia's Rangelands* (eds. G. N. Harrington, A. D. Wilson and M. D. Young) pp. 141–156. (CSIRO: Melbourne.)

Hodgkinson, K. C., Terpstra, J. W., and Müller, W. J. (1995). Spatial and temporal pattern in the grazing of grasses by sheep within a semi-arid wooded landscape. *Rangeland Journal* **17**, 154–165.

Holling, C. S. (1973). Resilience and stability of ecological systems. *Annual Review of Ecology and Systematics* **4**, 1–23.

Holling, C. S. (ed.) (1978). *Adaptive Environmental Assessment and Management*. (Wiley: Chichester, UK.)

Holling, C. S. (1995). What barriers? What bridges? In *Barriers and Bridges to the Renewal of Ecosystems and Institutions* (eds. L. H. Gunderson, C. S. Holling and S. S. Light) pp. 3–36. (Columbia University Press: New York.)

Holling, C. S., Schindler, D. W., Walker, B. H., and Roughgarden, J. (1995). Biodiversity in the functioning of ecosystems: an ecological synthesis. In *Biodiversity Loss: Ecological and Economic Issues* (eds. C. A. Perrings, K. G. Mäler, C. Folke, C. S. Holling and B. O. Jansson) pp. 44–83. (Cambridge University Press: Cambridge.)

Hoser, R. T. (1991). Burrowing bettong *Bettongia lesueur* Quoy and Gaimard 1824. In *Endangered Animals of Australia*, p. 201. (Pierson: Sydney.)

Kimber, R. (1983). Black lightning: Aborigines and fire in central Australia and the Western Desert. *Archaeology in Oceania* **18**, 38–45.

Landsberg, J., and Stol, J. (1996). Spatial distribution of sheep, feral goats and kangaroos in woody rangeland paddocks. *Rangeland Journal* **18**, 270–291.

Landsberg, J., James, C. D., Morton, S. R., Hobbs, T. J., Stol, J., Drew, A., and Tongway, H. (1997). *The Effects of Artificial Sources of Water on Rangeland Biodiversity*. (Environment Australia/CSIRO Wildlife and Ecology: Canberra.)

Langton, M. (1998). *Burning Questions: Emerging Environmental Issues for Indigenous Peoples in Northern Australia*. (Centre for Indigenous Natural and Cultural Resource Management, Northern Territory University: Darwin.)

Latz, P. (1995). Fire in the desert: increasing biodiversity in the short term, decreasing in the long term. In *Country in Flames*, Proceedings of the 1994 symposium on biodiversity and fire in northern Australia (ed. D. Bird Rose) pp. 77–86. (Department of Environment, Sport and Territories: Canberra/North Australia Research Unit, Australian National University: Canberra.)

Latz, P. K., and Griffin, G. F. (1978). Changes in Aboriginal land management in relation to fire and to food plants in central Australia. In *The Nutrition of Aborigines in Relation to the Ecosystem of Central Australia* (eds. B. S. Hetzel and H. J. Frith) pp. 77–85. (CSIRO: Melbourne.)

Leigh, J. H., and Noble, J. C. (1981). The role of fire in the management of rangelands in Australia. In *Fire and the Australian Biota* (eds. A. M. Gill, R. H. Groves and I. R. Noble) pp. 471–495. (Australian Academy of Science: Canberra.)

Leigh, J. H., Wood, D. H., Holgate, M. D., Slee, A., and Stanger, M. G. (1989). Effects of rabbit and kangaroo grazing on two semi-arid grassland communities in central–western New South Wales. *Australian Journal of Botany* **37**, 375–396.

Ludwig, J., Tongway, D., Freudenberger, D., Noble, J., and Hodgkinson, K. (eds.) (1997). *Landscape Ecology, Function and Management: Principles from Australia's Rangelands*. (CSIRO: Melbourne.)

Main, A. R. (1992). Management to retain biodiversity in the face of uncertainty. In *Biodiversity in Mediterranean Ecosystems in Australia* (ed. R. J. Hobbs) pp. 193–209. (Surrey Beatty: Chipping Norton, NSW.)

Masters, P. (1993). The effects of fire-driven succession and rainfall on small mammals in spinifex grasslands at Uluru National Park, Northern Territory. *Wildlife Research* **20**, 803–813.

Masters, P. (1997). The mulgara *Dasycercus cristicauda* (Krefft) (Marsupialia: Dasyuridae): ecological characteristics, implications for management and future research. In *Back to the Future*, Proceedings of a Natural Resources Research Workshop, Uluru-Kata Tjuta National Park (ed. L. G. Woodcock) pp. 62–67. (Australian Geological Survey Organisation: Canberra.)

McAlpin, S. (1997). Tjakura *Egernia kintorei*: The desert skink. In *Back to the Future*, Proceedings of a Natural Resources Research Workshop, Uluru-Kata Tjuta National Park (ed. L. G. Woodcock) pp. 57–61. (Australian Geological Survey Organisation: Canberra.)

Meggy, P. R. (1885). *From Sydney to Silverton*. (H. Solomon: Sydney.)

Moore, J. L., Howden, S. M., McKeown, G. M., Carter, J. O., and Scanlan, J. C. (1997). A method to evaluate greenhouse gas emissions from sheep grazed rangelands in south west Queensland. In *Proceedings of Modsim '97*, International Congress on Modelling and Simulation, University of Tasmania, Hobart (eds. D. A. McDonald and M. McAleer) pp. 137–142. (Modelling and Simulation Society of Australia: Canberra.)

Morton, S. R. (1990). The impact of European settlement on the vertebrate animals of arid Australia: a conceptual model. *Proceedings of the Ecological Society of Australia* 16, 201–213.

Morton, S. R., Doherty, M. D., and Barker, R. D. (1995a). *Natural Heritage Values of the Lake Eyre Basin in South Australia*. (CSIRO Division of Wildlife and Ecology: Canberra.)

Morton, S. R., Short, J., and Barker, R. D. (1995b). Refugia for biological diversity in arid and semi-arid Australia. Biodiversity Unit, Department of Environment, Sport and Territories, Biodiversity Series Paper no. 4, Canberra.

Morton, S. R., Stafford Smith, D. M., Friedel, M. H., Griffin, G. F., and Pickup, G. (1995c). The stewardship of arid Australia: ecology and landscape management. *Journal of Environmental Management* 43, 195–217.

Morton, S. R., Andersen, A. N., Cork, S. J., Hobbs, R. J., Noble, J. C., Saunders, D. A., Stafford Smith, D. M., and Young, M. D. (1996). *Looking After Our Land: A Future for Australia's Biological Diversity*. (CSIRO: Canberra.)

Myers, K., and Parker, B. S. (1965). A study of the biology of the wild rabbit in climatically different regions in eastern Australia. *CSIRO Wildlife Research* 10, 1–32.

Nicholson, A. T. (1993). Landholders' perspective: salinity management (past, present and future) for Barney's Reef Landcare Group. In *Land Management for Dryland Salinity Control*, Proceedings of a national conference, La Trobe University, Bendigo, pp. 158–162. (Land and Water Resources Research and Development Corporation and Victorian Salinity Program: Canberra.)

Noble, J. C. (1986a). Plant population ecology and clonal growth in arid rangeland ecosystems. In *Rangelands: A Resource under Siege*, Proceedings of the 2nd International Rangeland Congress, Adelaide (eds. P. J.

Joss, P. W. Lynch and O. B. Williams) pp. 16–19. (Australian Academy of Science: Canberra/CambridgeUniversity Press: Cambridge.)

Noble, J. C. (1986b). Prescribed fire in mallee rangelands and the potential role of aerial ignition. *Australian Rangeland Journal* 8, 118–130.

Noble, J. C. (1989a). Fire studies in mallee (*Eucalyptus* spp.) communities of western New South Wales: the effects of fires applied in different seasons on herbage productivity and their implications for management. *Australian Journal of Ecology* 14, 169–187.

Noble, J. C. (1989b). Fire regimes and their influence on herbage and mallee coppice dynamics. In *Mediterranean Landscapes in Australia: Mallee Ecosystems and their Management* (eds. J. C. Noble and R. A. Bradstock) pp. 168–180. (CSIRO: Melbourne.)

Noble, J. C. (1991). Behaviour of a very fast grassland fire on the Riverine Plain of southeastern Australia. *International Journal of Wildland Fire* 1, 189–196.

Noble, J. C. (1993a). Relict surface-soil features in semi-arid mulga (*Acacia aneura*) woodlands. *Rangeland Journal* 15, 48–70.

Noble, J. C. (1993b). Termites have a minor role in the decomposition of senescent grass tussocks in a semi-arid woodland in eastern Australia. In *Proceedings of the 17th International Grassland Congress*, Palmerston North, Hamilton and Lincoln, New Zealand and Rockhampton, Australia, pp. 58–59. (Tropical Grassland Society of Australia: Brisbane/New Zealand Grassland Association: Auckland.)

Noble, J. C. (1995). Bettong (*Bettongia* spp.) biocontrol of shrubs in semi-arid mulga (*Acacia aneura*) rangelands: an hypothesis. In *Ecological Research and Management in the Mulga Lands* (eds. M. J. Page and T. S. Beutel) pp. 139–145. (University of Queensland Gatton College: Gatton, Qld.)

Noble, J. C. (1996a). Mesomarsupial ecology in Australian rangelands: burrows, bettongs (*Bettongia* spp.) and biocontrol of shrubs. In *Rangelands in a Sustainable Biosphere*, Proceedings of the 5th International Rangeland Congress, Salt Lake City, Utah (ed. N. E. West) vol.1, pp. 395–396. (Society for Range Management: Denver.)

Noble, J. C. (1996b). Shrub population regulation in semi-arid woodlands before and after European settlement. In *Proceedings of the 9th Biennial Australian Rangeland*

Conference, Australian Rangeland Society, Port Augusta (eds. L. P. Hunt and R. Sinclair) pp. 62–65. (Australian Rangeland Society: Port Augusta, SA.)

Noble, J. C. (1997a). *The Delicate and Noxious Scrub: CSIRO Studies on Native Tree and Shrub Proliferation in the Semi-Arid Woodlands of Eastern Australia* (CSIRO Wildlife and Ecology: Canberra.)

Noble, J. C. (1997b). The potential for using fire in northern Australian pastoral lands. In *Fire in the Management of Northern Australian Pastoral Lands* (eds. A. C. Grice and S. M. Slatter) pp. 41–47. (Tropical Grassland Society of Australia: St Lucia, Qld.)

Noble, J. C., and Brown, J. R. (1997). A landscape perspective on rangeland management. In *Landscape Ecology, Function and Management: Principles from Australia's Rangelands* (eds. J. Ludwig, D. Tongway, D. Freudenberger, J. Noble and K. Hodgkinson) pp. 79–92. (CSIRO: Collingwood.)

Noble, J. C., and Tongway, D. J. (1986a). Pastoral settlement in arid and semi-arid rangelands. In *Australian Soils: The Human Impact* (eds. J. S. Russell and R. F. Isbell) pp. 217–242. (University of Queensland Press: St Lucia.)

Noble, J. C., and Tongway, D. J. (1986b). Herbivores in arid and semi-arid rangelands. In *Australian Soils: The Human Impact* (eds. J. S. Russell and R. F. Isbell) pp. 243–270. (University of Queensland Press: St Lucia.)

Noble, J. C., and Vines, R. G. (1993). Fire studies in mallee (*Eucalyptus* spp.) communities of western New South Wales: grass fuel dynamics and associated weather patterns. *Rangeland Journal* 15, 270–297.

Noble, J. C., Smith, A. W., and Leslie, H. W. (1980). Fire in the mallee shrublands of western New South Wales. *Australian Rangeland Journal* 2, 104–114.

Noble, J. C., Harrington, G. N., and Hodgkinson, K. C. (1986). The ecological significance of irregular fire in Australian rangelands. In *Rangelands: A Resource under Siege*, Proceedings of the 2nd International Rangeland Congress, Adelaide (eds. P. J. Joss, P. W. Lynch and O. B. Williams) pp. 577–580. (Australian Academy of Science: Canberra/Cambridge University Press: Sydney.)

Noble, J. C., Diggle, P. J., and Whitford, W. G. (1989). The spatial distributions of termite pavements and hummock feeding sites in a semi-arid woodland in eastern Australia. *Acta Oecologica/Oecologia Generalis* 10, 355–376.

Noble, J. C., MacLeod, N. D., Ludwig, J. A., and Grice, A. C. (1991). Integrated shrub control strategies in Australian semi-arid woodlands. In *Proceedings of the 4th International Rangeland Congress*, Montpellier (eds. A. Gaston, M. Kernick and H. Le Houérou) vol. 2, pp. 846–849. (Service Central d'Information Scientifique et Technique: Montpellier.)

Noble, J. C., Grice, A. C., MacLeod, N. D., and Müller, W. J. (1992). Integration of prescribed fire and sub-lethal chemical defoliation for controlling shrub populations in Australian semi-arid rangelands. In *Proceedings of the 1st International Weed Control Congress*, Monash University, pp. 362–364. (Weed Science Society of Victoria: Melbourne.)

Noble, J. C., Tongway, D. J., Roper, M. M., and Whitford, W. G. (1996a). Fire studies in mallee (*Eucalyptus* spp.) communities of western New South Wales: spatial and temporal fluxes in soil chemistry and soil biology following prescribed fire. *Pacific Conservation Biology* 2, 398–413.

Noble, J. C., Whitford, W. G., and Kaliszweski, M. (1996b). Soil and litter microarthropod populations from two contrasting ecosystems in semi-arid eastern Australia. *Journal of Arid Environments* 32, 329–346.

Noble J., MacLeod, N., and Griffin, G. (1997). The rehabilitation of landscape function in rangelands. In *Landscape Ecology, Function and Management: Principles from Australia's Rangelands* (eds. J. Ludwig, D. Tongway, D. Freudenberger, J. Noble and K. Hodgkinson) pp. 107–120. (CSIRO: Collingwood.)

Noble, J. C., Greene, R. S. B., and Müller, W. J. (1998a). Herbage production following rainfall redistribution in a semi-arid mulga (*Acacia aneura*) woodland in western New South Wales. *Rangeland Journal* 20, 206–225.

Noble, J. C., Habermehl, M. A., James, C. D., Landsberg, J., Langston, A. C., and Morton, S. R. (1998b). Biodiversity implications of water management in the Great Artesian Basin. *Rangeland Journal* 20, 275–300.

Noble, J. C., Detling J., Hik, D., and Whitford, W. G. (1999). Soil biodiversity and desertification in *Acacia aneura* woodlands. In *People and Rangelands: Building the Future*, Proceedings of the VI International Rangeland Congress, Townsville (eds. D. Eldridge and D. Freudenberger) vol. 1, pp. 108–109. (VI International Rangeland Congress, Inc.: Aitkenvale, Qld.)

Orr, D. M., and Holmes, W. E. (1984). Mitchell grasslands. In *Management of Australia's Rangelands* (eds. G. N. Harrington, A. D. Wilson and M. D. Young) pp. 241–254. (CSIRO: Melbourne.)

Orr, D. M., and Paton, C. J. (1993). Fire and grazing to interact to manipulate pasture composition in *Heteropogon contortus* (black speargrass) pastures. In *Proceedings of the 17th International Grassland Congress, Palmerston North, Hamilton and Lincoln, New Zealand and Rockhampton, Australia*, pp. 1910–1911. (Tropical Grassland Society of Australia: Brisbane/New Zealand Grassland Association: Auckland.)

Orr, D. M., McKeown, G. M., and Day, K. A. (1991). Burning and exclosure can rehabilitate degraded black spear grass (*Heteropogon contortus*) pastures. *Tropical Grasslands* 25, 333–336.

Paton, C. J., and Rickert, K. G. (1989). Burning, then resting reduces wiregrass (*Aristida* spp.) in black spear grass pastures. *Tropical Grasslands* 23, 211–218.

Peacock, R. W. (1900). Our western lands: their deterioration and possible improvement. *Agricultural Gazette of New South Wales* 11, 652–657.

Pedler, L. (1997). Fire and striated grasswrens in Uluru–Kata Tjuta National Park. In *Back to the Future*, Proceedings of a Natural Resources Research Workshop, Uluru–Kata Tjuta National Park (ed. L. G. Woodcock) pp. 91–93. (AGSO Record 1997/55.)

Perrings, C., and Walker, B. (1997). Biodiversity, resilience and the control of ecological–economic systems: the case of fire-driven rangelands. *Ecological Economics* 22, 73–83.

Perry, R. A. (1967). The need for rangeland research in Australia. *Proceedings of the Ecological Society of Australia* 2, 1–14.

Perry, R. A. (1977). The evaluation and exploitation of semi-arid lands: Australian experience. *Philosophical Transactions of the Royal Society of London, Series B*. 278, 493–505.

Pianka, E. R. (1979). Diversity and niche structure in desert communities. In *Arid-Land Ecosystems: Structure, Functioning and Management* (eds. D. W. Goodall and R. A. Perry) vol. 1, pp. 321–341. (Cambridge University Press: Cambridge.)

Powell, J. M. (1988). *An Historical Geography of Modern Australia: The Restive Fringe*. (Cambridge University Press: Cambridge.)

Priddel, D. (1989). Conservation of rare fauna: the regent parrot and the malleefowl. In *Mediterranean Landscapes in Australia: Mallee Ecosystems and their Management* (eds. J. C. Noble and R. A. Bradstock) pp. 243–249. (CSIRO: Melbourne.)

Ramson, W. S. (1988). *The Australian National Dictionary: A Dictionary of Australianisms on Historical Principles*. (Oxford University Press: Melbourne.)

Reid, J. R. W., Kerle, J. A., and Baker, L. (1993a). Mammals. In *Uluru Fauna: The Distribution and Abundance of Vertebrate Fauna of Uluru (Ayers Rock–Mt Olga) National Park, NT* (eds. J. R. W. Reid, J. A. Kerle and S. R. Morton) *Kowari* 4, 69–78. (Australian National Parks and Wildlife Service: Canberra.)

Reid, J. R. W., Kerle, J. A., Baker, L., and Jones, K. R. (1993b). Reptiles and frogs. In *Uluru Fauna: The Distribution and Abundance of Vertebrate Fauna of Uluru (Ayers Rock–Mt Olga) National Park, N.T* (eds. J. R. W. Reid, J. A. Kerle and S. R. Morton) *Kowari* 4, 58–68. (Australian National Parks and Wildlife Service: Canberra.)

Reid, J. R. W., Kerle, J. A., and Morton, S. R. (eds.) (1993c). *Uluru Fauna: The Distribution and Abundance of Vertebrate Fauna of Uluru (Ayers Rock–Mt Olga) National Park, NT, Kowari* 4. (Australian National Parks and Wildlife Service: Canberra.)

Remmert, H. (1991). The mosaic-cycle concept of ecosystems: an overview. In *The Mosaic-Cycle Concept of Ecosystems* (ed. H. Remmert) pp. 1–21. (Springer-Verlag: New York.)

Robson, A. D. (1995). The effects of grazing exclusion and blade-ploughing on semi-arid woodland vegetation in north-western New South Wales over 30 months. *Rangeland Journal* 17, 111–127.

Rolls, E. C. (1981). *A Million Wild Acres: 200 Years of Man and Australian Forest*. (Thomas Nelson: Melbourne.)

Royal Commission (1901). Royal Commission to inquire into the condition of the crown tenants of the Western Division of New South Wales. Part II. Minutes of evidence, appendices and returns. In *Votes and Proceedings of the New South Wales Legislative Assembly during the Session of 1901, with the Various Documents Connected Therewith* (6 vols.) vol. 4, pp. 1– 853. (Government Printer: Sydney.) (Mitchell Library Reference Q342.91/3.)

Saunders, D., and Walker, B. (1998). Biodiversity and agriculture. *Reform* Spring Issue 6, 11–16.

Scanlan, J. C. (1980). Effects of spring wildfires on *Astrebla* (Mitchell grass) grasslands in north west Queensland under varying levels of growing season rainfall. *Australian Rangeland Journal* 2, 162–168.

Scanlan, J. C. (1983). Changes in tiller and tussock characteristics of *Astrebla lappacea* (curly Mitchell grass) after burning. *Australian Rangeland Journal* 5, 13–19.

Schlesinger, C. A., Noble, J. C., and Weir, T. (1997). Fire studies in mallee (*Eucalyptus* spp.) communities of western New South Wales: reptile and beetle populations in sites of differing fire history. *Rangeland Journal* 19, 190–205.

Scifres, C. J., and Hamilton, W. T. (1993). *Prescribed Burning for Brushland Management: The South Texas Example*. (Texas A&M University Press: College Station.)

Spain, A. V., and McIvor, J. G. (1988). The nature of herbaceous vegetation associated with termitaria in north-eastern Australia. *Journal of Ecology* 76, 181–191.

Spain, A. V., Okello-Oloya, T., and Brown, A. J. (1983). Abundances, above-ground masses and basal areas of termite mounds at six locations in tropical north-eastern Australia. *Revue d'Écologie et de Biologie du Sol* 20, 547–566.

Spain, A. V., Sinclair, D. F., and Diggle, P. J. (1986). Spatial distribution of the mounds of harvester and forager termites (Isoptera: Termitidae) at four locations in tropical north-eastern Australia. *Acta Oecologica/Oecologia Generalis* 7, 333–352.

Stafford Smith, M., and Pickup, G. (1993). Out of Africa, looking in: understanding change. In *Range Ecology at Disequilibrium* (eds. R. H. Behnke Jr, I. Scoones and C. Kervan) pp. 196–226. (Overseas Development Institute: London.)

State of the Environment Advisory Council (1996). *Australia: State of the Environment*. (CSIRO: Melbourne.)

Suijdendorp, H. (1969). Deferred grazing improves soft spinifex association. *Journal of Agriculture of Western Australia* 10, 487–488.

Suijdendorp, H. (1981). Responses of the hummock grasslands of northwestern Australia to fire. In *Fire and the Australian Biota* (eds. A. M. Gill, R. H. Groves and I. R. Noble) pp. 417–424. (Australian Academy of Science: Canberra.)

Tongway, D. J., and Ludwig, J. A. (1990). Vegetation and soil patterning in semi-arid mulga lands of eastern Australia. *Australian Journal of Ecology* 15, 23–34.

Tongway, D. J., and Ludwig, J. A. (1996). Rehabilitation of semiarid landscapes in Australia: restoring productive soil patches. *Restoration Ecology* 4, 388–397.

Tongway, D. J., and Ludwig, J. A. (1997). The conservation of water and nutrients within landscapes. In *Landscape Ecology, Function and Management: Principles from Australia's Rangelands* (eds. J. Ludwig, D. Tongway, D. Freudenberger, J. Noble and K. Hodgkinson) pp. 13–22. (CSIRO: Melbourne.)

Tongway, D. J., Ludwig, J. A., and Whitford, W. G. (1989). Mulga log-mounds: fertile patches in the semi-arid woodlands of eastern Australia. *Australian Journal of Ecology* 14, 263–268.

Tothill, J. C. (1971). A review of fire in the management of native pasture with particular reference to north-eastern Australia. *Tropical Grasslands* 5, 1–10.

Tothill, J. C. (1976). Effect of grazing, burning and fertilising on the composition of a native pasture in the sub-tropics of south eastern Queensland. In *Proceedings of the 12th International Grassland Congress*, Moscow, vol. 2, pp. 515–521.

Tothill, J. C., and Gillies, C. (1992). *The Pasture Lands of Northern Australia: Their Condition, Productivity and Sustainability*. (Tropical Grassland Society of Australia: St Lucia, Qld.)

Turnbull, J.W. (1972). Dry country seeds. In *The Use of Trees and Shrubs in the Dry Country of Australia* (ed. N. Hall.) pp. 532–536. (Australian Government Publishing Service: Canberra.)

UNESCO (1979). *Map of the World Distribution of Arid Regions*. (UNESCO: Paris.)

Walker, B. H. (1985). A general model of savanna structure and function. In *Determinants of Tropical Savannas* (ed. B. H. Walker) pp. 1–12. (IRL Press: Oxford.)

Walker, B. H. (1992). Biodiversity and ecological redundancy. *Conservation Biology* 6, 18–23.

Walker, B. H. (1993). Rangeland ecology: understanding and managing change. *Ambio* 22, 80–87.

Walker, B. H. (1994). Landscape to regional-scale responses of terrestrial ecosystems to global change. *Ambio* 23, 67–73.

Walker, B. (1995). Conserving biological diversity through ecosystem resilience. *Conservation Biology* 9, 747–752.

Walker, B. H. (1998). The art and science of wildlife management. *Wildlife Research* 25, 1–9.

Walker, B., and Nix, H. (1993). Managing Australia's biological diversity. *Search* **24**, 173–178.

Walker, B., and Steffen, W. (1997). An overview of the implications of global change for natural and managed terrestrial ecosystems. *Conservation Ecology* [online] **1**, 2. Available from the Internet: http://www.consecol.org/vol 1/iss2/art2.

Walters, C. J., and Holling, C. S. (1990). Large-scale management experiments and learning by doing. *Ecology* **71**, 2060–2068.

Wellington, A. B. (1989). Seedling regeneration and the population dynamics of eucalypts. In *Mediterranean Landscapes in Australia: Mallee Ecosystems and their Management* (eds. J. C. Noble and R. A. Bradstock) pp. 155–167. (CSIRO: Melbourne.)

Westoby, M., Walker, B., and Noy-Meir, I. (1989). Opportunistic management for rangelands not at equilibrium. *Journal of Range Management* **4**, 266–274.

Williams, R. J. (1995). Tree mortality in relation to fire intensity in a tropical savanna of the Kakadu region, Northern Territory, Australia. *CALMScience Supplement* **4**, 77–82.

Wilson, A. D. (1990). The effect of grazing on Australian ecosystems. *Proceedings of the Ecological Society of Australia* **16**, 235–244.

Wilson, A. D., Hodgkinson, K. C., and Noble, J. C. (1988). Vegetation attributes and their application to the management of Australian rangelands. In *Vegetation Science Applications for Rangeland Analysis and Management* (ed. P. T. Tueller) pp. 253–294. (Kluwer: London.)

Wilson, A. D., and Mulham, W. E. (1979). A survey of the regeneration of some problem shrubs and trees after wildfire in western New South Wales. *Australian Rangeland Journal* **1**, 363–368.

Winter, W. H. (1987). Using fire and supplements to improve cattle production from monsoon tallgrass pastures. *Tropical Grasslands* **21**, 71–80.

Woinarski, J. C. Z. (1990). Effects of fire on the bird communities of tropical woodlands and open-forests in northern Australia. *Australian Journal of Ecology* **15**, 1–22.

Woinarski, J. C. Z. (1999). Fire and Australian birds: a review. In *Australia's Biodiversity: Responses to Fire* (eds. A. M. Gill, J. C. Z. Woinarski and A. York) pp. 55–112. (Department of Environment and Heritage: Canberra.)

Woinarski, J. C. Z., and Recher, H. F. (1997). Impact and response: a review of the effects of fire on the Australian avifauna. *Pacific Conservation Biology* **3**, 183–205.

Woinarski, J. C. Z., Brock, C., Fisher, A., Milne, D., and Oliver, B. (1999). Response of birds and reptiles to fire regimes on pastoral land in the Victoria River District, Northern Territory. *Rangeland Journal* **21**, 24–38.

Yen, A. L. (1989). Overstorey invertebrates in the Big Desert, Victoria. In *Mediterranean Landscapes in Australia: Mallee Ecosystems and their Management* (eds. J. C. Noble and R. A. Bradstock) pp. 285–299. (CSIRO: Melbourne.)

17

Fire management and biodiversity conservation: key approaches and principles

DAVID A. KEITH, JANN E. WILLIAMS AND JOHN C. Z. WOINARSKI

Abstract

Fire regimes have been a major driving force in the maintenance of biodiversity in many Australian ecosystems. Currently fire regimes are manipulated across much of the continent and affect, in often complex ways, the biodiversity of both relatively natural systems and those that are converted, fragmented, invaded or otherwise modified. This chapter focusses on the use of fire for managing biodiversity, drawing on the considerable advances in knowledge that have occurred in fire ecology and management in the last two decades. The policy and legislative environment within which fire managers operate is briefly addressed, but the main discussion considers a number of principles and approaches for the use of fire as a management tool including use of local populations and functional groups and variability within critical fire regime thresholds. These are applied in an adaptive approach to management that recognises the importance of setting explicit goals, precautionary management practices, experimentation, risk assessment and the need to implement monitoring so that management practices can be evaluated and potentially modified. Part of this approach will be to try to include the unpredictable (especially unplanned fires) in management planning. An ecosystem-based, landscape-scale approach to planning and management is advocated. To help achieve biodiversity conservation goals, spatial and temporal variability in fire regimes should be promoted, fire management targets should be defined as ranges rather than optima, and focus should especially be directed at groups of species with ecological traits which render them most susceptible to decline under different fire regimes.

Introduction

Within the last decade the conservation of biodiversity, in the broader context of regional planning and community involvement, has become the principal focus of natural resource management from the local to the international level. However, management of biodiversity is complex. For example, individual species can show considerable variability in growth form, fecundity, life history strategies and longevity. Also, biodiversity encompasses the compositional, structural and functional components of ecosystems (Mooney 1997). The spatially variable and temporally dynamic nature of biodiversity makes management particularly challenging. The overlying mosaic of human systems of tenure and governance and the broad range of values associated with biodiversity add further complexity. Australia has a particularly important stewardship role in terms of biodiversity conservation (Common and Norton 1992). As well as high levels of biodiversity, Australia's geological and evolutionary history has led to high levels of endemism, with around 85% of flowering plants, 82% of terrestrial mammals and 89% of reptiles found in Australia only occurring there.

Current fire regimes are manipulated across much of the continent and affect, in often complex ways, the biodiversity of both relatively natural systems (Gill 1996; R. J. Williams et al., this volume; Allan and Southgate, this volume) and those that are converted, fragmented, invaded or otherwise modified (Saunders et al. 1993; Fensham et al. 1994; James 1994; Gill and Williams 1996). Much of Australia has been substantially altered in the last 200 years (Purdie 1995; Commonwealth of Australia

1996a; Madden and Hayes 2001) and the original inhabitants of this country dispossessed over most of their lands. These two upheavals have broken the chain of millennia of fire management and, in many areas, prevented the new managers from documenting and understanding the formerly prevalent fire regimes. This loss has been to the detriment of much of the Australian biota, as new fire regimes, often poorly considered or short-term in their purpose, are imposed or flow from neglect. However, we argue here that rather than try to turn back the clock, the solution is to move forward into new management approaches based on sound and innovative fire science.

The focus of this chapter is on managing fire regimes for biodiversity conservation, which will be discussed within the current environmental, policy and legislative framework, concluding with some thoughts on future directions. Emphasis will be given to the key management approaches and principles including mosaic burning of landscapes, risk assessment and adaptive management. The discussion will range across a number of scales, from the use of fires in small, isolated remnants to regional approaches to conservation and management. Approaches used to address these issues in northern and southeastern Australia will be described and contrasted.

Legislative and policy framework for biodiversity conservation

Historically, much of the legislation concerning fire management in Australia has been about fire prevention and suppression, aimed at minimisation of loss to property and life. Legislators respond to deaths in bushfires by attempting to impose control on fire. In many cases this legislation narrows the range of acceptable or achievable fire regimes, and in some cases outlaws regimes that are required for the conservation of some biota (Hughes 1995). For fire management to be recognised and appropriately authorised, it is critical that the existing conservation conventions and agreements are translated to specific legislation which recognises that fire regimes play a fundamental but complex ecological role rather than being a simple destructive agent.

At a broad level, several policy and legal instruments exist with the overall goal of maintaining ecological processes and preventing the loss of indigenous biodiversity. These provide an overarching framework for biodiversity conservation, and hence fire management, in which land managers must operate. At the international level, a driving force has been the Convention on Biological Diversity (CBD) which arose out of the 1992 Rio Earth Summit. Along with Australia, around 170 countries have ratified this legally binding Convention which brings with it a number of responsibilities. For example, signatories to the Convention have to regulate or manage the relevant processes and activities where a significant adverse effect on biological diversity has been determined. In Australia, a national strategy for the conservation of biodiversity (Commonwealth of Australia 1996b), several state-based strategies and a biodiversity strategy for local government have all been developed in the last decade. To put policies into practice, these strategies need political support to ensure that they are successfully implemented. However it is widely agreed that at least the goals of the national biodiversity strategy will be difficult to achieve at the present rate of implementation and resourcing (S. Dovers and J. Williams unpublished data).

There is a substantial body of legislation of relevance to biodiversity conservation (e.g. the Commonwealth Environment Protection and Biodiversity Conservation Act 1999), endangered species protection and in particular the use of fire. Recent legal recognition of Aboriginal occupation prior to European settlement (Hughes 1995) also has implications, both for the use of fire and biodiversity conservation. At the state level, there are some quite strong legislative responsibilities related to fire planning and management for biodiversity conservation. For example, the fire management plan for Tarawi Nature Reserve, discussed later in this chapter, operates under the New South Wales Rural Fires Act 1997. This legislation defines the statutory obligations of the land manager and provides for establishment of District Bushfire Management

Committees as a means of integrating fire management across landscapes that comprise multiple managers with varying goals. The plan also operates under the New South Wales National Parks and Wildlife Act 1974, which defines the role of nature reserves and requires that fire management is not in conflict with the Plan of Management adopted for the reserve, and the New South Wales Threatened Species Conservation Act 1995, which defines requirements for impact assessment, planning and implementation of recovery for listed species. Other jurisdictions have different frameworks for fire planning and management, which incorporate biodiversity conservation to varying degrees.

It is critical that all levels of government meet their responsibilities defined by the CBD, particularly state and local governments, which have primary responsibility for land management. As important is the development of a framework for conserving biodiversity across a range of land tenures, particularly when a comprehensive, adequate and representative reserve system is currently not in place. Even if such a system existed, off-reserve management of indigenous ecosystems is increasingly recognised as a critical component of landscape management (Hale and Lamb 1997). Thus the management of indigenous biota on leasehold, freehold and Aboriginal-owned lands becomes particularly important for biodiversity conservation.

Fire regimes and biodiversity

Fires play an important role in shaping landscapes and their biota across almost all of Australia (Gill and Catling, this volume; Hodgkinson, this volume; Keith et al., this volume). Across much of northern and central Australia, topographic, edaphic and climatic variation is generally subdued and gradual, resulting in vast areas of relatively homogeneous environments, punctuated by very localised areas of contrast. Fires have long been instrumental in promoting diversity in this system, being a significant source of textural variation within the broad environments, and in mediating the occurrence of the relatively small patches of contrasted environments within the broader matrix. For example, the euca-

lypt open-forests of northern Australia are vast and relatively homogeneous, but over a decadal scale the prevailing fire regimes (and presumably those maintained by traditional Aboriginal management) impose a spatially complicated mosaic upon this relatively featureless environment (Fig. 17. 1) (Russell-Smith et al. 1997b). On an annual scale, the timing (and its correlates, intensity and extent) of fires, provides a major source of landscape variation, providing much of the spatial variability in habitat suitability which is probably necessary to maintain the constituent biota in this system (Braithwaite 1987; Woinarski 1990, 1999b). Nestled within the dominant savanna and open-forest environments of northern Australia are smaller areas of contrasting environments, most conspicuously including monsoon rainforests. Management of fire regimes by Aboriginal people has been a major determinant of the distribution, indeed existence, of these landscape features, and their maintenance within the dominant landscape matrix is now jeopardised where those regimes are disrupted (Russell-Smith and Bowman 1992; Harrington and Sanderson 1994; Russell-Smith et al. 1997a).

In southeastern Australia, the landscape is innately more diverse than northern Australia in terms of its physical characteristics and in terms of turnover in plant and animal communities across landscape gradients of varying spatial scales and pattern (Keith and Sanders 1990). Rainforests, eucalypt forests and woodlands, heathlands, grasslands, wetlands and alpine herbfields may thus be juxtaposed in complex mosaics within relatively small regions. The physical complexity and prominent meso-scale patterning of vegetation associated with the Great Dividing Range has an influence on local weather conditions and bushfire fuels, and all of these factors influence natural patterns of ignition, fire spread and behaviour, producing more variable fire regimes than occur in the tropics and central deserts. The importance of variability in fire regimes for the maintenance of biodiversity will be discussed below. Jackson (1968) first proposed a landscape model that linked vegetation dynamics with fire regimes over time-scales of centuries in a region of Tasmania containing a complex of

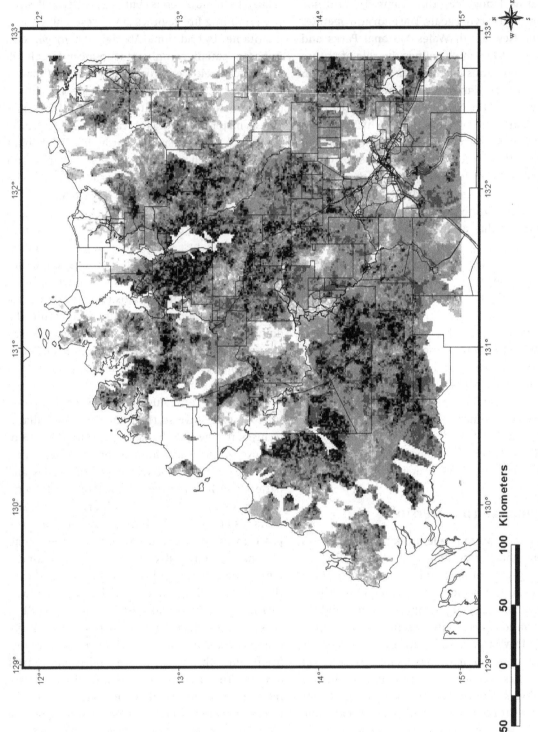

Fig. 17.1. Example of the heterogeneity imposed by fire upon an otherwise relatively featureless environment: fire history in eucalypt open-forests in part of coastal Northern Territory. Toning represents the number of fires recorded over a 6-year period (1993–98), with six increasingly dark tones representing frequencies from 0 to 6. White areas signify vegetation types other than eucalypt open-forests.

rainforest, eucalypt forest, scrub and sedgeland. Similarly, Ashton (1981) proposed multiple successional pathways in the dynamics of cool temperate wet sclerophyll forests driven by fire regimes that include infrequent high-intensity stand-replacing events. The vegetation and fire dynamics of warm temperate landscapes is potentially more complex (Keith and Bradstock 1994; Keith 1995). The role of Aboriginal burning in all of these temperate landscapes, although potentially significant, is largely speculative because of the earlier demise of their traditional practices in the south (Benson and Redpath 1997). However, new techniques are bringing insights into the nature of fires in southern Australia prior to the arrival of Europeans (Kershaw et al., this volume).

Examples of species declines associated with fires

Fire regimes have been implicated in local extinctions of several vascular plant species across Australia (Gill and Bradstock 1995; Keith 1996) and Leigh and Briggs (1992) list 19 plant species threatened with extinction at the state or federal level due to current inappropriate fire regimes. At the population level, fire-driven plant extinctions are likely to be widespread. Siddiqi et al. (1976) were amongst the first to document the local elimination of serotinous obligate seeders by high frequency fire regimes. This group of species is particularly prone to extinction under frequent fires (Cowling et al. 1990; Morrison et al. 1996; Bradstock et al. 1997b), although Gill and Bradstock (1995) list 19 examples of local extinctions that span a wide range of plant life histories, habitats and locations. Even some woody resprouters are prone to extinction under high-frequency fire regimes, although their rates of decline are generally slower than comparable obligate seeders (Bradstock and Myerscough 1988; Keith and Tozer 1997). The changes in habitat structure that come with the decline and elimination of woody plant species under frequent fire regimes have demonstrable implications for the persistence of other groups of biota (Catling 1991; York 1999). Other kinds of fire regimes have also been implicated in declines and extinctions of plants. For example, Keith and Bradstock (1994) described the decline of woody heathland understorey species under low-frequency fires due to competitive exclusion, while Kirkpatrick and Dickinson (1984) described the elimination of alpine conifers by single high-intensity fire events.

The involvement of fire regimes in the decline of many Australian birds has long been recognised. For example, Ashby (1924) considered that 'the indiscriminate burning of bush, which is the concomitant of all farming and grazing operations, is by a long way the major cause of the disappearance of many of our rarer birds'. Inappropriate fire regimes were implicated in the decline of 51 threatened bird taxa in Australia (Garnett 1992). In most of these cases, the threatened birds require long-unburnt vegetation and intervals between fires longer than those which have been imposed since European settlement (Woinarski 1999a). This is especially so for birds of heathlands, mallee and coastal 'scrub', and birds reliant on hollows for nesting or roosting. Alarmingly, it does not need a major change in fire regimes to trigger biodiversity loss: even minor changes in fire regimes may be critical for some bird species and can lead to almost imperceptibly gradual, but inexorable, decline, especially where habitats have been extensively fragmented (Brooker and Brooker 1994).

The complex interplay of fire, vegetation and land use in the decline of birds is exemplified in recent studies on the endangered granivorous golden-shouldered parrot (Psephotus chrysopterygius) on Cape York Peninsula. The development of extensive pastoralism, and a consequent abandonment or forcible replacement of traditional Aboriginal burning practices (Fensham 1997; Crowley and Garnett 2000), has led to some major changes in the relative extent of vegetation types. Most notably large areas of lowland grassland, the preferred breeding habitat of this species, have been replaced by Melaleuca thickets and woodlands (Stanton 1992; Crowley and Garnett 1998). In response, the parrot's range and abundance has diminished markedly and is likely to continue to extinction unless an appropriate fire regime (more frequent burning in the early wet season and late dry season) is reimposed

(Garnett and Crowley 1995). But the relationship of the golden-shouldered parrot to fire involves more than this longer-term vegetation change, as the annual pattern of fire extent and timing influences the spatial and temporal variability in the availability of its food resource, the seeds of particular grasses and herbs. In this highly seasonal environment, fires at different intensities and times alter the productivity and timing of this seed resource, and more extensive fires can homogenise the landscape, reducing the probability of the parrots being able to exploit landscape patchiness to maintain access to resources throughout the year. Potential changes to the current fire regime to help maintain parrot habitat are explored in the section below on managing for multiple values.

Managing for biodiversity conservation: key approaches and principles

The linkages between fire regimes and biodiversity conservation are undeniable, but how can this understanding be translated into on-ground management? A suitable management framework includes the following elements: (1) a means of reducing the complexity of biodiversity conservation to simple and measurable goals; (2) flexibility to deal with stochastic environments and uncertain knowledge; (3) a means of resolving conflicts to meet multiple management goals; and (4) a means of assessing perfomance, obtaining new knowledge and incorporating this knowledge into management practice. In this section, we outline key approaches and principles relevant to these aspects of fire management for biodiversity conservation.

Reducing complexity: local populations, functional groups and management thresholds

The full spectrum of biodiversity cannot be maintained if fire management only addresses single species in isolation. Species vary in the amount of genetic variation partitioned within and between their populations. Their roles within ecosystems will also vary from place to place depending on associated biota. Thus, in certain cases it may be desirable to address particular populations or even

individuals with special conservation measures (e.g. propagation, captive breeding). For a landscape process such as fire, maintenance of genetic, species and ecosystem diversity may be addressed in simple terms by focussing management on the conservation of populations (i.e. with the goal of avoiding local extinctions). This may seem a vexatious task, given that existing knowledge and available resources limit tangible action to only a fraction of the species within any management area. However, a strategy to conserve all extant populations, by focussing strategically on a few, should minimise losses of genetic and ecosystem diversity, even if these kinds of diversity are highly partitioned between populations. A good strategy to conserve *all* populations would be based on detailed knowledge, management action and monitoring of a *few*, particularly those species or groups of species which have traits that render them most susceptible to decline across any of the possible fire regimes. The complexity of managing biodiversity may thus be reduced by considering functional similarities among species (Keith 1996).

In plants, life history characteristics are important determinants of community dynamics in the face of recurring disturbances (Drury and Nisbett 1973; Noble and Slatyer 1980; Pickett and White 1985; Boutin and Keddy 1993). In Australia, functional classifications have been developed to describe fire-driven dynamics of plant communities (e.g. Noble and Slatyer 1980; Keith and Bradstock 1994). These schemes group species whose populations have similar methods of persistence or re-establishment after fire. Hence the members of a functional group share critical factors that mediate the response of their populations to a given fire regime.

The approach is illustrated by Noble and Slatyer's (1980) Vital Attribute scheme. For example, species in functional group CI have standing plants that are killed by fire. Re-establishment occurs from a short-lived seed bank that is exhausted by a single fire and only available when live adults are present at a site. Seedlings are only capable of establishing in the immediate post-fire period. Given high-frequency fire regimes, the critical factor mediating the persistence of CI species is their rate of seed bank accumulation after fire. The critical factor mediating their persistence under low-frequency fire regimes is the longevity of standing plants. The timing of seed bank establishment and senescence may be used to identify management thresholds for fire frequency. If fires are timed to occur only between these two life history events, then populations of CI species should have a better chance of persistence than if fires occur at shorter or longer intervals.

Fig. 17. 2a shows how a single management strategy for the conservation of all CI species could be developed by setting the fire interval thresholds using the CI species that are most sensitive to short and long fire intervals (i.e. those with slowest rate of seed bank accumulation and shortest life span, respectively). The minimum threshold assumes at least three reproductive seasons are required for adequate seedbank accumulation after maturation. Similar strategies could be devised for other functional groups and other components of the fire regime using the relevant critical parameters. Fig. 17. 2b, for example, shows how thresholds in soil heating (related to levels of ground fuel consumption) may be used to define a management strategy for conservation of a group of SI and ΣI species based on the highest temperature required to break dormancy among all species and the lowest lethal temperature for any species.

The approach is less developed for animals, although simple functional classifications have been proposed for ants (Andersen 1995), other invertebrates (Friend 1995; York 1999) and vertebrates (Keith *et al.*, this volume). Unlike plants in which life history characteristics exert critical influence over population dynamics, the principal functional attributes for animals relate to dispersal, behaviour and resource use, while life history tends to be less variable, at least among vertebrate species. For example, Fig. 17. 2c shows minimum fire frequency thresholds for conservation of a group of sedentary bird species, assuming at least five reproductive seasons are required to maintain populations over a long sequence of fires. The recovery of plants after fire is therefore often vitally important for other biota but the relationships described above may be unhinged by peculiarities of different dispersal patterns, idiosyncratic resource requirements and so

on. The simplifications and assumptions implicit in these examples will be discussed further below in relation to adaptive management approaches.

Flexibility: defining fire management targets as ranges rather than optima

The above examples show how management targets for each component of the fire regime may be defined as a range bounded by thresholds derived from functional characteristics of organisms that management is aiming to conserve. Alternatively, fire management targets could be defined as optimum settings in frequency, intensity and season which seek to maximise some ecological attribute, such as the density of a key species. Bradstock *et al.* (1995) give an example of a fire management strategy that sought to maximise the density of ground parrots (*Pezoporus wallicus*) by burning at a time since fire when peak density is reached. The strategy failed because it did not accommodate unplanned fires and because it ignored the fire ecology of plant species that provided essential food and shelter resources for the animal. The failure carries the following lessons:

- management targets defined as ranges, rather than optima, offer greater flexibility to deal with uncertainty and resolve potential management conflicts;
- targets defined narrowly by optima, while appearing conceptually simple, may be operationally difficult and expensive to achieve, given the stochastic imposition of unplanned fires (Bradstock *et al.* 1995), and the relative unpredictability of weather conditions during fire events.

Optimisation strategies lack the flexibility to deal with different species that may require different optima. By placing emphasis on trends in density that follow a single fire, they may ignore critical processes that operate over multiple fire intervals, as well as interactions with other species or physical factors that influence essential resources for the target species. Finally, the goal of density maximisation is often confused with that of conservation. In *Banksia cuneata*, the optimal fire frequency for maximising population size is likely to entail much

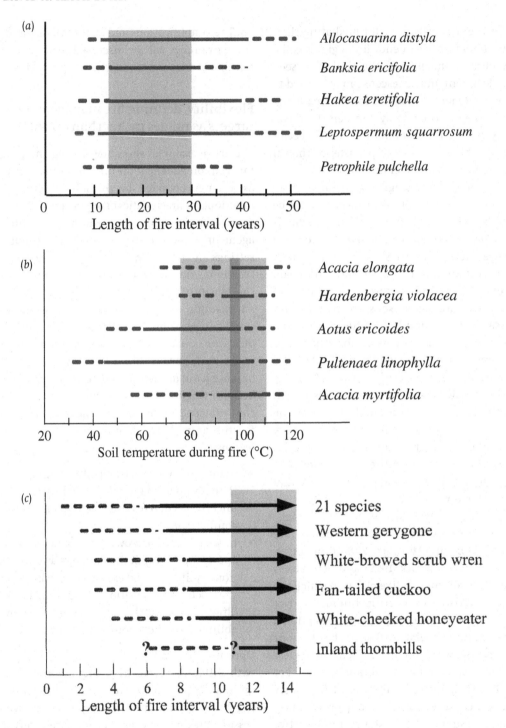

Fig. 17.2. (a) Fire interval thresholds for five bradysporous obligate seeding dominant shrubs of coastal heathland near Sydney representing Vital Attribute functional group CI. Unbroken horizontal lines represent life history stage between late maturation and early senescence. Broken lines represent variability in rates of maturation and senescence related to the influence of site quality, rainfall and population density on growth. The time of maturation is taken as the third season of fruit production after fire to allow for seed bank accumulation. Shaded vertical bar represents the minimal range of fire frequency that should ensure population persistence of all five species. Data compiled from Benson (1985), Bradstock and O'Connell (1988) and Keith (1991 and unpublished data).

shorter intervals between fires than those asso-ciated with minimal extinction risk for the species, primarily because of the unreliability of rainfall events that are essential for successful recruitment (Burgman and Lamont 1992).

Resolving conflicts between multiple management goals: temporal and spatial options

Conflicts often arise between fire management goals. Species or groups of species within a manage-ment area may have requirements for quite differ-ent fire regimes. Even when these requirements are expressed as ranges there may be a lack of overlap. The potential conflicts between the general goals of conservation, protection of life and property and sustainable production are well known (e.g. Morrison *et al.* 1996) and there may also be conflicts between conservation and some cultural roles of burning (Langton 1998). Traditionally, managers deal with conflicts by setting priorities. The goals perceived as least important may thus rarely be achieved. However, in fire management there are temporal and spatial options for achieving different goals that may only be in conflict at some times and some places.

Temporal variability in fire regimes

Consider the example of species-rich heathland near Sydney where conservation of different func-tional groups appears to be in conflict (Keith and Bradstock 1994). The dominant shrubs of this com-munity are obligate seeders with bradysporous seed banks (CI species of Fig. 17.2a). They represent a rela-tively small fraction of total floristic richness, but as larger shrubs they contribute to structural diversity and their nectar produced during winter flowering provides an important source of food for honeyeat-ers and small mammals at a low-resource time of year. Establishment of a seed bank takes 6–13 years after fire, depending on site quality and rainfall, while senescence occurs between ages 30–50 years (Bradstock and O'Connell 1988). Thus, we may plan to conserve populations of these important species by maintaining fire frequency within the range 13–30 years (Fig. 17. 2a). Understorey species make up the majority of floristic diversity within the com-munity. Most species have persistent soil seed banks or vegetative means of post-fire recovery, or both. The understorey includes woody resprouters that may take up to 15 years to develop fire-resistant organs (Bradstock and Myerscough 1988) and have a life span of at least 50 years. A feasible fire frequency range for conservation of this group would thus be 15–50 years, although fires at shorter intervals would not have an immediate impact due to the sur-vival of some adult plants after each fire.

There is considerable overlap in the fire fre-quency range required to conserve both groups of heathland plants (15–30 years). However, a more complex problem presents itself when interactions between the two groups are considered. The domi-nant species may form high-density thickets when burnt at 15–30 year intervals because seed banks

Fig. 17.2. (*cont.*)

(b) Germination soil temperature thresholds for five selected species of legume representing variation among 35 southeast Australian species in Vital Attribute functional groups SI and ΣI. Data adapted from Auld and O'Connell (1991) for seeds exposed to heating for 10 min duration. Unbroken horizontal lines represent the range of temperatures stimulating maximum levels of germination, broken lines represent temperatures stimulating suboptimal germination, while temperatures outside range spanned by horizontal lines stimulate negligible germination. Dark vertical band represents the range of temperatures (95–100 °C) that will stimulate maximum germination in all 35 species. Pale band represents range of temperatures (80–110 °C) that will stimulate more than negligible germination in all species. Fires that consume 0.6–2.0 kg m^{-2} fine ground fuel are required to generate this latter range of temperatures at soil depths of 1–3 cm where most seeds are thought to be stored (Bradstock and Auld 1995).

(c) Fire interval thresholds for 26 locally nesting bird species in heath near Perth. Broken lines represent the first five breeding seasons after a wildfire in 1985. Broken line represents period when we arbitrarily assume bird populations to be established in the post-fire community. Question marks indicate uncertainty in the recommencement of nesting in inland thornbills, which were not observed nesting in the 5-year post fire observation period, despite being common before the fire. Shaded vertical bar represents part of the range of fire frequency that may be required to ensure population persistence of all 26 species, assuming a sequence of five reproductive seasons is adequate for persistence. Data adapted from Brooker and Rowley (1991).

accumulate rapidly once plants are mature (Bradstock and O'Connell 1988; Morris and Myerscough 1988). There is evidence that competition from the dominants reduces the survival and fecundity of some understorey shrubs (Keith and Bradstock 1994). Floristic diversity may be expected to decline if overstorey thickets are maintained over successive long fire intervals due to the loss of understorey species. Therefore maintenance of full species diversity would seem unlikely under any regime of constant fire frequency. Full diversity could only be maintained by allowing fire frequency to vary in a way that includes occasional short intervals very close to the lower threshold for the dominant species (6 years). This would substantially reduce the density of overstorey species over the subsequent fire interval, allowing a period of reduced competition and enhanced survival and reproduction of understorey shrubs. Thus, by promoting variability in fire frequency between certain thresholds, apparent conflicts may be resolved and full diversity may be maintained.

The heathland example probably demonstrates a general axiom that maintenance of diversity requires variability in fire regimes (Keith and Bradstock 1994; van Wilgen *et al.* 1994; Morrison *et al.* 1995; Bradstock *et al.* 1995), which is consistent with disequilibrium theories of diversity (Connell 1978). Variability in disturbance regimes allows ecosystems to fluctuate between alternative states, without reaching 'equilibrium' whereby any one group of species dominates over long periods while others decline to eventual elimination. The rates of such declines are critical because they determine the timing of fire management actions required to reverse the trend before extinctions occur. It follows that only certain kinds of variability in fire regimes may promote maintenance of biodiversity. Appropriate kinds of variability are constrained by thresholds relevant to the life histories, behaviour and resource use of the organisms present and are further defined by the state and trends of the community at particular points in time. Management must therefore be adaptive if it is to implement variable fire regimes to achieve conservation of full diversity.

Fire-landscape mosaics

Spatial variability in fire regimes is considered important for the maintenance of biodiversity, particularly animal diversity (Saxon 1984). The notion that fire-related habitat mosaics provide refugia and a variety a successional stages necessary for the persistence of animal populations derives from a translocation into space of habitat relationships observed in relay succession after single fires (e.g. Fox and McKay 1981). However, many practical applications of this principle – the 'mosaic paradigm' (Bradstock *et al.* 1995) – result in rigid rotational block burning programs that are ill-conceived for their intended purpose (Williams *et al.* 1994; McCarthy and Burgman 1995). In these cases decisions about the scale, boundaries, order and timing of planned fires are made with little reference to ecological criteria relevant to the target organisms and reflect more the logistics of controlling fires. Although there is a scarcity of data on which to base the development of such criteria, some generalisations are emerging from available research. These are listed in Box 17. 1, with supporting evidence provided below.

Box 17.1. *Generalisations about fire-generated patches emerging from the literature (Supporting evidence follows in the text)*

1 Many animals are known to draw on multiple resources from spatial mosaics of vegetation that are dependent on recurring fire.

2 The effect of fire-generated patchiness upon animal populations is difficult to measure.

3 Animals may have varying levels of plasticity in resource use that affect their responses to fires and the scale and strength of their dependence on habitat patchiness.

4 Forms of spatial variability that are independent of fire may be important in meeting the requirements for food and shelter of many animals.

5 Patchiness generated by fire and other disturbances at the wrong scale can be detrimental to biodiversity conservation

(Williams *et al.* 1994; McCarthy and Burgman 1995.)

(1) A commonly used example to illustrate the role of fire mosaics in fauna conservation is that of the mala (*Lagorchestes hirsutus*), a small wallaby of central Australian deserts (Lundie-Jenkins 1993). These animals draw food and shelter from patches of different post-fire age. With the withdrawal of traditional Aboriginal burning regimes across the Tanami Desert over the last 50 years, the vegetation mosaic that mala relied on became coarser and more simple, reducing the range of resources locally available (Bolton and Latz 1978), and apparently precipitating declines in local populations. As the next point illustrates however, other factors have been implicated in the decline of these mammals, demonstrating the often complex nature of biodiversity loss. The golden-shouldered parrot (*Psephotus chrysopterygius*) of tropical Queensland savanna and mycophagous mammals such as potoroos and bettongs of temperate Australian forests (Lamont 1995) also draw on multiple resources from spatial mosaics of vegetation that are dependent on recurring fire.

(2) The effects upon animals of fire-generated patchiness have rarely been tested, perhaps partly because this may require large areas (for vertebrates at least), and because fire-related variability is easily confounded with other environmental variability at these scales. For example, the decline and extinction of many arid mammals cannot be attributed unambiguously to the sequential breakdown of Aboriginal burning regimes across central Australia (Burbidge *et al.* 1988), as distinct from other causes such as competition with introduced stock and herbivores and predation by introduced carnivores (Morton 1994). A study of three critical weight-range mammals on Barrow Island (Short and Turner 1994) illustrates the difficulty facing one of the few experimental studies on the subject. No differences were detected between two mosaic scales on the population density, condition and reproductive status of the animals, suggesting a lack of sensitivity to the observed scales of spatial variation. However, interpretation in terms of fire mosaics depends on assumptions that measurements of patchiness were relevant to resources used by the animals and that the patchiness observed in mapped vegetation and small-scale mining activities was a reasonable surrogate for variation that may be generated by fire.

(3) Brooker and Rowley (1991) found that 81% of bird species present in a west Australian heathland continued to breed after a 'severe' fire. Many of these altered their patterns of nest placement to suit the new pattern of habitat availability, suggesting some degree of plasticity in habitat requirements. However, the breeding success of some of these species was reduced because of the difficulty of attaching nests to the available substrates, while species that failed to continue breeding in the area did not resume nesting for 2 to more than 5 years after fire. Whelan *et al.* (1996) observed a fire-related winter dietary shift in the small nectarivorous–insectivorous dasyurid mammal *Antechinus stuartii* in an east Australian heathland. In an unburnt area scats of this species contained only *Banksia* pollen, whereas in a recently burnt area they contained pollen of both *Banksia* and *Xanthorrhoea*. *Banksia* flowers were rare and confined to small unburnt patches, whereas *Xanthorrhoea* flowers were abundant, but only present in the burnt area, their flowering being stimulated exclusively by fire (Keith *et al.*, this volume). The temporary availability of *Xanthorrhoea* pollen in burnt areas and the ability of *A. stuartii* to shift diet and utilise this resource when *Banksia* pollen was in short supply appears to be one of the critical factors in the persistence of the animal's populations through the first post-fire winter (Whelan *et al.* 1996). In contrast, post-fire records of another *Banksia*-feeding nectarivore, *Cercartetus nanus*, were rare and confined to the small unburnt patches.

(4) In the above example, Whelan *et al.* (1996) found that the rocky outcrops offered the most important shelter sites for *A. stuartii* and that the animals used these as a base to forage in both burnt and unburnt vegetation. In fragmented landscapes it is important to ensure that critical

habitat features are maintained, especially where options for fire management are limited.

(5) Rowley and Brooker (1987) observed that a sequence of patchy low-intensity fires was associated with the decline of wren populations in western Australian heath. There is apparently a relationship between the size of burnt areas and the impacts of herbivory and granivory. Dickinson and Kirkpatrick (1986) suggested that herbivorous macropods had a greater impact on eucalypt regeneration in small logging coups relative to larger coups. Whelan and Main (1979) found that the impact of invertebrate herbivores was greater in small burnt areas and near burnt edges than within larger areas. Post-fire mammal granivory had a major detrimental effect on seed bank accumulation in the endangered shrub *Grevillea caleyi* (Auld 1994). If foraging is concentrated in small burnt areas, as expected, then patch burning strategies substantially increase the risk of population extinctions (H. Regan, T. D. Auld, D. A. Keith and M. A. Burgman unpublished data). The effect of fire size on predation is likely to be greatest in habitats where ubiquitous mobile predators already exert a significant influence on ecosystem dynamics, as in temperate grassy woodlands and semi-arid regions with an abundance of feral or domestic herbivores.

While these generalisations reinforce the importance of spatial variation in fire regimes for biodiversity, more importantly they suggest a complex relationship in need of more research to support the development of management strategies for conservation. Even in northern Australia, few conservation reserves are large enough to maintain a fire mosaic over the long term which is sufficiently complex and dynamic to maintain all their constituent conservation values. Management of fire regimes for conservation must integrate fire management across tenure and broad regions. This has proven to be a difficult task, with the main impediments being diffuse or competing responsibilities, lack of knowledge, conflicting goals, lack of comprehension about the ecological role of fire regimes and fear of change.

Managing for multiple values

The case of the golden-shouldered parrot illustrates that it may be possible to define and establish fire regimes that promote conservation, production and other values. The current vegetation change associated with the decline of the parrot is not in the interests of the pastoralists whose management precipitated it. Hence there is scope for devising a more appropriate fire regime linked to changed grazing management, which can serve longer-term pastoral and conservation interests. Garnett and Crowley (1997) describe a trial on one pastoral property of such a regime to control invasion by *Melaleuca*, which involves spelling paddocks from grazing to allow fuel loads to build up sufficiently to carry intense fires in the early wet season ('storm burns'). High-intensity fire is also being used in nearby conservation reserves to return the balance of grasslands and *Melaleuca* woodlands in order to favour the parrots.

Spatial variability in fire regimes can help resolve other apparent conflicts in fire management, particularly between conservation and protection. Assets that require protection from fire, such as settlements or farmland, tourist sites etc., are not distributed randomly throughout landscapes. Furthermore, fires that threaten assets often follow well-known pathways defined by patterns in topography, fuels and weather conditions. In these cases the strategic location of protection measures such as fuel reduction fires, access trails, fire breaks etc. will return more effective and lower-cost protection than measures directed less discriminantly over the entire landscape. Conroy (1996) cites several examples where strategically placed prescribed fires were effective in containing the spread of wildfires under extreme weather conditions, although exceptions can occur (Bradstock and Scott 1995). A strategic approach to protection also offers a greater range of options for addressing other fire management goals, particularly in areas not directly relevant to the protection of assets. The current fire management plan for Crowdy Bay National Park (New South Wales National Parks and Wildlife Service 1997) shows an example of how strategic zonation of a management area, based on the location of identified assets, safe access routes and historical patterns in fire ignitions, movement, behaviour and weather, can be applied to resolve potential conflicts between conservation and protection (Fig. 17. 3).

Fire regimes and biodiversity conservation in fragmented landscapes

Extensive urban and agricultural land uses have imposed a further layer of pattern on temperate fireprone landscapes (Saunders *et al.* 1993; Commonwealth of Australia 1996a). Consequences can include interruption of fire spread patterns, increased frequency and more localised patterns of planned and unplanned ignitions, more active suppression of wildfires, and changed fuels and fire responses due to invasion of alien species. Overall, this translates to spatial and temporal simplification of fire landscape variability, which is critical for the maintenance of biodiversity, as previously discussed. As well, spatial options for fire management will be more limited if management areas are small or in landscapes fragmented by land uses with fire management goals in conflict with those of conservation. Budgetary constraints may further reduce management options. The compromises sometimes taken between fire protection and maintenance of conservation values can be particularly detrimental to remnant areas such as roadsides and may end up resulting in no management objectives being met (Milberg and Lamont 1995; Simmons and Adams 1997). Box 17. 2 contains two scenarios describing fire regimes in forest remnants, respectively adjacent to urban and agricultural areas, that were tested by drawing on the literature and expert knowledge (Gill and Williams 1996).

Box 17. 2. *Scenarios developed for fire regimes in remnants adjacent to urban and agricultural areas, and results from research used to test them.*
Urban interface scenario: There is a low frequency of unplanned fire in forest remnants. To prevent losses of life and property in adjacent urban areas, regular frequent prescribed burning is practised. Regular frequent prescribed burning reduces biodiversity.

Test of scenario: Support for the first two parts of this scenario was strong although the frequency of fires, prescribed or unplanned, may be a function of distance from the urban edge, the size of management unit, and the nature of the fuels. Urban predators may be expected to reduce vertebrate biodiversity, especially after fires.
Agricultural interface scenario: Clearing for agriculture leaves only small forest remnants which become fire free. Fire-free fragments eventually decline in plant species biodiversity.
Test of scenario: Forest fragments in rural areas vary widely in size and occur as roadside remnants, farm woodlots, Travelling Stock Reserves, State Forests and designated conservation reserves. The circumstances of burning vary widely. Grazing by domestic stock may negatively affect biodiversity, especially when combined with fire. Absence of fire may also reduce biodiversity.
Gill and Williams (1996).

This and other studies (for example, Hobbs, this volume; Lunt and Morgan, this volume) demonstrate the complex changes in fire regimes in fragmented native woodland and grassland systems, which are mainly found in southern and eastern Australia. The effective exclusion of fire from some remnants in rural landscapes, and the need for some form of disturbance in these systems, has lead some to recommend controlled use of grazing (Williams 2000). However, the impact of these practices on biodiversity conservation is practically unknown (Morgan 1998) and burning may end up being the most favoured management tool, at least in large remnants.

Management guidelines for fragmented systems, such as those developed for privately owned remnants on the New England tablelands in northeastern NSW (Clarke *et al.* 1998), recognise that fire is only one of several management tools available to tackle a suite of conservation issues, which may vary in combination from patch to patch. These issues may include weed invasion, lack of native regeneration, over-grazing, lack of downed woody

Table 17.1. *Applications of risk assessment to examine the merits and limitations of alternative fire management scenarios*

Study	Species	Fire management problem examined
Burgman and Lamont (1992)	*Banksia cuneata* (Matchstick Banksia)	Viability of populations under alternative fire management strategies directed at density maximisation and extinction risk minimisation
Brooker and Brooker (1994)	*Malurus splendens* (Splendid Fairy Wren)	Combined effects of fire and fragmentation on the viability of populations
Lindenmayer and Possingham (1995)	*Gymnobelideus leadbeateri* (Leadbeater's Possum)	Risks of population extinction under alternative management scenarios in timber production forest
Bradstock *et al.* (1997a)	*Banksia ericifolia* (Heath Banksia)	Viability of populations subject to fires of varying frequency and extent
H. Regan, T. D. Auld, D. A. Keith and M. A. Burgman (unpublished data)	*Grevillea caleyi* (Caley's Grevillea)	Effects of patch burning, fire frequency and seed predator control on population viability

material for fauna etc. The guidelines promote the development of management goals that are tailored to local management issues, as well as an adaptive approach to planning that incorporates monitoring and review (see below). Where a number of fragments of vegetation occur within a region, it is important to address complementarity in their management. This might entail differing but co-ordinated actions in individual remnants to address goals applicable to larger spatial scales, such as the provision of habitat diversity for wide-ranging predatory birds.

Accommodating uncertainty and new knowledge: risk assessment, adaptive management and monitoring

Several new approaches to conservation management have developed and been applied in the two decades since publication of *Fire and the Australian Biota* (Gill *et al.* 1981). One of the most powerful decision tools to emerge is that of risk assessment, whereby the probability of an adverse event (e.g. extinction) is estimated and compared between alternative management scenarios (Burgman *et al.* 1993). The process of estimating risks involves building models of biological populations or communities that may be conceptual, qualitative or quantitative (Burgman and Lindenmayer 1998), but nevertheless state explicit assumptions about the structure and dynamics of the system under consideration. The models incorporate stochasticity to account for uncertainty associated with unpredictable events and incomplete knowledge. There have been several recent studies applying these techniques to Australian fire management and conservation problems (Table 17. 1).

Adaptive management is an approach to implementing conservation of biodiversity that recognises our rudimentary understanding of its components, processes that underpin its function and its response to perturbations (Holling 1978). It also recognises the changing economic, social and political climate in which decisions are made (Salwasser 1993). Burgman and Lindenmayer (1998) list the principles of adaptive management as follows:

- consider alternative strategies to meet an explicit management goal;
- favour actions that are informative, reversible and robust to uncertainty;
- experiment on alternative options;
- validate assumptions underlying management decisions;
- monitor the results of management; and
- modify decisions and practices in the light of new information.

Although some of these elements have made their way into formal planning processes for fire management, the full extent of the adaptive approach is yet to be applied (Bradstock *et al.* 1995). The strategies applied in Fig. 17. 2 are consistent with an adaptive approach so long as research is focussed on their assumptions, monitoring is focussed to test biotic outcomes and management is adjusted accordingly. For example, the management strategy in Fig. 17. 2c assumes that five post-fire reproductive seasons yield sufficient recruitment to compensate mortality in all bird populations over an appropriate time frame. The fire frequency thresholds were based on presence/absence observations of nest building after a single fire at one location (Brooker and Rowley 1991), but without reference to nesting success and juvenile survival. If research on these parameters showed that recruitment rates were underestimated, the minimum fire frequency thresholds for bird conservation in Fig. 17. 2c would need to be lengthened. Similarly, research on the effects of fire patchiness may suggest that different frequency thresholds should apply to fires of different patchiness.

Monitoring is a critical element of adaptive management that has received some attention in fire policy, science and management. The ongoing collection of information on the biota and fire regimes and its evaluation in relation to management goals is increasingly a focus in strategic planning for fire management (Bradstock *et al.* 1995). The recent development of geographic information systems (GIS) has already had a major impact on the recording, assessment and communication of fire history data, as well as fire planning (e.g. Conroy 1996; Garvey 1996). Remote sensing techniques such as satellite imagery will also allow long-term monitoring of fire regimes, particularly in northern (O'Neill *et al.* 1993; Russell-Smith *et al.* 1997b) and central Australia where the fire scars are easier to interpret.

Monitoring of biodiversity is more complex and consequently less developed. Despite wide agreement on the importance of monitoring in biodiversity conservation (Commonwealth of Australia 1996b) and the development of viable and relevant techniques (e.g. Gill and Nicholls 1989), most monitoring projects fail on the ground either because of insufficient resourcing, flawed design and lack of problem focus, or a lack of commitment to define and implement recommendations for management from the results (Keith and Tozer 1997).

Fire as a restoration tool

As the extent of the modification of natural systems in Australia becomes more widely appreciated, attempts to restore or revegetate landscapes are becoming increasingly common. In particular, the impact of broad-scale land degradation on the productive potential of Australian landscapes (Graetz *et al.* 1995; Commonwealth of Australia 1996a; Murray Darling Basin Commission 1999; Madden and Hayes 2001), and the focus on revegetation as one means of meeting Australia's commitments to greenhouse targets, has seen considerable resources go into restoration/revegetation projects. This can range from the localised restoration of endangered plant and animal species or communities, to large-scale restoration for production and/or conservation objectives. Hobbs and Norton (1996) emphasised the need for guidelines and methodologies for restoration at a number of scales, but particularly at the landscape scale. A number of key processes were identified in restoration ecology by these authors that were considered essential for the successful integration of restoration into land management. These included the need to identify the processes leading to degradation or decline, and the development of methods to reverse or ameliorate these processes. The development of practical techniques for implementing restoration goals at a scale commensurate with the problem was also seen as essential (Hobbs and Norton 1996).

As fire regimes have played an important role in the evolution of most Australian ecosystems, it would be expected that they could also play a role when the restoration of biodiversity is the principal goal. The development of practical techniques can be complicated however, especially given the range of potentially degrading factors and the interactions between them. Even when these factors are identified and their impacts mediated, in many instances the legacy of these degrading processes is the loss or reduction of the number of plant and animal species. In order to reintroduce these biota into the landscape, an understanding of their biology and habitat requirements is essential. This is well illustrated by the inability, until quite recently, to germinate a number of plant species for restoration projects. While it was known that many species require heat associated with fires to germinate (Auld and O'Connell 1991), this by no means applied universally. Over the last decade however, the importance of smoke as a trigger for germination has been increasingly appreciated. Smoke is only one element of the range of changes induced by fire, but in situations where fire cannot be directly applied, then the ability to germinate plants using other techniques is an essential first step. In Australia, the pioneering work on smoke as a restoration tool has been by Kingsley Dixon and his colleagues at the Kings Park and Botanic Gardens in Perth (see Dixon *et al.* 1995; Roche *et al.* 1997). This group has developed a practical technique using smoke water for restoration, which is now widely used at a number of scales.

These researchers, however, are the first to point out that being able to germinate seeds is only one, albeit critical, factor that needs to be considered when restoring sites. Environmental weeds are a major threat to nature conservation in Australia (Commonwealth of Australia 1999; Williams and West 2001) and world-wide and, where present, need to be managed when attempting to restore systems. Fire has been used as a management tool in several systems to try to control weeds (Grice 1997; Downey 1999), often in combination with other management techniques such as the use of herbicides or mechanical treatment (Miller and Lonsdale 1992; Vranjic and Groves 1999). Sometimes however the use of fire can exacerbate the weed problem, and follow-up work is essential (Downey and Smith 2001). It is also important to ensure that the fire regime that is effective against the target weed is not detrimental to the community that is being restored.

The restoration of mine sites, such as when bauxite is mined in the *Eucalyptus marginata* forests of southwestern Western Australia, is another example of the use of fire as a restoration tool at the site level (Grant *et al.* 1997, 1998). Most managers work at this level, but fires also lend themselves to broad-scale landscape management, as exemplified by the use of aerial incendiaries to reduce fuels in forested areas (Williams and Gill 1995). Use of fire regimes as a restoration tool at the landscape scale however is in its infancy in Australia and guiding principles still need to be developed.

Aboriginal fire regimes

The impact of landscape burning by Aboriginals has been the subject of considerable recent investigation and comment, heated by ongoing debate and polemic about the historic and continuing role of Aboriginal proprietorship of this continent (Flannery 1990, 1994; Latz 1995; Kohen 1996; Benson and Redpath 1997; Bowman 1998; Langton 1998; Andersen 1999). The main relevant aspects of this argument concern the original impact of Aboriginal colonisation (through hunting and changed fire regimes), how (and why) Aboriginal people burnt their lands in the period immediately prior to European settlement, whether such a regime can and should again be implemented for conservation purposes, and, more recently, the recognition that there may be conflict between conservation goals and the 'traditional' use of fire on Aboriginal lands. Following Aboriginal entry to Australia, there was a substantial environmental rearrangement caused by (and/or leading to) radical change from 'natural' fire regimes, climatic fluctuations and a major extinction event. The sequence and relative significance of these factors is debatable, and attempts at such resolution probably futile (Choquenot and Bowman 1998). Regardless of that debate, in the tens of thousands of years since that entry, Aboriginal land management has left a major signature on Australian environments. Whatever the nature of Aboriginal fire regimes, a large complement of Australian biota persisted through them for tens of thousands of years. The last two centuries have brought a second transformation of fire regimes in Australia which, in tandem with habitat fragmentation and introduction of alien species, have led to major environmental changes (Price and Bowman 1994).

Some managers have seen the reinstitution of Aboriginal fire regimes either as the means to achieving conservation goals or as the goal itself. But such mimicry may be far too simplistic for the complicated task of conserving biodiversity because: (1) the fire regimes prevailing under the former Aboriginal management may be difficult to determine precisely, especially in the vast areas of the country where that regime has long lapsed; (2) traditional fire regimes may be impossible to re-establish or may lead to very different consequences, given the environmental changes since European colonisation (notably habitat fragmentation, and the spread of weeds, livestock and feral animals); and (3) Aboriginal fire regimes were instituted largely to maintain access to food resources, whereas conservation management may involve very different goals, such as the protection of a rare plant species which may have been of no value to Aboriginal people and indeed may have been disadvantaged by Aboriginal fire regimes. The reintroduction of Aboriginal fire regimes may also be too expensive to maintain across broad regions given the fine spatial scales over which they were apparently applied.

Aboriginal knowledge of landscapes and the role of fire in their management, where available, provides an invaluable guide for the ongoing management of biodiversity. However, the potential conflicts between conservation and Aboriginal cultural purposes is evident in fire management even in the best-resourced conservation reserve in northern Australia, Kakadu National Park. Here, recent dispute has arisen between Aboriginal owners of the Park and its conservation management agency, mostly centred around the damage to 'fire-sensitive' vegetation from fires lit by Aboriginal owners (KRSIS 1997) and concerns that traditional Aboriginal owners may be disempowered (Langton 1998). While this represents a complex social as well as environmental issue, which will no doubt be the subject of continuing debate (also see Bradstock and Cohn, this volume), some of the fire management approaches described above may have roles in the resolution of biodiversity and cultural goals where these conflict.

Putting science into practice: the Tarawi Nature Reserve fire management plan

A fire management plan recently developed for Tarawi Nature Reserve (Willson 1999), a 33,600 ha area of semi-arid mallee in south-west NSW, illustrates some of the legislative and policy instruments and scientific approaches to management discussed in this chapter. Key elements of the plan are outlined in Box 17.3.

Box 17.3. *Key elements of the Tarawi Nature Reserve fire management plan.*

General conservation goal

To manage fire to minimise the risk of any native species becoming extinct from the reserve.

Specific objectives

- To ensure consecutive fires are at least 20 years apart in any one area;
- To ensure a range of post-fire ages are present in the reserve (at least 50% of each mallee community should be >40 years old);
- To promote patchiness in wildfires, although a scale of patchiness is not specified due to lack of ecological knowledge; and
- To determine the biodiversity values of long-unburnt (>70 years) mallee.

Key strategy for meeting objectives

Strategic prescribed burning to contain the spread of wildfires that may violate the thresholds for fire frequency, post-fire age diversity or requirements for patchiness.

Management actions proposed to implement the strategy

- to develop fire management zones (asset protection, strategic fire management and heritage conservation; see Fig. 17.2 for an example);
- to define blocks (management units) for wildfire containment;
- to maintain designated fire trails;
- to carry out prescribed burning in strategic fire management zones targeting areas of high fire potential under 'normal' and 'speargrass' fuel years;
- to monitor fire regimes, maintain and refine fire history spatial database for the reserve and extend to adjacent lands;
- to provide and promote use of resource maps in fire planning and operations on neighbouring lands;
- to use the prescribed burning program to implement research and monitoring on key species, long-unburnt mallee and the impacts of feral herbivores.

The Tarawi fire management plan illustrates how many of the principles developed recently in conservation science may be put into practice. Local populations are used as operational surrogates for all levels of biodiversity, as discussed above. Management targets for frequency and spatial variability are expressed as ranges bounded by thresholds, rather than optima in fire regime parameters, thereby offering flexibility to deal with unplanned fires and resolve conflicting management goals. The thresholds are derived from an understanding of the life history of sensitive species belonging to extinction-prone functional groups. For example, the 20-year fire frequency threshold derives from the timing of seed bank accumulation in *Callitris verrucosa*, the slowest-maturing obligate-seedling plant species in the reserve. As well as temporal variability, the plan also places management emphasis on spatial variability through the promotion of patchiness in wildfires and the development of fire management zones with different purposes. The plan also addresses landscape-level fire management

through active participation in the District Bushfire Management Committee and by planning the extension of resource inventories off-reserve. The plan also has objectives and actions for management of cultural heritage sites and operational guidelines for wildfire suppression or backburning, use of chemicals, use of heavy equipment, use of water, etc. Resource inventory systems supporting the plan include a GIS with data layers for plant communities and fuel types, populations of significant species, cultural heritage and visitor assets, extent, date and ignition sources of previous fires, as well as infrastructure for fire operations such as access trails and water supply points.

An adaptive approach to management is demonstrated in the plan because reversible actions are to be implemented while research and monitoring are carried out to improve understanding of poorly known components of the system. Most of the reserve (>85%) has not been burnt since 1917, considerably exceeding management thresholds for fire frequency and age-class distribution. The strategic, rather than broad-area, use of prescribed fire and containment of wildfire within large blocks is a precautionary reversible strategy that will be applied so long as the effects of long fire intervals remain unknown. If long fire intervals are shown to be important for biodiversity, then a large area of the reserve has a chance of being maintained in that state. On the other hand, if long fire intervals are shown to be detrimental to biodiversity, fires could be lit in the long-unburnt areas, returning them immediately to a younger state. This flexibility to reverse fire management actions would not be available if broad-area prescribed fires were applied before the effects of long fire intervals were known; once burnt, vegetation cannot be returned to a long-unburnt state through short-term actions.

The Tarawi plan is indicative of fire management trends elsewhere in Australia. In Victoria for example, workshops held in 1998 on the management of fire for the conservation of biodiversity (Friend *et al.* 1999) will lead to the development of a draft set of *Guidelines for Ecological Burning*. These will set out the basic principles, procedures and responsibilities for planning and undertaking ecological burning, and could help further refine the integration of ecological and protection burning practices in Fire Operations Plans at the local level.

Conclusions

Biodiversity conservation is now the primary objective of a number of policies, strategies and Conventions from the local to international level. These are increasingly supported by legislation that relates to fire planning and management for biodiversity conservation, although historically the emphasis of fire legislation, policy and management has fallen exclusively on the protection of human life and property. This chapter focusses on the use of fire regimes for managing biodiversity, but recognises that fire management across landscapes has to integrate multiple managers with varying goals. Temporal and spatial variability in fire regimes has a critical role in reconciling potential conflicts in management goals, as well as in the conservation of biodiversity itself. For fire management to be sustainable in the longer term, social and economic issues must also be examined, especially when considering biodiversity conservation on private land, and management should be co-ordinated across tenures at a landscape (bioregional) scale.

Considerable advances in knowledge have occurred in fire ecology and management since the publication of *Fire and the Australian Biota* (Gill *et al.* 1981). The following recommendations emerge from principles and approaches outlined in this chapter:

- Adopt an adaptive approach to management – one that is precautionary and allows for the unpredictable (especially unplanned fires) in management planning.
- Use an ecosystem-based, landscape-scale approach to planning and management, i.e. emphasise the spatial context of any management unit.
- Recognise risk assessment as a powerful tool which allows exploration of alternative management options under explicit assumptions.
- Set explicit conservation management goals

that derive directly from critical ecological processes and lend themselves to evaluation through monitoring. Define fire management targets as ranges rather than optima.
- Aim to conserve the full range of biodiversity, from genes to ecosystems by applying strategies that reduce the complexity of biodiversity conservation and focus research and management on critical elements (e.g. local populations of the most sensitive species in functionally important groups, populations of threatened species).
- Incorporate Aboriginal knowledge of landscapes and the role of fire regimes into management objectives, where available and appropriate.
- Promote variability in fire regimes across space and time in ways that maintain biodiversity and resolve potentially conflicting management objectives.
- Monitor the impacts of fire regimes on biodiversity conservation to evaluate and potentially modify management practices.

Acknowledgements

Our thanks go to Grant Allan who provided the map for Fig. 17. 1 and to Ross Bradstock, Gordon Friend and Malcolm Gill for their comments on the chapter.

References

Allan, G., and Southgate. R. (2001). Fire regimes in the spinifex landscapes of Australia. In *Flammable Australia: The Fire Regimes and Biodiversity of a Continent* (eds. R. A. Bradstock, J. E. Williams and A. M. Gill) pp. 145–176. (Cambridge University Press: Cambridge.)

Andersen, A. N. (1995). A classification of Australian ant communities, based on functional groups which parallel plant life-forms in relation to stress and disturbance. *Journal of Biogeography* **22**, 15–29.

Andersen, A. N. (1999). Fire management in northern Australia: beyond command-and-control. *Australian Biologist* **12**, 63–70.

Ashby, E. (1924). Notes on extinct or rare Australian birds, with suggestions as to some of the causes of their disappearance. Part II. *Emu* **23**, 294–298.

Ashton, D. H. (1981). Fire in tall open-forests (wet sclerophyll forests). In *Fire and the Australian Biota* (eds. A. M. Gill, R. H. Groves and I. R. Noble) pp. 339–366. (Australian Academy of Science: Canberra.)

Auld, T. D. (1994). The role of soil seedbanks in maintaining plant communities in fire-prone habitats. In *Proceedings of the Second International Conference on Forest Fire Research*, Coimbra, Portugal, pp. 1069–1078.

Auld, T. D., and O'Connell, M. A. (1991). Predicting patterns of post-fire germination in 35 eastern Australian Fabaceae. *Australian Journal of Ecology* **16**, 53–70.

Benson, D. H. (1985). Maturation periods for fire-sensitive shrub species in Hawkesbury sandstone vegetation. *Cunninghamia* **1**, 339–349.

Benson, J. S., and Redpath, P. A. (1997). The nature of pre-European native vegetation in south-eastern Australia: a critique of Ryan, D. G., Ryan, J. R., and Starr, B. J. (1995) *The Australian Landscape: Observations of Explorers and Early Settlers*. *Cunninghamia* **5**, 285–328.

Bolton, B. L., and Latz, P. K. (1978). The western hare-wallaby *Lagorchestes hirsutus* (Gould) (Macropodidae) in the Tanami Desert. *Australian Wildlife Research* **5**, 285–293.

Bowman, D. M. J. S. (1998). The impact of Aboriginal landscape burning on the Australian biota. *New Phytologist* **140**, 385–410.

Boutin, C., and Keddy, P. A. (1993). A functional classification of wetland plants. *Journal of Vegetation Science* **4**, 591–600.

Bradstock, R. A., and Auld, T. D. (1995). Soil temperatures during experimental bushfires in relation to fire intensity: consequences for legume germination and fire management in south-eastern Australia. *Journal of Applied Ecology* **32**, 76–84.

Bradstock, R. A., and Cohn, J. S. (2001) Fire regimes and biodiversity in semi-arid mallee ecosystems. In *Flammable Australia: The Fire Regimes and Biodiversity of a Continent* (eds. R. A. Bradstock, J. E. Williams and A. M. Gill) pp. 238–258. (Cambridge University Press: Cambridge.)

Bradstock, R. A., and Myerscough, P. J. (1988). The survival and population response to frequent fires of two woody resprouters *Banksia serrata* and *Isopogon anemonifolius*. *Australian Journal of Botany* **36**, 415–431.

Bradstock, R. A., and O'Connell, M. A. (1988). Demography of woody plants in relation to fire: *Banksia ericifolia* L.f. and *Petrophile pulchella* (Schrad) R.Br. *Australian Journal of Ecology* **13**, 505–518.

Bradstock, R. A., and Scott, J. (1995). A basis for planning fire to achieve conservation and protection objectives adjacent to the urban interface. *CALMScience Supplement* **4**, 109–116.

Bradstock, R. A., Keith, D. A., and Auld, T. D. (1995). Managing fire for conservation: imperatives and constraints. In *Conserving Biodiversity: Threats and Solutions* (eds. R. Bradstock, T. D. Auld, D. A. Keith, R. T. Kingsford, D. Lunney and D. P. Sivertsen) pp. 323–333. (Surrey Beatty: Sydney.)

Bradstock, R. A., Bedward, M., Scott, J., and Keith, D. A. (1997a). Simulation of the effect of temporal and spatial variation in fire regimes on the population viability of a *Banksia* species. *Conservation Biology* **10**, 776–784.

Bradstock, R. A., Tozer, M. G., and Keith, D. A. (1997b). Effects of high frequency fire on floristic composition and abundance in a fire-prone heathland near Sydney. *Australian Journal of Botany* **45**, 641–655.

Braithwaite, R. W. (1987). Effects of fire regimes on lizards in the wet–dry tropics of Australia. *Journal of Tropical Ecology* **3**, 265–275.

Brooker, L. C., and Brooker, M. G. (1994). A model of the effects of fire and fragmentation on the population viability of the splendid fairy-wren. *Pacific Conservation Biology* **1**, 344–358.

Brooker, M. G., and Rowley, I. (1991). Impact of wildfire on the nesting behaviour of birds in heathland. *Wildlife Research* **18**, 249–263.

Burbidge, A. A., Johnson, K. A., Fuller, P. J., and Southgate, R. I. (1988). Aboriginal knowledge of the mammals of the central deserts of Australia. *Australian Wildlife Research* **15**, 9–39.

Burgman, M. A., and Lamont, B. B. (1992). A stochastic model for the viability of *Banksia cuneata* populations: environmental demographic and genetic effects. *Journal of Applied Ecology* **29**, 719–727.

Burgman, M. A., and Lindenmayer, D. B. (1998). *Conservation Biology for the Australian Environment.* (Surrey Beatty: Sydney.)

Burgman, M. A., Ferson, S., and Akcakaya, H. R. (1993). *Risk Assessment in Conservation Biology*. (Chapman and Hall: London.)

Catling, P. (1991). Ecological effects of prescribed burning practices on the mammals of south-eastern Australia. In *Conservation of Australia's Forest Fauna* (ed. D. Lunney) pp. 353–363. (Surrey Beatty: Sydney.)

Choquenot, D., and Bowman, D. M. J. S. (1998). Marsupial megafauna, Aborigines and the overkill hypothesis: application of predator–prey models to the question of Pleistocene extinction in Australia. *Global Ecology and Biogeography Letters* **7**, 167–180.

Clarke, P. J., Davison, E. A., and Trémont, R. M. (1998). *Your bushland: tips for managing bush plants in the New England Region.* (University of New England: Armidale, NSW.)

Common, M. S., and Norton, T. W. (1992). Biodiversity: its conservation in Australia. *Ambio* **21**, 258–265.

Commonwealth of Australia (1996a). *Australia: State of the Environment 1996.* (CSIRO: Melbourne.)

Commonwealth of Australia (1996b). *National Strategy for the Conservation of Australia's Biodiversity.* (Australian Government Publishing Service: Canberra.)

Commonwealth of Australia (1999). *National Weeds Strategy: A Strategic Approach to Weed Problems of National Significance*, revised edn. (Agriculture and Resource Management Council of Australia and New Zealand, Australian and New Zealand Environment and Conservation Council and Forestry Ministers, Canberra.)

Connell, J. H. (1978). Diversity in tropical rainforests and coral reefs. *Science* **199**, 1302–1310.

Conroy, R. J. (1996). To burn or not to burn? A description of the history, nature and management of bushfires within Ku-ring-gai Chase National Park. *Proceedings of the Linnean Society of New South Wales* **116**, 79–95.

Cowling, R. M., Lamont, B. B., and Enright, N. J. (1990). Fire and management of south-western Australian banksias. *Proceedings of the Ecological Society of Australia* **16**, 177–183.

Crowley, G. M., and Garnett, S. T. (1998). Vegetation change in the grasslands and grassy woodlands of east–central Cape York Peninsula, Australia. *Pacific Conservation Biology* **4**, 132–148.

Crowley, G. M., and Garnett, S. T. (2000). Use of fire by pastoralists in Cape York Peninsula: current practices and historical perspective. *Australian Geographer* **38**, 10–26.

Dale, A., and Bellamy, J. (1998). Regional Resource Use Planning in Rangelands: An Australian Perspective, LWRRDC Occasional Paper 06/98. (Land and Water Resources Research and Development Corporation: Canberra.)

Dickinson, K. J. M., and Kirkpatrick, J. B. (1986). The impact of grazing pressure in clearfelled, burned and undisturbed eucalypt forest. *Vegetatio* **66**, 133–136.

Dixon, K. W., Roche, S., and Pate, J. S. (1995). The promotive effect of smoke derived from burnt native vegetation on seed germination of Western Australian plants. *Oecologia* **101**, 185–192.

Downey, P. O. (1999). Fire and weeds: a management option or Pandora's box? In *Proceedings of Bushfire 99*, pp. 111–118. (Charles Sturt University: Albury, NSW.)

Downey, P. O. and Smith, J. M. B. (2001). Demography of the invasive shrub Scotch broom (*Cytisus scoparius* (L.) Link) at Barrington Tops, NSW: insights for management. *Austral Ecology* **25**.

Drury, W. H., and Nisbett, I. C. T. (1973). Succession. *Journal of the Arnold Arboretum* **54**, 331–368.

Fensham, R. J. (1997). Aboriginal fire regimes in Queensland, Australia: analysis of the explorers' record. *Journal of Biogeography* **24**, 11–22.

Fensham, R. J., Fairfax, R. J., and Cannell, R. J. (1994). The invasion of *Lantana camara* L. in Forty Mile Scrub National Park, north Queensland. *Australian Journal of Ecology* **19**, 297–305.

Flannery, T. F. (1990). Pleistocene faunal loss: implications of the aftershock for Australia's past and future. *Archaeology in Ocean* **25**, 45–55.

Flannery, T. F. (1994). *The Future Eaters: An Ecological History of the Australian Lands and People.* (Reed Books: Sydney.)

Fox, B. J., and McKay, G. M. (1981). Small mammal responses to successional changes in eucalypt forest. *Australian Journal of Ecology* **6**, 29–41.

Friend, G. (1995). Fire and invertebrates: a review of research methodology and the predictability of post-fire response patterns. *CALMScience Supplement* **4**, 165–174.

Friend, G., Leonard, M., MacLean, A., and Sieler, I. (1999). *Management of Fire for the Conservation of Biodiversity: Workshop proceedings.* (Department of Natural Resources and Environment: Melbourne.)

Garnett, S. (1992). *Threatened and Extinct Birds of Australia.* (Royal Australian Ornithologists Union: Melbourne.)

Garnett, S. T., and Crowley, G. M. (1995). Golden-shouldered parrot: options for management. Report to

Queensland Department of Environment and Heritage, Cairns.

Garnett, S. T., and Crowley, G. M. (1997). The golden-shouldered parrot of Cape York Peninsula: the importance of cups of tea to effective conservation. In *Conservation outside Nature Reserves* (eds P. Hale and D. Lamb) pp. 201–205. (University of Queensland: Brisbane.)

Garvey, M. (1996). Disseminating knowledge of wildfire using a geographic information system: three case studies. *Proceedings of the Linnean Society of New South Wales* 116, 109–114.

Gill, A. M. (1996). How fires affect biodiversity. In *Fire and Biodiversity: The Effects and Effectiveness of Fire Management*, Biodiversity Series Paper no. 8, pp. 47–55. (Commonwealth of Australia: Canberra.)

Gill, A. M., and Bradstock, R. (1995). Extinction of biota by fires. In *Conserving Biodiversity: Threats and Solutions* (eds. R. Bradstock, T. D. Auld, D. A. Keith, R. T. Kingsford, D. Lunney and D. P. Sivertsen) pp. 309–322. (Surrey Beatty: Sydney.)

Gill, A. M., and Catling, P. (2001). Fire regimes and biodiversity of forested landscapes of southern Australia. In *Flammable Australia: The Fire Regimes and Biodiversity of a Continent* (eds. R. A. Bradstock, J. E. Williams and A. M. Gill) pp. 351–369. (Cambridge University Press: Cambridge.)

Gill, A. M., and Nicholls, A. O. (1989). Monitoring fire-prone flora in reserves for nature conservation. In *Fire Management on Nature Conservation Lands* (eds. N. Burrows, L. McCaw and G. Friend) pp. 137–151. (Western Australian Department of Conservation and Land Management: Perth.)

Gill, A. M., and Williams, J. E. (1996). Fire regimes and biodiversity: the effects of fragmentation of southeastern Australian eucalypt forests by urbanisation, agriculture and pine plantations. *Forest Ecology and Management* 85, 261–278.

Gill, A. M., Groves, R. H., and Noble, I. R. (eds.) (1981). *Fire and the Australian Biota*. (Australian Academy of Science: Canberra.)

Graetz, R. D., Wilson, M. A., and Campbell, S. K. (1995). Landcover disturbance over the Australian continent: a contemporary assessment. Biodiversity Unit, Biodiversity Series Paper no. 7, Department of the Environment, Sport and Heritage, Canberra.

Grant, C. D., Loneragan, W. A., Koch, J. M., and Bell, D. T.

(1997). Fuel characteristics, vegetation structure and fire behaviour of 11–15-year-old rehabilitated bauxite mines in Western Australia. *Australian Forestry* 60, 16–23.

Grant, C. D., Koch, J. M., Smith, R. D., and Collins, S. J. (1998). A review of prescription burning in rehabilitated bauxite mines in Western Australia. *CALMScience* 2, 357–371.

Grice, A. C. (1997). Post-fire regrowth and survival of the invasive tropical shrubs *Cryptostegia grandiflora* and *Ziziphus mauritiana*. *Australian Journal of Ecology* 22, 467–474.

Hale, P., and Lamb, D. (eds.) (1997). *Conservation outside Nature Reserves*. (University of Queensland: Brisbane.)

Harrington, G. N., and Sanderson, K. D. (1994). Recent contraction of wet sclerophyll forest in the wet tropics of Queensland due to invasion by rainforest. *Pacific Conservation Biology* 1, 319–327.

Hobbs, R. (2001). Fire regimes and their effects in Australian temperate woodlands. In *Flammable Australia: The Fire Regimes and Biodiversity of a Continent* (eds. R. A. Bradstock, J. E. Williams and A. M. Gill) pp. 305–326. (Cambridge University Press: Cambridge.)

Hobbs, R. J., and Norton, D. A. (1996). Towards a conceptual framework for restoration ecology. *Restoration Ecology* 4, 93–110.

Hodgkinson, K. C. (2001). Fire regimes in *Acacia* wooded landscapes: effects on functional processes and biological diversity. In *Flammable Australia: The Fire Regimes and Biodiversity of a Continent* (eds. R. A. Bradstock, J. E. Williams and A. M. Gill) pp. 259–277. (Cambridge University Press: Cambridge.)

Holling, C. S. (ed.) (1978) *Adaptive Environmental Assessment and Management*, International Series on Applied Systems Analysis no. 3, International Institute for Applied Systems Analysis. (Wiley: Toronto.)

Hughes, C. J. (1995). One land: two laws – Aboriginal fire management. *Environmental and Planning Law Journal* 12, 37–49.

Jackson, W. D. (1968). Fire, air, water and earth: an elemental ecology of Tasmania. *Proceedings of the Ecological Society of Australia* 3, 9–16.

James, S. (1994). Symposium summary: Plant diseases in ecosystems: threats and impacts in south-western Australia'. *Journal of the Royal Society of Western Australia* 77, 99–100.

Keith, D. A. (1991). Coexistence and species diversity in

upland swamp vegetation: the roles of an environmental gradient and recurring fires. PhD thesis, University of Sydney.

Keith, D. A. (1995). Mosaics in Sydney heathland vegetation: the roles of fire, competition and soils. *CALMScience Supplement* 4, 199–206.

Keith, D. (1996). Fire driven extinction of plant populations: a synthesis of theory and review of evidence from Australian vegetation. *Proceedings of the Linnean Society of New South Wales* 116, 37–78.

Keith, D. A., and Bradstock, R. A. (1994). Fire and competition in Australian heath: a conceptual model and field investigations. *Journal of Vegetation Science* 5, 347–354.

Keith, D. A., and Sanders, J. A. (1990). Vegetation of the Eden region, south-eastern Australia: species composition, diversity and structure. *Journal of Vegetation Science* 1, 203–232.

Keith, D. A., and Tozer, M. G. (1997). Experimental design and resource requirements for monitoring flora in relation to fire. In *Bushfire '97* (eds. B. J. McKaige, R. J. Williams and W. M. Waggitt) pp. 274–179. (CSIRO: Darwin.)

Keith, D. A., McCaw, W. L., and Whelan, R. J. (2001). Fire regimes in Australian heathlands and their effects on plants and animals. In *Flammable Australia: The Fire Regimes and Biodiversity of a Continent* (eds. R. A. Bradstock, J. E. Williams and A. M. Gill) pp. 199–237. (Cambridge University Press: Cambridge.)

Kershaw, P., Clarke, J. S., Gill, A. M. and D'Costa, D. M. D. (2001). A history of fire in Australia. In *Flammable Australia: The Fire Regimes and Biodiversity of a Continent* (eds. R. A. Bradstock, J. E. Williams and A. M. Gill) pp. 3–25. (Cambridge University Press: Cambridge.)

Kirkpatrick, J. B., and Dickinson, K. J. M. (1984). The impact of fire on Tasmanian alpine vegetation and soils. *Australian Journal of Botany* 32, 613–629.

Kohen, J. L. (1996). Aboriginal use of fire in southeastern Australia. *Proceedings of the Linnean Society of New South Wales* 116, 19–26.

KRSIS (Kakadu Region Social Impact Study) (1997). *Community Action Plan*, Report of the Aboriginal Project Committee, July 1997. (Australian Government Publishing Service: Canberra.)

Lambert, J. A., Elix, J. K., Chenowith, A., Sole, C., Craig, D., and Fourmile, H. L. (1995). Approaches to bioregional planning. Part 2. Department of Environment, Sport and Territories, Biodiversity Series Paper no. 10, Canberra.

Lamont, B. B. (1995). Interdependence of woody plants, higher fungi and small marsupials in the context of fire. *CALMScience Supplement* 4, 151–158.

Langton, M. (1998). *Burning Questions: Emerging Environmental Issues for Indigenous Peoples in Northern Australia*. (Centre for Indigenous Natural and Cultural Resource Management, Northern Territory University: Darwin.)

Latz, P. (1995). *Bushfires and Bushtucker: Aboriginal Plant Use in Central Australia*. (IAD Press: Alice Springs.)

Leigh, J. H., and Briggs, J. D. (eds.) (1992). *Threatened Australian Plants: Overview and Case Studies*. (Australian National Parks and Wildlife Service: Canberra.)

Lindenmayer, D. B., and Possingham, H. P. (1995). *The Risk of Extinction: Ranking Management Options for Leadbeater's Possum*. (Australian National University and Australian Nature Conservation Agency: Canberra.)

Lundie-Jenkins, G. (1993). Ecology of the rufous hare-wallaby, *Lagorchestes hirsutus* Gould (Marsupialia: Macropodidae), in the Tanami Desert, Northern Territory. I. Patterns of habitat use. *Wildlife Research* 20, 457–476.

Lunt, I. D., and Morgan, J. W. (2001). The role of fire regimes in temperate lowland grasslands of southeastern Australia. In *Flammable Australia: The Fire Regimes and Biodiversity of a Continent*. (eds. R. A. Bradstock, J. E. Williams and A. M. Gill) pp. 177–196. (Cambridge University Press: Cambridge.)

Madden, B., and Hayes, G. (2001). *National Investments in Rural Landscapes: An Investment Scenario for National Farmers Federation and Australian Conservation Foundation with the assistance of Land and Water Resources Research and Development Corporation*. (The Virtual Consulting Group and Griffin nim Pty Ltd: Albany, NSW.)

McCarthy, M. A., and Burgman, M. A. (1995). Coping with uncertainty in forest management. *Forest Ecology and Management* 74, 23–36.

Meredith, C. W. (1988). *Fire in the Victorian Environment: A Discussion Paper*. (Conservation Council of Victoria: Melbourne.)

Milberg, P., and Lamont, B. B. (1995). Fire enhances weed invasion of roadside vegetation in southwestern Australia. *Biological Conservation* 73, 45–49.

Miller, I. L., and Lonsdale, W. M. (1992). Ecological management of *Mimosa pigra*: use of fire and competitive pastures. In *A Guide to the Management of Mimosa pigra* (ed. K. L. S. Harley) pp. 104–106. (CSIRO: Melbourne.)

Mooney, H. A. (1997). Ecosystem function of biodiversity: the basis of the viewpoint. In *Plant Functional Types* (eds. T. M. Smith, H. H. Shugart and F. I. Woodward) pp. 341–354. (Cambridge University Press: Cambridge.)

Morgan, J. W. (1998). Importance of canopy gaps for recruitment of some forbs in *Themeda triandra*-dominated grasslands in south-eastern Australia. *Australian Journal of Botany* **46**, 609–627.

Morris, E. C., and Myerscough, P. J. (1988). Survivorship, growth and self-thinning in *Banksia ericifolia*. *Australian Journal of Ecology* **13**, 181–189.

Morrison, D. A., Cary, G. J., Penjelly, S. M., Ross, D. G., Mullins, B. G., Thomas, C. R., and Anderson, T. S. (1995). Effects of fire frequency on plant species composition of sandstone communities in the Sydney region: inter-fire interval and time-since-fire. *Australian Journal of Ecology* **20**, 239–247.

Morrison, D. A., Buckney, R. T., and Bewick, B. J. (1996). Conservation conflicts over burning bush in south-eastern Australia. *Conservation Biology* **76**, 167–175.

Morton, S. (1994). European settlement and the mammals of arid Australia. In *Australian Environmental History: Essays and Cases* (ed. S. Dovers) pp. 141–166. (Oxford University Press: Melbourne.)

Murray Darling Basin Commission (1999). *The Salinity Audit of the Murray Darling Basin: A 100-Year Perspective, 1999.* (Murray Darling Basin Commission: Canberra.)

New South Wales National Parks and Wildlife Service (1997). *Fire management plan for Crowdy Bay National Park.* (New South Wales National Parks and Wildlife Service: Sydney.)

Noble, I. R., and Slatyer, R. O. (1980). The use of vital attributes to predict successional changes in plant communities subject recurrent disturbances. *Vegetatio* **43**, 5–21.

O'Neill, A. L., Head, L. M., and Marthick, J. K. (1993). Integrating remote sensing and spatial analysis techniques to compare Aboriginal and pastoral fire patterns in the east Kimberley, Australia. *Applied Geography* **13**, 67–85.

Pickett, S. T. A., and White, P. S. (1985). *The Ecology of Natural Disturbance and Patch Dynamics.* (Academic Press: New York.)

Purdie, R. (1995). Conserving biodiversity: threats and solutions – overview. In *Conserving Biodiversity: Threats and Solutions* (eds. R. Bradstock, T. D. Auld, D. A. Keith, R. T. Kingsford, D. Lunney and D. P. Sivertsen) pp. 410–416. (Surrey Beatty: Sydney.)

Price, O., and Bowman, D. M. J. S. (1994). Fire-stick forestry: a matrix model in support of skillful fire management of *Callitris intratropica* R.T. Baker. *Journal of Biogeography* **21**, 573–580.

Roche, S., Koch, J. M., and Dixon, K. W. (1997). Smoke enhanced seed germination for mine rehabilitation in the southwest of Western Australia. *Restoration Ecology* **5**, 191–203.

Rowley, I., and Brooker, M. (1987). The response of a small insectivorous bird to fire in heathlands. In *Nature Conservation: The Role of Remnants of Native Vegetation* (eds. D. A. Saunders, G. A. Arnold, A. A. Burbidge and A. J. M. Hopkins) pp. 211–218. (Surrey Beatty: Chipping Norton, NSW.)

Russell-Smith, J., and Bowman, D. M. J. S. (1992). Conservation of monsoon rainforest isolates in the Northern Territory, Australia. *Biological Conservation* **59**, 51–63.

Russell-Smith, J., Lucas, D., Gapindi, M., Gunbunuka, B., Kapirigi, N., Namingum, G., Lucas, K., Giuliani, P., and Chaloupka, G. (1997a). Aboriginal resource utilization and fire management practice in western Arnhem Land, monsoonal northern Australia: notes for prehistory, lessons for the future. *Human Ecology* **25**, 159–195.

Russell-Smith, J., Ryan, P. G., and DuRieu, R. (1997b). A Landsat MSS-derived fire history of Kakadu National Park, monsoonal northern Australia, 1980–1994: seasonal extent, frequency and patchiness. *Journal of Applied Ecology* **34**, 748–766.

Russell-Smith, J., Ryan, P. G., Klessa, D., Waight, G., and Harwood, R. (1998). Fire regimes, fire-sensitive vegetation, and fire management of the sandstone Arnhem Plateau, monsoonal northern Australia. *Journal of Applied Ecology* **35**, 829–846.

Salwasser, H. (1993). Sustainability needs more than better science. *Ecological Applications* **3**, 587–589.

Saunders, D. A., Hobbs, R. J., and Ehrlich, P. R. (eds.) (1993). *Nature Conservation*, vol. 3, *Reconstruction of Fragmented Ecosystems.* (Surrey Beatty: Sydney.)

Saxon, E. C. (1984). *Anticipating the Inevitable: A Patch-Burn Strategy for Fire Management at Uluru (Ayers Rock–Mt Olga) National Park*. (CSIRO: Melbourne.)

Short, P. S., and Turner, B. (1994). A test of the vegetation mosaic hypothesis: a hypothesis to explain the decline and extinction of Australian mammals. *Conservation Biology* **8**, 439–449.

Siddiqi, M. Y., Carolin, R. C., and Myerscough, P. J. (1976). Studies in the ecology of coastal heath in NSW. III. Regrowth of vegetation after fire. *Proceedings of the Linnean Society of New South Wales* **101**, 53–63.

Simmons, D., and Adams, R. (1997). Conflict or compromise? Impacts of roadside slashing for fire management on roadside heathland, Victoria. In *Bushfire '97* (eds. B. J. McKaige, R. J. Williams and W. M. Waggitt) looseleaf addition. (CSIRO: Darwin.)

Stanton, J. P. (1992). J. P. Thomson oration. The neglected lands: recent changes in the ecosystems of Cape York Peninsula and the challenge of their management. *Journal of the Queensland Geographical Society* **7**, 1–18.

van Wilgen, B. W., Richardson, D. M., and Seydack, A. H. W. (1994). Managing fynbos for biodiversity: constraints and options in a fire-prone environment. *South African Journal of Science* **90**, 322–329.

Vranjic, J., and Groves, R. (1999). 'Best-practice' management strategies for the South African weed, bitou bush (*Chrysanthemoides monilifera* subsp. *rotundata*). In *Proceedings of the 12th Australian Weeds Conference* (eds. A. C. Bishop, M. Boersma and C. D. Barnes) pp. 288–293. (Tasmanian Weed Society: Hobart.)

Whelan, R. J., and Main, A. R. (1979). Insect grazing and post-fire plant succession in south-west Australian woodland. *Australian Journal of Ecology* **4**, 387–389.

Whelan, R. J., Ward, S., Hogbin, P., and Wasley, J. (1996). Responses of heathland *Antechinus stuartii* to the Royal National Park wildfire in 1994. *Proceedings of the Linnean Society of New South Wales* **116**, 97–108.

Williams, J. E., (2000). Managing the Bush: recent research findings from the EA/LWRRDC National Remnant Vegetation R&D Program. National Research and Development Program on Rehabilitation, Management and Conservation of Remnant Vegetation, Research Report 4/00, Canberra.

Williams, J. E., and Gill, A. M. (1995). *Forest Issues 1: The Impact of Fire Regimes on Native Forests in Eastern NSW*,

Environmental Heritage Monograph Series no. 2. (New South Wales National Parks and Wildlife Service: Sydney.)

Williams, J. E. and West, C. (2000). Environmental weeds in Australia and New Zealand: issues and approaches to management. *Austral Ecology* **25**, 425–444.

Williams, J. E., Whelan, R. J., and Gill, A. M. (1994). Fire and environmental heterogeneity in southern temperate forest ecosystems: implications for management. *Australian Journal of Botany* **42**, 125–137.

Williams, R. J., Griffiths, A. D., and Allan, G. (2001). Fire regimes and biodiversity in the savannas of northern Australia. In *Flammable Australia: The Fire Regimes and Biodiversity of a Continent* (eds. R. A. Bradstock, J. E. Williams and A. M. Gill) pp. 281–304. (Cambridge University Press: Cambridge.)

Willson, A. (1999). *Tarawi Nature Reserve Fire Management Plan*. (New South Wales National Parks and Wildlife Service: Hurstville).

Woinarski, J. C. Z. (1990). Effects of fire on bird communities of tropical woodlands and open-forests in northern Australia. *Australian Journal of Ecology* **15**, 1–22.

Woinarski, J. C. Z. (1999a). Fire and Australian birds: a review. In *Australia's Biodiversity: Responses to Fire – Plants, Birds and Invertebrates*, pp. 55–180. (Environment Australia: Canberra.)

Woinarski, J. C. Z. (1999b). Prognosis and framework for the consideration of biodiversity in rangelands: building on the north Australian experience. In *Proceedings VI International Rangelands Congress* (eds. D. Eldridge and D. Freudenberger) pp. 639–645. (VI International Rangelands Congress, Inc: Aitkenvale, Qld.)

Woinarski, J. C. Z., Eckert, H. J., and Menkhorst, P. W. (1988). A review of the distribution, habitat and conservation status of the western whipbird *Psophodes nigrogularis leucogaster* in the Murray mallee. *South Australian Ornithologist* **30**, 146–153.

York, A. (1999). Long-term effects of repeated prescribed burning on forest invertebrates: management implications for the conservation of biodiversity. In *Australia's Biodiversity: Responses to Fire – Plants, Birds and Invertebrates*, pp. 181–266. (Environment Australia: Canberra.)

Part VIII

Final

18

Fire regimes and biodiversity: legacy and vision

A. MALCOLM GILL, ROSS A. BRADSTOCK AND JANN E. WILLIAMS

Abstract

Fires have occurred across Australian landscapes for millions of years and human use of fire has probably been a feature there for tens of thousands of years. Scientific knowledge of fires and their effects has risen steeply in the last 50 years but manipulation of landscapes using fire for the conservation of biodiversity is largely a phenomenon of the last 20 to 30 years. Two important concepts guiding fire management for biodiversity are those of the fire regime and of the functional group. Statistical variation in the components of the regime, especially fire interval, may be necessary for maintaining biodiversity in some cases. Fire regimes can be designed but, in practice, the outcome will be influenced by patterns in the weather, the incidence of unplanned fires (including arson) and fuel dynamics. Knowledge of fires and their effects resides in the oral traditions of Aboriginal people as well as in 'Western' society. Further cross-referencing of knowledge would be valuable. There is a need for the development of a nationally co-ordinated programme of research that seeks an understanding of fire regimes, biodiversity and management systems across Australia's diverse climates, land uses and tenures.

Introduction

While the occurrence of fire in Australia undoubtedly goes back millions of years (Kershaw *et al.*, this volume) the antiquity of the manipulation of fire by people in Australia is probably equal to the antiquity of the Aboriginal people – tens of thousands of years (see Miller *et al.* 1999). In the last two centuries, i.e. since white settlement, the landscapes on which contemporary fire management is carried out has changed, sometimes little apparently, some-times grossly. In the last two or three decades, the nature of human interactions with landscape fire in Australian landscapes has changed dramatically due to the application of science and modern technology to fire management.

Fire ecology is a modern discipline of 'Western' science. It has its roots both in academia and in practical land management. The scientific literature reveals the development of academic thought, and sometimes reports on the practices of land managers, but the human attitudes, values and ideas affecting fire management usually remain undocumented. However, the divergent views of various sectors of the Australian populace underlie debates on the problems challenging fire managers today.

Maintaining the full complement of native biota has been a major aim of many managers in conservation reserves, state forests and even on private land. Today, this aim is often expressed as the 'conservation of biodiversity', the 'native' qualifier being implicit. 'Biodiversity', a new word used internationally, encompasses 'ecosystems', 'species' and 'genes' (see Keith *et al.*, this volume b). The last of these components is hardly mentioned in this book but, in the decades ahead, we anticipate that it will be a component that will be addressed in relation to the inheritance and evolution of traits that allow adaptation, or not, to particular fire regimes. We anticipate that associated with this development will be the adoption of the techniques of gene technology to help explore the effects of fire regimes on micro-organisms, for example.

In this final chapter we briefly respond to the following questions as a precursor to outlining our view of future directions in fire ecology and the management of fire regimes for the conservation of biodiversity in Australia:

- What are the historical roots of fire ecology and the fire-management of biodiversity in Australia?
- Which are the key concepts created in the study and application of fire ecology for biodiversity conservation?
- What are the predominant fire-management ideas, their apparent origins and their value for biodiversity conservation?

An historical sketch from the mid 19th century to the 1960s

From the mid 19th century came the first recorded ecological observations on the interplay between fires and the biota. The explorer Mitchell (1848) noted that 'Fire, grass, kangaroos, and [original] human inhabitants, seem all dependent on each other for existence in Australia.' Soon after Major Mitchell's report were the memorable 1851 fires in southeastern Australia. Kiddle (1967), from source material, mentioned that dead and dying native animals followed these fires and that 'birds had dropped from the trees killed by heat rather than fire'. Other birds 'flew desperately out to sea; a few found refuge on ships'. Only a few decades later an example of astute ecological observation and deduction came on the effects of fires on the flora of Kangaroo Island, South Australia. There, Tate (1883) noted the occurrence of: fire ephemerals (plants stimulated to appear after fire but 'lost' before the next), 'seeders' and 'sprouters' (see below); post-fire germination of soil-stored seed; repeated fires causing plant species' extinction; and the effects of fire intensity on seed dispersal. It appears that, despite Tate's insights, his observations had little or no effect on the development of fire ecology as it was only much later that these concepts were further considered.

As pastoralists spread across Australia with their flocks and herds from the early 19th century onwards, they soon recognised the importance of fires to their livelihood (Noble and Grice, this volume). Low-intensity fires could be used to protect property by removing grassy fuels under safe burning conditions (Howitt 1856) and woody plants could be subdued, and palatable herbage encouraged, by the use of fire (Tate 1883). Such activities have probably been carried out in many parts of the world for centuries and even millennia and represent the deep roots of modern prescribed burning, a practice based on mathematical fire-spread models.

Observations on the effects of fires on plants and animals continued sporadically post Tate (1883) but it was not until the mid 20th century that scientific activity increased substantially. A harbinger of change was the research of Jarrett and Petrie (1929) who examined the effects of the 1926 fire in Central Victorian forests and noted the different responses of trees and shrubs, particularly, to the one fire. Beadle (1940) took measurements of the soil temperatures reached under burning material and noted the difference in responses of shrubs – easily killed, and relatively resistant – to fire. Henry (1961) set up the first experiments in Australia on the effects of fires, in 1952, in Queensland. Specht *et al.* (1958) showed how plant-species composition changed with increasing time after fire in South Australian heathlands. Unprecedented in geographic scope was the study of the effects of fires on vegetation in the Northern Territory (Stocker 1966). Soon after, Jackson (1968) proposed an innovative scheme of vegetation dynamics for plant communities in Tasmania based on overlapping statistical distributions in fire intervals for different vegetation types. Jackson's study followed the work of Gilbert (1959) who pointed out the relationships between 'rainforest' (closed forest) and tall open-forest as affected by fire interval and fire intensity. The modelling of fire behaviour in southeastern and southwestern Australia in the 1960s (McArthur 1962, 1966, 1967; Peet 1965), an important step, enabled safer prescribed burning – and therefore the protection of life and economic assets.

The 50s and 60s of the 20th century saw the beginnings of a theoretical base to fire ecology in Australia and the beginnings of an exponential rise in the volume of scientific literature (after Rose 1993).

Key concepts and their legacy

The fire regime

Fires occur as discrete events but their effects depend on the history of those events, the seasons in which the fires occurred and their properties. The sequence of fires is known as a 'regime' consisting of the components 'intensity', 'type', 'between-fire interval' and 'season' (Gill 1975). The regime concept provides a welcome shift beyond phrases such as, 'fire as a factor', 'adaptation to fire' and 'fire-sensitive species' (see below) although these phrases are still extensively used as shorthand through tradition or habit, or because of a lack of appreciation of the regime concept. It is a concept that implies which variables can be important in affecting ecological outcomes. It allows description of how fires can affect ecosystems without necessarily knowing all the details. It indicates which fire variables may have been important in the development of current ecosystems.

Knowledge of the effects of individual components of the fire regime has grown significantly during the last 20 years. Thus, it may be expected that the way in which effects of fire regimes may be addressed have changed too. A greater refinement in the understanding of the nature of these variables and their effects has resulted. The components, to some extent, have been surrogates for other variables or complexes of variables (see below).

Because the effects of repeated fires on long-lived plants requires a 'point' base, fire regimes are considered as acting at 'points' on the landscape. This is not to say that fire area, patch size or within-fire heterogeneity, are unimportant to certain ecological problems (Gill 1998). Rather, the point-based system allows maximum flexibility in application. Thus 'area' can be considered by aggregation of 'points' when the effects of a single fire are considered (e.g. maximum or mean intensity) but when several overlapping fires occur over time in the same region it is difficult to retain an 'area' focus in describing their effects. Fire characteristics vary within a single fire, even on flat terrain (Catchpole et al. 1992), so that a point-based system enables that variation to be considered.

For any one point on the land surface, one measurement of fire intensity (Byram 1959) is possible for any one fire. Effects at any point vary with height above the surface and the depth below it. Intensity can be a poor predictor of subterranean effects (Bradstock and Auld 1995) while its aerial effects can be moderated by other variables such as wind (Mercer and Weber 1994). In the future, 'intensity' may be portrayed as a frequency or probability distribution for different parts of landscapes (e.g. Cary, this volume). Fire characteristics like 'residence time' – the time of flaming combustion – may be more useful in the prediction of below-ground effects but require measurement at the time of fire which can be difficult. Residence time may be modelled by using a correlation with fuel loading. In such circumstances, 'residence time' may be seen as a component of the fire regime replacing 'intensity'.

'Intensity' is of little help in describing the nature or the effects of fires burning below ground. Below-ground fires tend to smoulder rather than flame and persist at points much longer than above-ground fires. Because of this, whether a fire burns above, or below, ground has been considered as the dichotomous choice within the component 'type of fire'. 'Burnout time' – the time of combustion at a point – may be a measure of such fires but, because the effects may be influenced by the depth of the burning material, no single measure is likely to satisfy all applications. There is a need for much greater understanding of these fires and their effects on ecosystems. While there is relatively little area affected by peat fires in Australia their effects can be very significant (Gill 1996).

'Between-fire interval' or 'fire interval' is the time in years between fires. The fire interval constitutes one way of measuring fire frequency. The fire interval is perhaps the most appropriate measure of fire frequency in an ecological sense because it allows direct insight into consequences for time-dependent life history processes, such as maturation in plants and animals. Where once attention was focussed on the ecological significance of a particular interval between fires, an emerging view is that both the mean and the variance of fire intervals are significant, along with temporal fluctuations in these parameters (see Clark *et al.*, this volume; McCarthy and Cary, this volume).

'Season of burn' remains the least understood of the fire regime components. 'Season' seems to have been used as a surrogate for seasonal changes in fire intensity in some ecosystems (associated with weather and fuel and within-fire patchiness in tropical savanna; e.g. R. J. Williams *et al.*, this volume) and weather. Season of fire may be important in terms of the timing of various seasonal plant-phenological or physiological processes with ramifications for demographic change (e.g. Bradstock and Cohn, this volume; R. J. Williams *et al.*, this volume). As knowledge grows greater clarity and sophistication can be expected in the expression and understanding of the 'season' component and its ecological effects.

Functional types

The idea of grouping plants according to their behaviour rather than their taxonomy is long-standing (e.g. Raunkiaer's 1934 life form concept). With regard to fire, the 'fire-adapted' and 'fire-sensitive' categorisation has a significant legacy of use. The concept of adaptation to fire (and its converse) relies on the assumption that particular traits are adaptive in an evolutionary sense (Gill 1975, 1981). Thus, in plants, traits such as thick bark, insulated woody fruits or fire stimulated germination of seeds were regarded as adaptive. Species in possession of such traits can cope with the passage of a fire and were apparently considered to reflect selective pressure of fire in both evolutionary and ecological time.

The limitations of such a view have been extensively discussed (Gill 1975, 1981) and there is much debate as to whether features such as serotiny or resprouting in plants reflect selective pressures of fire or other factors such as predation, herbivory or nutrient limitations (Bond and van Wilgen 1996). Most importantly, a classification based on traits in isolation, gives a very limited view of the influence of fire regimes (i.e. predictive power is low). The notion that 'fire-adapted' vegetation can withstand any fire or cycle of fire persists in many minds. The dichotomous split of vegetation into 'fire-adapted' and 'fire-sensitive' classes reinforces a mind-set that fires can be managed in relative isolation as 'events'

(see below). When used in this way, the dichotomy essentially deals with the event-related components of the fire regime: type, intensity and season. A management focus on control of fire in general, and fire intensity in particular, follows readily (see below).

Terms such as 'fire-adapted' and 'fire-sensitive' are used in a variety of ways, often without consistency or explicit definition, with consequent incongruities. For example, many 'fire-sensitive' rainforest species have the capability to resprout following a single fire (Russell-Smith and Stanton, this volume) whereas 'fire-adapted' mallee eucalypts can suffer high mortality when exposed to repeated annual fires in autumn (Bradstock and Cohn, this volume). Hence the dichotomy is often inappropriately used to infer syndromes of interval-related behaviour of populations, species or communities (i.e. the presence or otherwise of an event-related, 'fire-adaptive' trait is wrongly assumed to indicate a response to fire interval).

The significance of traits requires evaluation within the context of life history (Whelan *et al.*, this volume) as well as fire regime. There has been considerable development in the last two to three decades of functional classifications that partially resolve this problem and offer the potential to integrate and predict responses to both event- and interval-related components of fire regimes. Much of the impetus for this comes from another dichotomy, the separation of plant species in relation to their characteristic survival response to individual fires: i.e. 'seeder' versus 'sprouter' (Whelan *et al.*, this volume).

Gill (1981) provided a formal definition of the term 'fire-sensitive' in relation to the effect of a single fire on plants (death when leaf scorch from fire is 100%). He developed a species classification in which fire intensity and plant maturity were standardised, therefore allowing comparisons between species to be made. The assumption in schemes like that of Gill (1981) is that surface, rather than belowground (peat) fires are involved. Gill's (1981) classification involves a hierarchy of categories, based on the nature of seed storages and resprouting modes nested within the original dichotomy. While remaining a system intended to indicate responses to a single fire it recognises the presence of life-history attributes that require explicit time frames, e.g. time to produce seeds.

Concurrently the vital attribute system of Noble and Slatyer (1980) explicitly recognised the importance of attributes such as life span and the dependence or otherwise of seedling establishment on fire. It has a modelling orientation with fire interval as the main fire-regime component to which it is readily related. The influence, development and range of applications of this system has been wide (i.e. its application to explain vegetation dynamics in a number of chapters in this volume), but its practical application in fire management systems as a decision-support tool, is only now being realised (Keith *et al.*, this volume b).

While there have been attempts at definition of functional classifications for animals in relation to fires and fire regimes (Whelan *et al.*, this volume) there has been less momentum and unanimity in their uptake. In part this could reflect the different attributes that are important for animals compared with plants, particularly in relation to vegetative/non-vegetative and by inference, flammable/non-flammable components of habitat. Further developments are required.

A progression in thinking has occurred from relatively simple dichotomies to a more complex set of hierarchical categories and choices. This has occurred in direct response to changes in the way in which we view fires and their ecological effects. That species and communities respond to disturbance regimes, and thus both event and interval dependent processes, is axiomatic (Bond and van Wilgen 1996). The fire-regime requirements of species may be wide-ranging yet subtle. Multi-faceted classifications, such as vital attributes, offer a powerful means of compressing a potentially bewildering spectrum of requirements of an assemblage of species into a tractable form. None the less they are not perfect and further developments and refinement may be expected. Our knowledge and use of functional types reflects biases in the state of knowledge of fire ecology in general. Knowledge of the fire regime responses of vascular plants and some higher vertebrates (mammals), while incom-plete, is sufficient to permit the development of functional classifications. As yet however, little is known about the basic responses of lower plants and invertebrates. Equally it should be emphasised that the use of predictions based on classification systems, when applied in management, requires verification through observation and measurement in the field (see below).

Management legacies

Use of planned fire and suppression: the taming of 'wildfire'

The technology associated with fire-suppression operations has advanced very rapidly since the Second World War. The list of innovations continues to grow but mention can be made of the four-wheel-drive vehicle, aircraft (both fixed wing and helicopters), global positioning systems, computers and communication systems. However, even as the technology used in suppression operations has improved and proliferated, there has been a shift from a strong emphasis on suppression of fires in 'wildlands' and commercial eucalypt forests in southern Australia to the imposition of planned fire regimes using low-intensity fires (e.g. see Pyne 1991). While there may be a number of reasons for the use of prescribed fire in particular circumstances, such as in the pastoral industry (Noble and Grice, this volume), in many landscapes a continuing reduction in the area affected by unplanned fires is keenly sought. On most farms, in orchards and in pine plantations, the aim has been the complete exclusion of fire except for the dispatch of debris after the harvesting of graminoid crops or forest trees. With the twin approaches of prescribed fire (low intensity) and the effective suppression of unplanned fires, a 'taming' of fire has been implicitly sought.

It has been stated that Aboriginal people universally and perfectly controlled the incidence of unplanned fire by their planned use of fire (Langton 2000) and it would appear that much contemporary use of planned fire is promulgated on this idea. However, the validity of this idea is not universally accepted (Allan and Southgate, this volume; Gill and Catling, this volume). In all circumstances of human intervention, in the past and present, unplanned fires are likely to have occurred, whether these were triggered by arsonists, human carelessness, lightning or even volcanic activity. In particular, large fires may result when unplanned ignitions coincide with extreme fire-weather conditions thereby burning younger fuels than those prescribed for burning under mild weather. Fuel reduction by prescribed fire can affect the extent of unplanned fires but the relationship between the two varies from place to place (e.g. Gill and Moore 1997; Gill *et al.* 2001).

The use of prescribed fires carries an expectation that the fires will not spread beyond the target area or jurisdiction of interest. However, use of planned fire has led to the inevitable fire escapes and the associated unintended damage and suppression costs (Meredith 1996). There are risks as well as potential benefits from prescribed fires.

Exclusion of fire for a critical period of time across significant portions of landscapes may be required for ecological reasons (e.g. to allow maturation of 'seeder' species or the development of animal habitats). Such circumstances can result in accumulations of fuel sufficient to enable fire intensities beyond controllable levels when weather is severe (Morrison *et al.* 1996; Williams and Bradstock 2000). Comprehensive, landscape-level analyses of the interaction between planned and unplanned fires and resultant effects on fire regimes are awaited, though a number of recent modelling studies (Gill and Bradstock 1999; Bradstock *et al.* 1998; Gill *et al.* 2001; McCarthy and Cary, this volume) have explored the nature of these interactions.

Modern resource management by government agencies is predicated on fiscal stability, commensurate with finite economic costs and returns. Consequently, budgeted expenditure for prescribed fires and suppression costs can be fixed. Thus, the extent of prescribed burning each year may be 'set' thereby leading to a more-or-less constant interval between fires if a regular sequence of block burning is followed. Such a trend toward a constant interval between fires was detected in southwestern

Australian eucalypt forests by Lang (1997). However, the aim of burning a constant proportion of landscape each year as part of a fixed rotation may become obsolete in some cases. This does not preclude a contrived, but random, pattern of burning with a fixed target for area burnt by prescription each year; this may be desirable for the conservation of biodiversity assuming that the mean values are appropriate. In Kakadu National Park in northern Australia, deliberate burning, along with some unplanned fire, has created a random pattern of fire intervals (Gill *et al.* 2001).

The planned use of fire has not resulted in a complete taming of 'wildfire'. 'Taming' of unplanned fires has been partial rather than complete.

Events versus regimes

The desire for control of the unplanned and the unexpected has had an understandably strong influence on research and management cultures. Traditionally, fire behaviour research has been focussed on predicting the characteristics and spread of fire so that the firefighter, and the suppression operation, can be supported. The legacy is an emphasis on the 'event' – the weather at the time, the fuels present, the nature of the terrain, the firefighting resources available. It is, by nature, location-specific and mechanistic. To the people using this information, such as the fire management and suppression agencies, an event orientation was usual. Response to the event was seen as their most important role. Fire has been seen as an 'emergency' to be controlled and managed through suppression and/or manipulation of fuel.

Ecological outcomes are the result of a series of fires and their inherent characteristics (i.e. the fire regime), not just the results of the last event. The emphasis on the event, which is common in the field of emergency management, has been moderated by the recognition of the need for preparedness and recovery phases as well as the response (event) phase. When seen in this light, the event is followed by a recovery (or repair) phase which merges into a preparedness phase prior to the next event; a cycle of activity is involved. The post-event phase depends then on the type and magnitude of the event itself.

Its season of occurrence and temporal separation from the last event could also be important. A regime is involved. Thus a regime orientation may be applicable even to organisations traditionally concerned only with events (Bradstock and Gill 1999).

The differences between 'regime' and 'event' approaches reflect long-standing differences in outlook and concepts. For example, Andersen (1999) argued that many problems in fire management were due to an inherent focus on actions (e.g. lighting fires) and their immediate effects, rather than the wider question of the cumulative effect of actions and other factors (regimes). Debate and acrimony over fire management issues is often due to an underlying or unconscious clash between these paradigms. A greater understanding of these differing perspectives may help to overcome such problems in the future. A regime perspective arises if fire management is seen primarily as fire-interval management (after Gill and McCarthy 1998) or fuel-age management – whether the objective is the conservation of biodiversity or protection of socio-economic assets – and secondarily as fire-attribute (season and intensity) management.

'Event' approaches tend to be reinforced when the assets threatened are obvious and reflect social and economic values. The media reinforce event orientations by sensational coverage of fire outbreaks and by noting the presumed 'destruction' of the vegetation. An orientation toward economic, demographic or silvicultural performance often seems to encourage an event perspective too. Perhaps surprisingly, an event orientation exists in management for biodiversity conservation in some circles. In part this is cultural, reflecting the training, practical orientation and generational status of many managers. In equal measure it is due to the difficulty of comprehending the nature of regimes and their effects, and the lack of planning systems that compel such comprehension. Thus the discovery of a rare species may still trigger a ban on prescribed burning even though it is possible that the species is rare because fires have been too infrequent. The declaration of a small reserve may elicit a strong fire-protection response even though fires were a part of the ecosystem before reservation. Such approaches are often deemed precautionary, but it is difficult to see how this is justified without compilation of information on prevailing fire regimes as a precursor to 'precaution'.

Management philosophies

The use of philosophical argument (such as 'the flora is adapted to fire'), has great appeal because it may allow management to take place in the absence of relevant knowledge and/or constraints imposed by context. A 'recipe' or philosophical formula obviates the need to change and adapt – management becomes unerring and fixed. Lack of knowledge is relative: it has expanded rapidly in the past few decades but overall our knowledge of how biodiversity responds to fire regimes is meagre. As a result philosophies and recipes are used as surrogates for informed management (Gill 1977, Gill and Bradstock 1995) but, in the context of biodiversity conservation, most are deficient in some way.

Gill (1977) defined three management options for biodiversity conservation.
(1) *Laissez-faire* management with no intervention – on the presumption that 'natural' fire regimes will prevail in a beneficial way; (2) recreation of the fire regimes of Aboriginal people; or, (3) the creation of new or synthetic fire regimes to suit the explicit needs of species within a contemporary context. A choice between these options presents a dilemma, in part because it is difficult to tease out what each entails on the ground, or because each option is open to a considerable range of interpretation and opinion. It is evident today that the dilemma remains. Adherents of the re-creation of natural or anthropogenic fire regimes can be readily found, often on the basis of complex motives and ideals rather than quantitative information about the nature of prehistoric regimes. Often, the motivation is to reinstitute particular actions often without any clear evidence or understanding of what will ensue in contemporary circumstances, nor evidence about the degree and context in which such actions took place in the past.

The third option has received support in parallel with the growth of research on the relationship between biodiversity and fire regimes, and the additional realization that the contemporary condition and context of biodiversity has been substantially altered. Thus fire regimes, even if known perfectly from some point in the past, may not result in the same effect in a contemporary or future context. The problem of lack of knowledge bedevils this option. Implicit within it is the idea that knowledge of plants and animals themselves can be captured in such a way as to conserve them. It could be said that the plants and animals themselves contain the crucial knowledge needed for their conservation. Knowledge grows but remains limited – how do we cope given this imperfection?

Where to with management of fire regimes for biodiversity?

Some challenges

Many challenging questions await answers. Prominent examples include:

- How much variation in fire regimes (and in which components) is desirable for the maintenance of the native biota in fragmented landscapes?
- How can the nature and effects of variation in fire regimes be further uncovered?
- To what extent will managers be able to control fire regimes where ignition of fires through carelessness and/or arson is common?
- To what extent will the spread of exotic plants, particularly grasses, alter fire regimes and affect biodiversity?
- How to enhance the dialogue between indigenous and non-indigenous land managers to improve fire management?
- To what extent will Aboriginal people be involved in fire management as a function of restitution of land rights and custodianship?
- What will be the consequences of a changing climate?
- What management systems will best cope with these problems and make the best use of knowledge and resources?

These issues are addressed below and suggestions for future activities identified in the 'Conclusions'.

Variation in fire regimes?

Within a landscape there will be a range of fire regimes. This variation may be due to heterogeneity of the vegetation or of topography or of management action but there may be random variation as well. Most scientific attention has centred on variation in the between-fire interval (e.g. Allan and Southgate, this volume; Clark *et al.*, this volume; McCarthy and Cary, this volume) but there is undoubtedly variation, in intensity (e.g. Cary, this volume), season and type of fire. The extent of such variation, and the resilience of biodiversity to it, are important issues.

Clark (1996) showed how the statistical probability of a fire interval could be related to changes in the abundance of prominent plant species in North American forests using analysis of pollen in sediments and fire scars on trees. His work provided a validation of predictions from life-history models (Clark 1991). He noted that the nature of the variance in fire intervals was of equivalent importance to the mean fire interval in terms of its influence on particular plant life histories. Clark *et al.* (this volume) have indicated how the presence of a mix of species with fundamentally different life history characteristics is mediated by the set of fire regimes in a landscape. Ultimately the set of regimes is important, there being no invariable interval, for example, that is likely to serve the needs of a diverse assemblage of species.

McCarthy *et al.* (1999), using simple probabilistic models of fire interval, showed that local losses in seeder-species populations, stemming from the adverse 'tails' of the fire-interval distribution, may be counterbalanced by dispersal from places subjected to favourable fire intervals. A shift in the nature of fire recurrence may shift the distribution of fire intervals and in turn the balance between persistence, extinction and recolonisation. A critical domain of variation in fire intervals, where persistence outweighs local extinction, can therefore be defined as a function of the life histories of the species present in an area. Modelling studies dealing with Australian examples have attempted to quantify this domain, in terms of probability of persistence of both plant and animal species (Bradstock *et al.* 1998; McCarthy *et al.* 1999). Such modeling studies indicate the importance of dispersal in mediating the persistence of species in landscapes containing a dynamic range of fire regimes. Research on dispersal characteristics of plants and animals will therefore be of considerable importance to elucidating the landscape-level dynamics of species in relation to fire regimes.

Beyond the species level, there is recognition that the interplay between different functional groups of plants may be mediated by variation in fire regimes. Evidence suggests that floristic diversity in temperate shrublands and grasslands is a positive function of the variance of fire intervals (Keith *et al.*, this volume a; Lunt and Morgan, this volume). Bradstock *et al.* (1995) attempted to quantify the domain of fire-regime variation for shrubland communities in a way that could be used for decision support in an adaptive management system. In Victoria, idealised fire age-class distributions, selected with reference to plant life histories, are being used in the development of management plans for biodiversity conservation (Tolhurst 1999).

Rapid developments of modelling on the above themes is anticipated. However, the real challenge will lie in empirical scrutiny of the problem. How can such models be tested in reality? How can experiments and correlative studies be designed to explore the relationship between variability in regimes and biodiversity? Inherently it would seem a move toward observations and experiments that document fire regimes and ecological responses at a landscape scale may have to be made. Equally a move must be made beyond experiments that administer unvarying treatments of fire-regime components, however valuable these have proved in the past. Concepts of fire-regime variability not only will affect the way that research and management is conducted in the future but also the way the past is investigated and interpreted. Techniques are required that enable quantitative estimation of past fire regimes and the nature of their variation.

Studies of this kind offer much potential for the verification and calibration of models (see above) and a basis for formal linkage to contemporary empirical studies.

Parallel to these developments in research, there lies an immense challenge in creating awareness of the importance of variation of fire regimes in a management sense. The growing familiarity of managers with geographic information systems (GIS) and their routine use in decision-making may ease the way. As databases containing fire-history information improve in resolution and temporal scope, managers will become more familiar with the broad patterns that emerge from consideration of multiple 'layers' of individual fires. Adaptive management systems that stimulate managers, to update, scrutinise and interpret fire history information in this way are required to enhance this process (see below).

Can fire regimes be tailor-made?

Moving beyond the management of fires as isolated events will require detailed knowledge about how management actions will affect fire regimes. At the most fundamental level, fire regimes will be a function of rates of fire incidence (numbers of fires). As the rate of fire incidence increases the average between-fire interval at a point in the landscape will decrease if fire size is unaffected. However, with an increase in fire incidence due to human causes, there may be a distinct bias in their locations. Various parts of a complex terrain may be affected differently and the nature of the vegetation and its role as fuel will all be influential (McCarthy and Cary, this volume). It may be expected, for example, that a more diverse range of fire intervals will occur in topographically complex, compared with uniform, landscapes with a resultant effect on diversity of biota (Clark *et al.*, this volume).

A key influence on individual fires and resultant fire regimes is weather. All landscapes exist within a climatic context that is inherently variable at a hierarchy of temporal scales. Fluctuations in weather variables at diurnal and seasonal scales are apparent but changes may occur over longer, irregular, time-scales too. The weather has a great bearing on where a fire burns and with what characteristics (Catchpole, this volume). The shape, size, internal patchiness and distribution of intensities will be some function of weather in interaction with other factors, such as fuel type and quantity. Such effects of weather are complex. Fuel availability, for example, is limited by weather in a number of ways. At a continental scale, the degree of spatial continuity of surface fuels composed of the litter and live foliage of perennial vegetation is a positive function of average annual rainfall. In temperate forests, woodlands and heaths surface fuels are sufficiently continuous to carry a fire soon after a prior fire (2–5 years: Hobbs, this volume; Gill and Catling, this volume; Keith *et al.* this volume a; McCarthy and Cary, this volume). In semi-arid and arid shrublands and grasslands such levels of continuity may not be achieved until 10–15 years post-fire under average rainfall conditions (Allan and Southgate, this volume; Bradstock and Cohn, this volume; Hodgkinson, this volume; Noble and Grice, this volume).

Large fires can occur when there is a high degree of connectivity of fuel at a landscape scale. Such 'high connectivity' events occur in all Australian ecosystems, but their frequency and the nature of weather phenomena that drive them differ between regions and ecosystems. In temperate, mesic systems where accumulation of surface fuels is regular and the spatial cover of vegetation is relatively high, high-connectivity events occur when there is a coincidence of drought and extreme ambient weather, enhancing flammability. In contrast, in arid and semi-arid systems where spatial cover of litter and other fuels is relatively sparse, such events occur under extreme ambient conditions following periods of above-average rainfall which enhance accumulations of ephemeral grassy fuels. It can be argued that fire regimes are strongly structured by fires that coincide with these high-connectivity events. Not only the recurrence of such conditions, but the relative level of saturation of such events with ignitions will determine the frequency of large fires. Fires in the intervening period between high-connectivity events may interact and modify the nature of fires during such events, but

the nature of such interactions is currently speculative. Much research is needed to document the nature of recurrence of high-connectivity conditions, the rate of incidence of ignition relative to such conditions, the interaction with fires in 'normal conditions' and overall fire-regime outcomes.

The wet–dry tropical savanna provides an interesting convergence of these trends. On the one hand, high rainfall ensures recurrence of high fuel connectivity through growth of grasses on an annual scale. On the other hand, the conditions of the late dry season are conducive to large fires through extreme fire-weather and the effects of seasonal drought on curing (R. J. Williams *et al.*, this volume). Current management is predicated on using fire in the early dry season (i.e. 'normal' conditions) to restrict the extent of late dry-season fires (high-connectivity conditions). Management is shifting the season of fire and the resultant distribution of intensities, but the resultant interval remains close to maximal with an annual to biennial mean (Russell-Smith *et al.* 1997b; Gill *et al.* 2001). In other parts of the country the recurrence cycle of high-connectivity events may be measured in decades, at least, rather than annually, while 'normal' conditions recur more or less on annual cycles. Resultant effects of this disparity between rates are not well understood in terms of the range of fire regimes that may be possible under differing rates of ignition.

Management has the potential to influence not only the rate of fire incidence but also the weather with which fires are associated. A prescribed burning programme for example may not only increase the overall rate of fire incidence in a landscape but also skew the associated distribution of weather toward the moderate end of the spectrum. In an opposite way, the action of a malicious arsonist could increase the rate of fire incidence associated with severe weather. By implication, any management action that seeks to affect the behaviour of the arsonist will affect the fire regime. While both prescribed fires and arson fires will affect the set of fire regimes in a landscape, there are few studies that shed quantitative insight on the problem. Cary (this volume) gives some indication of the sensitivity of fire regimes to alteration in weather. In that case, climate change was predicted to increase mean fire intensity and decrease interval as a result of an increase in average ambient temperature and decrease in relative humidity. Those results not only are useful for indicating trends in response to climate change, they may also indicate how a shift in aberrant human behaviour (e.g. an increase in arson) could alter fire regimes under a constant climate.

A crucial influence on the make-up of fire regimes is the size of the landscape unit being considered. Modelling (Wimberley et al. 2000) indicates that the random spatial variability in fire regimes in a relatively small sample portion of an extensive forested landscape – say in an area smaller than the largest fire area – will be greater than that in a larger portion. This occurs even without isolating the target portion from ignitions in the 'complete' landscape. The end result is that the proportions of 'old growth' and populations of the associated biota in relatively small areas can fluctuate widely.

The consequences for biodiversity of fire-regime change due to fragmentation require exploration. Clark et al. (this volume) have indicated how the presence of a mix of species may be mediated by the set of fire regimes in a landscape. A narrower range of species may be expected to co-exist in small remnants compared with a large non-fragmented landscape (Hobbs, this volume; Lunt and Morgan, this volume). The contemporary situation of fragmented landscapes with discontinuities in ecological processes and possible constraints on fire regimes provides a context for fire management that is unique (Hobbs, this volume).

Currently, a cohesive landscape-level paradigm or set of hypotheses concerning the nature of fire regimes and their sensitivity to fragmentation and management is lacking. It is envisaged that models, theories and practices that account for the interaction of planned and unplanned events, patterns of recurrence of normal and extreme weather and resultant effects on fuels and fire spread will be at the core of this paradigm. Development of such a paradigm should provide a systematic understanding of how different fire regimes arise and how management options can shape outcomes.

Use of indigenous knowledge and involvement of indigenous people

The focus of this book reflects the perspective of 'Western' science. However, a considerable body of literature has developed on the use and understanding of fires in the landscape by the indigenous people of Australia, especially over the last few decades (e.g. Hallam 1975; Kimber 1983; Griffin and Allan 1986; Burrows and Christensen 1990; Baker and Mutitjulu Community 1992; Russell-Smith et al. 1997a; Bowman 1998). There has been a general shift from reports by anthropologists on the use of fire by an apparently dying culture, to papers written by non-indigenous people working closely with Aboriginal communities, to papers authored jointly or solely by indigenous authors (e.g. Russell-Smith et al. 1997a).

Discussion concerning the use of indigenous knowledge and the involvement of indigenous people in fire management is essential for many reasons. For example, large areas of Australia are under direct ownership by Aboriginal people (AUSLIG 1993) and an increasing number of conservation reserves are being jointly managed (Langton 2000). Further, indigenous and non-indigenous managers can learn from sharing their knowledge (Reid et al. 1992). The use of fire by indigenous peoples has been to create and maintain conditions necessary for survival – particularly the supply of plant food and of vegetation which supported animal food – and for ceremonial, spiritual and communication purposes. Whatever the underlying purpose, there can be little doubt that Aboriginal burning played a central role in the dynamics of landscapes subsequently colonised by Europeans (Bowman 1998). Thus, while the maintenance of biodiversity is considered a 'Western' concept that may be outside traditional Aboriginal experience (Rose 1995; Morrison 2000), it is likely that the widespread use of fire by indigenous peoples influenced the distribution and abundance of biota of Australia. Traditional knowledge is increasingly being identified as an important element in land management in general, and biodiversity conservation in particular. Bowman (1998) identified an urgent need to directly involve Aboriginal people, especially the older ones, in collaborative research on fire ecology in environments where Aborigines maintain close links to the land.

The challenge for contemporary land managers – both indigenous and non-indigenous – is to communicate effectively despite their different world views and their use of different languages (including different Aboriginal languages). This dialogue will challenge both groups, especially because of the different perspectives and philosophies they bring to the table (Rose 1995; Andersen 1999). For example, the spiritual perception of landscape brought by indigenous peoples, such as the protection of sacred sites from fire, is something outside the experience of most non-indigenous peoples. However, these perspectives are important and need to be considered in fire-management planning. The locations of these sites are generally known only by particular members of the Aboriginal community, so their explicit consideration in management plans is difficult. Like most natural resource issues, fire management is not only about science – it is as much a social and cultural issue.

Adaptive management of fire regimes and biodiversity

Management of fire regimes and biodiversity must become more systematic and adaptive, because fire regimes are inherently labile and the requirements of species are broad-ranging, variable and not well known. Yet, it is these fundamental characteristics of fire regimes and biodiversity that sit uncomfortably with individuals, institutions and bureaucracies whose inherent inclination is toward planning, prescription and control and whose focus is restricted to regard fires as discrete events. New management systems are needed to cope with the nature of fire regimes and biodiversity. We consider the following elements are fundamental to such systems.

- Clear objectives are required. A *biodiversity* objective needs to be framed in a way that informs all

management strategies and actions no matter how trivial. When a manager either lights or suppresses a fire, that person needs to know clearly how that action will help achieve an overarching biodiversity objective for the landscape. If management objectives are obscure, then the motivation for action is lost and the process itself becomes a surrogate, or perhaps false objective. In landscapes where biodiversity conservation is a management priority, avoidance of the loss of species is a primary goal that has immediacy and relevance. Articulation of such a goal in clear language, plus reinforcement of the idea that fire regimes and the actions that shape them are the means to an end, not the end in themselves, are keys to good fire management.

- Given a diverse range of fire-regime requirements of diverse species, decision-making needs to be based explicitly on a knowledge of key life-history attributes, habitat requirements and the condition of species and communities in the managed landscape. Thus the status of species' populations or knowledge of them has to be the trigger for decision-making. To use the well-worked example of maturation periods in obligate-seeder plant species, data on the time to first flowering or seed set can be used to specify a minimum interval (or range thereof) between fires. Data can be obtained first-hand through ongoing monitoring of plants in the field (e.g. Gill and Nicholls 1989) or from pre-existing studies or from some combination of them.

- The outcomes from a managed fire regime will vary from place to place and time to time as a result of the condition of the species present. Intuitively it seems that an assessment of measures of ecological condition needs to be made at least annually. The passage of a year can alter life-history stages in ways that are ecologically significant even if no fire occurs. Processes to institutionalise continual evaluation and decision-making are required.

- The condition of the landscape in terms of its fire regime is the complementary ingredient to the condition of its biodiversity. Decision-making can be informed if a reasonable knowledge of fire regimes is attained by continually updating an inventory of fires and their individual characteristics. A commitment to regular record-keeping is a critical and basic part of management.

- Management is undertaken at a landscape scale. Thus the condition of biodiversity is assessed with reference to a landscape-level objective. Implicitly this involves consideration of a set of fire regimes. Plant and animal populations may wax and wane with local extinctions being an inevitable consequence within portions of a 'global' landscape exposed to 'natural' fire regimes. Because of landscape fragmentation and its effects, however, there may be inherent limits on the temporal and spatial scales that any local extinction will be tolerated. These may prompt interventions to minimise the chances of even local extinction.

- A complete management system requires measures that can be used to evaluate its performance and to serve as a safeguard against unexpected consequences that could arise through the use of incomplete or imperfect knowledge. Given the elements outlined above, performance evaluation measures must determine whether management actions resulted in the anticipated set of fire regimes and the desired responses from biodiversity. Measurement of the fire regimes that result from actions is essential but, on its own, insufficient. Biodiversity is the ultimate outcome and success should be evaluated accordingly. A potential spin-off of such a process is that it leads to a continual upgrading of the knowledge base upon which future iterations of the management cycle will be based.

Keith *et al.* (this volume b) outlined instances where elements of an adaptive system such as that described above have been implemented. However, much more needs to be done to develop and implement systems which embody these principles. An adaptive system with biodiversity and fire regimes at its focus has the potential to solve the ongoing challenge to create a set of fire regimes suited to enhance the contemporary status of biodiversity. It also offers a basis for rational evaluation of management options in landscapes where the imperative for the protection of life and property or other goals may compete or conflict with biodiversity conservation. Contemporary landscapes, no matter how seemingly pristine, present unique challenges in terms of altered biodiversity and ecosystem processes, alien species and human activity. An ideal management system must ultimately deliver relevant solutions to both contemporary and future problems.

Conclusions

While detailed answers to many of the issues raised above are not yet available, the way forward involves building upon current knowledge and concepts to achieve an integrated body of concepts that link fire regimes, biodiversity and management systems in ways that are applicable to all Australian ecosystems. The underlying links may be substantially similar in all environments although they may differ quantitatively. There is much to be learnt by examining trends in key processes at large scales, across gradients in climate and ecosystems as we have attempted to begin in this chapter. Similarly there is much to be learnt from comparisons of management experiences across diverse tenures and landscapes. Both research and management could benefit from a national focus. This does not mean that we advocate centralisation of these disciplines but we do encourage a view beyond the local patch. Indeed, our capital of knowledge may expand more rapidly if limited national resources can be harnessed to study more effectively the problems of fire-regime effects on ecosystems and to disseminate such knowledge.

Treaties, agreements, performance indicators, legislation and moral imperatives at international, national, state and local levels reflect concerns about the ecological impacts of fire regimes – including the generation of globally significant 'greenhouse gases' and losses to biodiversity. We suggest that national co-ordination is needed to make significant progress on key themes which emerge from the overview contained in this book:

- measurement and description of past, present and future fire regimes across Australia using remote sensing, archives of fire records and maps, traditional knowledge and biological data;
- the study of fuel dynamics, including fuel modelling suited to geographic extrapolation or interpolation, for fuel systems dominated by grasses, litter and shrubs;
- the development of systems to support the involvement of all key land managers, including Aboriginal people, in fire management;
- experimentation and modelling on the ecological effects of fire regimes as dictated by the interaction between planned and unplanned sources of fire and other sources of disturbance;
- experimentation and modelling to specifically address the effects of fragmentation, urbanisation, resource extraction, exotic plants and animals and global warming on fire regimes and their interactions with biodiversity;
- further investigation of the role of fire regimes in the rehabilitation of biodiversity in landscapes significantly altered or degraded by land use and resource extraction and incorporation of such knowledge into practical management;
- creation of a general system for the determination of indicator species sensitive to changes in fire regimes at a local level;
- creation of an official national register, or coordinated regional databases, of biotic responses to fires as a way of harnessing and stimulating interest in fire-related aspects of natural history;
- creation of a national education programme regarding fire regimes and biodiversity for professionals and volunteers involved in fire management, tertiary and secondary students in relevant disciplines and the wider public; and
- creation of a national award for fire-regime management, in one of the categories of 'pastoral property', 'production forest' or 'conservation reserve', which explicitly includes the conservation of biodiversity.

We highlight these themes as those in particular need of development in order to advance prospects for future management. Knowledge of fire regimes and the Australian biota has changed substantially in the recent past, as reflected in the chapters of this volume. While concepts and knowledge are expanding, native landscapes and the distributions of native biota are shrinking. Many challenges remain for scientists and managers attempting to discover and deliver appropriate fire regimes for biodiversity in these changing landscapes.

References

Allan, G., and Southgate, R. (2001). Fire regimes in the spinifex landscapes of Australia. In *Flammable Australia: The Fire Regimes and Biodiversity of a Continent* (eds. R. A. Bradstock, J. E. Williams and A. M. Gill) pp. 145–176. (Cambridge University Press: Cambridge.)

Andersen, A. (1999). Cross-cultural conflicts in fire management in Northern Australia: not so black and white. *Conservation Ecology* 3: 6. URL: http//www.consecol.org/vol3/iss1/art6

AUSLIG (1993). Map 93/020. In *Australia: Land Tenure*, 1st Edn 1. (Commonwealth of Australia: Canberra.)

Baker, L. M., and Mutitjulu Community (1992). Comparing two views of the landscape: Aboriginal traditional ecological knowledge and modern scientific knowledge. *Rangeland Journal* 14, 174–189.

Beadle, N. C. W. (1940). Soil temperatures during forest fires and their effect on the survival of vegetation. *Journal of Ecology* 28, 180–192.

Bond, W. J., and van Wilgen, B. W. (1996). *Fire and Plants*. (Chapman and Hall: London.)

Bowman, D. J. M. S. (1998). The impact of Aboriginal landscape burning on the Australian biota. *New Phytologist* 140, 385–410.

Bradstock, R. A., and Auld, T. D. (1995). Soil temperatures during experimental bushfires in relation to fire intensity: consequences for legume germination and fire management in south-eastern Australia. *Journal of Applied Ecology* 32, 76–84.

Bradstock, R. A., and Cohn, J. S. (2001). Fire regimes and biodiversity in semi-arid mallee ecosystems. In *Flammable Australia: The Fire Regimes and Biodiversity of a Continent* (eds. R. A. Bradstock, J. E. Williams and A. M. Gill) pp. 238–258. (Cambridge University Press: Cambridge.)

Bradstock, R. A., and Gill, A. M. (1999). When is a fire an ecological emergency? *Australian Journal of Emergency Management* **14**, 6–8.

Bradstock, R. A., Keith, D. A., and Auld, T. D. (1995). Fire and conservation: imperatives and constraints on managing for diversity. In *Conserving Biodiversity: Threats and Solutions* (eds. R. A. Bradstock, T. D. Auld, D. A. Keith, R. Kingsford, D. Lunney and D. Sivertsen), pp. 323–333. (Surrey Beatty: Chipping Norton, NSW.)

Bradstock, R. A., Bedward, M., Kenny, B. J., and Scott, J. (1998). Spatially explicit simulation of the effect of prescribed burning on fire regimes and plant extinctions in shrublands typical of south-eastern Australia. *Biological Conservation* **86**, 83–95.

Burrows, N. D., and Christensen, P. E. S. (1990). A survey of Aboriginal fire patterns in the Western Desert of Australia. In *Fire and the Environment: Ecological and Cultural Perspectives*, pp. 297–305. United States Department of Agriculture, Forest Service, General Technical Report SE-69.

Byram, G. M. (1959). Combustion of forest fuels. In *Forest Fires: Control and Use* (ed. K. P. Davis) pp. 61–89. (McGraw-Hill: New York.)

Cary, G. J. (2001). Importance of a changing climate for fire regimes in Australia. In *Flammable Australia: The Fire Regimes and Biodiversity of a Continent* (eds. R. A. Bradstock, J. E. Williams and A. M. Gill) pp. 26–46. (Cambridge University Press: Cambridge.)

Catchpole, E. A., Alexander, M. E., and Gill, A. M. (1992). Elliptical fire perimeter and area intensity distributions. *Canadian Journal of Forest Research* **22**, 968–972.

Catchpole, W. (2001). Fire properties and burn patterns in heterogeneous landscapes. In *Flammable Australia: The Fire Regimes and Biodiversity of a Continent* (eds. R. A. Bradstock, J. E. Williams and A. M. Gill) pp. 49–76. (Cambridge University Press: Cambridge.)

Clark, J. S. (1991). Disturbance and tree life history on the shifting mosaic landscape. *Ecology* **72**, 1102–1118.

Clark, J. S. (1996). Testing disturbance theory with long-term data: alternative life-history solutions to the distribution of events. *American Naturalist* **148**, 976–996.

Clark, J. S., Gill, A. M., and Kershaw, A. P. (2001). Spatial variability in fire regimes: its effects on recent and past vegetation. In *Flammable Australia: The Fire Regimes and Biodiversity of a Continent* (eds. R. A. Bradstock, J. E.

Williams and A. M. Gill) pp. 125–141. (Cambridge University Press: Cambridge.)

Gilbert, J. M. (1959). Forest succession in the Florentine Valley, Tasmania. *Papers and Proceedings of the Royal Society of Tasmania* **93**, 129–151.

Gill, A. M. (1975). Fire and the Australian flora: a review. *Australian Forestry* **38**, 4–25.

Gill, A. M. (1977). Management of fire-prone vegetation for plant species conservation in Australia. *Search* **8**, 20–26.

Gill, A. M. (1981). Adaptive responses of Australian vascular plant species to fires. In *Fire and the Australian Biota* (eds. A. M. Gill, R. H. Groves and I. R. Noble) pp. 243–272. (Australian Academy of Science: Canberra.)

Gill, A. M. (1996). How fires affect biodiversity. Commonwealth of Australia, Department of Environment, Sports and Territories, Biodiversity Series, Paper no. 8, pp. 47–55. Canberra.

Gill, A. M. (1998). An hierarchy of fire effects: impact of fire regimes on landscapes. In *Proceedings of the 3rd International Conference on Forest Fire Research and the 14th Conference on Fire and Forest Meteorology*, November 1998, vol. 1, pp. 129–144. (Luso: Portugal.)

Gill, A. M., and Bradstock, R. A. (1994). The prescribed burning debate in temperate Australian forests: towards a resolution. In *Proceedings of the 2nd International Forest Fire Research Conference*, pp. 703–712. (Coimbra: Portugal.)

Gill, A. M., and Bradstock, R. A. (1995). Extinctions of biota by fires. In *Conserving Biodiversity: Threats and Solutions* (eds. R. A. Bradstock, T. D. Auld, D. A. Keith, R. Kingsford, D. Lunney and D. Sivertsen) pp. 309–322. (Surrey Beatty: Chipping Norton, NSW.)

Gill, A. M., and Bradstock, R. A. (1999). Prescribed burning: patterns and strategies. In *Proceedings of the 13th International Conference on Fire and Forest Meteorology*, October 1996, Lorne, Victoria, Australia, pp. 3–6. (International Association of Wildland Fire: Fairfield, WA.)

Gill, A. M., and Catling, P. C. (2001). Fire regimes and biodiversity of forested landscapes of southern Australia. In *Flammable Australia: The Fire Regimes and Biodiversity of a Continent* (eds. R. A. Bradstock, J. E. Williams and A. M. Gill) pp. 351–369. (Cambridge University Press: Cambridge.)

Gill, A. M., and McCarthy, M. A. (1998). Intervals between

prescribed fires in Australia: what intrinsic variation should apply? *Biological Conservation* **85**, 161–169.

Gill, A. M., and Moore, P. H. R. (1997). Contemporary fire regimes in the forests of southwestern Australia. Contract Report to Environment Australia. (CSIRO Plant Industry: Canberra.)

Gill, A. M., and Nicholls, A. O. (1989). Monitoring fire-prone flora in reserves for nature conservation. In *Fire Management on Nature Conservation Lands* (eds N. Burrows , L. McCaw and G. Friend) pp. 137–151. (Western Australian Department of Conservation and Land Management: Perth.)

Gill, A. M., Ryan, P. G., Moore, P. H. R., and Gibson, M. (2001). Fire regimes of World Heritage Kakadu National Park, Australia. *Austral Ecology* **25**, 616–625.

Griffin, G., and Allan, G. (1986). Fire and the management of Aboriginal owned lands in Central Australia. In *Science and Technology for Aboriginal Development*, Project Report no. 3. (eds. B. Foran and B. Walker). (CSIRO Division of Wildlife and Rangelands Research, Canberra.)

Hallam S. J. (1975). *Fire and Hearth: a Study of Aboriginal Usage and European Usurpation in South-Western Australia*. (Australian Institute of Aboriginal Studies: Canberra.)

Henry, N. B. (1961). Complete protection versus prescribed burning in the Maryborough hardwoods. *Queensland Forest Service Research Notes* **13**, 33pp.

Hobbs, R. (2001). Fire regimes and their effects in Australian temperate woodlands. In *Flammable Australia: The Fire Regimes and Biodiversity of a Continent* (eds. R. A. Bradstock, J. E. Williams and A. M. Gill) pp. 305–326. (Cambridge University Press: Cambridge.)

Hodgkinson, K. C. (2001). Fire regimes in *Acacia* wooded landscapes: effects on functional processes and biological diversity. In *Flammable Australia: the Fire Regimes and Biodiversity of a Continent* (eds. R. A. Bradstock, J. E. Williams and A. M. Gill) pp. 259–277. (Cambridge University Press: Cambridge.)

Howitt, W. (1856). Black Thursday. *Household Words*, 10 May 1856, 388–395.

Jackson, W. D. (1968). Fire, air, water and earth: an elemental ecology of Tasmania. *Proceedings of the Ecological Society of Australia* **3**, 9–16.

Jarrett, P. H., and Petrie, A. H. K. (1929). The vegetation of the Black's Spur region: a study of the ecology of some

Australian mountain *Eucalyptus* forests. II. Pyric succession. *Journal of Ecology* **17**, 249–281.

Keith, D. A., McCaw, W. L., and Whelan, R. J. (2001a). Fire regimes in Australian heathlands and their effects on plants and animals. In *Flammable Australia: The Fire Regimes and Biodiversity of a Continent* (eds. R. A. Bradstock, J. E. Williams and A. M. Gill) pp. 199–237. (Cambridge University Press: Cambridge.)

Keith, D. A., Williams, J. E., and Woinarski, J. C. Z. (2001b). Biodiversity conservation – key approaches and principles. In *Flammable Australia: The Fire Regimes and Biodiversity of a Continent* (eds. R. A. Bradstock, J. E. Williams and A. M. Gill) pp. 401–425. (Cambridge University Press: Cambridge.)

Kershaw, A. P, Clark, J. S., Gill, A. M., and D'Costa, D. M. (2001). A history of fire in Australia. In *Flammable Australia: The Fire Regimes and Biodiversity of a Continent* (eds. R. A. Bradstock, J. E. Williams and A. M. Gill) pp. 3–25. (Cambridge University Press: Cambridge.)

Kiddle, M. (1967). *Men of Yesterday*. (Melbourne University Press: Melbourne.)

Kimber, R. (1983). Black lightning: Aborigines and fire in Central Australia and the Western Desert. *Archaeology in Oceania* **18**, 38–45.

Lang, S. (1997). Burning the bush: a spatio-temporal analysis of jarrah forest fire regimes. BSc thesis, Australian National University, Canberra.

Langton, M. (2000). 'The fire at the centre of each family': Aboriginal traditional fire regimes and the challenges for reproducing ancient fire management in the protected areas of northern Australia. In *Fire! The Australian Experience*, Proceedings from the National Academies Forum seminar, 30 September–1 October 1999, University of Adelaide, South Australia, pp. 3–32. (Australian Academy of Technological Sciences and Engineering Limited: Melbourne.)

Lunt, I. D., and Morgan, J. W. (2001). The role of fire regimes in temperate lowland grasslands of southeastern Australia. In *Flammable Australia: The Fire Regimes and Biodiversity of a Continent* (eds. R. A. Bradstock, J. E. Williams and A. M. Gill) pp. 177–196. (Cambridge University Press: Cambridge.)

McArthur, A. G. (1962). Control burning in eucalypt forests. Commonwealth of Australia, Forestry and Timber Bureau Leaflet no. 80, Canberra.

McArthur, A. G. (1966). Weather and grassland fire

behaviour. Commonwealth of Australia, Forestry and Timber Bureau Leaflet no. 100, Canberra.

McArthur, A. G. (1967). Fire behaviour in eucalypt forests. Commonwealth of Australia, Forestry and Timber Bureau Leaflet no. 107, Canberra.

McCarthy, M. A., and Cary, G. J. (2001). Fire regimes in landscapes: models and realities. In *Flammable Australia: The Fire Regimes and Biodiversity of a Continent* (eds. R. A. Bradstock, J. E. Williams and A. M. Gill) pp. 72–93. (Cambridge University Press: Cambridge.)

McCarthy, M. A., Gill, A. M., and Lindenmayer, D. B. (1999). Fire regimes in mountain ash forest: evidence from forest age structure, extinction models and wildlife habitat. *Forest Ecology and Management* 124, 193–203.

Mercer, G. N., and Weber, R. O. (1994). Plumes above line fires in a cross wind. *International Journal of Wildland Fire* 4, 201–207.

Meredith, C. W. (1996). Is fire management effective? Commonwealth of Australia, Department of Environment, Sports and Territories, Biodiversity Series, Paper no. 8, pp. 227–231. Canberra.

Miller, G. H., Magee, J. W., Johnson, B. J., Fogel, M. L., Spooner, N. A., McCulloch, M. T., and Aycliffe, L. K. (1999). Pleistocene extinction of *Genyornis newtoni*: human impact on Australian megafauna. *Science* 283, 205–208.

Mitchell, T. L. (1848). *Journal of an Expedition into the Interior of Tropical Australia, in Search of a Route from Sydney to the Gulf of Carpentaria.* (Longmans: London.)

Morrison, D. A., Buckney, R. T., Buick, B. J., and Cary, C. J. (1996). Conservation conflicts over burning bush in south-eastern Australia. *Biological Conservation* 76, 167–175.

Morrison, J. (2000) A case study: indigenous land management and capacity building in south-east Arnhem land and the Gulf of Carpentaria. In *Native Solutions: Indigenous Knowledge and Today's Fire Management*, an International Symposium abstract URL: http://www.parks.tas.gov.au/manage/conferences/abstracts.html

Noble, I. R., and Slatyer, R. O. (1980). The use of vital attributes to predict successional changes in plant communities subject to recurrent disturbances. *Vegetatio* 43, 5–21.

Noble, J. C., and Grice, A. C. (2001). Fire regimes in semi-arid and tropical pasture lands: managing biological diversity and ecosystem function. In *Flammable Australia: The Fire Regimes and Biodiversity of a Continent* (eds. R. A. Bradstock, J. E. Williams and A. M. Gill) pp. 373–400. (Cambridge University Press: Cambridge.)

Peet, G. B. (1965). A fire danger rating and controlled burning guide for the northern Jarrah (*Euc. marginata* Sm) forest, of Western Australia. Forests Department of Western Australia Bulletin no. 74, Perth.

Pyne, S. (1991). *Burning Bush: A Fire History of Australia.* (Henry Holt: New York.)

Raunkiaer, C. (1934). *The Life Forms of Plants and Statistical Plant Geography.* (Clarendon Press: Oxford.)

Reid, J., Baker, L., Morton, S. R., and Muṯitjulu Community (1992). Traditional knowledge + ecological survey = better land management. *Search* 23, 249–251.

Rose, B. (1995). *Land Management Issues: Attitudes and Perceptions amongst Aboriginal People of Central Australia.* (Central Land Council: Alice Springs.)

Rose, R. (1993). Bush fire: research, policies and practitioners. In *The Burning Question: Fire Management in NSW* (ed. J. Ross) pp. 5–9. (University of New England: Armidale, NSW.)

Russell-Smith, J., and Stanton, P. (2001). Fire regimes and fire management of rainforest communities across northern Australia. In *Flammable Australia: The Fire Regimes and Biodiversity of a Continent* (eds. R. A. Bradstock, J. E. Williams and A. M. Gill) pp. 329–350. (Cambridge University Press: Cambridge.)

Russell-Smith, J., Gapindi, M., Gunbunuka, B., Kapirigi, N., Namingum, G., Lucas, K., and Chaloupka, G. (1997a). Aboriginal resource utilization and fire management practice in western Arnhem Land, monsoonal northern Australia: notes for prehistory, lessons for the future. *Human Ecology* 25, 159–195.

Russell-Smith, J., Ryan, P. G., and Durieu, R. (1997b). A LANDSAT MSS-derived fire history of Kakadu National Park, monsoonal northern Australia, 1980–94: seasonal extent, frequency and patchiness. *Journal of Applied Ecology* 34, 748–766.

Specht, R. L., Rayson, P., and Jackman, M. E. (1958). Dark Island Heath (Ninety-Mile Plain, South Australia). VI. Pyric succession: changes in composition, coverage, dry weight, and mineral nutrient status. *Australian Journal of Botany* 6, 59–88.

Stocker, G. C. (1966). Effects of fires on vegetation in the Northern Territory. *Australian Forestry* 30, 223–230.

Tate, R. (1883). The botany of Kangaroo Island. *Transactions and Proceedings and Report of the Royal Society of South Australia* 6, 116–171.

Tolhurst, K. (1999). Towards the implementation of ecologically based fire regimes in the Grampians National Park. In *Management of Fire for the Conservation of Biodiversity*, Workshop Proceedings (eds. G. Friend, M. Leonard, A. MacLean and I. Sieler) pp. 30–38. (Victorian Department of Natural Resources and Environment: Melbourne.)

Whelan, R. J., Rodgerson, L., Dickman, C. R., and Sutherland, E. F. (2001). Critical life cycles of plants and animals: developing a process-based understanding of population changes in fire-prone landscapes. In *Flammable Australia: The Fire Regimes and Biodiversity of a Continent* (eds. R. A. Bradstock, J. E. Williams and A. M. Gill) pp. 94–124. (Cambridge University Press: Cambridge.)

Williams, R. J., and Bradstock, R. A. (2000). Fire regimes and the management of biodiversity in temperate and tropical *Eucalyptus* forest landscapes in Australia. In *Fire and Forest Ecology: Innovative Silviculture and Vegetation Management*, Tall Timbers Fire Ecology Conference Proceedings, no. 21 (eds. W. K. Moser and C. F. Moser.) pp. 139–150. (Tall Timbers Research Station: Tallahassee, FL.)

Williams, R. J., Griffiths, A. D., and Allan, G. (2001). Fire regimes and biodiversity in the savannas of northern Australia. In *Flammable Australia: The Fire Regimes and Biodiversity of a Continent* (eds. R. A. Bradstock, J. E. Williams and A. M. Gill) pp. 281–304. (Cambridge University Press: Cambridge.)

Wimberley, M. C., Spies, T. A., Long, C. J., and Whitlock, C. (2000). Simulating historical variability in the amount of old forests in the Oregon Coast Range. *Conservation Biology* 14, 167–180.

Taxonomic index

General index

Page numbers in bold indicate where the indexed term appears in a figure.